Lecture Notes in Networks and Systems 724

The series "Lecture Notes in Networks and Systems" publishes the latest developments in Networks and Systems—quickly, informally and with high quality. Original research reported in proceedings and post-proceedings represents the core of LNNS.

Volumes published in LNNS embrace all aspects and subfields of, as well as new challenges in, Networks and Systems.

The series contains proceedings and edited volumes in systems and networks, spanning the areas of Cyber-Physical Systems, Autonomous Systems, Sensor Networks, Control Systems, Energy Systems, Automotive Systems, Biological Systems, Vehicular Networking and Connected Vehicles, Aerospace Systems, Automation, Manufacturing, Smart Grids, Nonlinear Systems, Power Systems, Robotics, Social Systems, Economic Systems and other. Of particular value to both the contributors and the readership are the short publication timeframe and the world-wide distribution and exposure which enable both a wide and rapid dissemination of research output.

The series covers the theory, applications, and perspectives on the state of the art and future developments relevant to systems and networks, decision making, control, complex processes and related areas, as embedded in the fields of interdisciplinary and applied sciences, engineering, computer science, physics, economics, social, and life sciences, as well as the paradigms and methodologies behind them.

Indexed by SCOPUS, INSPEC, WTI Frankfurt eG, zbMATH, SCImago.

All books published in the series are submitted for consideration in Web of Science.

For proposals from Asia please contact Aninda Bose (aninda.bose@springer.com).

Radek Silhavy · Petr Silhavy
Editors

Artificial Intelligence Application in Networks and Systems

Proceedings of 12th Computer Science On-line Conference 2023, Volume 3

Springer

Editors
Radek Silhavy
Faculty of Applied Informatics
Tomas Bata University in Zlin
Zlin, Czech Republic

Petr Silhavy
Faculty of Applied Informatics
Tomas Bata University in Zlin
Zlin, Czech Republic

ISSN 2367-3370 ISSN 2367-3389 (electronic)
Lecture Notes in Networks and Systems
ISBN 978-3-031-35313-0 ISBN 978-3-031-35314-7 (eBook)
https://doi.org/10.1007/978-3-031-35314-7

This Springer imprint is published by the registered company Springer Nature Switzerland AG
The registered company address is: Gewerbestrasse 11, 6330 Cham, Switzerland

Preface

We are honored to present the refereed proceedings of the 12th Computer Science Online Conference 2023 (CSOC 2023), composed of three volumes: Software Engineering Perspectives, Artificial Intelligence Trends, and Cybernetics Perspectives in Systems. This Preface is intended to introduce and provide context for the three volumes of the proceedings.

CSOC 2023 is a prominent international forum designed to facilitate the exchange of ideas and knowledge on various topics related to computer science. The conference was held online in April 2023, using modern communication technologies to provide researchers worldwide with equal participation opportunities.

The first volume, Software Engineering Research in System Science, encompasses papers that discuss software engineering topics related to software analysis, design, and the application of intelligent algorithms, machine, and statistical learning in software engineering research. These papers provide valuable insights into the latest advances and innovative approaches in software engineering research.

The second volume, Networks and Systems in Cybernetics, presents papers that examine theoretical and practical aspects of cybernetics and control theory in systems or software. These papers provide a deeper understanding of cybernetics and control theory and demonstrate how they can be applied in the design and development of software systems.

The third volume, Artificial Intelligence Application in Networks and Systems, is dedicated to presenting the latest trends in artificial intelligence in the scope of systems, systems engineering, and software engineering domains. The papers in this volume cover various aspects of artificial intelligence, including machine learning, natural language processing, and computer vision.

In summary, the proceedings of CSOC 2023 represents a significant contribution to the field of computer science, and they will be an excellent resource for researchers and practitioners alike. The papers included in these volumes will inspire new ideas, encourage further research, and lead to the development of novel and innovative approaches in computer science.

April 2023

Radek Silhavy
Petr Silhavy

Organization

Program Committee

Program Committee Chairs

Petr Silhavy	Tomas Bata University in Zlin, Faculty of Applied Informatics
Radek Silhavy	Tomas Bata University in Zlin, Faculty of Applied Informatics
Zdenka Prokopova	Tomas Bata University in Zlin, Faculty of Applied Informatics
Roman Senkerik	Tomas Bata University in Zlin, Faculty of Applied Informatics
Roman Prokop	Tomas Bata University in Zlin, Faculty of Applied Informatics
Viacheslav Zelentsov	Doctor of Engineering Sciences, Chief Researcher of St. Petersburg Institute for Informatics and Automation of Russian Academy of Sciences (SPIIRAS)
Roman Tsarev	Department of Information Technology, International Academy of Science and Technologies, Moscow, Russia
Stefano Cirillo	Department of Computer Science, University of Salerno, Fisciano (SA), Italy

Program Committee Members

Juraj Dudak	Faculty of Materials Science and Technology in Trnava, Slovak University of Technology, Bratislava, Slovak Republic
Gabriel Gaspar	Research Centre, University of Zilina, Zilina, Slovak Republic
Boguslaw Cyganek	Department of Computer Science, University of Science and Technology, Krakow, Poland
Krzysztof Okarma	Faculty of Electrical Engineering, West Pomeranian University of Technology, Szczecin, Poland

Monika Bakosova	Institute of Information Engineering, Automation and Mathematics, Slovak University of Technology, Bratislava, Slovak Republic
Pavel Vaclavek	Faculty of Electrical Engineering and Communication, Brno University of Technology, Brno, Czech Republic
Miroslaw Ochodek	Faculty of Computing, Poznan University of Technology, Poznan, Poland
Olga Brovkina	Global Change Research Centre Academy of Science of the Czech Republic, Brno, Czech Republic and Mendel University of Brno, Czech Republic
Elarbi Badidi	College of Information Technology, United Arab Emirates University, Al Ain, United Arab Emirates
Luis Alberto Morales Rosales	Head of the Master Program in Computer Science, Superior Technological Institute of Misantla, Mexico
Mariana Lobato Baes	Research-Professor, Superior Technological of Libres, Mexico
Abdessattar Chaâri	Laboratory of Sciences and Techniques of Automatic control & Computer engineering, University of Sfax, Tunisian Republic
Gopal Sakarkar	Shri. Ramdeobaba College of Engineering and Management, Republic of India
V. V. Krishna Maddinala	GD Rungta College of Engineering & Technology, Republic of India
Anand N Khobragade (Scientist)	Maharashtra Remote Sensing Applications Centre, Republic of India
Abdallah Handoura	Computer and Communication Laboratory, Telecom Bretagne, France
Almaz Mobil Mehdiyeva	Department of Electronics and Automation, Azerbaijan State Oil and Industry University, Azerbaijan

Technical Program Committee Members

Ivo Bukovsky, Czech Republic
Maciej Majewski, Poland
Miroslaw Ochodek, Poland
Bronislav Chramcov, Czech Republic
Eric Afful Dazie, Ghana
Michal Bliznak, Czech Republic

Donald Davendra, Czech Republic
Radim Farana, Czech Republic
Martin Kotyrba, Czech Republic
Erik Kral, Czech Republic
David Malanik, Czech Republic
Michal Pluhacek, Czech Republic
Zdenka Prokopova, Czech Republic
Martin Sysel, Czech Republic
Roman Senkerik, Czech Republic
Petr Silhavy, Czech Republic
Radek Silhavy, Czech Republic
Jiri Vojtesek, Czech Republic
Eva Volna, Czech Republic
Janez Brest, Slovenia
Ales Zamuda, Slovenia
Roman Prokop, Czech Republic
Boguslaw Cyganek, Poland
Krzysztof Okarma, Poland
Monika Bakosova, Slovak Republic
Pavel Vaclavek, Czech Republic
Olga Brovkina, Czech Republic
Elarbi Badidi, United Arab Emirates

Organizing Committee Chair

Radek Silhavy — Tomas Bata University in Zlin, Faculty of Applied Informatics
radek@silhavy.cz

Conference Organizer (Production)

Silhavy s.r.o.
Website: https://www.openpublish.eu
Email: csoc@openpublish.eu

Conference Website, Call for Papers

https://www.openpublish.eu

Contents

Prediction Model for Tax Assessments Using Data Mining and Machine Learning

Anthony Willa Sampa[1](✉) and Jackson Phiri[2]

[1] University of Zambia, Lusaka, Zambia
anthonysampa@yahoo.com
[2] Department of Computer Science Lusaka, University of Zambia, Lusaka, Zambia
jackson.phiri@cs.unza.zm

Abstract. Tax administration remains an integral part of a country's economic growth. Most tax administrations across the world face similar challenges in the collection process, the most common of which is the compliance. It is important therefore to be able to detect revenue leakages as much as possible in order to increase overall collection. In this research we reviewed the tax assessment process that attempts to detect revenue leakages due to under declarations, fraud and declaration errors. We developed a machine learning model using supervised learning to detect declaration audit and assessment selections that are likely to lead to a significantly high revenue collection from these leakages. Due to the large volumes of audit cases generated by audit selection methods, some of which yield very little collections after the audits, it was important to intelligently separate cases that are likely lead to smaller insignificant revenue collection from the cases that yield significant revenue against in comparison with resources used to perform audits and assessments. The test results from the model we created produced positive results with an accuracy of 83%. The results showed that the model would help the revenue authority effectively increase revenue collections with the same amount of resources.

Keywords: Compliance · Audit · Assessment · Machine Learning · Tax

1 Introduction

Taxation plays an integral role in the economy of any country in the world. Tax collection is the way in which citizens contribute to the overall development of a country's economy and economic growth in their individual capacity. The taxation system is designed to fairly and equitably collect contributions from citizens earning an income depending on their source of income. From the perspective of economics, tax is well thought out as a form of income redistribution tool from the well-to-do in society to the underprivileged members of society as well as generating revenue for the government expenditure [1]. The collections go towards national development activities such as infrastructure, heath care and education that are aimed at improving the lives of citizens and overall development of the country. The taxation system in Zambia is a self-declaration type of system. It

R. Silhavy and P. Silhavy (Eds.): CSOC 2023, LNNS 724, pp. 1–14, 2023.
https://doi.org/10.1007/978-3-031-35314-7_1

is therefore important to efficiently and effectively assess taxpayer's declarations and calculate the correct taxes the taxpayers are supposed to pay through an assessment. The tax collections to GDP ratio is typically much lower in developing countries than in developed countries [2]. Tax revenues above 15% of a country's gross domestic product are considered good indicators of economic growth according to the World Bank [3]. Generally, difference in tax revenues between the poorest and the richest nations of the world are almost exclusively explained by the weakness of direct taxation in developing countries [2]. In this research, we developed a prediction model for the prediction of taxpayer assessments using data mining and supervised machine learning in order to automatically detect tax declarations that lead to significant positive assessments thereby increasing revenue collection and making the process more effective and efficient.

2 Literature Review and Related works

Taxes can be used to redistribute income in a country's economy in order to reduce inequality or as an instrument for regulation to encourage or discourage particular activities in order to enhance social welfare [4]. The wealth of any nation is assessed by its performance in infrastructure provision through its construction industry. The construction industry is large, unpredictable, and requires large capital outlays [5]. Generally, the objective of tax collection is the same throughout the world, but the solution to a good tax administration system is most of the time specific to a country. How the taxes and tax administration systems are structured is influenced by the specific country's policy and needs, strengths and weaknesses in order to make the collection equitable, fair, and beneficial to the majority of the country's citizens. In developing countries like Zambia, there is a high need for government spending due to the many social and economic areas that require attention. In developing countries, the challenges faced by citizens are common. The common challenges in the economy include poor health facilities that cannot satisfy the population, poor school systems and infrastructure and poor roads required to move essential goods and services effectively. In developing countries, there is a high dependency on tax collections in order for the economy to thrive. Revenue authorities are constantly seeking to ensure they are collecting the right amount of taxes by identifying, fixing revenue leakages and increase compliance levels of the tax population.

In Zambia, taxpayers self-declare their business activity to the Zambia revenue authority. The taxpayer has two alternatives: (1) the taxpayer may report actual income and pay the necessary taxes on it immediately; or (2) the taxpayer may report less than their actual income. Should they chooses the latter, their payoff will depend in part on the likelihood of undergoing audit and the effectiveness of the audit if it is performed [6]. When tax reporting is voluntary as in the other voluntary tax systems, enforcement of the tax collection is carried out mainly through irregular audits, with penalties often assessed if the taxpayer is discovered to have under-declared their taxable income [7]. Taxpayers reporting less than the actual income may not always be intentional on the taxpayer's part. At times, it may be as a result accounting errors, incomplete information while at times it may be because of ignorance from the taxpayer in understanding the tax systems. Is some cases the taxpayer may over declare their income and pay more than they actually need to, in which case they will be eligible for a tax refund. The main objective of tax audits is to investigate the credibility of the declared or assessed tax [8].

Femidah and Phiri [9] described tax compliance as the act of adhering to or conforming with laws or rules. James and Alley [10] described tax non-compliance as a failure of taxpayer to accommodate tax responsibilities, whether they performed unintentionally or intentionally. A compliant taxpayer is one who has met all their tax obligations according to the laws of the country. This includes ensuring all the required tax declarations as submitted on or before their due dates. The taxpayer is also expected to make all necessary payments towards their obligations in time as stipulated in the law.

Tax evasion is payment of the payment of a suppressed amount or non-payment at all of required amount required by law from one's taxable income. Evasion usually takes the form of a taxpayer purposefully deciding to under declare their income or claim an excessive amount as a refund [11]. In his research, Wallschutzky [11] reviewed two groups of evader from the general populace. From the responses received from the evaders it was found that one of the primary reasons people evade taxes is they felt they were not getting their money's worth for their taxes, that the taxes were also too high. It was also found that the evaders felt the government did not spend the taxpayer's taxes prudently and the burden of taxes fell disproportionately on low-income earners.

Most taxpayer will evade taxes if they expect many others to do the same. In the same vein, most taxpayer will pay taxes if they expect few others to evade. These behaviors exist both among individual taxpayers and among business taxpayers even if, all else being equal, the latter are more likely to evade than the former [12]. In a research done by Bruno [12] on tax enforcement, compliance and morale, it was found that high uncertainty and sunk costs of tax enforcements had a negative effect on tax compliance as opposed to a higher value for tax revenues needed for public goods and services has a positive impact. It was also found that citizens who use third party reporting such as fiscalization systems tend to evade taxes less than those who self-declare. Evasion is done in many different ways with aim to go undetected by the revenue authority. Some fraud schemes tend to be very simple and straightforward while others are very complex and may involve many different legitimate processes within the system and other external processes. One such type of scheme is done by a series of independent transactions and processes, which by themselves do not amount to fraud but when used together, they show that the taxpayer's objective was evasion [13].

Intelligent behavioral change is very critical to achieving success in dealing with taxpayer compliance. The goal of data mining, machine learning and artificial intelligence taxation is to be able to achieve the intelligent behavior from computers, accurately to an acceptable level. Hoglund [14] proposed a decision support tool for the prediction of tax payment defaulters using genetic algorithm-based variable selection. His dataset consisted of Finnish limited liability firms that have defaulted on employer contribution taxes or on value added taxes. He found that variables measuring solvency, liquidity and payment period of trade payables were important variables in predicting tax defaults. The research focusses on taxpayers who have defaulted, but in reality defaulters may come from even those that have been paying taxes thus cannot give an accurate prediction. Taxpayers may be paying and declaring regularly but an inconsistent or on downward trend. Such trends may suggest that their business is stressed or a form of tax evasion. Furthermore, taxpayer may file nil return simply for the sake of meeting their obligations or ignorance and such would be left out. A study was conducted to demonstrate how data mining and machine learning could help to advance the effectiveness of public schemes and inform policy decisions in Italy. The study concentrated on a tax rebate system that was introduced in that country in 2014. A tax rebate is a form of relief given to a taxpayer upon request that reduces their tax liability if the taxpayer incurred too much taxes during their business process such as manufacturing. The main purpose of tax a rebate policy is to enhance a country's competitiveness in foreign markets and avoid double taxation on export goods in some cases [15]. The study was aimed at showing the impact of machine learning algorithm and how it would help in increasing the effectiveness of the process of selecting the beneficiaries of the scheme. It would help in transparency, as it was to use an easily interpretable machine-learning algorithm to determine the beneficiaries [16]. The results showed that 29.5% of the funds earmarked to the policy (about 2 billion euro) could have been saved without reducing the overall consumption expenditure, as they were channeled to households not targeted by ML [16]. Sabbeh [17] compared and analyzed the performance of different machine-learning techniques that are used for the prediction of customers that are likely to stop doing business with a particular entity. They explored different Machine learning systems models to analyze customers' personal data to give the organization a competitive advantage by increasing their customer retention rate. The dataset that was used for the experiments of this study was a customer database of a telecommunication company. The dataset contained customers' statistical data, which included 17 explanatory features related to customers' service usage during day, international calls, customer service calls. They used ten analytical techniques that belong to different categories of learning. The chosen techniques include Discriminant Analysis, Decision Trees (CART), instance-based learning (k-nearest neighbors), Support Vector Machines, Logistic Regression, ensemble–based learning techniques (Random Forest, Ada Boosting trees and Stochastic Gradient Boosting), Naïve Bayesian, and Multi-layer perceptron. The classifiers were evaluated and it was found that the random forest and ADA Boost gave the best accuracy results of 96%. It was also found that Multi-layer perceptron and Support vector machine could also be recommended with 94% accuracy. Yilmazer and Kocaman [18] tried to solve the highly complex problem of the development by using of new assessment approaches for mass appraisal of real estate using machine learning and artificial intelligence to help municipalities in Turkey calculate the tax property values accurately against the taxpayer's declarations. They studied a solution for mass appraisal in urban residential areas where commercial properties were also available. They used Linear multiple regression and random forest to develop the model. For the random forest model, 1162 cases were split as for training (80%) and validation (20%) randomly [18]. The out of bag error (OOB), which is an indicator and a major tool of measuring was used to determine the number of trees to be used in the model. In their conclusion, it was found that the random forest algorithm was capable of explaining the dependent variable accurately and produced slightly better results than the linear multiple regression method for mass appraisal. They also concluded that results obtained with the residential appraisal approach could be used for the correction of declared tax values in Turkey as at the time municipalities computed the property tax values based on taxpayers' declarations. A study to predict tax avoidance by means of social network analytics was carried out by Lismont [19]. They developed a predictive model by building a network of firms connected through shared board membership.

They then, applied three analytical techniques, logistic regression, decision trees, and random forests to create five models using either firm characteristics, network characteristics or different combinations of both. It was found that the random forest with firm characteristics, network characteristics of firms and network characteristics of board members provided the best performance of the model, which had predictive ability of tax avoidance. They concluded that financial analysts and regulatory agencies could use their insights to predict which firms are likely to be low-tax and potentially at risk. Pal [20] carried out a to study is to show results obtained with the random forest classifier and to compare its performance with the support vector machines (SVMs) in terms of classification accuracy, training time and user defined parameters. They used Landsat Enhanced Thematic Mapper Plus data of an area in the UK with seven different land covers for the training and testing data for the model creation. Results from this study suggests that the random forest classifier performed equally well to SVMs in terms of classification accuracy and training time. This study also concludes that the number of user-defined parameters required by random forest classifiers is less than the number required for SVMs and easier to define.

3 Methodology

In this section, we looked at how we collected data, prepared it and how we used it to create a model for the identification of tax declarations for tax assessments using the Cross Industry Standard Process for Data Mining (CRISP-DM) methodology. CRISP-DM is considered the standard and an industry-independent process model for applying data mining projects [21]. The methodology defines a non-rigid sequence of six phases, which allow the building and implementation of a data-mining model to use in a real environment [22]. The CRISP-DM reference model for data mining provides an overview of the life cycle of a data-mining project. It contains the phases of a project, their respective tasks, and their outputs [23]. CRISP-DM has several characteristics that render it useful for evidence mining. It provides a generic process model that holds the overarching structure and dimensions of the methodology [24]. The cycle of the project consists of six stages namely business understanding, data understanding, data preparation, modelling, evaluation and deployment. Based on the understanding we got in the business understanding stage, we were able to extract the required dataset for our model. We collected sample data from the tax administration system database to be used for training and testing with a total of 199,999 records in the year 2020. From the data extraction, we were able to identify and extract nine features we were to use in the machine learning process.

We performed data preparation by checking and ensuring we had no null values in our data using specific python libraries. We were also able to create new required variables, convert and clean data that was incorrectly formatted into the correct usable format acceptable by the machine-learning algorithm.

The machine-learning model was developed using taxpayer declaration data containing information with variances between input and output declarations and categories into nine categories, which made up our model features. The first of these is called invoice reduction: These are transactions where a reduction in a seller's sales invoices was detected on the taxpayer's declaration in comparison with the purchaser's declaration for the same invoice. The second category is called nil value on invoice category: These are transactions where it was detected that the seller's invoice was declared as zero while the purchaser's invoice contained a positive value. The third category is the nil declaration category: this category contains cases where the entire declaration was declared as zero but transactions were declared from other purchaser's declarations. The fourth category called no declaration category: This is where the seller did not submit a declaration for a particular period and the purchaser declared purchases from them. The fifth category is the six months payment compliance, which shows whether the taxpayer was compliant in terms of making payments for their tax obligations in the last six months. The sixth category is the one-year payment compliance, which shows how compliant the taxpayer was compliant in terms of making payments for their tax obligations in the last one year. The seventh category is the six months declaration compliance, which shows the taxpayer's declaration compliance in the last six months. The eighth category is the one-year declaration compliance, which shows how compliant the taxpayer has been in making their tax declarations in the last one year. The final category is the assessments in the last six months, which shows if the taxpayers were assessed in the last six months of the given period. Table 1 and Fig. 1 below illustrate the feature distribution.

Table 1. Table showing the feature distribution.

Feature	Positive total	Negative total	Total flagged for Audit
Invoice-Reduction	28898	171101	2724
Nil Value	20435	179564	14680
Nil Declaration	12671	187328	6826
No Declaration	122419	77580	72998
Payment Compliance (6months)	99621	100378	37838
Payment Compliance (1year)	92662	107337	50195
Declaration Compliance (6months)	100001	99998	38039
Declaration Compliance (1year)	95000	104999	46398
Assessments (6months)	88916	111083	55134

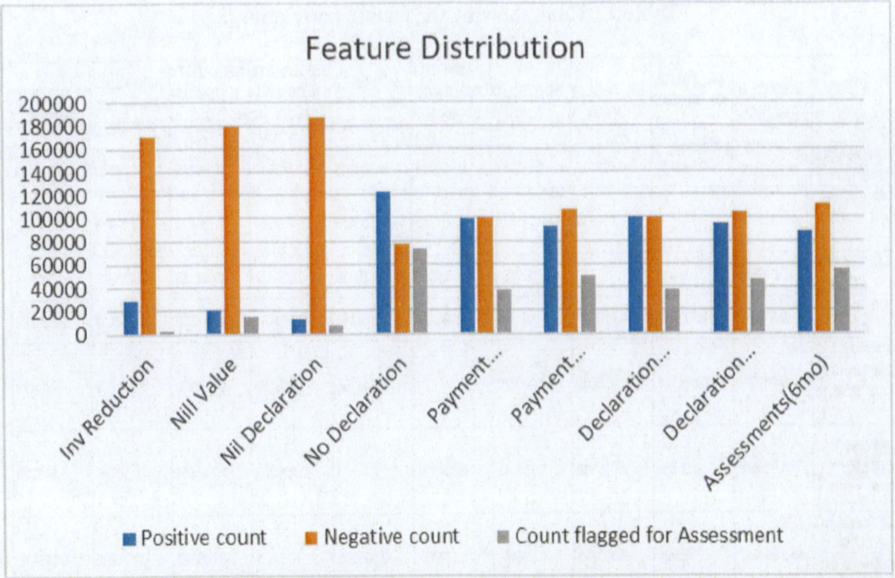

Fig. 1. Chart illustrating the feature distribution.

3.1 Model Design and Implementation

Using the features identified and analyzed in Table 1, we were able to develop a prediction model using the Random Forest algorithm. The data was cleaned, prepared, and converted to a usable format for the machine-learning engine. We converted categorical data in the form of "Y" and "N" to "1" and "0". "1" to represent "Y" and "0" to represent "N". To achieve this, we used the python library One Hot Encoding that converts categorical data to numeric data as per mapping. We then used the correlation heatmap to observe correlations between and among variables. The table below illustrates these correlations (Table 2 and Fig. 2).

We used the Random forest algorithm to create the prediction model. The random forest algorithm is a supervised learning algorithm that creates decision trees on data samples, then gets the prediction from each of them and finally selects the best solution by means of voting [25]. The random forest classifier uses the Gini Index as an attribute selection measure, which measures the impurity of an attribute with respect to the classes [20]. For a given training set T, selecting one case at random and saying that it belongs to some class, the Gini index is written as follows:

$$\sum\sum_{j\neq i}(f(C_i,T)/|T|)(f(C_j,T)/|T|) \tag{1}$$

Table 2. Table showing the feature correlations.

	INVOICE_REDUCTION_1	NIL_VALUE_ON_INVOICE_1	NIL_DECLARATION_1	NO_DECLARATION_1	PAYMENT_COMPLIANCE_SIX_MONTHS_1	PAYMENT_COMPLIANCE_ONE_YEAR_1	DECLARATION_COMPLIANCE_SIX_MONTHS_1	DECLARATION_COMPLIANCE_ONE_YEAR_1	ASSESSMENTS_IN_SIX_MONTHS_1	FLAG_ASSESSMENT_1
INVOICE_REDUCTION_1	1	-0.063137	-0.002559	-0.516247	-0.000236	-0.084925	-0.000547	-0.049399	-0.10539	-0.243464
NIL_VALUE_ON_INVOICE_1	-0.063137	1	-0.016919	0.111319	-0.001183	0.03639	-0.002201	0.031813	0.34188	0.233449
NIL_DECLARATION_1	-0.002559	-0.016919	1	0.020561	0.0008	0.015039	-0.000722	-0.002168	0.286606	0.083813
NO_DECLARATION_1	-0.516247	0.111319	0.020561	1	-0.000222	0.184642	-0.000481	0.114036	0.238209	0.553688
PAYMENT_COMPLIANCE_SIX_MONTHS_1	-0.000236	-0.001183	0.0008	-0.000222	1	-0.004104	0.156837	-0.002668	-0.001705	-0.00472
PAYMENT_COMPLIANCE_ONE_YEAR_1	-0.084925	0.03639	0.015039	0.184642	-0.004104	1	-0.004205	0.17556	0.10711	0.305138
DECLARATION_COMPLIANCE_SIX_MONTHS_1	-0.000547	-0.002201	-0.000722	-0.000481	0.156837	-0.004205	1	-0.000975	-0.001321	-0.003572
DECLARATION_COMPLIANCE_ONE_YEAR_1	-0.049399	0.031813	-0.002168	0.114036	-0.002668	0.17556	-0.000975	1	0.059394	0.208047
ASSESSMENTS_IN_SIX_MONTHS_1	-0.10539	0.34188	0.286606	0.238209	-0.001705	0.10711	-0.001321	0.059394	1	0.43812
FLAG_ASSESSMENT_1	-0.243464	0.233449	0.083813	0.553688	-0.00472	0.305138	-0.003572	0.208047	0.43812	1

The random forest classifier therefore consists of N trees, where N is the number of trees to be grown, which can be any value defined by the user. To classify a new dataset, each instance of the datasets is handed down to each of the N trees. The algorithm selects a class having the most out of N votes, for that case [2]. The model was developed using python and python libraries. The data was split into training and testing datasets with 80% for the training set and 20% for the testing set. We setup the random forest classifier and set the n_estimators to 200. n_estimators are the number of trees you want to build before taking the maximum voting or averages of predictions [26]. This is one of the main parameters that can be used to tune the model performance. Generally, a higher n_estimator value may lead to better results but will make the code run slower and we used this feature to tune the model until we could not get any further improvement in the performance.

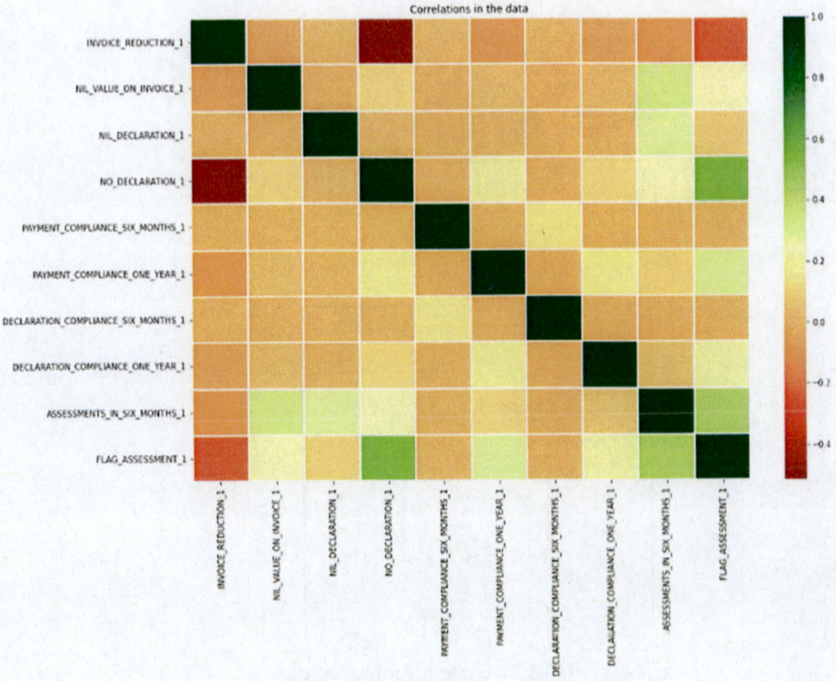

Fig. 2. Chart illustrating the feature correlations.

3.2 Web Application Design and Implementation

The web application was developed with two sub functions both of which use the machine learning prediction model to make predictions based on the input. The first was the web UI application, which has a graphical user interface to allow users to interact with the system. The second sub function is an API web service for integration with other already existing systems to allow them to use the prediction engine as an extension of their existing processes. Both the web application and the web service applications were developed using python language and running on flask framework as the web application host. The illustration below depicts the system setup (Fig. 3).

The system was designed and developed to use MySQL database for the relational database management system. The user management data and transactional data were stored on the database. As taxpayers make their declarations on the main tax administration system, the data is replicated on the database we implemented. We developed a database procedure and scheduled job to run at a specified time of the day to collect the required transaction data relating to the tax declarations and prepares it for feeding into the prediction engine. On user request, the data is pushed to the model to make predictions.

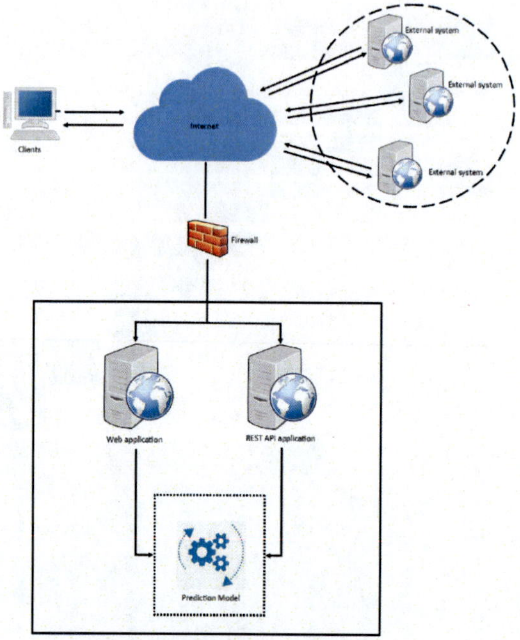

Fig. 3. System Architecture.

4 Results

The study collected 200,001 records from taxpayer tax declarations which was used for developing the supervised learning prediction model using the random forest machine learning algorithm. The model was tuned to get the most optimal results and we used a number of evaluations methods and techniques to validate the performance of the model. The following are the methods that were used:

4.1 Random Forest Score Classifier

The random forest score method measures the model's performance from an accuracy point of view by measuring how many labels the classifier got right during the prediction process. This gives a fairly good first indication of how our model performed. We evaluated our model using the score method of the model and it gave us a score of 0.839525 which is about 83% accuracy.

4.2 Confusion Matrix

We also used the confusion matrix for analysis. The confusion matrix is a common and popular method used to evaluate a classification problem. The illustration below shows our results and showed us statistics such as where the model predicted incorrectly and where it predicted correctly (Fig. 4).

The following table shows the detailed results of the confusion matrix evaluation (Table 3):

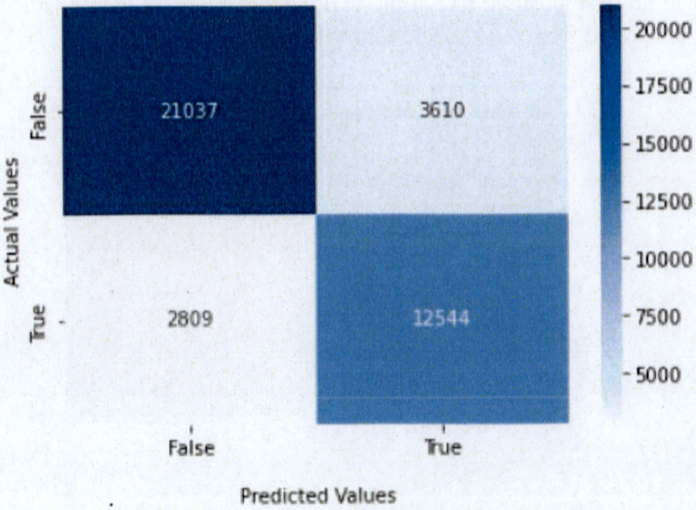

Fig. 4. Confusion Matrix.

Table 3. Table detailed results of the confusion matrix.

Clasification	Description	Calculation	Rate
Accuracy	Shows us overall, how often is the classifier correct	(TP+TN)/total = (12439+ 21041)/40000	**0.83**
Precision	Shows us how often it is correct when it predicts yes	TP/predicted yes = 12439/16187	**0.77**
Sensitivity	Shows us how often it predicts yes when it is actually yes	TP/actual yes = 12439/15211	**0.82**
False Positive Rate	Shows us how often it predicts yes when it is actually no	FP/actual no = 3748/24789	**0.15**
Specificity	Shows us how often it predicts no when it is actually no	TN/actual no = 21041/24789	**0.84**
Error Rate	Shows us overall, how often is it wrong	(FP+FN)/total = (3748+2772)/ 40000	**0.16**

4.3 ROC Curve

The Receiver operating characteristic Curve is another tool we used to measure the performance of our model. The ROC curve measures the performance of a classification model at a number of threshold settings. It is therefore plots the true positive rate or recall against the false positive rate at the various thresholds of the classification. The ROC curve generated from our model is show below: (Fig. 5)

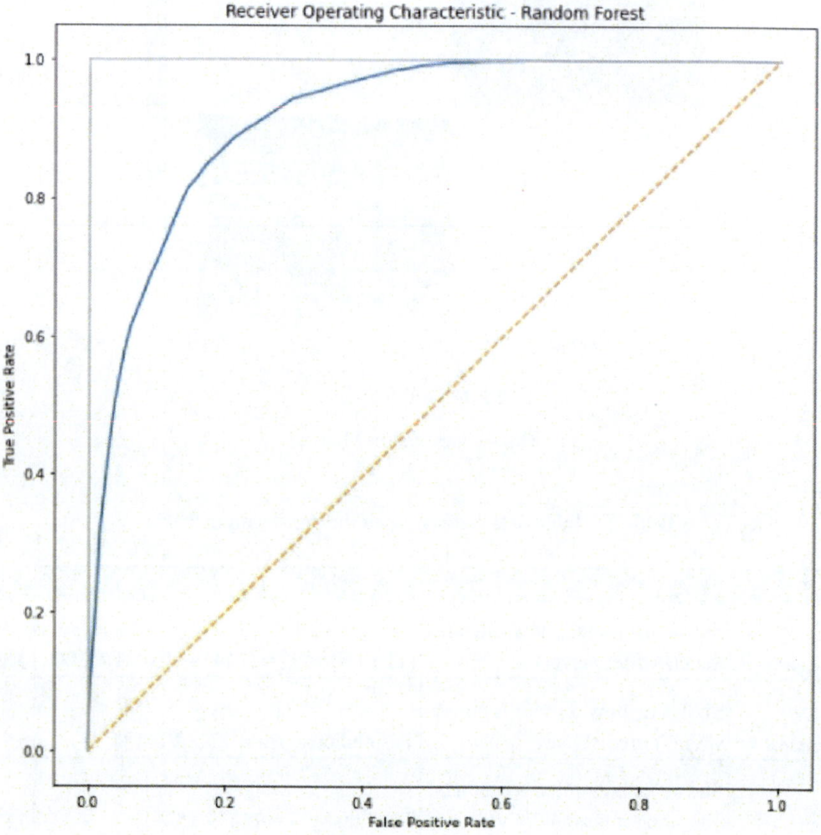

Fig. 5. ROC Curve

The AUC shows us how the model is able to distinguish between different classes. We aimed to get a high AUC score from our model evaluation. Our model scored 0.9166820855361439 on the AUC score.

5 Discussion and Conclusion

We developed and implanted a web based tax assessment prediction system using data mining and machine learning. The system comprises of two sub modules, a web based application with a graphical user interface and an API for integration with already existing systems. We developed and prediction model based on data mining and machine learning as a prediction engine for the two modules. We used the random forest algorithm to develop the model using training and test data extracted from the tax administration system and was split by 20% to 80%. We evaluated our model using the random forest classifier, confusion matrix, receiver operator characteristic (ROC) curve and the Area under curve (AUC). From the evaluation results, we were able to get an accuracy of 83%, precision of 77%, sensitivity of 81%, specificity of 85%, false positive rate of 14% and an error rate of 16%. We were also able to get an AUC score of 0.9166820855361439, which indicated to us that the model was able to reasonably distinguish between two classes of assessments with a significant tax amount and the others.

We also found that there was a strong relationship between no declaration transactions and the cases that were flagged for assessment, which could indicate that taxpayers who do not file their declarations regularly are often audited and assessed regularly. We were also able to see a similar relationship for taxpayers who had undergone previous assessment tend to be audited and assessed again. This could be an indication taxpayers who are found to have under-declared through an audit and assessment are usually repeat offenders. This information could help the revenue authority to investigate these trends further and put in place policies and strategies to improve compliance among taxpayers.

References

1. Saez, E.: Reported incomes and marginal tax rates, 1960–2000: evidence and policy implications (2004). Accessed 06 Feb 2022
2. Auriol, E., Warlters, M.: Taxation base in developing countries. J. Public Econ. **89**(4), 625–646 (2005). https://doi.org/10.1016/j.jpubeco.2004.04.008
3. Junquera-Varela, R., Haven, B.: Getting to 15 percent: addressing the largest tax gaps. https://blogs.worldbank.org/governance/getting-15-percent-addressing-largest-tax-gaps. Accessed 15 Sept 2022
4. Nhekairo, W.: The taxation system in Zambia. https://www.taxjustice-and-poverty.org/fileadmin/Dateien/Taxjustice_and_Poverty/Zambia/JCTR/JCTR_2014_taxstudy.pdf. Accessed 29 Jan 2022
5. Kaliba, C., Muya, M., Mumba, K.: Cost escalation and schedule delays in road construction projects in Zambia. Int. J. Proj. Manag. **27**(5), 522–531 (2009). https://doi.org/10.1016/j.ijproman.2008.07.003
6. Chang, O.H., Nichols, D.R., Schultz, J.J.: Taxpayer attitudes toward tax audit risk. J. Econ. Psychol. **8**(3), 299–309 (1987). https://doi.org/10.1016/0167-4870(87)90025-0
7. Scotchmer, S., Slemrod, J.: Randomness in tax enforcement. J. Public Econ. **38**(1), 17–32 (1989). https://doi.org/10.1016/0047-2727(89)90009-1
8. General Tax Information – Zambia Revenue Authority. https://www.zra.org.zm/tax-information-details/. Accessed 30 Jan 2022
9. Phiri, F., Ndlovu, N.: Tax Compliance - SA Institute of Taxation. https://www.thesait.org.za/news/524096/Tax-Compliance.htm. Accessed 20 May 2022
10. James, S., Alley, C.: Tax compliance, self-assessment and tax administration, University Library of Munich, Germany, 26906 (2002). https://ideas.repec.org/p/pra/mprapa/26906.html. Accessed 06 Feb 2022
11. Wallschutzky, I.G.: Possible causes of tax evasion. J. Econ. Psychol. **5**(4), 371–384 (1984). https://doi.org/10.1016/0167-4870(84)90034-5
12. Bruno, R.L.: Tax enforcement, tax compliance and tax morale in transition economies: a theoretical model, p. 39
13. Hemberg, E., Rosen, J., Warner, G., Wijesinghe, S., O'Reilly, U.-M.: Tax non-compliance detection using co-evolution of tax evasion risk and audit likelihood. In: Proceedings of the 15th International Conference on Artificial Intelligence and Law, San Diego California, June 2015, pp. 79–88 (2015). https://doi.org/10.1145/2746090.2746099

14. Höglund, H.: Tax payment default prediction using genetic algorithm-based variable selection. Expert Syst. Appl. **88**, 368–375 (2017). https://doi.org/10.1016/j.eswa.2017.07.027
15. Cui, Z.: 'China's export tax rebate policy. China Int. J. **1**(2), 339–349 (2003). https://doi.org/10.1353/chn.2005.0035
16. Andini, M., Ciani, E., de Blasio, G., D'Ignazio, A., Salvestrini, V.: Targeting with machine learning: an application to a tax rebate program in Italy. J. Econ. Behav. Organ. **156**, 86–102 (2018). https://doi.org/10.1016/j.jebo.2018.09.010
17. Sabbeh, S.F.: Machine-learning techniques for customer retention: a comparative study. ijacsa **9**(2) (2018). https://doi.org/10.14569/IJACSA.2018.090238
18. Yilmazer, S., Kocaman, S.: A mass appraisal assessment study using machine learning based on multiple regression and random forest. Land Use Policy **99**, 104889 (2020). https://doi.org/10.1016/j.landusepol.2020.104889
19. Lismont, J., et al.: Predicting tax avoidance by means of social network analytics. Decis. Support Syst. **108**, 13–24 (2018). https://doi.org/10.1016/j.dss.2018.02.001
20. Pal, M.: Random forest classifier for remote sensing classification. Int. J. Remote Sens. **26**(1), 217–222 (2005)
21. Schröer, C., Kruse, F., Gómez, J.M.: A systematic literature review on applying CRISP-DM process model. Procedia Comput. Sci. **181**, 526–534 (2021). https://doi.org/10.1016/j.procs.2021.01.199
22. Moro, S., Laureano, R.M.S., Cortez, P.: Using data mining for bank direct marketing: an application of the CRISP-DM methodology, p. 5
23. Wirth, R., Hipp, J.: CRISP-DM: towards a standard process model for data mining, p. 11
24. Solano, J.A., Lancheros Cuesta, D.J., Umaña Ibáñez, S.F., Coronado-Hernández, J.R.: Predictive models assessment based on CRISP-DM methodology for students performance in Colombia - Saber 11 Test. Procedia Comput. Sci. **198**, 512–517 (2022). https://doi.org/10.1016/j.procs.2021.12.278
25. Classification Algorithms - Random Forest. https://www.tutorialspoint.com/machine_learning_with_python/machine_learning_with_python_classification_algorithms_random_forest.htm. Accessed 10 May 2022
26. Random Forest Parameter Tuning | Tuning Random Forest. Analytics Vidhya, 09 June 2015. https://www.analyticsvidhya.com/blog/2015/06/tuning-random-forest-model/. Accessed 17 May 2022

A Review of Evaluation Metrics in Machine Learning Algorithms

Gireen Naidu(✉), Tranos Zuva, and Elias Mmbongeni Sibanda

Vaal University of Technology, Vanderbijlpark, Gauteng, South Africa
gireen.naidu1@gmail.com, {tranosz,eliass}@vut.co.za

Abstract. With the increase in the adoption rate of machine learning algorithms in multiple sectors, the need for accurate measurement and assessment is imperative, especially when classifiers are applied to real world applications. Determining which are the most appropriate evaluation metrics to effectively assess and evaluate the performance of a binary, multi-class and multi-labelled classifier needs to be further understood. Another significant challenge impacting research is that results from models that are similar in nature cannot be adequately compared if the criteria for the measurement and evaluation of these models are not standardized. This review paper aims at highlighting the various evaluation metrics being applied in research and the non-standardization of evaluation metrics to measure the classification results of the model. Although Accuracy, Precision, Recall and F1-Score are the most applied evaluation metrics, there are certain limitations when considering these metrics in isolation. Other metrics such as ROC\AUC and Kappa statistics have proven to provide additional insightful into the effectiveness of an algorithms adequacy and should also be considered when evaluating the effectiveness of binary, multi-class and multi-labelled classifiers. The adoption of a standardized and consistent evaluation methodology should be explored as an area of future work.

Keywords: Evaluation metrics · machine learning · accuracy · ROC · AUC

1 Introduction

Machine learning has gained significant acceptance in various industries [1]. Machine learning is a branch of data modelling which assists in analytical prediction building. The machine learning model studies from the data, classify general patterns and conceptualize decisions with nominal human interaction. Machine learning is primarily utilized when there is a complicated business problem that requires data analysis to solve. Depending on the computational requirements, machine learning algorithms can be an extremely efficient and effective mechanism for organizations to resolve complex business problems [2]. As highlighted in Fig. 1, machine learning mainly uses two types of learning techniques:

1) Supervised Machine Learning

R. Silhavy and P. Silhavy (Eds.): CSOC 2023, LNNS 724, pp. 15–25, 2023.
https://doi.org/10.1007/978-3-031-35314-7_2

2) Unsupervised Machine Learning

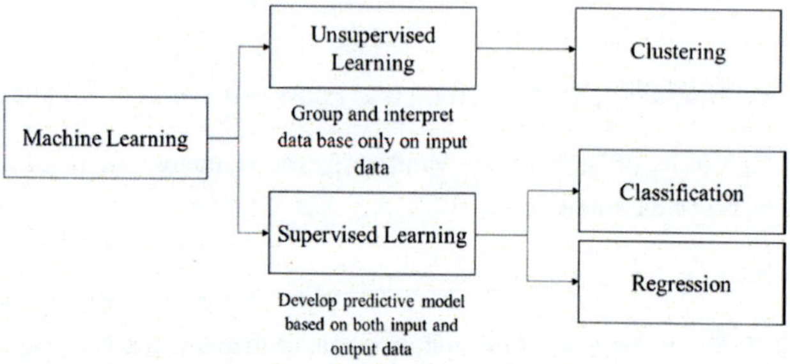

Fig. 1. Machine learning techniques

Supervised machine learning is the computative analytical capability of studying correlations between observations in training datasets and then executing this behavioural pattern in generating a predictive model capable of deducing observations of unseen data [3]. In supervised machine learning, there is an input predictor variable (X) and a target output variable (Y), and we use an algorithm as shown in Eq. 1 to understand the association between the input to the output [4]:

$$Y = f(X) \tag{1}$$

The objective is to effectively calculate the association function that when the new source data (X) is introduced the model predicts the target variable (Y) for that data. The learning is called supervised learning when instances are given with known labels. The features can be continuous, categorical, or binary [5]. Supervised learning problems can be grouped into regression and classification. Popular machine learning algorithms include Logistic Regression (LR), K-Nearest Neighbours (KNN), Decision Trees (DT), Support Vector Machine (SVM), Naïve Bayes (NB), Neural Networks (NN) and XGBoost (XGB).

Evaluation metrics are a set of statistical indicators that will measure and determine the effectiveness and adequacy of the binary, multi-class or multi-labelled classifier in relation to the classification data being modelled. The evaluation metric is used to quantify and aggregate the quality of the trained classifier when validated with the unseen data. The results of these evaluation metrics will determine if the classifier has performed optimally, or further refinement of the classifier is required.

This review paper focused on highlighting the various evaluation metrics being applied in machine learning algorithms. Identified challenges and issues are also discussed. This paper is organized as follows: Related Work, Evaluation Metrics, Results and Discussion, Conclusion and Future Work.

2 Related Work

Many researchers have produced machine learning models in different industries and numerous studies have made comparisons of the classifiers that were applied. Various experiments were conducted with different types of information to achieve the most adequate classification and discriminant analysis techniques for various classification mechanisms to solve relevant business problems. This section will demonstrate the variation of evaluation metrics used in measurement of machine learning algorithms.

A study conducted by [6] utilized Accuracy, Precision, Recall, F-Measure and ROC. Other metrics were also utilized. The logistic regression and logit boost models were developed to measure customer churn. Since most companies cannot provide privileged customer information, the researcher sought data from data repository sites containing a replica of data similar to those of telecoms companies. The logistic regression had an 85.2% accuracy rate while the logit boost had 85.17%.

A study conducted by [7] utilized Accuracy, Error Rate and F-Measure to evaluate their machine learning model. Bayesian boosting was introduced with logistic regression to improve prediction accuracy. It was noted that Bayesian boosting extended the processing time of the prediction algorithm. [8] applied Confusion Matrix and Accuracy metrics to evaluate the accuracy of the prediction model. Backwards regression was applied on a dataset of 22 variables with over 2000 customer observations. Using their experiment resulted in an accuracy rate of 80%. [9] utilized Accuracy as the evaluation criteria. Logistic regression was applied on a publicly available dataset. Feature engineering was executed to select the most important variables contributing to prediction accuracy. The study was able to achieve a 79% accuracy rate.

[10] applied Confusion Matrix, Accuracy, Recall and F1-Score to evaluate the models. Logistic regression, Random Forest Classifier and XGBoost were used to predict customer churn. Data filtering, noise removal and feature selection mechanisms were employed to clean the dataset. The dataset was split into 80% for training and 20% for testing.

Research by [11] only applied Accuracy to evaluate the model. A churn prediction model was built using decision tree classification on a dataset of 33 000 observations that contained customer demographic as well as location and usage patterns. The model utilized the different variations of the decision tree algorithm i.e., Chi-squared Automatic Interaction Detection, Classification and Regression Trees, Quick, Unbiased, Efficient Statistical Tree and Exhaustive CHAID whereby the Exhaustive CHAID technique proved to have the highest level of accuracy.

Research by [12] used Accuracy, Precision, Recall, F-Score to evaluate the models. Since real data from telecoms companies are not readily available, the research they carried out utilized the IBM Watson data set that was released in 2015. Using the publicly available dataset, they identified that the rate of churners was 26% based on the data available. However, since the research utilized three Customer Churn Models, namely K-NN, Random forests and XGBoost, the conclusion was that XGBoost performs the best of the three denoting that the Random Forest model is not as accurate and effective in predicting customer churn rates.

In the next section we summarize the various machine learning studies and the respective evaluation metrics that were applied to measure the effectiveness and success in fulfilling its intended application. The various evaluation metrics were used to evaluate the generalization capability of the trained classifier. Accuracy is one of the most common metrics in practice used by many researchers to evaluate the generalization ability of classifiers. Through Accuracy, the trained algorithm is evaluated based on complete correctness which refers to the total of instances that are positively predicted as true by the trained algorithm when tested with the new data [13]. Table 1 demonstrates the various evaluation metrics being applied in research:

Table 1. Evaluation Metrics comparison in studies reviewed

Reference	*Year*	*Algorithms*	*Accuracy*	*Precision*	*Recall*	*F1 Score*	*AUC \ ROC*	*Confustion Matrix*
[14]	2022	DT, NB, LR, NN, SVM	**Yes**	**Yes**	**Yes**	**Yes**	No	No
[15]	2018	NN	**Yes**	**Yes**	**Yes**	No	**Yes**	**Yes**
[16]	2019	XGB, RF, DT	No	No	No	No	**Yes**	No
[12]	2019	KNN, RF, XGB	**Yes**	No	No	Yes	No	No
[17]	2019	LR, RF, Perceptron,NB, DT	No	**Yes**	**Yes**	**Yes**	No	**Yes**
[18]	2019	NN	**Yes**	No	No	No	No	No
[6]	2020	LR and Logit Boost	**Yes**	**Yes**	**Yes**	**Yes**	**Yes**	**Yes**
[19]	2020	LR, KNN	**Yes**	**Yes**	**Yes**	**Yes**	No	No
[7]	2020	LR	**Yes**	No	No	**Yes**	No	No
[10]	2020	RF, LR, XGB	No	**Yes**	**Yes**	**Yes**	No	No
[20]	2020	LR, RF, NN	No	No	No	No	No	**Yes**
[9]	2021	XGB Classifier & LR	**Yes**	No	No	No	No	No
[21]	2021	LR, RF, KNN	**Yes**	**Yes**	**Yes**	**Yes**	No	No
[22]	2021	Gradient Boosting, RF, LR, XGB , DT, KNN	**Yes**	No	No	No	No	No
[23]	2021	LR, DT, SVM, Logit Boost, RF	**Yes**	**Yes**	**Yes**	**Yes**	No	No
[24]	2021	XGB , LR, DT, and NB	**Yes**	**Yes**	**Yes**	**Yes**	No	No
[25]	2019	LR, NB, KNN, RF, MLP	**Yes**	No	No	**Yes**	No	No
[26]	2019	RF, DT	No	No	No	No	**Yes**	No

From the information depicted in Table 1, it is evident that various evaluation metrics are being applied in machine learning models. Accuracy (25%) is the preferred evaluation metric, followed by Recall (20%), F1-Score (20%) and Precision (18%).

3 Evaluation Metrics

In general, the evaluation metric can be described as the measurement tool that measures the performance of classification and discriminant analysis techniques for binary and multi-class classification. Different metrics evaluate various features of the classifier induced by the classification algorithm. Various evaluation metrics were used to measure the accuracy and effectiveness of the prediction model. Below is a summary of the evaluation metrics that can be applied to machine learning classifiers.

Confusion Matrix
The confusion matrix is one of the easiest and most intuitive metrics used for finding the precision of a classification model [27].

There are four categories in 2 × 2 confusion matrix:

Predicted Values \ Actual Values	Positive (1)	Negative (0)
Positive (1)	***True Positive (TP)***	***False Positive (FP)***
Negative (0)	***False Negative (FN)***	***True Negative (TN)***

Precision (P)
Precision indicates the ratio of correct predictions, where the model predicts true positives. It is computed as true positives divided by sum of true positives and false positives as shown in Eq. 2:

$$\mathrm{P} = \frac{TP}{TP + FP} \tag{2}$$

Sensitivity (S)
Sensitivity measures the model's ability to predict true positives of each available category and how effective the model was able to predict them. Formally it is calculated as the ratio of true positives and the sum of true positives and false negatives as shown in Eq. 3:

$$\mathrm{S} = \frac{TP}{TP + FN} \tag{3}$$

Accuracy (A)
Accuracy is defined as the percentage of correct predictions for the test data. Formally it is calculated as the ratio of true positives and true negatives and sum of the total number of predictions as shown in Eq. 4:

$$\mathrm{A} = \frac{(TP + TN)}{(TP + FP + TN + FN)} \tag{4}$$

F1 Score
F1 Score is a function of Precision and Recall and is useful when measuring the effectiveness of unbalanced datasets [28]. The formulae for F1 Score can be seen in Eq. 5:

$$\text{F1 Score} = \frac{(P * R)}{(P + R)} \tag{5}$$

Specificity (SP)

Specificity itself can be described as the algorithm/model's ability to predict a true negative of each category available. In literature, it is also known simply as the true negative rate. The formulae for SP can be seen in Eq. 6:

$$\mathrm{SP} = \frac{TN}{TN + FP} \tag{6}$$

Error Rate (ER)
The Error Rate measures the ratio of incorrect predictions over the total number of instances evaluated. The formulae for ER can be seen in Eq. 7:

$$\mathrm{ER} = \frac{(FP + FN)}{(TP + FP + TN + FN)} \tag{7}$$

Geometric-Mean (GM)
This metric is used to obtain the maximum tp rate and tn rate, and simultaneously keeping both rates fairly balanced. The formulae for GM can be seen in Eq. 8:

$$\mathrm{GM} = \sqrt{TP * TN} \tag{8}$$

Mean Square Error (MSE)
MSE is formulated by applying the mean of the square of the difference between the first set of inputted data and predicted observations of the dataset. The formulae for MSE can be seen in Eq. 9:

$$MSE = \frac{1}{n}\sum\nolimits_{i=1}^{\mathrm{n}}\left(Y_i - \tilde{Y}_i\right)^2 \tag{9}$$

Here n is the number of records in the dataset. The sigma symbol on the Y co-ordinates denotes that the difference between actual and predicted observations taken on every i value ranging from 1 to n.

Kappa Statistic (K)
Essentially, the kappa statistic is a measurement of how directly the observations classified by the machine learning algorithm compared against the data labelled as true positive, influencing the accuracy of a machine learning classifier as evaluated by the anticipated accuracy. The formulae for K can be seen in Eq. 10:

$$\mathrm{K} = \frac{(P_0 + P_e)}{(1 - P_e)} \tag{10}$$

where P_0 is the observed outcome class, and P_e is the expected outcome class. It effectively implies how effective the model is performing over the outcome of a model that just randomly guesses according to the number of observations of each class.

Receiver Operating Characteristic Curve (ROC) and Area Under Curve (AUC)
Receiver operating characteristic (ROC) curve is a graph indicating a classification model's performance at all thresholds of classification. Two variables are plotted along the ROC curve:

$$\text{True positive rate (TPR)} = \frac{TP}{TP + FN} \tag{11}$$

$$\text{False positive rate (FPR)} = \frac{FP}{FP + TN} \tag{12}$$

TPR and FPR are plotted at different thresholds of classification. When the classification threshold is increased, more objects are classified as positive which results in higher true positives and false positives [29]. Area under the ROC curve (AUC) is the measurement of the area below the ROC curve. AUC-ROC enables the visualization of a machine learning classifier's performance. The ROC curve essentially removes the noise from the signal. The AUC curve, on the other hand, summarizes the ROC curve and indicates the classifier's ability to distinguish distinct classes. The performance of the machine learning model is considered performing well in discriminating between positive classes and negative classes as the AUC increases.

A classifier is said to be capable of correctly distinguishing between entire positive and negative classes when the value of AUC is 1. The classifier, however, will tend to read all positives as negatives and all negatives as positives. If AUC is higher than 0.5 but lower than 1, the classifier has a tendency to identify positive from negative classes since it is capable of identifying true positives and true negatives. The classifier cannot ascertain between negative and positive classes if the value of AUC is 0.5 [13]. The figures below highlight the ROC\AUC graphically (Figs. 2 and 3):

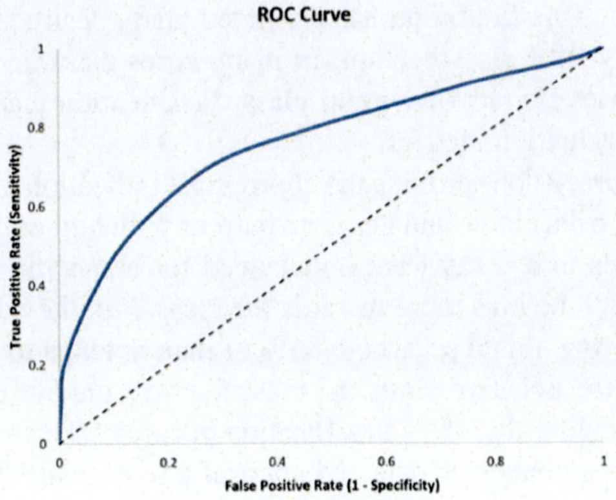

Fig. 2. ROC Curve

Applications of ROC analysis includes model comparison and evaluation, model selection, model presentation, model construction, and model combination. As measurement of classification performance, AUC-ROC has been found to possess several advantages such as: high analysis of variance sensitivity, independence from selected threshold, indicates positive and negative class separation, and unaffected by previous class probabilities [30].

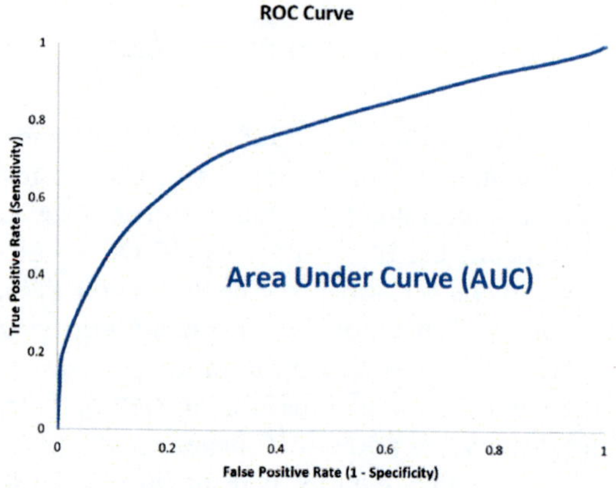

Fig. 3. ROC Curve showing AUC

4 Results and Discussion

Most of the metrics being reviewed were originally established and primarily applicable for binary classification. With the increasing complexity of non-binary classification models, the current evaluation metrics are limited when it comes to multi-class and multi-labelled classification. This limitation has restricted many desired metrics for scalable usability in varying type of classification. In many cases classifiers are not limited to two-classes. Many cases also involve multi-class classification that possibly requires a more flexible set of evaluation metrics.

Even though Accuracy does provide the appropriate indication of prediction reliability, it sometimes can be deceptive and generate outputs with higher than actual accuracy. When the observations of a dataset are imbalanced for a specific variable, the resulting accuracy scores will be bias more towards the class with the maximum imbalance. For example, considering a data set where 85% of data belongs to a positive class and only 15% represents the negative class, the classifier will predict most of the samples in the positive class giving the 85% classification accuracy. However, practically this value is misleading for imbalanced data and not useful to evaluate the performance of a prediction model.

The research indicated that even though Accuracy was utilized for majority of the studies as the statistical metric for the evaluation, Accuracy must not be utilized in isolation for the evaluation of model effectiveness.

Furthermore, the F1 Score metric applies equal importance to Precision and Recall. It is not always appropriate to equate a false positive to a false negative. In different real-life situations, the cost of false positives and false negatives are different. For example, it is much worse to classify a sick person and healthy (FN), than classify a healthy person as sick (FP). Recall percentage can prove useful in these situations as it will measure the high number of FNs.

5 Conclusion

With the increase in the adoption rate of machine learning algorithms in multiple sectors including healthcare, financial, transportation, retail, manufacturing, etc., emphasis has been placed on classification and discriminant analysis techniques for binary and multi-class classification to solve relevant business problems.

Present challenges with classification or regression supervised learning algorithms are determining which are the most appropriate evaluation metrics to effectively assess and evaluate the performance of the algorithm. Furthermore, the need for proper benchmarking and baselining across the various algorithms has become of vital importance. The effectiveness of various algorithms cannot be adequately compared if varying evaluation metrics are used.

Another challenge is what techniques can be adopted to improve the effectiveness of these models.

From the various articles reviewed, it was evident that numerous evaluation metrics are being applied across the various machine learning algorithms. Accuracy, Precision, Recall and F1 Score are the most applied evaluation metrics however there are limitations when considering these metrics in isolation. However, other metrics such as ROC\AUC have proven to be more insightful into the effectiveness of an algorithm adequacy and should be also considered when evaluating the effectiveness of machine learning algorithms.

The adoption of a standardized and consistent evaluation methodology should be explored as an area of future work. This methodology should also define the thresholds that classify a machine learning algorithm as effective or not. Additionally, a systematic methodology for model evaluation will ensure that existing and future models can be adequately benchmarked. This methodology could even be isolated to the type of machine learning application such as natural language processing, information retrieval, image recognition, etc.

References

1. Wehle, H.: Machine learning, deep learning, and AI: what's the difference. TY - BOOK AU - Wehle, Hans-Dieter PY, 2017/07/24 SP - T1, ER (2017)
2. Sayed, H., Abdel-Fattah, M.A., Kholief, S.: Predicting potential banking customer churn using apache spark ML and MLlib packages: a comparative study. Int. J. Adv. Comput. Sci. Appl. **9**, 674–677 (2018). https://doi.org/10.14569/ijacsa.2018.091196
3. Fabris, F., de Magalhães, J.P., Freitas, A.A.: A review of supervised machine learning applied to ageing research. Biogerontology **18**(2), 171–188 (2017). https://doi.org/10.1007/s10522-017-9683-y
4. Mahbobi, Tiemann: Regression Basics 2015. https://opentextbc.ca/introductorybusinessstatistics/chapter/regression-basics-2/. Accessed 6 June 2022
5. Kotsiantis, S., Kanellopoulos, D., Pintelas, P.: Handling imbalanced datasets: a review. Science **2006**(30), 25–36 (1979)
6. Jain, H., Khunteta, A., Srivastava, S., Jain, H., Khunteta, A., Srivastava, S.: Churn prediction in telecommunication using logistic regression and logit boost. Procedia Comput. Sci. **167**, 101–112 (2020). https://doi.org/10.1016/j.procs.2020.03.187

7. Arivazhagan, B., Sankara Subramanian, D.R.S., Scholar, R.: Customer churn prediction model using regression with Bayesian boosting technique in data mining. In: IjaemaCom 2020, vol. XII, pp. 1096–104 (2020)
8. Sebastian, H.T., Wagh, R.: Churn analysis in telecommunication using logistic regression. Orient. J. Comput. Sci. Technol. **10**, 207–212 (2017)
9. Parmar, P.: Telecom churn prediction model using XgBoost classifier and logistic regression algorithm. Int. Res. J. Eng. Technol. (IRJET) **08**, 1100–1105 (2021)
10. Kavitha, V., Kumar, H., Kumar, M., Harish, M.: Churn prediction of customer in telecom industry using machine learning algorithms. Int. J. Eng. Res. Technol. (IJERT) **9**, 181–184 (2020). https://doi.org/10.17577/ijertv9is050022
11. Nisha, S., Garg, K.: Churn prediction in telecommunication industry using decision tree. Int. J. Eng. Res. **6**, 439–443 (2017). https://doi.org/10.17577/ijertv6is040379
12. Pamina, J., Beschi Raja, J., Sathya Bama, S., Soundarya, S., Sruthi, M.S., Kiruthika, S., et al.: An effective classifier for predicting churn in telecommunication. J. Adv. Rese. Dyn. Control Syst. **11**, 221–229 (2019)
13. Hossin, M., Sulaiman, M.N.: A review on evaluation metrics for data classification evaluations. Int. J. Data Min. Knowl. Manag. Process **5**, 01–11 (2015). https://doi.org/10.5121/ijdkp.2015.5201
14. Lalwani, P., Mishra, M.K., Chadha, J.S., Sethi, P.: Customer churn prediction system: a machine learning approach. Computing **104**, 271–294 (2022). https://doi.org/10.1007/s00607-021-00908-y
15. Karanovic, M., Popovac, M., Sladojevic, S., Arsenovic, M., Stefanovic, D.: Telecommunication services churn prediction - deep learning approach. In: 2018 26th Telecommunications Forum, TELFOR 2018 - Proceedings, Institute of Electrical and Electronics Engineers Inc. (2018). https://doi.org/10.1109/TELFOR.2018.8612067
16. Ahmad, A.K., Jafar, A., Aljoumaa, K.: Customer churn prediction in telecom using machine learning in big data platform. J. Big Data **6**(1), 1–24 (2019). https://doi.org/10.1186/s40537-019-0191-6
17. Ullah, I., Raza, B., Malik, A.K., Imran, M., Islam, S.U., Kim, S.W.: A churn prediction model using random forest: analysis of machine learning techniques for churn prediction and factor identification in telecom sector. IEEE Access **7**, 60134–60149 (2019). https://doi.org/10.1109/ACCESS.2019.2914999
18. Cao, S., Liu, W., Chen, Y., Zhu, X.: Deep learning based customer churn analysis. n.d
19. Joolfoo, M., Jugurnauth, R., Joofloo, K.: Customer churn prediction in telecom using big data analytics. IOP Conf. Ser. Mater. Sci. Eng. **768** (2020). https://doi.org/10.1088/1757-899X/768/5/052070
20. Kavita, M., Sharma, N., Aggarwal, G.: Churn prediction of customer in telecommunications and e-commerce industry using machine learning. Palarch's J. Archaeol. Egypt Egyptol. **17**, 6–15 (2020)
21. Senthilnayaki, B.: Customer churn prediction. Iarjset **8**, 527–531 (2021). https://doi.org/10.17148/iarjset.2021.8692
22. Singh, D., Jatana, V., Kanchana, M.: Survey paper on churn prediction on telecom. SSRN Electron. J. **27**, 395–403 (2021). https://doi.org/10.2139/ssrn.3849664
23. Jain, H., Khunteta, A., Srivastava, S.: Telecom churn prediction using seven machine learning experiments integrating features engineering and normalisation. Comput. Sci. Sch. Basic Appl. Sci. (2021). Poornima University
24. Xu, T., Ma, Y., Kim, K.: Telecom churn prediction system based on ensemble learning using feature grouping. Appl. Sci. **11** (2021). https://doi.org/10.3390/app11114742
25. Baldominos, A., Cervantes, A., Saez, Y., Isasi, P.: A comparison of machine learning and deep learning techniques for activity recognition using mobile devices. Sensors **19** (2019). https://doi.org/10.3390/s19030521

26. Almaguer-Angeles, F., Murphy, J., Murphy, L., Portillo-Dominguez, A.O.: Choosing machine learning algorithms for anomaly detection in smart building IoT scenarios. In: IEEE 5th World Forum on Internet of Things, WF-IoT 2019 - Conference Proceedings, Institute of Electrical and Electronics Engineers Inc., pp. 491–495 (2019). https://doi.org/10.1109/WF-IoT.2019.8767357
27. Lantz, B.: Machine Learning with R: Expert Techniques for Predictive Modeling. Packt Publishing Ltd. (2019)
28. Vafeiadis, T., Diamantaras, K.I., Sarigiannidis, G., Chatzisavvas, K.C.: A comparison of machine learning techniques for customer churn prediction. Simul. Model. Pract. Theory **55**, 1–9 (2015). https://doi.org/10.1016/j.simpat.2015.03.003
29. Vakili, M., Ghamsari, M., Rezaei, M.: Performance analysis and comparison of machine and deep learning algorithms for IoT data classification. n.d
30. Majnik, M., Bosnić, Z.: ROC analysis of classifiers in machine learning: a survey. Intell. Data Anal. **17**, 531–558 (2013). https://doi.org/10.3233/IDA-130592

Hardware Implementation of IoT Enabled Real-Time Air Quality Monitoring for Low- and Middle-Income Countries

Calorine Katushabe[1,2](✉), Santhi Kumaran[3], and Emmanuel Masabo[4]

[1] African Center of Excellence in Internet of Things (ACEIoT), College of Science and Technology (C.S.T.), University of Rwanda, Nyarugenge, P.O. Box 3900, Kigali, Rwanda

[2] Department of Computer Science & Information Technology, Faculty of Computing, Library and Information Science, Kabale University, P.O Box 317, Kabale, Uganda
calorinekatushabe@gmail.com,ckatushabe@kab.ac.ug

[3] School of ICT, Copperbelt University, P.O Box: 21692, Kitwe, Zambia

[4] African Center of Excellence in Data Science (ACEDS), College of Business and Economics (CBE), University of Rwanda, Kigali, Rwanda

Abstract. The environment and human health are both impacted by air quality. In Africa, where air quality monitoring systems are rare or nonexistent, poor air quality has caused far more deaths and environmental damage than anywhere else in the world. Air pollution in Africa is a result of the continent's growing urbanization, industrialization, road traffic, and air travel. Particularly in Africa, air pollution continues to be a silent killer, and if it is not addressed, it will continue to cause fatal health disorders like heart disease, stroke, and chronic respiratory organ disease. In this study, the potential of IoT is greatly exploited to measure air pollution levels in real time. An Arduino Uno microcontroller board based on the ATmega328P integrated with the Arduino Integrated Development Environment is used to build the prototype. The designed prototype consists of the different sensors that capture air pollutant concentration levels from the environment. All the data pertaining to air quality are monitored in real-time using Thingspeak, an IoT-based platform. The monitoring results are visible through the mobile application developed; as a result, this creates awareness to the public and the concerned policy makers can make well informed decisions.

Keywords: Air Pollution · Internet of Things · Air Quality Monitoring · ThingSpeak · Low- and middle-income countries

1 Introduction

As per the State of Global Air 2021 report, air pollution is a major global risk factor for diseases. It is reported that air pollution claims over 6.5 million

R. Silhavy and P. Silhavy (Eds.): CSOC 2023, LNNS 724, pp. 26–36, 2023.
https://doi.org/10.1007/978-3-031-35314-7_3

people annually [1]. The effects of polluted air are felt both globally and locally; if ignored, they pose big a threat to all living creatures, including humans. According to research published by the WHO, about 90% of the world's population still lack access to environments that meet good air quality criteria particularly in the low- and middle-income countries [2].

Although most developed countries have attempted to develop strategies and methods to improve air quality using rapidly developing technologies and, in the long run, mitigate the problem of air pollution, most low- and middle-income countries still lag behind on this matter and lack air pollution monitoring systems or management strategies in place despite the widespread understanding of the negative effects of air pollution on productivity and human health [3–5].

In fact, in more than 100 countries in the low- and middle-income countries in the world do not even have air quality monitoring systems, and in many cases, a few that have air quality monitoring programs, data generated is not made available to the public [6,7].

Health issues are growing at quicker rate particularly in urban areas of developing countries where industry and growing range of vehicles ends up in unharness of ton of various pollutants. In rural areas, people are just unaware of the terrible hazard posed by kerosene, which is used to cook food and light homes across the continent [8]. Harmful effects of pollution embody delicate sensitivity like irritation of the throat, eyes and nose and more serious issues like respiratory illness, heart diseases, pneumonia, respiratory organ and aggravated bronchial asthma [9].

Various research conducted in a number of cities across the globe indicate that when air pollution goes high, also a number of lives being lost increase [10].

There is an urgent need to address air pollution issues in low- and middle-income countries, particularly in Africa, where urbanization and industrialization are increasing along with population density otherwise people in these regions have a higher risk of developing lung cancer, heart disease, and stroke as a result of air pollution.

This study focuses on the urban cities of Uganda, where large number of used cars and motorbikes, which are the main forms of transportation, enter these cities at rising rates each year, along with the opening of new industries in every part of these urban centers. Health is greatly impacted by air pollution, which alone results in millions of hospitalizations each year.

Some of the major sources of air pollution in these regions include dust from unpaved roads and open burning of garbage by individuals as a method of managing uncollected waste. More so, open burning of waste releases dangerously high levels of air pollution into the air via combustion. Additionally, there is a great deal of polluted air originating from numerous factories and power plants in these cities. Another great significant source of air pollution are the vehicle emissions which are produced by numerous imported used cars [11].

1.1 Lack of Air Quality Monitory Systems in Africa

Although air pollution is a serious silent killer in Africa, the scale of this problem is little known because of the absence of air quality monitoring systems on the on the ground in several African countries. In low-and middle-income countries dirt air is an under-recognized threat to children's health [12,13].

During the WHOs Public Health, Environment and Social Determinants of Health department presentation at a conference sponsored by the Novartis Foundation on air pollution estimates around the world in 2019, the figures concerning Africa were not indicated rather there was "NA" which showed that little on air pollution was known on the entire continent though air pollution is one of the greatest threats of any urbanization [14,15]. It is noted that air pollution remains a big silent killer of about 712,000 infants annually more than the toll of unsafe water, malnutrition and poor sanitation; this costs the continent over $200 billion economically.

Therefore, it's time for Africa to take advantage for new technologies and monitor air quality because monitoring is key to document impacts and progress of cubing the dirty air problem; this creates awareness to the public and the concerned policy makers can make well informed decisions.

Therefore, in this work a hardware implementation of Internet of Things (IoT) based real-time air quality monitoring is designed. IoT is a revolutionary innovation in technology that can enable air quality monitoring systems. By integrating intelligent sensing technologies like sensors into a "physical entity," the IoT enables the linking of a network of items to obtain information about the purpose at any time.

Traditional methods for estimating air quality such as variance analysis, clustering analysis and regression analysis and, have been used; however, these methods have limitations due to the non-linear relationship between pollutant datasets [16]. More so, the IoT based air quality monitoring systems that are already in place, seem to be established in the already developed countries. Although some of the big cities in Africa have invested in monitoring networks that measure air pollution levels, there are still thousands of cities in the continent with limited or no air quality measurements at all; even those few cases where air quality is measured, the information may not be shared widely with the public, policy and advisory boards responsible may not be notified.

According to WHO Global Burden of Disease report 2020 in Fig. 1 it shows that low-and middle-income countries more exposed to dirt air particularly Sub-saharan Africa with more than 80
Therefore, its high time air quality monitoring techniques and strategies be transferred to the low- and middle-income countries to assist and avoid fatal repercussions of air pollution.

Air quality monitoring is an essential IoT application that monitors the air pollutant concentrations levels. Therefore this study shows the realization of IoT utilization in air quality monitoring.

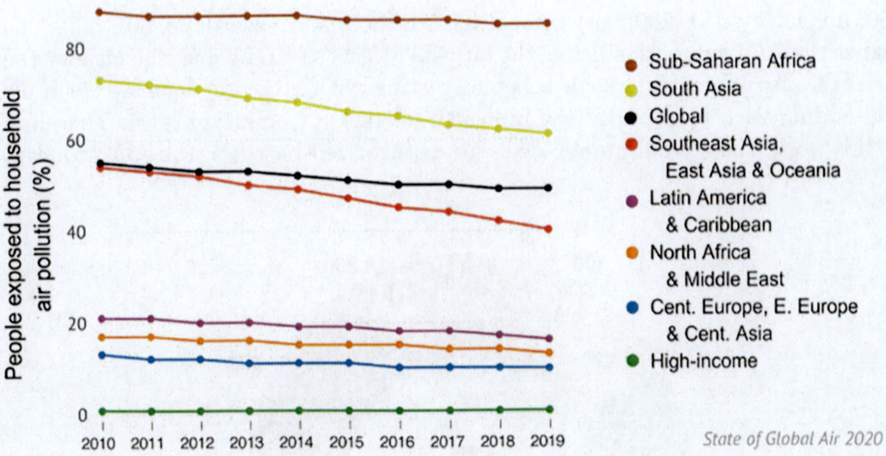

Fig. 1. Number of people exposed to air pollution per region in percentages [17]

1.2 Air Quality Index in Air Quality Monitoring

The Air Quality Index (AQI) is typically used to present information on air quality concentrations in a specific region and their health effects. AQI enables Air quality levels to be presented to the public in a more understandable manner. It is a method of generalizing how the quality of air is described around the world. It serves as an informational tool for individuals in a given society to better understand the effects of air pollution on both environment and health. It provides information on how clean or dirt the air is as well as any potential health implications. The AQI equation follows the linear interpolation formula; therefore, to measure the air pollutants values based on the AQI, the formula in Eq. 1 is followed.

$$\text{Ip} = \frac{Ihigh - ILow}{BPhigh - BPlow}(Cp - BPlow) + ILow \tag{1}$$

Where,

Ip = the index for pollutant p
Cp = is the monitored concentration of pollutant p
BPHigh = the breakpoint that is greater than or equal to Cp
BPLow = the breakpoint that is less than or equal to Cp
IHigh = the AQI value corresponding to BPHigh
ILow = the AQI value corresponding to BPLow

AQI values used in this work follow Environmental Protection Agency (EPA) standards. This is an international agency that describes guidelines and standards in relation to the air pollution concentration levels and its effect as elaborated in Fig. 2 [18]. AQI values range from 0 to 500 and is categorized into five categories: good, 0–50; moderate, 51–100; unhealthy for sensitive groups, 101–

150; unhealthy, 151–200; very unhealthy, 201–300; and hazardous, 301–500. The higher the AQI value, the dirtier the air; the lower the AQI value, the cleaner the air. AQI also indicates a color scheme starting with green for healthy clean air and ending with maroon for the unhealthiest dirt air pollutant level. Therefore in this work, the air pollutants levels are categorized based on the AQI values.

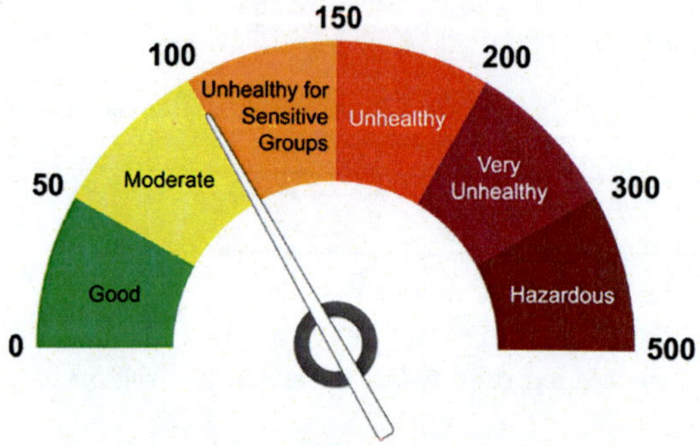

Fig. 2. AQI standards and guidelines

2 Methods and Materials

2.1 Proposed Hardware Model and Work Flow

This section presents an IoT-based architecture concept for hardware implementation of IoT enabled real-time air quality monitoring. The overall architecture is described in Fig. 3. It starts with a sensor node consisting of MQ-135 for CO_2, SO_2 and NO_2, MQ7 for CO and DHT11 composite sensor for both temperature and humidity. These sensors are connected to ATmega328P microcontroller; the Arduino IDE is used to create the logical code to calculate the AQI values from the air pollutant concentrations, which are then uploaded to the microcontroller via the Arduino UNO. The results are then sent to the thingspeak server via a Wi-Fi signal using the NodeMCU module ESP8266 that contains the Wi-Fi module in it. Then, sensor calibrated AQI values from thingspeak server are visualised through a mobile application and more so, to the thingSpeak dashboard inform of graphs upon the user's query from anywhere through the internet. Data displayed via a mobile application and thingSpeak platform is the result of air quality monitoring that has been done in the real-time. The Fig. 4 shows the flowchart on how the system functions flow.

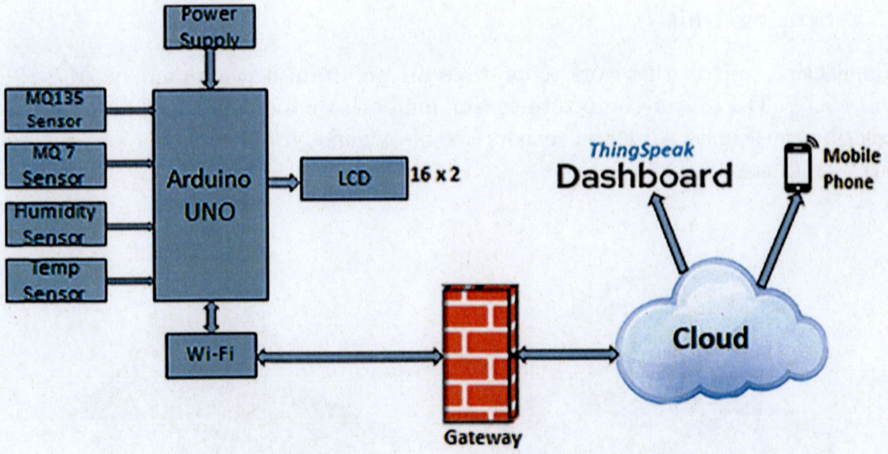

Fig. 3. Block Diagram for the proposed system

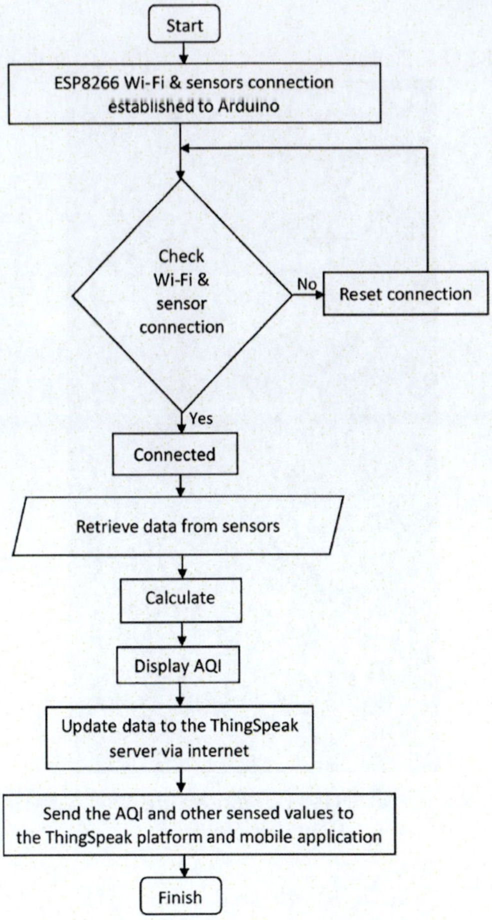

Fig. 4. Data flow on the software Implementation

2.2 Sensing Unit

The sensing unit in this work, comprises air pollutant sensors namely: MQ-135 and MQ 7, DHT11; a composite sensor for both temperature and humidity as described in Fig. 5. All these sensors are of extremely low cost, low power, and ultra-small size.

(a) MQ-135 Gas Sensor

(b) MQ-7 Gas Sensor

(c) DHT11 Temperature & Humidity Sensor

Fig. 5. Sensors used in this work

MQ-135 Gas Sensor
This is a gas sensor that detects gases like smoke, carbon dioxide, aromatic chemicals, ammonia, nitrogen, and oxygen. MQ-135 gas sensor can be implemented to detect different harmful gases. It is affordable and especially suited for applications involving the monitoring of air quality. In this work, we used MQ-135 to measure carbon dioxide concentrations. It measures between 10 and 1000 PPM. The output from this sensor is voltage levels, hence it must be converted to PPM. This is accomplished by writing a proper code. The output it gives is analog, therefore on the ESP8266, it is connected to the analog pin.

MQ-7 Gas Sensor
MQ-7 is high sensitivity to gas to carbon monoxide gas. In this work, it was used to detect concentration levels of carbon monoxide. This sensor can detect concentrations ranging from 10 to 10,000 PPM. It has a stable and long life.

DHT11 Temperature and Humidity Sensor
The DHT11 is a low-cost, basic digital temperature and humidity sensor. It produces digital output. It measures temperature in range of 0–50° C and humidity from 20–90 percent. It measures temperature and humidity with an accuracy of ±1°C and ±1% respectively.

3 Experimental Setup

Using the USB cable, Arduino Uno is connected to the PC to both upload the code to the board and also supply the power. We connect the output of the sensors to the analog input pins of the Arduino Uno board. Also power and ground are connected to the power and ground pins of the board respectively. For the ESP8266 Wi-Fi module, which operates on 3.3 V power instead of 5 V, it is necessary to construct a voltage divider circuit using resistors of certain values to provide 3.3 V across its ends before connecting the Wi-Fi module's Vcc and ground across the ends of the resistor. With this model, the Arduino Uno's receive and transmit pins are connected to the Wi-Fi module's transmitter (Tx) and receiver (Rx), respectively. Fig. 6 shows the hardware setup of various components to measure the value of AQI.

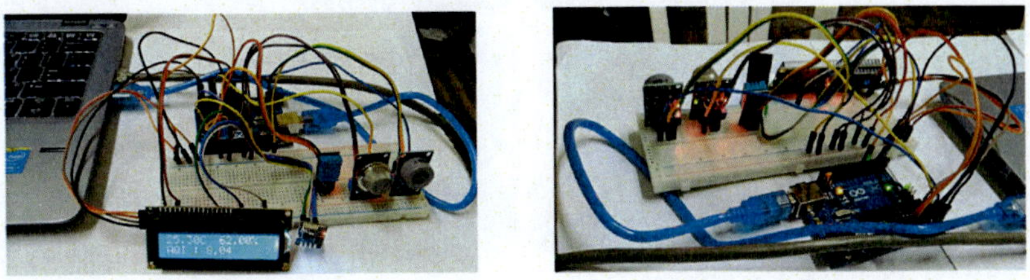

(a) (b)

Fig. 6. Hardware Setup of the componets

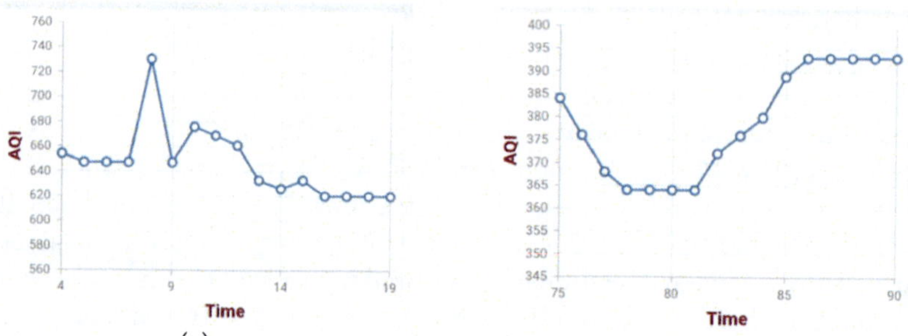

(a) (b)

Fig. 7. AQI values on thingSpeak dashboard

4 Results and Discussion

The designed sensor node can be deployed to monitor both outdoor and indoor air quality levels. Data obtained from the sensor node is archived to the thingSpeak cloud database and the results are displayed to the thingSpeak dashboard and mobile application. Data is received by thingSpeak platform and can be monitored on the channel pages provided. The data is on the thingSpeak platform in form of graphs as in indicated in Fig. 7 The X-axis and Y-axis in the graphs represent time and different parameters respectively. The obtained values of AQI measured in parts per million (ppm)

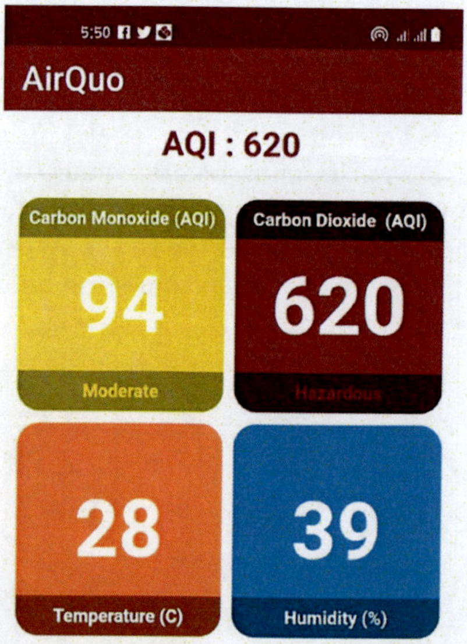

Fig. 8. Visualization of AQI values via mobile application

Furthermore values of AQI can be accessed via a mobile application. This is possible because thingSpeak server provides an API to access data that has been collected and stored on thingSpeak server. The mobile application designed in this work is based on android operating system. It is developed using Kotlin, an open source programming language used in android based mobile application development. Fig. 8 shows a screenshot of the air pollutant concentrations captured on the mobile application. As indicated in the screeshot, AQI values as well as temperature and humidity values are captured. Amongest the air pollutants captured, the highest value is picked in the line with AQI calculation algorithm and displayed as the AQI value on the platform as well as its category.

5 Conclusion

In this work, we present hardware implementation of IoT based real-time air quality Monitoring. It is observed that sensors integrated with IoT systems are such a valuable resource in monitoring air pollutant concentrations in real-time. These sensors are inexpensive, low power consuming and accurate capable of capturing air pollutant concentration levels thus can be utilized by the low- and middle-income countries particularly in Africa to monitor the quality of air. The continuous updating of data on the thingSpeak server enables the users to take well informed and timely decisions whenever is it necessary. Therefore, this system can be deployed in industries, factories, schools, homes, parking lots and other busy areas to monitor the state of air pollution from anywhere by anyone using the mobile application or a computer and in turn air pollution can be curbed to save the entire ecosystem.

References

1. Organization, W.: Others WHO global air quality guidelines: particulate matter (PM2. 5 and PM10), ozone, nitrogen dioxide, sulfur dioxide and carbon monoxide. World Health Organization (2021)
2. Organization, W.: WHO guidelines for indoor air quality: household fuel combustion. World Health Organization (2014)
3. Marécal, V., et al.: Others A regional air quality forecasting system over Europe: the MACC-II daily ensemble production. European Geosciences Union (2015)
4. Suk, W., et al.: Others Environmental pollution: an under-recognized threat to children's health, especially in low-and middle-income countries. Environ. Health Perspect. **124**, A41–A45 (2016)
5. Pinder, R., Klopp, J., Kleiman, G., Hagler, G., Awe, Y., Terry, S.: Opportunities and challenges for filling the air quality data gap in low-and middle-income countries. Atmos. Environ. **215**, 116794 (2019)
6. Gani, S., et al.: Systematizing the approach to air quality measurement and analysis in low and middle income countries. Environ. Res. Lett. **17**(2), 021004 (2022)
7. Amegah, A., Agyei-Mensah, S.: Urban air pollution in Sub-Saharan Africa: time for action. Environ. Pollut. **220**, 738–743 (2017)
8. Organization, W.: Others WHO housing and health guidelines. World Health Organization (2018)

9. Katushabe, C., Kumaran, S., Masabo, E.: Fuzzy based prediction model for air quality monitoring for Kampala city in East Africa. Appl. Syst. Innov. **4**, 44 (2021)
10. IHME state of Global air report in collaboration between the Institute for Health Metrics and Evaluation's Global Burden of Disease Project and the Health Effects Institute. GlobalBurden (2017)
11. Fayiga, A., Ipinmoroti, M., Chirenje, T.: Environmental pollution in Africa. Environ. Dev. Sustain. **20**, 41–73 (2018)
12. Adaji, E., Ekezie, W., Clifford, M., Phalkey, R.: Understanding the effect of indoor air pollution on pneumonia in children under 5 in low-and middle-income countries: a systematic review of evidence. Environ. Sci. Pollut. Res. **26**, 3208–3225 (2019)
13. Perera, F.: Pollution from fossil-fuel combustion is the leading environmental threat to global pediatric health and equity: Solutions exist. Int. J. Environ. Res. Publ. Health. **15**, 16 (2018)
14. Chutel, L.: Scientists aren't sure exactly how bad air pollution is in Africa but think it's worse than we thought. https://qz.com/africa/905656/air-pollution-is-increasing-in-africa-along-with-rapid-urbanization/. Accessed 20 Jul 2022
15. Shi, X.: Environmental health perspectives for low-and middle-income countries. Glob. Health J. **6**, 35–37 (2022)
16. Jovanelly, T., Okot-Okumu, J., Nyenje, R., Namaganda, E.: Comparative assessment of ambient air standards in rural areas to Uganda city centers. J. Publ. Health Dev. Countries **3**, 371–380 (2017)
17. Hei, I.: State of global air 2020: a special report on global exposure to air pollution and its health impacts. Institute For Health Metrics And Evaluation, And Health Effects Institute. https://www.stateofglobalair.org (2020)
18. Plaia, A., Ruggieri, M.: Air quality indices: a review. Rev. Environ. Sci. Bio/Technol. **10**, 165–179 (2011)

Predicting the Specific Student Major Depending on the STEAM Academic Performance Using Back-Propagation Learning Algorithm

Nibras Othman Abdulwahid[1(✉)], Sana Fakhfakh[2,3], and Ikram Amous[3]

[1] Miracl Laboratory, Enet'com,Technopole of Sfax, Sfax University, Road Tunis km 10, 3018 Sfax, Tunisia
nebrasse.toubane@enetcom.u-sfax.tn

[2] Department of information systems, College of Computer Engineering and Sciences, Prince Sattam bin Abdulaziz University, Al-Kharj 11942, Saudi Arabia
s.fakhfakh@psau.edu.sa

[3] Miracl Laboratory, Enet'com, Technopole of Sfax, Sfax University, Road Tunis km 10, 3018 Sfax, Tunisia
ikram.amous@enetcom.usf.tn

Abstract. The classical educational system in some countries, such as the Arabic countries, depends on the final year's scores to predict academic performance. In contrast, the STEAM educational system considers the students' scores for all the studying years alongside the students' skills and interests to predict academic performance. However, the STEAM educational system predicts the academic performance of students in five to seven general majors regardless of the variant factors that may affect the students' future careers. Hence, in this research, a seven majors and five factors (SAF) model has been proposed to assign a specific major to every student based on their academic background, interests and skills, and the main influencing factors. The SAF model uses a supervised back-propagation artificial neural network and is trained by a scale-conjugate learning algorithm. The SAF model has the capability to predict a specific major among 17 different majors for every student with a high learning performance (1.4147), plausible error value (-0.1211), and rational number of learning epochs (223).

Keywords: artificial neural network · predicting student's academic performance · steam education · back-propagation · scale-conjugate learning algorithm

1 Introduction

Predicting student performance has become a tough issue, as the amount of data in educational institutions has grown and data-mining tools have advanced. As a result, it has become vital to discover the elements that influence student success in higher education: specifically, utilizing predictive data-mining approaches [1].

R. Silhavy and P. Silhavy (Eds.): CSOC 2023, LNNS 724, pp. 37–54, 2023.
https://doi.org/10.1007/978-3-031-35314-7_4

Higher education institutions can be judged on their ability to help their students make informed decisions and prepare for the events and circumstances that affect their academic outcomes. Defining academic achievement is a necessary first step before exploring the characteristics and contexts that may inspire students' performance. However, most studies conclude that academic success is contingent exclusively on performance within the assessment window. However, in the present era, there are several widely disseminated viewpoints that argue that many important aspects influence a student's academic achievement .

The purpose of this study is to employ the most extensively studied elements that impact students' academic performance in higher education institutions. This includes pre-academic achievement, student demographics, e-learning activity, psychological characteristics, and environments [1].

The aforementioned factors are the most prevalent. The two most essential criteria are prior academic success and student demographics, which were addressed in 69 of previous publications with a very high prediction probability [2]. Meanwhile, prior academic accomplishment is the most essential component by 40, since it is the student's historical base, which is established by the grades (or any other academic performance indicators) achieved by the student in the past (pre-university and university data) [2].

The back-propagation (BP) algorithm is used to correct errors in the neural network through feed-forward. It is one of the ways to learn the networks that transmit information through the reverse propagation of the original direction of the information. This method is based on supervised learning and needs detailed data from which the network learns during the training phase [3].

We selected the BP method instead of [4] feed-forward algorithm because it delivers superior results in classification and regression (prediction) issues when the output layer of the neural network has a single cell.

Then comes the BP stage, where the network recalculates the error value in each neuron from the hidden networks. Finally, the new weights are updated as the network recalculates all the weights and replaces them with new values [3].

A scaled conjugate gradient is a supervised-learning technique for neurofeedback networks that eliminates the need for time-consuming linear searches in the conjugate direction in the remainder of the algorithms. The fundamental concept is to integrate two methodologies (one of which uses the Levenberg-Marquardt algorithm with the conjugate gradient approach). Furthermore, if the network's functions, weight, net income, and transmission are all useful derivative functions, the method trains it. BP is also utilized for weight and simulation factors [5].

The neural network technology used is a three-layer feed-forward network (FFNN). The three layers are an input layer that takes input variables, a hidden layer that captures connections between variables, and an output layer that predicts the outcome. [4]; They showed the effect of internal representations transmitted from one neuron to another in directing behavior, accelerating the learning process. Reorientation of internal models may reduce the number of periods required for optimal performance [6]. These representations can be used to reflect students' talents and interests while simulating a student's academic performance system.

Many studies have attempted to develop intelligent classifiers to predict student achievement and ignore the importance of identifying the main factors that lead to predicting final performance. Moreover, this is necessary to enable leaders to identify the strengths and weaknesses of their academic programs and thus improve student achievement [7].

The main contributions of this paper are:

1- We have built an intelligent model that predicts the acceptance of students according to the STEAM educational system with five factors (prior academic achievements, student demographic, e-learning activity, physiological attributes, and environments) that affect student acceptance.
2- Our research is the first to use a STEAM prediction technique that uses a conjugate gradient-scale supervised-learning algorithm.
3- The results we obtained are very close to reality.
4- The size of the data used is enormous compared to previous works, and most of its limitations were due to the sample size.
5- Our research contains a complete part of recent previous works that shed light on the use of factors that affect student acceptance.

2 Related Works and Research Gap

Several studies have attempted to develop intelligent classifiers to predict student performance by focusing on identifying the main influencing factors that improve the quality of predicting students' academic performance. We reviewed the relevant works that we characterized as identifying the most critical factors that affect students' academic performance. In higher education, predicting student performance is a useful endeavor, since it achieves tactical advantages that include the creation of early warning and course-suggestion systems, the identification of negative student behaviors, and the automation of course evaluation. The precise prediction of student academic progress, however, is a challenging research project that necessitates a thorough comprehension of all factors and conditions affecting the students and their learning [7].

Nasri Harb and Ahmed El Shaarawi investigated the socio-economic characteristics of students at the College of Business and Economics (CBE), United Arab Emirates University UAEU . They used a dataset with a sample of 864 CBE students. Also, they used the students' socio-economic characteristics factor and the regression-analysis method to predict this factor. Consequently, the result showed the most important factor affecting student performance was the student's competence in English and that non-national students outperformed national students and female students outperformed their male counterparts [5].

Secil Bal et al. investigated the impacts of teacher efficacy and motivation on students' academic achievement in science education in secondary and high schools located in Iran and Russia. They used a dataset representing 790 students and 350 teachers drawn from 15 schools in Iran and Russia. Also, they used these factors: teacher self-efficacy, student-learning motivation, and academic achievement. Meanwhile, they used these techniques: correlation analysis, regression analysis, and t-test analysis. Consequently, the findings of this study will provide the basis for future research on this topic of growing scholarly and practical importance. Nonetheless, this research has limitations, such as the sample size and the focus area of the research [8].

Moruzzella Rossi analyzed the factors that affect academic achievement in higher education. The dataset used represented 808 students from the Faculty of Economics and Business Administration at Universidad Andres Bello. Rossi used six factors, which included socio-demographic, economic, and cognitive student characteristics such as age, gender, parents' education, parents' income, and high school grades. He used tow-technique OLS regression and logistic regression. Consequently, the result was that family socio-economic status indicators, such as parental education and parents' income, do not have incidence on the academic performance of evening undergraduate students [9].

Elizabeth Thiele et al. used a hauntological technique to examine if students in mixed courses obtained higher grades than those in fully online courses and to discover demographic indicators of academic achievement in Web-based college courses. The dataset used represented 2,174 students (M 27.6, SD 9.5 years; range: 17–68 years) enrolled in Web-based courses at the University of Southern Maine (USM) during the Fall semester of 2011. Also, they used eight factors: gender, age, academic plan, college affiliation, course, full-time vs. part-time status, graduate vs. undergraduate enrollment, and course instructional mode (blended vs. online). They used the regression technique. The results showed that the current study expands knowledge on the veracity of blended learning vs. online education, especially for older nontraditional students. Also, this research had some limitations that made it incapable of addressing differences in online instructional styles and student motivation factors [10].

A study by Tamara Thiele et al. supports this, by examining the links between school grades, school type, school performance, socioeconomic disadvantage, neighborhood participation, gender, and academic achievement at a British university. The data were from a British university, one of the six original "red brick" civic universities and a founding member of the Russell Group. They used six factors (school grades, school type, school performance, socioeconomic disadvantage, neighborhood participation, and gender) as well as academic achievement. They also used the logistic regression (LR) technique. The results show that students from the most deprived areas performed less well than students from the richest. Asian and black students performed less well than white students. Female students performed better than their male counterparts. Nonetheless, this research had its limitations. First, it was not possible to control for all factors that affected university attainment. Second, this study included only those students who successfully completed their degrees and not those who failed or did not complete their course. A final and common limitation relevant to the present study lay in the high proportion of missing data, as this could significantly bias analyses and results [11].

Amjed Abu Saa et al. aimed to identify the most studied factors that affect students' performance as well as the most common data-mining techniques

Table 1. Related works overview.

Reference	Proposed	Data set	Factors	AI Technique	Results	Limitation
[12]	Surveyed the current literature regarding the ANN methods used in predicting students' academic performance		1- sample size. 2- level. 3- field of education or the study context. 4- GPA. 5- demographic variables.	artificial neural network	ANN and EDM are works together to get student prediction	limited application of the ANN in a real-life situation
[7]	Proposed the hybrid regression model (HRM).	1- survey 357-related articles (29 features). 2- survey 71-related articles (70% student performance prediction, 21% student dropout). 3- public datasets (Kaggle).	1- previous academic achievements of students (student grades obtained both pre-university [e.g., high school] and during university) 2- student demographics (gender, age, and ethnicity) 3- student psychological traits 4- learning environment 5- e-learning activity	1- collaborative filtering 2- fuzzy rules 3- LASSO linear regression	Demonstrated that integrating different predictive modeling techniques leads to high-accuracy predictions compared to using a single approach.	1- Limited datasets do not include all possible factors that may impact student achievements. 2- Future work will explore the mentioned limitations in more inclusive detail using real academic datasets.
[2]	Provided educators with easier access to data mining techniques, enabling all the potential of their application to the field of education.	1- SIS. 2- survey logs.	1- prior academic achievement. 2- student demographics. 3- student environment. 4- psychological. 5- student e-learning activity	1- classification. 2- regression. 3- clustering.	Showed that prior academic achievement, student demographics, e-learning activity, and psychological traits were the most reported factors	
[13]	Explored factors influencing the use of e-learning by students in private HEIs in Nigeria using a technology organization environment (TOE).	Use a data collection method encompassing semi-structured interviews with 15 students from L-University drawn purposefully from the Landmark directory and a hybrid thematic analysis to analyze the data.	1- technology-related factors, 2- organization-related factors. 3- environment-related factors, 4- impact-related factors.	Data analysis	100% of the participants revealed that technology-related factors such as ease of use, speed and accessibility, and service delivery influence their usage of the e-learning facilities, while 93% of the participants indicated that organizational factors such as training support and diversity shape their use of e-learning facilities.	One of the challenges of qualitative research is normally the size of data and sample used
[14]	Proposed the BCEP prediction framework and proposed the PBC model based on a summary of existing e-learning behavior classification methods.	The Open University Learning Analytics Dataset (OULAD) [illegible] is one of the most comprehensive international open datasets in terms of e-learning data diversity, including student demographic data and interaction data between students and VLE.	student demographic	1- SVC (R). 2- SVC (L). 3- naïve Bayes (NB). 4- KNN (U). 5- KNN (D). 6- sofmax10.	Showed that the BCEP prediction framework performed markedly better than the traditional learning-performance predictor construction method.	1- Behavior is frequently limited to independent e-learning behaviors. 2- The classification of e-learning behaviors is generally limited to theoretical research.
[15]	The study suggested the contribution of "sense of coherence" and "need for cognition" to students' learning and supports recent suggestions about the crucial role of trait characteristics and mental health in learning.	A convenience sample of 406 (13.7% male and 86.3% female) undergraduates studying in a social science department in Greece is taken into account.	1- motivational factors including mental-health variables. 2- cognitive-psychological factors related to success and quality of learning give a "false picture of real learning" and study success.	TThey used CFA, cluster analysis, MANOVA, discriminant analysis, and the decision tree model.	The findings reveal four student profiles: surface-unorganised students, deep-organized students, high-dissonant students with a low sense of coherence, and moderate-dissonant students with low need for cognition.	1- It should be emphasized that the results cannot be generalized to other university disciplines. 2- The cross-sectional design does not allow us to identify changes but only differences between the years of study. 3- The sample suffered from a high gender imbalance. 4- More information is needed from other instruments to reach a sufficient understanding of the web of individual variables that contribute to students' learning.
[16]	[illegible] an investigation of the application of C5.0, J48, CART, NB, K-nearest neighbor (KNN), random forest and support vector machine for prediction of students' performances.	Three datasets from school level, college level, and e-learning platforms.	1- demographic. 2- academic 3- behavioral attribute (namely, number of raised hands and visited resources have high correlation). 4- internal assessment, 5- attendance. 6- factors such as age, gender, annual income, parent's occupation, and parent's education included	C5.0, J48, CART, NB, KNN, random forest and support vector machine	Results show that the performances of random forest and C5.0 are better than J48, CART, NB, KNN, and SVM.	
[17]	This study was more suited to examine the critical role of internal protective factors in students' academic resilience.	Data on the linguistic equivalence of the Turkish and English forms were obtained from 56 English language teaching (ELT) students (50 females, six males).	1- external factors (child-rearing attitudes or parenting style and ecological-education-value perception). 2- internal factors (academic self-efficacy and academic motivation)	1-expectation-maximization (EM) algorithm. 2-Skewness kurtosis coefficients and normality tests were used to examine the normality. 3-covariance matrix.	The findings of the study were presented in three stages: 1- adaptation of ARS into Turkish culture. 2- development of EEVPS. 3- testing hypothesized model.	1- A limited number of internal and external resources were discussed. 2- Parenting-style measures were discussed in which only the maternal style was included in this study.
[18]	Research work conducted between 2010 and November 2020 was surveyed to present a fundamental understanding of the intelligent techniques used for the prediction of student performance.	We synthesized and analyzed a total of 62 relevant papers with a focus on three perspectives: 1- the forms in which the learning outcomes are predicted. 2- the predictive analytics models developed to forecast student learning. 3- the dominant factors impacting student outcomes.	1- student online learning activities. 2- term assessment grades. 3- student academic emotions	1- regression. 2- supervised machine-learning models.	They called upon the research community to implement the recommendations concerning 1- the prediction of program-level outcomes and. 2- validation of the predictive models using multiple datasets from different majors and disciplines.	1- Due to the lack of datasets and the variety of approaches employed to predict student outcomes, a meta-analysis of the prior findings was not possible. 2- They deliberately restricted our search of the intelligent predictive models of learning outcomes to the last decade only (i.e., 2010–2020). 3- This search was limited to peer-reviewed journals and conference publications, which may have missed dissertations and unpublished literature.

applied to identify these factors. In this study, 36 research articles out of a total of 420 from 2009 to 2018 were critically reviewed and analyzed by applying a systematic literature review approach. Also, they used five factors: students' previous grades, class performance, e-learning activity, demographics, and social information. Three of the techniques used were decision trees, naïve Bayes (NB), and artificial neural networks. The results showed that the most common factors were grouped under four main categories: students' previous grades and class performance, e-learning activity, demographics, and social information [1]. see Table 1 for a list of other related studies that used many factors to predict student academic performance.

3 The Proposed Approach

Our proposed approach uses the seven majors and five factors (SAF) model, a new intelligent approach for predicting students' academic achievements due to many occupational criteria and factors. It is complementary to the previously proposed system-the seven-criteria, twelve-year, seven-group (STS) model-in principle, objectives, and dataset [19]. But there are some additions to the modern approach that we will discuss in detail as shown in Fig. 1. This figure has three basic components: 1- a description of the dataset, 2- the STS model, and 3- the SAF model. Below, more details about each region will be provided.

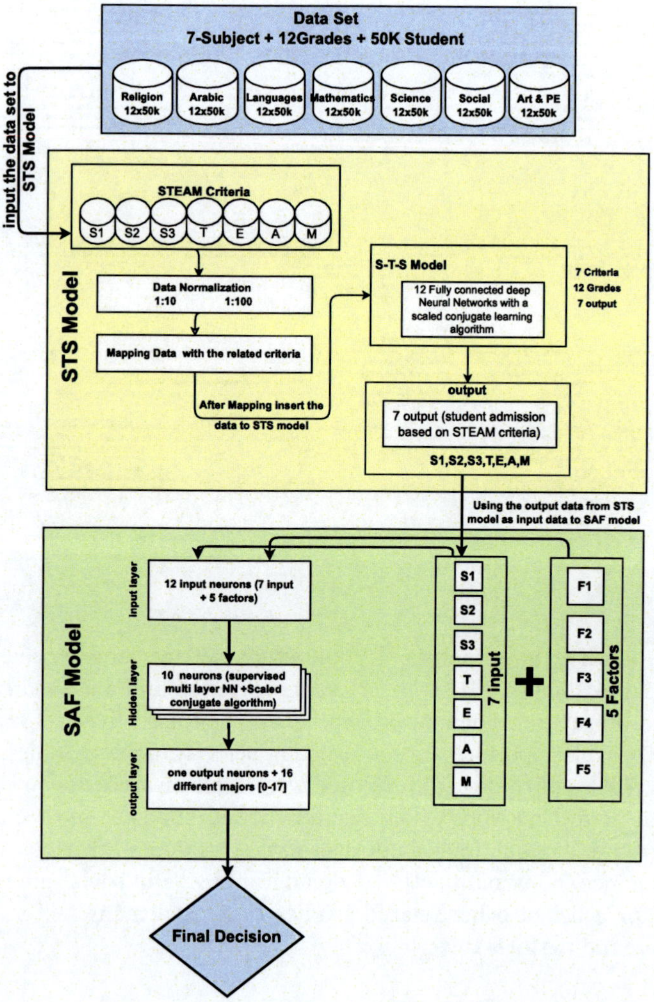

Fig. 1. STS and SAF models' frameworks.

3.1 The Dataset Description

This research was based on the proposed scores of 50,000 students from random schools in Baghdad, Iraq. The dataset is a three-dimensional matrix that consists of 50,000 students' seven scores for 12 grades. Figure 2 depicts a histogram representation of the 50,000 students with scores ranging from 50–100 for all stages. The dataset has been divided into 70 for training, 15 for testing, and 15 for validation. Initially, the dataset was a three-layer matrix consisting of 12 grades, 7 subjects,-and 50,000 students (12×7×50,000); the 12 is the number of classes (6 primary + 6 secondary), the 7 is the number of subjects (religion, languages, mathematics, science, history/geography, art, and sport), and the 50,000 is the number of students. Later, the three-layer matrix was reshaped to a two-layer matrix to obtain 7-subject and 50,000-student dimensions.

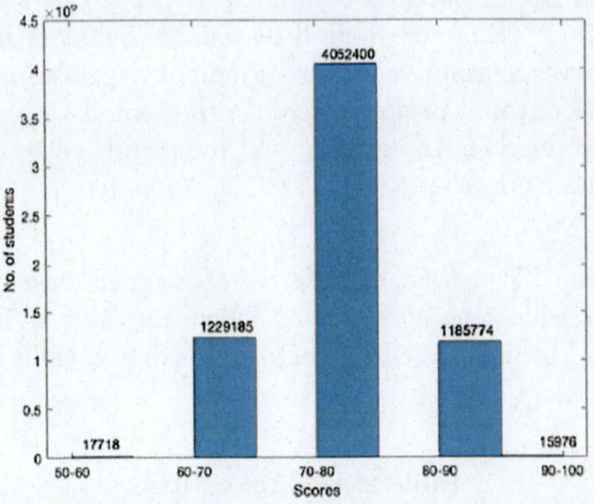

Fig. 2. Input dataset histogram.

3.2 The STS Model

The STS model is a fully connected deep neural network model, consisting of seven parallel deep neural networks, each with twelve fully connected networks. The seven lines represent the seven criteria, and the twelve fully connected networks signify the 12 grades as shown in Fig. 3.

Previously, the researchers proposed the STS model. Basically, the input dataset was a three-layer matrix consisting of seven subjects, 12 grades, and 50,000 students. However, the output of this model expresses that one of the seven criteria related to the interests and skills of the student is not his/her specific major. In any case, the STS model rearranged the input dataset due to the STEAM criteria. Different criteria represented different subjects. Thus, each criterion had a different row number but fixed columns. The criteria matrices were then normalized in the 1–100 range before being mapped to the related subjects.

Fig. 3. STS model framework.

Afterward, seven parallel fully connected neural networks (FCNN) predicted the STEAM criteria outputs. Those seven criteria should be forwarded as an input to another artificial neural network combined with five other inputs.

The results of the STS model showed reasonable behavior in terms of performance, error, fit curves, training-condition gradient, regression, and best number of epochs to achieve optimal performance. It represented the expected academic results according to the STEAM system, which depends entirely on the interests and skills of the entry criteria.

STEAM Criteria. The seven criteria were assigned based on the subjects related to STEAM education concepts as shown in Table 2. The subjects were assigned to STEAM groups based on their interests and skills in line with their academic destination.

Table 2. STEAM criteria

Criteria	Subject
S1	Religion and Arabic language
S2	Science, English and foreign languages
S3	History and Geography
T	Biology, Physics, Chemistry, English and foreign languages
E	Mathematics, Physics, Chemistry, English and foreign languages
A	Physical education and Art
M	Mathematics

Data Normalization. At this stage, two categories of data had been normalized: 1- Normalization of grades (average student grades), in which the average

Table 3. Dataset setup.

Step	Procedure
1	Set a three-dimensional array with 7 rows, 12 columns, and 50,000 layers
2	Set the rows as the subjects of every grade, Set the subjects to 7 only, as follows: (Religion, Arabic, English and Foreign languages, Mathematics, Science, Social, and Art and Physical education).
3	Subjects normalization: If there are one or more foreign languages beside the English language subjects, then get the average score for all of them After grade 4, the social subject contains the average scores of: Geography, History, and Nationality subjects Before grade 4, the scores are in range 1–10, thus they have been normalized to be in range 1–100 After grade 7, the science subject contains the average scores of: Biology, Chemistry, and Physics subjects
4	Set the columns as the grades numbers: Grade 1 – Grade 6 stand for primary school, Grade 7 – Grade 9 stand for middle school, Grade 10 – Grade12 stand for high school
5	Set the third dimension layer to the number of students
6	All the scores for all the grades should be equal or grater than 50 (success threshold)

values of grades ranging from 0–10 for the primary level and 10–100 for the intermediate levels were normalized by multiplication. They were multiplied by 10 to standardize all values on a standard scale without altering the range of machine-learning values. 2- Normalization of the topic (average of three issue grades), in which, according to the educational system in Iraq, the standard of lessons (chemistry, physics, and biology) for the intermediate and preparatory stages is determined by adding the grades and dividing by three, which reflects the total number as shown in Table 3.

Data Mapping. At this stage, STEAM-based options are selected. The STEAM subject sequence can be used to designate each group of subjects that represents the name of a group of STEAM aggregates (S1, S2, S3, T, E, A, M). For example, S1 stands for "religion and Arabic", S2 stands for "sciences, English, and other languages" and displays in Table 2, etc., and mapping is done by associating the matrix for each part with the target matrix (1 × 50,000). Additionally, Objective 1 suggests a mean of more than 90 while Objective 0 indicates a criterion less than 90. As noted in Table 4, the score must be either 0 or 1.

Table 4. STS model input initialization and setup

Step	Procedure
1	Mapping the input matrix regarding the criteria mentioned in Table 2, as follows: S1 input is a 2×50000 matrix, S2 input is a 2×50000 matrix, S3 input is a 1×50000 matrix, T input is a 2×50000 matrix, E input is a 3×50000 matrix, A input is a 1×50000 matrix, M input is a 1×50000 matrix,
2	The Target is a 1×50000 matrix, Map the target matrix for every input, as follows: Target = 1 when the related score has an average value greater than 90, Target = 0 when the related score has an average value less than 90,
3	The output values should be either 0 or 1, as follows: **If** S1 = 1, **then** the related student eligible to S1 academic group, **If** S2 = 1, **then** the related student eligible to S2 academic group, **If** S3 = 1, **then** the related student eligible to S3 academic group, **If** T = 1, **then** the related student eligible to T academic group, **If** E = 1, **then** the related student eligible to E academic group, **If** A = 1, **then** the related student eligible to A academic group, **If** M = 1, **then** the related student eligible to M academic group, **else**, the related student is not eligible to the related criteria

3.3 The (SAF) Model

This proposed model forwards the outputs of the STS model with new added inputs to get more specific outputs. The input of this model is a combination of some proposed factors besides the outputs of the STS model. The ANN of the model is a supervised neural network trained by a scale-conjugate back-propagation learning algorithm. The output of the SAF model is mapped to institutes and colleges to represent the deserved specific school of the student upon his/her STEAM educational background based on their skills and interests. Table 5 shows the workflow procedure of the SAF model. The SAF model consists of three main parts: the input part (input layer SAF model), the ANN part (hidden layer SAF model), and the mapping part (output layer SAF model).

Table 5. Dataset setup

Step	Procedure (SAF Model)
1	**Forward** the input dataset to the STS model
2	**Get** the seven-group outputs
3	**Forward** the outputs to the SAF model
4	**Combine** the outputs of the STS model with the factors of the SAF model
5	**Train** the multi-layer neural networks using the scale conjugate learning algorithm
6	**Get** the output results with a value: [0–17]
7	**map** the output value due to Table 7

3.4 Factors Impacting Student Performance

The five new inputs represented the following factors, as shown in Table 6, after getting the seven-group outputs from the STS model. These inputs can be considered the input of a new proposed supervised artificial neural network named "SAF model." The SAF model consists of three main parts: the input part, the ANN part, and the mapping part. Figure 1 shows the combination framework of the STS and SAF models. In our study, we looked at the five most important and common factors that affect how well students do in school (previous academic achievements, student demographics, e-learning activity, psychological traits, and environments). The first factor (F1) shows the student's previous academic achievements, which are given a value of 1 or 0 depending on how good or bad they were. The second factor (F2) shows the student's background. Its value, either 1 or 0, shows how close or far away the student is from the college or university. The third factor (F3) shows how much the student has done with e-learning. The value is 1 or 0 depending on how good or bad the student is at e-learning. The student's psychological traits are shown by the fourth factor (F4), which has a value of 1 or 0 based on how good or bad the traits are. The fifth factor (F5) is about the student's study environment. The value of this factor is 1 or 0 depending on whether the student's study environment is good or bad. Moreover, understanding the factors and their impact on students will help reveal key strengths and weaknesses in achieving student learning outcomes.

Table 6. Additional factors

Item	Name	Value
F1	Prior academic achievements	[1=Good , 0=Bad]
F2	Student demographics	[1=Near , 0=Far]
F3	E-Learning activity	[1=Good , 0=Bad]
F4	Psychological attributes	[1=Good , 0=Bad]
F5	Environments	[1=Good , 0=Bad]

3.5 Input Layer (SAF) Model

This part combines the seven-group results of the STS model and the five assigned factors in one column vector to make it 12 inputs. The seven group is the seven STEAM criteria (S1, S2, S3, T, E, A, M) that relate to the interests and skills of the student. The seven criteria have been assigned based on the subjects related to STEAM education concepts as shown in Table 2. The students' subjects were assigned to STEAM groups based on their interests and skills aligned with their academic destination. Those seven criteria should be forwarded as input to another artificial neural network combined with five other inputs (factors). The five most important and common factors affect how well students do in school (previous academic achievements, student demographics, e-learning activity, psychological traits, and environments).

3.6 Hidden Layer (SAF) Model

The ANN of the model is a supervised multi-layer neural network trained by a scale-conjugate learning algorithm. A scaled conjugate gradient is a supervised learning algorithm for neurofeedback networks that avoids the time-consuming linear search in the conjugate direction used by other algorithms as shown in Table 9. The basic idea is to use two methods together (one of which uses the Levenberg-Marquardt algorithm with the conjugate-gradient algorithm). Also, the algorithm trains the network if its functions, weight, net income, and transmission are all useful derivative functions. Back-propagation is also used to change the variables for weight and simulation.

3.7 Output Layer (SAF) Model

The output part, which is the final decision to accept a student based on the STEAM system (i.e., the student's talents and preferences), and the existence of five factors for each student are used to predict his/her admission. Moreover, the output part of the model maps the one-output neuron at the output layer to the 17 different majors. The result of this neuron has 18 values (0–17), of which the 0 refers to the institute and (1–17) refer to the college's group in different academic majors as shown in Table 7.

4 Final Decision

The SAF model showed an excellent response when estimating predicted output based on the actual result. The final decision to accept a student is based on the STEAM system, i.e., the student's talents, and preferences, and the existence of five factors for each student is used to predict his/her admission as shown in Table 7.

Table 7. SAF model - Output part mapping

Output	Criteria	No. of factors	Major
0	–	–	Institute
1	S1	>4	College of Medicine
2	S1	>3	College of Dentistry
3	S1	>2	College of Pharmacy
4	S1	>1	College of Veterinary Medicine
5	S1	0	College of Nursing
6	T	>2	College of Engineering
7	T	0	College of Agricultural Engineering Sciences
8	S2	0	College of Science
9	S3	>4	College of Management and Economics
10	A	>2	College of Physical Education and Sports Sciences
11	M	>3	College of Education
12	A	>2	College of Fine Arts
13	S3	>2	College of Law
14	S3	0	College of Political Sciences
15	S3	>1	College of Islamic Sciences
16	A	0	College of Media
17	A	1	College of Languages

5 Optimization Procedure

A scaled conjugate gradient is a supervised-learning algorithm for neurofeedback networks that avoids the time-consuming linear search in the conjugate direction used by other algorithms as shown in Table 9. The basic idea is to use two methods together (one of which uses the Levenberg-Marquardt algorithm with the conjugate gradient approach). Also, the algorithm trains the network if its functions, weight, net income, and transmission are all useful derivative functions. Back-propagation is also used to change the variables for weight and simulation.

Most methods to optimize for function minimization use the same strategy. Minimization is a local, iterative method that finds the closest approximation of a function to the current point in weight space that is as close to the function as possible. People often use a first-order or second-order Taylor expansion of the function to get a close approximation of the function. The idea behind the strategy is shown by the pseudo method below, which minimizes the error function E(w)

$$E(w) = \left(..., \sum_{p=1}^{P} \frac{dE_p}{dw_{ij}}, ... \right) \tag{1}$$

Table 8. Optimization procedure

	Optimization procedure
1	Choose initial weight vector w_1 and set $k = 1$
2	Determine a search direction p_k and a step size α_k so that $E(w_k + \alpha_k p_k) < E(w_k)$
3	Update vector: $w_{k+1} = w_k + \alpha_k p_k$
4	**IF** $E(w_k) \neq 0$ **then** set $k = k + 1$ and go to 2, **else** return w_{k+1} as the desired minimum

Table 9. Scaled conjugate algorithm

Algorithm 1: scaled conjugate gradient	
step 1	Choose weight vector w_1 and scalars $\sigma > 0, \lambda_1 > 0$ and $\bar{\lambda_1} > 0$. Set $p_1 = r_1 = -E(w_1), k = 1$ and $success = true.$
step 2	**If:** $success = true$ **then** calculate second order information: $\sigma_k = \frac{\sigma}{\lvert p_k \rvert}$, $s_k = \frac{E(w_k+\sigma_k p_k)-E(w_k)}{\sigma_k}$, $\delta_k = p_k^T s_k$.
step 3	Scale s_k: $s_k = s_k + (\lambda_k - \bar{\lambda_k})p_k$, $\delta_k = \delta_k + (\lambda_k - \bar{\lambda_k}) \lvert p_k^2 \rvert$,
step 4	**If** $\delta_k \leq 0$ **then** make the Hessian matrix positive definite: $s_k = s_k + (\lambda_k - 2\frac{\delta_k}{\lvert p_k \rvert^2})p_k$, $\bar{\lambda_k} = 2(\lambda_k - \frac{\delta_k}{\lvert p_k \rvert^2})$, $\delta_k = -\delta k + \lambda_k \lvert p_k \rvert^2$, $\lambda_k = \bar{\lambda_k}$.
step 5	Calculate step size: $\mu_k = p_k^T r_k$, $\alpha_k = \frac{\mu_k}{\delta k}$.
step 6	Calculate the comparison parameter: $\Delta_k = \frac{2\delta_k(E(w_k)-E(w_k+\alpha_k p_k))}{\mu_k^2}$.
step 7	**If** $\Delta_k \geq 0$ **then** a successful reduction in error can be made: $w_{k+1} = w_k + \alpha_k p_k$, $r_{k+1} = -E(w_{k+1})$, $\bar{\lambda_k} = 0$, $success = true$, **If** $kmodN = 0$ **then** restart algorithms: $p_{k+1} = r_{k+1}$ **else** create new conjugate direction: $\beta_k = \frac{\lvert r_{k+1} \rvert^2 - r_{k+1} r_k}{\mu_k}$, $p_{k+1} = r_{k+1} + \beta_k p_k$. **If** $\Delta_k \geq 0.75$ **then** reduce the scale parameter: $\lambda_k = \frac{1}{2}\lambda_k$, **else** a reduction in error is not possible: $\bar{\lambda_k} = \lambda_k$, $success = flase$.
step 8	**If** $\Delta_k\ l0.25$ **then** increase the scale parameter: $\lambda_k = 4\lambda_k$.
step 9	**If** the steepest descent direction $r_k \neq 0$ **then** set $k = k + 1$ and go to 2, **else** terminate and return w_{k+1} as the desired minimum.

where P is the number of patterns presented to the network during training and E_p is the error associated with pattern p [20], as shown in Table 8.

6 Results

The SAF model effectively responded to estimate the predicted outputs according to the actual results (confusion matrix) as shown in Fig. 4. The confusion matrix represents the number of predicted outcomes in the rows as compared to the actual data in the columns. The confusion matrix shows the 16 majors only, regardless of the institute major number 0. For every column index, the number of successful predictions should be located at the same row index. And the other misses may appear around the related index (the closest index with the lowest error and vice versa). For example, column index 2 represents the college of dentistry and the number of successful predictions is 7,479 as a dental student, while 130 has been predicted in row index 10 (mistakenly) for the college of physical education and sports sciences. Consider another example: all the students at the college of media (index [16, 16]) have been successfully predicted without any other misses. It is worth mentioning that the sum of the predicted values in the confusion matrix equals the number of input students (50,000). However, the scale-conjugate learning algorithm showed an optimized gradient (0.3956) after 229 epochs as shown in Fig. 5. Meanwhile, the best validation per-

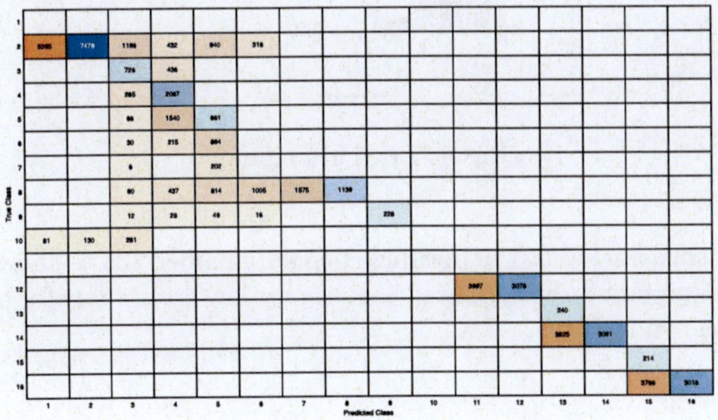

Fig. 4. Actual and predicted outputs .

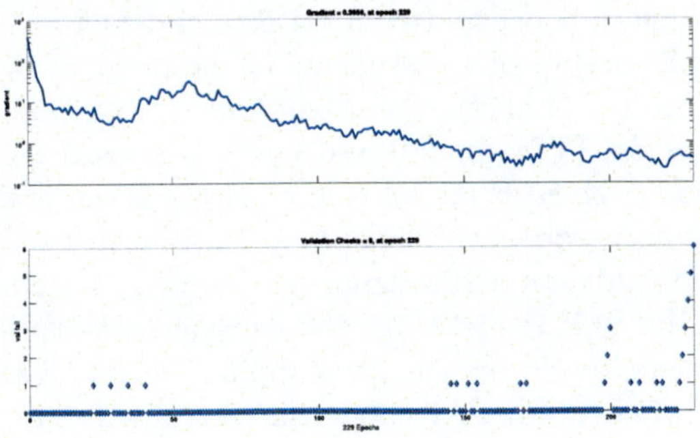

Fig. 5. Training steps.

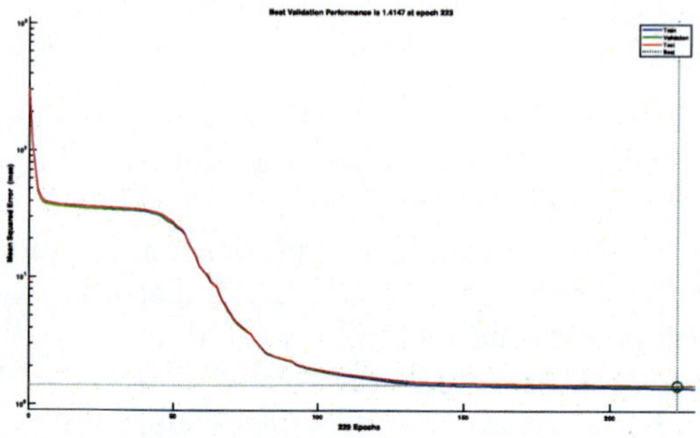

Fig. 6. Optimised performance.

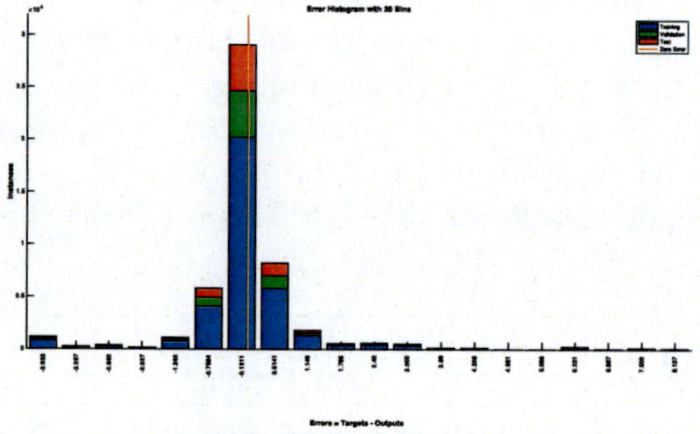

Fig. 7. Error histogram.

formance reached 1.4147 at the optimized epoch number 223 as shown in Fig. 6. Also, most histogram error instances have a very low error value (approximately -0.1211) as shown in Fig. 7.

7 Conclusion

A new smart approach has been proposed to predict students' academic achievement according to several occupational criteria and factors. The SAF model predicted the specific majors of the students in the range of 16 colleges and one general institute. Such a prediction was based on the output criteria of a previously proposed model (STS model) and newly proposed factors that mainly affect the prediction. The SAF model combined the seven criteria and the five factors all together and propagated them to a multi-layer supervised neural network to predict the student's performance in 17 different academic majors.

The results of the SAF model show the considerable effect of the proposed factors on the predicted performance. By mapping the actual and predicted outputs in a confusion matrix, the SAF model successfully distributed the students to their well-deserved academic majors after 223 epochs and significant validation performance reached 1.4147 with an error value of -0.1211.

Although the SAF model has successfully predicted the academic majors of the students on a very wide scale, it has correlated the majors with a random factor of the proposed factors. Such a limitation can be overcome in future work by assigning a specific factor(s) for every individual academic major.

References

1. Abu Saa, A., Al-Emran, M., Shaalan, K.: Factors affecting students' performance in higher education: a systematic review of predictive data mining techniques. Technol. Knowl. Learn. **24**(4), 567–598 (2019)
2. Alyahyan, E., Düştegör, D.: Predicting academic success in higher education: literature review and best practices. Int. J. Educ. Technol. High. Educ. **17**(1), 1–21 (2020)
3. Ginting, S.L.B., Fathur, M.A.: Data mining, neural network algorithm to predict students grade point average: backpropagation algorithm. J. Eng. Sci. Technol. **16**(3), 2028–2037 (2021)
4. Khalid, M., et al.: Cortico-hippocampal computational modeling using quantum neural networks to simulate classical conditioning paradigms. Brain Sci. **10**(7), 431 (2020)
5. Aich, A., Dutta, A., Chakraborty, A.: A scaled conjugate gradient backpropagation algorithm for keyword extraction. In: Bhateja, V., Nguyen, B.L., Nguyen, N.G., Satapathy, S.C., Le, D.-N. (eds.) Information Systems Design and Intelligent Applications. AISC, vol. 672, pp. 674–684. Springer, Singapore (2018). https://doi.org/10.1007/978-981-10-7512-4_67
6. Khalid, M., Wu, J., Ali, T.M., Moustafa, A.A., Zhu, Q., Xiong, R.: Green model to adapt classical conditioning learning in the hippocampus. Neuroscience **426**, 201–219 (2020)
7. Alshanqiti, A., Namoun, A.: Predicting student performance and its influential factors using hybrid regression and multi-label classification. IEEE Access **8**, 203827–203844 (2020)
8. Charandabi, S.E., Kamyar, K.: Using a feed forward neural network algorithm to predict prices of multiple cryptocurrencies. Eur. J. Bus. Manage. Res. **6**(5), 15–19 (2021)
9. Rossi, M.: Factors affecting academic performance of university evening students. J. Educ. Hum. Dev. **6**(1), 96–102 (2017)
10. Vella, E.J., Turesky, E.F., Hebert, J.: Predictors of academic success in web-based courses: Age, GPA, and instruction mode. Qual. Assur. Educ. **24**(4), 586–600 (2016)
11. Thiele, T., Singleton, A., Pope, D., Stanistreet, D.: Predicting students' academic performance based on school and socio-demographic characteristics. Stud. High. Educ. **41**(8), 1424–1446 (2016)
12. Baashar, Y., et al.: Toward predicting student's academic performance using artificial neural networks (ANNs). Appl. Sci. **12**(3), 1289 (2022)
13. Eze, S.C., Chinedu-Eze, V.C., Okike, C.K., Bello, A.O.: Factors influencing the use of e-learning facilities by students in a private higher education institution (HEI) in a developing economy. Humanit. Soc. Sci. Commun. **7**(1), 1–15 (2020)
14. Qiu, F., et al.: Predicting students' performance in e-learning using learning process and behaviour data. Sci. Rep. **12**(1), 1–15 (2022)

15. Karagiannopoulou, E., Milienos, F.S., Rentzios, C.: Grouping learning approaches and emotional factors to predict students' academic progress. Int. J. Sch. Educ. Psychol. **10**(2), 258–275 (2022)
16. Sathe, M.T., Adamuthe, A.C.: Comparative study of supervised algorithms for prediction of students' performance. Int. J. Mod. Educ. Comput. Sci. **13**(1) (2021)
17. Aliyev, R., Akbaş, U., Özbay, Y.: Mediating role of internal factors in predicting academic resilience. Int. J. Sch. Educ. Psychol. **9**(3), 236–251 (2021)
18. Namoun, A., Alshanqiti, A.: Predicting student performance using data mining and learning analytics techniques: a systematic literature review. Appl. Sci. **11**(1), 237 (2020)
19. Abdulwahid, N.O., Fakhfakh, S., Amous, I.: Simulating and predicting students' academic performance using a new approach based on steam education. J. Univ. Comput. Sci. **28**(12), 1252–1281 (2022)
20. Møller, M.F.: A scaled conjugate gradient algorithm for fast supervised learning. Neural Netw. **6**(4), 525–533 (1993)

Data Mining, Natural Language Processing and Sentiment Analysis in Vietnamese Stock Market

Cuong Bui Van, Trung Do Cao(✉), Hoang Do Minh, Ha Pham Dung, Le Nguyen Ha An, and Kien Do Trung

DATX Vietnam, Hanoi, Vietnam
trung.do@datx.com.vn

Abstract. Stock investors have many channels to exchange information, social platforms such as Facebook, Twitter or Telegram are popular information channels used. In Vietnam, one of the most popular channels is Zalo platform, where investors exchange information in daily chat rooms. This study presents a method of crawling data from Zalo chat rooms and evaluating the sentiment of these comments about stocks or the market in general. The status is rated as positive, negative or neutral. The technique such as Vietnamese language processing, machine learning (ML) or neural networks (NN) will be used.

Keywords: Vietnamese stock market · Data mining · Natural language processing (NLP) · PhoBert · Deep learning · Sentiment analysis

1 Introduction

The stock market is a financial playground that attracts a large number of investors. Vietnam's stock market has the index named VN-Index [1] which is considered an attractive and promising investment channel, attracting more and more domestic and international investors [2]. The growth margin of the VN-Index is still exceptionally large when Vietnam's stock market has attracted about 5% of the population to participate [3], this figure is only about 10% compared to developed countries such as the US or Europe [4]. With the economic growth rate among the fastest countries in the world, stable political and investment environment, the attraction to investors of Vietnam's stock market is huge.

One of the most valuable information exchange platforms that Vietnamese stock investors often use is Zalo [5], on this platform chat rooms will be established to gather investors with the number of 100 to 1000 people, especially there are many rooms that attract tens of thousands of traders to participate. The rooms are used for traders to share opinions, discuss on stock codes, general market movements. Those comments directly reflect the trader's views and the movements of VN-Index and stock codes. On the contrary, the psychological state and opinions of investors will also directly affect the movement of stocks or the general market. Therefore, if the information in these chat

R. Silhavy and P. Silhavy (Eds.): CSOC 2023, LNNS 724, pp. 55–76, 2023.
https://doi.org/10.1007/978-3-031-35314-7_5

rooms is collected and the status of comments are evaluated, it will be extremely helpful to get the market movements as well as forecast the trend of the market and stocks.

This study proposes a method for crawling chat texts from Zalo stock rooms and then evaluating those comments. The texts will be evaluated basing on positive, negative and neutral statuses which are defined as follow,

- Positive: Traders express positive views about a stock code or markets (VN-Index). For example, the text chat expresses to buy or continue to hold a certain stock code. This information will be rated as 1.
- Negative: It is the opposite of the "Positive" when traders give negative reviews of stocks. For example, trader advises to sell the stock or not buy a stock code. This information will be set as -1.
- Neutral: Traders do not have a clear statement about the stock codes, the market or the comment that is not related to the stock. This information will be marked as 0.

2 Related Work

There is a lot of research on stock price forecasts or stock market trends, where forecasts based on media information are especially popular. Since the time that Bidirectional Encoder Representations from Transformers (BERT) was established, there has been a lot of research to develop BERT for natural language processing (NLP), one of which is the sentence embedding technique. For example, Nils Reimers et al. [6] developed a Sentence-BERT technique called Siamese BERT-networks for sentence embedding. The study continued to develop by the author in [7]. Muxi [8] developed a model to forecast market movement based on text data from Reddit. The study used sentence embedding based on the BERT approach by Nils Reimers et al. before applying the sentiment analysis in work no. [9], after that, it developed a Neural Network model (CNN-based) for stock market prediction. Sanjeev Arora, et al. [10] apply weighted average of vectors for sentence embedding. Ryan Kiros et al. [11] propose an unsupervised learning of a generic, distributed sentence encoder in which an encoder-decoder model is trained by using the continuity of text from books. The Recurrent neural networks (RNN) encoder-decoder and GRU activation used in the study bring the remarkable result.

BERT has been applied and developed for many languages around the world. This is the basic foundation for research on stock market assessment and forecast based on information from media in different countries. Peerapon Vateekul et al. [12] propose a method including LSTM (Long Short-Term Memory Network) and BERT by using news headlines in Thai for prediction of the stock market movement. Mingzheng Li et al. [13] submit an approach of sentiment analysis model for Chinese stock reviews based on BERT.

Basing on BERT, the PhoBERT researched and developed for the Vietnamese [14] making it more convenient to study information from Vietnamese media. This study presents the performance of a software system for collection data from Zalo rooms which work real-time with movements of the Vietnamese stock market (VN-Index) and sentiment analysis. Specific work performed includes,

- Crawl Vietnamese text chat from Zalo platform

- Vietnamese language processing for purely sentences
- Model training and sentence labeling by using PhoBERT and DNN (Deep-learning Neural Networks)

3 Approach

3.1 General

The main functional blocks of the program are depicted in the Fig. 1, including.

- Firstly, data from Zalo chat rooms is collected real time through the UC01-Data Crawler module. This data is stored in real-time into the Data Storage layer. This data is text chat in Vietnamese of stock investors in the Zalo rooms
- Secondly, texts are processed in UC02 module consisting works of
  - Remove links, tag names, icons, newline characters, special characters…
  - Perform word segmentation in sentences (compound morphemes)
  - Remove data with empty content and unlabeled data
- Finally, the sentences are used for sentiment analysis by three states of "Positive", "Negative" and "Neutral" which are labeled by scores of 1, −1 and 0, respectively. The model for sentence evaluating is DNN which is trained by a set of crawled sentences labeled by experts.

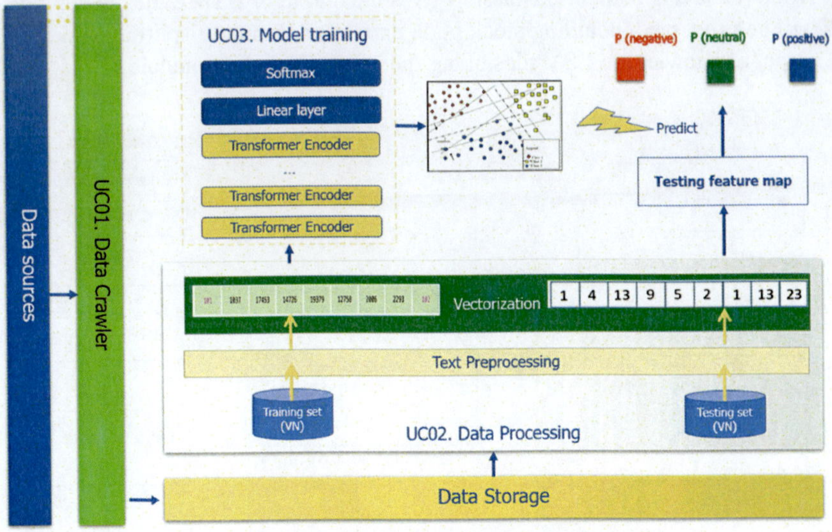

Fig. 1. Overall structure of the system

The general flowchart of the programme is shown in Fig. 2.

The chart expresses steps of cookie saving, html saving, data saving belonging to the crawling work. While sentence labeling and updating refer to processing, training and marking of Vietnamese sentences.

Fig. 2. General flowchart of program

3.2 Data Crawling

The crawling work is implemented by UC01- Data Crawler with the general flowchart is presented in Fig. 3. At first, the Crawler logins the account in the website to scan the chat group for saving contents. This step gives the output of HTML files. After that, the files will be processed including stock code extraction, sentence splitting and reunion. The output of Crawler are CSV files being the input of labeling module.

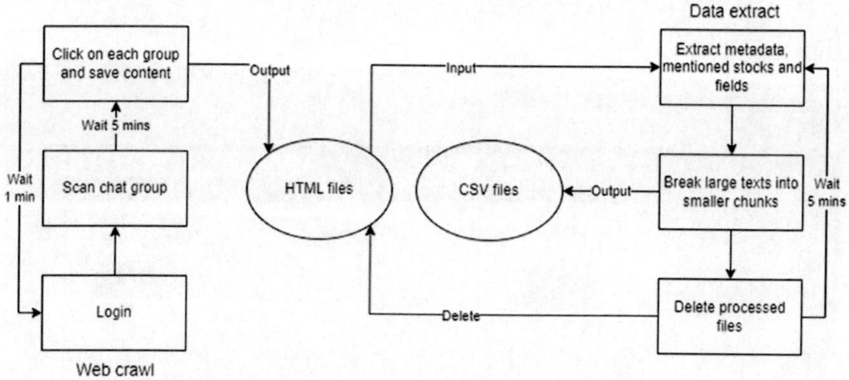

Fig. 3. General flowchart of data crawling program

UC01 collects data in real time including variables shown in the Table 1.

Table 1. Collected real-time data.

Name	Description
Group_name	Name of zalo chatting group of stock
Member_length	The number of member of a group
Send_time	The time when the message is sent
Sender_name	Name of member sending the message
Contents_original	Original message, has not splitted
Contents	Splitted message and unsplitable message
Stock_code	Stock code is mentioned in the message
Field	Field is mentioned in message
IsAdmin	Is the message sender an administrator of a group? 1: Yes; 0: No
Chat_id	Private code of a message
Account	Account used for data crawling
Platform	Platform used for data crawling (google.com)
Label	Labels: 1: Positive; 0: Neutral; -1: Negative
IsParent	1: splitable sentence; 0: unsplitable sentence
IsQuote	1: message containing reply; 0: message not containing reply

All above information will be included in the output crawled CSV files. The detailed explanation of some main variables as below,

- Stock_code: Each enterprise on the Vietnamese stock market has its own symbol code, the vast majority of these codes consist of three letters that are abbreviations of the company name.

  For example: VIC is stock code of Vingroup Joint Stock Company

- Field: The variable indicates the enterprise's field of activity on the stock exchange. This information is used for the collection of codes in the same field.

  For example: Field of VIC is "Construction and Real Estate Industry"

- Account: Zalo account used to crawl data, each Zalo account corresponds to one and only one mobile phone number, the name of the account is set by the user. The phone number and password of the account are declared in the software for the software to log into the rooms. The more chat rooms Zalo can join, the more data the software crawls.
- IsAdmin: Indicates whether the sender of the message is the administrator of that room or not. Administrators are usually brockers or people who recommend buying or selling stocks or are actively consulted by room members when decisions need to be made. Since the opinion of the administrator is dominant in the room, it should

be distinguished from the opinions of other members. The software automatically determines which member is the administrator of that room.

Cookie Saving

The purpose of saving cookies is to help login to zalo accounts without checkpoints (account verification). The algorithm is shown in the flowchart of Fig. 4.

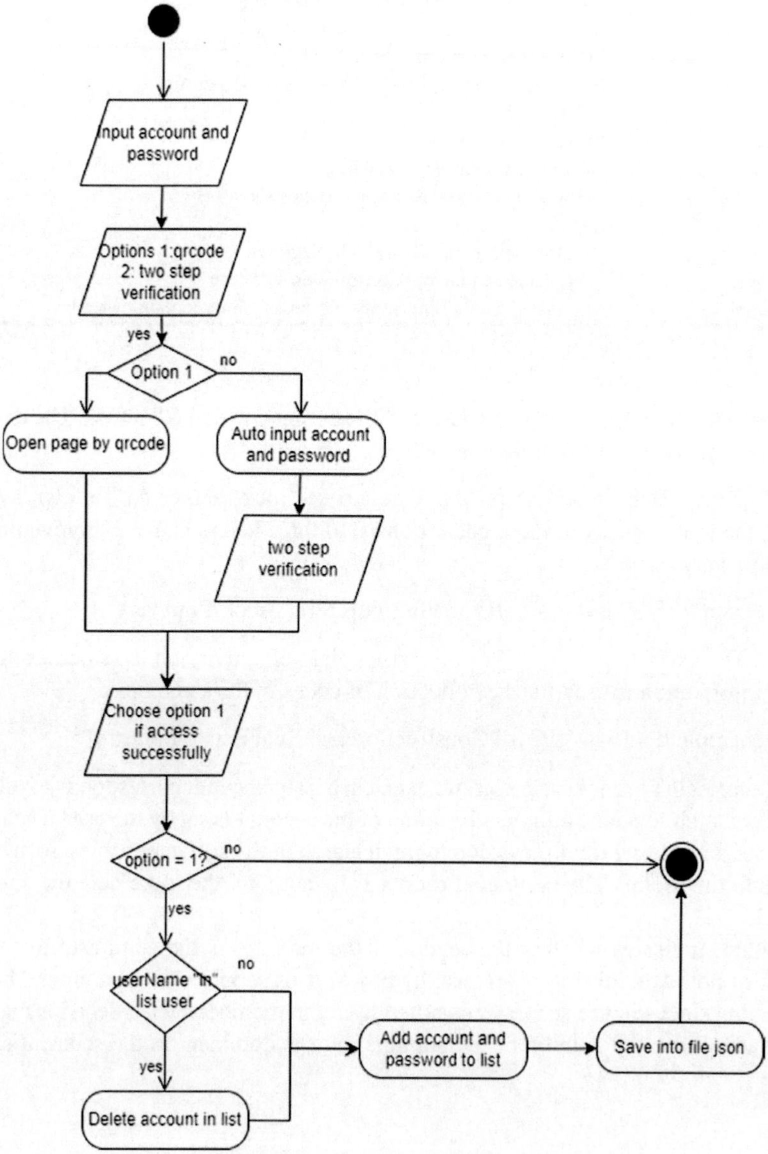

Fig. 4. Cookie saving flowchart

HTML Saving

The HTML saving of zalo groups is to increase crawl processing time. Instead of going to each group to read and process the data, the system will save all the data in HTML form and process it later. The algorithm is shown in Fig. 5.

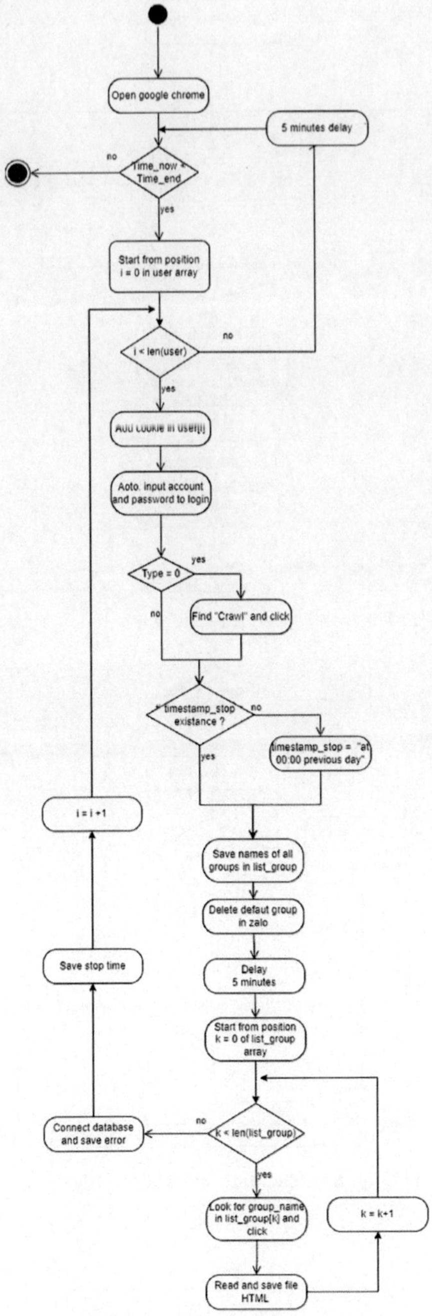

Fig. 5. HTML saving flowchart

In the Fig. 5, the step of "Read and save file HTML" is carried out by the algorithm shown in Fig. 6.

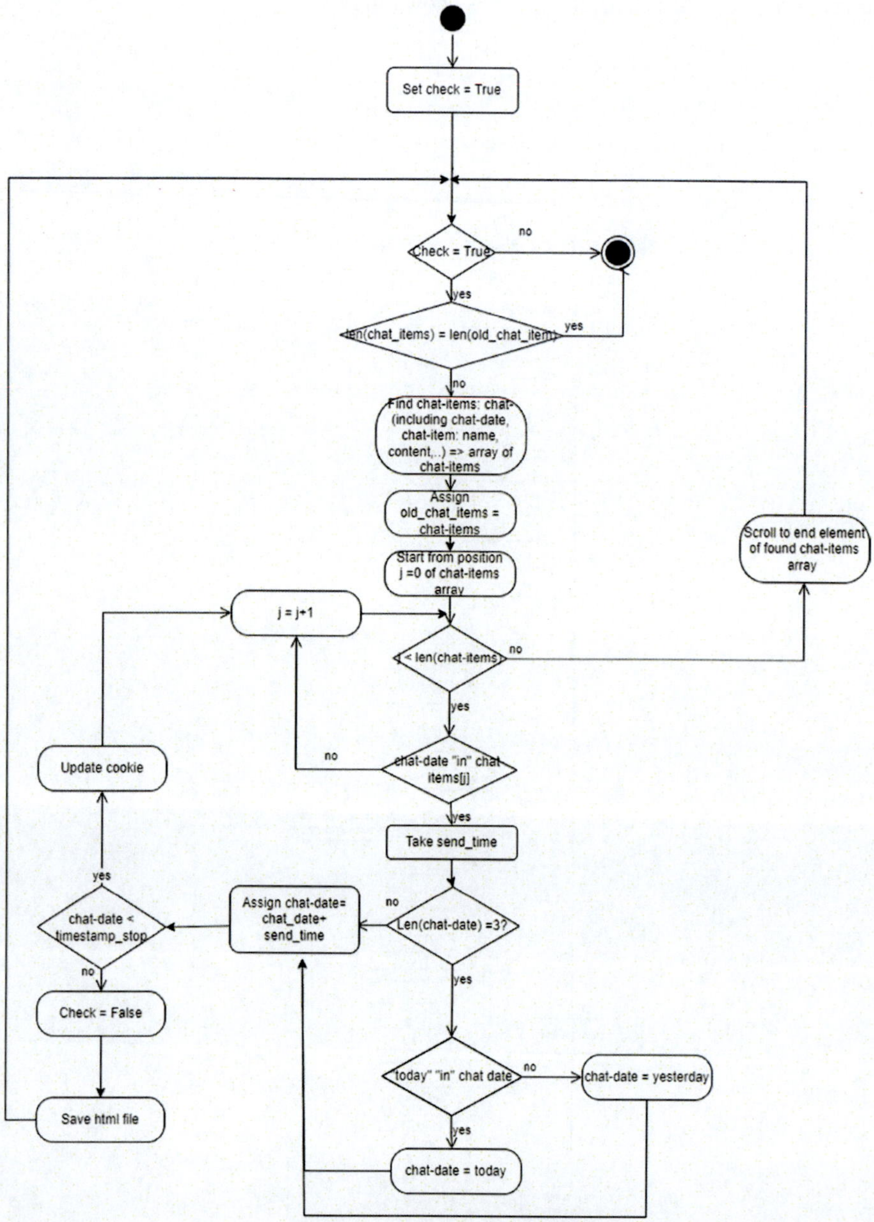

Fig. 6. Read and save HTML file

3.3 Labeling System

The labeling system is designed including steps of Text processing and Modeling. In which Text processing includes Pre-processing and Embedding as shown in Fig. 7. This system consists of two modules being UC02 and UC03 which presented in the Fig. 1 above.

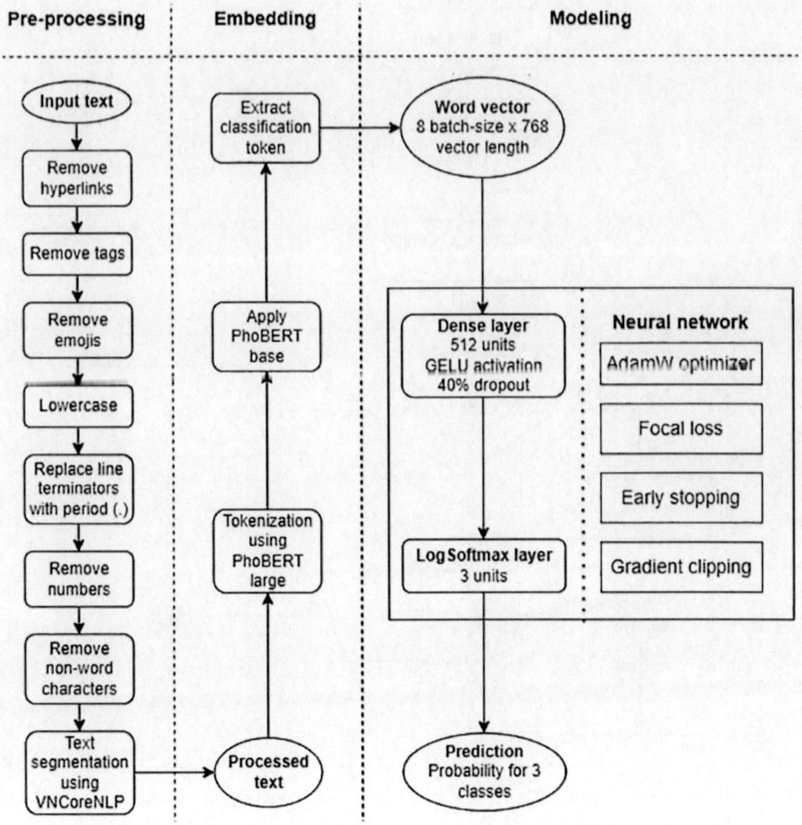

Fig. 7. General structure of labeling system

Text Preprocessing

Data normalization, the step that provides input to model training and prediction. Data retrieved from the database is normalized by several steps:

- Delete duplicates
- Data filtering: the rows of data are error, empty is removed
- Remove special characters, characters that are not useful during Model training and Prediction

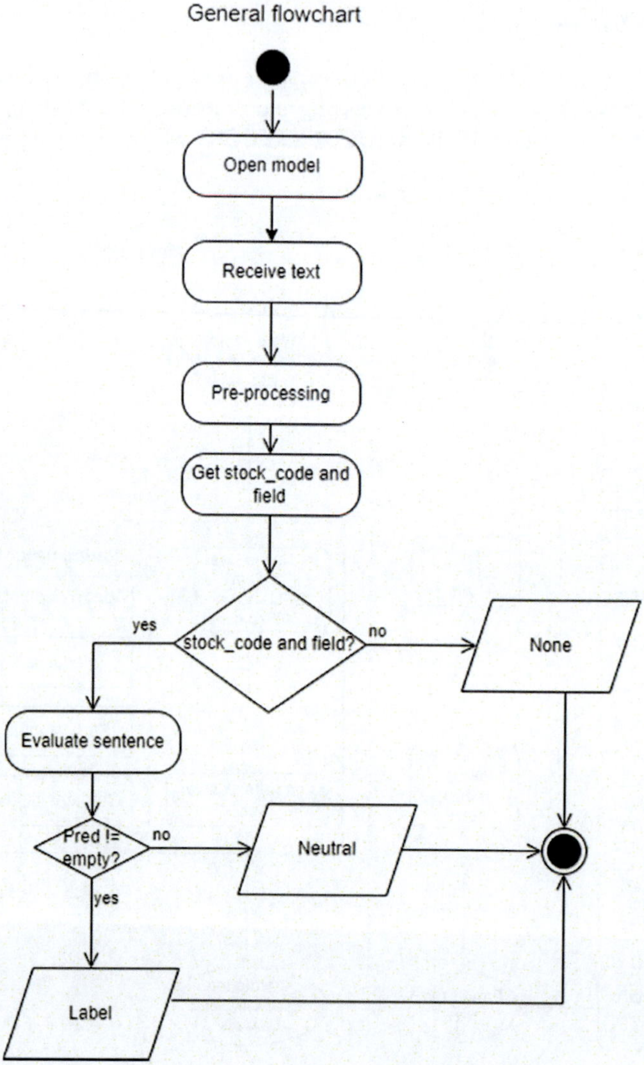

Fig. 8. Data processing and normalization

- Convert text data into vector data. In this step the technique applied is Tokenization.

  Data processing and normalization is shown by the Fig. 8.
  The logic of text pre-processing is shown in Fig. 9

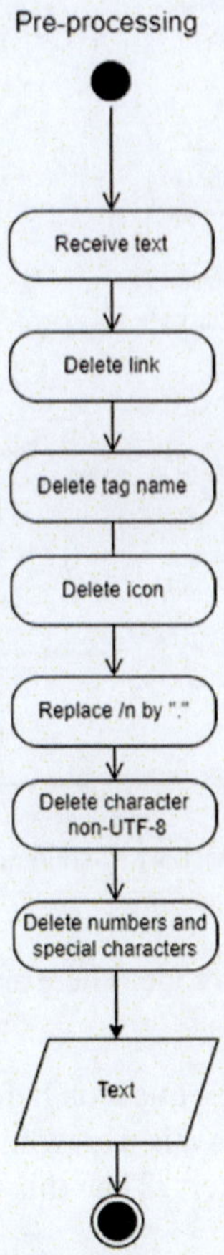

Fig. 9. Flowchart of text pre-processing

The pre-processing is designed to fulfill following tasks:

- Delete all the links
- Delete the tag name in the sentences
- Delete all the icon, e.g. the emotions
- Replace "/n" by comma (",")
- Delete character non-UTF-8
- Delete numbers and special characters

The output of pre-processing is pure chat text sentences. The pre-processed sentences will be the input of the next step of stock code and field obtaining as shown in Fig. 8. This step is built to extract the stock code mentioned in the sentences as well as the field of that ticker.

The logic of sentence labeling is shown in Fig. 10.

In the process of data processing, chat sentences containing more than one stock code will be separated so that when conducting labeling, there will be a specific and

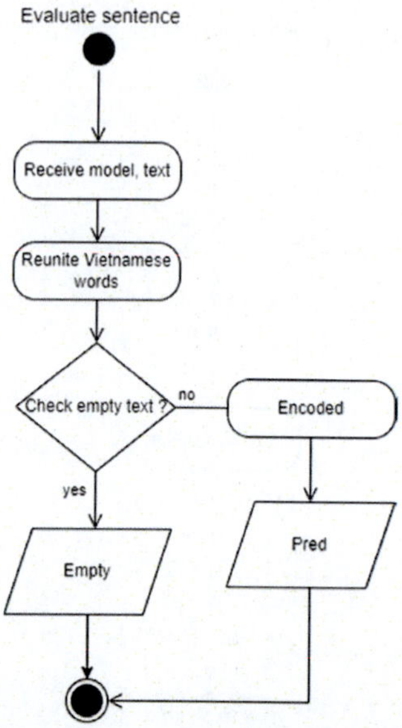

Fig. 10. Flowchart of sentence labeling

accurate assessment of each securities code. There are two cases where the chat contains more than one ticker:

- Respondent attaches someone else's question before, and the answer mentions a different stock code than the question. When crawling, the software will get a composite text with at least two different tickers. Then this text will be separated question and answer by using curly braces "{".
- Chat sentences (of a member) containing many different securities codes will be separated into smaller sentences containing only one stock code by using a period ("." or a comma (",") in the sentence (Figs. 11 and 12).

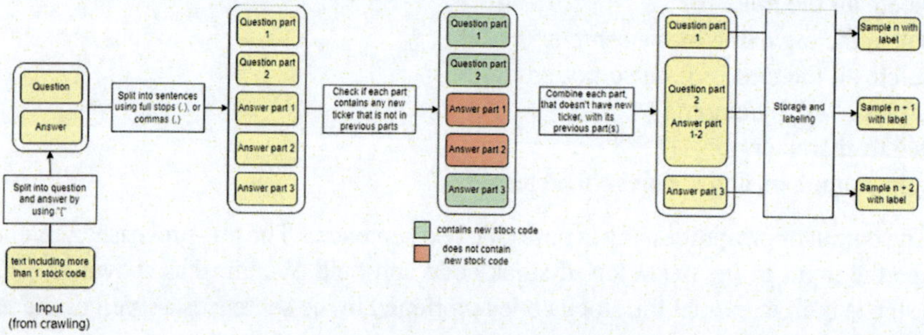

Fig. 11. Multi-sticker text processing

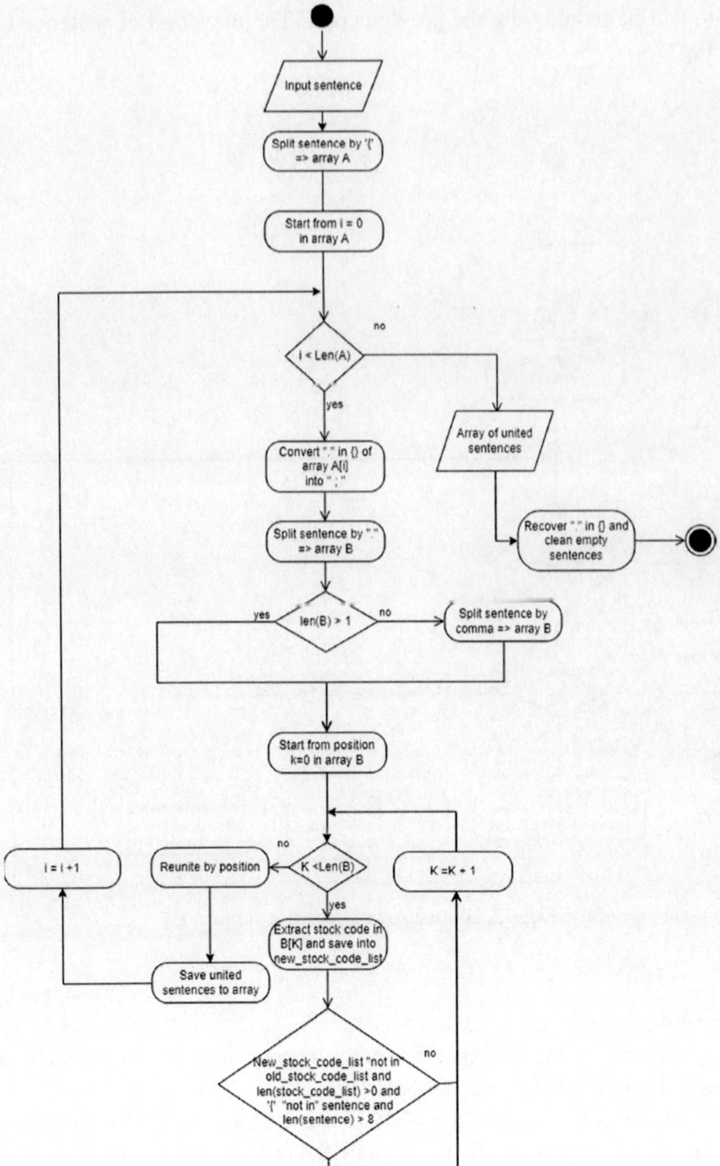

Fig. 12. Flowchart of sentence splitting

After the sentence separating work, there are cases where it is necessary to combine some text pieces to form a complete sentence in which paragraphs that do not contain a new code will be paired with the previous part. The flowchart of sentence reunion is shown in Fig. 13.

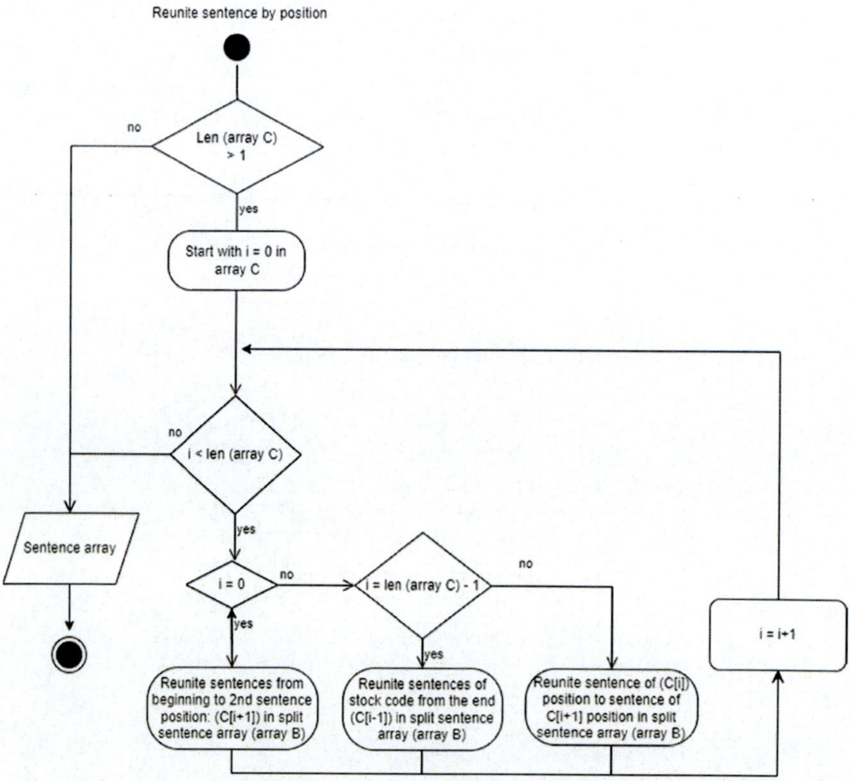

Fig. 13. Flowchart of sentence reuniting

Model Training

The Module training has input of specific vectors representing of text data. These vectors are processed through transformer encoder layers.

The training model is built on the BERT-Base architecture [14] with 12 overlapped layers of Encoder. After the 12th floor, the data is fed through the Liner Layer and Softmax Layer. The Softmax Layer serves as a classifier.

The model after being trained will be stored in a ".pt" file and used to label new data patterns.

The processing flow diagram is given in the Fig. 14. The steps are described as follows.

- Data is used for model training being sentences of Vietnamese chat text which is labelled manually by expert. The data will be preprocessed at UC02 (text processing module) to eliminate noise and invalid information, then divided into subsets of Training, Val and Testing with the rate of 60%, 20% and 20% respectively.

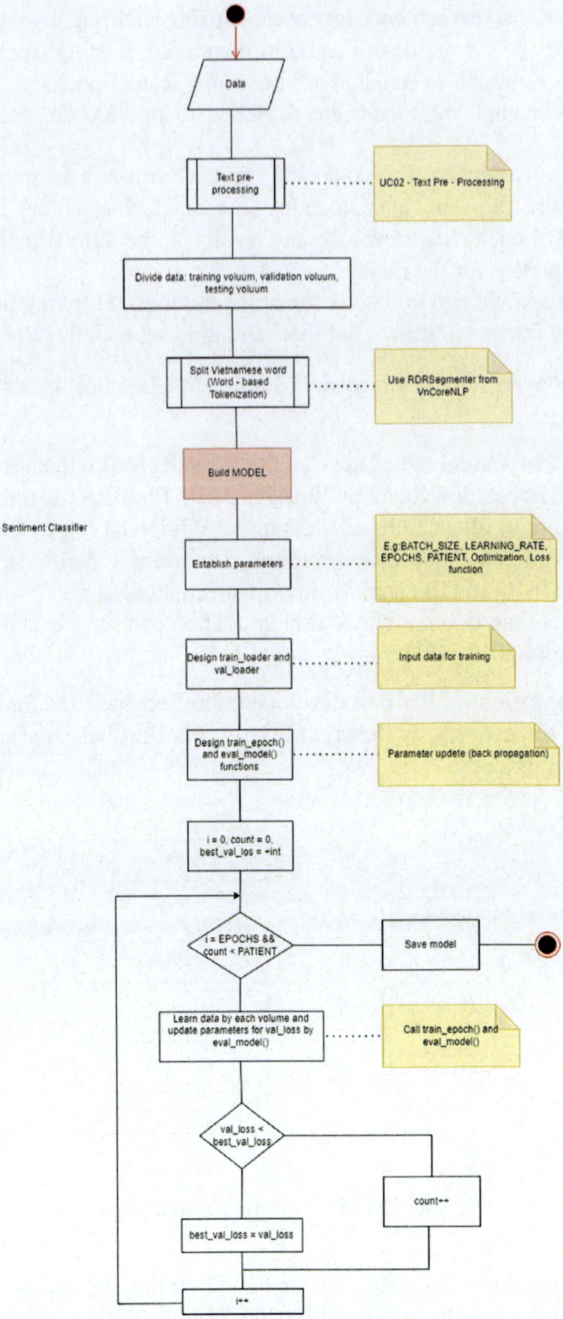

Fig. 14. Training structure

- The Vietnamese chat text sentence is processed using RDRSegmenter of VnCoreNPL [15]. After that, the system begins to build the model, starting from the declaration of hyperparameters such as Batch_Size, Learning_Rate, Epochs.
- The Train_loader and Val_loader are designed to provide the text input data for training.
- Two functions of train_epoch() and eval_model() are built to train the model. In both of them, the train_epoch() calculates accuracy as loss on the Training dataset. While the eval_model() helps evaluate the results on the Validation set to update the parameters (weights) for the model.
- The training work will end in both cases of the number of running times equaling to the set value of Epochs or the loss of the Validation set reaching the minimum.

Details of the layers in the built model (Sentiment Classifier) are as follows:
PhoBert Layers:

- The PhoBERT [14] model is the latest language model for Vietnamese, trained exclusively for Vietnamese, developed by VinAI in 2020. PhoBERT is trained on about 20 GB of data including about 1 GB of Vietnamese Wikipedia corpus and the remaining 19 GB taken from Vietnamese news corpus. The training period is 8 weeks with 4 Nvidia V100 GPUs (16 GB each) to support speed increases.
- The training is based on architecture and approach of Facebook's RoBERTa introduced in mid-2019 [16].

There are two versions of PhoBert developed, PhoBert-base and PhoBert-large with 12 and 24 layers, respectively, as shown in Fig. 15. PhoBert-base, the simpler structure will be applied in this study.

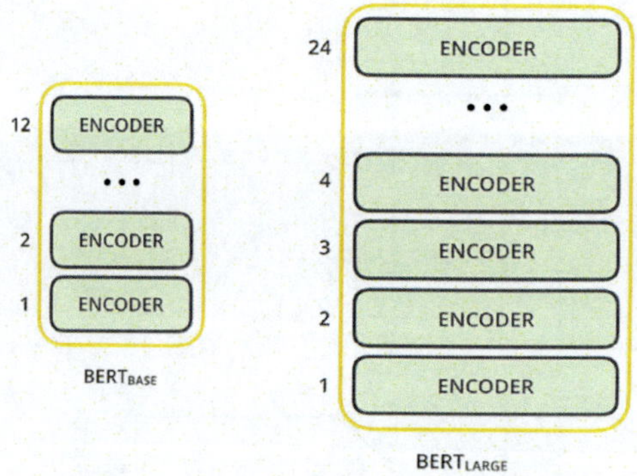

Fig. 15. PhoBert-base and PhoBert-large

The input of the model is a token vector of 512. While the output will be the first output vector being the specific token position [CLS] with the dimension of 768 to handle the classification problem.

Dense Layer: deeply connected to its previous layer, this layer works to resize the output by performing matrix and vector multiplication to create a vector of the new dimension (Fig. 16).

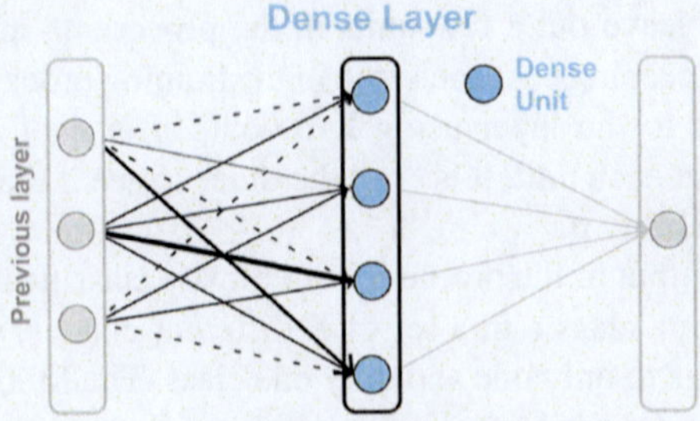

Fig. 16. Dense layer

GELU: the nonlinear activation function with the full name is Gaussian error linear unit. The task of the function is generally to deal with complex nonlinear relationships between data features and model outputs. The GELU function will help the model converge faster than the Sigmoid or RELU functions. The formula of the GELU in number (1) function and the schematic diagram are shown in Fig. 17.

$$g(x) = 0.5x\left(1 + \frac{2}{\sqrt{\pi}}\int_0^{\frac{x}{\sqrt{2}}} e^{-t^2}\,dt\right) \tag{1}$$

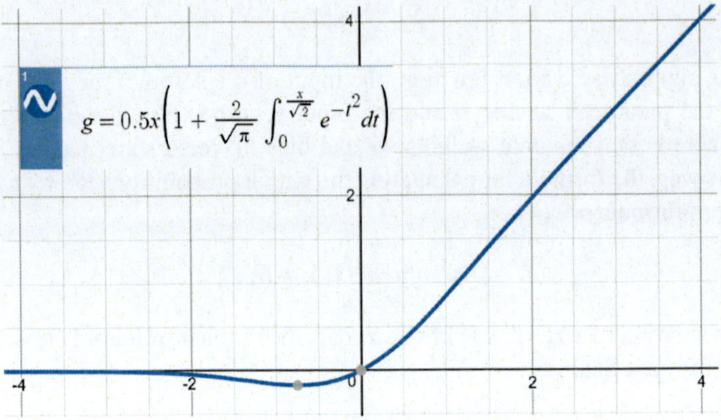

Fig. 17. GELU function

The data crawled from stock Zalo rooms is a complex data set because it contains the opinions of millions of traders on thousands of stock codes. Therefore, applying a nonlinear activation function instead of a linear function will help the neural network evaluate the data more accurately.

- Dropout: A common problem with model training is the problem of overfitting the training data. Overfitting, if it occurs, will make the training quality high, but the actual prediction quality on the test data is significantly low (in other words, it is only learned from the training data). To overcome this problem, the Dropout technique is used to intentionally leave out a few units in the process of spreading information (avoiding the case of peeping). At each step in the training process, when performing Forward Propagation to the layer using Drop-out, instead of calculating all units present on the layer, at each unit, it is "roll the dice" to see if that unit has calculated or not based on probability p.

- Softmax layer: The output of the problem is multiclass classification, that is, labeling each observation with a class c in a set of K different classes, while assuming that an observation belongs to only one and only one class. The following representation will be used: the output y for each input x will be a vector of length K. If class c is the correct class, then set $y_c = 1$ and set all other elements of y to 0, ie $y_c = 1$ and $y_j = 0 \; \forall \, j \neq c$. Such a vector y, with one value = 1 and other values = 0 is called a ***one-hot-vector***. The task of the classifier now becomes to provide a prediction vector $\hat{y}$. For each class k, the value of $\hat{y}$ is estimated by the probability $P(\hat{y}_k{=}1 \mid \mathbf{x})$.
- To calculate $P(\hat{y}_k{=}1 \mid \mathbf{x})$ the softmax function receives the input of vector $\mathbf{z} = [\mathbf{z}_1, \mathbf{z}_2, \ldots, \mathbf{z}_K]$ with z_i $(i = \overline{1, n})$ being the distribution probability. The value of z_i is in the range 0 to 1 and the sum of all elements z_i $(i = \overline{1, n})$ of vector z is 1, meaning,

$$z_1 + z_2 + \cdots + z_K = 1 \tag{2}$$

- With the vector **z** of K dimension, the softmax is defined in calculation no. (3)

$$softmax(z_i) = \frac{\exp(z_i)}{\sum_{j=1}^{K} \exp(z_j)} (1 \leq i \leq K) \tag{3}$$

Like the sigmoid activation function, the input of the softmax function will be the product of the parameter vector **w** and the input vector **x**, plus the bias term b. The difference here is that separate weights w_k and bias b_k vectors are needed for each k class. In this way, the formula for calculating the output probability vector $\hat{y}$ can be built as showing in formula no. (4).

$$\hat{y} = softmax(\boldsymbol{Wx} + \boldsymbol{b}) \tag{4}$$

where $\boldsymbol{W}$ is a matrix of size [K x f] in which K is the number of output classes, and f is the number of input features.

4 Experiments

The experiments are implemented with two datasets which were obtained in two different periods. The first one is February to March 2022 and the second one is May to July 2022.

4.1 Dataset Number One

Data Description

– 8016 samples of crawled data.
– Allocate labels in specific input datasets as shown in Fig. 18.

This dataset was labeled manually by experts to compare with output of the software.

– The data is randomly divided in a 6:2:2 ratio. Concrete,

- Training data is: 4809 samples
- Training model verification data: 1604 samples
- Test result data: 1603 samples

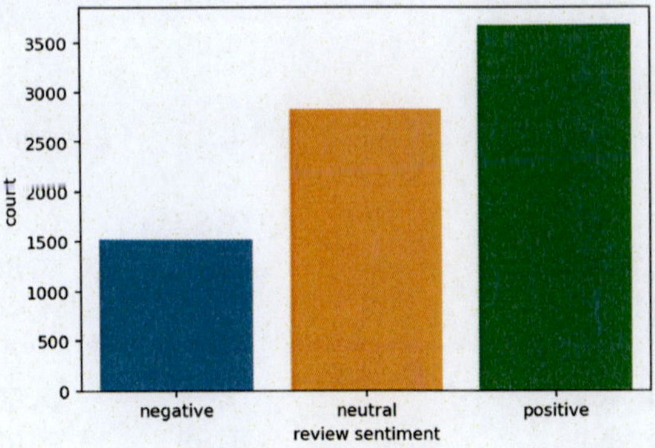

Fig. 18. Data distribution

Results

The target function: Self adjusting Dice Loss [17].

- Test no.1:

	precision	recall	f1-score	support
negative	0.86	0.86	0.86	303
neutral	0.88	0.86	0.87	566
positive	0.90	0.92	0.91	734
accuracy			0.88	1603
macro avg	0.88	0.88	0.88	1603
weighted avg	0.88	0.88	0.88	1603

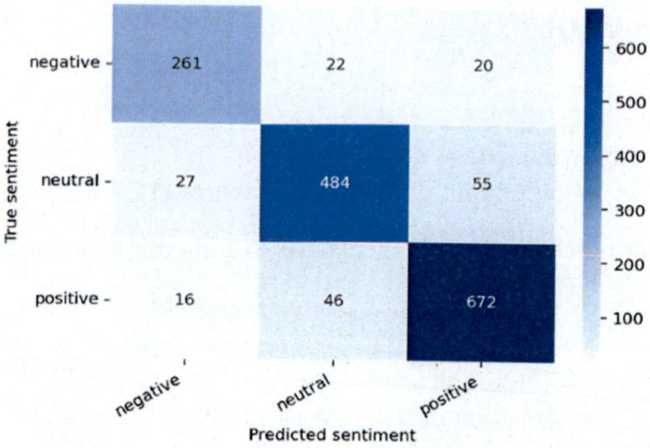

Fig. 19. Result of test no. 1

- Test 2:

	precision	recall	f1-score	support
negative	0.87	0.87	0.87	303
neutral	0.87	0.84	0.86	566
positive	0.89	0.91	0.90	734
accuracy			0.88	1603
macro avg	0.88	0.88	0.88	1603
weighted avg	0.88	0.88	0.88	1603

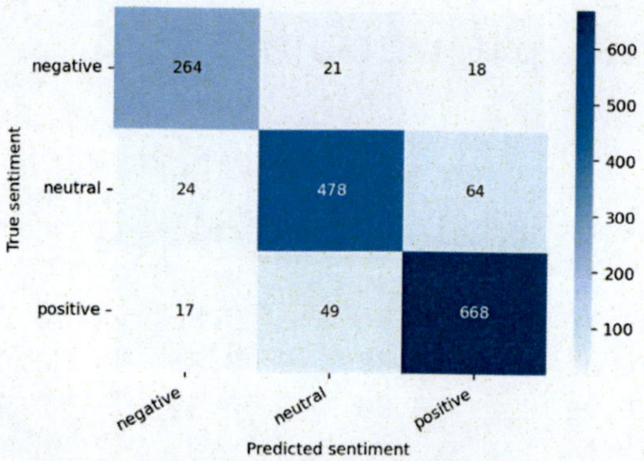

Fig. 20. Result of test no. 2

The precision reaches from 85% to 90% and is relatively uniform in all three cases "Positive", "Negative" and "Neutral".

4.2 Dataset Number Two

Data Description

- The data includes 9017 samples were collected randomly (devided in the rate of 6:2:2). In which, 1785 samples were used for verification of result.
- Target function: Self adjusting Dice Loss [16]

The result is,

	precision	recall	f1-score	support
negative	0.71	0.76	0.74	214
neutral	0.87	0.83	0.85	921
positive	0.83	0.86	0.84	650
accuracy			0.83	1785
macro avg	0.80	0.82	0.81	1785
weighted avg	0.83	0.83	0.83	1785

The precision varies from 70% to 85% and averages around 80% where sentences with "Negative" labels achieve the lowest accuracy at only 70%.

4.3 Comments

- The results of the tests confirm the initial success of the software being built.
- Labeling results depend on the state of the market, when the market is in a good state (uptrend) at a certain stage (2–3) months, then positive comments appear a lot and the percentage of label 1 ("Positive") accounts for quite high. In contrast, in a downtrend, the proportion of -1 ("Negative") labels is high, while when the market is sideways, the ratio of labels 1 and -1 is more balanced.
- The number of labels 0 (neutral) usually accounts for a higher proportion (about 40% to 50%) of labels 1 and -1, which is a normal proportion in chat rooms because much of the information given is not related to the stock market or does not mention a specific ticker. This is also the basic characteristic of a free chat room.

5 Conclusions

The study presented a complete solution for crawling and analyzing data from a popular social networking platform in Vietnam. This information reflects the sentiment of investors according to the real-time movement of the VN-Index. This method can also be applied to other social networking platforms, making the collected information much more diverse and richer.

The work mainly concentrates on data crawling and language processing while labeling job also get encouraging achievement. The study will be able to promote by developing the model. The output of this software also applied for prediction of stock codes or the market. These are the goals of continued studies by authors.

References

1. https://vn.investing.com/indicies/vn
2. Vietnam stock market remains attractive to foreign investors | Business | Vietnam+ (VietnamPlus)
3. 5 pct of Vietnam's population stock investors - VnExpress International
4. Share of Americans investing in stocks 2022 | Statista
5. Zalo – Apps on Google Play
6. Reimers, N., Gurevych, I.: Sentence-BERT: sentence embeddings using Siamese BERT-networks. In: Proceedings of the 2019 Conference on Empirical Methods in Natural Language Processing and the 9th International Joint Conference on Natural Language Processing (EMNLP-IJCNLP), Hong Kong, China, November 2019, pp. 3982–3992 (2019)
7. Reimers, N.: Sentence transformers: Multilingual sentence embeddings using bert/roberta/xlm-roberta & co. with pytorch, March 2021. https://github.com/UKPLab/sentence-transformers
8. Xu, M.: NLP for Stock Market Prediction with Reddit Data. Stanford. https://web.stanford.edu/class/cs224n/reports/final_reports/report030.pdf
9. Bollen, J., Mao, H., Zeng, X.: Twitter mood predicts the stock market. J. Comput. Sci. **2**(1), 1–8 (2011)
10. Arora, S., Liang, Y., Ma, T.: A simple but tough-to-beat baseline for sentence embeddings. In: ICLR (2017)
11. Kiros, R., Zhu, Y., Salakhutdinov, R., Zemel, R.S., Torralba, A., Urtasun, R., Fidler, S.: Skip-thought vectors. CoRR, abs/1506.06726 (2015)
12. Prachyachuwong, K., Vateekul, P.: Stock trend prediction using deep learning approach on technical indicator and industrial specific information. Information **12**, 250 (2021). https://doi.org/10.3390/info12060250(2021)
13. Mingzheng, L., Lei, C., Jing, Z., Qiang, L.: A Chinese stock reviews sentiment analysis based on BERT model. Res. Sq. 2020 (2020, in preprint)
14. Nguyen, D.Q., Nguyen, A.T.: PhoBERT: pre-trained language models for Vietnamese (2020)
15. Vu, T., Nguyen, D.Q., Nguyen, D.Q., Dras, M., Johnson, M.: VnCoreNLP: a Vietnamese natural language processing toolkit. In: Proceedings of NAACL-HLT 2018: Demonstrations, New Orleans, Louisiana, 2–4 June 2018, pp. 56–60. Association for Computational Linguistics (2018)
16. Liu, Y., et al.: RoBERTa: a robustly optimized BERT pretraining approach (2019)
17. Li, X., Sun, X., Meng, Y., Liang, J., Wu, F., Li, J.: Dice loss for data-imbalanced NLP tasks, department of computer science and technology. Zhejiang University (2020)

Two Approaches to E-Book Content Classification

Alexey V. Bosov(✉) and Alexey V. Ivanov

Federal Research Center "Computer Science and Control" of the Russian Academy of Sciences , Vavilova 44/2, Moscow 119333, Russia
ABosov@frccsc.ru, AIvanov@ipiran.ru

Abstract. The problem of automatic assessment of the quality of educational content, in particular, the content of an electronic textbook, in e-learning systems is considered. One of the components of such an assessment is the classification of tasks for independent work or practical examples. The purpose of the classification is to determine the section of the discipline being studied in accordance with the formulation of the task. The problem at hand is complicated by the presence of mathematical expressions in the text content for classification. In this case, mathematical expressions were presented in the TeX scientific markup language. Two approaches have been tried for classification. The first one is to use a direct encoding of content at the character level to extract features and apply image classification methods. The second is to adapt natural language processing algorithms to formulas. In the practical part of the study, we demonstrate both approaches using simple feed-forward neural networks and draw conclusions about the prospects for the practical implementation of the educational content classifier.

Keywords: E-learning system · Learning content · Classification algorithms · Content quality assessment · Machine learning

1 Introduction

An electronic textbook (digital textbook, e-textbook as a special type of e-book) [1] has long been an integral part of the learning process [2], as well as an indispensable tool for distance learning [3].

Computer interactivity is an important part of the electronic textbook and its main feature. In particular, the well-formed content of an electronic textbook can be used to automate the learning process. For example, task solving, testing, and quizzes can be automated. In addition, this content can be utilized to form an individual learning trajectory by automating the selection of tasks, composite tasks, tests, and exam tickets. To some extent, all these functions are implemented in the existing e-learning systems or can be implemented in the near future. The development of such systems includes various stages, such as choosing a data storage platform (DBMS), choosing a learning content management system (LCMS), or solving auxiliary mathematical problems. Examples of such problems are academic performance evaluation and prediction, test generation, test complexity levels evaluation, and more [4–9].

R. Silhavy and P. Silhavy (Eds.): CSOC 2023, LNNS 724, pp. 77–87, 2023.
https://doi.org/10.1007/978-3-031-35314-7_6

Our attention is focused on one of these problems – the problem of assessing the quality of educational content. The need for such an assessment certainly arises when using e-learning systems. Automation tools have not yet been proposed for this fundamental problem. At present, only professional and social communities discuss the quality of traditional textbooks or the educational plans underlying these textbooks. At the same time, the content used in e-learning systems is much more diverse due to its computer interactivity. This content is electronic, so automating its processing and analysis, in particular, seems like a natural task.

Although at present we cannot propose a universal and complete methodology for automating the assessment of the quality of educational content, we have managed to solve one of the tasks that make up such a methodology quite successfully. It is the task of classifying the elements of educational content according to the topics of the discipline being studied. Such elements may include a task for independent work, a test, a practical example, or an examination paper question. Since we can automatically determine the topic of the course being studied, and in the near future the complexity of the task, this provides an almost ready-made tool for quality analysis. Namely, a tool used to assess the completeness and sufficiency of electronic content, determine the complexity of a set of tasks (test, question of an examination paper), generate tasks of a given level of complexity and/or thematic focus, and so on.

Our approach is based on the classification of text content. We use an electronic textbook on the theory of functions of a complex variable. E-book sections serve as categories for classification, and tasks from each section are used to form a data set for training the classifier. If the tasks had exclusively textual formulations in natural language, then the problem would be reduced to an ordinary text classification. The problem at hand is complicated by the presence of mathematical expressions in the text content. We see two possible approaches to consider this.

The first approach is to model content elements as images, that is, directly encoding them at the character level. Partly, attention to this approach is inspired by the most well-known means of recognizing typeset and handwritten mathematical expressions in this area [10–12]. It should be noted that the recognition result in such projects is presented in the formats of electronic editors, in particular LaTeX. Could we look for knowledge about the subject of a content element in the "image" of its formulation, including formulas? We note right away that the TeX scientific markup language [13], which has become the lingua franca of the scientific world, was used as the preferred way to represent mathematical expressions in the content.

The second approach is to use text classification techniques, that is, direct encoding content elements at the word level. There are many projects on the subject of mathematical knowledge search [14], in particular, full-text search using mathematical expressions. Two kinds of mathematical search implementation strategies can be identified. The first strategy is to extend the existing text search system by converting formulas to strings (LeActiveMath [15, 16], MathDex [17], EgoMath [18], DLMFSearch [19], and MIaS [20]). The second strategy, which can be called ontological, is to implement some kind of knowledge extraction model for mathematical expressions (MathWebSearch in [21], WikiMirs in [22], hierarchies in [23, 24]). The mentioned results cannot be directly used in the problem at hand; nevertheless, they encouraged us to choose the natural language processing approach. It is possible to apply procedures such as word segmentation (tokenization), and lemmatization (obtaining the basic dictionary form of a word) to formulas, and then, having formed dictionaries and frequency statistics, apply standard classification algorithms. Expectedly, procedures suitable for natural languages do not work well for formulas presented in TeX. Therefore, our task is to adapt text-processing algorithms to formulas.

This paper is organized as follows. First, it describes the preparation of content for the algorithm of the first approach. Next, an adaptation of text processing algorithms to formulas is presented (Sect. 3). The calculus performed with the textbook at our disposal is discussed in Sect. 4.

2 An Initial Data and an "Image" Model

The textbook on the discipline "Theory of functions of a complex variable" provided by its authors [25] was used as the initial data. The content of the textbook is grouped into 9 sections (chapters). These sections were used as categories for classification. Tasks for classroom work and tasks for independent work were taken from each section. We have translated these tasks into English. In total, 174 tasks from [25] were used. The number of tasks in sections ranged from 6 to 42. In order to make the distribution more uniform, tasks from other textbooks in the same discipline were added to the sections with the least number of tasks. As a result, a training block of 200 tasks was formed, distributed over the existing 9 sections as follows (Table 1).

Table 1. Distribution of tasks by sections.

Section number	Category name	Number of tasks
1	Complex numbers and operations on them	42
2	Functions of a complex variable	31
3	Differentiability. Analytic functions	18
4	Integration of complex variable functions	31
5	Series in complex area	15
6	Function zeros. Isolated Singular Points	16
7	Residues	16
8	Laplace transform	15
9	Application of the operating method	16

The task blocks from each section were saved as separate TeX documents. Each task contains a textual formulation and several formulas, so the description of each task in the final data set has two parts. The first is formed from the text formulation of the task.

The second contains formulas. The procedure for generating these descriptions includes the following steps:

- splitting the document into separate tasks;
- extracting the text part of the task (all formulas are deleted);
- extracting and merging all the formulas related to the task.

Regular expressions were used as the main tool to implement these steps. The procedures for generating the descriptions were written in Python programming language. The step of splitting the text into tasks requires identifying the distinctive features of the text, which determine whether a piece of text is a task number, a formula, or a description. For example, the unique feature of the task number is the presence of the character sequence "No.". As a rule, these characters are bolded using the "\textbf" tag. In a number of tasks, there are references to other task numbers, which should be excluded. The resulting regular expression looks as follows:

```
[^ ](\\textbf{)?(No.)(.*?\})?
```

The following regular expressions were used to extract or delete formulas:

```
\\\(([\s\S]+?)\\\)
\\\[([\s\S]+?)\\\]
\$([\s\S]+?)\$
```

These expressions make it possible to extract formulas in TeX format written using the delimiters "\(" and "\)", "\[" and "\]", "$" and "$".

In addition, when forming task descriptions, it was necessary to take into account the possibility of having several versions for one formulation. Such versions may be designated as "a)", "b)", "c)" or "1)", "2)", "3)". The structure of some tasks has two levels. For example, the student needs to calculate several versions of the integral for several areas. Several tasks have all versions combined into one formula and separated by the character ";". From such tasks, as many tasks were formed as the number of versions actually presented. The final description of each task includes all the formulas belonging to each version plus the formulas available in the shared part of the task formulation. Extracting all versions from a task is done with the following regular expressions:

1. extraction of versions designated as "1)", "2)", "3)"

```
\s\d+\)(?=[\s\*]*\\\(|[\s\*]*\\\[|[\s\*]*\$)
```

2. extraction of versions designated as "a)", "b)", "c)"

```
\s[a-z]\)(?=[\s\*~]*\\\(|[\s\*~]*\\\[|[\s\*~]*\$)
```

In addition, a formula splitting procedure was written for the case when all versions are combined into one formula and separated by the ";" symbol.

Each formed task description also includes a label that contains the number of the textbook section from which the task was extracted. The described process is illustrated in Fig. 1. The source text of the task with the formulas included in it corresponds to block A. Blocks B and C show the steps for forming the "image" model of the task. In this case, the transition to step B means the formation of two vectors filled with characters. The first vector contains the characters of the text part of the task description; the second contains the characters of the formula part. In step C, the final model of the problem is formed. The lengths of the vectors are set as the maximum lengths of the text and formula parts, respectively, out of all 200 tasks available. Characters are replaced with their UTF-16 encoded numeric values. The parts of the vectors that were not filled with characters are filled with zeros. Finally, the numeric values are normalized: since most character codes are in the range 0…127, each character code is divided by 128, and the quotients greater than 1 are limited to 1.

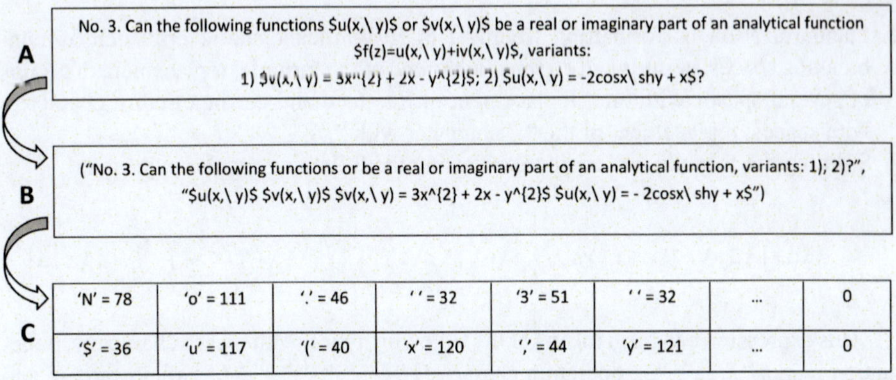

Fig. 1. The "image" model of the task.

In this study, we tried to apply various classification algorithms to the training data set thus formed. We did not get significantly different results. At the same time, the best result for the mixed text-formula model, which is discussed in the next section, was obtained for a multilayer feed-forward neural network [26]. Therefore, we decided to apply such a network to both models. For the model in this section, the network had two hidden layers with 270 and 32 neurons and an activation function $f(x) = \max(0, x)$. The choice of hyperparameters was made by grid search with cross-validation. The network for the model from the next section, which also has two input vectors corresponding to the text and formula parts of the task description, has two hidden layers with 50 and 16 neurons and an activation function $f(x) = \max(0, x)$. We used the implementation of the multilayer neural network given in https://scikit-learn.org/stable/modules/neural_networks_supervised.html#neural-networks-supervised.

3 Classification Based on a Mixed Text-Formula Model

3.1 Data Preparation

Unlike the previous model, the preparation of task descriptions for a mixed text-formula model contains an additional step that eliminates the ambiguity of the notation of mathematical expressions in TeX. For example, the cosine function can be written as the tag "\cos" or the character sequence "cos". The same is true for other functions. Another example is the possible omission of some operations, such as multiplication, and parentheses around function arguments. If such situations are not excluded when preparing task descriptions, this will lead to ambiguities at the stage of formula tokenization. In particular, operation signs, TeX tags, sequences of Latin letters, and numbers are considered tokens. As a result, such tokens as "xe", "iy", "zdz", "cost" may appear, which should have been presented as two separate tokens. This is why it was necessary to prepare the formulas before they were included in the task descriptions. A procedure has been developed for taking into account the ambiguity of mathematical expressions notation that performs:

1. Standardization of formatting (removal of insignificant characters, such as non-breaking spaces, periods and commas at the end of the formula; replacement of groups of several spaces with one; replacement of line feed and carriage return characters with spaces; replacement of the ";" character with ",");
2. Splitting the formula into basic tokens using the following regular expression

```
(\d+|\d+\.?\d+|[\s,/\+\-\|\:\.\_\^\(\)\{\}<>]|\\\{|\\[a-
zA-Z\-\.]+)
```

This expression can form tokens of the following types: whitespace characters, punctuation marks ",", ".", ":", operation characters "+", "-", "/", "<", ">", lowercase and superscript "_", "^", brackets " (",")", "{","}", "|", TeX tags, integers and real numbers;

3. Decomposition of tokens of the "sequence of Latin letters" type by recognizing and standardizing the notation of special names, including trigonometric, hyperbolic, and logarithmic function names. For example, the token "coshy" is replaced by "\cosh y". The special names "Re", "Im", "matrix", "res" are processed using the following regular expression:

```
^(d[a-z]|Re|Im|matrix|res)
```

This expression is used by a procedure that allows replacing character sequences of the form "xdx" with "x dx", "Imz" with "Im z", etc. The rest of the character sequence, which does not contain special names, is split up letter-by-letter and forms separate tokens.

In the illustration for the mixed text-formula model of the task (Fig. 2), this stage corresponds to block E. Pay attention to the difference with the block B from Fig. 1, which is illustrated by the elements "\cos x" and "\sh y".

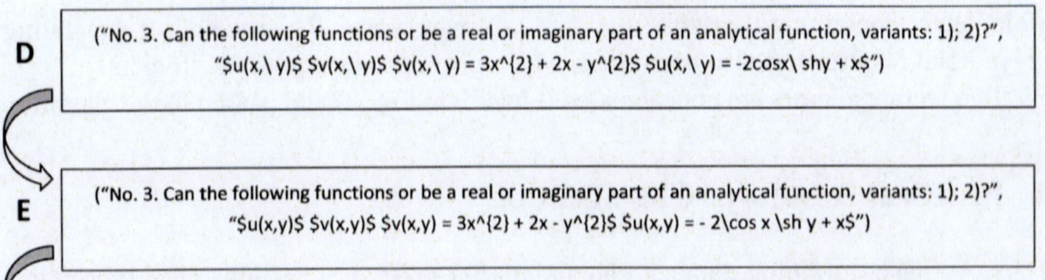

1	2	3	...	...	...	162	1	2	3	4	...	...	70
Can following	Following function	real or	...	...	...	...	u	x	y	\cos	...	...	...
0.3	0.5	0.3	...	...	...	...	0.3	0.5	0.4	0.2	...	...	...

Fig. 2. The text-formula model of the task.

3.2 Feature Extraction in a Text-Formula Model

Next, two feature vectors were formed from the text and formula parts of the task descriptions. The feature vector of the text description of the task was formed using the "bag of words" model [27], which is typical for natural language processing. Task texts are split into tokens, then a dictionary is formed from the tokens, then the text model is formed as a histogram – a vector of weights of all elements of the dictionary in the task description. The word weights were calculated using the tf-idf measure [28], which is the frequency of a word in a description multiplied by the reciprocal of the frequency of a word in all descriptions (in the corpus). The spaCy library, https://spacy.io/, was used to perform text tokenization (lexical analysis) and lemmatization (word form normalization). The scikit-learn library, https://scikit-learn.org/stable/, was used to generate the dictionary and vectors. Spaces, stop words, punctuation marks, and numbers were excluded from the generated list of tokens. Canonical forms of words (lemmas) were used as tokens. Words found in more than 70% of task descriptions were not included in the dictionary. Since the "bag of words" representation is not sensitive to the order of words in a sentence, the n-gram mechanism was additionally involved. Bigrams gave the best result, that is, the dictionary was made up of pairs of adjacent words.

The feature vector from the formula description of the problem was also formed using the "bag of words" model, but most of the operations with tokens required significant customization. The dictionary and the vector with the measure tf-idf were formed in the same way as the text part.

To split the text of formulas into basic tokens, the same regular expression was used as in the data preparation stage (see Sect. 3.1). Next, whitespace characters, "." character, TeX tags "\left", "\right", "\left.", "\right.", which only affect the visual representation of the formula, are excluded from the resulting list of tokens. In addition, during the formula tokenization process, all tokens corresponding to integer and real numbers are replaced with the "[number]" token, which reduces the size of the dictionary and improves classification accuracy.

All tokens, regardless of the frequency of their appearance in the descriptions, were included in the dictionary. An attempt to apply the mechanism of n-grams was also made for the formula part of the task, but it did not lead to success: the classification result with a dictionary containing n-grams (n = 2, …, 9) was worse. The method of generating polynomial features has also been tried, which proved to be equally inefficient.

Two feature vectors are concatenated (block F in Fig. 2) and passed to the classifier.

4 Practical Results of Classification

Thus, we have a training data set containing 200 tasks in 9 sections. The procedures described in the previous section have been applied to task descriptions; as a result, two sets of input vectors were formed: I) using an "image" model of the task, II) using a text-formula model. We applied the previously selected feed-forward neural networks to each set and analyzed the classification quality. To do this, the existing set of task descriptions was randomly split into two parts: the first part of the set was used for training; the remaining tasks were used to evaluate the classification quality. Three ratios of training and control parts were used: a) 100–100, b) 125–75, c) 150–50. Thus, for example, the result for case IIb in the analysis results table below means a mixed model and a network trained on 125 examples and tested on 75.

It should also be noted that the lengths of the vectors for model I) were 536 for the text part and 302 for the formula part, and the mixed model II) gave a dictionary size of 162 for the text part and 70 for the formula part.

The quality of classification is assessed by the common metric "precision-recall" [29] that is, the harmonic mean of the ratios of the results of correct classifications to their sum with the number of false positives (accuracy) and the ratios of the results of correct classifications to their sum with the number of false negatives (recall). This metric, known as the f1-score, is calculated for all categories and all test cases together. In the case of IIc, this metric reached 1.0. However, this result raises obvious doubts about the reliability of the quality assessment and requires a significant increase in the size of the training data, which was impossible to achieve within the framework of this study (Table 2).

Table 2. Classification quality analysis.

Model and amount of training	F1-score
Ia	0.71
Ib	0.80
Ic	0.86
IIa	0.93
IIb	0.96
IIc	1.00

5 Conclusions

We have interpreted the results as follows. First, the text-formula model shows the best results. It is the expected result since feature extraction in this model is performed by methods, while empirical, but highly adapted to the language specifics. That is, even though the language of formulas is fundamentally different from the natural language of people, it is still a language, and natural language processing methods can be successfully applied to it. Secondly, the "image" model showed a rather weak result, somewhat unexpected. At the beginning of the study, we assumed that in such a simple formulation, the network would be able to extract, if not the mathematical knowledge itself, then its features that are significant in the problem at hand. These expectations were not met for this model.

To identify a possible cause within the framework of this study, we carried out several additional computer experiments. In particular, we tried to classify tasks using only the formula part. As a result, the f1-score value for the text-formula model decreased to the same order of magnitude that the "image" model gives on the full data set (Ia, Ib, Ic values in the table). We interpreted this result as a significantly worse ability of the "image" model to extract features from natural language text, while both models perform approximately the same when extracting features from formula parts.

A number of experiments were also carried out with other classification algorithms and other neural network architectures. They gave the following results. For the "image" model, the naive Bayes classifier shows the worst result, the other algorithms do not show significant differences in classification quality. For the text-formula model, all algorithms show acceptable classification quality, but the multilayer feed-forward network gives the best result.

Finally, it should be noted that the quality of the classification shown by the text-formula model (IIa, IIb, IIc in the table) allows us to talk about the possibility of its practical application in existing e-learning systems.

Acknowledgments. The study was supported by the Russian Science Foundation Grant No. 22–28-00588, https://rscf.ru/project/22-28-00588/. The research was carried out using the infrastructure of the Shared Research Facilities «High Performance Computing and Big Data» (CKP «Informatics») of FRC CSC RAS (Moscow).

References

1. Suarez, S.J., Michael, F., Woudhuysen, H.R.: The Oxford Companion to the Book. Oxford University Press, Oxford (2010)
2. Baek, E.-O., Monaghan, J.: Journey to textbook affordability: an investigation of students' use of eTextbooks at multiple campuses. Int. Rev. Res. Open Distrib. Learn. **14**(3), 1–26 (2013)
3. Garrison, D.R.: E-Learning in the 21st Century: A Framework for Research and Practice. Taylor & Francis (2003)
4. Van der Linden, W.J., Scrams, D.J., Schnipke, D.L., et al.: Using response-time constraints to control for differential speededness in computerized adaptive testing. Appl. Psychol. Meas. **23**(3), 195–210 (1999)

5. Rasch, G.: Probabilistic Models for Some Intelligence and Attainment Tests. The University of Chicago Press, Chicago (1980)
6. Kibzun, A.I., Inozemtsev, A.O.: Using the maximum likelihood method to estimate test complexity levels. Autom. Remote Control **75**(4), 607–621 (2014)
7. Naumov, A.V., Mkhitaryan, G.A.: On the problem of probabilistic optimization of time-limited testing. Autom. Remote Control **77**(9), 1612–1621 (2016)
8. Kuravsky, L.S., Margolis, A.A., Marmalyuk, P.A., Panfilova, A.S., Yuryev, G.A., Dumin, P.N.: A probabilistic model of adaptive training. Appl. Math. Sci. **10**(48), 2369–2380 (2016)
9. Bosov, A.V.: Adaptation of Kohonen's self-organizing map to the task of constructing an individual user trajectory in an e-learning system. In: Silhavy, R., Silhavy, P., Prokopova, Z. (eds.) CoMeSySo 2022. LNNS, vol. 597, pp. 554–564. Springer, Cham (2023). https://doi.org/10.1007/978-3-031-21438-7_44
10. Zanibbi, R., Blostein, D., Cordy, J.R.: Recognizing mathematical expressions using tree transformation. IEEE Trans. Pattern Anal. Mach. Intell. **24**(11), 1455–1467 (2002)
11. Blostein, D., Grbavec, A.: Recognition of mathematical notation. In: Wang, P.S.P., Bunke, H. (eds.) Handbook of Character Recognition and Document Image Analysis, pp. 557–582. World Scientific Publishing Company (1997)
12. Chan, K.F., Yeung, D.Y.: Mathematical expression recognition: a survey. Int. J. Doc. Anal. Recognit. **3**(1), 3–15 (2000)
13. Knuth, D.E.: The TeXbook. Addison-Wesley, Reading (1984)
14. Guidi, F., Coen, C.S.: A survey on retrieval of mathematical knowledge. Math. Comput. Sci. **10**(4), 409–427 (2016)
15. Libbrecht, P., Melis, E.: Methods to access and retrieve mathematical content in ACTIVE-MATH . In: Iglesias, A., Takayama, N. (eds.) ICMS 2006. LNCS, vol. 4151, pp. 331–342. Springer, Heidelberg (2006). https://doi.org/10.1007/11832225_33
16. Libbrecht, P., Melis, E.: Semantic search in leactivemath. In: Proceedings of the First WebALT Conference and Exhibition, Eindhoven, Hollan, pp. 97–109 (2006)
17. Miner, R., Munavalli, R.: An approach to mathematical search through query formulation and data normalization. In: Kauers, M., Kerber, M., Miner, R., Windsteiger, W. (eds.) MKM Calculemus 2007 2007. LNCS, vol. 4573, pp. 342–355. Springer, Heidelberg (2007). https://doi.org/10.1007/978-3-540-73086-6_27
18. Mišutka, J., Galamboš, L.: System description: EgoMath2 as a tool for mathematical searching on Wikipedia.org. In: Davenport, J.H., Farmer, W.M., Urban, J., Rabe, F. (eds.) CICM 2011. LNCS, vol. 6824, pp 307–309. Springer, Heidelberg (2011). https://doi.org/10.1007/978-3-642-22673-1_30
19. Miller, B.R., Youssef, A.: Technical aspects of the digital library of mathematical functions. Ann. Math. Artif. Intell. **38**(1), 121–136 (2003)
20. Sojka, P., Líška, M.: Indexing and searching mathematics in digital libraries. In: Davenport, J.H., Farmer, W.M., Urban, J., Rabe, F. (eds.) CICM 2011. LNCS, vol. 6824, pp. 228–243. Springer, Heidelberg (2011). https://doi.org/10.1007/978-3-642-22673-1_16
21. Kohlhase, M., Anca, S., Jucovschi, C., Palomo, A.G., Sucan, I.A.: MathWebSearch 0.4, a semantic search engine for mathematics (manuscript 2008). http://mathweb.org/projects/mws/pubs/mkm08.pdf
22. Hu, X., Gao, L.C., Lin, X.Y., Zhi, T., Lin, X.F., Baker, J.B.: Wikimirs: a mathematical information retrieval system for Wikipedia. In: Proceedings of the 13th ACM/IEEE-CS Joint Conference on Digital Libraries, Indiana, pp. 11–20, ACM (2013)
23. Tian, X.: A mathematical indexing method based on the hierarchical features of operators in formulae. In: Proceedings of the 2017 2nd International Conference on Automatic Control and Information Engineering (ICACIE 2017), pp. 49–52. Atlantis Press (2017)

24. Liu, H., Tian, X., Tian, B., Yang, F., Li, X.: An improved indexing and matching method for mathematical expressions based on inter-relevant successive tree. J. Comput. Commun. **4**(15), 63–78 (2016)
25. Bityukov, Yu.I., Martyushova, Ya.G.: Solving Problems in the Theory of Functions of a Complex Variable. MAI Publishing House (2022)
26. Haykin, S.: Neural Networks and Learning Machines, 3rd edn. Pearson Education, New Jersey (2009)
27. McTear, M.F., Callejas, Z., Griol, D.: The Conversational Interface: Talking to Smart Devices. Springer, Cham (2016). https://doi.org/10.1007/978-3-319-32967-3
28. Salton, G., McGill, M.J.: Introduction to Modern Information Retrieval. McGraw-Hill, New York (1983)
29. Van Rijsbergen, C.J.: Information Retrieval, 2nd edn. Butterworth-Heinemann (1979)

On the Geometry of the Orbits of Killing Vector Fields

A. Ya. Narmanov(✉) and J. O. Aslonov

National University of Uzbekistan, Universitet Str., Tashkent 100174, Uzbekistan
narmanov@yandex.ru

Abstract. The study of transformations that preserve the space-time metric plays an extremely important role in mathematical physics. It's sufficient to say that the most important conservation laws are associated with such transformations. These transformations generate the so-called Killing vector field. Killing vector fields in physics indicate the symmetry of the physical model and help find conserved quantities such as energy and momentum. In this paper, we study the question about the classification of the geometry of orbits of Killing vector fields #CSOC1120.

Keywords: Killing Vector Fields · Orbit of the Family of Vector Fields · Foliation

1 Introduction

The geometry of Killing vector fields was studied in the works of W. Killing [1], V. N. Berestovsky [2, 3], Yu. G. Nikonorova [2, 3], M.O. Katanaev [4], A.Narmanov [5, 6] and other authors.

As noted above, in a number of areas of physics, for example, in the theory of the electromagnetic field, in the theory of heat, in static physics and in the theory of optimal control, it is necessary to consider not only vector fields, but a family of vector fields. In this case, the main object of research is the orbit of the system of vector fields.

At present, the geometry of the orbits is one of the important problems of modern geometry, which has been studied by many mathematicians due to its importance in optimal control theory, differential games, and the geometry of singular foliations [5, 7–13].

Definition 1. If the infinitesimal transformations $x \to X^t(x)$ of the field X on the Riemannian manifold M preserves the distance between points X is called a Killing vector field.

Example 1. On the Euclidean space $R^3(x, y, z)$, we have Killing fields:

$$X_1 = \frac{\partial}{\partial x},\ X_2 = \frac{\partial}{\partial y},\ X_3 = \frac{\partial}{\partial z},\quad X_4 = -y\frac{\partial}{\partial z} + z\frac{\partial}{\partial y},$$

R. Silhavy and P. Silhavy (Eds.): CSOC 2023, LNNS 724, pp. 88–94, 2023.
https://doi.org/10.1007/978-3-031-35314-7_7

$$X_4 = -y\frac{\partial}{\partial z} + z\frac{\partial}{\partial y},\ X_5 = -z\frac{\partial}{\partial x} + x\frac{\partial}{\partial z},\ X_6 = -x\frac{\partial}{\partial y} + y\frac{\partial}{\partial x}.$$

The infinitesimal transformations of fields $X_1,\ X_2,\ X_3$ are parallel transfers in the direction of the coordinate axes and the fields $X_4,\ X_5,\ X_6$ generate rotations.

Example 2. The Killing vector field on $R^4(x_1, x_2, x_3, x_4)$

$$X = -x_2\frac{\partial}{\partial x_1} + x_1\frac{\partial}{\partial x_2} - x_4\frac{\partial}{\partial x_3} + x_3\frac{\partial}{\partial x_4}.$$

is tangent to the three dimensional sphere S^3. The vector lines of this vector field generate a smooth bundle called Hopf bundle.

We need the following proposition [2, 9].

Proposition 1. For a vector field $X = \sum_{i=1}^{n} \xi_i \frac{\partial}{\partial x_i}$ the condition $\frac{\partial \xi_i}{\partial x_j} + \frac{\partial \xi_j}{\partial x_i} = 0\ i \neq j,\ \frac{\partial \xi_i}{\partial x_i} = 0,\ i = 1, ..., n.$ is necessary and sufficient for it to be a Killing field.

2 The Geometry of Killing Vector Fields

In this section we consider Killing vector fields on two-dimensional surfaces.

Theorem 1. Every Killing vector field on the two-dimensional cylinder is vector field of constant length.

Proof. Suppose the considered surface M is parameterized as follows.

$$\begin{cases} x = Rsinu, \\ y = Rcosu, \\ z = v. \end{cases}$$

The following fields

$$X_1 = -x\frac{\partial}{\partial y} + y\frac{\partial}{\partial x}, X_2 = \frac{\partial}{\partial z}.$$

are Killing vector fields on this manifold.

The vector lines of these fields are, respectively, circles and straight lines, which are parallel to the generatrix of the cylinder. It is known that these lines are geodesic on the cylinder.

For every Killing vector field X on M there are smooth functions $\lambda_1(x, y, z)$ and $\lambda_2(x, y, z)$ such that $X = \lambda_1(x, y, z)X_1 + \lambda_2(x, y, z)X_2$. From proposition 1we obtain the following equality: $\lambda_1(x, y, z) = \lambda_1(z),\quad \lambda_2(x, y, z) = \lambda_2(x, y)$ and

$$\frac{\partial \lambda_1}{\partial z} + y\frac{\partial \lambda_2}{\partial x} = 0, \quad \frac{\partial \lambda_1}{\partial z} - x\frac{\partial \lambda_2}{\partial y} = 0. \tag{1}$$

Lie bracket $[X, X_2]$ has the following form

$$[X, X_2] = \lambda_1[X_1, X_2] + X_2(\lambda_1)X_1 + \lambda_2[X_2, X_2] + X_2(\lambda_2)X_2.$$

A simple calculation shows $[X_1, X_2] = 0$. By $[X_2, X_2] = 0$ we have $[X, X_2] = X_2(\lambda_1)X_1 + X_2(\lambda_2)X_2$ and $\frac{\partial \lambda_1}{\partial z} = 0$. It follows from (1) $\frac{\partial \lambda_2}{\partial x} = \frac{\partial \lambda_2}{\partial y} = 0$. Consequently, $\lambda_1(x, y, z)$ and $\lambda_2(x, y, z)$ are constant.

For a vector line of the field X we have the following system:

$$\begin{cases} \frac{\partial x}{\partial t} = \lambda_1 y \\ \frac{\partial y}{\partial t} = -\lambda_1 x \\ \frac{\partial z}{\partial t} = \lambda_2 \end{cases} \tag{2}$$

The integral lines of the vector field is a helix, if $\lambda_1 \neq 0, \lambda_2 \neq 0$. If $\lambda_1 = 0, \lambda_2 \neq 0$ it is a straight line, if $\lambda_1 \neq 0, \lambda_2 = 0$ it is a circle.

Example 3. Let $M = S^2 \times R^1$ be embedded in R^4 using the following parametric equations

$$\begin{cases} x = \cos u \sin v \\ y = \cos u \cos v \\ z = \sin u \\ w = t. \end{cases}$$

The field $X = y\frac{\partial}{\partial x} - x\frac{\partial}{\partial y}$ is the Killing field on the M. The vector lines of the field X passing with starting point (x_0, y_0, z_0, w_0) has the form

$$\begin{cases} x(t) = x_0 cost + y_0 sint \\ y(t) = -x_0 sint + y_0 cost \\ z(t) = z_0 \\ w(t) = w_0 \end{cases}$$

If $z_0 \neq 0$, then the trajectory of this system is not a great circle on S^2, cconsequently it is not a geodesic.

Definition 2. The orbit $L(x)$ of the system D of fields with starting point x is defined (see [8]) as the set of points $y \in M$ for which there exist real numbers $t_1, t_2, \dots, t_k$ and vector fields $X_{i_1}, X_{i_2}, \dots, X_{i_k}$ from D (where k is positive integer) such that

$$y = X_{i_k}^{t_k}(X_{i_k-1}^{t_k-1} \dots (X_{i_1}^{t_1}(x)))$$

Example 4. We consider fields $X = 2x\frac{\partial}{\partial y} + y\frac{\partial}{\partial z}$ and $Y = 2z\frac{\partial}{\partial y} + y\frac{\partial}{\partial x}$ on three-dimensional Euclidean space. Their Lie bracket is $[X, Y] = 2x\frac{\partial}{\partial x} - 2z\frac{\partial}{\partial z}$. The first integral of these vector fields is the surface $y^2 - 2xz = 0$. The orbit of the system D is surfaces that are defined using the equations $y^2 - 2xz = C$, where C is a constant.

3 The Classification of Geometry of Orbits

We have the following classification theorem.

Theorem 2. The geometry of orbits of Killing vector fields on R^3 have one of the following types:

1) All orbits are parallel straight lines;

2) The orbits are concentric circles and centers of these concentric circles.
3) The orbits are helical lines;
4) The orbits are parallel planes;
5) The orbits are concentric spheres and a point;
6) The orbits are concentric cylinders and axis of cylinders;
7) Every orbit coincides with R^3.

For the proof of the Theorem 2 one needs the following lemma from [14].

Lemma. Let F be a partition of the complete Riemannian manifold M into orbits of Killing fields. Let γ_0 is geodesic with starting point x_0 to the point y_0, orthogonal to the orbit. Then for every $x \in L(x_0)$ there $\exists$ a geodesic γ with starting point x to the point of the orbit $L(y_0)$ of length γ_0 and orthogonal to the all orbits.

Proof of the Theorem 2. 1) We assume $L(p_0) = p_0$ for some a unique point p_0. In this case the set D must contain more than one vector field. Let S_r^2 be a sphere with the radius $r > 0$ centered at p_0. Then the infinitesimal transformations of the vector fields from D take the sphere S_r^2 into itself. It follows the orbit $L(q)$ for the point $q \in S_r^2$ is S_r^2. Consequently the orbits are concentric spheres and a point p_0.

2) Let us assume $\exists$ p_1 and p_2 such that $L(p_i) = \{p_i\}, i = 1, 2$. Hence it follows that the straight line p_1p_2 is fixed. For other points not on this line, the orbit is a circle with center on this line.
3) Assume $dimL(p) = 1$ for all $p \in R^3$. Then the orbits generate one dimensional foliation.By Lemma leaves of this foliation are helical lines with a common axis or all leaves of the foliation are parallel lines.
4) We assume $dimL(p) \geq 1$ for all points and $dimL(p) = 2$ for some point p and $dimL(q) = 1$ for some point q.

Let $dimL(q) = 1$ for a point q. We have possibilities:

a) the orbit $L(q)$ be a straight line, there are rotations around $L(q)$. It follows the orbits are concentric cylinders with common axis $L(q)$.
b) If orbit $L(q)$ is a helix, then the case is similar to the previous case.

5) If $dimL(p) = 2$ for all points p it follows from [7] that the leaves are parallel planes.
6) There is the point q such that $\dim L(q) = 3$. In this case, from [8] and [9], we have $L(q)$ is a open subset of R^3. It follows from the paper [6] $L(p_0)$ is a closed set. Consequently $L(q) = R^3$ Theorem is proved.

4 On the Compactness of the Orbits

This part of the paper devoted to the geometry of Killing fields under the condition that the vector fields and the Riemannian metric are connected by the condition.

Assume D be a system of smooth vector fields defined on a smooth manifold M of dimension n. Let $A(D)$ denote the minimal Lie subalgebra that contains D, $A_x(D)$ the vector space, consisting of all vectors $\{X(x) : X \in A(D)\}$.

If $dimA_x(D) = k$ for all $x \in M$, where $0 < k < n$, orbits of the system D generate a k dimensional foliation [8]. If we suppose $dimA_x(D) = n - 1$ for all $x \in M$, then the orbits of the system are submanifolds of dimension $n - 1$ [8].

Theorem 3. If $Xg(Y, Z) = g([X, Y], Z) + g(Y, [X, Z])$ for $X \in A(D)$ and for every fields $Y, Z \in V(M)$ on the complete manifold M, then if some orbit is compact, then all orbits are compact.

Proof. It was proved in [14] that under the condition of the theorem, which connects vector fields and the Riemannian metric g, the orbits of the system D generates a Riemannian foliation of codimension one.

We denote by $F(x)$ the tangent space of the orbit $L(x)$ at the point x, $H(x)$ the orthogonal complement $F(x)$ in T_xM, $x \in M$, where T_xM is the tangent space at x. Two subbundles appear $TF = \{F(x) : x \in M\}$ and $H = \{H(x) : x \in M\}$, where H is the orthogonal complement of TF.

A line $\gamma : [0, 1] \to M$ is called horizontal if $\dot{\gamma}(t) \in H(\gamma(t))$ for each $t \in [0, 1]$. A line that lies in the leaf of the foliation F is called vertical.

We suppose the orbit $L_0 = L(x_0)$ is a compact set. Let L be some orbit other than L_0. For an arbitrary point $x \in L$ by $d(x, L_0)$ we denote the distance from the point x to the orbit L_0. Due to the fact that L_0 is a compact orbit, there is a point $p_x \in L_0$ such that $d(x, L_0) = d(x, p_x)$, where $d(x, p_x)$ is the distance. Since the Riemannian manifold M is complete, there is a geodesic line γ_x that realizes this distance [14].

This means that the geodesic γ_x connects the points x, p_x and its length is equal to the distance $d(x, p_x)$. Note that, in addition, this geodesic is orthogonal to the leaf L_0 [14] and, since the foliation is Riemannian, it is orthogonal to all leaves.

Let $z \in L$ be a point of the orbit other than the point x, $\nu : [0, 1] \to L$ a curve in L connecting the points x and z: $\nu(0) = x$, $\nu(1) = z$. As the foliation F is Riemannian and the M is complete, for each pair of vertical and horizontal lines $\nu, h : I \to M$ with $h(0) = \nu(0)$, there is a piecewise smooth mapping $P : I \times I \to M$ such that the line $t \to P(t, s)$ is a vertical line for every $s \in I$, and the line $s \to P(t, s)$ is horizontal for every $t \in I$, where $P(t, 0) = \nu(t)$ for every $t \in I$ and $P(0, s) = h(s)$ for every $s \in I$. This homotopy is called the vertical-horizontal homotopy.

By the theorem in [8], there exists a homotopy $P : I \times I \to M$ for the curves ν, γ_x such that $P(0, s) = \gamma_x(s)$ for $s \in I$ and for each $t \in I$ the curve $s \to P(t, s)$ is a horizontal geodesic of length $d(x, p_x)$. Thus, the length of the geodesic $\gamma_z(s) = P(1, s)$ is equal to $d(x, p_x)$.

Thus, from each point z of the orbit L comes the geodesic $\gamma_z : [0, 1] \to M$, the length of which is equal to the distance from the point $z \in L$ to L_0, and the lengths of all of geodesics γ_z are equal to $d = d(x, p_x)$.

Now we will show that, the orbit L is compact. Since the orbit L has dimension $n - 1$, only one geodesic passes through each point $z \in L$. Therefore, the geodesic flow $z \to exp_z(v)$ transfers L to L_0, where $v \in T_zM$ is the horizontal vector and $|v| = d$, Since the geodesic smoothly depends on the initial points, this map is bijective, and therefore a diffeomorphism. In particular, as the image of a compact set under a diffeomorphism, the orbit L is compact. The theorem is proved.

5 Applications in Partial Differential Equations

Definition 4. [10] A group G of transformations acting on the set M is called the symmetry group of differential equation if every element $g \in G$ transforms a solution of the equation to a solution.

Let us to bring simple example. For the heat equation, $u_t = u_{xx}$ Killing vector field $X = a\frac{\partial}{\partial t} + b\frac{\partial}{\partial x}$ is infinitesimal generator of symmetry group of the heat equation. The vector field X generates the group of following transformations. We can check that the function $f = bt - ax$ is invariant function of the transformation group. This invariant function allows us to search solution of heat equation in the form $u(t, x) = V(\xi)$, where $\xi = bt - ax$. The function $u(t, x) = V(\xi)$ is a solution of second-order differential equation: $a^2V'' - bV' = 0$. As result, we obtain large class of solutions of the heat equation:

$$u(t, x) = C_1\frac{a}{b}e^{bt-ax} + C_2,$$

where C_1, C_2 are arbitrary constants.

The geometry of Killing fields is used for the study of partial differential equations [10, 15].

6 Conclusion

The topology of the orbits of vector fields on a connected Riemannian manifold is studied when there is a given connection between the vector fields and the Riemannian metric. Classification of the orbits is obtained. In the last section, it is given some applications of the geometry of vector fields in in theory of partial differential equations.

References

1. Killing, W.: Ueber die Grundlagen der geometrie. J. Reine Angew. Math. **109**, 121–186 (1892)
2. Berestovskii, V.N., Nikonorov, Y.: Killing vector fields of constant length on Riemannian manifolds. Siber. Math. J. **49**(3), 395–407 (2008)
3. Berestovskii, V.N., Nikitenko, E.V., Nikonorov, Y.: Classification of generalized normal homogeneous Riemannian manifolds of positive Euler characteristic. Differ. Geom. Appl. **29**(4), 533–546 (2011)
4. Katanaev, M.O.: Geometrical methods in mathematical physics. Applications in quantum mechanics. Part 1. Lektsionnye Kursy NOC 25, pp. 3–174 (2015). https://doi.org/10.4213/lkn25
5. Azamov, A., Narmanov, A.: On the limit sets of orbits of systems of vector fields. Differ. Equ. **2**, 271–275 (2004)
6. Narmanov, A.Ya., Saitova, S.S.: On the geometry of orbits of killing vector fields. Differ. Equ. **50**(12), 1584–1591 (2014)
7. Narmanov, A., Qosimov, O.: Geometry of singular Riemannian foliations. Uzbek Math. J. **3**, 129–135 (2011)

8. Sussman, H.: Orbits of families of vector fields and integrability of distributions. Transl. AMS **180**, 171–188 (1973)
9. Stefan, P.: Accessible sets, orbits and foliations with singularities. Bull. AMS **80**(6), 699–713 (1974)
10. Olver, P.: Applications of Lie Groups to Differential Equation. Springer, New York (1989). https://doi.org/10.1007/978-1-4684-0274-2
11. Sussmann, H.: Differ. Equ. **20**, 292–315 (1976)
12. Lovric, M.: Rocky Mt. J. Math. **30**, 315–323 (2000)
13. Sachkov, Y.: Rus. Math. Surv. **77**(1), 99–163 (2022)
14. Narmanov, A.: On the transversal structure of controllability sets of symmetric control systems. Differ. Equ. **32**(6), 780–787 (1996)
15. Narmanov, O.: Invariant solutions of two dimensional heat equation. Vestnik Udmurtskogo Universiteta. Matematika. Mekhanika. Komp'yuternye Nauki **29**(1), 52–60 (2019)

The Classification of Vegetations Based on Share Reflectance at Spectral Bands

S. Kerimkhulle(✉), Z. Kerimkulov, Z. Aitkozha, A. Saliyeva, R. Taberkhan, and A. Adalbek

L.N. Gumilyov Eurasian National University, 2, Satpayev Street, Astana 010008, Kazakhstan
kerimkhulle@gmail.com

Abstract. This paper studies the problem of classification for the vegetation based on the share reflectance at spectral bands. For this, the theory and methodology of regression and data analysis, algorithms and technologies of remote sensing, several modern scientific literatures are involved. As a result, a system of regression models was built with factor variables Barley, Corn, and Wheat vegetation in the share reflectance at bands of the spectral space and the parameters based on ordinary least squares method were estimated #CSOC1120.

Keywords: Data Analysis · Remote Sensing · Least Squares Method

1 Introduction

Today, the problem of using remote sensing technology is relevant and a huge amount of scientific literature is devoted to this topic, some of which are given in this study.

Known that [1] precision agriculture is used for mapping, monitoring and analysis to changes in vegetations. In particular, the paper proposes a fast and reliable semi-automated workflow implemented for processing unmanned aerial vehicle multispectral images and aimed at detecting and extracting crowns of olive and citrus trees located in the region of Calabria (Italy) to obtain energy maps within the framework of precision agriculture. Also note that the multi-resolution segmentation task was implemented using layers of spectral and topographic channels, the classification stage was implemented as a process tree.

Further, we note that the work [2] was devoted to the survey of agricultural crops using the technology of remote sensing of the earth. To obtain information about the state of agrobiologically, agrochemical and agrophysical characteristics of crops, IoT sensors and multispectral images of crop areas were used to create agrotechnological maps of crops. We also note that here a deep neural network with two hidden layers was used for data processing.

Also, in work [3], it is intended to determine the hidden dependences arising from the seasonal cycles of the productivity of agricultural crops in the irrigated and rainfed land in the central region of Myanmar, and the system solutions for the assessment and metric measurement of plant cover and vegetation indices of this condition. The results

R. Silhavy and P. Silhavy (Eds.): CSOC 2023, LNNS 724, pp. 95–100, 2023.
https://doi.org/10.1007/978-3-031-35314-7_8

of the research are based on the scientific point of view that it is possible to create a land plot map taking into account the systemic, temporal and spatial risks of agricultural production.

2 Materials, Data and Methods

The study of the problem on the classification of vegetation based on the share reflectance in the bands is carried out by the theory and methodology of regression and data analysis [4], algorithms and technologies of remote sensing [5], a several of scientific literature in [6].

In this work, the center of the vegetation mass in the spectral space for the three classes of vegetation will be used as the arithmetic mean of the weighted values [7]:

$$C_V = \frac{\sum(\text{Barley} + \text{Corn} + \text{Wheat}) \times (\text{Barley} + \text{Corn} + \text{Wheat})}{\sum(\text{Barley} + \text{Corn} + \text{Wheat})} \quad (1)$$

And it will be used the correlation-regression equation for three classes of vegetation: Barley, Corn, Wheat, and least squares to estimate the dependency parameters reflectance at spectral bands according to the following equation system specification [4]:

$$\text{Barley} = \beta_{0,\text{Corn}} + \beta_{1,\text{Corn}} \times \text{Corn} + \varepsilon_{\text{Corn}}, \quad (2)$$

$$\text{Corn} = \beta_{0,\text{Wheat}} + \beta_{1,\text{Wheat}} \times \text{Wheat} + \varepsilon_{\text{Wheat}}, \quad (3)$$

$$\text{Wheat} = \beta_{0,\text{Barley}} + \beta_{1,\text{Barley}} \times \text{Barley} + \varepsilon_{\text{Barley}}, \quad (4)$$

$$\begin{aligned} &E(\varepsilon_{\text{V}}(\cdot)|\text{V}) = 0, \\ &Var(\varepsilon_{\text{V}}(\cdot)|\text{V}) = \sigma^2 I, \\ &Cov(\varepsilon_{\text{V}}(\cdot), \varepsilon_{\text{V}}(\cdot\cdot)|\text{V}) = 0, \\ &\varepsilon_{\text{V}}(\cdot)|\text{V} \approx N\big(0, \sigma^2 I\big), \end{aligned} \quad (5)$$

where $\varepsilon_{\text{V}}(\cdot)$ – the random model of error; $\text{V} \stackrel{\text{def}}{=}$ Corn, Wheat, Barley – a matrix of the observations compiled by of Barley, Corn, Wheat; $E(\cdot)$ – a expectation; $Cov(\cdot)$ – a covariance; $Var(\cdot)$ – a variance; I – a identity matrix; $N(0, 1)$ – the standard normal distribution.

3 Results

3.1 Preparation Data of Share Reflection for Vegetations

We choose three classes of vegetation: Barley, Corn, Wheat, we prepare of data of share reflectance at all spectral bands for examined vegetations, the source of these data is "Egistic" LLP [8], the city of Astana in the Republic of Kazakhstan, obtained based on a fielded survey of agricultural enterprises in the Akmola region. Then we get data of share reflectance of all spectral bands for the three classes of vegetations: Barley, Corn, Wheat, where the horizontal axis of the spectral space contains Wavelength, nm, and the vertical axis – Share reflectance at all spectral bands for the vegetations: Barley, Corn, Wheat (Fig. 1).

3.2 Classification of Vegetations Based on Share Reflectance at Bands

The solution of the problem on the classification of vegetation based on the share reflectance at spectral bands is implemented in the following sequence:

- it was chosen as an indicator of the share reflectance at band 8 – NIR as factorial feature of the spectral space (Fig. 2);
- it was chosen as an indicator of the share reflectance at band 11 – SWIR as result feature of the spectral space (Fig. 2);
- it was built a regression model at variable Corn vegetation in share reflectance of spectral space based on ordinary least squares estimator:

$$\text{Barley} = 0.1102 + 0.7013 \times \text{Corn},\ R^2 = 0.9763\ (0.001)\ (0.007) \tag{6}$$

where dependent variable is Barley vegetation in share reflectance of spectral space (Table 1, Fig. 2);

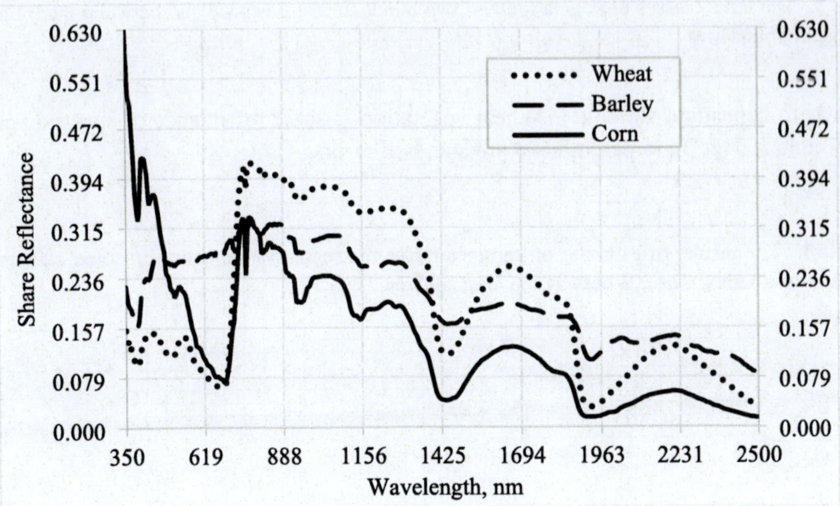

Fig. 1. Data in spectral space for the three classes of vegetation: Wheat, Corn, and Barley

- it was built a regression model at variable Wheat vegetation in share reflectance of spectral space based on ordinary least squares estimator:

$$\text{Corn} = 0.5102 - 0.9325 \times \text{Wheat},\ R^2 = 0.9746\ (0.002)\ (0.009) \tag{7}$$

where dependent variable is Corn vegetation in share reflectance of spectral space (Table 1, Fig. 2);

- it was built a regression model at variable Barley vegetation in share reflectance of spectral space based on ordinary least squares estimator:

$$\text{Wheat} = -0.0153 = 1.3220 \times \text{Barley},\ R^2 = 0.9868\ (0.002)\ (0.009) \tag{8}$$

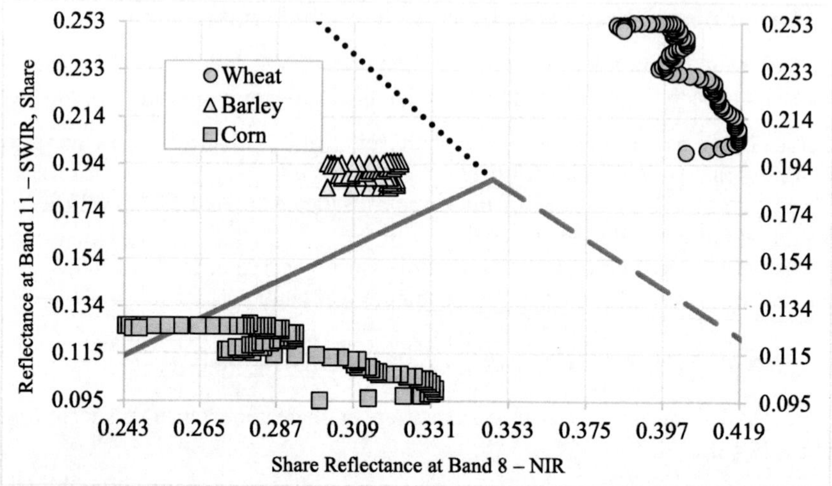

Fig. 2. The three classes of vegetation and two bands in feature space and between them linear decision boundaries

where dependent variable is Wheat vegetation in share reflectance of spectral space (Table 1, Fig. 2), in parentheses are standard errors.

Table 1. The results of estimate of parameters for the regression equation to three classes of vegetation: Barley, Corn, Wheat by the least squares

	Barley	Corn	Wheat
$\beta_{1,\mathrm{Corn}}$	0.7013*** (0.007)		
$\beta_{1,\mathrm{Wheat}}$		– 0.9325*** (0.009)	
$\beta_{1,\mathrm{Barley}}$			1.3220*** (0.009)
Constant	0.1102*** (0.001)	0.5102*** (0.002)	– 0.0153*** (0.002)
Numb of obs	280	280	280
R-square	0.9763	0.9746	0.9868

Note. The dependent variable – Barley, Wheat, Corn in share. In parentheses are standard errors. *, **, *** – estimate is significant at 10%, 5%, 1% level

4 Discussion

Further development of the problem of using remote sensing can be directed both in terms of diversity and in depth of theory, methodology and technology.

In particular, the study [9] noted that to improve the problems of agricultural land cover segmentation, it is proposed to use an indicator, the generalized vegetation index, which can be connected to many architectures of neural networks, as an information input of satellite sensors. The results show a 0.9–1.3% improvement in the intersection-over-union of the vegetation-related classes and consistently improve the overall mean intersection-over-union by 2% over the baseline.

At the same time, a new application of the technology of remote sensing of the earth in [10] is the research to determine the degree of fire consumption of leaves and crowns of pear trees in orchards. Indeed, using machine learning technology to construct a database developed from multispectral response coefficients using aerial photographs of healthy leaves, asymptomatic diseased leaves, and symptomatic diseased leaves, it was able to determine with 95.0% accuracy the degree of fire blight damage.

In work [11], based on linear regression using the least squares, a model was developed to predict the yield of potatoes in the state of Idaho. With the help of remote sensing, for cloudless days of the growing season in 2017, data were obtained on various vegetation indices. The yield prediction model was cross validated. The results of the study showed that the spectral indices, together with the topography of the field, made it possible to predict yields based on crop type and variety.

5 Conclusions

Thus, this paper studies the problem of classification for the vegetation based on the share reflectance at spectral bands. For this, the theory and methodology of regression and data analysis, algorithms and technologies of remote sensing, several modern scientific literatures are involved. As a result, a system of regression models was built with factor variables Barley, Corn, and Wheat vegetation in the share reflectance at bands of the spectral space and the parameters based on ordinary least squares method were estimated.

Acknowledgments. This research was funded by the Science Committee of the Ministry of Science and of Higher Education the Republic of Kazakhstan (Grant No. AP09259435).

References

1. Modica, G., Messina, G., De Luca, G., Fiozzo, V., Praticò, S.: Monitoring the vegetation vigor in heterogeneous citrus and olive orchards. A multiscale object-based approach to extract trees' crowns from UAV multispectral imagery. Comput. Electron. Agric. **175**, 105500 (2020)
2. Shafi, U., et al.: A multi-modal approach for crop health mapping using low altitude remote sensing, Internet of Things (IoT) and machine learning. IEEE Access **8**, 112708–112724 (2020)

3. Feyisa, G.L., et al.: Characterizing and mapping cropping patterns in a complex agro-ecosystem: an iterative participatory mapping procedure using machine learning algorithms and MODIS vegetation indices. Comput. Electron. Agric. **175**, 105595 (2020)
4. Greene, W.H.: Econometric Analysis. Pearson (2011)
5. Landgrebe, D.A.: Signal Theory Methods in Multispectral Remote Sensing. Wiley, Hoboken (2003). https://doi.org/10.1002/0471723800
6. Richards, J.A., Jia, X.: Remote Sensing Digital Image Analysis. An Introduction. Springer, Heidelberg (2006). https://doi.org/10.1007/3-540-29711-1
7. ISO 2854:1976. Statistical interpretation of data – Techniques of estimation and tests relating to means and variances. www.iso.org/standard/7854.html. Accessed 12 Dec 2022
8. Egistic LLP. https://egistic.kz. Accessed 12 Dec 2022
9. Sheng, H., Chen, X., Su, J., Rajagopal, R., Ng, A.: Effective data fusion with generalized vegetation index: Evidence from land cover segmentation in agriculture. In: Proceedings of IEEE Computer Society Conference on Computer Vision and Pattern Recognition Workshops, Seattle, WA, USA, pp. 267–276. IEEE (2020). https://doi.org/10.1109/CVPRW50498.2020.00038
10. Bagheri, N.: Application of aerial remote sensing technology for detection of fire blight infected pear trees. Comput. Electron. Agric. **168**, 105147 (2020)
11. Abou Ali, H., Delparte, D., Griffel, L.M.: From pixel to yield: forecasting potato productivity in Lebanon and Idaho Int. Arch. Photogramm. Remote Sens. Spat. Inf. Sci. ISPRS Arch. **XLII-3/W11**(ISPRS), 1–7 (2020). https://doi.org/10.5194/isprs-archives-XLII-3-W11-1-2020

The Problem of Information Singularity in the Storage of Digital Data

Alexander V. Solovyev(✉)

Federal Research Center "Computer Science and Control" of Russian Academy of Sciences, 44/2 Vavilova Street, Moscow 119333, Russia
soloviev@isa.ru

Abstract. This article analyzes the possible emergence of the problem of information singularity in the organization of digital data storage, including long-term storage. The information singularity is understood as the state of the information system, when the rate of new digital data appearance per unit of time exceeds the rate of their processing and interpretation. The study deals with the problems of information singularity in the processing of information of multimedia data, streaming videos, in the selection of digital data for long-term storage, including the examination of value. The problem of the emergence of information singularity when searching for related documents and managing distributed registries is also considered. In addition, the problem of digitizing paper archival documents is considered as a possible example of a local information singularity. The article proposes a number of approaches to solving the problem of information singularity. The conclusion is made about the advantages and disadvantages of using artificial intelligence to solve the problem. The advantages and disadvantages of other solutions to the problem under consideration are given. #CSOC1120.

Keywords: Digital Data · Information Singularity · Data Storage

1 Introduction

Obviously, the organization of long-term storage of digital data makes sense only if this data is of great value. Digital data in a changing digital environment becomes the object of management. The task of organizing storage is to ensure the stability of all parameters of digital data, such as authenticity, reliability, interpretability, security, stability (including catastrophic impacts).

The degree of study of these problems, especially reliability problems, is quite high [1], which makes it possible to formulate the main postulates of the technology for ensuring long-term preservation and to implement many of these provisions [2].

However, recently many researchers have begun to talk about the problem of the onset of information singularity in connection with the digitalization of society and the economy. This can lead to a significant transformation of society and cause many negative consequences [3, 4].

R. Silhavy and P. Silhavy (Eds.): CSOC 2023, LNNS 724, pp. 101–105, 2023.
https://doi.org/10.1007/978-3-031-35314-7_9

According to many statistical studies, the doubling rate of digital information in the world is estimated from 6 to 12 months. At the same time, the doubling of the speed of computers, according to Moore's law, occurs every 18 months. Or rather, not even speed, but the number of transistors per unit area of a chip chip, which may imply the onset of doubling limits due to the achievement of the molecular level of transistor sizes.

It can be assumed that the point in time is not so far away when the rate of receipt of new digital information will exceed the rate of its processing. Such a state of information systems, when the rate of appearance of new digital data per unit of time exceeds the rate of their processing, we will call the information singularity.

Let's consider this problem in more detail.

2 Overview of Information Singularity Issues

2.1 Storing and Interpreting Streaming Videos

Recently, due to the improvement of technology, more and more devices appear that conduct continuous video recording.

The appearance of cameras on almost every corner has not surprised anyone for a long time. Almost all areas of activity from traffic and access control to premises to observations in experimental physics and astronomy are captured by streaming video.

Digital observational data of experiments are of high value and should certainly be preserved. At the same time, it is important not only to save such data, which in itself requires large information resources, but also to be able to process and interpret in the future.

2.2 Expertise in the Value of Digital Data

Another problem is the problem of selecting digital data for long-term storage, including their examination of value.

The same streaming videos of the progress of experiments can be part of the information subject to long-term storage. More and more of this information can be created over time. Assessing the value of such information is always difficult due to the long time required for the interpretation of digital data by a person or even a computer.

2.3 Linked Data Search and Distributed Digital Data

Another problem that leads to information singularity is the search for related digital data. For example, digital data that contains a link to other digital data, including regulatory documents that were relevant at the time of the experiment.

Of course, we are talking about electronic documents. In this case, the problem lies in the need to search for missing data among a huge amount of digital data, including geographically distributed ones.

The second problem is the problem of managing distributed registries in general.

Today, blockchain technology is used to solve the problem. However, the growth of chains under conditions of long-term storage creates the problem of parsing (checking) constantly growing chains of blocks, which will require huge capacities. As a result, during the execution of the next operation, the following may happen: it will take more time to check the chain than is allowed for the execution of the next operation.

As a result, a denial of service may occur due to the impossibility of performing an operation to add a new portion of digital data, build a connection with existing data, interpret and verify data, etc.

2.4 Digitization of Documents

An example of a local information singularity can be a situation when it is required to digitize huge volumes of paper documents, for example, documents on the progress and results of physical experiments conducted in the pre-digital era, and there are very few resources for digitization.

A similar situation may be typical in the transition of large organizations to a digital platform with large archives of paper documents.

In the absence of the necessary amount of resources to digitize the existing archival fund of paper documents, digitization can be extended in time for many years, which can significantly depreciate the transition to a digital platform.

3 Possible Solution to the Information Singularity Problem

Based on the analysis of the growth rate of digital information, given the almost inevitable slowdown in the growth of computer performance, the onset of the information singularity is almost inevitable.

And this will lead to significant socio-economic consequences [6], the essence of which can be formulated as the separation of the digital world from society and the inability to control the digital environment.

In this regard, it is necessary to determine possible ways to solve the problem of information singularity.

The first, most radical solution to the problem is the search for fundamentally new ways of processing information at the technical level.

The solution here seems to be the development of quantum computers that allow solving the problem of NP-problems, therefore, when implemented, leading to the possibility of almost unlimited parallel processing of information.

Parallel computing remains a definite solution to the problem in the absence of a full-fledged quantum computer, however, as in the case of processing distributed registries, it requires increasing energy costs. This is contrary to both the installation of green energy and common sense.

The second is the use of artificial intelligence for operations such as pattern recognition and interpretation of streaming videos and other digital data of large volumes.

The third solution related to artificial intelligence is the use of various forms of machine learning.

4 The Problem of Solving the Information Singularity Problem

However, all of the above approaches to solving the problem of information singularity are not without fundamental shortcomings. And the creation of specific technical solutions based on these approaches is associated with enormous difficulties.

So, for example, a quantum computer is still only a hypothetical device. So far, only limited prototypes have been implemented. Full implementation requires not only an industrial revolution that affects the most diverse sectors of the economy, such as the production of superconductors, but also the improvement of the physical principles of its operation.

Parallel computing requires more and more resources, and, consequently, energy costs. So, for example, mining farms for cryptocurrencies require energy costs comparable to the energy costs of cities with a population of tens, if not hundreds of thousands of people.

Given the problem of increasing the speed of information, energy costs will skyrocket in parallel with the growth of digital data.

Also, artificial intelligence is not without flaws, which is seen as a promising approach to solving the problem of information singularity.

The first disadvantage is that the rules for artificial intelligence systems are created by people and cannot be perfect.

As a consequence, this may lead to the fact that the initial training samples for machine learning may be inadequate to the conditions for solving the set information processing tasks, and therefore give incorrect results.

The uncontrolled development of artificial intelligence tools can lead to a technological singularity, during which artificial intelligence will develop uncontrollably and lead to completely different goals that were set when creating artificial intelligence systems for processing digital data.

5 Conclusion

This article provides an overview of the problem of information singularity in the organization of digital data storage.

Information singularity is understood as the state of the information system, when the rate of new digital data appearance per unit of time exceeds the rate of their processing and interpretation.

Typical problems of information singularity are considered in the processing of multimedia data, streaming videos, selection of digital data for long-term storage, including examination of value, digitization of paper documents, search for related data and management of distributed registries.

The article proposes approaches to solving the problem of information singularity and makes a brief overview of the advantages and disadvantages of the proposed solutions.

In future studies, it is planned to consider in more detail each approach to solving the research problem, to analyze their risks and advantages.

References

1. Akimova, G.P., Pashkin, M.A., Soloviev, A.V., Tarkhanov, I.A.: Modeling the methodology to assess the effectiveness of distributed information systems. Adv. Sci. Technol. Eng. Syst. **5**, 86–92 (2020). https://doi.org/10.25046/aj050110
2. Solovyev, A.V.: Long-term digital documents storage technology. In: Radionov, A.A., Karandaev, A.S. (eds.) RusAutoCon 2019. LNEE, vol. 641, pp. 901–911. Springer, Cham (2020). https://doi.org/10.1007/978-3-030-39225-3_97
3. Reingold, L.A., Solovyev, A.V.: Singularity problems in the context of digitalization. In: Proceedings of the International scientific conference "Co-evolution of technology and society in the context of the digital age", pp. 15–17 (2020)
4. Klychikhina, O.V., Reingold, L.A., Solovyev, A.V.: On the socio-economic consequences of the introduction of promising digital technologies. In: Proceedings of the XII International Scientific and Practical Conference "Regions of Russia: Development Strategies and Mechanisms for the Implementation of Priority National Projects and Programs" Part II, pp. 367–374 (2021)
5. Lyman, P., Varian, H.R.: How much information. Release of the University of California. UC Berkeley, Berkeley, US (2003). http://www.sims.berkeley.edu/research/projects/how-much-info-2003/. Accessed 12 Feb 2022
6. Reingold, L.A., Reingold, E.A., Solovyev, A.V.: Socioeconomic technologies – development trends in the age of digitalization. CEUR Workshop Proc. **2763**, 122–127 (2020). https://doi.org/10.30987/conferencearticle_5fce27706c3e14.59191136

Autonomous System for Locating the Maize Plant Infected by Fall Armyworm

Farian S. Ishengoma[1,3](✉), Idris A. Rai[2], and Ignace Gatare[1]

[1] African Center of Excellence in the Internet of Things (ACEIoT), College of Science Technology, University of Rwanda, P.O. Box 3900, Kigali, Rwanda
farishen@sua.ac.tz

[2] Department of Computer Science, School of Computing, Communications and Media, The State University of Zanzibar, P.O. Box 146, Zanzibar, Tanzania
rai@suza.ac.tz

[3] Department of Informatics and Information Technology, College of Natural and Applied Sciences, Sokoine University of Agriculture, P.O. Box 3000, Morogoro, Tanzania

Abstract. Digital agriculture helps farmers collect, analyze, and monitor farms while in remote locations, thus spending less time and money while getting more yield. Digital agriculture employs a variety of techniques, including image processing and machine learning. Previous studies have proposed various systems that use unmanned aerial vehicles (UAV) and machine learning techniques to precisely predict maize plants infested by various diseases. In this paper, we propose an extension of the system to locate the exact position of infected maize plants, as well as estimate the size of the infected area. This enables farmers to take appropriate action based on the precise location of the infection, rather than the entire farm. The proposed system performs three tasks: first, it crops UAV images into smaller images and transplants the Global Positioning System coordinates (GPSc) from UAV images into cropped images; second, it extracts the coordinates of the infested maize plant and counts similar GPSc that determine the size of the infected area on every UAV image, and finally, it sends a report to the farmer indicating the infested plants and the size covered by infection on every UAV image. The report helps farmers act quickly and only spray pesticides on infected areas, which saves them time and money. Using a dataset from a maize farm infected by fall armyworm, we show that the system is effective in locating the infected plants and areas.

Keywords: Maize · fall armyworms · unmanned aerial vehicle · GPS coordinates · convolution neural network

1 Introduction

Precision agriculture is the science of using high-tech sensors and analysis tools to improve crop yields and assist management decisions. It increases production, reduces operating costs, and ensures effective fertilizer and irrigation management [1]. Farmers, for example, can monitor the behavior of their plants at any time and from any location by

R. Silhavy and P. Silhavy (Eds.): CSOC 2023, LNNS 724, pp. 106–113, 2023.
https://doi.org/10.1007/978-3-031-35314-7_10

employing three techniques: unmanned aerial vehicles (UAV), machine-learning algorithms (ML), and the internet of things (IoT) [2]. Using these three techniques, various methods have been proposed to automate the farming process and reduce farmers' cost and time.

For instance, Bhoi et al. proposed an IoT-assisted UAV-based rice pest detection model that uses the Image cloud to identify pests in rice during field production. The model is able to detect and identify pests and send the information to the owner for further action [2]. Moreover, Selvaraj et al. proposed a method for categorizing bananas under mixed-complex Africa landscape using pixel-based classifications and ML models derived from multi-level satellite images (Sentinel 2, Planet Scope, and WorldView-2) and UAV (Mica Sense Red Edge) platforms. Their pixel-based banana classification method, which combined features of vegetation indices and principal component analysis, achieved up to 97% overall accuracy [3]. Wu et al. proposed a method that employs two advanced deep learning algorithms, namely, the Faster Region-based Convolutional Neural Network (Faster RCNN) and You Only Look Once version 3 (YOLOv3). Model performance was analysed using precision (mAP), size, and processing speed. The results show that all four models (YOLOv3 MobileNet, YOLOv3 Draknet53, Faster RCNN ResNet101, and Faster RCNN ResNet50) had similar precision (0.602–0.64), but the YOLO-based models were smaller and faster than the Faster R-CNN-adapted models. The authors used an UAV to collect a large number of images from a pine tree canopy during an early stage of infection in order to create a training dataset [4]. Furthermore, Yadav et al. proposed a method for early detection of bacteriosis disease to reduce pesticide use and crop loss. Deep learning method and imaging preprocessing algorithms were used to develop convolutional neural network (CNN) models for bacteriosis detection from peach leaf images. The proposed solution compares the outcomes of the imaging and CNN methods. The model architectures created using different deep learning (DL) algorithms performed the best, with an accuracy of 98.75% in identifying the corresponding peach leaf (bacterial and healthy) in 0.185 s per image [5].

Several other systems have recently been developed that make use of UAVs, IoT devices, ML, and DL [6]. For data collection, UAVs equipped with cameras and other IoT devices are used, while ML and DL algorithms are used to create models that can learn the behavior of data. In this paper, we propose a system that uses UAV and ML techniques to precisely locate the position of an infected maize plant, as well as estimate the size of the infected area. The system performs three tasks: firstly, it crops UAV images of size 5472×3080 into 150×150 pixels and transplants the Global Positioning System coordinates (GPSc) from UAV images into cropped images, secondly, it extracts the coordinates of the infested maize plant and counts similar GPSc that determine the size of the infected area on every UAV image, and finally, it sends a report to the farmer indicating the infested plants and the size covered by infection on every UAV image.

The rest of the paper is organized as follows. The proposed solution is presented in the following section, and the experimental results are presented and discussed in Sect. 3. Finally, in Sect. 4, we conclude the paper.

2 Proposed Solution

This section explains the proposed method flow chart, evaluation parameters, and proposed algorithms. Our solution is based on an extension of the previous method proposed by Ishengoma et al. [7], and the dotted blocks shown in Fig. 1 indicate the extended solution. Our aim is to locate the position of infected maize plants, as well as estimate the size of the infected area using images captured by UAV and classified by ML techniques. Specifically, the hybrid convolution neural network (HCNN) model was used to classify images into four classes, i.e., healthy, infected, weed, and redundant. The n-infected images are shown in the healthy class, while the infected images are shown in the infected class. The weed class, on the other hand, identifies the various types of weeds, while the redundant class includes maize tassels, maize stems, and soil [7].

In this study, the data was collected using a quadcopter drone, the Phantom 4 Pro v2, which flew 5 m above the ground. It has an integrated camera with an 8.6mm standard focal length and a sensor width of 12.8 3mm. The UAV captured images on a farm in Morogoro, Tanzania, at Mikese (6°47′7.6″S 37°54′44.3″ E). Field observations were conducted from March 15 to March 21, 2020.

The captured images are uploaded to the cloud via a fourth-generation network, then downloaded and cropped into 150 × 150 pixels, as shown in Fig. 1. Since cropping removes GPS coordinates (GPSc) from images, Algorithm 1 is used to add this metadata to cropped images. The cropped images are then run through the contrast enhancement block to remove background noise and reduce misclassification. Using the HCNN model, these images are divided into four categories, namely health, infected, weed, and redundant. Finally, the infected image lists are sent to a pathologist and a farmer. The pathologist list contains only infected images, whereas the farmer list contains the names of infected images, locations of infected images, the size of each infected UAV image, and the number of infected images on each UAV image. The involvement of pathologists is critical for confirming plant disease and recommending pesticides for use.

2.1 Evaluation Parameter

The size of the UAV image, on the other hand, is calculated by multiplying the ground sample distance (GSD) by the width (I_w) and the length (I_l) of the image to get the actual width (W_{gnd}) and length (L_{gnd}) on the ground, respectively. The GSD is the amount of surface area covered by a single image in flight. The size of the UAV image on the ground assists the farmer in identifying the surface area containing the diseases. Equations 1–3 explain how to calculate the GSD, the W_{gnd}, and the L_{gnd} of the image. It should be noted that the smaller the SGD, the clearer the picture.

$$GSD(m) = \frac{S_w h_{gnd}}{F_c I_w} \tag{1}$$

$$W_{gnd} = I_w GSD(m) \tag{2}$$

$$L_{gnd} = I_l GSD(m) \tag{3}$$

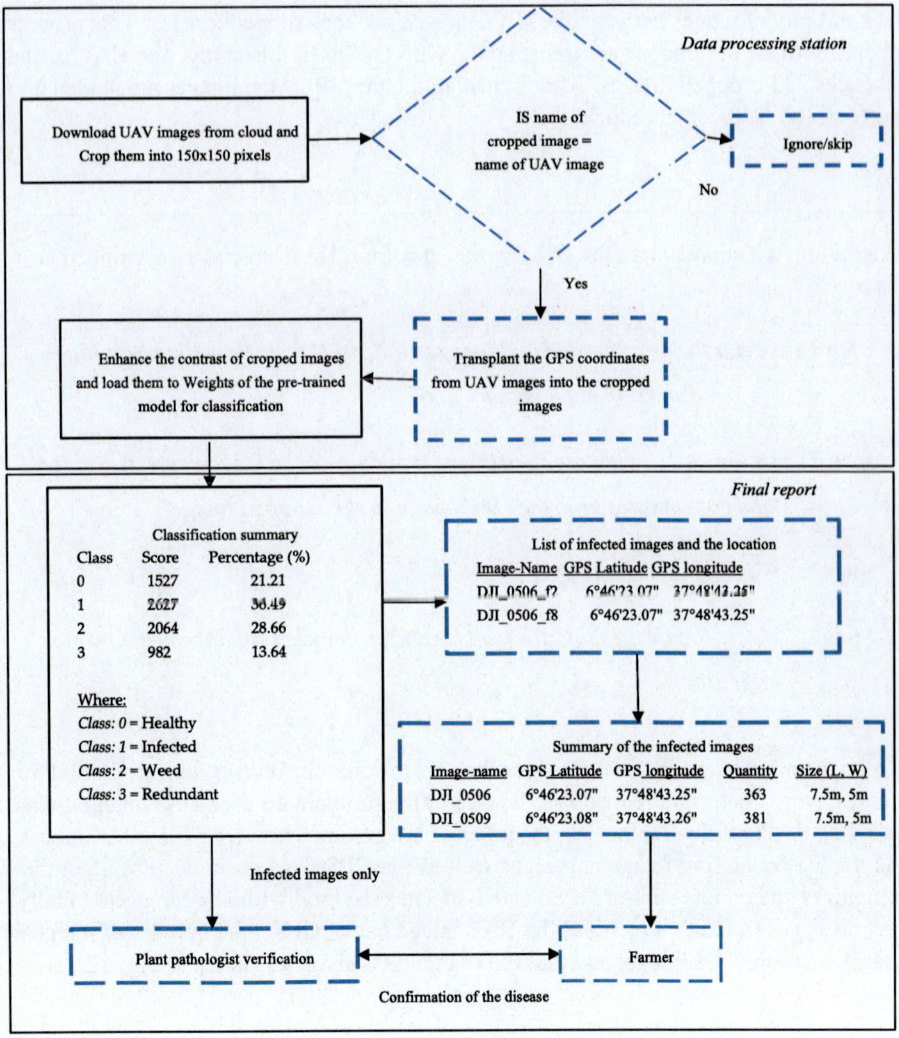

Fig. 1. The proposed system workflow

Where S_w is the width of the UAV sensor, while F_c is the length of the UAV camera's focus. The h_{gnd} is the height used during image capture to represent the distance from the ground to the UAV.

2.2 Proposed Algorithm

In this section, we describe two proposed algorithms for transplanting GPSc into cropped images and counting infected cropped images. Algorithm 1 is used to transplant the GPSc from the UAV image into the cropped images as shown in Fig. 1. The system first downloads the UAV images of size 5472 × 3080 pixels and then crops them into 150 × 150 pixels. After cropping, the system transplants the GPSc onto the images by checking the similarity of names between the UAV images and the cropped images. This process repeats until all the images are transplanted with GPSc. In this study, one UAV image produces 720-cropped images. After transplanting the GPSc, the images are loaded into the contrast enhancement block.

Algorithm 1: Transplanting the GPS coordinates from UAV images into cropped images.

Step 1: *Crop the downloaded UAV images to 150x150 pixels while retaining the source names in each image.*

Step 2: *If the name of the cropped image is similar to the UAV image, transplant the coordinates from the UAV image to the cropped image.*

Step 3: *Else, ignore the image.*

Step 4: *Load the images into the next block, for contrast enhancement and classification.*

On the other hand, the goal of Algorithm 2 is to count the images with similar GPSc, thus allowing the farmer to evaluate the size of the problem on each UAV image. After receiving the classification output, the infected images are extracted from other images, and the list containing image names, as well as the GPSc, is generated. The algorithm recognizes and counts similar GPSc and associates the total with UAV images. Finally, a report is generated that includes the UAV image name, GPSc, the number of infected images on each UAV image, and the size of the UAV image as shown in Fig. 1.

Algorithm 2: Counting infected images on UAV images

Step 1: *Separate the infected class from the other classes.*

Step 2: *Extract the image names and GPSc from the infected class.*

Step 3: *Append the names linked with their GPSc to the new list.*

Step 4: *Count the number of similar GPSc on the list and link the total to UAV images.*

---

3 Experimental Results and Discussion

To evaluate the system's performance, 10 UAV images are cropped into 150 × 150 pixels, yielding 7200 images. Each UAV image produces 720 cropped images. The cropped images are classified by using the HCNN model, which achieved an accuracy of 96.98%. According to the testing results, 21.21% of 1527 images are classified as healthy, 36.49% of 2627 images are infected, 28.66% of 2064 images are weed, and 13.64% of 982 images are redundant. The infected images are then counted using Algorithm 2 based on image name and coordinate similarity to assist the farmer in determining the magnitude of infection on each UAV image, as shown in Table 1.

Table 1 provides an overview of the farmer's report. The report includes a list of infected images with the corresponding GPSc, the number of infected images on each UAV image, and the ground distance covered by each UAV image. The GPSc and ground distance covered by the UAV image help the farmer apply pesticides to specific areas infested by fall armyworms (faw) more efficiently. Furthermore, the number of infected images on each UAV image directs the farmer to the areas that require more attention. To improve farm management, the infection rate is divided into three stages: above 50% indicates more infection, 25% to 40% indicates average infection, and less than 25% indicates less infection. The number of infected images on the DJI_0506 and DJI_0509 are 363 (50.42%) and 381 (52.92%) respectively, implying that farmers should focus their efforts on these areas because more than half of the images are infected. Moreover, the infection rate on DJI_0547, DJI_0551, DJI_0564, DJI_0567, and DJI_0579 images is 237 (32.92%), 276 (38.33%), 278 (38.61%), 272 (37.78%), and 275 (38.19%), respectively, which is considered moderate because less than half of the images are infected. The DJI_0532 and DJI_0576 images, on the other hand, have lower infection rates of 170 (23.61%) and 171 (23.75%), respectively. It can also be seen that the DJI_0509 image has the highest percentage of infected rate of 52.92%, while the DJI_0532 image has the least percentage of infected rate of 23.61%. It should be noted that images were captured at a distance of 5m from the ground using a UAV equipped with an integrated camera with an 8.6 mm standard F_c and a S_w of 12.83mm. As a result, using Eqs. 1–3, the SGD, W_{gnd}, and L_{gnd} in this study are 0.00136m, 7.5m, and 5m, respectively.

Table 1. Final report received by farmer

UAV Image name	GPSc		Number of infected images on every UAV image		Size (m)	
	Latitude	Longitude	Quantity	Percentage (%)	W_{gnd}	L_{dgn}
DJI_0506	6°46′23.07″	37°48′43.25 ″	363	50.42	7.5	5
DJI_0509	6°46′23.08 ″	37°48′43.26 ″	381	52.92		
DJI_0532	6°46′22.61 ″	37°48′43.38 ″	170	23.61		

(continued)

Table 1. (*continued*)

UAV Image name	GPSc		Number of infected images on every UAV image		Size (m)	
	Latitude	Longitude	Quantity	Percentage (%)	W_{gnd}	L_{dgn}
DJI_0537	6°46′22.64 ″	37°48′43.00 ″	204	28.33		
DJI_0547	6°46′23.03 ″	37°48′43.06 ″	237	32.92		
DJI_0551	6°46′23.20 ″	37°48′43.09”	276	38.33		
DJI_0564	6°46′23.57 ″	37°48′43.18”	278	38.61		
DJI_0567	6°46′23.42 ″	37°48′43.15 ″	272	37.78		
DJI_0576	6°46′23.34 ″	37°48′42.99 ″	171	23.75		
DJI_0579	6°46′23.02 ″	37°48′43.26 ″	275	38.19		

4 Conclusion

In this study the goal was to propose a system that uses UAV and ML methods to locate an infected maize plant, calculate the size of the infected area, and report this information to the farmer. Because cropping removes metadata, Algorithm 1 is used to transplant coordinates from UAV images into cropped images by associating their names, while algorithm 2 is used to extract infected images from other classes, count them based on GPSc similarities, and finally link them to UAV images. This data enables farmers to calculate the degree of infection per UAV image and prioritize inspection and pesticide application. To put the system to the test, 10 UAV images were cropped to 150 × 150 pixels, yielding 7,200 images that were then classified into four categories using the HCNN model: healthy, infected, weed, and redundant. The infected images were extracted and counted using an algorithm 2, and the results show that DJI 0509 has higher infection rate of 52.92%, while DJI 0532 has lower infection rate of 23.61%. However, because all images were captured at a height of 5 m, the distance covered on the ground by each image is consistent across all images.

Funding. The African Center of Excellence on the Internet of Things funded this research (ACEIoT).

Conflicts of Interest. The authors claim to have no known conflicts of interest.

References

1. Singh, P., et al.: Hyperspectral remote sensing in precision agriculture: Present status, challenges, and future trends. LTD (2020)
2. Bhoi, S.K., et al.: An Internet of Things assisted Unmanned Aerial Vehicle based artificial intelligence model for rice pest detection. Microprocess. Microsyst. **80**, 103607 (2021)
3. Gomez Selvaraj, M., et al.: Detection of banana plants and their major diseases through aerial images and machine learning methods: a case study in DR Congo and Republic of Benin. ISPRS J. Photogramm. Remote Sens. **169**, 110–124 (2020)

4. Wu, B., et al.: Application of conventional UAV-based high-throughput object detection to the early diagnosis of pine wilt disease by deep learning. For. Ecol. Manage. **486**, 118986 (2021)
5. Yadav, S., Sengar, N., Singh, A., Singh, A., Dutta, M.K.: Identification of disease using deep learning and evaluation of bacteriosis in peach leaf. Ecol. Inform. **61**, 101247 (2021)
6. Tsouros, D.C., Bibi, S., Sarigiannidis, P.G.: A review on UAV-based applications for precision agriculture. Information **10**(11), 349 (2019). https://doi.org/10.3390/info10110349
7. Ishengoma, F.S., Rai, I.A., Ngoga, S.R.: Hybrid convolution neural network model for a quicker detection of infested maize plants with fall armyworms using UAV-based images. Ecol. Inform. **67**, 101502 (2022)

The Best Model for Determinants Impacting Employee Loyalty

Dam Tri Cuong(✉)

Industrial University of Ho Chi Minh City, Ho Chi Minh City, Vietnam
damtricuong@iuh.edu.vn

Abstract. Employees are essential to a firm and play a significant role in its success. High levels of employee loyalty are essential for the growth of the company. It will motivate employees that have a high level of loyalty to the company to work hard and put up their best effort. Moreover, employee retention and developing employee loyalty are becoming increasingly vital in the present era of globalization and a dynamic labor market since employees are crucial to every part of the company's operations. Therefore, the goal of this study is to use the AIC technique to identify the best model for the variables that influence employee loyalty. A Google form was used to collect study data from 225 employees in Ho Chi Minh City, Vietnam. The non-probability approach was used to collect the sample. The research's conclusions showed that the best model for factors influencing employee loyalty included five elements - compensation, work environment, relationships with coworkers, training and development, and job satisfaction - that had a positive impact on employee loyalty, with compensation serving as the most important factor.

Keywords: The best model · Determinants · Employee loyalty · Vietnam

1 Introduction

Employees have always been valuable assets for any firm. They might be considered the lifeblood of an organization because of their vital nature. The majority of businesses are becoming more and more technology-driven because of technological advancement. However, because the technology requires human resources to function, this circumstance does not lessen the worth of employees in an enterprise. In most businesses, competition is getting more intense because of factors like globalization. This condition also has an impact on the labor market because businesses need more human resources to stay competitive in their particular industries [1]. Besides, as employees contribute to the organization's competitive edge, they are noteworthy assets. Competent workers can boost productivity and performance at work, whereas less competent workers can make businesses less effective at attaining their objectives. The company will survive and succeed through the efficient and effective use of its human resources. Because employees play an important role in every aspect of the company's operations, keeping good employees on board and fostering employee loyalty are becoming more and more

R. Silhavy and P. Silhavy (Eds.): CSOC 2023, LNNS 724, pp. 114–124, 2023.
https://doi.org/10.1007/978-3-031-35314-7_11

crucial in the current era of globalization and a dynamic labor market [2]. Similarly, personnel are important elements of a business and have a significant impact on how successful a firm is. Employee loyalty at high levels is essential to the growth of the company. Employees will work harder and give the company their best effort if there is a high level of loyalty among them [3]. Additionally, it has become harder and harder for businesses to keep their personnel. Employee demands are rising, and they have very high expectations for their positions. However, losing important personnel can have detrimental effects on businesses [4, 5]. The U.S. Department of Labor estimates that replacing an employee costs a business one-third of the annual compensation of a new hire [5]. It is crucial for businesses to understand how to keep qualified staff in these situations [6]. As a result, for businesses to remain competitive, they must not only hire the best individuals but also maintain their employment for a long time. Keeping employees motivated and engaged while allowing them to work for as long as workable is currently the most challenging issue that businesses must solve [1]. Employee loyalty, which refers to employers' efforts to keep employees to achieve corporate objectives, has always been a hot issue of discussion and research in the field of human resource management. In actuality, employees are the ones who decide whether to go or stay. Employee loyalty is a sign of long-term dedication by staff to the company [7]. Employee loyalty results from how a person is treated, encouraged and valued while performing his tasks [2]. Therefore, to investigate the impacting factors, we must adhere to the definition of employee loyalty [7, 8].

On the other side, employee loyalty research has been the attention of current researchers and practitioners [see 2, 7, 9–11]. Prior research has shown that factors including compensation [2, 12], work environment [9, 13], coworker relationships [11, 14], training and development [15–17], and job satisfaction [18, 19] all impact employee loyalty. There are, however, few studies that pinpoint the optimal model for factors that affect employee loyalty by using the AIC technique, particularly in Vietnam. Therefore, this study aims to close this gap.

2 Literature Review

2.1 Employee Loyalty

Employee loyalty is the degree to which employees feel devoted to the company, committed to it, included in it, care about it, and responsible for it. Another way to describe employee loyalty is the level of overall employee desire to contribute personally to the success of the company [20]. Employee loyalty refers to a person's desire and commitment to remain with a company and actively take part in its operations. When an employee sees himself as the core of the organization and an indispensable component of it, he is showing dedication and voluntary participation toward the company [21]. Employee loyalty is described as employees who have strong feelings for the organization, will grow together with it, feel a sense of duty and purpose at work, give their knowledge and experience to further the company's objectives, and play their part in assisting the company in achieving its strategic goals [22]. Likewise, being loyal means being sincere, trustworthy, devoted, and attached to a person, location, or group of people. Employee loyalty is a quality that motivates individuals to work hard and is a driving force that encourages them to use most of their time, expertise, and abilities to accomplish the objectives of the business. Employee loyalty is a company's intangible asset that helps with long-term success. Loyalty among employees has a direct influence on an organization's growth and success [10].

2.2 Compensation

Compensation is known as income received besides base salary [23]. Compensation refers to compensation packages that a company offers to its employees, where the compensation is received directly or indirectly by the employees as money, commodities, or other facilities for the contribution the employees make to the firm [24]. Compensation, as known, is the number of packages that an employer offers employees in exchange for the labor they provide. Employees receive the sum of all compensation in exchange for their services [2]. There are two types of compensation: direct compensation and indirect compensation. Wages, salaries, bonuses, and incentives are examples of direct compensation given to employees. In contrast, indirect compensation is a reward given by the company to its employees as benefits like allowances, health insurance, and other benefits [24]. Moreover, the compensation system is a key concern in every commercial organization. This is most likely because it is seen as a significant predictor of employee success and loyalty [12]. Numerous studies have demonstrated that compensation has a positive and considerable impact on employee loyalty [2, 24, 25].

2.3 Work Environment

Work environment refers to everything that surrounds employees and has the potential to influence how well they do their given responsibilities [26]. Besides, the work environment comprises all the equipment and supplies a person uses, as well as the working practices and arrangements used by both individuals and organizations. The work environment is the area where people do tasks that have the potential to impact how those tasks are carried out [9]. Employees, on the other hand, want to continue working in an environment where they have good relationships with their coworkers and the opportunity to grow within the company [11]. Prior scholars disclosed that the work environment is a predictor of employee loyalty, and there is a positive link between work environment and employee loyalty [9, 13, 16].

2.4 Coworker Relationships

Coworker relationships are often built on two ideas: the leader-member connection and coworkers' interactions. Coworkers' relationships are a sort of interpersonal relationship that exists inside an organization [27]. Relationships with coworkers inside the company indicate that superiors frequently inspire and motivate workers to produce a cordial environment between managers and staff, leading to the perception among staff that superiors are family. To match the attention and help of my superiors, employees will attempt to work even more challenging. As a result, employees must have the support and help of their peers when needed [28]. They will cooperate well at work and the task will be completed efficiently if they find a welcoming, comfortable workplace or if the relationships between employees are always open and warm [29]. Moreover, to carry out their duties and pursue prospects for advancement within the company, employees would prefer to continue working in an environment where everyone gets along well. The aspects of a congruous purpose or growth, work environment, and connections with coworkers are widely mentioned in employee loyalty [11]. Previous studies proposed that coworker relationships have a predictor of employee loyalty and impact employee loyalty [11, 29].

2.5 Training and Development

Employees consider training and development as a key factor in building employee loyalty as it is each employee's capacity for personal growth and self-realization. Employees report higher work satisfaction and loyalty when they have more opportunities to enhance their skills and sense of self [30]. Besides, one of the human resources department's most obvious duties is training. The chance to learn new skills measures employee evaluation of the company's training program. For workers to perform their responsibilities to the standards of the organization, training entails supplying them with the fundamental information and abilities they require. Companies that offer high-quality training help employees feel more emotionally connected to the company, which eventually leads to a desire to stay put [15]. Feelings of competence that might come from taking part in training programs increase loyalty among employees [15]. The previous research revealed that training and development have a predictor of employee loyalty and impact employee loyalty [15–17].

2.6 Job Satisfaction

Job satisfaction is referred to as an enjoyable or favorable emotional state that results from work experiences. It is characterized as a function of how well an employee's demands are met in a certain workplace [19]. Job satisfaction is a transitory attribute that responds to both internal and external influences. Job satisfaction is an optimistic attitude concerning one's work performance [31]. The degree to which an employee is happy with the benefits of their job, particularly in terms of intrinsic motivation, is another way to describe job satisfaction [32]. Job satisfaction is one metric used to assess how satisfied an employee is with the tasks assigned as well as the money received. Job satisfaction is a crucial metric that helps the management of a firm determine whether or not its employees are happy in their roles because, if they are not, it is predicted that employee motivation will be low and that employee performance will not be at its best [33]. Job satisfaction is one of the crucial elements of employee loyalty. According to the definition of job satisfaction, it is a pleasing merging of psychological, physiological, and situational states toward the workplace that are brought on by experiences or performance appraisals. It is directly correlated with organizational performance variables including productivity, loyalty, and retention [34]. Former scientists demonstrated that job satisfaction is an antecedent of employee loyalty and impacts favorably on employee loyalty [18, 19, 34].

Considering the aforementioned assessment, we have applied the AIC technique to empirical research in this study to select the optimum model for factors influencing employee loyalty in Vietnam. The equation of the optimum model for employee loyalty is proposed:

$$Y = \beta 0 + \beta 1COM + \beta 2WEN + \beta 3CRE + \beta 4TDE + \beta 5JSA + \varepsilon \quad (1)$$

Where:

Y = ELO: Employee loyalty, COM: Compensation, WEN: Work environment, CRE: Coworker relationships, TDE: Training and development,

JSA: Job satisfaction, $\mathcal{E}$: Error

3 Methodology

3.1 Analytical Technique

Akaike [35] introduced the AIC or Akaike information criterion. The Kullback-Leibler divergence method used by AIC is an information theory technique for the choice of the model (variable). One of the most often used strategies for model selection in linear regression analysis is AIC [36]. AIC's model selection index is another useful technique that is frequently used with regression models. This is due to the scientist's ability to precisely use the model selection criteria while deciding between many models [37]. The model with the lowest AIC value qualifies as the ideal one.

The AIC index provided by the RStudio program is used in this study to determine the ideal model. The candidate model's AIC is determined using the formula below [38]:

$$AIC = 2k - 2\ln(L) \quad (2)$$

Where: The number of the model's parameters is k, and the log-likelihood of the candidate model is ln(L)

3.2 Data and Sample

225 employees at companies in Ho Chi Minh City, Vietnam, participated in an online survey that was used to collect study data using a Google form. The population sample was selected using a non-probability approach. The criteria were assessed using a five-point Likert scale (1 being fully opposed to 5 being completely in agreement). The following items from [16, 22, 24, 29, 39] were used in the questionnaire to measure the research's variables: compensation (four items), work environment (four items), coworker relationships (four items), training and developing (three indicators), job satisfaction (four indicators), and employee loyalty (four indicators).

4 Results and Discussion

4.1 Results

Descriptive Statistics

The characteristics of the population are shown in Table 1.

Table 1. Population characteristics

Attribute	Category	Frequency	Percent
Gender	Female	137	60.9
	Male	88	39.1
	Total	225	100.0
Age	18–25	61	27.1
	26–40	82	36.4
	41–55	71	31.6
	>55	11	4.9
	Total	225	100.0

According to Table 1, there are 137 female employees in the sample, who make up 60.9% of the total, and 88 male employees, who make up 39.1%. Regarding age, 61 employees between the ages of 18 and 25 make up 27.1% of the workforce, 82 between the ages of 26 and 40 make up 36.4%, 71 between the ages of 41 and 55 make up 31.6%, and 11 above the age of 55 make up 4.9%.

AIC (Akaike Information Criterion)
The AIC finding can be compiled in Table 2.

The AIC findings are shown in Table 2 for each stage leading up to the final model search. The model with predictor variable (COM) and AIC = −48.54 is the first stage. AIC = −75.24 and two predictor variables (COM + WEN) are used in the second phase. In the third phase, AIC = −115.38, three predictor variables (COM + WEN + CRE). The fourth step has an AIC of −125.36 and four predictor variables (COM + WEN + CRE + TDE). The model with five independent variables (COM + WEN + CRE + TDE + JSA) with an AIC value of -140.8 is the fifth and final phase.

Table 2. AIC

Model	AIC
ELO ~ COM	−48.54
ELO ~ COM + WEN	−75.24
ELO ~ COM + WEN + CRE	−115.38
ELO ~ COM + WEN + CRE + TDE	−125.36
ELO ~ COM + WEN + CRE + TDE + JSA	−140.80

RStudio ends at the final model, which includes five independent variables (COM + WEN + CRE + TDE + JSA), after exploring the best model in five steps. This research showed that the model with five independent variables is the best one since it has the lowest AIC index.

Model' Predictor Coefficients Testing
The outcome of predictor coefficients in the final model is shown in Table 3.

Table 3. The final model's predictor coefficients

	Estimate	Std. Error	t-value	p-value
(Intercept)	-0.96721	0.28057	-3.447	0.0000
COM	0.40029	0.06008	6.663	0.0000
WEN	0.18981	0.05913	3.210	0.0000
CRE	0.21974	0.05770	3.808	0.0000
TDE	0.16732	0.05739	2.916	0.0000
JSA	0.26402	0.06286	4.200	0.0000

Residual standard error: 0.7218 on 219 degrees of freedom
Multiple R-squared: 0.5703 Adjusted R-squared: 0.5605
F-statistic: 58.14 on 5 and 219DF, p-value: < 0.0000

Table 3 reveals that the final model's independent variable coefficients have a p-value of 0.05 or below and are statistically significant. As a result, the final model's assumption regarding predictor variables is confirmed. As a result, the following model is the best one:

$$Y = -3.447 + 0.40\ \text{COM} + 0.19\ \text{WEN} + 0.22\ \text{CRE} + 0.17\ \text{TDE} + 0.26\ \text{JSA} + \varepsilon \quad (3)$$

Additionally, the R-squared for the best model is 0.5703, which indicates that 5 independent factors variables (COM + WEN + CRE + TDE + JSA) account for 57.03% of the variation in employee loyalty.

Testing of Variance Inflation Factor (VIF)
Table 4 shows the results of VIF testing for the best model.

Table 4. The VIF

	COM	WEN	CRE	TDE	JSA
VIF	1.259947	1.260628	1.449943	1.256900	1.722057

Since the predictor variables' VIF criteria are less than 5, as shown in Table 4, there is no multicollinearity between the predictor factors.

4.2 Discussion

The results showed that the final model created by applying the AIC technique consists of the dependent variable (Employee loyalty - LOY) and 5 antecedent variables (compensation - COM, work environment - WEN, colleague relationships - CRE, training and development - TDE, and job satisfaction - JSA). According to the final model, five independent variables - compensation (COM), work environment (WEN), colleague relationships (CRE), training and development (TDE), and job satisfaction (JSA) - all had a positive impact on employee loyalty (LOY) and were statistically significant.

Additionally, as seen in Table 3, compensation had the greatest effect on employee loyalty ($\beta = 0.40$). Besides, if compensation (COM) was raised by 1 point, employee loyalty would also rise by 0.40 points. This conclusion is in line with certain research [2, 23, 24], and it also shows that employee loyalty was the strongest antecedent. Managers should thus enhance employee compensation, they should focus on some attributes of compensation such as wages, salaries, bonuses, incentives, allowances, health insurance, and other benefits if they want to promote employee loyalty. The second most important factor affecting employee loyalty was job satisfaction ($\beta = 0.26$). Therefore, if job satisfaction climbed by one point, employee loyalty would have grown by 0.26. Scholars have endorsed this result (e.g. [18, 19, 34]). Managers can take note of several strategies to boost job satisfaction, such as offering workers exciting, fulfilling work to do and allowing for flexible working hours so that they have a feeling of well-deserved success. Employee loyalty is increased if employers provide a job they are happy with. Relationships between coworkers came in third as a significant factor affecting employee loyalty ($\beta = 0.22$). The result was a 0.22 increase in employee loyalty for every 1-point improvement in coworker relationships. The researchers have demonstrated this result (e.g. [11, 29]). Therefore, managers offer experience, technical expertise, and passion to help employees to improve the connections between coworkers at work. Employees may communicate more effectively and recognize their specific contributions to the business thanks to this strategy. To foster harmony between people, families, and the firm, the business might make an effort to comprehend the lives of families of its employees. Additionally, the human resources management division might plan some fun events to celebrate milestones and foster a welcoming atmosphere to increase employee loyalty [11]. The fourth significant factor affecting employee loyalty was the work environment ($\beta = 0.19$), which was followed. The researchers have demonstrated this result (see [9, 13, 16]). Therefore, managers should develop a productive workplace that promotes happiness, confidence, and cleanliness to improve employee loyalty. Finally, the fifth critical factor affecting employee loyalty was training and development ($\beta = 0.17$). Employee loyalty rose by 0.17 if training and development improved by 1 point. The researchers have confirmed this result (e.g. [15–17]). Therefore, managers should establish policies that promote equity, offer chances for professional and personal growth, and support training. As a result, employee loyalty is increased.

5 Conclusions and Limitations

By using the AIC method, this finding confirms the ideal model for the variables affecting employee loyalty. Five factors - compensation (COM), work environment (WEN), coworker relationships (CRE), training and development (TDE), and job satisfaction (JSA) - were shown to have a favorable impact on employee loyalty (ELO) in the best model. Therefore, managers who wish to increase the number of loyal employees should take these things into account.

The best model used in this study, however, only adequately described 57.03% of the variation in employee loyalty by 5 independent factors (COM, WEN, CRE, TDE, and JSA). Therefore, other factors should be included in the model in this future research to improve the degree to which employee loyalty is described.

References

1. Kossivi, B., Xu, M., Kalgora, B.: Study on determining factors of employee retention. Open J. Soc. Sci. **04**, 261–268 (2016). https://doi.org/10.4236/jss.2016.45029
2. Putra, B.N., Jodi, I.W.G.A., Prayoga, I.M.: Compensation, organizational culture and job satisfaction in affecting employee loyalty. J. Int. Conf. Proc. **12**, 11–15 (2019)
3. Sutanto, E.M., Perdana, M.: Antecedents variable of employees loyalty. Jurnal Manajemen dan Kewirausahaan **18**, 111–118 (2016). https://doi.org/10.9744/jmk.18.2.111
4. Stroh, L., Reilly, A.H.: Loyalty in the age of downsizing. Sloan Manage. Rev. **38**, 83–88 (1997)
5. Michaud, L.: 5 keys to maximum employee retention. Natl. Public Account. **47**, 36–37 (2002)
6. Martensen, A., Grønholdt, L.: Internal marketing: a study of employee loyalty, its determinants and consequences. Innov. Mark. **2**, 92–116 (2006)
7. Zhong, X., Zhang, Y.X., Li, S., Liu, Y.: A multilevel research on the factors influencing employee loyalty under the new employer economics. Bus. Manage. Res. **9**, 1–8 (2020). https://doi.org/10.5430/bmr.v9n2p1
8. Frank, F.D., Finnegan, R.P., Taylor, C.R.: The race for talent: retaining and engaging workers in the 21st century. Hum. Resour. Plan. **27**, 12–25 (2004)
9. Ena, Z., Sjioen, A.E., Riwudjami, A.M.: The effect of work environment on employee loyalty with work stress as an intervening variable at Bella Vita Hotel - Kota Kupang. Quant. Econ. Manage. Stud. **3**, 65–76 (2022). https://doi.org/10.35877/454ri.qems865
10. Angayarkanni, R., Shobaba, K.: Factors influencing the loyalty of employees: a study with reference to employees in Chennai. J. Critic. Rev. **7**, 5915–5921 (2020)
11. Lai, C.: Factors affecting employee loyalty of organizations in Vietnam. Int. J. Organ. Innov. **14**, 115–127 (2021)
12. Akhigbe, O.J., Ifeyinwa, E.E.: Compensation and employee loyalty among health workers in Nigeria. Arch. Bus. Res. **5** (2017). https://doi.org/10.14738/abr.511.3778
13. Sukawati, T.B., Suwandana, I.G.: Effect of physical work environment, workload, and compensation on employee loyalty at visesa ubud resort. Am. J. Human. Soc. Sci. Res. **5**, 399–408 (2021)
14. Nguyen, H.H., Nguyen, T.T., Nguyen, P.T.: Factors affecting employee loyalty: a case of small and medium enterprises in tra Vinh Province, Viet Nam. J. Asian Finan. Econ. Bus. **7**, 153–158 (2020). https://doi.org/10.13106/jafeb.2020.vol7.no1.153
15. Costen, W.M., Salazar, J.: the impact of training and development on employee job satisfaction, loyalty, and intent to stay in the lodging industry. J. Hum. Resourc. Hospital. Tourism **10**, 273–284 (2011). https://doi.org/10.1080/15332845.2011.555734

16. Thuong, V.K., Huy, V.K.: Factors affecting the staff loyalty at private companies in can tho city. Int. J. Small Bus. Entrepren. Res. **7**, 15–22 (2019)
17. Ismail, H., Puteh, F.: Factors that influence employee loyalty: a study at manufacturing sector in Klang and Shah Alam industrial zone. In: E-Proceeding 8th International Conference On Public Policy And Social Science (ICoPS), pp. 539–543 (2021)
18. Jigjiddorj, S., Tsogbadrakh, T., Choijil, E., Zanabazar, A.: The mediating effect of employee loyalty on the relationship between job satisfaction and organizational performance. Adv. Econ. Bus. Manage. Res. **78**, 197–202 (2019). https://doi.org/10.2991/emt-19.2019.37
19. Yousaf, I., Nisa, S., Nasir, N., Batool, M., Gulfam, R.: Factors affecting the job satisfaction: implications of employee loyalty and employee turnover. Int. J. Acad. Res. Bus. Arts Sci. **2**, 16–27 (2020). https://doi.org/10.5281/zenodo.3745567
20. Upasana, K.: Influence of compensation on employee loyalty to organization. Asian J. Multidiscipl. Stud. **3**, 195–200 (2015)
21. Sharma, M.: Job Satisfaction and Employee Loyalty: a study of working professionals in Noida NCR. Gurukul Bus. Rev. **15**, 36–43 (2019). https://doi.org/10.5958/2321-5763.2016.00015.9
22. Chen, S., Xu, K., Yao, X.: Empirical study of employee loyalty and satisfaction in the mining industry using structural equation modeling. Sci. Rep. **12**, 1–15 (2022). https://doi.org/10.1038/s41598-022-05182-2
23. Wulandari, N., Arifin, A., Khoiriyah, M., Pujiningtiyas, R.A.I., Arifin, M.: Effect of empowerment and compensation on employee loyalty. In: Proceedings of the International Conference on Health Informatics, Medical, Biological Engineering, and Pharmaceutical (HIMBEP), pp. 259–263 (2021). https://doi.org/10.5220/0010330902590263
24. Manurung, P..: The effect of direct and indirect compensation to employee's loyalty: case study at directorate of human resources in Pt Pos Indonesia. J. Indonesian Appl. Econ. **7**, 84–102 (2017). https://doi.org/10.21776/ub.jiae.2017.007.01.6
25. Sumaryathi, N.K.D., Dewi, I.G.A.: The effect of compensation on employee loyalty with job satisfaction as a mediator. Am. J. Human. Soc. Sci. Res. **4**, 367–373 (2020)
26. Ramadhanty, D.P., Saragih, E.H., Aryanto, R.: The influence of the work environment on the loyalty of millennial employees. Adv. Econ. Bus. Manage. Res. **149**, 264–271 (2020). https://doi.org/10.2991/aebmr.k.200812.046
27. Lin, S.-C., Shu, J., Lin, J.: Impacts of coworkers relationships on organizational commitment-and intervening effects of job satisfaction. Afr. J. Bus. Manage. **5**, 3396–3409 (2011). https://doi.org/10.5897/AJBM10.1558
28. Matzler, K., Fuchs, M., Schubert, A.K.: Employee satisfaction: does Kano's model apply? Total Qual. Manag. Bus. Excell. **15**, 1179–1198 (2004). https://doi.org/10.1080/1478336042000255569
29. Antoncic, J.A., Antoncic, B.: Employee satisfaction, intrapreneurship and firm growth: a model. Ind. Manag. Data Syst. **111**, 599–607 (2011). https://doi.org/10.1108/02635571111133560
30. Kumar, D.N.S., Shekhar, N.: Perspectives envisaging employee loyalty. J. Manag. Res. **12**, 100–112 (2012)
31. Robbins, S., Timothy, A.: Organizational Behavior. Pearson Education Inc, New Jersey (2013)
32. Statt, D.A.: The Routledge Dictionary of Business Management. Routledge, London, U.K (2004)
33. Nurhasan, R., Ahman, E., Suryadi, E., Setiawan, R.: Generation Y behavior: employee loyalty based on job satisfaction and workplace spirituality. J. Int. Conf. Proc. **4**, 1–6 (2021)
34. Zanabazar, A., Jigjiddorj, S.: Impact of employee's satisfaction in employee loyalty, retention and organization performance. Int. J. Manage. Appl. Sci. **4**, 51–55 (2018)
35. Akaike, H.: A new look at the statistical model identification. IEEE Trans. Autom. Control **19**, 716–723 (1974). https://doi.org/10.1109/TAC.1974.1100705

36. Chaurasia, A., Harel, O.: Model selection rates of information based criteria. Electron. J. Stat. **7**, 2762–2793 (2013). https://doi.org/10.1214/13-EJS861
37. Karlsson, P.S., Behrenz, L., Shukur, G.: Performances of model selection criteria when variables are Ill conditioned. Comput. Econ. **54**(1), 77–98 (2017). https://doi.org/10.1007/s10614-017-9682-8
38. Acquah, H.D.: Comparison of Akaike information criterion (AIC) and Bayesian information criterion (BIC) in selection of an asymmetric price relationship. J. Develop. Agric. Econ. **2**, 001–006 (2010)
39. Coughlan, R.: Employee loyalty as adherence to shared moral values. J. Manag. Issues **17**, 43–57 (2005)

Intrusion Detection with Supervised and Unsupervised Learning Using PyCaret Over CICIDS 2017 Dataset

Stefan Krsteski, Matea Tashkovska, Borjan Sazdov, Luka Radojichikj, Ana Cholakoska, and Danijela Efnusheva(✉)

Faculty of Electrical Engineering and Information Technologies, Ss. Cyril and Methodius University, Skopje 1000, North Macedonia
danijela@feit.ukim.edu.mk

Abstract. The most crucial security mechanisms against the complex and expanding network threats are intrusion detection systems. Assuming that researchers have been working on building effective machine learning models that address the problem of network attacks in recent years, this study focuses on building a machine learning model, based on the CICIDS 2017 dataset, for intrusion detection using PyCaret – an open-source library that automates machine learning workflows. Detailed exploratory data analysis was performed, which led to replacing unnecessary data. Mainly, this research approaches the challenge as a classification problem. Different classification algorithms were used for this purpose such as: Random Forest, Decision Tree, SVM – Linear Kernel, k-NN classifier and Naïve Bayes classifier. Random Forest was the best performing classifier with 99.6% accuracy and F1-Macro score of 0.917. Besides classification, clustering and anomaly detection were implemented. PyCaret clustering with two different algorithms, Agglomerative Clustering and K-Means Clustering, identified two clusters. Our aim was to distribute the data into two clusters, benign data cluster and network attacks (malign) cluster. Agglomerative Clustering achieved 0.90 silhouette score and 0.54% accuracy, while K-Means clustering achieved 0.56 silhouette score and 0.75% accuracy. The accuracy was calculated by comparing the cluster of each record with the actual class, benign or malign. Classification proved to be highly efficient, while clustering and anomaly detection were less efficient and have room for improvement.

Keywords: Machine Learning · Intrusion Detection · PyCaret

1 Introduction

Approximately two-thirds of the global population will have Internet access by 2023. It is estimated that there will be 5.3 billion users (66% of global population) in comparison to 3.9 billion (51% of global population) in 2018 [1]. With the growth of Internet traffic, the number of attacks rises. That is why system security is crucial, which is achieved with Intrusion Detection System (IDS). IDS technology has been attempting to provide

R. Silhavy and P. Silhavy (Eds.): CSOC 2023, LNNS 724, pp. 125–132, 2023.
https://doi.org/10.1007/978-3-031-35314-7_12

effective, high quality, intrusion monitoring for decades. IDSs can be implemented as software or hardware systems, which monitor the events occurring in a computer system or network. These systems analyze events for signs of security issues. Intrusions are defined as attempts to compromise the confidentiality, integrity and availability, or to bypass the security mechanisms of a computer or network. In recent years, with the development of machine learning (ML) and artificial intelligence (AI), researchers have been working on building effective machine learning models that address the problem of network attacks. Machine learning was used for this task, because it can identify various trends and patterns with a huge amount of data. Furthermore, machine learning is able to automate numerous decision-making tasks, reduce false positive rates and produce accurate IDS. Finding a high-quality solution is greatly aided by the numerous publicly accessible datasets. CICIDS 2017 [2], a dataset published by the "Canadian Institute of Cybersecurity" at the "University of New Brunswick" is one of the most up to date datasets that contains the most common latest attacks and in this paper the focus is given on this dataset. The objective of this paper is to explore the performance of supervised and unsupervised learning in detecting network environment anomalies by using PyCaret [3] - an open-source, low-code machine learning library in Python, that automates machine learning workflows. The rest of the paper is organized as follows: Sect. 2 provides some related work using intrusion detection. Section 3 describes the analysis and preprocessing of the dataset. Section 4 presents the different algorithms used. Section 5 gives the results from the analysis. Section 6 ends the paper with the conclusion.

2 Related Work

With the exponential development of technology, the intruders are also advancing their intrusion methods. This increases the demand for sophisticated intrusion detection systems. In recent years, machine learning-based intrusion detection systems have gained a lot of attention. There are various online available datasets that can be used to create effective machine learning models. The "Canadian Institute for Cybersecurity" (CIC) has issued eleven datasets since 1998, the most recent dataset being CICIDS 2017 [2], which contains the most common contemporary attacks. Several attempts have been made to develop a machine learning model based on CICIDS 2017. Yulianto et al. [4] used the CICIDS 2017 dataset to improve an Ada-boost intrusion detection system (IDS) performance. They focus on surpassing the problem of imbalanced training data. Synthetic Minority Oversampling Technique (SMOTE), Principal Component Analysis (PCA), and Ensemble Feature Selection (EFS) improved the model and it achieved Area Under the Receiver Operating Characteristic curve (AUROC) of 92%. Some of the papers focus on detailed analysis of the CICIDS 2017 dataset. Ranjit Panigrahi and Samarjeet Borah in [5] examined the specific features of the CICIDS2017 dataset and identified its problems. They concluded that the major issue was class imbalance, regarding classification, and as a solution they proposed class relabeling. Another dataset that was used in building machine learning intrusion detection systems is NSL-KDD [6]. This dataset was also published by the "Canadian Institute of Cybersecurity", and it suggested solving the inherent problems of the KDD '99 dataset. Cholakoska et al. analyzed this dataset – NSL-KDD and studied the effectiveness of various classification algorithms for anomaly detection [7]. They managed to build a model that detects anomalies with 99% accuracy, using the Random Forest method for classification. Intrusion detection is in the focus of many researchers, since it addresses one of the biggest problems of the Internet.

3 Dataset Analysis and Preprocessing

As mentioned above, we used the CICIDS 2017 dataset for this research. The findings of network traffic analysis performed using CICFlowMeter are included in CICIDS 2017 and are labeled flows based on timestamp, source and destination IP addresses, source and destination ports, protocols, and attacks [2]. The background traffic was created by using the abstract behavior of 25 users based on the following protocols: HTTP (Hypertext Transfer Protocol), HTTPS (Hypertext Transfer Protocol Secure), FTP (File Transfer Protocol), SSH (Secure Shell), and email protocols. The data capturing lasted five days, from Monday to Friday. This dataset consists of 8 different network attacks: Brute Force FTP, Brute Force SSH, DoS (Denial of Service), Heartbleed, Web Attack, Infiltration, Botnet and DDoS (Distributed Denial of Service). In this dataset 11 criteria, necessary for building a reliable model, were identified [2]: Complete Network configuration, Complete Traffic Labeled Dataset, Complete Interaction, Complete Capture, Available Protocols, Attack Diversity, Heterogeneity, Feature Set and Meta Data. More than 80 network flow features were extracted from the network traffic measured by CICFlowMeter. The size of the whole dataset is 51.1 GB. We used an abbreviated version of the dataset which included 20% of the data of the original CICIDS2017 dataset. As a key step in developing a successful machine learning algorithm, detailed exploratory data analysis was performed. A conclusion was drawn, that the classes were highly imbalanced, which is shown in Fig. 1, and this is the main downside of this dataset.

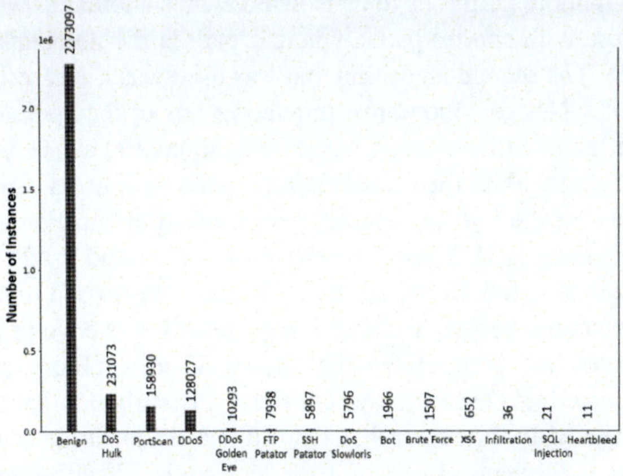

Fig. 1. Class distribution of CICIDS 2017

The picture above shows that aside from DoS Hulk, PortScan and DDoS the other attacks are much less common compared to the benign class, especially Heartbleed, SQL

Injection, and Infiltration web attacks. Further, this analysis brought us to the realization that some of the features contained only zero values, so in terms of information gain, they were useless. WEKA [8] software was used to detect these features and later they were removed. The next step of the study was checking for NaN and null values. The NaN values found in some of the features were replaced by the mean of the feature vector. The dataset also contained infinite values, which were dropped. After the analysis of the data, the dataset was split into 80% for the training and 20% for the test set and the study proceeded on training the model via PyCaret, a machine learning library in Python that automates workflows.

4 Methodology

The features are used as input to the classification algorithms in order to learn a classification model that will correctly classify the anomaly. The target variable consists of the most up-to-date common attacks. There are fourteen types of attacks, including: DoS Hulk, PortScan, DDoS, DDoS GoldenEye, FTP-Patator, SSH-Patator, DoS slowloris, DoS Slowhttptest, Bot, Web Attack - Brute Force, Web Attack - XSS, Infiltration, Web Attack - Sql Injection, Heartbleed. For the development of the model, the PyCaret library was used [3], which contains several machine learning libraries and frameworks, such as scikit-learn [9], XGBoost [10], LightGBM [11], CatBoost, spaCy, Optuna, Hyperopt, Ray, and a few more. The different types of algorithms that were used are anomaly detection, classification, and clustering. For anomaly detection two outlier detection techniques were used, that identify anomalies instead of normal observations. The first algorithm used was Isolation Forest [12]. The Isolation Forest algorithm is based on the principle that anomalies are observations that are few and different. By randomly selecting a feature and subsequently a split value for the feature, Isolation Forest builds partitions on the dataset. Evidently, compared to normal points in the dataset, anomalies require fewer random partitions to be isolated, hence anomalies will be represented as points in the tree with shorter paths, where a path is the number of edges crossed from the root node. The second algorithm that was used was k-Nearest Neighbors [13]. The capabilities of k-NN go beyond just predicting sets of data points or their values. Anomalies can be found with it as well. Identifying anomalies can be the end goal, such as in intrusion detection. With the classification approach, four machine learning algorithms implemented in the PyCaret toolkit were used: Random Forest Classifier [14], Decision Tree Classifier [15], SVM - Linear Kernel [16] and k-NN. Contrary to the supervised approaches stated above, clustering is an unsupervised method that utilizes datasets with no outcome (target) variable and no prior knowledge of the relationships between the observations, or simply unlabeled data. There are four types of clustering: Centroid-based Clustering, Density-based Clustering, Distribution-based Clustering and Hierarchical Clustering. In this research, Centroid-based Clustering is used, specifically K-means Clustering. K-means clustering uses "centroids", K different randomly initiated points in the data, and assigns every data point to the nearest centroid. The centroid is shifted to the average of all points assigned to it, after each point has been assigned. The other type of clustering that was tested is Hierarchical Clustering, more precisely Agglomerative Clustering. Agglomerative Clustering is a method of clustering where each data point starts in its own cluster. Then, these clusters are recursively combined by merging the two clusters that are the most similar to one another. The goal is to conduct a detailed research of multiple techniques for supervised and unsupervised learning in order to conclude which one delivers better results for intrusion detection. Different metrics were used in order to evaluate all the algorithms and methods mentioned previously. To evaluate the anomaly detection method, only accuracy was used as a metric, which represents the number of correctly predicted data points out of all the data points. Evaluation metrics for classification included accuracy (1), precision, recall and F1-score. Recall is calculated by dividing the true positives by anything that should have been predicted as positive (2). Precision is calculated by dividing the true positives by anything that was predicted as a positive (3). F1-score is defined as the harmonic mean between precision and recall (4). These are commonly used evaluation techniques for classification. Because it is crucial to correctly identify the intrusions (classes not labeled 'BENIGN'), greater weight is added to precision, recall, and F1-score over accuracy.

$$\text{Accuracy} = \text{Number of correct predictions} \,/\, \text{Total number of predictions} \tag{1}$$

$$\text{Recall} = \text{True Positive} \,/\, (\text{True Positive} + \text{False Negative}) \tag{2}$$

$$\text{Precision} = \text{True Positive} \,/\, (\text{True Positive} + \text{False Positive}) \tag{3}$$

$$\text{F1} = 2 * \text{Precision} * \text{Recall} \,/\, (\text{Precision} + \text{Recall}) \tag{4}$$

In contrast to supervised learning, where the model's performance can be assessed using the ground truth, clustering analysis lacks a reliable evaluation measure that we can use to assess the effectiveness of various clustering techniques. For evaluating the clustering algorithm, Silhouette score [17] was used, which is a metric used to calculate the performance of a clustering technique (5). Its value ranges from -1 to 1.

1: Means clusters are well apart from each other and clearly distinguished.

0: Means clusters are indifferent

-1: Means that data belonging to clusters may be wrong/incorrect

$$\text{Silhouette Score} = (\text{b} - \text{a}) \,/\, \max(\text{a}, \text{b}) \tag{5}$$

where, a is average distance between each point within a cluster and b is average distance between all clusters.

5 Experimental Results

The results obtained from anomaly detection, more accurately from Isolation Forest, and k-NN are shown in Table 1. The results are poor with both algorithms. The reason for this is the distribution of the target variable - 75% are benign and 25% are anomalies. This is not a sufficient enough percentage to be classified as an anomaly.

In comparison to the algorithms for detecting anomalies, classification algorithms produce substantially better outcomes. The results are shown in Table 2 (Accuracy,

Table 1. Anomaly Detection results.

Algorithm	Accuracy
K-nearest neighbors	0.706
Isolation Forest	0.740

Precision, Recall and F1-score). From the results it can be seen that all of the models have high accuracy, that is mainly because of the class 'BENIGN' which represents 75% of the total instances. However, the F1-score was similarly high, indicating that the models are reliable.

Table 2. Classification results.

Classifier	Accuracy	Recall	Precision	F1-score
Random Forest	0.996	0.849	0.996	0.917
Decision Tree	0.995	0.863	0.995	0.924
k-NN	0.993	0.783	0.993	0.876
SVM - Linear	0.963	0.564	0.957	0.710

As for the clustering, the results are shown in Table 3 (Silhouette score and Accuracy). Agglomerative Clustering separates the dataset into two clusters well apart from each other, as seen from the Silhouette score. However, based on the accuracy we can conclude that it incorrectly isolates the class "BENIGN" from the anomalies.

Compared to the agglomerative clustering, the K-means algorithm does not provide two clear clusters, but has a better accuracy. Even so, this does not imply that the K-means clustering is effective.

Table 3. Clustering results

Algorithm	Silhouette score	Accuracy
K-nearest neighbors	0.90	0.54
Isolation Forest	0.54	0.75

The accuracy of the top-performing method for each of the three separate approaches—classification, anomaly detection, and clustering—is displayed in Fig. 2. The highest accuracy was achieved by approaching the problem as a classification issue, using the Random Forest Classifier. On the other side, the performance of anomaly detection with Isolation Forest, and k-Means Clustering was significantly lower achieving accuracies of 74% and 75% respectively. This is the case because of the high class imbalance.

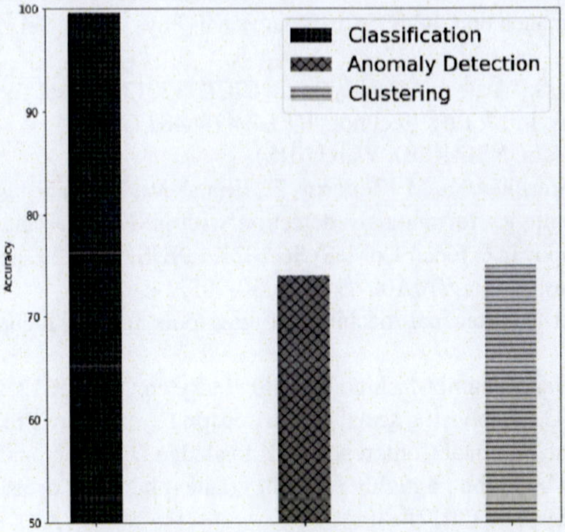

Fig. 2. Comparison of models

6 Conclusion

In this study using the PyCaret library, we compared two machine learning approaches, supervised and unsupervised learning, for detecting intrusions. This shows that machine learning methods are effective for detection of attacks, and therefore for security in systems. The implementation of machine learning algorithms is even easier with the PyCaret tool, which is simple to use for developing models and their evaluation. The effectiveness of the methods mentioned above is proven by the achieved results. The best performing model is the Random Forest Classifier, which achieved the highest F1-score (91%), as well as the highest accuracy (99%). These results were achieved using classification, however the other two methods (anomaly detection and clustering, with PyCaret) did not prove to be helpful. The ineffectiveness of anomaly detection is due to the distribution of classes (25% are attacks), which is a large percentage to be counted as an anomaly. Clustering has not shown to be efficient, and similar to anomaly detection, it has to be improved. In our future work, we intend to implement deep learning [18] to identify network intrusions and enhance clustering to boost performance.

References

1. Cisco, U.: Cisco annual internet report (2018–2023) white paper. Cisco: San Jose, CA, USA (2020)

2. Sharafaldin, I., Lashkari, A.H., Ghorbani, A.A.: Toward generating a new intrusion detection dataset and intrusion traffic characterization. In: 4th International Conference on Information Systems Security and Privacy (ICISSP), Portugal (2018)
3. pycaret.org. PyCaret, April 2020
4. Arif, Y., Sukarno, P., Suwastika, NA.: Improving adaboost-based intrusion detection system (IDS) performance on CIC IDS 2017 dataset. J. Phys. Conf. Ser. **1192**(1) (2019). (IOP Publishing)
5. Ranjit, P., Borah, S.: A detailed analysis of CICIDS2017 dataset for designing Intrusion Detection Systems. Int. J. Eng. Technol. 7.3.24, 479–482 (2018)
6. Ghulam Mohi-ud-din. NSL-KDD. Web (2018)
7. Cholakoska, A., Shushlevska, M., Todorov, Z., Efnusheva, D.: Analysis of machine learning classification techniques for anomaly detection with NSL-KDD data set. In: Silhavy, R., Silhavy, P., Prokopova, Z. (eds.) CoMeSySo 2021. LNNS, vol. 231, pp. 258–267. Springer, Cham (2021). https://doi.org/10.1007/978-3-030-90321-3_21
8. Witten, Ian, H., et al.: Practical machine learning tools and techniques. Data Mining **2**(4) (2005)
9. Pedregosa et al.: Scikit-learn: Machine Learning in Python. JMLR **12**, 2825–2830 (2011)
10. Chen, T., Carlos, G.: Xgboost: a scalable tree boosting system. In: Proceedings of the 22nd ACM SIGKDD International Conference on Knowledge Discovery and Data Mining (2016)
11. Guolin, K., et al.: Lightgbm: a highly efficient gradient boosting decision tree. Adv. Neural Inform. Process. Syst. **30** (2017)
12. Liu, F.T., Ting, K.M., Zhou, Z.H.: Isolation forest. In: 2008 Eighth IEEE International Conference on Data Mining (ICDM), pp. 413–422. Pisa, Italy (2008)
13. Mucherino, A., Papajorgji, P.J., Pardalos, P.M.: k-nearest neighbor classification In: Data Mining in Agriculture. Springer Optimization and Its Applications, vol. 34. Springer, New York (2009)
14. Liaw, A., Wiener, M.: Classification and regression by randomForest. R news **2**(3), 18–22 (2002)
15. Wu, X., et al.: Top 10 algorithms in data mining. Knowl. Inform. Syst. **14.1**, 1–37 (2008)
16. Cortes, C., Vapnik, V.: Support-vector networks. Mach. Learn. **20**(3), 273–297 (1995)
17. Shahapure, K.R., Nicholas, C.: Cluster quality analysis using silhouette score. In: 2020 IEEE 7th International Conference on Data Science and Advanced Analytics (DSAA). IEEE (2020)
18. Lansky, J., et al.: Deep learning-based intrusion detection systems: a systematic review. IEEE Access **9**, 101574–101599 (2021). https://doi.org/10.1109/ACCESS.2021.3097247

A Web Application for Moving from One Spot to Another by Using Different Public Transport - a Logical Model

Kamelia Shoilekova(✉)

Angel Kanchev University of Ruse, 8 Studentska Str, 7000 Ruse, Bulgaria
kshoylekova@uni-ruse.bg

Abstract. This document describes the logical model of a Web application for moving from one spot to another by using different means of transport (trains or buses transport). To create this application, several different technologies are used: big data, cloud technologies, architectural pattern MVC-aiming to optimize time for movement and several cloud services.

Keywords: application · MVC · Big data · services · cloud technologies

The development of the information technologies (IT) has brought huge, revolutionary changes in all spheres of life. IT are not a factor that influence only the computer systems and webs or the mobile phones communications. Digital technologies are used everywhere with their possibilities to render and transfer data from every object. This fact has brought about a new discipline and direction, which is known as Internet of Things - interconnected heterogeneous objects in Internet.

Data sources of this paradigm can be the means of transport and systems as well, that render an account and track activities in that sphere. From a global point of view, the digital data created a world, where enormous and various amount of data is created in high speed.

This fact issued the appearance of a new discipline or direction known as Big Data. The importance of Big Data is huge, because that can help creating different analyses and optimizations of events and processes and that could be the base on which different models could be built. The data could help businesses to raise effectiveness of processes and to find new tendencies [10].

The idea for developing this research appeared, when the movement from spot A to spot B happened to be a difficult task, because:

- Part of the roads web is closed, because of road reconstruction processes;
- There are accidents, which causes a slow traffic or/and a bypass must be found;
- A connection to another means of transport must be made;
- There is no (bus/train) route, which offers movement from A to B spot.

R. Silhavy and P. Silhavy (Eds.): CSOC 2023, LNNS 724, pp. 133–138, 2023.
https://doi.org/10.1007/978-3-031-35314-7_13

1 MVC

The architectural pattern for design is used, based on the separating of business logic of graphic interface and the data in an application. It "separates" the application into three interconnected parts. This is done in order to separate the internal presentation of the information from the ways, by which the information is presented and accepted by the user (Fig. 1).

Model-the core of the application, which is beforehand defined by the field for which it is designed. The model manages the data, the logic and the application rules.

The view is the exiting flow of information (the one that the application sends as an answer to the user, in response to his request). Several possible views are available for one and the same information.

The Controller is the third part of the pattern. It accepts the user's entrance (i.e. the data, that the user inputs with the request, etc.) and transfers them into commands to the model of the view [1, 2, 4, 9].

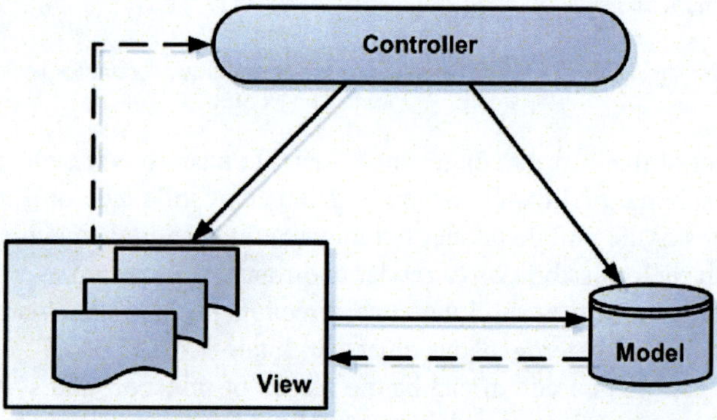

Fig. 1. Architecture of MVC

2 Model of Application

In order to describe maximum precisely and clearly the created application we are going to divide it into four layers (Fig. 2).

Physical Layer:
At this layer we have the possibility to feed the data and to extract information connected with traffic, being late or accidents.

Description Layer:
At this layer the user's access and the different services are defined. Every user can give current grading of the current itinerary and can interfere for signaling about a slow down or removing of the reason, that has caused the slow down, or he can add information

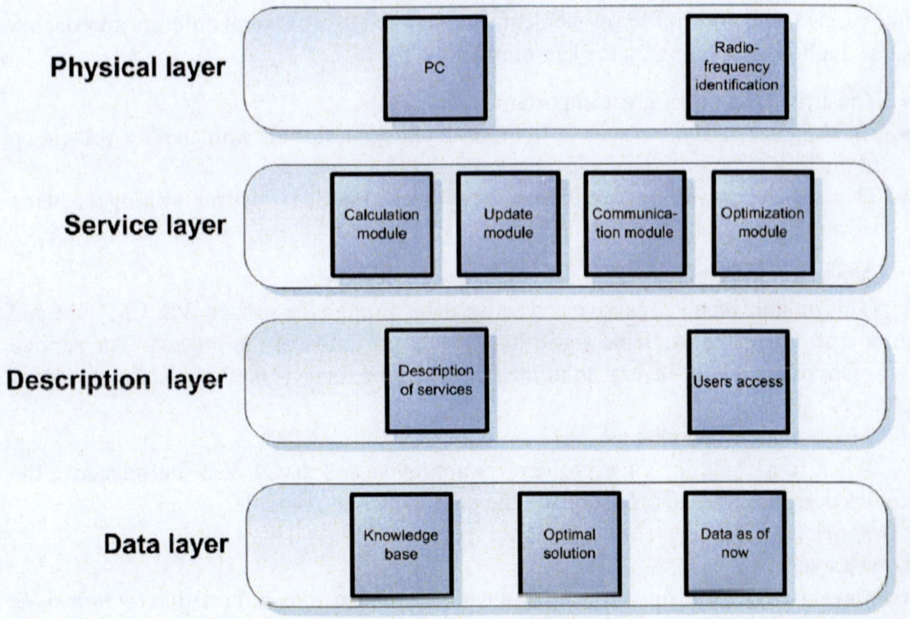

Fig. 2. Layers of the application

to the data. At that layer another important peculiarity is the communication between the different modules of the application, aiming to receive new information, in order to move from A to B spot.

Service Layer:
The various services are defined in the layer of the description, which allows the users to follow the execution of the process in a beforehand input range, to get the data for different calculations. That layer is composed out of three modules:

Communication module-this module executes the communication between different users and the system and the modules. When we define the route from spot A to spot B, it is necessary the system to take us out - is there a need to change transport or is there a slow down, how long is the time for travelling, how long a stay will take etc., i.e. every existing problem that may appear in a real situation, when travelling. In order to get the whole information, it is necessary to collect information for the a process in real time, it is necessary to record different problems that may appear and then to make calculations and measurements, that will give the opportunity to adjust the time for travelling [3, 5, 8].

All data is transferred to the digital module through the service WS_CM_CalM (it extracts data from the communication module and transfers it to the calculation module), measuring the time for the execution of the process and the module for optimization through the service WS_CM_OM (it extracts data from the communication module and transfers it to the optimization module), which aims to offer an optimal solution.

The calculation module - this module calculates the time for travelling. The Dijkstra algorhythm, for finding the shortest track for movement, is used, but in case of an unexpected situation-traffic, an accident, etc., it is necessary several different approaches to be applied, depending on the key moment:

- The arrival time is of great importance;
- An obligatory arrival to spot C, in order to change transport with another vehicle up to the final spot;
- The time for travelling - no matter how much time it is lost for waiting or being slowed down.

Module for optimization:

This module receives passive and active data, through the service WS_CM_OM and after that it uses the available algorithms for optimization of the process. The service WS_Opt offers one or several solutions, and then the focus is on the module of current change.

Module of current change:

After some solution of a problem is approved and saved, it is introduced I the knowledge base. It extracts data from the communication module.

Data Layer:

This layer is accessible for verification or restoring of data like: information for processes that are started everyday, new knowledge data base (current information for slow downs, change of timetables, etc.) and an optimal solution [5–7].

Figure 3 represents the communication between the MVC and the business logic layer.

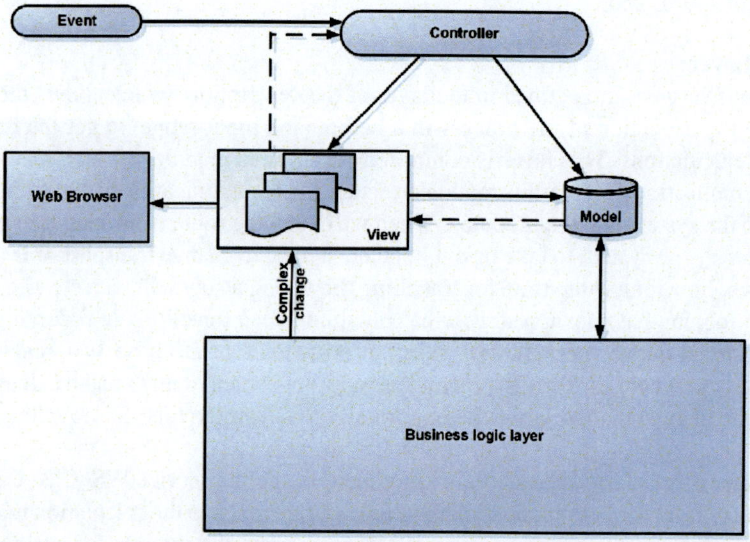

Fig. 3. Logical model of the application

3 Conclusion

To reach its purpose, the research uses different services, in order to extract the timetables of different transport firms.

We have described the moments, in which the offered approach is the most effective, we have described possible situations, when optimization of time is needed or optimization of the whole movement process is needed. Big Data is used in the research, though not demonstrably. The use of Big Data for optimizing and analysis of transport systems leads to increasing of their autonomy, intelligence, interconnection and to new possibilities for overcoming roads problems in urban areas [11]. The intelligent transport systems provide effective integration and cooperation between different technologies and services. Processing of huge volume of data, offers effective, safe and useful solutions for a lot of activities in transport.

Acknowledgements. This publication reflects research from the scientific project "Investigation of effective knowledge management mechanisms applied in software engineering when creating projects with Agile methodologies" - of the "Scientific Research" fund of Ruse University "Angel Kanchev", 2023.

References

1. Bansiya, J., Davis, C.G.: A hierarchical model for object-oriented design quality assessment. IEEE Trans. Software Eng. **28**(1), 4–17 (2002)
2. Chiarelli, A. (2016). Mastering JavaScript ObjectOriented Programming. Packt Publishing Ltd.
3. Moreri, Kealeboga & Maphale, Lopang. (2014). A Conceptual Participatory Framework for Road Accidents Location using Free Open Source GIS Tools
4. G Krastev, V Voinohovska, Smart Mobile Application for Public Transport Schedules - Logical Model, TEM Journal, 2020, ISSN 2217-8309, DOI: https://doi.org/10.18421/TEM92-15
5. Benotmane, Z., Belalem, G., Neki, A.: A Cloud Computing Model for Optimization of Transport Logistics Process. Transport and Telecommunication Journal. **18** (2017). https://doi.org/10.1515/ttj-2017-0017
6. Boyanov, Luben, System for Big Data processing in Transport (August 19, 2021). Mechanics Transport Communications, vol. 19, No 2, 2021, ISSN 1312–3823 (print), ISSN 2367–6620 (online), pp. XII-9 - XII-14, Available at SSRN: https://ssrn.com/abstract=3950641
7. I. Rusev, R. Rusev, B. Ivanova (2021) TRAINING OF MEDICAL STAFF FOR USING DATA – USAGE OF THE FULL CAPABILITY OF SMART DEVICES, EDULEARN21 Proceedings, pp. 8804–8810
8. Maphale, Lopang & Sereetsi, Lone & Moreri, Kealeboga & Manisa, Michael. (2015). Design of a Public Transport Web Map Application (WMA) for the city of Gaborone
9. Krastev, G., Voinohovska, V., "Smart mobile application for public transport schedules – data organization and program realization,",: International Congress on Human-Computer Interaction. Optimization and Robotic Applications (HORA) **2020**, 1–4 (2020). https://doi.org/10.1109/HORA49412.2020.9152884

10. Shoilekova, K., Ivanova, B. (2022). The Necessity of Information Extraction from Big Data Systems for the Purpose of Business Process Optimization. In: Silhavy, R. (eds) Software Engineering Perspectives in Systems. CSOC 2022. Lecture Notes in Networks and Systems, vol 501. Springer, Cham. https://doi.org/10.1007/978-3-031-09070-7_5
11. Atanasova, D., Venelinova, N. (2022). Comparative Analysis of Decision-Making Models in National Healthcare Systems of EU Member-States: Change-Drivers' Identification Contemporary Methods in Bioinformatics and Biomedicine and Their Applications. BioInfoMed 2020. Lecture Notes in Networks and Systems, vol 374. Springer, Cham. https://doi.org/10.1007/978-3-030-96638-6_1

Traffic Prediction for VRP in Intelligent Transportation Systems

Piotr Opioła(✉), Piotr Jasiński, Igor Witkowski, Katarzyna Stec, Bazyli Reps, and Katarzyna Marczuk

Aleet, Wólczańska 125, 90-521 Łódź, Poland
p.opiola@aleet.co

Abstract. One of the biggest challenges in modern urban cities is how to deliver goods and services efficiently, that is to reduce congestion on roads and pollution while improving life comfort with less time spent in traffic or waiting for deliveries. In this paper, we study how machine learning methods, in particular artificial neural networks, can help with solving this problem by predicting the traffic flow, which in turn may improve performance of the vehicle routing methods, especially wait and ride times. Once we have accurate traffic speed predictions, we can use it as the cost matrix to find the optimal travel times between nodes in a road network graph. We focus mostly on convolutional networks, including graph convolutions, which are suitable for representing the graph-like structure of the analyzed data. All the models are tested on GPS data from Warsaw and compared to baseline models. The best performing model is then tested with a VRP algorithm for ride-sharing taxi. Our results suggest that artificial neural networks can reduce the wait times significantly compared to simple statistical estimation, specifically our test shows a reduction by 18%.

Keywords: Machine Learning · Traffic Prediction · Graph Convolutional Networks · Vehicle Routing Problem · Traffic Simulation

1 Introduction

With e-commerce booming, consumers expect their parcels to arrive in the minimum time, at the minimum cost. At the same time, new travel modes develop rapidly, especially in the area of shared mobility, including bike-sharing, e-scooter sharing, car-sharing, and ride hailing [1]. Furthermore, all of the transportation services are expected to be designed and managed with the awareness of sustainability and reduce carbon emissions [2]. In order to achieve this, logistics needs to be aided with advanced technology that helps making the right decisions by providing all the necessary information, analysis and recommendations. One example of such technology are Vehicle Routing Problem (VRP) solutions, which attempt to find the optimal resource assignment as well as the optimal order of destinations to visit, resulting in faster delivery, lower fuel consumption and lower emissions [3].

The VRP methods traditionally minimize the distance or the travel time. While the distance can be easily obtained from the road network structure, the accurate travel times

R. Silhavy and P. Silhavy (Eds.): CSOC 2023, LNNS 724, pp. 139–147, 2023.
https://doi.org/10.1007/978-3-031-35314-7_14

are unknown and they can change dynamically. A usual approach is to approximate them with simple statistical models [4, 5]. We study an approach which predicts the travel times from real-time traffic data using advanced machine learning methods, specifically artificial neural networks with focus on taking advantage of the graph structure of the data. The current state of knowledge for such application of ML is limited. Several examples of recently undertaken experiments are listed in [6], which shows that the most common methods are decision trees and basic ANN's. An example of using more complex convolutional neural networks (CNN) models for traffic prediction can be found in [7].

We have decided to focus on methods that take advantage of the graph-like structure of the data, especially Temporal Graph Convolutional Network (TGCN) [8] and Adaptive Graph Convolutional Recurrent Network (AGCRN) [9]. The former model introduces an interesting concept of a graph convolution defined by the adjacency matrix rather than by a static kernel. The later method enhances the idea by making the adjacency matrix trainable. These methods proved to be accurate on the Performance Measurement System (PeMS) data, commonly used as a traffic prediction benchmark [10]. Our experiments confirmed them to be beneficial on aggregated FCD data as well. We also compared them to a model based on traditional kernel convolutions.

2 Real-Time Prediction System Design

One of the crucial parts of a prediction system is the input data, and in the case of a system which is expected to respond to dynamic changes on the roads it is equally important to have data processing pipelines in place to be able to update the model frequently and calculate the most current predictions in real-time [11]. We will discuss briefly the software architecture needed for this task as well as the types of input datasets.

2.1 Dataset

The first problem to solve is the choice of the data source. The common solutions include road sensors [12], camera images [13], images from the route planners' websites [14], mobile phones localization [15], meteorological information [16], and GPS devices [17]. Road sensors have the advantage of easy and accurate counting of all the traffic on a single road segment. However, the number of road segments equipped with such sensors is limited, and thus the method has little value for VRP problems. The second serious candidate is the data from GPS devices. This can be divided into devices installed in vehicles and mobile phones carried by passengers. The problem with mobile phones is that they are also used by pedestrians and so such data is too noisy if we are only interested in road traffic. At the same time, the GPS data from the vehicle devices has the disadvantage of a relatively small percentage of vehicles having such devices. Also, we need to distinguish between vehicles stopped due to congestion and vehicles stopped for any other reason, such as picking up passengers. Other data sources can be considered a potentially valuable supplement, but they are not sufficient as the main source. The most reasonable trade-off for us is the GPS data from vehicle devices.

Despite the disadvantages, it is the only source which contains an acceptable quantity of measurements from all the road segments and is not as noisy as the mobile phone data.

Processing the data through all the stages from the source to the model input is complicated and requires a well-defined architecture to be reliable and computed efficiently within expected time intervals in an automated manner. The GPS location data, also called Floating Car Data (FCD) is usually provided as a CSV file with timestamps and geographic coordinates. Those locations need to be matched to the road network, consisting of road segments. We define a road segment as a part of a road between intersections. The dataset we used for testing our models covers the area of the city of Warsaw, Poland, which is an urban area with a population of approximately 2 million inhabitants and population density of approximately 3500 per km^2. The number of registered vehicles is over 2 million while the number of cars driving into the city center during the day can be estimated to be half a million [20].

2.2 Data Processing Pipelines

Matching the FCD data to the road network is not a trivial task, but it has been covered well in literature, along with concepts such as Kalman filter and local path searching [18]. We have used an implementation of the method described in [19], which integrates a Markov model with precomputed shortest paths. In addition to map matching, it is important to identify anomalies in the data, such as cars that are not moving for too long and could falsely be interpreted as stuck in traffic while they have been simply parked.

The next step is aggregation, both by time and space. We group the data into 15-min time bins, for each of the road segments. With over 80 thousand segments, it becomes troublesome to use that data for model training as it doesn't fit in the memory. For that reason, we use an iterable dataset and load the records into memory in batches. The whole processing pipeline requires significant resources to deal with a year-long dataset, which contains around a billion records. Cloud platforms along with Apache Spark come in handy and guarantee scalability. In order to process the historical data and build a model, we run around a hundred parallel workers.

The real-time continuous prediction is less demanding in terms of resources, but requires more attention to single-process performance. Within several minutes, we download the most current data from the provider, go through map matching and aggregation, apply preprocessing transforms, predict the traffic speed for each segment and update the cost matrix used by a ride-sharing VRP algorithm [21]. This way we always have the most up to date input to the model. Also, the model itself is up to date thanks to being rebuilt once a day. The new rebuilt version is compared to the currently deployed model. If the metrics of the newly trained model are better, it is being deployed in place of the previous one. This is a fully automated process, which can be easily monitored thanks to logs and alerts.

Developed prediction models are deployed in Aseem, our in-house built mesoscopic traffic simulator, to compare the impact of various traffic prediction models on an actual transportation system.

2.3 Missing Values

One more important problem to solve in our methodology is missing data. Although this is a common issue, it requires special attention in our case. This is because of the quantity of the missing data, which is around 90%. Thus, the usual approach of replacing empty values with mean or median is not applicable here. Such a large ratio of missing values is caused by the fact that each of the 80 thousand road segments is a separate feature. Considering that only a small percentage of all vehicles are tracked, it is understood that many segments are not covered within a 15-min time interval. This raises the need for an accurate approximation. We have dealt with this issue by applying the adjacency matrix, that is a matrix which has a positive value on the *(i, j)* position if a car can move directly from the *i*-th segment to the *j*-th segment, and zero otherwise. This kind of matrices are used in Graph Convolutional Networks. An interesting feature of an adjacency matrix is that when multiplied by a vector of segment-wise features, it propagates the features to adjacent segments. The values in the matrix can be limited to 0 and 1, to only indicate the existence of a connection, or they can be real values from 0 to 1, to indicate how much of the traffic moves from one segment to another on average. If we normalize the columns of the matrix by dividing by their sums, then multiplying by the matrix is in fact an approximation of each input value with its adjacent values. The procedure is repeated until non-empty values get propagated to all the segments. The traffic speed is first normalized with respect to the speed limit. This technique proved to give satisfactory results.

3 Models

There are many possible approaches to traffic prediction. First, we have to choose the properties that will describe the traffic characteristics well, and that will suit our purpose, which in this case is solving a VRP problem. The most commonly used parameters are volume and average speed. Both of them are useful as the model input, but for the output we are interested mostly in the average speed, as it can be used directly to calculate the travel times. While the average speed can be easily obtained from aggregated FCD data, we don't know the overall number of vehicles that went through the street segment, only the number of the tracked vehicles. However, this value is good enough for the model input, as it still allows tracking volume changes and differences between segments. In order to make the speed property comparable between segments, we divide it by the max speed limit, which results in a value related to the congestion index.

Traffic prediction falls into the category of time series prediction problems, which suggests using specific models, such as ARIMA or LSTM, which place emphasis on the time dependency. On the other hand, our data is not a single sequence of values changing in time, but a whole network of such sequences, which in turn indicates the need for modeling space dependency. This can be done with a basic dense layer, but it is more efficient to identify local space dependencies with convolutional layers. The regular convolutional layer assumes the input to have a grid structure, which is useful in image processing where the image is a grid of pixels, but it is less applicable when the input is a graph. We can, however, use a dense layer to map the graph into a grid first and then apply the convolution. The grid representing a single graph will be 2-dimensional, but we can also include the time dimension in the grid and then it will become 3-dimensional, which implies the use of a 3D convolutional layer. The nodes in the grid can be interpreted as clustering of the city map which aggregates highly correlated graph nodes. We will refer to this model as CNN.

Another approach to modeling space dependency is the GCN method described in [8]. It takes advantage of the adjacency matrix, described in the previous section. Based on the matrix, the convolution layer passes information only between adjacent nodes. The matrix is not trainable, so it has to be calculated a priori. The GCN model is enhanced with the GRU layers for tracking time dependency - in this version it is called Temporal GCN (TGCN). Another iteration of GCN models is the AGCRN model [9]. The main difference from TGCN is that the adjacency matrix is trainable. More specifically, it is calculated from a trainable matrix called embedding, which has two purposes: Node Adaptive Parameter Learning (NAPL) and Data Adaptive Graph Generation (DAGG). The NAPL module is responsible for generating a trainable adjacency matrix, while the DAGG module generates a new, smaller graph, which aggregates correlated nodes from the original graph (Figs. 1 and 2).

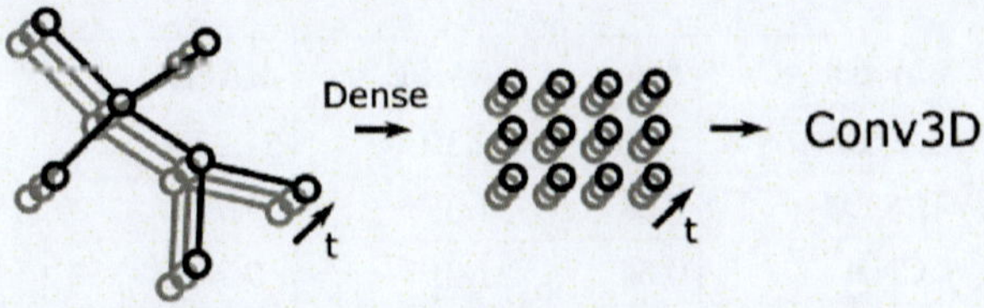

Fig. 1. The CNN model maps a sequence of road networks into a sequence of grids and applies a three-dimensional convolution.

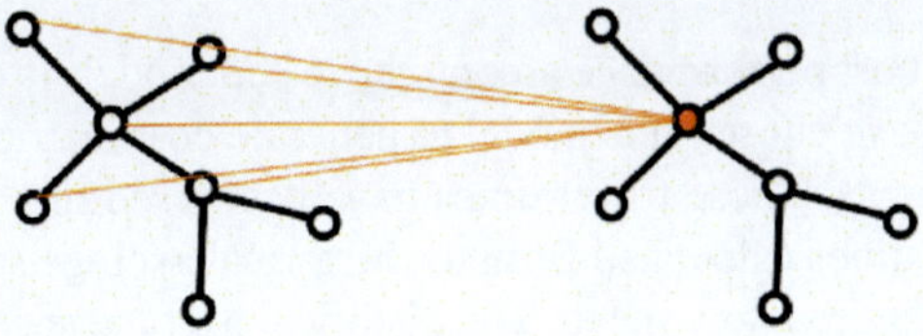

Fig. 2. The GCN model layer connects the input and output nodes with respect to their topological relationship in the road network.

4 Results

We compare the models by measuring their performance with the following metrics: Mean Absolute Error (MAE), Root Mean Square Error (RMSE) and Mean Absolute Percentage Error (MAPE). The models to compare are: ARIMA, LSTM, CNN, TGCN and AGCRN. The training time is limited to 1 h and runs on a single GPU. The parameters for each model are listed in Table 1.

Table 1. Model parameters.

Model name	Parameters
ARIMA	p = 4, d = 1, q = 4
LSTM	Hidden layer size: 20
CNN	Hidden layer size: 200, Conv3D layers: 2
TGCN	GCN layers: 2
AGCRN	Embedding size: 10, Hidden layer size: 64

For the input we take 16 time steps (4 h), aggregate them into 4 groups, each having 4 original time steps, and calculate the mean value for each group. For the output we take the mean value of 4 time steps (1 h). The results for each model are presented in Table 2.

Table 2. Model performance comparison for 1 h prediction horizon.

Model	MAE	RMSE	MAPE
ARIMA	1.70	2.30	21.23%
LSTM	0.68	1.06	7.79%
CNN	0.64	1.01	7.24%
TGCN	1.23	1.75	14.71%
AGCRN	0.49	0.89	5.75%

It is clear that the best performance among the tested models is demonstrated by the AGCRN model. We have chosen this model to generate cost matrices with travel times for each pair of segments. These travel times have been used in a VRP algorithm and compared with travel times calculated from the historical average speed and with travel times obtained from the free flow speed. The historical average speed is calculated over the last 6 weeks for the given day of the week and the given hour. The simulation covers a period of 17 h, from 7am to midnight, with 210408 requests made within the analyzed time. The simulated fleet represents a cab-sharing taxi company with 10,000 vehicles. The results are presented in Table 3.

The difference in results obtained with each travel time estimation method is mostly emphasized in the wait times. With convolutional graphs we were able to reduce the wait times by 18% compared to historical averages, and 24% compared to free flow-based estimation. Due to inaccurate travel times in the baseline prediction methods the VRP algorithm was choosing slower routes which propagated to non-optimal assignment of vehicles to passengers. The high passenger occupancy rate in the free flow method is caused by the lack of information about congestion which leads to optimistic assignments. The highest advantage of AGCRN is clearly visible in Figs. 3, 4 and 5, between 9 and 11 am.

Table 3. VRP simulation: AGCRN vs historical average vs free flow.

Metric	AGCRN	Historical	Free flow
Percentage of completed requests	69.38%	67.07%	68.86%
Average wait time	00:23:27	00:28:29	00:30:47
Average trip time	01:28:33	01:28:40	01:31:03
Average distance driven per completed request (km)	15.41	15.96	15.50
Percentage of empty km	11.59%	12.69%	11.96%
Average passenger occupancy in nonempty vehicles	1.67	1.60	1.72

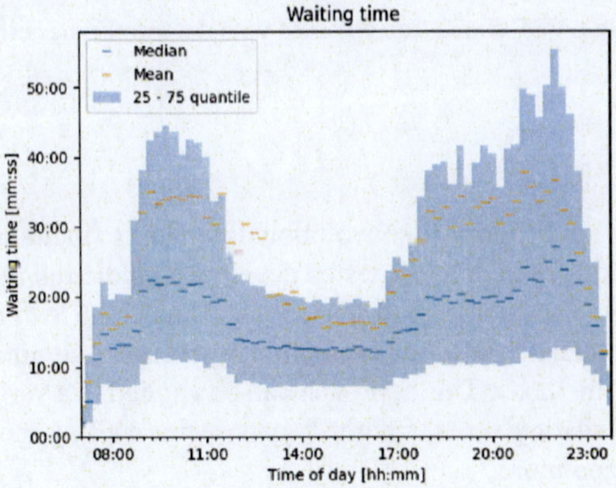

Fig. 3. Waiting times in a VRP simulation obtained with the travel times estimated by the AGCRN model.

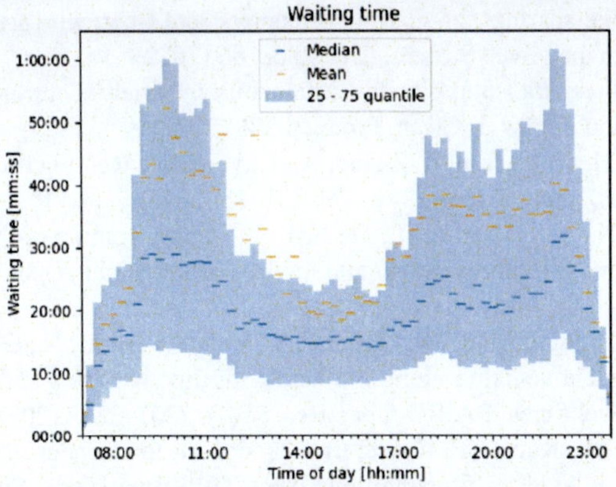

Fig. 4. Waiting times in a VRP simulation obtained with the travel times estimated by the historical averages.

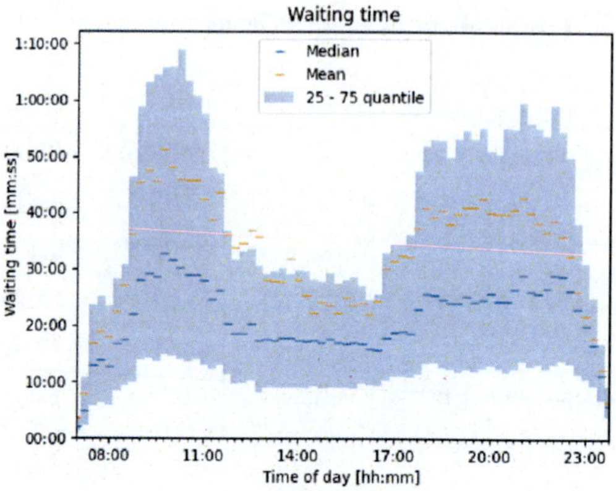

Fig. 5. Waiting times in a VRP simulation obtained with the travel times estimated from the free flow speed.

4.1 Conclusion

This paper evaluates application of convolutional networks (including graph convolutions) models for traffic prediction. We tested developed prediction models on FCD data for the city of Warsaw. The results show that advanced machine learning methods, especially GCN models, in real-time traffic prediction systems can significantly increase the performance of VRP methods. This approach can be applied to a variety of real-life use cases, including cab-sharing fleets, freight logistics, ride-hailing, postal services, bike sharing rebalancing and more.

References

1. Mouratidis, K.: Bike-sharing, car-sharing, e-scooters, and Uber: who are the shared mobility users and where do they live? Sustain. Cities Soc. **86** (2022)
2. Shah, K. J., et al.: Green transportation for sustainability: Review of current barriers, strategies, and innovative technologies. J. Clean. Product. **326** (2021)
3. Kim, G., Ong, Y., Heng, C., Tan, P., Zhang, A.: City Vehicle Routing Problem (City VRP): a review. IEEE Trans. Intell. Transp. Syst. **16**(4), 1–13 (2015)
4. Tas, D., Dellaert, N., Van Woensel, T., De Kok, T.: Vehicle routing problem with stochastic travel times including soft time windows and service costs. Comput. Oper. Res. **40**(1), 214–224 (2013)
5. Khanchehzarrin, S., Shahmizad, M., Mahdavi, I., Mahdavi-Amiri, N., Ghasemi, P.: A model for the time dependent vehicle routing problem with time windows under traffic conditions with intelligent travel times. RAIRO-Oper. Res. **55**(4), 2203–2222 (2021)
6. Sihag, G., Parida, M., Kumar, P.: Travel time prediction for traveler information system in heterogeneous disordered traffic conditions using GPS trajectories. Sustainability **14**(16), 10070 (2022)
7. Hou, Y., Edara, P.: Network scale travel time prediction using deep learning. Transp. Res. Rec. **2672**(45), 115–123 (2018)
8. Zhao, L., et al.: T-GCN: A Temporal Graph Convolutional Network for Traffic Prediction. arXiv:1811.05320 [cs.LG] (2021)

9. Bai, L., Yao, L., Li, C., Wang, X., Wang, C.: Adaptive Graph Convolutional Recurrent Network for Traffic Forecasting. arXiv:2007.02842 [cs.LG] (2020)
10. Performance Measurement System (PeMS) Data Source. https://dot.ca.gov/programs/traffic-operations/mpr/pems-source
11. Yu, J.J.Q., Gu, J.: Real-time traffic speed estimation with graph convolutional generative autoencoder. IEEE Trans. Intell. Transp. Syst. **20**(10), 3940–3951 (2019)
12. Polson, N., Sokolov, V.: Deep learning for short-term traffic flow prediction. Transp. Res. Part C: Emerg. Technol. **79**, 1–17 (2017)
13. Fredianelli, L., et al.: Traffic flow detection using camera images and machine learning methods in ITS for noise map and action plan optimization. Sensors (Basel) **22**(5), 1929 (2022)
14. Wang, T., Hussain, A., Sun, Q., Li, S.E., Jiahua, C.: The prediction of urban road traffic congestion by using a deep stacked long short-term memory network. IEEE Intell. Transp. Syst. Mag. **14**(4), 102–120 (2022)
15. Tettamanti, T., Varga, I.: Mobile phone location area based traffic flow estimation in urban road traffic. Adv. Civil Environ. Eng. **1**(1), 1–15 (2014)
16. Braz, F.J., et al.: Road traffic forecast based on meteorological information through deep learning methods. Sensors **22**(12), 4485 (2022)
17. Guo, J., Liu, Y., Yang, Q., Wang, Y., Fang, S.: GPS-based citywide traffic congestion. Transp. A Transp. Sci. **17**(2), 190–211 (2021)
18. Chen, F., Shen, M., Tang, Y.: Local path searching based map matching algorithm for floating car data. Procedia Environ. Sci. **10**, 576–582 (2011)
19. Yang, C., Gidófalvi, G.: Fast map matching, an algorithm integrating hidden Markov model with precomputation. Int. J. Geogr. Inf. Sci. **32**(3), 547–570 (2018)
20. Warsaw Traffic Survey. https://en.um.warszawa.pl/-/warsaw-travels-1, City of Warsaw (2015)
21. Alonso-Mora, J., Samaranayake, S., Wallar, A., Frazzoli, E., Rus, D.: On-demand high-capacity ride-sharing via dynamic trip-vehicle assignment. Proc. Natl. Acad. Sci. **114**(3), 462–467 (2017)

Enhancing Monte-Carlo SLAM Algorithm to Overcome the Issue of Illumination Variation and Kidnapping in Application to Unmanned Vehicle

Agunbiade Olusanya Yinka(✉) and Avuyile NaKi

Department of Information Systems, University of the Western-Cape, Bellville, Cape Town, South Africa
agunbiadeolusanya@yahoo.com

Abstract. A research problem that has been thoroughly investigated by researchers is the Simultaneous Localization and Mapping (SLAM). Its potential ability to solve self-navigation in robot is the center of attraction for most researchers studying it and numerous SLAM approaches have been developed throughout the years with great success. However, there are problems such as kidnapping and Illumination variation (Shadow and Light variation) limiting the acceptance of SLAM. These problems complicate image interpretation which might have a negative impact on robot trajectory and or, in worst case scenario, the robot inability to recovery from kidnapping. In this study, the original Monte Carlo algorithm will be upgraded to overcome these challenges. The filtering algorithms will be introduced to overcome issue of illumination variation, while the Initialize localization and grid base mapping was employed to overcome kidnapping. In this experiment, the enhanced MCL was compared with the original MCL, and Matlab was employed for the purpose of simulation. Evaluation was based on quantitative and qualitative approach and result obtained shows better performance of the enhanced MCL over the original MCL. This achievement by the enhanced MCL for attaining better robot trajectory, in real life could support self-exploratory into unknown area, auto-driving and route planning while reducing injuries and accident for people employed in hazardous surroundings.

Keywords: Autonomous Navigation · Monte Carlo Algorithm · Kidnapping · Illumination variation and robot trajectory

1 Introduction

An essential method for autonomous guidance is simultaneous localization and mapping. Its popularity amongst many researchers is backed up by its ability to contribute to successful navigation. Thus, a number of SLAM strategies have been suggested with important advancements, although they are still susceptible to certain issues. Among the issues that many scholars complained about seemed to be environmental noises (shadow and light intensity). The impact of environmental noises makes it challenging

R. Silhavy and P. Silhavy (Eds.): CSOC 2023, LNNS 724, pp. 148–159, 2023.
https://doi.org/10.1007/978-3-031-35314-7_15

to attain visual feature interpretation [1, 2]. This fundamental problem led some scholars to suggest using sensors such as sonar and laser range for deploying their own SLAM technique [3, 4]. Thus, the appearance of varying illumination in the image, on the other hand, may lead to a kidnapping problem, which represents a scenario that occurs when the robot is unaware of its new location [5]. This occurs whenever an autonomous vehicle moves without being aware of its odometry measurement, and the outcome may result in system failure with no chance of recovery [5]. Furthermore, kidnapping can occur due to a variety of factors, including failing sensors, robots teleporting to some other location, and an increase in measurement noise [6]. The result of this constraint is that a robot is unable to determine its pose. This situation violates the principle of resolving the SLAM problem, since position estimation is mandatory for map creation [5]. However, these problems were researched and the Monte Carlo (MCL) algorithm would be modified to overcome these challenges. The deployment of this study findings will boost productivity and safety while also advancing SLAM research. The paper structure is as follows: The literature review conducted for this study was presented in Sect. 2, the modified MCL algorithm was presented in Sect. 3, the dataset, experiment, and results were discussed in Sect. 4, and the study's conclusion was covered in Sect. 5.

2 Literature Review

Numerous approaches to SLAM have been presented in the recent years, with impressive results. Furthermore, due to their advantages over one another, various types of algorithms had been used in SLAM studies [16]. This segment discusses the different SLAM algorithms, their benefits, and their drawbacks. Additionally, it discusses the study direction for future SLAM.

In the review study of [7], they discuss a SLAM technique for autonomous mobile robot navigation that employs an active stereo camera for vision. The stereo images are analyzed using the MCL algorithm. The operation is split into two phases: prediction and updating. The collection of chosen particles estimated either from previous state is introduced to the motion model at the prediction step, and the observation estimation as well as the sample weight are considered at the update stages. During this level, resampling prioritizes particles by a high probability of being linked with a real object/feature to create a sample of new set. The discussed vision framework involves feature/object detection as well as avoidance in addition to SLAM technique presented. The model for stereo vision that is utilized for various purposes is shown in Fig. 1.

The presented method has been put to the test and the outcome of the trajectory error compared to the real measurement of the ground-truth is impressive. Additionally, it demonstrates its capacity to maneuver in a dynamic setting in an unstructured and complex environment. However, due to variance in the stereo camera sensors' observing positions, data association is difficult whenever an image scene has the characteristics of light change, specular reflection, and inconsistent point clouds. In order to enhance the technique's efficiency during the data association stage, the researchers intend to address illumination variance in their upcoming research.

In the research of [8], the unmanned vehicle created must travel in an unfamiliar environment by creating a map of its surroundings using the features gathered from the

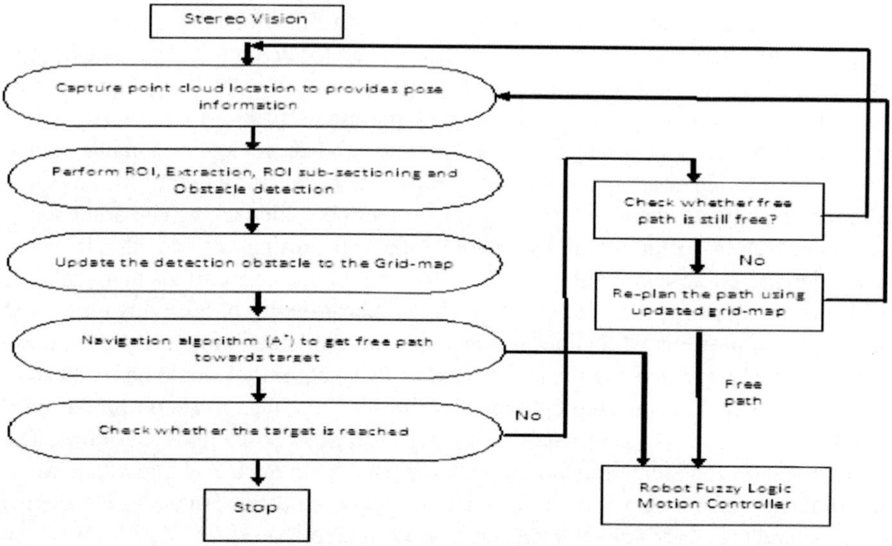

Fig. 1. A Framework for stereo vision multi-purpose [7].

environment and utilizing the same generated map to localize itself with minimal error. In order to create the map, the data was analysed using the occupancy grid mapping technique and the sonar sensor. The relative positioning methods based on the Kalman filter are also supported by wheel encoders and inertial measurement units, which help to determine the precise vehicle pose. Additionally, its navigation was also guided by the discrete step-wise modelling. All presented techniques were carried out on a Raspberry Pi, that functions as the robot's main computational processor. The Raspberry Pi 3 model has a 1.2 GHz 64-bit quad-core Advanced RISC Machine (ARM) V8 CPU with 1GB RAM. The suggested SLAM technique was capable of addressing a number of problems, including the dynamic environment, actuator defects, computing complexity, and trajectory error. The technique, however, can only be used in an indoor setting; attempts to test it in an outside setting result in system failure. In addition, using ultrasonic range sensors, because they are costlier than most varieties of sensors like Lidar and cameras, could be another disadvantage. Finally, the problem of robot kidnapping, in which the robot lost position as a result of build-up of inconsistencies, also restricts the effectiveness of the suggested SLAM system. Thus, in their future SLAM technique all current limitation is planned to be addressed.

Considering the review conducted for this study, it is obvious the illumination variation and kidnapping are still persisting till date. In this research, the illumination variation and kidnapping are taking into consideration and the Monte-Carlo algorithm will be modified to overcome these challenges.

3 Methodology

Five stages make up the modified MCL-SLAM procedures. The Image acquisition stage employs sensors to gather environmental data. The feature extraction stage is where the features or landmarks are extracted and used in the subsequent processing in upcoming phases. The Filtering step identifies and eliminates illumination variation, while the extracted features are employed for map conception at SLAM phase. The Navigation stage is responsible for controlling the robot's motion by transmitting signal components to the actuator. The technical flows of the Enhanced MCL-SLAM technique based on the research of [9] are illustrated in Fig. 2.

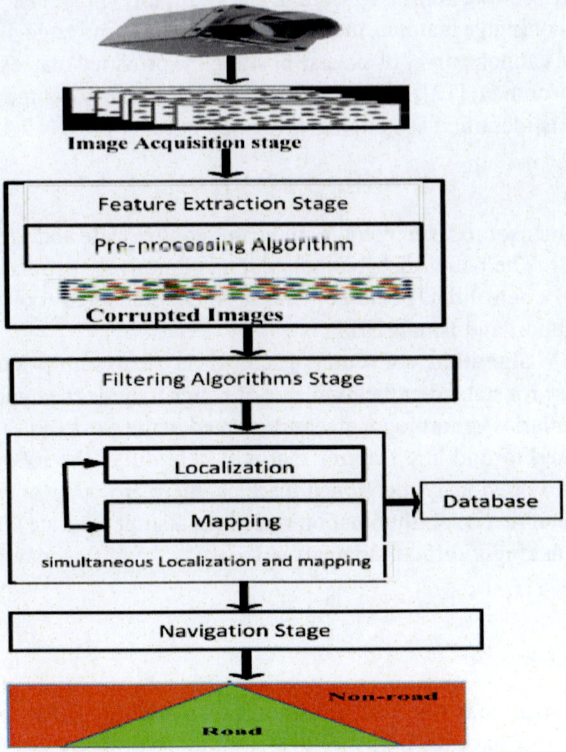

Fig. 2. The modified MCL framework. [9].

3.1 Image Acquisition Stage

During this process, the image acquisition phase converts captured visual features to a stream of electrical indication that can be interpreted as a form of digital image [10]. Digital images are produced by a variety of sensors, including range sensors, radar, tomography, and cameras, among others. Thus, the camera was utilized in this phase for transforming captured featured into object that can be readable or analyzed. This sensor was preferred because it has the benefit of acquiring more data than some other sensors [11], which will assist further during object detection. The camera captures image sequences and transferred them to the SLAM method, which processes and builds a map which will be utilized for vehicle movement [12]. Thus, in the process of acquiring image features from the environment using camera is the source for which illumination variation and kidnapping appears in the image.

3.2 Feature Extraction Stage

The 2nd phase for the enhanced MCL-SLAM algorithm is the Feature extraction. This stage uses statistical techniques and a variety of filters to extract image features from various regions (drive section, non-drive section, and ambiguity) [13]. The SLAM approach supports a variety of image features, including color, texture, and edge. In general SLAM concept, the robot cannot extract its precise position the provided map except if it gathers data from its environment [12]. This extracted data from the environment that can assist with the position is identified as belief as presented in Eq. 1 [14].

$$bel(s_t) = p(s_t \backslash z_t, u_t) \tag{1}$$

where z_t signifies sensor measurement, u_t indicates control state and s_t denotes the state sample at a time t. The belief distribution, which is iteratively estimated based on the measurement and control data, emerges into a potent statistical tool for tackling the SLAM issue (Bukhori and Ismail 2017).

In MCL-SLAM algorithm, collection of samples is use to signify subsequent belief. Samples are theory for state identification, one of which include the hypothesis that wall and obstacle boundaries are orthogonal was presented in the work of [17]. These sample collections are used to find key features that will help direct the robot's direction. In addition to the two previously mentioned theories, there are other ones. Thus, several samples under the influence of illumination variations disrupt the state identification theories, thereafter making it difficult to analyze the image, and this may result in system failure [15].

3.3 Filtering Stage

This is the third stage that was added to modified the MCL-SLAM procedure since illumination variances have the influence to distort image properties, reduce visual perception, and have a major impact that can minimize the performance of SLAM technique [16]. Furthermore, illumination variations under intense scenario can result in kidnapping because they alter the appearance of the image and make the robot unfamiliar with its environment [6, 16]. This step was designed to minimize the impact of illumination variance on feature attributes, which is crucial for creating maps for vehicle/robot localization. In this research, attention is focused on shadow and lighting condition. At this point, the region of the shadow in the captured frame is being focused on. It was computed using Eq. 2 utilizing the theory that shadows arise in images as a result of direct light being hindered by an opaque object.

$$l_i = (t_i \cos \theta_i l_d + I_e) R_i \tag{2}$$

the l_i represent the value for $i-$ th pixel in RGB space, l_d and l_e signifies the direct light intensity and ambient light intensity measured in RGB color space respectively, R_i signifies the reflectance surface of the pixel, θ_i signifies the angle between the surface norm and the direction of the direct light, t_i represent the attenuation factors of the direct light, when $t_i = 1$ implies that the object point is in a sunshine area and when $t_i = 0$ implies that the object points is in a shadow area. $K_i = t_i \cos\theta_i$ signifies shadow coefficient for the $i-$ th pixel and $r = {}^{l_d}/_{l_e}$ represent the ratio between the direct light and ambient light.

Giving that the model has assisted with shadow detection and to create a shadow-free region, the shadowed pixel is corrected using the pixel from the corresponding non-shadowed region that is close to the shadow region as illustrated in Eq. 3 [16].

$$I_i^{shadow\,free} = \frac{r+1}{k_i r + 1} l_i \tag{3}$$

Thereafter the dark channel technique is then used to target the areas of the images with the high light intensity. In order to identify light intensity, it is believed that, in the dark channel model shown in Eq. 4, regions which are affected by high light ought to have high intensity values, whereas regions which were not impacted will have low values.

$$I^{dark}(x) = \min_{y \in v(x)} \left(\min_{c \in (r,g,b)} (I^c(y)) \right) \tag{4}$$

the $v(x)$ signifies the local patch centered at x, x denotes the image coordinate and I^c signifies colour channel of I.

Thereafter, an improved version of OSTU automatic thresholding was employed for labelling the dark channel image affected with high light. The marked image is provided below in Eq. 5.

$$\text{mark}(x) = \begin{cases} 1 & \textit{if } I^{dark}(x) \rangle t^* \\ 0 & \textit{Otherwise} \end{cases} \tag{5}$$

After proper identification of the high light region, the removal method is founded on the specular to diffuse theory discussed by [16]. The outcome presents an image that is not impacted with light intensity effect (I^D) as presented in Eq. 6.

$$I^D(\wedge_{\max}) = I - \frac{\max_{u \in (r,g,b)} I_u - \wedge_{\max} \sum_{u \in (r,g,b)} I_u}{1 - 3\wedge_{\max}} \tag{6}$$

3.4 Simultaneous Localization and Mapping Stage

A robot's capacity for localization refers to its ability for determining its own pose and orientation inside an environment, whereas that of mapping refers to its capacity for creating a model representation of an unfamiliar location [19]. These are basic operations for SLAM. Localizing the robot and generating a map in robotics are two distinct tasks that are intimately connected because the autonomous vehicle requires the map to find its position (localize) while the vehicle localization is essential to generate the model/map of the location [18]. SLAM method has drawn many scientists to focus it and in effort to fix this connection between the two problems. Numerous SLAM systems have been suggested in the literature with impressive results achieved [19, 20]. However, we enhanced the MCL-SLAM technique to cope with kidnapping. Additionally, initializing localization, which will be used to address the kidnapping issue, will be carried out with the assistance of the database integrated in to SLAM strategy shown in Fig. 1. The original MCL algorithm described in Eq. 7 has been adjusted at this point to address the SLAM issue; this modification will be covered in Sect. 3.4.1.

$$p(S_{1:t}, m|Z_{1:t}, U_{1:t-1}) = p(m|S_{1:t}, Z_{1:t}).p(S_{1:t}|Z_{1:t}, U_{1:t-1}) \tag{7}$$

the $U_{1:t-1} = U_1, \ldots\ldots\ldots, U_{t-1}$ represent the odometry measurement and the $S_{1:t} = S_1, \ldots\ldots\ldots, S_t$ signifies the samples and the $Z_{1:t} = Z_1, \ldots\ldots\ldots, Z_t$ is denoted as the observation measurement provided in map m [16].

3.4.1 Simultaneous Localization and Mapping Stage

Kidnapping is a problem that occurs when a robot moves unexpectedly in an environment without the robot noticing the movement [5]. As a result of this constraint, a vehicle is unable to extract its position, which violates the SLAM problem-solving principle since position extraction is necessary for map construction or may cause robot failure with no chance of recovery [16]. The algorithm chosen to tackle this challenge has to be capable of achieving the following goals: pose measurement, kidnap problem detection, and global localization [5]. Given that a robot is using the map to navigate in an unfamiliar environment based on the enhanced MCL-SLAM method, how can we be sure that the vehicle/robot is kidnapped? After kidnapping, particles move from local samples to global samples, and then, after re-localization is vice-versa. An important component of robot kidnapping and recovery is the global samples. Additionally, consideration is given to the particle probabilities, such particle's with maximum probability lower than the threshold coefficient γ, triggers robot kidnapping [21]. The kidnapping illustration is described in Eq. 8.

$$\text{The } Kidnapping\ vehicle_t = \begin{cases} 1 & w_t^{max.} > \gamma \\ 0 & Otherwise \end{cases} \tag{8}$$

Thereafter Kidnapping detection, the scan to map matching algorithm known as the SIFT descriptor. In this method, the matching technique is applied for a comprehensive review of the recent observations of the vehicle/robot as related to reference map (pre-mapped environment) saved in the database to determine the robot's pose and orientation in relation to the chosen reference map found in the database. This reference map selected is what the robot will use to estimate its pose to continue with the procedure of simultaneous localization and mapping.

3.5 Simultaneous Localization and Mapping Stage

Autonomous vehicle depends on a precise position extraction and map of its surroundings that was resolved at the SLAM phase. Notwithstanding, the robot's route, that takes into account the maps created for a secure path for perfect movement from the starting location to the end, hasn't been resolved [16]. Furthermore, the issue becomes more complicated because the robot won't have a detailed map information of its surroundings, but rather later after receiving updates from other phases [22]. The SLAM system must therefore, whenever it obtains new map information, it will revise and re-plan its optimum trajectory course. There are numerous varieties of navigation algorithms that have been proposed in the literature, including the popular navigational algorithms D* and A* [22]. However, the D* algorithm was chosen for this study due to its efficient heuristic use and ability to handle more updated map data. The D* algorithm function based on the grid-based technique by splitting the map/observation area into $m \times m$ grid. Furthermore, the D* algorithm also employs the cost function for path planning. The robot navigates via cells with the least cost functions after calculating the cost functions for each and every cell [23]. Let X represent the area of observation that is divided into $m \times m$ grids, the cost function $f(R, X)$ for the trajectory of the robot form the current pose (R) through the area of observation to the final position (G) is estimated using Eq. 9 [22].

$$f(R, X) = g(X) + h(R, X) \tag{9}$$

the $g(X)$ signifies the minimum cost function from X to G while the $h(R, X)$ denotes the estimated cost function from R to X.

4 Experiment and Result

This study created a private dataset from the electrical building of a university in South Africa using a camera sensor in order to test the performance of the improved MCL-SLAM algorithm. In carrying out this experiment, our target is on the illumination variance because under high intensity of such noises, kidnapping can be trigger [5, 6, 16]. Since the MCL-SLAM is enhanced for eliminating the impact of Illumination variation and kidnap robot. We increased the illumination variation in the image by 5% to intensify the noises by increasing the contrast in such a way that the shadow in the image is darker and increase the brightness in such a way to severely increase the light intensity in the image using the Eqs. 10 and 11 respectively.

$$G(i, j) = F(i, j) * c \; \left({}^{c\,>0\,Contrast\ Increase}_{c\,<0\,Contrast\ Decrease} \right) \tag{10}$$

$$G(i, j) = F(i, j) + b \; \left({}^{b\,>0\,Brightness\ Increase}_{b\,<0\,Brightness\ Decrease} \right) \tag{11}$$

where $G(i,j)$ represent the image pixel in Gray Scale level and $F(i,j)$ is the pixel representation in the input image [28].

The modified MCL-SLAM algorithm and the original MCL-SLAM algorithm without any modifications were compared in the experiment using both qualitative and quantitative approaches. Increasing the noise image will show if the enhanced MCL is capable to overcoming the issue of illumination variation and we can be able to study the performance of the algorithm when kidnapping is triggered. The simulation was done using Matlab and the 'Imadjust' function of the Matlab was used to increase the illumination variation by 5% and the result obtained is shown in Fig. 3.

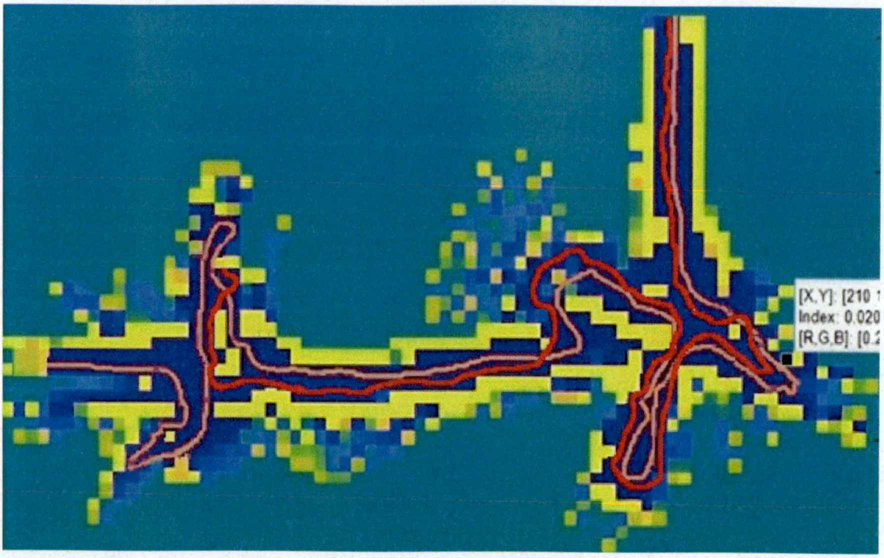

Fig. 3. The Qualitative Result.

The Fig. 3 shows two robot paths moving through a simulated environment as a qualitative outcome. The original MCL algorithm, which has not been modified, may be found here: https://ch.mathworks.com/matlabcentral/fileexchange/8068-montecarlo and its trajectory is displayed in red. The modified MCL technique is shown in orange. The virtual map's blue and yellow sections designate the segments that can be drivable and non-drivable respectively. Figure 3 shows the trajectories of both SLAM algorithms, but the enhanced MCL algorithm exhibits less miss-detection than the original MCL algorithm. Additionally, the original MCL-SLAM algorithm encountered kidnapping and couldn't recover, was unable to arrive at the appropriate location. When the virtual map at which kidnapping happened is compared to the actual image. It was observed that the image due to increase in noise level has caused a severe illumination variance, and the enhanced MCL algorithm only could endure it due to the presence of filters and its ability to re-localize from kidnapping. However, quantitative evaluation is then used to estimate trajectory performance in order to identify the SLAM procedure having the greatest efficiency in comparison to the ground truth trajectory [24, 25]. This ground truth measurement was obtained using a laser sensor-based method similar to the one presented [26]. There are different quantitative evaluations [27], however, the Absolute Trajectory Error (ATE), is utilized in this study and description is provided in equation [25].

$$\mathrm{F_k} = SP_k\left(Q_k^{-1}\right) \tag{12}$$

The $\mathrm{F_k}$ represent the ATE at the time k is used to find the euclidean transformation S corresponding to the least-square solution that maps the estimated trajectory $P_{1:m}$ onto the ground truth trajectory $Q_{1:m}$. Illustration of the qualitative result is presented in

Fig. 4. This evaluation was employed to calculate the inaccuracy/error in the trajectory obtained by the SLAM algorithms shown in Fig. 4.

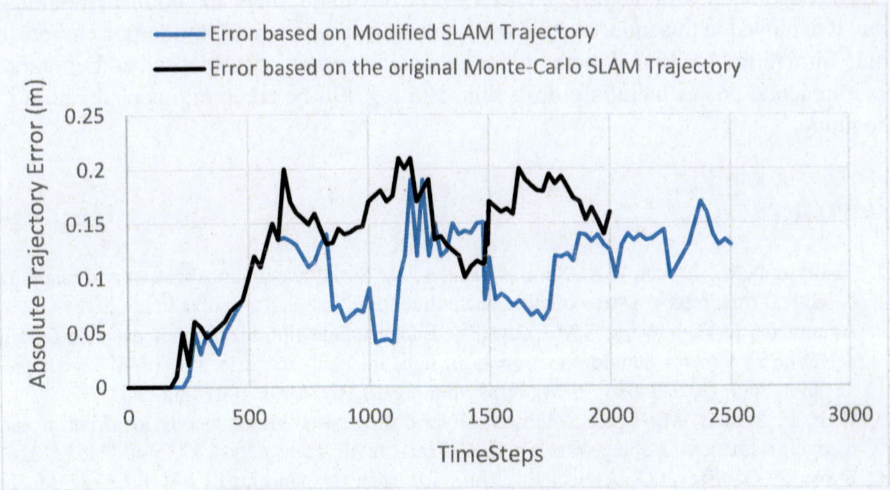

Fig. 4. Quantitative Absolute Trajectory Error.

The quantitative result used has represented in Fig. 4 shows the measured error in the robot's trajectory. The blue line in the graph represents the modified MCL-SLAM and the graph line in black color denotes the original MCL-SLAM with no modification. Considering the total distance covered, the modified MCL-SLAM achieved the minimum error for the majority of trajectories compared to the original MCL-SLAM with no modification. Additionally, it was found here that error difference between the modified MCL-SLAM and the original MCL-SLAM started increasing when more areas were covered into the challenging part of the image with severe illumination variations until it was observed that the original MCL-SLAM algorithm got kidnapped with no recovery. This explain why we have limited measurement in the black graph line which represent the original MCL-SLAM algorithm. The results shown in Fig. 4 backs the need for filters and in most location, they are typical issues that the robot may run into at any time.

5 Conclusion

SLAM study is evolving constantly because it allows the simultaneous execution of localization and mapping procedure. This is a crucial necessity for tackling the problem of self-driving in robots without human assistance. Thus, there are several challenges restricting the success of SLAM, the above success has yet to be fully realized. Furthermore, the literature review conducted in this study suggest that the SLAM systems are still suffering from the challenges of illumination variances and kidnapping [7, 8, 16]. As a result of filters and initialized localization introduced into the SLAM technique, the presented enhanced MCL-SLAM technique has the qualities of resolving the issue of illumination and kidnapping. The experimental findings confirm the enhanced MCL-SLAM algorithm has the capacity to deal with the study's target limitation. The research also demonstrates improved trajectory accuracy with lower Absolute Trajectory Error (ATE) once linked with the original MCL-SLAM with no filters and no form modification. If deployed in this industry, this will enhance the accuracy of autonomous robots in route formation. Additional methods for enhancing system robustness to address more environmental noises including mist, rain, and fog will be taken into consideration in the future.

References

1. Thamrin, N.M., Adnan, R.A., Sam, R., Razak, N.: Simultaneous localization and mapping based real-time inter-row tree tracking technique for unmanned aerial vehicle (2012)
2. Agunbiade, O.Y., Ngwira, S.M., Zuva, T.: Enhancement optimization of drivable terrain detection system for autonomous robots in light intensity scenario. Int. J. Adv. Manufac. Technol. **74**(5–8), 629–636 (2014). https://doi.org/10.1007/s00170-014-5972-7
3. Choi, J., Maurer, M.: Local volumetric hybrid-map-based simultaneous localization and mapping with moving object tracking. IEEE Trans. Intell. Transp. Syst. **17**, 2440–2455 (2016)
4. Byron, B., Geoffrey, J.G.: A spectral learning approach to range-only SLAM. Int. Conf. Mach. Learn. Atlanta, Georgia **28**, 1–19 (2013)
5. Guyonneau, R., Lagrange, S., Hardoun, L., Lucidarme, P.: The kidnapping problem of mobile robots: a set membership approach. In: 7th National Conference on Control Architectures of Robots (2012)
6. Negenborn, R., Johansen, P.P., Wiering, M.: Robot localization and kalman filter. Institute of Information and Computing Sciences (2003)
7. Makhubela, J.K., Zuva, T., Agunbiade, O.Y.: Framework for visual simultaneous localization and mapping in a noisy static environment. In: International Conference on Intelligent and Innovative Computing Applications (ICONIC), 6–7 Dec, pp. 1–6 (2018)
8. Khan, S.A., Chowdhury, S.S., Niloy, N.R., Aurin, F.T.Z., Ahmed, T., Mostakim, M.: Simultaneous localization and mapping (SLAM) with an autonomous robot, pp. 1–41. Department of Electrical and Electronic Engineering, School of Engineering and Computer Science, BRAC University (2018)
9. Fernández-Madrigal, J.A., Claraco, J.L.B.: Simultaneous localization and mapping for mobile robots: introduction and methods. In: Information Science Reference (2013)
10. Al-Amri, S.S., Kalyankar, N.V., Khamitkar, S.D.: A comparative study of removal noise from remote sensing image. Int. J. Comput. Sci. Issues (IJCS) **7**(1), 32–36 (2010)
11. Wen, Q., Yang, Z., Song, Y., Jia, P.: Road boundary detection in complex urban environment based on low-resolution vision. In: Proceedings of the 11th Joint Conference on Information Science, State Key Laboratory on Intelligent Technology and Systems, Tsinghua National Laboratory for Information Science and Technology, Department of Computer Science and Technology, Tsinghua University Beijing. 100084, China, pp. 1–7 (2008)
12. Luis, M., Pablo, B., Pilar, B., Jose, M.: Multi-cue visual obstacle detection for mobile robots. J. Phys. Agents **4**(1) (2010)
13. Yenikaya, S., Yenikaya, G., Duven, E.: Keeping the vehicle on the road- a survey on-road lane detection system. ACM Computing Surveys. Uludag University, Turkey, vol 46, pp. 1–43 (2013)
14. Bukhori, I., Ismail, Z.H.: Detection of kidnapped robot problem in Monte-Carlo localization based on the natural displacement of the robot. Int. J. Adv. Robot. Syst. SAGE, 1–6 (2017)

15. Agunbiade, O.Y., Zuva, T., Johnson, A.O., Zuva, K.: Enhancement performance of road recognition system of autonomous robots in shadow. Sign. Image Process. Int. J. **4**(6) (2013)
16. Agunbiade, O.Y., Zuva, T.: Image enhancement from illumination variation to improve the performance of simultaneous localization and mapping technique. In: 4th International Conference on Information and Computer Technologies (ICICT). IEEE Explore Digital Library, 11–14 March 2021. pp. 115–121. https://doi.org/10.1109/ICICT52872.2021.00027
17. Jean-Arcady, M., David, F.: Map-based navigation in mobile robots: a review of map-learning and path-planning strategies. Cogn. Syst. Res. **4**, 283–317 (2003)
18. Oh, S., Hahn, M., Kim, J.: Simultaneous localization and mapping for mobile robots in dynamic environments. In: International Conference on Information Science and Applications (ICISA), 24–26 June, pp. 1–4 (2013)
19. Clipp, B., Zach, C., Lim, J., Frahm, J.-M., Pollefeys, M.: Adaptive, Real-Time Visual Simultaneous Localization and Mapping, pp. 1–8. Department of computer Science, University of North Caroline (2009)
20. Fuentes-Pacheco, J., Ruiz-Ascencio, J., Rendón-Mancha, J.M.: Visual simultaneous localization and mapping: a survey. Artific. Intell. Rev. **43**(1), 55–81 (2012). https://doi.org/10.1007/s10462-012-9365-8
21. Chuho, Y., Byung-Uk, C.: Detection and Recovery for Kidnapped-Robot Problem Using Measurement Entropy, vol. 261, pp. 293–299. Springer-Verlag Berlin Heidelberg (2011)
22. Djekoune, O., Achour, K., Toumi, R.: A sensor based navigation algorithm for a mobile robot using the DVFF approach. Int. J. Adv. Robot. Syst. (2009)
23. Zhang, Z., Liu, S., Tsai, G., Hu, H., Chu, C.-C., Zheng, F.: PIRVS: An Advanced Visual-Inertial SLAM System with Flexible Sensor Fusion and Hardware Co-Design, p. 95054. PerceptInc, Santa Clara, CA, USA (2017)
24. Cras, J.L., Paxman, J., Saracik, B.: Vision based localization under dynamic illumination. In: 5th International Conference on Automation, Robotics and Applications (ICARA), pp. 453–458 (2011)
25. Lang, Z., Ma, X., Dai, X.: Simultaneous Planning Localization and Mapping in a Camera Network. vol. 24, pp. 1037–1058 (2010)
26. Ceriani, S., et al.: Rawseeds ground truth collection systems for indoor self-localization and mapping. Auton. Robots **27** (2009)
27. Shalal, N., Low, T., McCarthy, C., Hancock, N.: Orchard mapping and mobile robot localisation using on-board camera and laser scanner data fusion – part a: tree detection. Comput. Electron. Agric. **119**, 254–266 (2015)
28. Sinecen, M.: Digital image processing with MATLAB (2016)

A Neighborhood Overlap-Based Binary Search Algorithm for Edge Classification to Satisfy the Strong Triadic Closure Property in Complex Networks

Natarajan Meghanathan(✉)

Department of Electrical and Computer Engineering and Computer Science, Jackson State University, 1400 Lynch Street, Jackson, MS 39217, USA
natarajan.meghanathan@jsums.edu

Abstract. We use the neighborhood overlap (NOVER) scores of the edges as the topological basis for their classification to strong ties or weak ties. Edges with NOVER scores above a threshold are classified as strong ties; otherwise, weak ties. We propose a binary search-based algorithm to determine the minimum threshold NOVER score ($NOVER_{\min}^{threshold}$) that could be used to classify the edges as strong ties or weak ties such that the STC property (if a node A has strong ties with two nodes B and C, then B and C are expected to be connected with an edge) is satisfied for at most f_{STC}^{nodes} fraction of the nodes. We ran our algorithm on several benchmarking real-world network datasets: we observe the inverse relationship (with increase in f_{STC}^{nodes}) between $NOVER_{\min}^{threshold}$ and f_{st}^{edges} to shift from a linear decrease to a logarithmic decrease with increase in variation in node degree for the real-world networks.

Keywords: Strong Ties · Weak Ties · Tie Strength Analysis · Strong Triadic Closure Property · Binary Search Algorithm · Social Network Analysis · Neighborhood Overlap · Variation in Node Degree

1 Introduction

Social networks are characteristic of tightly-knit communities within each of which there are typically more than the minimum number of links required to connect the member nodes, whereas there are relatively very few edges between two communities. Edges connecting the member nodes within a community are more likely to have a significant amount of shared neighborhood such that messages forwarded within the community are more likely to reach the member nodes multiple times; such edges are referred to as strong ties [1] in the SNA (Social Network Analysis) literature. On the other hand, edges connecting two different communities are likely to have a sparse or no shared neighborhood; such edges are referred to as weak ties [1] in the SNA literature. Throughout the paper, the terms 'node' and 'vertex', 'edge' and 'link', 'network' and 'graph' are used interchangeably. They mean the same.

R. Silhavy and P. Silhavy (Eds.): CSOC 2023, LNNS 724, pp. 160–169, 2023.
https://doi.org/10.1007/978-3-031-35314-7_16

Tie strength analysis in the literature has often been conducted using offline information such as the number of interactions between the end vertices of the edges [2]. Tie strength analysis is a classical SNA problem in which edges are classified as strong ties and weak ties using either topological information only or offline information only (such as the number of messages shared through the edges [2]) or both. Tie strength analysis could improve the accuracy of link prediction [3, 4], enhance the spreading model of a disease and information as well as be useful for targeted marketing. The accuracy of the classification algorithm is evaluated by examining whether the strong triadic closure property (STC) is satisfied at each node. The STC property for social networks is stated as follows [1]: If a node A has strong ties with nodes B and C, then B and C are expected to have at least a weak tie. In other words, the STC property basically requires that any two strong tie neighbors of a node be connected with an edge. The STC property is typically expected to be satisfied at every node in order to consider the whole network to have satisfied the STC property [1].

In this paper, we do not use any offline information and choose to use the neighborhood overlap (NOVER) score [1] as the only topological basis to classify edges as strong ties and weak ties. Moreover, instead of expecting the STC property to be strictly satisfied at every node in the network, we setup a parameter f_{STC}^{nodes} ($0 < f_{STC}^{nodes} \leq 1$) that represents the fraction of nodes in the network that need to at most satisfy the STC property. The NOVER score [1] for an edge *u*-*v* is the ratio of the number of neighbors that are common to both *u* and *v* to the number of unique nodes that is at least a neighbor of *u* and/or *v*, excluding *u* and *v* themselves. We propose an optimization variable called the *threshold NOVER score* for a network such that edges whose NOVER scores are above the threshold NOVER score will be classified as strong ties and edges whose NOVER scores are less than or equal to the threshold NOVER score are classified as weak ties. We propose a binary search algorithm to determine the minimum value for the threshold NOVER score ($0 \leq NOVER_{\min}^{threshold} \leq 1$) to be used to classify the edges of a network such that at most f_{STC}^{nodes} of the nodes satisfy the STC property. We also determine the fraction of edges as strong ties (f_{st}^{edges}) in the network corresponding to the (f_{STC}^{nodes}, $NOVER_{\min}^{threshold}$) values for which the STC property is satisfied for the network.

The rest of the paper is organized as follows: Sect. 2 presents the proposed binary search algorithm to determine the minimum threshold NOVER score that could be used to classify the edges as strong ties or weak ties such that the STC property is satisfied for at most a certain fraction of nodes in the network. Section 3 presents the application of the binary search algorithm for a suite of 9 benchmarking real-world networks (whose variation in node degree ranges from that for random networks to scale-free networks) and the results obtained with respect to ($NOVER_{\min}^{threshold}$, f_{st}^{edges}) for f_{STC}^{nodes} values ranging from 0.1 to 1. Section 4 presents related work in the literature and highlights the contributions of this work. Section 5 concludes the paper.

2 Binary Search Algorithm to Determine the Threshold NOVER Score

Binary search is typically used to search for a key in a sorted array (a discrete search space, ranging from a left index: LI to a right index: RI) such that an invariant property is satisfied for the search space throughout the execution of the algorithm. For the classical key search problem, the invariant property to be satisfied is that if at all the search key is in the sorted array, its index of occurrence in the array has to be greater than or equal to LI and less than or equal to RI. The invariant property is to be maintained throughout the execution of the binary search algorithm.

For the problem considered in this research, we form a sorted NOVER array (referred to as *NOVER_Array*) comprising of the unique NOVER scores of the edges. As the NOVER scores of the edges could range from 0.0 to 1.0 (which could potentially be the initial values for the LI and RI of the search space), we include 0.0 and/or 1.0 at the first index and last index of *NOVER_Array* if the latter does not already comprise them. For a given network and f_{STC}^{nodes}, the search key is the minimum threshold NOVER score ($NOVER_{\min}^{threshold}$) that could be used to classify the edges as strong ties or weak ties such that the STC property is satisfied for at most f_{STC}^{nodes} of the nodes. In the classical key search problem, the search key is given along with the sorted array, and we need to determine whether or not it is in the array (if it is in the array, at what index?). For the search problem in hand, the search key is guaranteed to be in the array; but, we don't know what it is (that would correspond to the minimum threshold NOVER score) and its index of occurrence.

We first seek to setup the invariant property (that needs to be satisfied at the left index: LI and right index: RI of the search space) for our search problem. ***Left Index***: If the threshold NOVER score (above which the edges are classified as strong ties; otherwise, weak ties) is set to 0.0, then all the edges whose NOVER scores are above 0.0 will be classified as strong ties: the STC property is more likely to fail at several nodes in such cases. For a given f_{STC}^{nodes} value and a threshold NOVER score of 0.0, if the fraction of nodes satisfying the STC property is greater than or equal to f_{STC}^{nodes}, then 0.0 is the $NOVER_{\min}^{threshold}$ value we are looking for and we do not need to run binary search. For a given f_{STC}^{nodes}, we need to do binary search only if the STC property is not satisfied for at most f_{STC}^{nodes} of the nodes when the threshold NOVER score is 0.0, which could be the initial value for the NOVER score corresponding to left index (LI) of the search space. Hence, the invariant property for the left index (LI) of the search space is that the STC property is not satisfied for at most f_{STC}^{nodes} fraction of the nodes when the threshold NOVER score is less than or equal to the value at index LI in the *NOVER_Array*. ***Right Index***: If the threshold NOVER score is set to 1.0, then all the edges in the network will be classified as weak ties (there will be no edges classified as strong ties) and the STC property will be implicitly satisfied (i.e., there will not be any strong tie neighbors for any node) at all nodes in the network (irrespective of the f_{STC}^{nodes} value). In other words, for any f_{STC}^{nodes} value, we could set the initial value of the right index (RI) to be 1.0 and expect the STC property to be (implicitly) satisfied for at most f_{STC}^{nodes} fraction of the nodes. Hence, the invariant property for the right Index (RI) of the search space is that the STC

```
Inputs: Network Graph G = (V, E); NOVER_Array[0, ..., n-1], f_STC^nodes
        where 'n' is the number of unique NOVER scores of the edges, including 0.0 and 1.0
Auxiliary Variables: Left Index LI, Right Index RI and Middle Index MI

Begin Binary Search Algorithm

if STC_property_satisfied(threshold NOVER score = 0.0) then
        return NOVER_min^threshold = 0.0
end if

LI = 0          // NOVER_Array[0] = 0.0
RI = n-1        // NOVER_Array[n-1] = 1.0

while (RI - LI > 1) do
        MI = (LI + RI) / 2
        if STC_property_satisfied(threshold NOVER score = NOVER_Array[MI]) then
                RI = MI
        else
                LI = MI
end while
return NOVER_min^threshold =NOVER_Array[RI]
End Binary Search Algorithm
```

Fig. 1. Binary Search Algorithm to Determine Minimum Threshold NOVER Score for Classification of Edges (Strong Ties/Weak Ties) to Satisfy the Strong Triadic Closure Property

property is satisfied for at most f_{STC}^{nodes} fraction of the nodes when the threshold NOVER score is set to the value at index greater than or equal to RI in the *NOVER_Array*.

For a given network, a *NOVER_Array* of 'n' unique NOVER score values (with 0.0 and 1.0 as the values at index 0 and $n-1$ respectively) and f_{STC}^{nodes} value, we start with LI = 0 and RI = n-1. The binary search algorithm (see Fig. 1) proceeds iteratively as follows: At the beginning of each iteration, we determine the middle index MI as the average of LI and RI: i.e., MI = (LI + RI)/2. We now check whether the STC property is satisfied for at most f_{STC}^{nodes} fraction of the nodes if the NOVER score corresponding to index MI in the *NOVER_Array* is used as the threshold NOVER score to classify edges as strong ties and weak ties. If the STC property is satisfied, we set RI = MI; otherwise, we set LI = MI. This would maintain the invariant property that the STC property is not satisfied for index values less than or equal to LI and is satisfied at index values greater than or equal to RI. We continue to the next iteration as long as the latest values of the LI and RI are such that RI - LI > 1. The moment RI - LI equals 1 (i.e., the left index and right index are next to each other with regards to the *NOVER_Array*), we stop the algorithm and declare the NOVER score corresponding to the right index RI as the minimum threshold NOVER score ($NOVER_{\min}^{threshold}$) at which the STC property is satisfied for at most f_{STC}^{nodes} fraction of the nodes.

The number of iterations of the binary search algorithm is $\log_2(n)$, where n is the number of entries (i.e., unique NOVER scores of the edges along with 0.0 and 1.0) in the *NOVER_Array*. For each iteration, we need to check if the STC property is satisfied at each node in the network. If 'N' and 'M' are the number of nodes and edges in the network respectively, the average degree per node is $2M/N$. For the STC property to be satisfied at a node, any two strong tie neighbors of the node need to have an edge between them. There could be on average $2M/N$ strong tie neighbors for a node and it would take $2M/N * 2M/N = 4M^2/N^2$ time to check the presence of an edge between any two strong tie neighbors of a node. It would take $N * 4M^2/N^2 = 4M^2/N = \Theta(M^2/N)$ time to evaluate if the STC property is satisfied at all the N nodes in the network for a particular iteration of the algorithm. For $\log_2(n)$ iterations, the time complexity to check the satisfaction of the STC property at all the N nodes in the network would be $\Theta(\log n * M^2/N)$. On the other hand, if we were to follow a brute force approach for our search problem, the time complexity would be $\Theta(n * M^2/N)$.

We now present an example (Fig. 2) to illustrate the procedure adopted to check whether the STC property is satisfied or violated at each node in the network. The threshold NOVER score used in the example is 0.0 and the classification of the edges as strong ties or weak ties is accordingly shown. Then, for each node: we list its strong tie and weak tie neighbors. For the STC property to be satisfied at a node, any two strong tie neighbors of the node should have an edge between them. If a node has no (0) or just one (1) strong tie neighbor, the STC property is considered to be implicitly satisfied at the node. With a threshold NOVER score of 0.0, five of the nine nodes satisfy the STC property. If the f_{STC}^{nodes} value is 5/9, we could conclude that the STC property is satisfied for at most f_{STC}^{nodes} fraction of the nodes; otherwise, not.

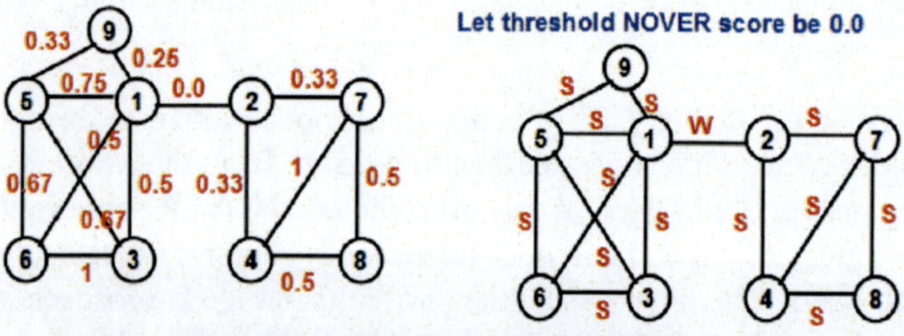

Node	Strong Tie Neighbors	Weak Tie Neighbors	Strong Triadic Closure Property
1	3, 5, 6, 9	2	VIOLATED (no edge: 3, 9; 6, 9)
2	4, 7	1	YES
3	1, 5, 6	-	YES
4	2, 7, 8	-	VIOLATED (no edge between 2, 8)
5	1, 3, 6, 9	-	VIOLATED (no edge: 3, 9; 6, 9)
6	1, 3, 5	-	YES
7	2, 4, 8	-	VIOLATED (no edge between 2, 8)
8	4, 7	-	YES
9	5, 1	-	YES

Fig. 2. Example Illustration to Check for the Satisfaction of the STC Property at each Node in a Network [Threshold NOVER Score = 0.0]

3 Execution of the Binary Search Algorithm on Real-World Networks

We now present the results of the execution of the binary search algorithm for a suite of 9 real-world networks (see Table 1) whose variation in node degree (captured in the form of the spectral radius ratio for node degree: λ_{sp}^{k}) ranges from that of random to scale-free networks. The real-world networks chosen span several domains (such as social networks, transportation networks, biological networks, etc.). We use the CINES cloud cyber infrastructure (https://net.science/) to extract the neighborhood information of nodes in the real-world networks. We ran the binary search algorithm of Sect. 2 on each of these networks and determined the ($NOVER_{\min}^{threshold}$, f_{st}^{edges}) values for the networks as the f_{STC}^{nodes} value is increased from 0.1 to 1, in increments of 0.1. As expected, the $NOVER_{\min}^{threshold}$ values increased and the f_{st}^{edges} values decreased with increase in f_{STC}^{nodes}. We plot the $NOVER_{\min}^{threshold}$ (X-axis) vs. f_{st}^{edges} (Y-axis) and observe that the f_{st}^{edges} values decrease with increase in $NOVER_{\min}^{threshold}$. We observe the tendency of this decrease to change from linear decline to a logarithmic decline as the λ_{sp}^{k} values of the networks get larger.

Table 1. Details of the Real-World Networks Analyzed

Net #	Network Name	# Nodes	# Edges	λ_{sp}^{k}	Ref.
Net 1	Football Network	115	613	1.011	[6]
Net 2	Cat Brain Network	65	730	1.197	[7]
Net 3	US Marine Highway Network	16	26	1.246	[8]
Net 4	UK Faculty Network	81	577	1.353	[9]
Net 5	Karate Network	34	78	1.466	[10]
Net 6	Adjacency Noun Network	112	425	1.733	[11]
Net 7	Anna Karnenina Network	138	493	2.475	[12]
Net 8	Celegans Metabolic Network	453	2025	2.943	[13]
Net 9	Perl Developers Network	839	2111	5.217	[14]

Figure 3 shows the $NOVER_{\min}^{threshold}$ vs. f_{st}^{edges} plots for the 9 real-world networks: We fit the $NOVER_{\min}^{threshold}$, f_{st}^{edges} values to a logarithmic curve, $f_{st}^{edges} = \beta * \ln(NOVER_{\min}^{threshold}) + c$ as well as a straight line $f_{st}^{edges} = \beta' * (NOVER_{\min}^{threshold}) + c'$. The parameters β and β' incur negative values as they capture the rate of decrease in the f_{st}^{edges} values with increase in $NOVER_{\min}^{threshold}$ and c, c' are proportionality constants.

The linear fit incurs a higher R^2 value only for the Football network (Net 1) whose $\lambda_{sp}^{k} = 1.01$, falling in the regime of random networks. For the rest of the 8 real-world networks (Net 2 through Net 8, whose λ_{sp}^{k} values range from 1.19 to 5.22), we observe the logarithmic curve fit to incur a higher R^2 value. For each network in Fig. 3, we show the fit that incurs the higher R^2 value along with the distribution of the ($NOVER_{\min}^{threshold}$, f_{st}^{edges}) values. Table 2 lists the values of the parameters β and c (for the logarithmic curve fit) as well as parameters β' and c' (for the linear fit) as well as the corresponding R^2 values of the two fits and the larger of the R^2 values is highlighted.

Table 2. Results of Empirical Analysis and Degree Variation Index

Net #	Logarithmic Curve Fit			Linear Fit			Degree Variation Index (DVI)
	β	C	R^2	β '	C'	R^2	
Net 1	−0.241	0.156	0.726	−1.387	0.888	**0.941**	N/A
Net 2	−0.906	−0.332	**0.985**	−1.786	1.204	0.954	−0.906
Net 3	−0.535	−0.378	**0.992**	−2.308	0.962	0.991	−0.535
Net 4	−0.621	−0.251	**0.989**	−1.794	1.081	0.983	−0.621
Net 5	−0.363	−0.227	**0.997**	−2.005	0.891	0.896	−0.363
Net 6	−0.273	−0.342	**0.978**	−3.108	0.713	0.859	−0.273
Net 7	−0.224	−0.017	**0.992**	−1.840	0.921	0.926	−0.224
Net 8	−0.205	−0.077	**0.961**	−1.365	0.709	0.614	−0.205
Net 9	−0.163	−0.210	**0.998**	−2.703	0.699	0.829	−0.163

We notice the λ_{sp}^{k} values to increase exponentially as β values increase (i.e., approach 0). We observe a positive correlation (see Fig. 4) between the parameter β and λ_{sp}^{k} values for the eight real-world networks Net 2... Net 9. We hence propose the logarithmic fit to be more appropriate to capture the extent of variation in node degree without actually determining the λ_{sp}^{k} value. We claim that the parameter β be considered as a measure of the Degree Variation Index (DVI) of the network; it is independent of the number of nodes and edges of the network and it can serve as a computationally-light alterative to the computationally-heavy spectral radius ratio for node degree (λ_{sp}^{k}) that needs to be determined using spectral analysis of the adjacency matrix of the network graph.

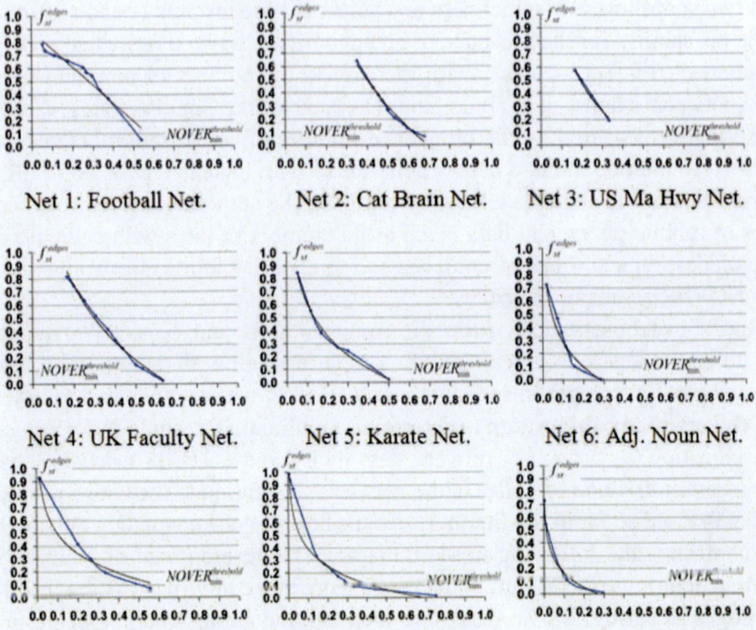

Net 1: Football Net. Net 2: Cat Brain Net. Net 3: US Ma Hwy Net.

Net 4: UK Faculty Net. Net 5: Karate Net. Net 6: Adj. Noun Net.

Net 7: Anna Karn Net. Net 8: Celegans M Net. Net 9: Perl Dev. Net.

Fig. 3. Plots of the ($NOVER_{\min}^{threshold}$, f_{st}^{edges}) Distribution for the Real-World Networks and their Linear/Logarithmic Curve Fit (fit with the higher R^2 value is shown)

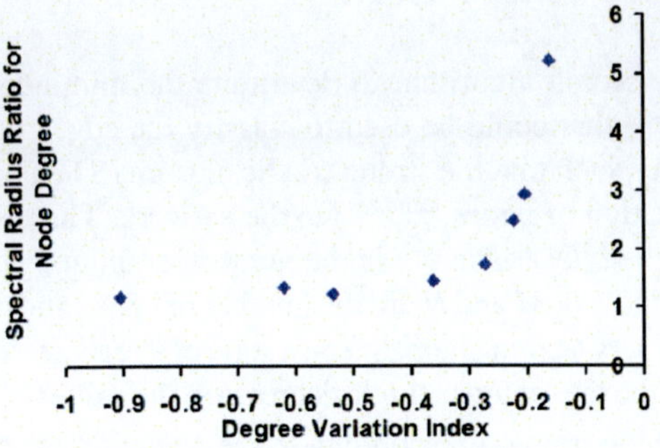

Fig. 4. Degree Variation Index vs. Spectral Radius Ratio for Node Degree

4 Related Work and Contributions

Granovetter was the first to propose the notion of strong ties vs. weak ties in his seminal work "The Strength of Weak Ties" [15]. His hypothesis is that larger the fraction of common friends (a measure of the NOVER score) between two people, the larger the strength of the tie (an edge) between them. In a classical empirical work [16] on the impact of the number of common neighbors on triadic closure, the authors observed that between two snapshots of a social network (taken at time instants t_1 and t_2 such that $t_1 < t_2$), the probability for the presence of an edge between two users x and y at t_2 (who are not connected at t_1) increases with increase in the number of common neighbors observed between x and y at t_1. In another classical work, Onnela et al. [17] observed that tie strength (measured by the number of minutes of conversation between mobile phone users) in a mobile phone network increased with increase in the NOVER scores of the end users of the edge. In a recent work [18], the authors observed four temporal attributes of mobile phone call data (such as the number of days with calls, number of bursty cascades, typical times of contacts, and temporal stability) to strongly correlate with the NOVER scores of the edges.

All the above observations form the motivation for our paper. We hypothesize NOVER to be the key topological feature correlating well with the temporal attributes that characterize tie strength in social networks. Hence, we are motivated to develop an efficient (binary search) algorithm to determine a minimum threshold NOVER score for edge classification (as strong ties or weak ties) such that the STC property is satisfied at most for a certain fraction of nodes in the network. A solution to the above problem has not been addressed so far in the literature. To the best of our knowledge, ours is the first work to contribute the following to the literature: (1) Development of a binary search algorithm to determine the minimum NOVER score that can be used as the threshold to classify edges as strong ties vs. weak ties such that the strong triadic closure property is satisfied at most for a certain fraction of nodes in the network; (2) Development of a computationally-light metric (Degree Variation Index) on the basis of tie strength to quantify the extent of variation of node degree in the network.

5 Conclusions and Future Work

We propose a binary search algorithm to determine the minimum threshold NOVER score ($NOVER_{\min}^{threshold}$) that could be used to classify the edges in a social network as strong ties or weak ties such that the strong triadic closure (STC) property is satisfied at most for a certain fraction of nodes (f_{STC}^{nodes}) in the network. The time complexity of the algorithm is $\Theta(\log n * M^2/N)$, where n is the number of unique NOVER scores of the edges, including 0.0 and 1.0, M and N are the number of edges and nodes in the network. After executing the binary search algorithm on a suite of 9 real-world networks of diverse domains, we observe the rate of logarithmic decrease in the values of the fraction of edges as strong ties (f_{st}^{edges}) incurred corresponding to the $NOVER_{\min}^{threshold}$ value at which the STC property is satisfied for at most f_{STC}^{nodes} fraction of the nodes to increase with increase in the spectral radius ratio for node degree. We hence propose the rate of logarithmic decrease (quantified using a measure called the Degree Variation Index, DVI) to be a computationally-light alternative to the computationally-heavy spectral radius ratio for node degree for quantifying the variation in node degree in a scale of [−1,..., 0] that is independent of the number of nodes and edges in the network. In future, we plan to analyze the correlation between DVI and the modularity of communities in social networks as well as extend the analysis for synthetic networks generated using theoretical network models.

Acknowledgement. The work leading to this paper was partly funded through the U.S. National Science Foundation (NSF) grant OAC-1835439 and through a grant received by Jackson State University as part of its partnership with the Maritime Transportation Research and Education Center (MarTREC) at The University of Arkansas. The views and conclusions contained in this paper are those of the authors and do not represent the official policies, either expressed or implied, of the funding agencies.

References

1. Easley, D., Kleinberg, J.: Networks, Crowds, and Markets: Reasoning about a Highly Connected World, 1st edn. Cambridge University Press (2010)
2. Pappalardo, L., Rossetti, G., Pedreschi, D.: How well do we know each other? Detecting tie strength in multidimensional social networks. In: Proceedings of the 2012 International Conference on Advances in Social Networks Analysis and Mining, pp. 1040–1045 (2012)
3. Lu, L., Zhou, T.: Link prediction in weighted networks: The role of weak ties. Europhys. Lett. **89**(1), 18001 (2010)
4. Li, N., Feng, X., Ji, S., Xu, K.: Modeling relationship strength for link prediction. In: Proceedings of the Pacific-Asia Workshop on Intelligence and Security Informatioo, pp. 62–74 (2013)
5. Meghanathan, N.: Spectral radius as a measure of variation in node degree for complex network graphs. In: Proceedings of the 7th International Conference on u- and e- Service, Science and Technology, pp. 30–33 (2014)
6. Girvan, M., Newman, M.: Community structure in social and biological networks. Proc. Natl. Acad. Sci. USA **99**(12), 7821–7826 (2002)
7. de Reus, M.A., van den Heuvel, M.P.: Rich club organization and intermodule communication in the cat connectome. J. Neurosci. **33**(32), 12929–12939 (2013)
8. https://www.maritime.dot.gov/grants/marine-highways/marine-highway
9. Nepusz, T., Petroczi, A., Negyessy, L., Bazso, F.: Fuzzy communities and the concept of bridgeness in complex networks. Phys. Rev. E **77**(1), 016107 (2008)
10. Zachary, W.W.: An information flow model for conflict and fission in small groups. J. Anthropol. Res. **33**(4), 452–473 (1977)
11. Newman, M.: Finding community structure in networks using the Eigenvectors of matrices. Phys. Rev. E **74**, 3, 036104 (2006)
12. Knuth, D.E.: The Stanford GraphBase: A Platform for Combinatorial Computing, 1st edn. Addison-Wesley (1993)
13. Duch, J., Arenas, A.: Communication detection in complex networks using extremal optimization. Phys. Rev. E **72**, 027104 (2005)
14. Heymann, S.: CPAN-Explorer, an interactive exploration of the Perl ecosystem. Gephi Blog (2009)
15. Granovetter, M.: The strength of weak ties. Am. J. Sociol. **78**(6), 1360–1380 (1973)
16. Kossinets, G., Watts, D.J.: Empirical analysis of an evolving social network. Science **311**(5757), 88–90 (2006)
17. Onnela, J.P., et al.: Structure and tie strengths in mobile communication networks. Appl. Phys. Sci. **14**(18), 7332–7336 (2007)
18. Ureña-Carrion, J., Saramäki, J., Kivelä, M.: Estimating tie strength in social networks using temporal communication data. EPJ Data Sci. **9**(1), 1–20 (2020). https://doi.org/10.1140/epjds/s13688-020-00256-5

Novel Framework for Potential Threat Identification in IoT Harnessing Machine Learning

A. Durga Bhavani[1](✉) and Neha Mangla[2]

[1] Department of Computer Science and Engineering, BMS Institute of Technology and Management, Bengaluru, India
durga842004@bmsit.in

[2] Department of Information Science and Engineering, Atria Institute of Technology, Bengaluru, India

Abstract. The applications and services running over Internet-of-Things (IoT) interacts with massive number of connections with larger devices in order to perform high rate of data transmission. However, this scale of vastness of network structure in IoT also invites various security threats. Review of existing approaches towards intrusion detection system showcase various beneficial outcomes as well as limitations too. The proposed scheme addresses this issue by harnessing the potential of machine learning for achieving higher detection capability towards intruders. An unsupervised learning scheme using autoencoder is applied in proposed scheme where clustering operation is carried out towards reducing the possible anomalies in identification. The study outcome exhibits that proposed scheme offers comparatively higher accuracy than existing multiple learning models used in intrusion detection in IoT.

Keywords: Internet-of-Things · Intrusion Detection System · Autoencoder · machine learning · clustering

1 Introduction

Cryptography is a good way to make security services better by making source authentication mechanisms work, but it's not very good at protecting against usability or availability attacks [1]. Studies on Intrusion Detection Systems (IDS) for the Internet of Things (IoT) have received a lot of attention due to the development of machine learning (ML) methods. In fact, IDS and cryptographic-based approaches work together to guarantee the highest levels of confidentiality, integrity, authenticity, and availability [2, 3]. Signature-based and anomaly-based methods, for example, are two types of network IDS [4]. Anomaly-based techniques typically take into account normal traffic patterns, whereas signature-based methods primarily take into account labeled attack traffic patterns. However, each of these two approaches has its own benefits and drawbacks. Real-time detection of new and complex attacks is limited because signature-based IDS

R. Silhavy and P. Silhavy (Eds.): CSOC 2023, LNNS 724, pp. 170–179, 2023.
https://doi.org/10.1007/978-3-031-35314-7_17

only uses known traffic patterns. Anomaly-based IDS, on the other hand, typically has a higher rate of false alarms. To eliminate the large number of alerts generated, this may necessitate additional time and resources. Additionally, the existing IDS are insufficient due to their design's disregard for resource efficiency and scalability [10]. There are many schemes, but IoT security is still a major concern, which calls for an effective IDS framework to combat various cyber threats and attacks. Apart from this, it is also noted that existing studies are soon adopting machine learning in order to understand the patterns and behaviors of the attack more closely [6–10]. Adoption of machine learning in encryption makes the operation of security further more intellectual and smart good enough for identification of intrusion. Therefore, the proposed scheme presents a novel machine learning approach towards improving the security features in IoT. Section 2 discusses about the existing research work followed by problem identification in Sect. 3. Section 4 discusses about proposed methodology followed by elaborated discussion of method implementation in Sect. 5. Comparative analysis of accomplished result is discussed under Sect. 6 followed by conclusion in Sect. 7.

2 Related Work

Wang et al. [11] offered an analytical framework for determining the circumstances in which malicious devices in the IDS attack-prevention is designed using game theory. Basset et al. [12] presented a semi-supervised method for IDS that takes into account sequential traffic flow features during the training phase. Dutta et al. [13] presented a sparse autoencoder-based anomaly detection method, and an LSTM with logistic regression is utilized to further identify malicious traffic. Aygun et al. [14] presented a comparison of two autoencoders for detecting new or unknown attacks is conducted. Shone et al. [15] presented an unsymmetric autoencoder system that provides improved dimensionality reduction and data generalization. The work that Tabassum et al. [16] recommended a privacy-conscious IDS using autoencoder-based preprocessing operations present precise dataset modeling and eliminate redundant data. The efforts made by Kaur et al. [17] identified and described various security attacks using a Convolution Neural Network (CNN). Ferrag et al. [18] developed a multiclass classification system for the Internet of Things network with CNN model with multiple layers and a programmable loss function has been used by the authors to reduce attack detection time. An IDS based on a fuzzy logic system was developed by the authors of the study by Haripriya and Kulothungan [19] to protect IoT devices from DoS attacks. Ciklabakkal et al. [20] have presented a mechanism to detect intrusions in the IoT network by combining the autoencoder with a number of machine learning methods. However, the types of attacks that the authors are considering have not been explicitly discussed. Faker and Dodge [21] used deep learning in the design of the IDS and evaluated how well it performed on the two datasets CIC-IDS2017 and UNSW-NB15. Alkadi et al. [22] made use of blockchain technology to guarantee the reliability and safety of distributed IoT-IDS. Bi-Long Short-Term Memory (LSTM) is used to create IDS that deal with sequential network flow, and Ethereum blockchain-based smart contracts are used to guarantee privacy features in this study. The UNSW-NB15 and BoT-IoT datasets are used to evaluate the effectiveness of the presented model. In the study by Fatani et al. [23], AI-enabled

IDS is designed using CNN to address feature engineering issues and extracted relevant attributes. The next section highlights about the identified research problems.

3 Problem Description

After reviewing the existing intrusion detection schemes, various limiting factors have been noted viz. i) existing security schemes are proven resistive for specific forms of intruders only or whose attack behavior is well defined and hence they are not applicable for dynamic attackers, ii) adoption of machine learning is one smart solution however, majority of them perform sophisticated operation with extensive response time which cannot be use for instantaneous detection of uncertain attackers, iii) effective classification towards suspicious behavior of an attacker prior to applying machine learning is not reported in any study which could have higher chances of offering higher accuracy, and iv) majority of existing approaches has higher dependencies of training dataset in order to furnish higher accuracy value.

4 Proposed Methodology

The proposed scheme is a continuation of our prior work [24, 25]. The primary objective of the proposed research project is to develop a novel Intrusion Detection System (IDS) that can detect potential cyberattacks with high accuracy and recover the Internet of Things application as soon as possible by initiating a suitable response.

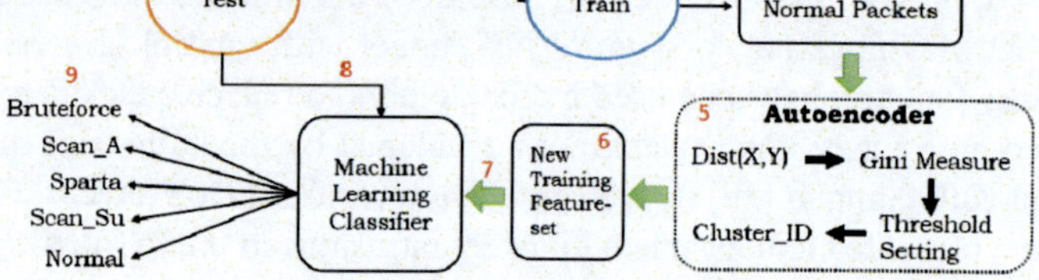

Fig. 1. Proposed Methodology

The study in this work (Fig. 1) examines security on all three layers: the transport layer (TCP and UDP analysis), the application layer, and the network layer using semi-supervised approach. In this method, an autoencoder and supervised learning methods are used to model features and detect intrusions. More specifically, there are three fundamental modules in the proposed system. The proposed system's constructs a data preprocessing module, which is meant to deal with missing and null values as well as data encoding. Autoencoder and distance function feature modeling and representation are emphasized in the second module of the proposed system. Implementing machine learning classifiers for multiclass intrusion detection and classification is the focus of the proposed system's third module.

5 Method Implementation

To begin, the study conducts dataset exploration and visualization to comprehend the dataset's characteristics and the need for appropriate data treatment. An appropriate data preprocessing is carried out to clean the data and make it suitable for further feature modeling. The idea of an auto-encoders-based clustering mechanism is used in the proposed study to learn hidden representations of various feature sets. The irrelevant features are eliminated when clustering mechanisms are used. Thus, the correlation measure from each cluster is used to select important and non-redundant features. By reconstructing data samples in a more suitable form, the auto-encoder addresses the issue of intra-variance in data distribution and maps the original data feature space into a new feature vector space. Clustering-based feature selection, in which each packet is given a distinct cluster identity, is used to improve performance. Since Autoencoder does not include a label in its output, it is suitable for unsupervised learning. When the autoencoder is fed data that is not the same as the training data, it will produce data that is completely different from the input. Consequently, the autoencoder can frequently be utilized as a filter for data clustering. The best features that were obtained for intrusion detection and classification will be evaluated in the following section of the proposed work. In order to accomplish this, the study employs a variety of Machine Learning (ML) strategies for attack classification.

5.1 Dataset Adopted for Study

The evaluation of the proposed study is carried out using latest MQTTIoT-IDS-2020 dataset [24]. The dataset was constructed by simulating an attack in a research-controlled environment where an attacker's IP address serves as a label for identifying the attacker is used to identify the attack packets. The attack simulation setup includes an attacker, a camera feed server, and a central communication module (broker) to resemble the 12 MQTT sensors in a real network. MQTT brute-force attack and generic networking scanning are included in this setup. The study is carried out considering samples in normal samples (Nor), Sparta (SPA) and MQTT Bruteforce (MBA) attacks is significantly higher than in other attack classes like scanUDP (SUA) and scanaggressive (SCA).

5.2 Autoencoder Design

A neural network with two core module encoders and a decoder is called an auto-encoder. After receiving the input data, the encoder module compresses it into a low-dimensional space. An autoencoder's decoder module reconstructs input data in a low-dimensional space using the compressed result from the encoder module. Additionally, it is able to reconstruct data into a decoder layer after encoding it into a hidden layer. The study considers the training dataset x, with n-dimensional vector, where n is the number of data samples. Through the use of the mapping function $f_x()$, the encoder module converts the input x_i into the hidden representation h_i:

$$h_i = f(x_i) = a((Wx_i + b)) \tag{1}$$

In the above expression, W denotes weight, b denotes bias, function $f()$ denotes the mapping function, and a denotes the activation function. The decoder module makes an effort to map the hidden representation back into the input layer, which then builds the new or reconstructed input data in the following manner:

$$y_i = g(h_i) = a((Wh_i + b)) \tag{2}$$

The primary objective here is to identify the parameter with the greatest potential to effectively reduce the difference between input and reconstructed data throughout the entire training process. Autoencoder is an unsupervised method that, in the end, gives each packet a cluster id. Classification methods use the cluster-id as a feature (or input). Using ANNs, this is an advanced method for feature engineering. Any classification method's performance will improve as a result of this. The autoencoder's training takes into account only normal traffic samples, not attacks samples. However, the input feature descriptor, also known as the predictor variables, is converted into numerical vectors using min-max normalization operations in accordance with equation three as follows:

$$x(i) = \frac{x(i) - min(x(i))}{max(x(i)) - min(x(i))} \tag{3}$$

The proposed scheme considers 1 input layer, 5 hidden layer, ReLu as activation function, and MSE as loss function. The mechanism of training the autoencoder is shown as follows:

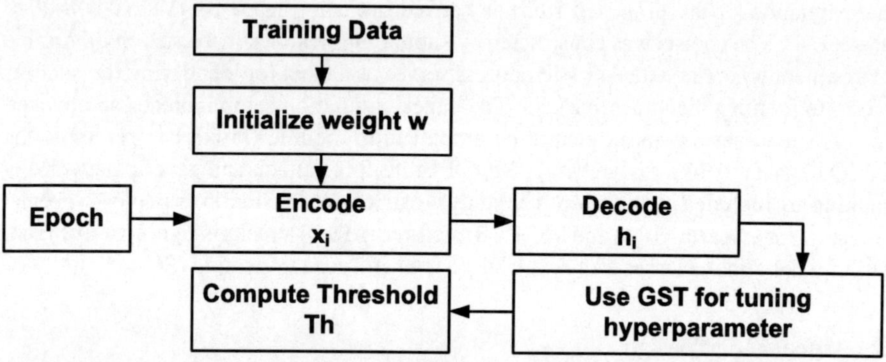

Fig. 2. Autoencoder Training

According to Fig. 2, the difference between the input and output of the network traffic data determines its category. When the difference is found to be higher than the threshold than it specifies an event of an attack. The empirical expression for threshold Th is as follows:

$$\text{Th} = \frac{1}{n}\sum\nolimits_{i=1}^{n}\left(x_i - x_i^{'}\right)^2 \tag{4}$$

In the above expression (4), the dependable parameters are input arguments x_i and x_i' to calculate a threshold value (Th). The computed threshold value (Th) is 0.001, and n is the total number of data samples in the training samples. Method 2 goes over the entire autoencoder-based clustering procedure. The proposed feature representation scheme is implemented in three phases by Method 2. The assignment of cluster ids, the calculation of distance, and the calculation of threshold. The method calls method-1, which is a trained autoencoder model, in the first step. The data sample of typical traffic patterns is used for the autoencoder training. However, before beginning a training procedure, the data set needs to be vectorized and normalized. The study first uses a pre-trained method to determine learning parameters W (weights) in order to improve model generalizability and avoid vanishing gradient issues during the learning process. Additionally, the optimal hyperparameter settings for tuning the model during the learning process are determined using the grid-search technique known as GST. The entire testing dataset is fed to the trained model in the following step of the method testing of the trained model, which produces reconstructed data after processing. Additionally, the following distance function, which makes use of the Euclidean distance formula, is used to calculate the distance vector:

$$\vec{D} = \sqrt{\sum\nolimits_{i=1}^{n}\left|x_i - x_i^{'}\right|^2} \tag{5}$$

The above expression (5) showcases vector Distance D between input data (x_i) and reconstructed data ($x_i^{'}$). The threshold values are compared to the distance value that is used to assign a cluster-id to groups with unique features that are unique to normal class attack classes (Fig. 3). The Gini index is used to calculate the threshold value. The Gini value is used to determine the threshold values for assigning cluster-ids because the autoencoder only specifies the reconstruction loss. The Gini value is simply a measure of how a set of similar attributes are distributed across a threshold.

In this study, the Gini measure is used to set the system's thresholds so that the best clustering occurs. The Gini value is used to determine the threshold values for assigning cluster-ids because the autoencoder only specifies the reconstruction loss. The Gini value is simply a measure of how a set of similar attributes are distributed across a threshold.

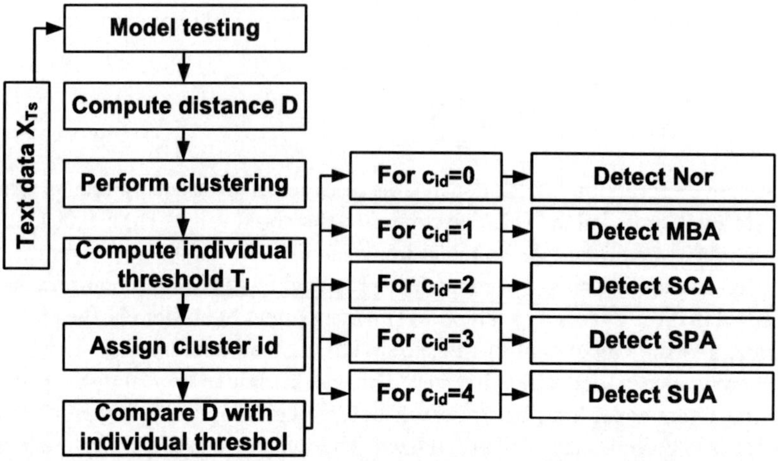

Fig. 3. Variable Attack Identification using machine learning

6 Results Discussion

On an Anaconda platform, the proposed system's design and development are carried out with python scripted in Jupyter notebook. The three shallow ML classifiers used in the proposed study are Naive Bayes (NB), K-Nearest Neighbor (KNN), and Random Forest (RF). These classifiers take the cluster id as their input and return observed data in normal and attack classes. Half of the dataset is used to train the classifier, and the other half is used to evaluate the model. This is because there are a lot of traffic samples in the dataset set, which is quite large and contains approximately 30 million of them. When processing more than 70% of a dataset, the error "Computing resource exhaustion" frequently occurs. Only an effective feature representation technique, which does not need to be implemented with 100% of the dataset but can provide sufficient feature code for training machine learning classifiers with just a few samples, is introduced in the proposed research to address this kind of issue. In order to demonstrate how the proposed feature representation technique improves the IDS's performance, outcome and performance analysis are presented in this section. In this regard, the proposed feature representation model (WFR) and the unproposed feature representation model (WoFR) are used to implement ML classifiers for performance evaluation. Precision, recall, and the F1 score are calculated for each data class because the proposed study focuses on multiclass intrusion detection. In addition, the classification performance of three ML classifiers is demonstrated against class imbalance issues and through extensive analysis.

As MQTT-based attacks are more dynamic and complex because they can easily mimic benign or normal behavior, the proposed work has taken into account the case study of MQTT enabled IoT networks in order to improve the performance of data driven or learning based IDS for the accurate detection and classification of attacks in dynamic networks such as IoT. Any kind of IDS will find it extremely difficult and challenging to precisely identify attacks or intrusions in these situations. As a result, the proposed study offers a feature representation framework that enables machine learning models to

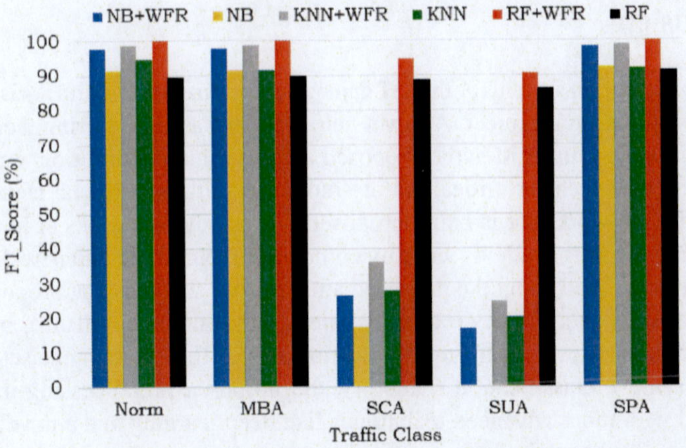

Fig. 4. Comparative analysis in terms of F1_score

learn the distribution of data features more precisely and better generalize latent features. With features enhanced by the proposed scheme, three supervised classifiers have been implemented and trained (WFR). Then, their results are compared to those of classifiers trained using normal data, or WoFR, without the proposed method (Fig. 4). The results achieved with the proposed feature representation (FR), as shown by the comparison, are promising. The superior classifier is one that was trained on data that were enhanced by the proposed scheme WFR. However, in the cases of WFR and WoFR, RF has outperformed the other two of the three classifiers. This is also shown in Fig. 4, which provides a comparison of each technique in terms of its F1_score.The RF is based on the decision tree (DT) method's design principle, which uses a similar approach to figure out how a dataset's features should split into nodes to form a tree. Euclidian distancing is the foundation of the KNN; The labels with lower support values should perform better. Based on the Bayes' Theorem, the NB classifier is a probabilistic model that makes the specific assumption that the data is a particular feature of the data that is independent of any other form of features. Because of this assumption, the NB classifier is unable to classify SUA attacks as SCA attacks that follow a similar pattern to normal classes. KNN and NB do not perform well in classification because the dataset is associated with an imbalance factor. It should be noted that the extensive result is not obtained because there are insufficient and pertinent studies on the comparable dataset used in the proposed study. Only 49 Google Scholar citations have been found for the dataset thus far. However, the study looks at handling data imbalance factors in future work to improve classification performance and make it more appropriate for the real-time scenario. By calibrating it to generalize what to learn and what not to learn, the proposed method will be extended with the deep learning approach in subsequent research.

7 Conclusion

The intrusion detection system for the IoT ecosystem against MQTT attacks is the subject of the proposed research project. A novel autoencoder-based clustering technique that gives each packet a cluster-id is the proposed study's main contribution. An advanced data treatment approach for critical feature modeling with an artificial neural network is provided by the work that is being proposed. The study also looks at how well different machine learning methods work by comparing them to old and new datasets that have been preprocessed using the proposed autoencoder scheme. The proposed feature representation method's efficacy is demonstrated by the outcome statistics, and comparative evaluation justifies its real-time application scope. To deal with dataset imbalance factors and achieve classification reliability, the proposed preprocessing method will be combined with more advanced techniques like deep learning and a novel regularize scheme in subsequent work.

References

1. Balogh, S., Gallo, O., Ploszek, R., Špaček, P., Zajac, P.: IoT security challenges: cloud and blockchain, postquantum cryptography, and evolutionary techniques. Electronics **10**(21), 2647 (2021). https://doi.org/10.3390/electronics10212647
2. Asharf, J., Moustafa, N., Khurshid, H., Debie, E., Haider, W., Wahab, A.: A review of intrusion detection systems using machine and deep learning in Internet of Things: challenges, solutions and future directions. Electronics **9**(7), 1177 (2020). https://doi.org/10.3390/electronics9071177
3. Vanin, P., et al.: A study of network intrusion detection systems using artificial intelligence/machine learning. Appl. Sci. **12**(22), 11752 (2022). https://doi.org/10.3390/app122211752
4. Díaz-Verdejo, J., Muñoz-Calle, J., Alonso, A.E., Alonso, R.E., Madinabeitia, G.: On the detection capabilities of signature-based intrusion detection systems in the context of web attacks. Appl. Sci. **12**(2), 852 (2022). https://doi.org/10.3390/app12020852
5. Thakkar, A., Lohiya, R.: A review on machine learning and deep learning perspectives of IDS for IoT: recent updates, security issues, and challenges. Arch. Comput. Meth. Eng. **28**(4), 3211–3243. (2021)
6. Liu, H., Lang, B.: Machine learning and deep learning methods for intrusion detection systems: a survey. Appl. Sci. **9**(20), 4396 (2019). https://doi.org/10.3390/app9204396
7. Fu, Y., Du, Y., Cao, Z., Li, Q., Xiang, W.: A deep learning model for network intrusion detection with imbalanced data. Electronics **11**(6), 898 (2022). https://doi.org/10.3390/electronics11060898
8. Banaamah, A.M., Ahmad, I.: Intrusion detection in IoT using deep learning. Sensors **22**(21), 8417 (2022). https://doi.org/10.3390/s22218417
9. Albulayhi, K., Al-Haija, Q.A., Alsuhibany, S.A., Jillepalli, A.A., Ashrafuzzaman, M., Sheldon, F.T.: IoT intrusion detection using machine learning with a novel high performing feature selection method. Appl. Sci. **12**(10), 5015 (2022). https://doi.org/10.3390/app12105015
10. Verma, P., et al.: A novel intrusion detection approach using machine learning ensemble for IoT environments. Appl. Sci. **11**(21), 10268 (2021). https://doi.org/10.3390/app112110268
11. Wang, D.-C., Chen, I.-R., Al-Hamadi, H.: Reliability of autonomous Internet of Things systems with intrusion detection attack-defense game design. IEEE Trans. Reliab. **70**(1), 188–199 (2021)

12. Abdel-Basset, M., Hawash, H., Chakrabortty, R.K., Ryan, M.J.: Semi-supervised spatiotemporal deep learning for intrusions detection in IoT networks. IEEE Internet Things J. **8**(15), 12251–12265 (2021)
13. Dutta, V., Chora's, M., Pawlicki, M., Kozik, R.: A deep learning ensemble for network anomaly and cyber-attack detection. Sensors **20**, 4583 (2020)
14. Aygun, R.C., Yavuz, A.G.: Network anomaly detection with stochastically improved autoencoder based models. In: Proceedings of the 2017 IEEE 4th International Conference on Cyber Security and Cloud Computing (CSCloud). New York, NY, USA, 26–28 June 2017, pp. 193–198 (2017)
15. Shone, N., Ngoc, T.N., Phai, V.D., Shi, Q.: A deep learning approach to network intrusion detection. IEEE Trans. Emerg. Top. Comput. Intell. **2**, 41–50 (2018)
16. Tabassum, A., Erbad, A., Mohamed, A., Guizani, M.: Privacy-preserving distributed IDS using incremental learning for IoT health systems. IEEE Access **9**, 14271–14283 (2021)
17. Kaur, G., Lashkari, A.H., Rahali, A.: Intrusion traffic detection and characterization using deep image learning In: IEEE International Symposium on Dependable, Autonomic and Secure Computing, International Conference Pervasive Intelligence Computing International Conference Cloud Big Data Computing, International Conference Cyber Science and Technology Congress (DASC/PiCom/CBDCom/CyberSciTech), pp. 55–62 (2020)
18. Ferrag, M.A., Maglaras, L., Moschoyiannis, S., Janicke, H.: Deep learning for cyber security intrusion detection: approaches, datasets, and comparative study. J. Inf. Secur. Appl. **50**, 102419 (2020). https://doi.org/10.1016/j.jisa.2019.102419
19. Haripriya, A., Kulothungan, K.: Secure-MQTT: an efficient fuzzy logic-based approach to detect DoS attack in MQTT protocol for internet of things. EURASIP J. Wirel. Commun. Netw. **2019**, 90 (2019)
20. Ciklabakkal, E., Donmez, A., Erdemir, M., Suren, E., Yilmaz, M.K., Angin, P.: ARTEMIS: an intrusion detection system for MQTT attacks in Internet of Things. In: Proceedings of the 2019 38th Symposium on Reliable Distributed Systems (SRDS), Lyon, France, 1–4 October 2019, pp. 369–3692
21. Faker, O., Dogdu, E.: Intrusion detection using big data and deep learning techniques. In: Proceedings of the 2019 ACM Southeast Conference, Kennesaw, GA, USA, 18–20 April 2019, pp. 86–93
22. Alkadi, O., Moustafa, N., Turnbull, B., Choo, K.-K.R.: A deep blockchain framework-enabled collaborative intrusion detection for protecting IoT and cloud networks. IEEE Internet Things J. **8**(12), 9463–9472 (2021)
23. Fatani, A., Abd Elaziz, M., Dahou, A., Al-Qaness, M.A.A., Lu, S.: IoT intrusion detection system using deep learning and enhanced transient search optimization. IEEE Access **9**, 123448–123464 (2021)
24. Hindy, H.: Hanan Hindy, IEEE DataPort, 23-Jun-2020. https://ieee-dataport.org/authors/hanan-hindy. Accessed 31 Oct 2022

Multivariate Statistical Techniques to Analyze Crime and Its Relationship with Unemployment and Poverty: A Case Study

Anthony Crespo(✉), Juan Brito, Santiago Ajala, Isidro R. Amaro, and Zenaida Castillo

School of Mathematical and Computational Sciences, Yachay Tech University, Hda. San José s/n y Proyecto Yachay, San Miguel de Urcuquí 100119, Ecuador
{brian.crespo,juan.brito,santiago.ajala,iamaro, zcastillo}@yachaytech.edu.ec

Abstract. Crime is a problem that affects people all over the world. This is the case in countries such as Ecuador, which has been affected by a recent increase in crime. The present work makes use of multivariate statistical tools such as clustering and HJ-Biplot to carry out a study of crime in Ecuador in the period corresponding from January 2021 to May 2022. This includes an analysis of the number of crimes at the monthly and provincial level. Additionally, to identify groups of provinces with similar characteristics in these variables, an analysis of the crime rate per 100,000 inhabitants is performed. The same techniques were applied to find possible correlations between poverty, unemployment, and crime. Even though the results do not show a strong correlation between poverty, unemployment, and crime variables, it is observed that the Amazon provinces, which have the highest rate of rapes, also present the highest rates in all variables of poverty for both years, 2019 and 2021.

Keywords: Crime · Unemployment and Poverty · Ecuador · Clustering · HJ-Biplot

1 Introduction

Like most of the countries in the region, Ecuador is exposed to crime every day. The numbers reflecting crime in Ecuador vary from year to year; thus, this is the subject of many investigations by government officials. In 2021, the government reinforced the presence of security forces in the streets because the number of homicides doubled compared to 2020 [3]. Additionally, in 2021 one of the biggest prison crisis in Ecuador was evidenced, and the crime rate increased [12]. As a solution to this event, the government implemented several states of exception during 2021 to try to reduce crime rates. Currently, Ecuador registers a rate of 15.48 violent deaths per 100,000 inhabitants [9].

R. Silhavy and P. Silhavy (Eds.): CSOC 2023, LNNS 724, pp. 180–192, 2023.
https://doi.org/10.1007/978-3-031-35314-7_18

Ecuador not only faces a fight against crime but unemployment and poverty also affect its citizens. Throughout history, poverty has undergone radical changes over the years. In December 2021, 27.7% of Ecuadorians were immersed in poverty, living with less than USD 2.85 per day, which contributes to increasing inequalities in the population[10]. On the other hand, unemployment is evident, not only in small towns but also in large provinces such as Guayas and Pichincha. According to unemployment indicators, published by the National Institute of Statistics and Census or INEC, by its acronym in Spanish, in April 2022, the unemployment rate reached 4.7%, which in comparison to the unemployment rate in April 2021 (5.1%) represents a decrease of 0.4% points, which is not a significant reduction according to INEC [1].

This article focuses on the analysis of crime in Ecuador in 2021 and 2022, as well as its relationship with unemployment and poverty during 2019 and 2021. Clustering and HJ-Biplot were used to analyze the data provided by the INEC. The document is organized as follows. Section 2 presents the materials and methods, which in this case are the data used and the details of its preprocessing. It also presents a description of the statistical tools employed in this research, clustering, and HJ-Biplot. Section 3 shows the graphic results obtained from the study, together with its discussion and analysis. Finally, the conclusions of the research are drawn in Sect. 4.

2 Materials and Methods

2.1 Data Description

Data on crime, unemployment, poverty levels, and a number of inhabitants per province, during the years covered by this study, were provided by the INEC [1].

Crime. The provided crime data covers the period from January 2021 to May 2022, and presents a classification by month and by province. From this data, nine variables were taken: homicides, personal robberies, auto parts theft, motorcycle theft, car theft, house theft, business robbery, rapes, and the number of inhabitants per province.

Unemployment and Poverty. The provided unemployment and poverty data are classified by month and province and cover the years 2018, 2019, and 2021. In this case, nine variables were selected: unemployment, adequate employment, unpaid employment, NEET people, poor by income, extreme poverty by income, basic needs poverty, multidimensional poverty, and extreme multidimensional poverty.

In order to avoid over-plotting and achieve adequate visualization of the graphs of each analysis, a code was assigned to each province and to each study variable. This codification is shown in Tables 1, 2, and 3. Additionally, to refer to the crime rates per 100,000 inhabitants corresponding to each variable, 100K was added as a suffix to its corresponding code.

Table 1. Codes for provinces

Province	Code	Province	Code	Province	Code
Azuay	A	Guayas	G	Pichincha	P
Bolívar	B	Imbabura	I	Tungurahua	T
Cañar	U	Loja	L	Zamora Chinchipe	Z
Carchi	C	Los Ríos	R	Galapagos	W
Cotopaxi	X	Manabí	M	Sucumbíos	K
Chimborazo	H	Morona Santiago	V	Orellana	Q
El Oro	O	Napo	N	Santo Domingo	J
Esmeraldas	E	Pastaza	S	Santa Elena	Y

Table 2. Codes for crime variables

Variable	Code	Variable	Code
Homicides	$X1$	House theft	$X6$
Personal robberies	$X2$	Business robbery	$X7$
Auto parts theft	$X3$	Rapes	$X8$
Motorcycle theft	$X4$	Number of inhabitants per province	$X9$
Car theft	$X5$		

Table 3. Codes for unemployment and poverty

Variable	Code	Variable	Code
Unemployment	$Y1$	Extreme poverty by income	$Y6$
Adequate employment	$Y2$	Basic needs poverty	$Y7$
Unpaid employment	$Y3$	Multidimensional poverty	$Y8$
NEET people	$Y4$	Extreme multidimensional poverty	$Y9$
Poor by income	$Y5$		

2.2 HJ-Biplot

Biplots are multivariate statistical techniques that allow us to visualize data, as a $n \times p$ matrix, with markers $j_1, ..., j_n$ for rows and $h_1, ..., h_p$ for columns, chosen so that both markers can be represented simultaneously, on the same Euclidean space, with optimal quality. In 1972, Gabriel proposed two kinds of Biplots: the JK-Biplop, representing the rows with the maximum quality, and the GH-Biplot, taking the columns with the maximum quality instead of the rows (see [5]).

On the other hand, in 1986, Galindo [6] proposed the HJ-Biplot that seeks to obtain maximum quality of representation in rows and columns simultaneously. This technique is used mainly to represent the data in a low-dimensional space for ease of interpretation. Usually, the matrix rows are represented by points x_i (row markers) and the columns by vectors y_i (column markers). The interpretation does not require specialized statistical knowledge; it is enough to know how to interpret the length of a vector, the angle between two vectors, and the distance between two points.

Mathematical Analysis of HJ-Biplot. The Biplots methods are based on the *SVD* decomposition of data matrix Y.

$$Y = UDV^T$$

where U is the matrix whose columns contain the eigenvectors of YY^T, D is a diagonal matrix with the singular values of Y sorted in descending order, and V is the matrix whose columns contain the eigenvectors of Y^TY. U and V must be orthonormal, that is, $U^TU = I$ and $V^TV = I$, to guarantee the uniqueness of the factorisation. One can express the entry y_{ij} of the data matrix Y, as:

$$y_{ij} = \sum_{s=1}^{min(n,p)} \lambda_s u_{ij} v_{js}$$

The process of obtaining a Biplot representation is based on dividing our $Y_{n,p}$ data into two different matrices, $A_{n,k}$, and $B_{k,n}$. When the n rows of A and the p columns of B are displayed in a single graphical representation. Depending on the type of Biplot, the definition of these matrices can change. For example, in the case of a GH-biplot, $A = U$ and $B^T = VD$.In the case of a JK-Biplot $A = UD$ and $B^t = V$. Finally in HJ-biplot $A = UD$ and $B^t = VD$. For a more extensive review see [2] and [5].

To take advantage of the two-dimensional HJ-Biplot representation, we must recognize the following geometric elements:

1. The orthogonal projection of the x_i individual onto the y_i vector indicates the order of the individual in that variable.
2. The distances between individuals represent how much they are similar to each other.
3. The length of the vector y_i indicates the variance of this variable.
4. The cosines of the angles between the column vectors approximate the correlations between said variables so that acute angles are associated with indicators with a high positive correlation; obtuse angles indicate a negative correlation, and right angles indicate uncorrelated variables.

Recent applications of the HJ-Biplot can be seen in [11] and [13].

2.3 Clustering

In statistical analysis, the need to treat data by groups instead of individually has been solved by clustering, a relatively traditional technique that is widely used nowadays [4]. Clustering takes the data and generates categories or clusters, placing the data that differ most significantly into different categories [7]; as a result, the data elements of each cluster are similar to each other but dissimilar to those of other groups. This facilitates the overview of the data at the cluster level and also allows us a closer analysis of the data in each cluster [15].

K-means. It is a technique that partitions a data set into k disjoint groups or clusters in such a way that the elements of the same cluster present similar characteristics. The objective of this non-hierarchy method is to minimize the intra-cluster distance:

$$\sum_{i=1}^{k}\sum_{\mathbf{x}\in C_i} \mathbf{d}^2(\mathbf{x}-\mathbf{u}_i)$$

Where a set of n observations $X = \{\mathbf{x}_1, \mathbf{x}_2, \ldots, \mathbf{x}_n\}$, with each $\mathbf{x} \in \mathbb{R}^d$, and a value $k \leq n \in \mathbb{N}$, the set $C = \{\mathbf{c}_1, \mathbf{c}_2, \mathbf{c}_3, \ldots, \mathbf{c}_k\}$ is the desired clusters, $U = \{\mathbf{u}_1, \mathbf{u}_2, \ldots, \mathbf{u}_k\}$ is a set with the means of each cluster and finally d denotes the distance employed.

The main steps are:

1. Choose the value of k.
2. Choose randomly as many centroids as the number of clusters defined. The centroids do not necessarily have to be observations from the data set. However, in the most straightforward implementation of the algorithm, the centroids are chosen to be observations from the data set.
3. For all $\mathbf{x}_i \in X$, find the closest centroid $\mathbf{c}_j \in C$.
4. Compute and assign the new centroid of each cluster.

Steps 3 and 4 are repeated until a stopping criterion has been met: the centroids stop changing, or the maximum number of iterations is reached.

2.4 Methodology

Two analyses were performed using clustering and HJ-Biplot. Specifically for clustering, K-means was applied, using the elbow method to determine k, the optimal number of groups for each data set.

First Analysis. This analysis covers the period corresponding to January 2021-May 2022, and it is developed in three parts: analysis of crime per month, per province, and per 100,000 inhabitants. Crime data corresponding to the 24 provinces and the time period of interest were filtered for the first two parts. Subsequently, the data was organized to obtain the number of crimes raised each month at the national and provincial levels. Finally, to analyze crime in each province regardless of its inhabitants, crime was expressed as a rate for every 100,000 inhabitants. For this last step, the data corresponding to the number of inhabitants per province was used.

Second Analysis. The second analysis combines the information given by variables of crime, poverty, and unemployment. Its objective is to determine if there is any correlation between the crimes in each province and their situation regarding employability and poverty. Unemployment and poverty data are published annually; however, data for the year 2020 are not available on INEC's website; therefore, data for 2019 and 2021 were taken.

As mentioned before, Clustering and HJ-Biplot techniques were used for the two proposed analyses, over previously normalized data. These statistical techniques were applied using modules from the R software and the MultBiplotR package by Vicente-Villardón [14]. The results are discussed in the next section.

3 Results and Discussion

As a previous step to the presentation of results and their discussion, it is necessary to consider the quality of representation of both variables and individuals. Variables and individuals can be interpreted if their quality of representation is greater than half of the highest representation quality among all the variables or individuals [8]. Thus, appendices contain details of the quality of representation of the variables. Those of the individuals are shown together with the HJ-Biplot plot. Additionally, the interpretations regarding clusters that contain non-interpretable individuals are supported by their means with respect to the study variables.

3.1 Analysis of Crime from January 2021 - May 2022

Crime per Month. K-means was applied for the months from January 2021 to May 2022, considering the number of crimes of each type raised in each month. Using the elbow method, it was found that the appropriate number of clusters is 3. Additionally, an HJ-Biplot was performed. Figure 1 shows the Biplot graph together with the clusters and the quality of representation of months.

As seen in Fig. 1, the months corresponding to the first semester of 2021, form a first group. July, August, September, October, November 2021, and February 2022 constitute the second group. The third group comprises December 2021, January, March, April, and May 2022.

From the Biplot, the groups were classified as the months with the lowest, medium and highest crime in the period analyzed. Additionally, it can be seen that the elements that make up each cluster correspond almost perfectly with the months of the first and second semesters of 2021 and the first semester of 2022, respectively. Therefore, there seems to be an increase in almost all variables of crime from one semester to another.

Crime per Province. In order to seek possible relationships between the number of crimes and each of the provinces, a clustering of the provinces was carried out. The number of clusters obtained from the elbow method was 3. An HJ-Biplot was also performed using the data. From the Biplot, in Fig. 2, it is

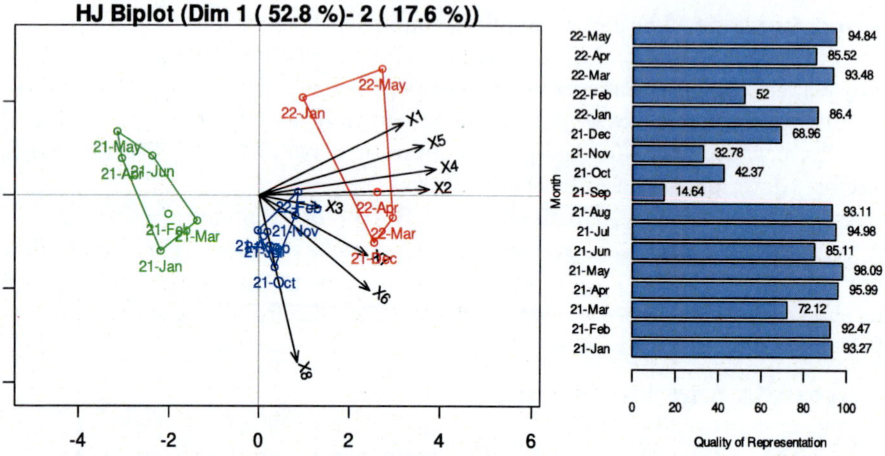

Fig. 1. HJ-Biplot for months and number of crimes.

observed that there is a high correlation of all variables from X1 to X8 with respect to variable X9 (Population), which means that the provinces with the highest number of inhabitants tend to have the highest number of crimes.

Analyzing the clustering and the Biplot results, Guayas and Pichincha form the group of provinces with the highest number of crimes. Likewise, the second group, made up of Azuay, El Oro, Esmeraldas, Los Ríos, Manabí, and Santo Domingo de los Tsáchilas represents the provinces with an intermediate number of crimes. Finally, the provinces Bolívar, Cañar, Carchi, Cotopaxi, Chimborazo, Imbabura, Loja, Morona Santiago, Napo, Pastaza, Tungurahua, Zamora Chinchipe, Galápagos, Sucumbíos, Orellana and Santa Elena can be characterized as the group of provinces with the lowest number of crimes in the period under study. Another interesting result from the Biplot is that the number of crimes in Guayas and Pichincha significantly exceeds the rest of the provinces.

Crime Rate per 100,000 Inhabitants. Clustering and HJ-Biplot techniques were also used to analyze the rate of each crime per 100,000 inhabitants. Figure 3 shows the results.

As can be seen in the figure, the number of groups formed is once again 3. Group 1, in red, comprises four provinces of the Coast region: El Oro, Esmeraldas, Guayas, Los Ríos, and one of the Interandean region: Pichincha. These provinces present the highest rates in the variables X1, X2, X3, X4, and X5. Group 2, in blue, is made up of Azuay, Bolívar, Cañar, Carchi, Cotopaxi, Chimborazo, Imbabura, Loja, Manabí, Napo, Tungurahua, Zamora Chinchipe, Galápagos, Santo Domingo, and Santa Elena. This group presents the lowest crime rates in most variables. An important thing to notice is that all the provinces of the Interandean region, except Pichincha, are part of this second group. To these are added three from the Coast: Manabí, Santo Domingo, and Santa Elena; two from the Amazon: Napo and Zamora Chinchipe; and the Insu-

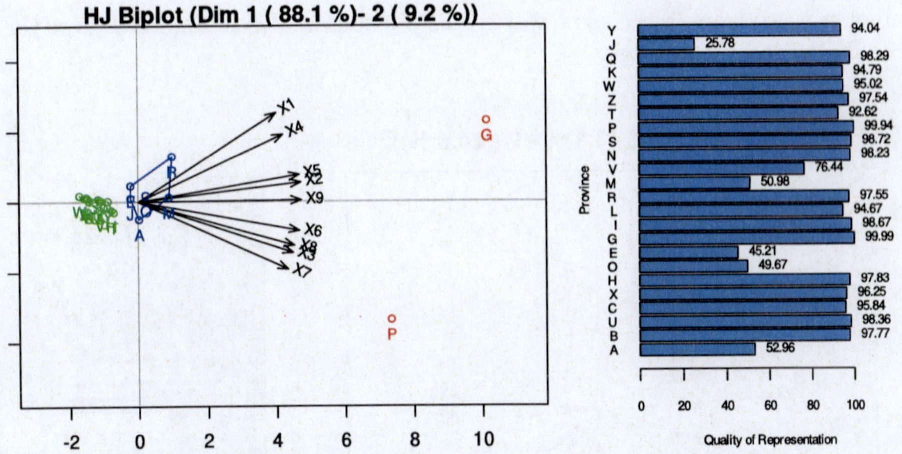

Fig. 2. HJ-Biplot for provinces and number of crimes.

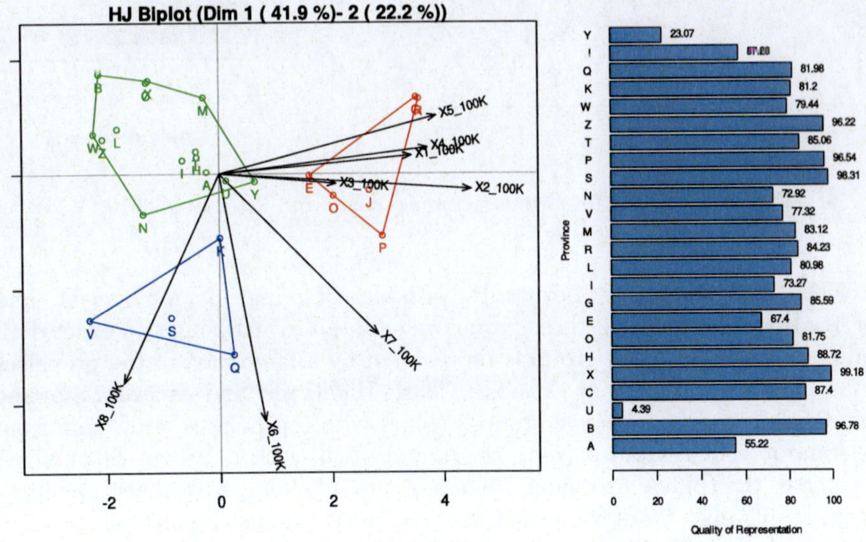

Fig. 3. HJ-Biplot for provinces and crime per 100,000 inhabitants.

lar region represented by Galápagos. Finally, group 3, in green, is formed by Morona Santiago, Pastaza, Sucumbíos, and Orellana. These provinces have the highest crime rates associated with variables X6 and X8. Interestingly, these are 4 of 6 provinces in the Amazon region.

3.2 Analysis of Crime, Unemployment, and Poverty from 2019-2021

Crime, Unemployment, and Poverty in 2019. This analysis uses all the data corresponding to crime, unemployment, and poverty in 2019. The cluster number was chosen in the same way as the previous analyses. Figure 4 shows the HJ-Biplot and the clusters.

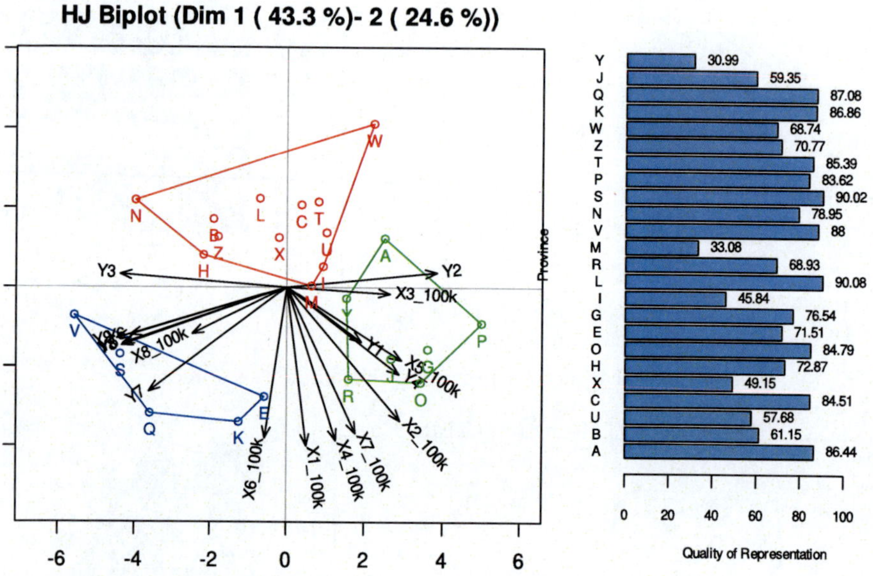

Fig. 4. HJ-Biplot for Crime, unemployment, and Poverty of 2019.

When analyzing 2019, both with variables of crime, unemployment, and poverty through K-means, three groups were obtained. Although an exact division does not occur, each group is represented by a large part of the provinces of the three main regions of Ecuador: Coast, Highlands, and Amazon. Starting from the left, one can observe group 2 (blue) with 5 provinces, 4 of them from the Amazon. Then, group 1 (red), the largest of all, with 12 provinces, of which 8 belong to the Highlands region. Finally, group 3 (green), with the remaining 7 provinces, of which 5 are from the Coast region. It can also be noticed that 5 of the 6 provinces with the largest population, i.e., Guayas, Pichincha, Los Ríos, Azuay, and El Oro, are in this last group.

The HJ-Biplot analysis showed a correlation between the variables Y5, Y6, Y8, Y9, and X8_100K, and it allows us to realize that the Y's variables correspond to different indices that measure poverty, while the variable X8_100K corresponds to the rate of rapes. This relates greater poverty to a higher number of rapes. Additionally, the provinces with the highest levels of poverty and rape belong to group 2 (blue), precisely the cluster in which most of the Amazon provinces are found.

In general, the provinces in the third group with the highest number of inhabitants are also those with adequate employment. Focusing on the variables corresponding to security, it can be observed a high correlation between the variables

X1_100K (Homicide), X4_100K (Motorcycle Theft), and X7_100K (Business robbery) as well and there is a high correlation between X5_100k (Car Theft) and Y4 (NEET people). Another interesting observation is when analyzing the variables Y3 (Unpaid employment) and Y2 (Adequate employment), where an opposite correlation can be seen. It can also be distinguished that the provinces of the Highlands, except Pichincha, generally have better security indices, being the provinces with fewer homicides and, in general, less of any kind of robberies. Finally, the correlation between poverty variables Y5, Y6, Y7, Y8, and Y9 and variables of crime rates X1_100K, X4_100K, X7_100K, and X2_100K, in general, is small, making the relationship between unemployment, poverty, and crime poor in Ecuador.

Crime, Unemployment, and Poverty in 2021. The same procedure for analyzing the 2019 data was followed with the 2021 data. Figure 5 shows the corresponding HJ-Biplot and clusters for 2021.

In contrast to the year 2019, for the year 2021, 3 provinces changed cluster. These changes further reinforce the trend of the clusters regarding the division of the three most important regions of the country. Azuay was incorporated into group 1 (red), monopolizing 9 of the 10 provinces of the Highlands region, one province more compared to 2019. Napo changed from group 1 to group 2 (blue). This group continues with 5 provinces in 2021, but now all are from the Amazon. Finally, Esmeraldas moved from group 2 to group 3 (green). From the results, it can be concluded that for both, unemployment and poverty and crime indices, there is a similar behavior between provinces of the same region, except for

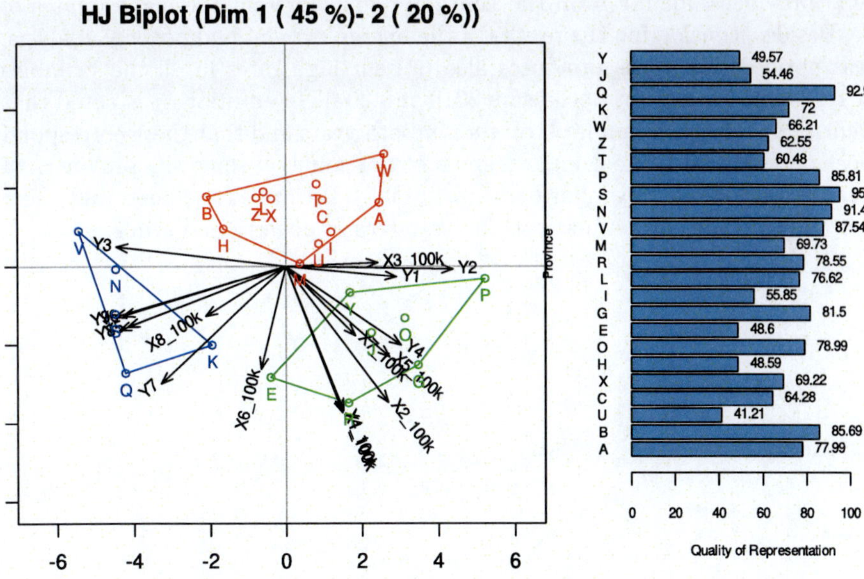

Fig. 5. HJ-Biplot for crime, unemployment, and poverty of 2021.

the Insular region, whose only province was grouped with the Highlands region. On the other hand, 2019 and 2021 maintained a high correlation between the variables representing poverty rates and the number of rapes.

When analyzing how the variables change over time, an increase in correlation between the crime rate variables is noted. For example, the correlation between X1_100K (Homicide) and X4_100K (Motorcycle theft) became almost one. Lastly, another change that can be seen from 2019 to 2021 is that, in general, all provinces shifted slightly downwards toward the direction of the homicide vector, which presented a slight increase in its length too. This suggests that there was an increase in the number of homicides in general. Focusing on the data, we noted that the number of homicides in the country stepped from 1177 in 2019 to 2460 in 2021.

4 Conclusion

Crimes are a cause for concern in many countries; hence the importance of investigating their relationship with the social environment. In the present work, an analysis of the data related to the topic, in Ecuador, was carried out, using multivariate statistical tools, such as clustering and HJ-Biplot. The analysis allowed us to draw conclusions about the crime in the country in the period January 2021 - May 2022 and its relationship to poverty and unemployment. From the study, it was possible to identify an evident increase in crime by semester. In particular, Guayas and Pichincha were found to be the provinces with a high number of crimes and crime rates, which was correlated to their high number of inhabitants compared with other provinces. It is important to highlight that several provinces, mainly from the Amazon region, present the highest rates of rape. Besides, considering the results of the unemployment and poverty analysis, it was noticed that these provinces also present high rates in all the variables that represent poverty for both years 2019 and 2021. In addition, it is noted that the clusters maintain a similar structure in both years and that these correspond almost perfectly with the 3 main regions of the country; thus, the provinces of the same region follow a similar behavior. Finally, it can be concluded that there is not a strong correlation between the variables of poverty and crime.

Appendix A - Quality of Representation of the variables

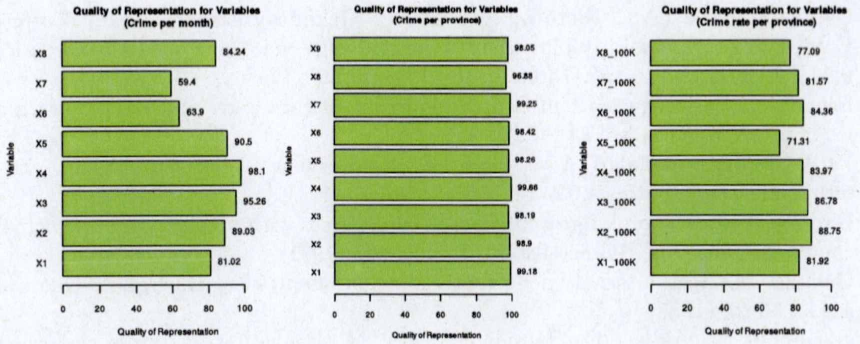

(a) Crime per month (b) Crime per province (c) Crime rate per province

Fig. 6. Qualities of Representation for the First Analysis

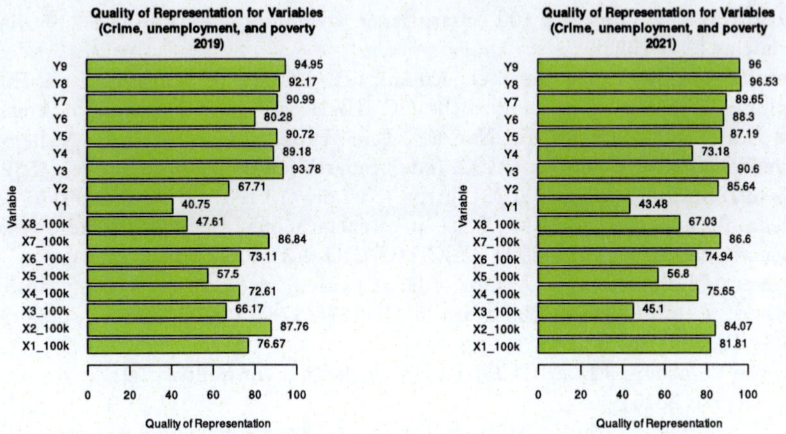

(a) Crime, unemployment, and poverty of 2019 (b) Crime, unemployment, and poverty of 2021

Fig. 7. Qualities of Representation for the Second Analysis

References

1. Instituto Nacional de Estadística y Censos. https://www.ecuadorencifras.gob.ec/institucional/home/
2. Cubilla-Montilla, M., Nieto-Librero, A.B., Galindo-Villardón, M.P., Torres-Cubilla, C.A.: Sparse hj biplot: A new methodology via elastic net. Mathematics **9**(11), 1298 (2021). https://doi.org/10.3390/math9111298
3. España, S.: La inseguridad en Ecuador encierra en casa a los ciudadanos y saca a los militares a las calles. El País (Feb 2022)
4. Fu, L., Lin, P., Vasilakos, A.V., Wang, S.: An overview of recent multi-view clustering. Neurocomputing **402**, 148–161 (2020)
5. Gabriel, K.R.: The biplot graphic display of matrices with application to principal component analysis. Biometrika **58**(3), 453–467 (1971)
6. Galindo, M.: Una alternativa de representación simultánea: HJ-Biplot. Qüestiió **10**(1), 13–23 (1986)
7. Golalipour, K., Akbari, E., Hamidi, S.S., Lee, M., Enayatifar, R.: From clustering to clustering ensemble selection: A review. Eng. Appl. Artif. Intell. **104**, 104388 (2021)
8. González, J.M., Fidalgo, M.R., Martín, E.J., Vicente, S.: Study of the evolution of air pollution in Salamanca (Spain) along a five-year period (1994–1998) using HJ-Biplot simultaneous representation analysis. Environm. Model. Softw. **21**(1), 61–68 (2006)
9. Mella, C.: Ecuador alcanza la tasa más alta de muertes violentas de la última década. Primicias (08 Feb 2022)
10. Orozco, M.: 27,7 de cada 100 ecuatorianos viven con menos de usd 2,85 diarios. Primicias (Jan 2022)
11. Riera-Segura, L., Tapia-Riera, G., Amaro, I.R., Infante, S., Marin-Calispa, H.: HJ-Biplot and clustering to analyze the COVID-19 vaccination process of American and European Countries. In: Narváez, F.R., Proaño, J., Morillo, P., Vallejo, D., González Montoya, D., Díaz, G.M. (eds.) SmartTech-IC 2021. CCIS, vol. 1532, pp. 383–397. Springer, Cham (2022). https://doi.org/10.1007/978-3-030-99170-8_28
12. Rueda, R.C.A., Ruiz, L.M.M.: Crisis penitenciaria en el Ecuador. Estudio casos de masacres carcelaria 2021–2022. RECIMUNDO **6**(3), 222–233 (2022)
13. Tenesaca-Chillogallo, F., Amaro, I.R.: Covid-19 data analysis using hj-biplot method: A study case. Bionatura **6**(2), 1778–1784 (2021). https://doi.org/10.21931/RB/2021.06.02.18
14. Vicente Villardon, J.L.: MULTBIPLOT: A package for multivariate analysis using biplots (Jan 2010)
15. Wu, X., Cheng, C., Zurita-Milla, R., Song, C.: An overview of clustering methods for geo-referenced time series: From one-way clustering to co-and tri-clustering. Int. J. Geogr. Inf. Sci. **34**(9), 1822–1848 (2020)

Bidirectional Recurrent Neural Network for Total Electron Content Forecasting

Artem Kharakhashyan and Olga Maltseva(✉)

Institute for Physics, Southern Federal University, Rostov-on-Don 344090, Russia
oamaltseva@sfedu.ru

Abstract. To ensure the operation of various terrestrial and space communication systems, it is necessary to know the total electronic content (TEC) of the ionosphere. At the same time, this parameter has become the testing ground for the development and testing of forecasting methods based on machine learning. In the previous paper of the authors, using the example of the Juliusruh reference station, the efficiency of the ARMA, LSTM, GRU methods was studied, including the use of solar and geomagnetic activity indices (Dst, IMF, Np, F10.7, 10Kp, AE). This new paper presents the results of two stages of modification of LSTM and GRU neural networks: (1) the implementation of multilayered LSTM and GRU architectures, which provided an improvement in the forecasting accuracy, while the 10Kp and Np indices turned out to be the most effective, (2) modifying these architectures to support bidirectional processing, greatly improving the efficiency of all methods. It is also shown that the addition of the Dst and F10.7 indices contributed to the improvement in efficiency. Since the state of the ionosphere described by the TEC parameter strongly depends on the coordinates of the observation point, we used a latitudinal chain of stations along the 30° E meridian and a set of high-latitude Russian stations included in the IGS system. An improvement in forecasting accuracy by a factor of two in comparison to the previous results and the results of the first stage of the current research was showcased. Without bidirectional processing, RMSE changes from 2.9 TECU to 1.5 TECU with latitude increase, MAPE changes from 8% to 16%. The use of the bidirectional architecture leads to the following estimates in dependence of latitude: RMSE changes from 1.4 TECU to 0.5 TECU, MAPE changes from 6% to 4%. With the longitude increase, the RMSE for the initial architectures changed from 1.6 TECU to 1.2 TECU, MAPE changed from 16% to 12%. Bidirectional architectures allowed to reduce RMSE to 0.6–0.8 TECU, MAPE to 4–6%. #CSOC1120.

Keywords: Deep Learning · Recurrent Neural Networks · Bidirectional · BiLSTM · BiGRU · Total Electron Content · Global Positioning System · Forecasting

1 Introduction

The total electron content (TEC) of the ionosphere is a powerful parameter that is necessary for providing global communications, so its study and forecasting is of great importance. In recent years, there has been a surge in publications on the TEC forecasting due to the use of a neural network approach, and each article contains certain

R. Silhavy and P. Silhavy (Eds.): CSOC 2023, LNNS 724, pp. 193–207, 2023.
https://doi.org/10.1007/978-3-031-35314-7_19

new elements. In this research, the TEC forecast is focused on providing information about the state of the ionosphere necessary for such technological systems as HF and satellite communication, satellite navigation, space-based radar and imaging, terrestrial radar surveillance and others [1]. Thus, from a wide range of applications of machine learning techniques for space weather prediction, the review of publications will be limited to works related to TEC forecasting. Reference [2] provides an overview of recent forecasting methods along with the quantitative assessment of the accuracy of previous methods for 1 h forecasting and up to 1-day and 2-day forecasting: the RMSE for 1 h TEC forecast in low-latitudes ranges from 2 to 5 TECU for different learning algorithms and different levels of solar activity. For the mid-latitude 1 h VTEC forecast, RMSE is about 1.5 TECU, for 1 day TEC forecast RMSE was 4 TECU in high solar activity and 2 TECU in low solar activity In [3], the forecast was made on the basis of GIM CODE maps using LSTM NN. The difference from other methods involved use of EUV flux and Dst-index. A comparison with the CODE GIM TEC showed that the first/second hour RMSE was 1.27/2.20 TECU during storm time and 0.86/1.51 TECU during quiet time.

In [4], a large overview of the use of LSTM methods for TEC forecasting was given and, to improve its accuracy, data of solar and geomagnetic activity were added in the form of parameters F10.7, Dst, and Ap. The value evaluation index was Mean Absolute Error (MAE), the activation function was Rectified Linear Unit (ReLU). For TEC prediction solution, the correlation coefficient was over 0.97 and RMSE was below 2.5 TECU, except in 2015, for 2 days ahead. The time spans variation showed that as the input time span increases, the quality of the predicted results may decrease. The RMSE range for all cases was 1.28–1.46 TECU, MAPE - in the range of 11.8–17%. Based on a large amount of data, an important conclusion was made is that due to the small proportion of geomagnetic storm time in the training data, the model may not capture the space "weathering", resulting in an unsatisfactory model prediction effect during geomagnetic storms. Article [5] points to the drawbacks of TEC prediction schemes based on RNN, which learn TEC representations from previous time steps, and each time-step made an equal contribution to a prediction. To overcome these drawbacks, in the paper [5] two improvements were proposed: (1) to predict TEC with both past and future time-step, Bidirectional Gated Recurrent Unit (BiGRU) was used, (2) to highlight critical time-step information, attention mechanism was used to provide weights to each time-step. The results of the method were compared with those of the following methods: Deep Neural Network (DNN), Artificial Neural Network (ANN), Recurrent Neural Network (RNN), Long Short-Term Memory (LSTM), Bidirectional Long Short-Term Term memory (BiLSTM), and Gated Recurrent Unit (GRU) - according to the CODE map for three latitude zones (low latitude, middle latitude, and high latitude) at longitude 100° E. The proposed by the authors Attentional BiGRU model was superior to the other models in the selected nine points. Experimental results, presented in the paper, show that the higher the latitude, the higher the prediction accuracy of the proposed model. Experimental results also show that in the middle latitude, the prediction accuracy of the model is less affected by solar activity, and in other areas, the model is greatly affected by solar activity.

In [6], the results of the ARMA, LSTM, GRU and some other methods were compared using Juliusruh station data for 2015. The ARMA method turned out to be the best for monthly and semi-annual forecasts, however, for individual days, the LSTM and GRU methods showed noticeably better results. In [6], the TEC forecast was also made taking into account space weather conditions using the indices Dst, IMF, Np, F10.7, AE for 12 months of 2015. To assess the effectiveness of the indices, linear correlation coefficients were calculated for TEC deviations from the median δTEC. The results showed that there is no clear advantage of any one coefficient. The deviation δTEC had the largest positive relationship with Dst in June, the largest negative - in December. The situation in May was characterized by a lack of correlation with any parameter. It is important to note that many articles showcase the year 2015 as the most difficult for analysis. For example, in the article [7], Fig. 3 gives a visual representation of the boundary TEC variations for 4 stations at different latitudes in the longitude range of 8–15° E during the quietest and disturbed day of each month 2015. In [4], the results for 2015 turned out to be much worse than for 2019, and the need to develop an appropriate model for disturbed conditions was noted.

In this paper, the neural networks presented in [6] are converted into the multilayer LSTM and GRU neural networks and then further modified into BiLSTM and BiGRU architectures. To ensure the continuity of the results in terms of comparison, these methods are considered for the Juliusruh station in 2015, and then the results for Russian stations are obtained. The paper presents 8 deep neural network architectures employed for TEC forecasting during single-step forecasts on limited training datasets.

2 Data and Methods

The paper proposes 8 new architectures of deep recurrent neural networks, which differ in their internal structure. Each neural network is independently trained on the experimental dataset and then used to forecast TEC values during the testing stage. Juliusruh was chosen as the reference station, where preliminary testing is carried out in order to determine proper layer composition and achieve greater accuracy compared to previously proposed regression methods and single-layer architectures. Seven stations located in Russia were selected for additional verification of the results.

2.1 Experimental Data

The values of global GIM-TEC maps were calculated from IONEX files with a time step of 2 h (ftp://cddis.gsfc.nasa.gov/pub/gps/products/ionex/) for Juliusruh (54.6° N, 14.6° E), Murmansk (69° N, 33° E), Petersburg (60° N, 30.7° E), Moscow (55.6° N, 37.2° E), Nicosia (35.1° N, 33.2° E), Norilsk (69.4° N, 88.1° E), Tixi (71.69° N, 128.86° E) and Bilibino (68.05° N, 166.44° E). The JPL map was chosen. The data on the indices of solar and geomagnetic activity Dst, F10.7, IMF, AE, Np and 10Kp, was taken from SPDF OMNIWeb Service (http://omniweb.gsfc.nasa.gov/form/dx1.html).

Processing using different indices (Dst, F10.7, IMF, AE, 10Kp, Np) took place in two stages. At the first stage, which included the study of four multilayer architectures, the best result was obtained using 10Kp, Np indices. The inclusion of other indices did not lead to a further increase in the accuracy of the forecast in that case. At the second stage, the transition to the bidirectional architecture made it possible to include two additional indices F10.7 and Dst in the calculation, which further enhanced the accuracy. Testing of the proposed architectures of neural networks is carried out on data for 2015, selected as a reference year. This year is notable for the presence of a significant number of geomagnetic storms of various duration and intensity, followed by irregular periods of quiet state; thus, this year is of particular interest for forecasting methods evaluation. The behavior of the F10.7 and Dst indices characterizing the solar and geomagnetic conditions in 2015 is shown in Fig. 1.

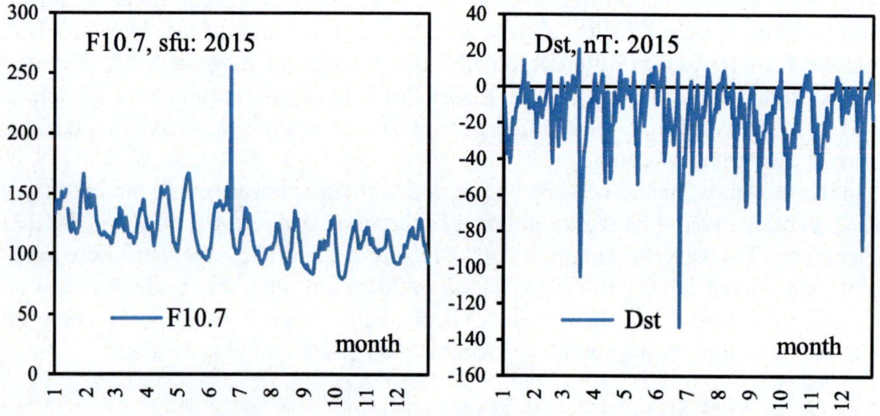

Fig. 1. Average daily values of F10.7 and Dst indices in 2015.

The behavior of median TEC values in 2015 is shown in Fig. 2 for stations along the 30° N meridian (left panel) and for Russian high-latitude stations included in the international IGS GPS base of stations (right panel). The values for Petersburg station, which is part of the meridional chain of stations, are very close to those for Moscow station.

The year 2015 belongs to the descending branch of the 24th solar cycle, therefore, in the behavior of the F10.7 index and the TEC parameter, there is a tendency for the values to decrease during this period. The latitudinal dependence of TEC is characterized by a significant increase in values from high to low latitudes; for the longitudinal dependence, the differences are not very large. The behavior of the Dst index indicates strong disturbed conditions.

2.2 Data Preprocessing

The dataset for year 2015 was used for training, verification and testing of the considered neural networks consisted of 4380 samples with a time step equal to 2 h. The first 40% part of samples was used for training, next 10% were used for validation and the last 50% corresponding to the second half of the year were used for testing. TEC samples are organized into subsequences formed by a sliding window with a width equal to 12

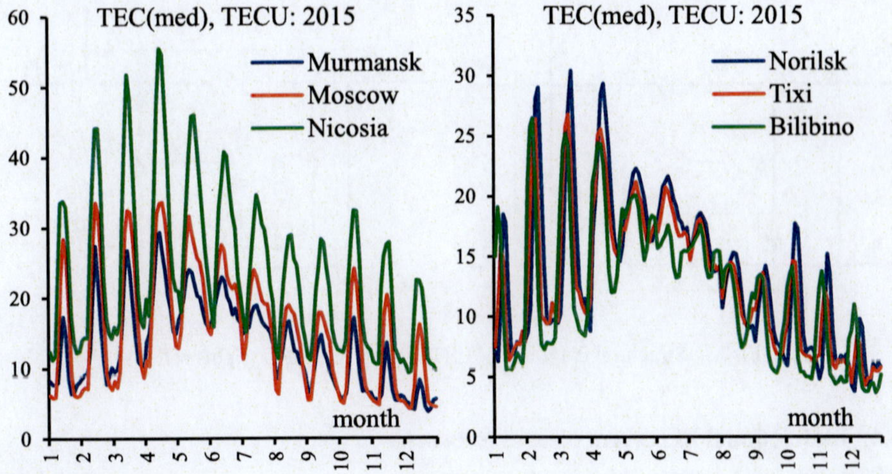

Fig. 2. TEC variations during 2015 for the latitudinal (left panel) and longitudinal (right panel) networks of stations.

samples. Np, 10Kp, Dst and F10.7 indices are organized into separate sequence features, but no sliding window is applied. The preliminary evaluation showed that other indices did not positively contribute to the results for the current dataset and architectures under consideration. Every index is organized into a separate sequence feature, but the selected indices depend on the architecture. Np and 10Kp indices are included in every case. Thereby, the TEC value at the time step t is predicted using 12 previous TEC values and a single previous value of the respective included indices. Data arrays are truncated appropriately to maintain the same array lengths for all features. Error values are replaced with the nearest previous value.

2.3 Recurrent Neural Networks

All architectures discussed in this article are based on the two most common architectures of recurrent neural networks, which have found extensive use in modeling and time series forecasting. These include Long Short-Term Memory (LSTM) and Gated Recurrent Unit (GRU). The structures of individual cells that make up the neural networks of the corresponding type are shown in Fig. 3.

Long short-term memory networks were introduced in [8]. This architecture stores the previously learned time dependencies in its internal cell state. The main advantage of LSTM is the ability to handle longer time sequences compared to a conventional recurrent neural network, which faces the vanishing gradient problem. LSTM allows to negate this drawback. Therefore, the inference accuracy of LSTM increases with the amount of historical information considered. Gated Recurrent Unit [9] is a simplified modification of LSTM, lacking cell state and having less gates.

The basic LSTM and GRU architectures were used in the first stage of the research to develop multilayer deep neural networks and evaluate their performance.

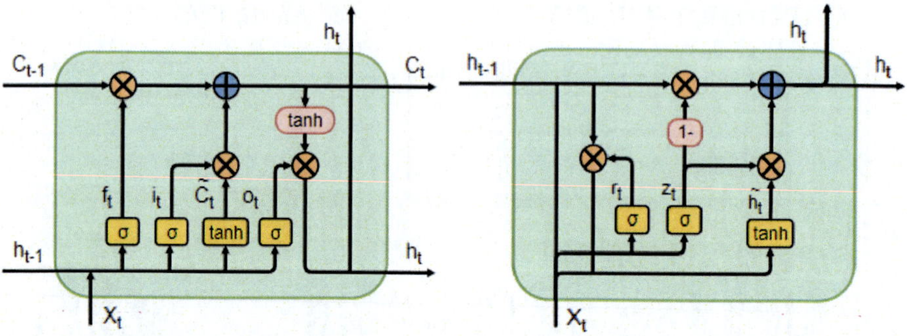

Fig. 3. Structural diagrams of LSTM (left) and GRU (right) cells.

2.4 Bidirectional Recurrent Neural Networks

While conventional LSTM and GRU neural networks are processing data in one direction, according to the order of a given input sample, their learning capabilities can be extended by introducing a bidirectional architecture. An overview of a bidirectional recurrent neural network is given in Fig. 4.

For a bidirectional architecture, the cells are arranged into two layers, one of which processes the original sample in the forward direction, and the other in the backward direction. The outputs of each of the layers are concatenated and form a common output. Bidirectional neural networks are capable of learning more complicated long-term dependencies between time steps compared to conventional architectures, and are especially useful when dealing with sequential events.

During the second stage of the research the developed at the first stage multilayer networks, including conventional LSTM and GRU, were modified to support bidirectional processing.

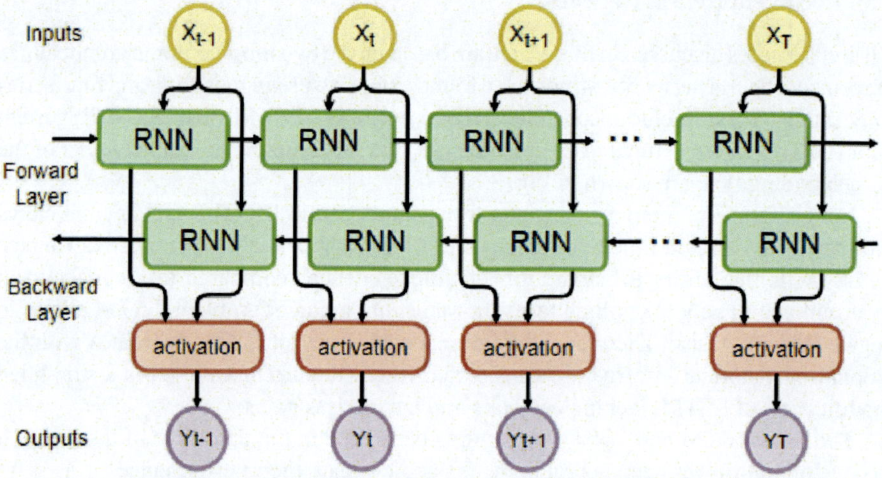

Fig. 4. Structural diagram of the bidirectional recurrent neural network.

2.5 The Proposed Deep Neural Networks for TEC Forecasting

In this paper, a comparison of a number of architectures of multilayer neural networks that differ in the composition of layers, and the principles used and the order of processing input data, is made. This section describes their features, commonalities and differences.

Activation Functions. Several different activation functions have been used. First of all, each of the considered architectures included Parametric Rectified Linear Unit (PReLU), which is a modification of the widespread RELU activation function.

PReLU activation function is defined similarly to ReLU, but it includes an arbitrary scaling factor C for negative inputs:

$$f(x) = \begin{cases} x, \ x \geq 0, \\ C \cdot x, \ x < 0. \end{cases} \quad (1)$$

PReLU activation function proved to be more flexible compared to ReLU during the early stages of the research, generally speeding up the learning process and increasing the prediction accuracy.

Two other common activation functions found in many LSTM and GRU implementations are the sigmoid and hyperbolic tangent (tanh) activation functions. The sigmoid activation function is defined as

$$f(x) = \frac{1}{1 + e^{-x}}, \quad (2)$$

and tanh activation function is

$$f(x) = \frac{e^{2x} - 1}{e^{2x} + 1}. \quad (3)$$

In this research, sigmoid activation function is used for each LSTM and GRU as the gate activation function, tanh is used as the state activation function, and each presented architecture includes a single PReLU activation function at a specific position in the layer stack.

RFF and FRF Multilayer Architectures. The Architectures Under Consideration can be Grouped According to 3 Attributes: The Order of Layers, the Principle of Processing, and the Indexes Used. The abbreviation ending "RFF" corresponds to the recurrent layer preceding two fully connected layers, while the ending "FRF" corresponds to the recurrent layer enclosed between two fully connected layers. Activation functions and normalization layers are omitted in abbreviations. Bidirectional architectures are marked by "Bi" prefix (BiLSTM, BiGRU). As mentioned earlier, all architectures under consideration include 10Kp and Np indices. Therefore, only the Dst and F10.7 indices are additionally indicated.

The structural diagrams of all architectures are shown in the Fig. 5. The value of N determines the number of cells in each respective layer. The recurrent layer was defined either as LSTM, or GRU, or BiLSTM, or BiGRU. This layer was followed by PReLU in every case. The batch normalization layer performs interlayer input normalization across mini-batches and was integrated into RFF architecture due unpredictable and unstable learning process. The same approach applied to FRF architecture provided no positive feedback, thus batch normalization layer is excluded there.

Each variable is normalized separately so that the minimum value was equal to zero and the maximum value was equal to unity. This approach to the input data normalization provided the highest prediction accuracy compared to other standard normalization methods for the architectures under consideration. Weights initialization in fully connected layers is performed using Glorot initializer [10]. Input and recurrent weights for recurrent layers were initialized using orthogonal matrix decomposition initializer [11]. Adaptive Moment Estimation (ADAM) method has been employed during the training process. The training was conducted using MATLAB environment and Deep Learning and Parallel Computing Toolboxes.

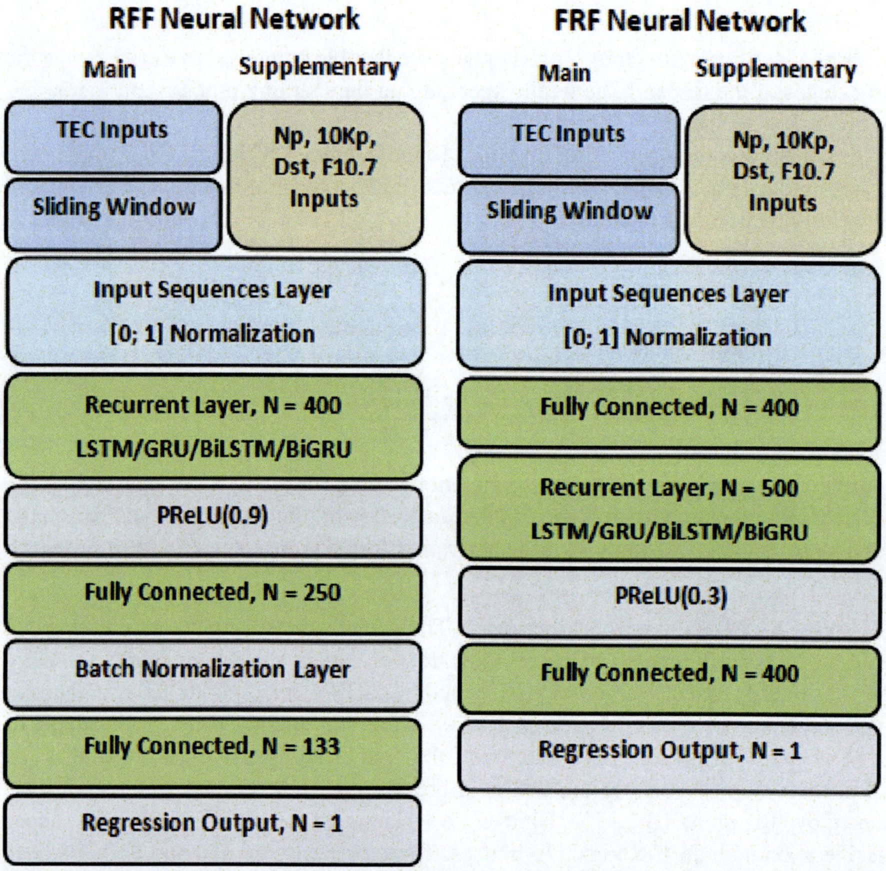

Fig. 5. Schematic diagrams of RFF and FRF deep neural network architectures. The selected recurrent layer and indices vary in each specific case.

Input parameters and training options for Adam optimizer have been selected as follows:

```
TrainingOptionsADAM with properties:
            GradientDecayFactor: 0.9000
     SquaredGradientDecayFactor: 0.9990
                        Epsilon: 1.0000e-08
               InitialLearnRate: 0.0030
              LearnRateSchedule: 'none'
            LearnRateDropFactor: 0.1000
            LearnRateDropPeriod: 10
               L2Regularization: 1.0000e-04
        GradientThresholdMethod: 'l2norm'
              GradientThreshold: Inf
                      MaxEpochs: 100
                  MiniBatchSize: 1
            ValidationFrequency: 20
             ValidationPatience: Inf
                        Shuffle: 'every-epoch'
           ExecutionEnvironment: 'gpu'
                 SequenceLength: 'longest'
           SequencePaddingValue: 0
       SequencePaddingDirection: 'left'
        ResetInputNormalization: 0
```

[ADAM optimizer parameters listing from MATLAB programming environment].

The analysis of the prediction accuracy for presented architectures is carried out on the basis of a comparison of statistical characteristics of the results: Mean Absolute Error (MAE), Mean Absolute Percentage Error (MAPE), root-mean-square error (RMSE) and correlation coefficient R.

3 Results and Discussion

A set of considered architectures and the corresponding results for the Juliusruh station in the form of statistical characteristics of MAE, MAPE, RMSE, R, averaged over half a year, are presented in Table 1.

The implementation of the RFF and FRF architectures (the first four rows of Table 1) improves the results compared to [6]. The additional modification of those architectures using of bidirectional processing leads to a 2-fold improvement in the forecast accuracy (the second four lines compared to the first ones). Supplementing the bidirectional RFF architecture with Dst and F10.7 indices (the last four lines) provides the best results. Results for bidirectional FRF architecture with Dst and F10.7 included are inconclusive.

The changes in TEC values, together with the behavior of the Dst index, the value of which is reduced by a factor of 10 for clarity, and the deviations δTEC, which determine the forecast error, are shown in Fig. 6 for Murmansk, Moscow, Nicosia stations along the 30° E meridian. The same variables for high-latitude stations(Norilsk, Bilibino) are given in Fig. 7. Figures are presented for Dst+F10.7+BiGRU-RFF architectures as they provided the best results. The plots for Murmansk and Moscow stations exclude several δTEC values reaching ~ 9 TECU and ~ 11 TECU, respectively, so as not to reduce

Table 1. Forecasting results for Juliusruh station in 2015.

Architecture	MAE	MAPE	RMSE	R
LSTM-RFF	1.11	11.54	1.57	0.97
GRU-RFF	1.16	11.52	1.62	0.97
LSTM-FRF	1.12	11.23	1.56	0.97
GRU-FRF	1.15	10.75	1.58	0.97
BiLSTM-FRF	1.07	10.34	1.44	0.978
BiGRU-FRF	0.85	9.30	1.08	0.990
BiLSTM-RFF	0.65	7.21	0.87	0.991
BiGRU-RFF	0.55	6.18	0.79	0.992
Dst+BiGRU-FRF	0.70	7.87	0.93	0.989
Dst+F10.7+BiGRU-FRF	0.62	6.95	0.80	0.992
Dst+BiLSTM-FRF	1.07	10.61	1.45	0.973
Dst+F10.7+BiLSTM-FRF	1.48	14.37	1.94	0.960
Dst+BiGRU-RFF	0.50	5.66	0.71	0.993
Dst+F10.7+BiGRU-RFF	0.47	5.23	0.74	0.993
Dst+BiLSTM-RFF	0.58	6.49	0.78	0.993
Dst+F10.7+ BiLSTM-RFF	0.56	6.48	0.92	0.990

the clarity of the figure. Those values are related to the conditions under which neural networks gave unsatisfactory results: a long recovery phase after a strong magnetic storm, against which the next magnetic storm occurs (left panel, in which the value of the Dst index is reduced by 10 times for clarity).

It can be seen that the forecast accuracy does not change significantly from station to station. Figures 8 and 9 show the latitudinal and longitude dependences of the statistical characteristics of RMSE and MARE for the architectures indicated inside the graphs. Recall that for all architectures, the indices 10Kp and Np were also used as input parameters.

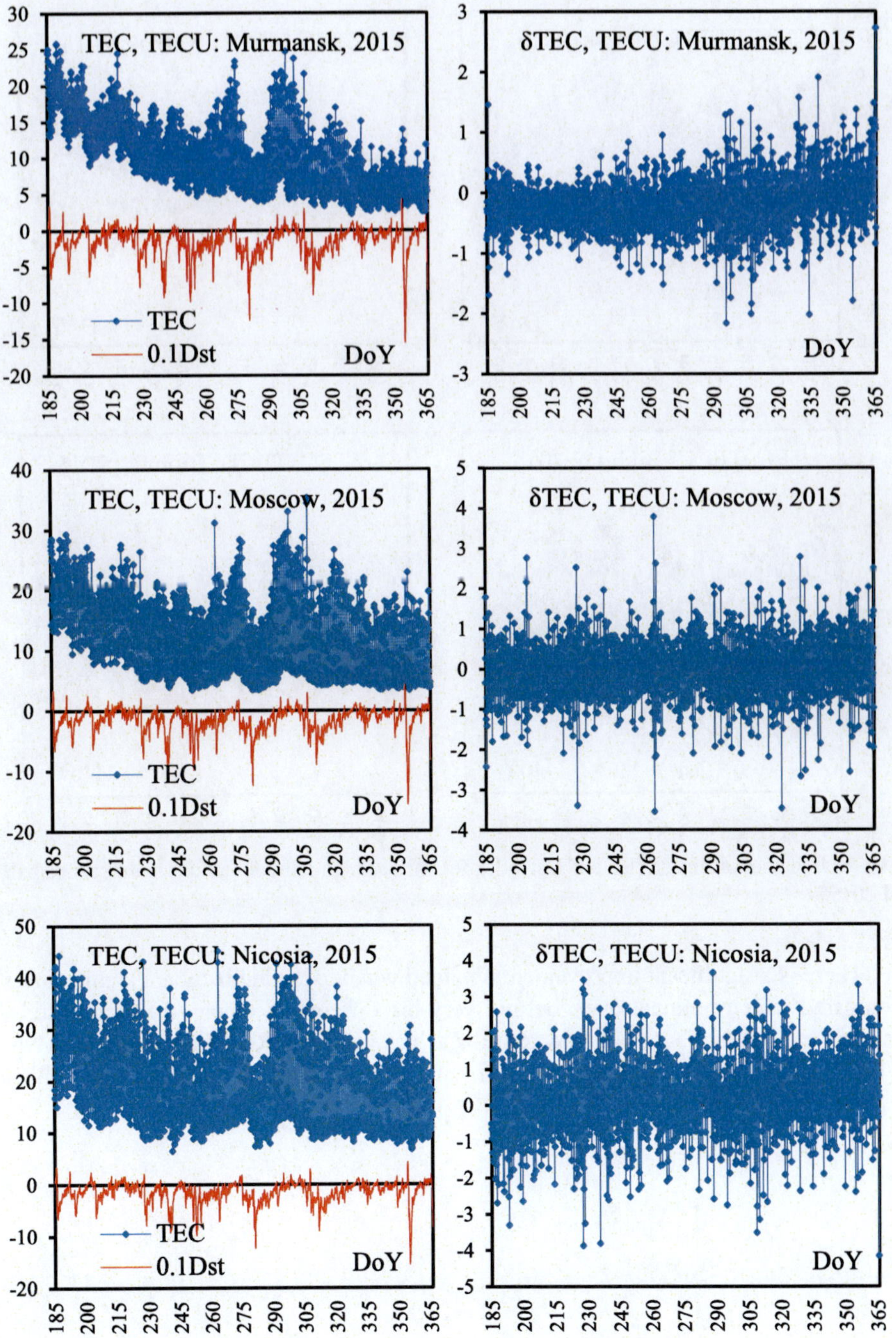

Fig. 6. Behavior of TEC, δTEC and Dst index for different stations along the 30° E meridian in the second half of 2015.

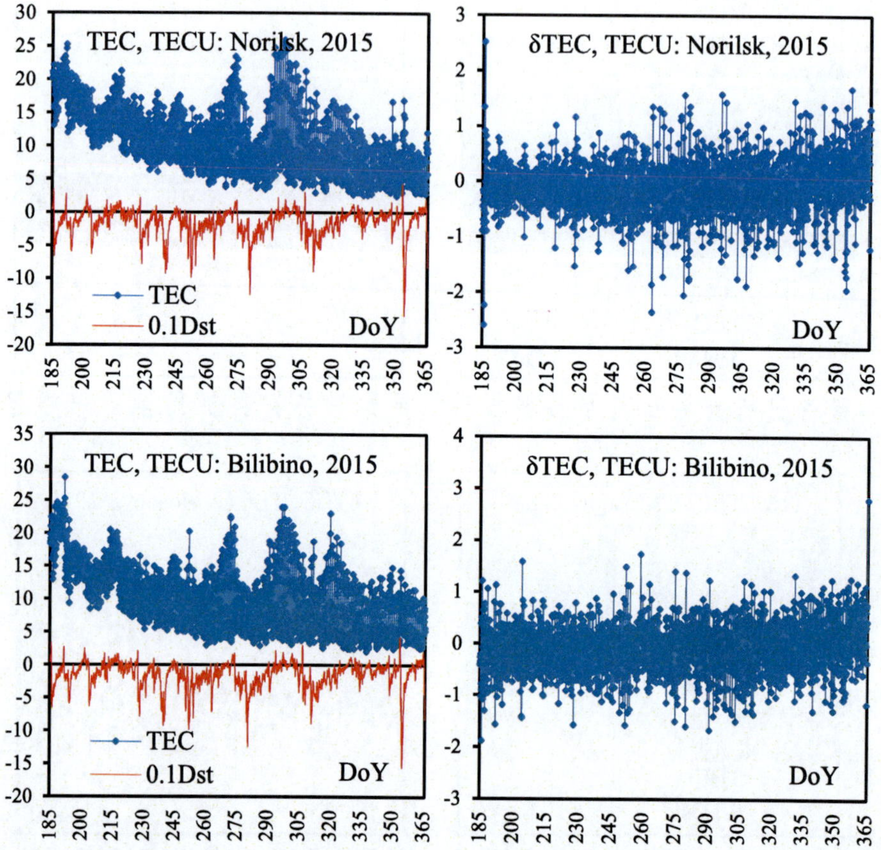

Fig. 7. Behavior of TEC, δTEC and Dst index for various high-latitude stations in the second half of 2015.

There is a significant improvement obtained when using bidirectional architectures compared to conventional ones, but not very big differences among the architectures in each respective group. The best result is achieved by BiGRU-RFF using 10Kp, Np, F10.7, Dst indexes as input parameters. In quantitative terms, there is an improvement by a factor of two.

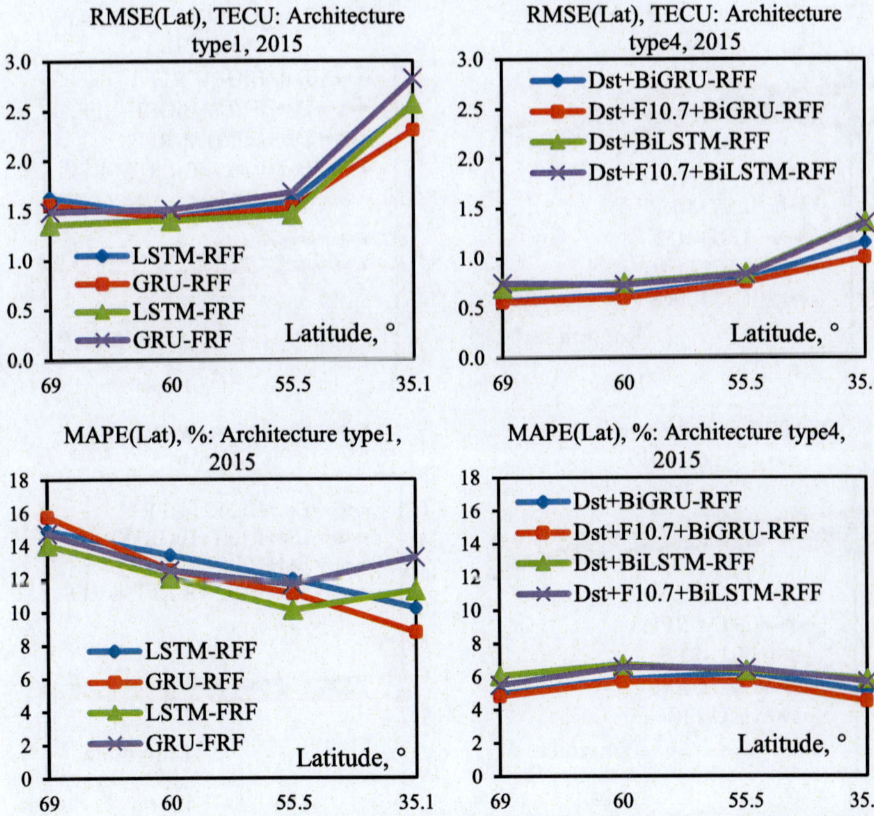

Fig. 8. Statistical characteristics of forecasting methods for stations along the meridian 30° E.

For conventional architectures RMSE in dependence of latitude varies from 2.9 TECU to 1.5 TECU, MAPE changes from 8% to 16%. The employment of bidirectional architecture leads to the following estimates: RMSE changes from 1.4 TECU to 0.5 TECU, MARE changes from 6% to 4%. For the longitudinal dependence - the initial RMSE varied from 1.6 TECU to 1.2 TECU, MAPE varied from 16% to 12%. Bidirectional architectures allowed to decrease values for RMSE to 0.6–0.8 TECU and for MAPE to 4–6%.

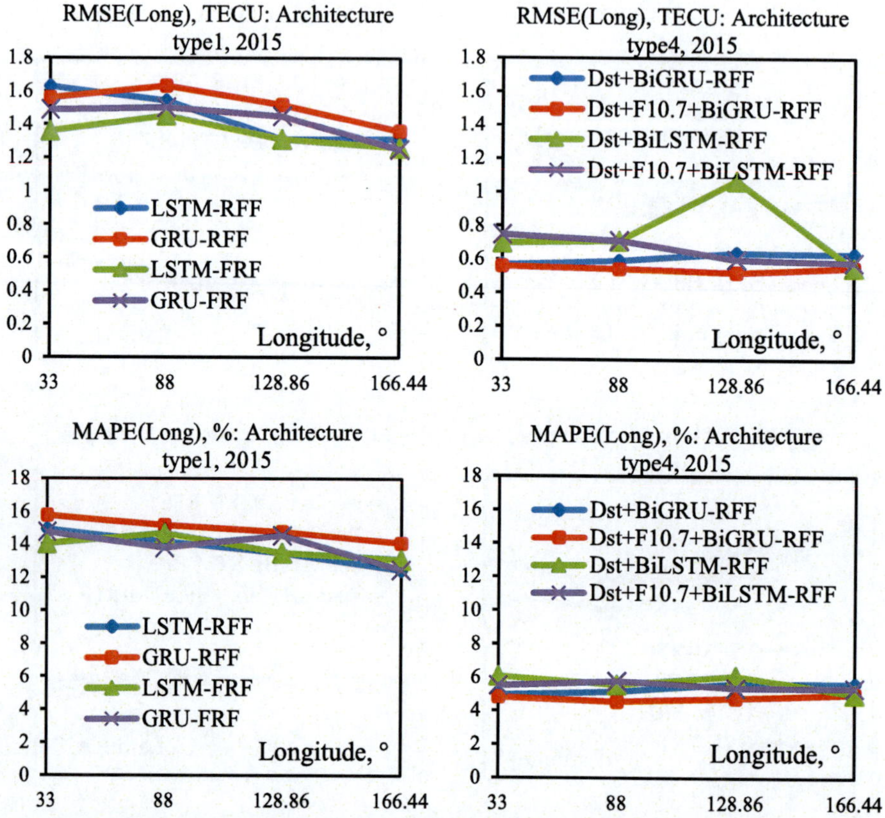

Fig. 9. Statistical characteristics of forecasting methods for high-latitude stations.

4 Conclusions

The total electron content (TEC) of the ionosphere is one of the most important parameters for ensuring the operation of various GPS-based technological systems. So far, only GIM IGS TEC maps can really provide a global scale of TEC variations over time with their continuity and independence from differences in TEC determination methods used at individual stations or regional networks of GPS receivers. Global coverage is their advantage, especially in places where there are no GPS receivers. Therefore, in this work, GIM JPL maps were used to predict TEC. The efficiency of 8 architectures of multilayer neural networks was studied with additional involvement of solar and geomagnetic activity indices. The graphs clearly illustrate the dependence of RMSE, MAPE of TEC forecasts for the latitudinal and longitude chains of Russian stations.

The best results were obtained for the BiGRU-RFF architecture using the indices 10Kp, Np, F10.7, Dst as the input parameters. In quantitative terms, there is an improvement by a factor of two. Without bidirectional processing, RMSE changes from 2.9 TECU to 1.5 TECU with latitude increase, MAPE changes from 8% to 16%. The use of the bidirectional architecture leads to the following estimates in dependence of latitude: RMSE changes from 1.4 TECU to 0.5 TECU, MAPE changes from 6% to 4%. With the longitude increase, the RMSE for the initial architectures changed from 1.6 TECU to 1.2 TECU, MAPE changed from 16% to 12%. Bidirectional architectures allowed to reduce RMSE to 0.6–0.8 TECU, MAPE to 4–6%.

Comparison of the results with literature data shows that, possibly, the bidirectional architecture acts as an arbiter in relation to the maximum TEC forecast accuracy achievable in practice due to the presence of unpredictable variations in the state of the ionosphere.

Acknowledgements. The TEC samples data was obtained from ftp://cddis.gsfc.nasa.gov/pub/gps/products/ionex/. Data on solar and geomagnetic activity was taken from SPDF OMNIWeb Service http://omniweb.gsfc.nasa.gov/form/dx1.html. The research was financially supported by Ministry of Science and Higher Education of the Russian Federation (State task in the field of scientific activity 2023).

References

1. Goodman, J.M.: Operational communication systems and relationships to the ionosphere and space weather. Adv. Space Res. **36**, 2241–2252 (2005). https://doi.org/10.1016/j.asr.2003.05.063
2. Natras, R., Soja, B., Schmidt, M.: Ensemble machine learning of random forest, AdaBoost and XGBoost for vertical total electron content forecasting. Remote Sens. **14**, 3547 (2022). https://doi.org/10.3390/rs14153547
3. Liu, L., Zou, S., Yao, Y., Wang, Z.: Forecasting global ionospheric TEC using deep learning approach. Space Weather **18**, e2020SW002501 (2020). https://doi.org/10.1029/2020SW002501
4. Ren, X., Yang, P., Liu, H., Chen, J., Liu, W.: Deep learning for global ionospheric TEC forecasting: different approaches and validation. Space Weather **20**, e2021SW003011 (2022). https://doi.org/10.1029/2021SW003011
5. Lei, D., et al.: Ionospheric TEC prediction base on attentional BiGRU. Atmosphere **13**, 1039 (2022). https://doi.org/10.3390/atmos13071039
6. Kharakhashyan, A., Maltseva, O., Glebova, G.: Forecasting the total electron content TEC of the ionosphere using space weather parameters. In: 2021 IEEE International Conference on Wireless for Space and Extreme Environments (WiSEE), pp. 31–36. (2021). https://doi.org/10.1109/WiSEE50203.2021.9613829
7. Rukundo, W.: Ionospheric electron density and electron content models for space weather monitoring. In: Magnetosphere and Solar Winds, Humans and Communication 2022, pp. 2–21 (2022). http://dx.doi.org/10.5772/intechopen.103079
8. Hochreiter, S., Schmidhuber, J.: Long short-term memory neural computation. Neural Comput. **9**(8), 1735–1780 (1997). https://doi.org/10.1162/neco.1997.9.8.1735
9. Cho, K., et al.: Learning phrase representations using RNN encoder-decoder for statistical machine translation. arXiv:1406.1078v3, [cs.CL] 3, (2014). https://arxiv.org/pdf/1406.1078.pdf
10. Glorot, X., Bengio, Y.: Understanding the difficulty of training deep feedforward neural networks. In: Proceedings of the Thirteenth International Conference on Artificial Intelligence and Statistics, Sardinia, Italy. AISTATS, pp. 249–356 (2010)
11. Saxe, A.M., McClelland, J.L., Ganguli, S.: Exact solutions to the nonlinear dynamics of learning in deep linear neural networks. arXiv preprint arXiv:1312.6120 (2013)

Convolutional Neural Network (CNN) of Resnet-50 with Inceptionv3 Architecture in Classification on X-Ray Image

Muhathir[1,2], Muhammad Farhan Dwi Ryandra[1,2], Rahmad B. Y. Syah[1,2(✉)], Nurul Khairina[1,2], and Rizki Muliono[1,2]

[1] Informatics Department, Faculty of Engineering, Universitas Medan Area, Medan, Indonesia
rahmadsyah@uma.ac.id

[2] Excellent Centre of Innovation and New Science, Universitas Medan Area, Medan, Indonesia

Abstract. The condition, which has recently become a hot topic of debate among people all over the world, was first discovered in the city of Wuhan, Hubei Province, China, and was later reported to the World Health Organization, namely the WHO (World Health Organization), on December 31, 2019 as an outbreak of a respiratory disease. It is caused by the virus SARS Corona Virus 2. (SARS-CoV-2). X-ray images can be used to detect COVID-19-infected lungs using a Convolutional Neural Network (CNN). The resnet-50 architecture and Inception v3 deep learning model were used in this research. This study employs several parameters, including epoch, batch size, and optimizer. The resnet-50 and inception v-3 architectures were utilized to optimize results. According to the results, the resnet-50 and inception v-3 architectures had the best epochs of 25, the best batch was 200, and the best optimizer was Adam. Overall, the most optimal hyperparameter is in epoch 25, batch 200 optimizer Adam with 99% accuracy and a computation time of 6 h for Inception v-3 and 9 h 21 min for resnet-50. The most optimal architecture in the classification of covid-19 based on x-ray images is the Inception v-3 architecture.

Keywords: Resnet-50 · Inception v3 · Classification · CNN · Covid

1 Introduction

The disease, which has recently become a hot topic of discussion for people all over the world, was first discovered in the city of Wuhan, Hubei Province, China, and was later reported to the World Health Organization, namely WHO (World Health Organization), on December 31, 2019 as an outbreak of a disease that infects the respiratory tract and is caused by a virus called Severe Acute Respiratory Syndrome Corona Virus 2 (SARS-CoV-2) [1, 2]. However, thousands of people in China, including many provinces (such as Hubei, Zhejiang, Guangdong, Henan, Hunan, and so on) died in the following month (January).

This virus can infect both humans and animals, with symptoms similar to MERS and SARS such as difficulty breathing, coughing, and fever, but Covid19 is spreading faster

R. Silhavy and P. Silhavy (Eds.): CSOC 2023, LNNS 724, pp. 208–221, 2023.
https://doi.org/10.1007/978-3-031-35314-7_20

[2, 3]. Corona Virus Disease 19, also known as COVID19, is a virus that has recently spread almost everywhere in the world, including Indonesia [4, 5]. Because of its rapid spread, this virus is known to be extremely dangerous [6].

The diagnosis of COVID19 is thought to be very similar to the diagnosis of SARS-CoV, because patients infected with COVID19 are likely to require hospitalization for isolation, and there is also a chance that patients infected with COVID19 will die. So far, COVID-19 testing has been done using a PCR (Polymerase Chain Reaction) test and a swab test on the respiratory tract [7], but this method takes time and does not provide an accurate diagnosis [7, 8].

Detection of covid-19 using x-rays or x-rays is a feasible solution, such as in research with deep residual networks [9] and also in research on the detection of the corona virus in X-Ray images [10]. This study demonstrated that classification of covid-19 using x-ray images is possible.

The CNN (Convolutional Neural Network) method will be used to classify covid-19 in this study. CNN is a type of neural network that is frequently used for image data and is frequently used to detect and recognize objects in images [11]. CNN is one of the deep learning algorithms that has been demonstrated in the pneumonia classification study [12] with an accuracy of 89.58%.

There are two architectures in the CNN (Convolutional Neural Network) method: ResNet and Inception. There are various types of resnet in the resnet architecture, ranging from 18, 34, 50, 101 to 152 layers. In this study, resnet-50 is used, and there are several types of inceptions in the inception architecture, including inception V-1, V-2, V-3, and V-4. Inception V-3 is used in this study because it is a popular transfer learning model [13, 19].

Previous research studies on the resnet-50 architecture, specifically fingerprint classification using resnet-50, discovered that the resnet-50 architecture has a 99% accuracy rate. This resnet-50 architecture has a 90% accuracy rate, and the pneumonia classification using resnet-50 has a 96% accuracy rate [13, 14].

Additionally, it was discovered that the v-3 inception architecture has an accuracy rate of 94% in earlier research investigations on the inception v3 architecture, namely the classification of cat races [12, 14]. Inception v3 architecture obtains an accuracy rate of 93% when used in animal categorization research and 93% when used in the performance of a deep learning model to detect the use of masks [14, 15].

This architecture was selected because it performed well in the ILSVRC competition [15]. The ILSVRC is an annual competition for image classification using various CNN architectures.

In order to assess performance, accuracy, and execution time between the two architectures, this study seeks to compare the two CNN deep learning models. This study should be able to demonstrate which of the two architectures is best for categorizing images [16–18].

2 State of the Art

The research methodology consists of gathering data, developing a classification model, training a classification model, testing, and performance evaluation. The first game explains the research methodology (Fig. 1).

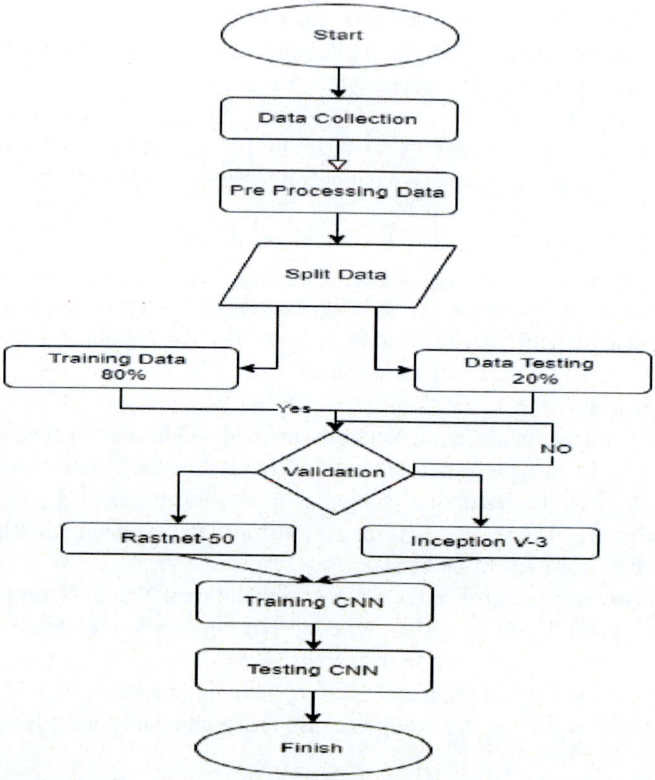

Fig. 1. Stage of the art Classification Model, Training Classification Model, Testing, And Performance Evaluation.

2.1 Dataset

The data for this study was obtained from an open source containing covid19 data via the https://www.kaggle.com/prashant268/chest-xray page from this source, and the combined dataset was then combined with two classes, namely covid19 and non-covid19 (Fig. 2).

The image is pre-processed before running the classification using the resnet-50 architecture and inception v3. The pixel sizes of the collected covid and negative covid images vary. As a result, the next step is to change the original image's pixel size so that each image is the same size, which is 224 224 pixels.

2.2 Model Deep Learning

The following step is to create a training model. In this study, I used the transfer learning method, which involved training a model that had previously been trained. An architectural overview of Inception v-3 and resnet-50 is provided below [3, 19].

Inception v-3 was built primarily on the Convolutional Neural Network (CNN) architecture. Various procedures/steps are performed in the Inception-V3 architecture shown

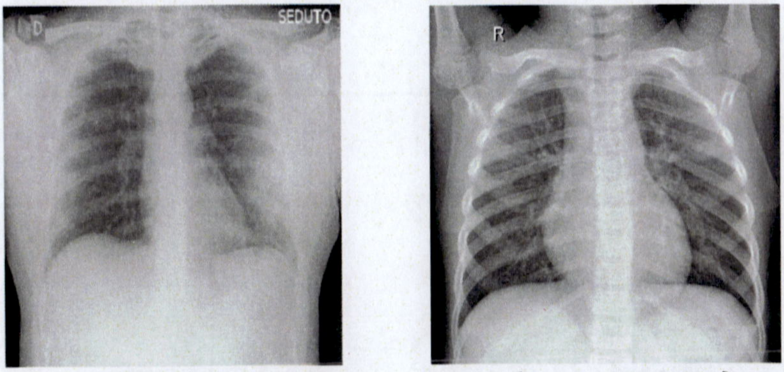

(1 Condition Covid) (2 condition normal)

Fig. 2. Condition figure show the dataset examples from each class.

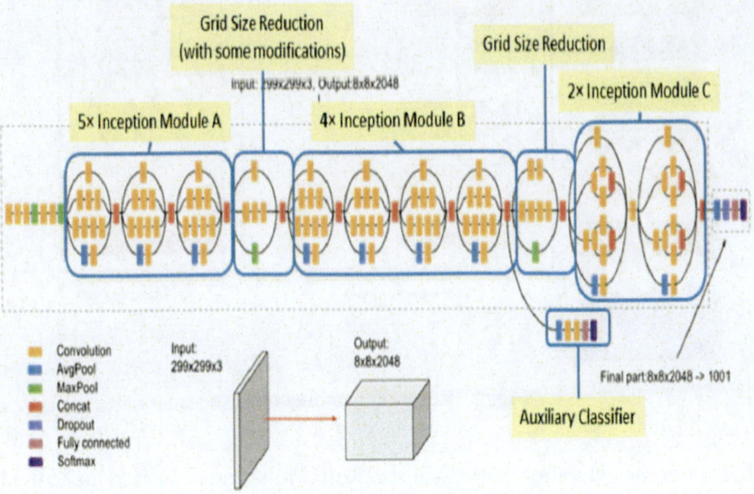

Fig. 3. Architecture Inception V-3

in Fig. 3, including convolution, Average Pool, MAXPOOL, dropout, fully connected, and SOFTMAX. The same steps were used in this study. The activation function, however, is SOFTMAX. Because the classification used in this study is a two-class classification, also known as a binary classification [1–3].

Following the pre-processing stage, the image is sent to the Inception v3 architecture for model training. This architecture is a 42-layer deep convolutional neural network architecture (Fig. 4).

After pre-processing, the image will be sent to the ResNet-50 architecture for model training. The 224 x 224-pixel image from the input layer is convoluted to the convolution layer in the first stage with a filter size of 7 7 and stride 2 [19]. Convolution results in a feature map, which is then normalized using Batch Normalization. The normalization

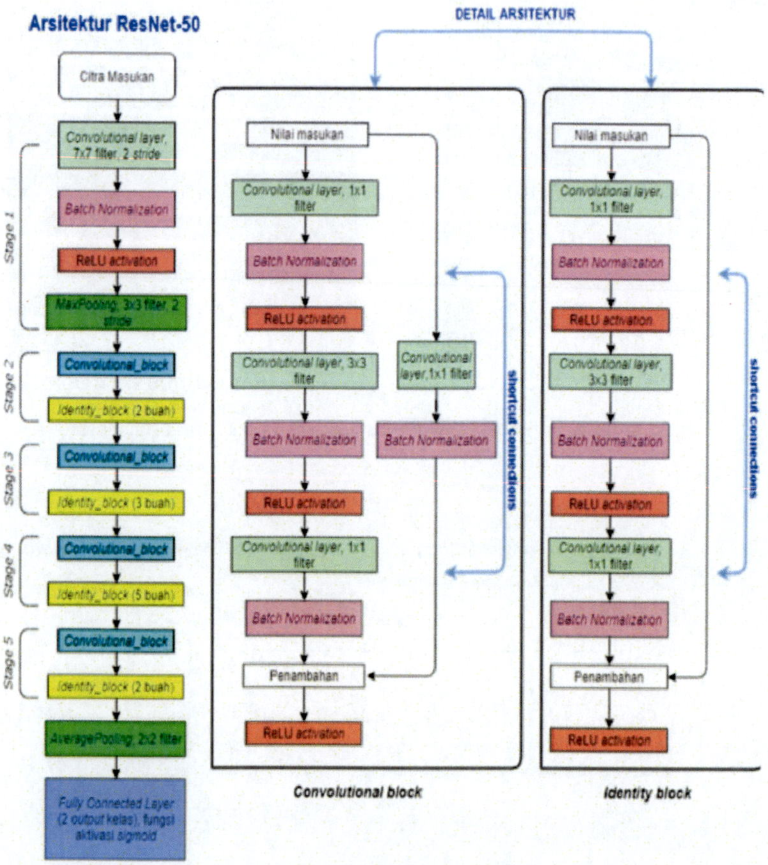

Fig. 4. Resnet-50 architecture.

results then enter the activation layer, where a ReLU function is used to make the feature extraction results non-linear. Several parameter values are initialized during the training process, including the number of epochs and batch size. After running the model through the training process, the model is tested by classifying the image to fit its class [2–4].

2.3 Evaluation Metrics

Accuracy, precision, recall, and F1-score are the metrics used to evaluate this model. To begin understanding the metrics, we will define TP (True Positive), TN (True Negative), FP (False Positive), and FN (False Negative), as shown in Table 1. Positive data that is predicted to be positive is referred to as TP, while negative data that is predicted to be negative is referred to as TN. While FN is the inverse of TP (positive data predicted to be negative), FP is the inverse of TN (negative data predicted to be positive) [4–6].

The formula for calculating accuracy, precision, recall, and F1 is shown below.

$$Accuracy = \frac{TP + TN}{TP + TN + FP + FN} \tag{1}$$

Table 1. Confusion Matrix

		Original Class	
		Positive	False Positive
	Positive	True Positive	False Positive
Prediction	Negative	False Negative	True Negative

$$Precision = \frac{TP}{TP + FP} \tag{2}$$

$$Recall = \frac{TP}{TP + FN} \tag{3}$$

$$F1Score = 2x\frac{PrecisionxRecall}{Precision + Recall} \tag{4}$$

Explanation:
TP = *true positive.*
FP = *false positive.*
FN = *false negative.*
TN = *true negative.*

2.4 Data Sharing and Hyperparameter

The data used in this study were 2159 lung x-ray images divided into two classes, Covid19 and normal. The Covid19 data set is 576, while the normal data set is 1583. There is training data and test data with a percentage of 80% and 20%, respectively. The Tables 2 and 3 below shows the distribution of the amount of image data in detail.

Table 2. Data Sharing

Category Data	Amount Data
Normal (Train)	1266
Normal (Test)	317
Covid (Train)	460
Covid (Test)	116

Table 3. Hyperparameter

Number	Name	Amount
1	Epoch	(5, 10, 25)
2	Batch	(100, 150, 200)
3	Optimizer	(Adam, RMSprop)

3 Result and Discussion

3.1 Result

3.1.1 Architecture Resnet-50

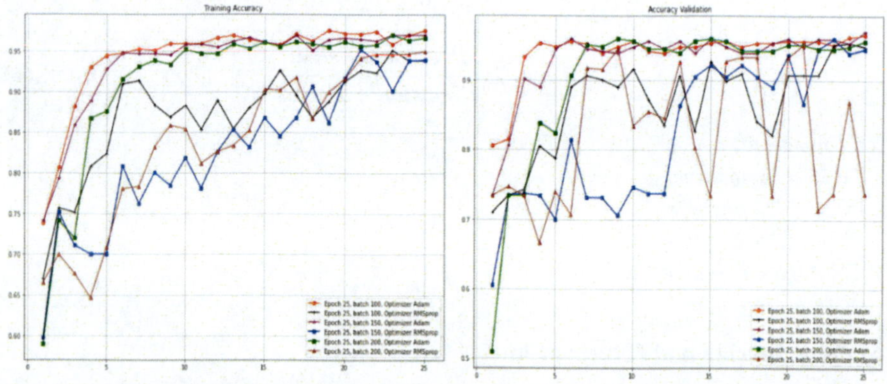

(1) **(2)**

Fig. 5. Resnet-50 Epoch 25 Training Accuracy (1) and Resnet-50 Epoch 25 Accuracy Validation (2)

The graph in **(1)** Fig. 5 above shows that the training accuracy of the resnet-50 architecture using the RMSprop optimizer increases despite fluctuations or fluctuations, with the best accuracy training in epoch 25 using the Adam optimizer, where the graph continues to rise throughout the epochs. And in **(2)** Fig. 5, this is the accuracy validation graph of the resnet-50 architecture, where the best graph is produced by the Adam optimizer, as seen in the graph above Fig. 5.

The loss accuracy of the resnet-50 architecture is depicted in Fig. 6. The graph above shows that the Adam optimizer is very stable, sloping down epoch by epoch, as opposed to the RMSprop optimizer, where the graph still looks up and down. And in Fig. 7, the best graph indicated by the Adam optimizer is a graph of the loss validation of the resnet-50 architecture with 25 epochs.

3.1.2 Inception V3 Architecture

The graph in **(1)** Fig. 8 shows that the training accuracy of the Inception v-3 architecture using the RMSprop optimizer increases despite fluctuations. The best accuracy training

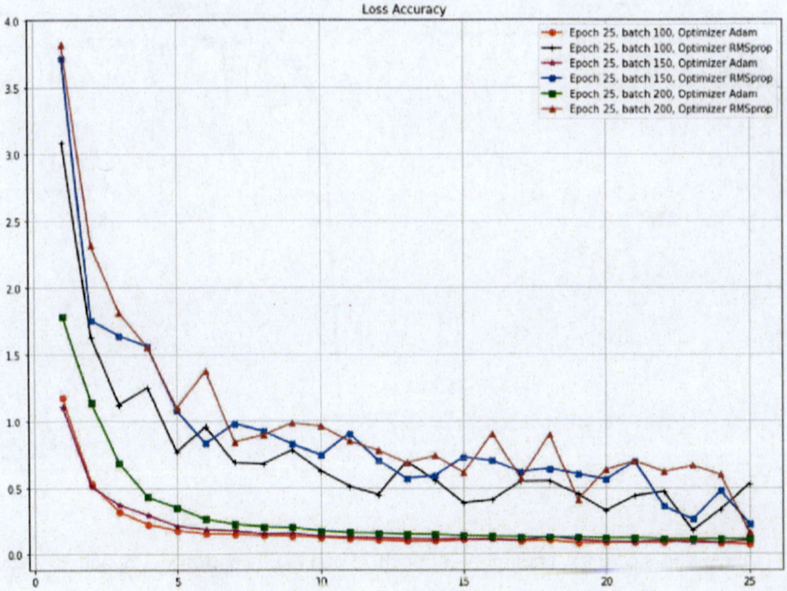

Fig. 6. Resnet Accuracy Loss -50 Epoch 25

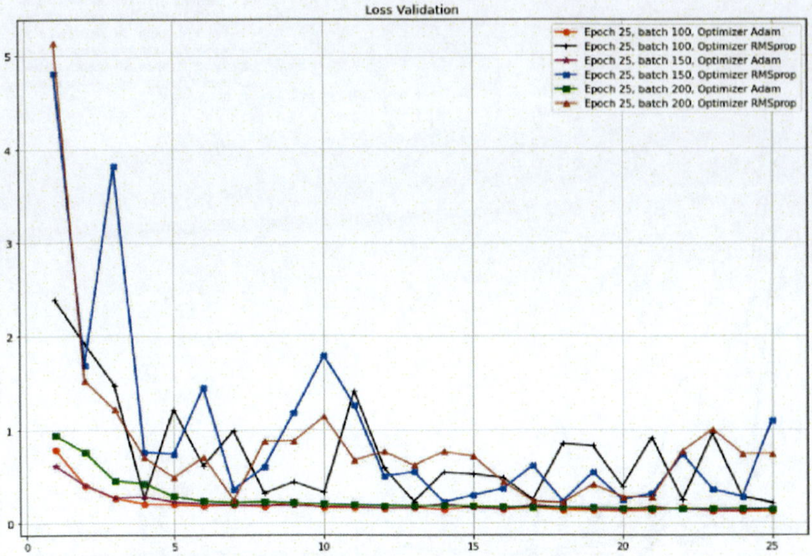

Fig. 7. Resnet-50 Loss Validation Epoch 25

in epoch 25 is using the Adam optimizer, where the stable graph continues to rise throughout the epoch. And shown in **(2)** Fig. 8 is the accuracy validation graph of the Inception v-3 architecture, where the best graph is produced by the Adam optimizer and the graph appears stable [19, 20].

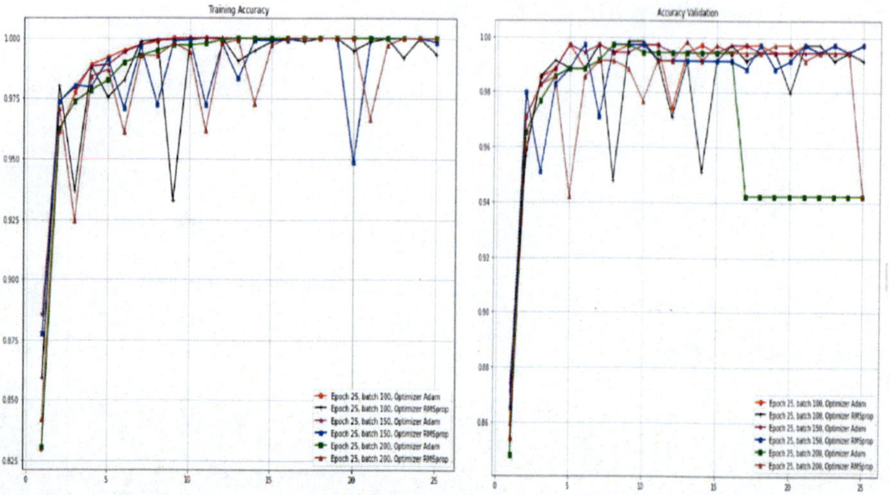

(1) **(2)**

Fig. 8. (1) Inception Accuracy Training v-3 Epoch 25 and (2) Inception v-3 Epoch 25 Accuracy Validation.

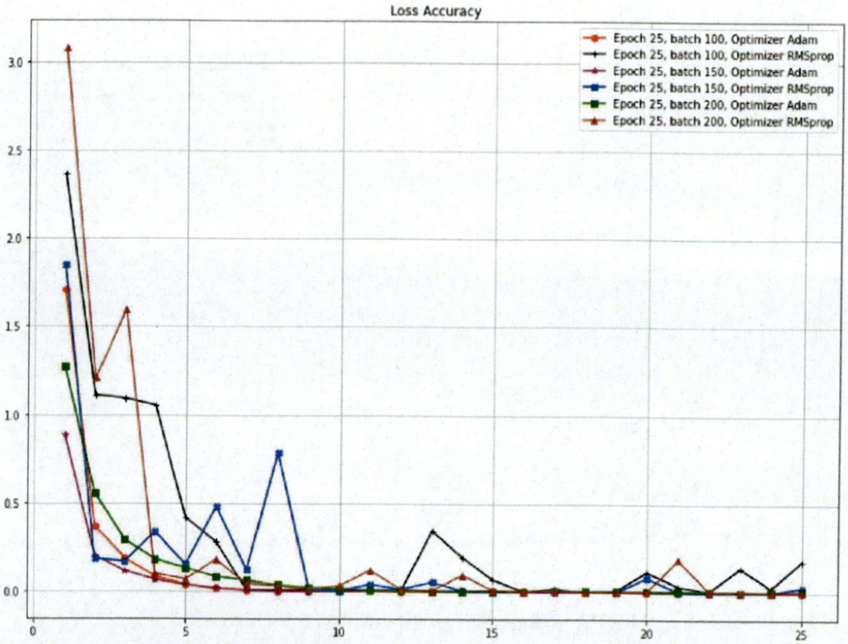

Fig. 9. Inception v-3 Loss Accuracy Epoch 25

Figure 9 shows a graph of the loss of accuracy of the Inception v-3 architecture with epoch 25. All graphs appear to be sloping down, but Adam's optimizer shows the best

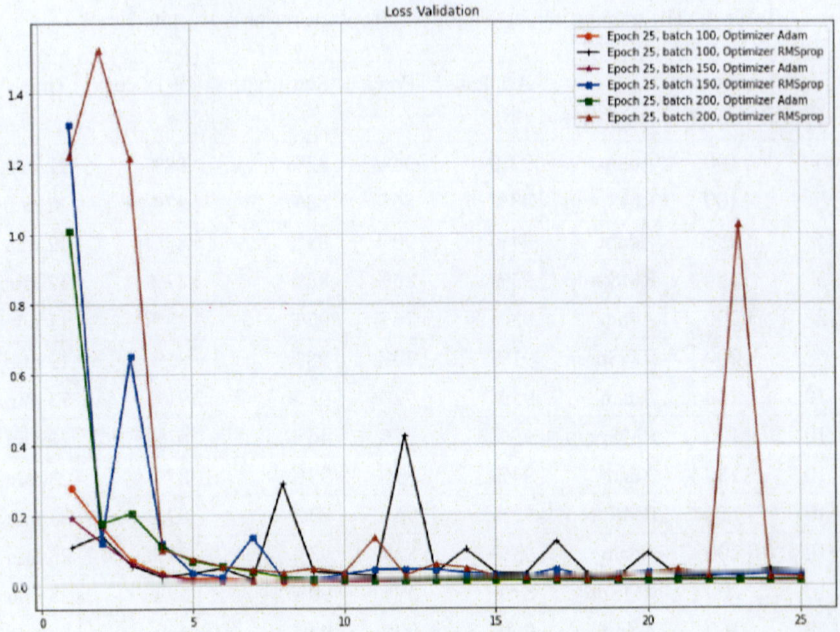

Fig. 10. Inception Validation Loss v-3 Epoch 25

graph. And in Fig. 10, the best graph is shown by the Adam optimizer, which is stable and decreasing, for the loss validation of the Inception v-3 architecture over 25 epochs.

3.1.3 Training Results and Accuracy for the Resnet-50 Architecture

Training results and accuracy as shown in the Table 4 below, the best epoch is 25, the best batch is 200, and the best optimizer is adam. Overall, the most optimal hyperparameter with 99% accuracy is in epoch 25 condition, batch 200 optimizer adam.

3.1.4 Training and Accuracy Results for the Inceptionv3 Architecture

Training and accuracy results As shown in the Table 5 below, the best epoch is 25, the best batch is 200, and the best optimizer is Adam. Overall, the most optimal hyperparameter with 99% accuracy is in the epoch 25 condition, batch 200 optimizer Adam.

The Table 6. Above shows the average classification of deep learning models, with the inception v-3 architecture having a 98% higher accuracy value than resnet-50 (92.3%). In terms of precision, the inception v-3 architecture outperforms resnet-50. This is 95.2%, while resnet-50 is 86.4%. The inception architecture also had higher sensitivity and f-1 scores, 97.3% and 96.1%, respectively, while resnet-50 was 88.7% and 86.6%, respectively. In terms of computing time, the Inception v-3 architecture outperforms the resnet-50 architecture, which takes up to 9 h and 21 min.

The Table 7. Above compares the best hyperparameters for conducting model training. The hyperparameter architecture used in Inception v-3 is epoch 25, batch 200 optimizer adam with 99% accuracy, 98% precision, 98% sensitivity, and 98% F-1 Score and

Table 4. The training and validation results of the resnet-50 architecture:

N	Hyperparameter			Accuracy	Precis	Sensitivities	F-1 Score	Time
	Epoch	Batch	Optimizer					
1	5	100	Adam	93%	94%	82%	88%	12 min
2	5	100	RMSprop	93%	86%	88%	87%	12 min
3	5	150	Adam	94%	90%	88%	89%	12 min
4	5	150	RMSprop	93%	86%	88%	87%	12 min
5	5	200	Adam	93%	76%	97%	85%	11 min
6	5	200	RMSprop	91%	70%	98%	85%	12 min
7	10	100	Adam	95%	92%	91%	91%	23 min
8	10	100	RMSprop	69%	97%	46%	62%	23 min
9	10	150	Adam	94%	83%	93%	87%	23 min
10	10	150	RMSprop	93%	96%	83%	89%	23 min
11	10	200	Adam	93%	82%	93%	87%	23 min
12	10	200	RMSprop	87%	57%	90%	69%	23 min
13	25	100	Adam	96%	92%	94%	96%	57 min
14	25	100	RMSprop	93%	96%	83%	89%	57 min
15	25	150	Adam	95%	88%	94%	90%	57 min
16	25	150	RMSprop	93%	79%	94%	85%	57 min
17	25	200	Adam	99%	98%	98%	98%	57 min
18	25	200	RMSprop	98%	94%	98%	96%	59 min

a computation time of 28 min. While in the resnet-50 hyperparameter architecture, epoch 25 is used, with 99% accuracy, 98% precision, 98% sensitivity, and 98% F-1 Score and a computation time of 57 min. Table 5 shows that the Inception v-3 architecture performs the best in classifying covid 19 based on x-ray images.

3.2 Discussion

The image data used in this study is the result of x-ray images in.jpg format, with 224 x 224-pixel x-ray images, and the deep learning model used is the Inception V-3 and Resnet-50 architecture. Contribution to research or renewal in comparison to other research is that several parameters are used, such as using three Epochs, namely 5, 10, and 25, batch sizes of three, namely 100, 150, and 200, and two optimizers, namely Adam and RMSPROP. Furthermore, it employs more image data than previous research.

The data used in this study is secondary data rather than primary data from the Kaggle platform, and the parameters used are three epochs, namely 5, 10, and 25. The batch sizes are 100, 150, and 200, and the parameters used are Adam and RMSPROP. Another flaw is the lengthy computation time required during the training process.

The implications for future research are as follows Table 8:

Table 5. The result of the Inception V-3 architecture's training and validation.

N	Hyperparameter			Accuracy	Hyperparameter	Sensitivities	F-1 Score	Time
	Epoch	Batch	Epoch					
1	5	100	Adam	98%	98%	96%	96%	8 min
2	5	100	RMSprop	98%	97%	97%	97%	8 min
3	5	150	Adam	98%	98%	96%	96%	8 min
4	5	150	RMSprop	98%	95%	98%	96%	8 min
5	5	200	Adam	98%	96%	96%	96%	8 min
6	5	200	RMSprop	95%	86%	96%	90%	8 min
7	10	100	Adam	98%	95%	98%	96%	15 min
8	10	100	RMSprop	98%	96%	92%	96%	16 min
9	10	150	Adam	98%	94%	98%	96%	15 min
10	10	150	RMSprop	98%	93%	100%	96%	15 min
11	10	200	Adam	97%	93%	98%	95%	17 min
12	10	200	RMSprop	98%	96%	96%	96%	15 min
13	25	100	Adam	98%	94%	98%	96%	40 min
14	25	100	RMSprop	99%	96%	100%	98%	40 min
15	25	150	Adam	99%	96%	100%	98%	39 min
16	25	150	RMSprop	98%	95%	98%	96%	38 min
17	25	200	Adam	99%	98%	98%	98%	28 min
18	25	200	RMSprop	99%	98%	98%	98%	28 min

Table 6. Average deep learning model classification comparison

Name Architectures	Accuracy	precision	Sensitivities	F-1 Score	Time
Inception v-3	98%	95,2%	97,3%	96,1%	6 h
Resnet-50	92,3%	86,4%	88,7%	86,6%	9 Jam 21 min

Table 7. Hyperparameter Optimal Training Model Comparison

Name Architectures	Accuracy	precision	Sensitivities	F-1 Score	Time
Inception v-3	99%	98%	98%	98%	28 min
Resnet-50	99%	98%	98%	98%	57 min

4 Conclusion

Based on the results of testing data on thousands of x-ray images of normal and covid-19 lungs with inception v-3 and resnet-50 architecture, it can be concluded that inception v-3 architecture has a 98% higher accuracy value, while resnet-50 has a 92.3% accuracy value. In terms of precision, the inception v-3 architecture outperforms resnet-50, scoring 95.2% versus 86.4%. The inception architecture score was also higher in terms of sensitivity and f-1, with 97.3% and 96.1%, respectively, while resnet-50 was 88.7% and 86.6%, respectively. In general, the two architectures perform very well in classifying lung x-ray images, but the best architecture in classifying covid-19 based on x-ray images is the inception v-3 architecture.

References

1. Koklu, M., Cinar, I., Taspinar, Y.S., Kursun, R.: Identification of Sheep Breeds by CNN-based pre-trained Inceptionv3 model. In: 2022 11th Mediterranean Conference on Embedded Computing (MECO), pp. 01–04. https://doi.org/10.1109/MECO55406.2022.9797214
2. Boldog, P., Tekeli, T., Vizi, Z., Dénes, A., Bartha, F.A., Röst, G.: Risk assessment of novel Coronavirus COVID-19 outbreaks outside China. J. Clin. Med. **9**(2), 571 (2020). https://doi.org/10.3390/jcm9020571
3. Çınar, A., Yıldırım, M., Eroğlu, Y.: Classification of pneumonia cell images using improved ResNet50 model. Traitement du Signal **38**(1), 165–173 (2021). https://doi.org/10.18280/ts.380117
4. Zhang, C., Li, J., Huang, J., Wu, S.: Computed tomography image under convolutional neural network deep learning algorithm in pulmonary nodule detection and lung function examination. J. Healthc. Eng. **2021**, 1–9 (2021). https://doi.org/10.1155/2021/3417285
5. Handayani, D., Hadi, D.R., Isbaniah, F., Burhan, E., Agustin, H.: Corona Virus disease 2019. Jurnal Respirologi Indonesia **40**(2), 119–129 (2020). https://doi.org/10.36497/jri.v40i2.101
6. Long, J., Rong, S.: Application of machine learning to badminton action decomposition teaching. Wirel. Commun. Mob. Comput. **2022**, 1 (2022). https://doi.org/10.1155/2022/3707407
7. Syah, R., Elveny, M., Nasution, M.K.M.: Clustering large dataset' to prediction business metrics. In: Silhavy, R., Silhavy, P., Prokopova, Z. (eds.) CoMeSySo 2020. AISC, vol. 1294, pp. 1117–1127. Springer, Cham (2020). https://doi.org/10.1007/978-3-030-63322-6_95
8. Hariyani, Y.S., Hadiyoso, S., Siadari, T.S.: "Deteksi Penyakit Covid-19 Berdasarkan Citra X-Ray Menggunakan deep residual network. ELKOMIKA: Jurnal Teknik Energi Elektrik Teknik Telekomunikasi, & Teknik Elektronika **8**(2), 443 (2020). https://doi.org/10.26760/elkomika.v8i2.443
9. Kusumawardani, R., Karningsih, P.D.: Detection and classification of canned packaging defects using convolutional neural network. PROZIMA (Prod. Optim. Manufact. Syst. Eng.) **4**(1), 1–11 (2021). https://doi.org/10.21070/prozima.v4i1.1280
10. Khairina, N., Sibarani, T.T.S., Muliono, R., Sembiring, Z., Muhathir, M.: Identification of pneumonia using the K-Nearest neighbors method using HOG Fitur feature extraction. J. Inform. Telecommun. Eng. **5**(2), 562–568 (2022). https://doi.org/10.31289/jite.v5i2.6216
11. Ula, M.-M., Sahputra, I.: Optimization of multilayer perceptron hyperparameter in classifying pneumonia disease through X-Ray images with speeded-up robust features extraction method. Int. J. Adv. Comput. Sci. Appl. **13**(10) (2022). https://doi.org/10.14569/IJACSA.2022.0131025

12. Li, X.-X., Li, D., Ren, W.-X., Zhang, J.-S.: Loosening identification of multi-bolt connections based on wavelet transform and ResNet-50 convolutional neural network. Sensors **22**(18), 6825 (2022). https://doi.org/10.3390/s22186825
13. Miranda, N.D., Novamizanti, L., Rizal, S.: Convolutional neural network Pada Klasifikasi Sidik Jari Menggunakan RESNET-50. Jurnal Teknik Informatika (Jutif) **1**(2), 61–68 (2020). https://doi.org/10.20884/1.jutif.2020.1.2.18
14. Syah, R., Al-Khowarizmi, A.-K.: Optimization of applied detection rate in the simple evolving connectionist system method for classification of images containing protein. Jurnal Ilmiah Teknik Elektro Komputer dan Informatika **7**(1), 154 (2021). https://doi.org/10.26555/jiteki.v7i1.20508
15. Nasution, M.K.M., Syah, R.: Data management as emerging problems of data science. In: Data Science with Semantic Technologies. Wiley, pp. 91–104 (2022). https://doi.org/10.1002/9781119865339.ch4
16. Muhathir, Al-Khowarizmi: Measuring the accuracy of SVM with varying Kernel function for classification of Indonesian Wayang on Images. In: 2020 International Conference on Decision Aid Sciences and Application (DASA), pp. 1190–1196, November 2020. https://doi.org/10.1109/DASA51403.2020.9317197
17. Ayan, E., Karabulut, B., Ünver, H.M.: Diagnosis of pediatric pneumonia with ensemble of deep convolutional neural networks in Chest X-Ray images. Arab. J. Sci. Eng. **47**(2), 2123–2139 (2022). https://doi.org/10.1007/s13369-021-06127-z
18. Sutrisno, S., Khairina, N., Syah, R.B.Y., Eftekhari-Zadeh, E., Amiri, S.: Improved artificial neural network with high precision for predicting burnout among managers and employees of start-ups during COVID-19 pandemic. Electronics **12**, 1109 (2023). https://doi.org/10.3390/electronics12051109
19. Guefrechi, S., Jabra, M.B., Ammar, A., Koubaa, A., Hamam, H.: Deep learning based detection of COVID-19 from chest X-ray images. Multimed. Tools Appl. **80**(21–23), 31803–31820 (2021). https://doi.org/10.1007/s11042-021-11192-5
20. Sahinbas, K., Ferhat, O.C.: Transfer learning-based convolutional neural network for COVID-19 detection with X-ray images. Data Science for COVID-19, pp. 451–466. Academic Press (2021). https://doi.org/10.1016/B978-0-12-824536-1.00003-4

Image Manipulation Using Korean Translation and CLIP: Ko-CLIP

Sieun Kim and Inwhee Joe(✉)

Department of Computer Science, Hanyang University, Seoul 04763, South Korea
{clown14,iwjoe}@hanyang.ac.kr
http://wm.hanyang.ac.kr/

Abstract. Deep Learning, a field of artistic intelligence (AI), is showing good results in natural language processing (NLP) and image processing classification. In the NLP field, in particular, the BERT-based model has become the main focus of the latest language model. It is a representative model that utilizes BERT pre-training and fine-tuning. Through the process of pre-training vast amounts of data and fine-tuning it, more natural NLP can be implemented. CLIP recently built a dataset with only web crawling without manual labeling to create a huge dataset that forms image-text pairs. With the CLIP Model, it tells you which image the input text is deeply related to. However, CLIP does not recognize Korean text when it is input, so it cannot accurately analyze it. In this paper, we propose to use the BERT Model of NLP and CLIP in the field of image processing to process images by receiving Korean text input. The Korean text is translated into English through the BERT Model and used as input text in the CLIP Model. The output that went through the two models reflected the contents of the Korean text. It can be seen that Output is related to the accuracy of Korean text.

Keywords: Computer Vision · Image Processing · Natural Language Processing · Machine Learning

1 Introduction

Among Deep Learning, an area of artistic intelligence (AI), Natural Language Processing (NLP) is a field that mainly processes and interprets text data using Machine Learning. NLP is used in Chatbot and translators. In particular, in recent years, automatic translator functions are sometimes added to messenger services to provide global services. In addition, the NLP Model is added to the Chatbot service so that the machine understands and understands the context of the user. Therefore, it is possible to provide a natural conversation between the user and the chatbot. Image Processing is the field that produces the most visual results in AI. In recent research, it has stood out in various fields by applying it to technologies such as drawing or distorting photographs using the GAN-based models. Among them, CLIP is a model that can distinguish images that are

R. Silhavy and P. Silhavy (Eds.): CSOC 2023, LNNS 724, pp. 222–230, 2023.
https://doi.org/10.1007/978-3-031-35314-7_21

most similar to the contents of the text by analyzing the input text. However, CLIP does not respond properly to input to Korean text and does not give a normal return. Therefore, in this paper, we would like to study the application of CLIP Model to sentences entered in Korean. In the process, the process of applying and translating the NLP model, the BERT Model, is included. I would like to conduct a study on the Korean text input in CLIP as follows:

- Since CLIP is a model that only supports English, it is not processed for Korean input, so it goes through the process of translating Korean into English through the BERT Model. Hangul should have contents about how to process images to be input.
- The Korean sentence input goes through tokenizing in the translation process. Since there is a difference in performance in the process of translating Hangul into English according to the tokenizing method, the translation is performed by adopting an appropriate tokenizing method.
- Put text translated from Korean to English as input value in CLIP Model. The CLIP Model analyzes the input English words or sentences. Identify the requirements or obejct of the sentence and return it.
- When the CLIP Model analyzes the text and returns the closest result, it reconstructs the requirement or object of the Korean text that was initially input on the input image through StyleGAN.

Sect. 2 of this paper explains the contents of existing studies, and Sect. 3 explains the process and method of the research. Subsequently, Sect. 4 analyzes the results generated in the research process. Finally, we discuss the conclusions in Sect. 5 and future research in Sect. 6.

2 Related Work

AI has recently become so common that it can be included in various ways. The field of language processing and image processing are being used in more fields, and the accuracy of the model has increased significantly. In addition, a larger data set is created and a pre-trained model that trains a vast amount of data sets is commonly used. A representative example is the BERT Model [3] in the NLP field. The BERT Model is a language model using Transformer. BERT first learns a vast amount of language data. Due to the vast amount of data, it is possible to perform better than the embeding methods used before the appearance of BERT. In addition, the BERT Model increases the probability value of the word through the Self-Attention operation.

In this paper, the requirements for deformation or distortion of images input in Korean are translated into English using the BERT Model. It is used as a Tokenizer for the translation module using the Tokenizer of the BERT Model. CLIP [5] is a model developed by OpenAI that is pre-trained with a large DataSet of imageNet [2], and can perform well by performing some additional learning like BERT. In this paper, we analyze the object and requirements for image transformation input into Korean and translated into English. CLIP recognizes

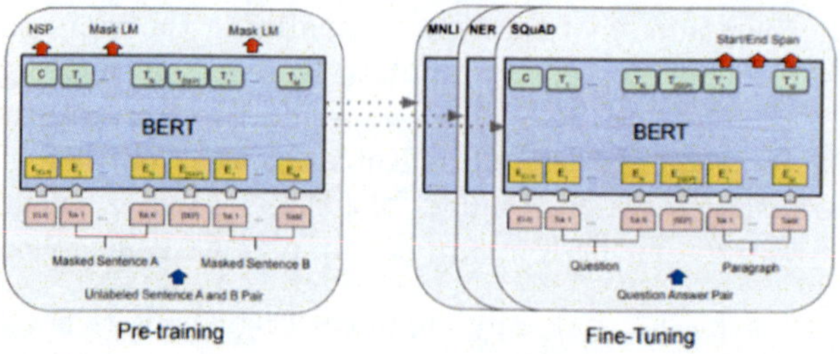

Fig. 1. BERT Modeling process. Pre-training and Fine-tuning [3]

in the image the object to be modified for the requirement. And the part that is highly related is recognized and returned. In the field of image processing, Style and GAN[1] have been actively studied recently. Image Style Transfer is a method of applying a style on top of an image and applying a painting style on top of a landscape photo as shown in the picture. Basically, we use the weights of CNN [4] to update the Noise image, creating a more similar style of painting. In this paper, Style performs the task of transforming the input image based on the result of the return from CLIP. CLIP does not support input to Hangul along with ImageNet. Therefore, there is a problem that the Korean input does not work properly when it comes in, and in this paper, we solved this part and conducted a study to respond to the Korean input.

3 Method

3.1 Structure

In this paper, a structure that reconstructs an image by processing a Korean text input is proposed as shown in Fig. 3. In Fig. 3, the Korean text, T, and image and I to be manipulated are received by input. The two go through different models. The Korean text is translated into English text, T' through the BERT model in order to be translated into English that can be recognized by the CLIP Model. Since the translated T' is an English text that the CLIP Model can recognize, it can be input together with Image and I. T' and I' are processed by the CLIP Model and return C, which is the return value of the CLIP Model. C contains information about under what conditions the image is transformed. C is once again entered as the input value of the StyleGAN Model along with Image I. It is changed according to the conditions of the Korean text and T received as input from the StyleGAN Model and is output as Return Image, I'.

3.2 BERT

The BERT Model is a model that pre-training a vast amount of data as described in Fig. 1. Additional learning can be done for Fine-tuning. In this paper, we intend to use BERT's Tokenizing method and translation module. The BERT

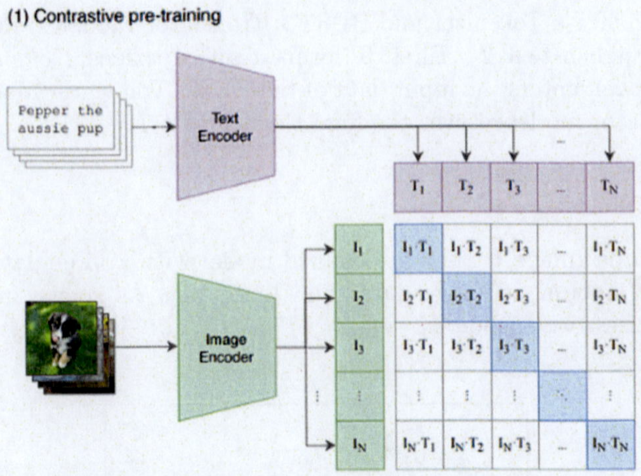

Fig. 2. CLIP Pre-training and Structure [5]

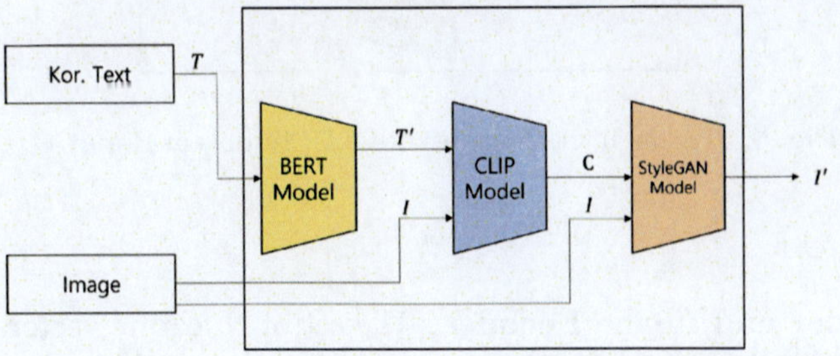

Fig. 3. CLIP Model processfor the Korean text input

Model can be said to be the sum of the three embedding values, Token Embedding, Segment Embedding, and Position Embedding, as described in Fig. 4. In this paper, we used BERT's Tokenizer and BERT's Translator Model to make Korean text T into English text $T^{'}$. First, T forms a set of tokens T_token via BERT Tokenizer. $T_tokenenters$ the input data of the BERT Translator Model, and the BERT Translator model returns the final English text $T^{'}$.

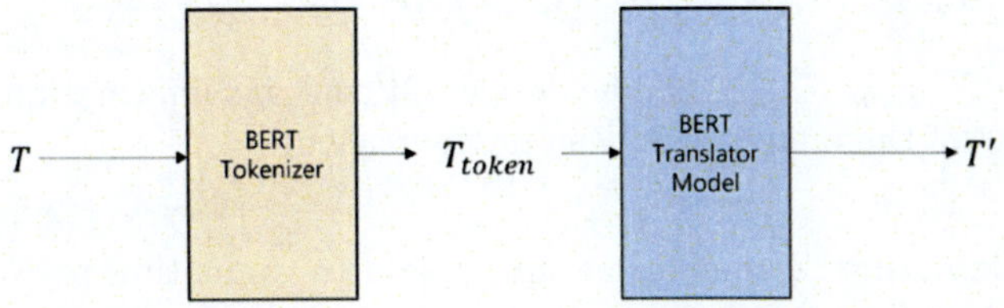

Fig. 4. Tokenizing and Translation process using BERT

We used BERT's Tokenizer and BERT's Translator Model to make Korean text T into English text $T^{'}$. First, T forms a set of tokens T_token via BERT Tokenizer. T_token enters the input data of the BERT Translator Model, and the BERT Translator model returns the final English text $T^{'}$.

3.3 CLIP

Existing research ImageNet is a pre-trained model with a large data set. CLIP used a larger amount of image data than ImageNet. As shown in Fig. 5 and Fig. 2, CLIP uses two encoders.

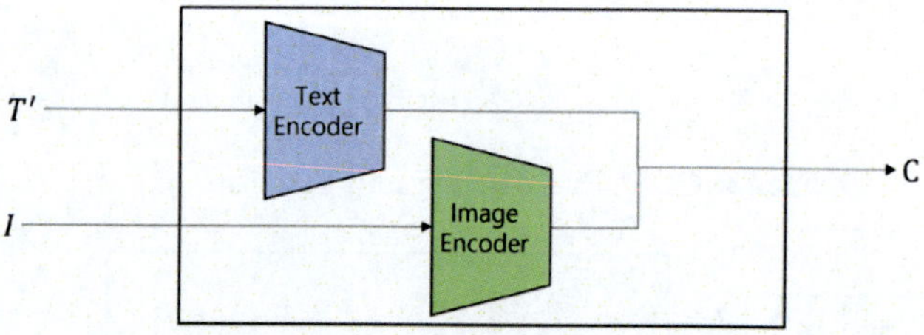

Fig. 5. Through the Encoder of CLIP to obtain Result C

Text Encoder and Image Encoder. The set of N feature vectors is derived by receiving the English text and $T^{'}$ as input. $T^{'}_{vector}$ may be expressed as:

$$T^{'}_{vector}=\{T^{'}_{1},T^{'}_{2},T^{'}_{3},\ldots,T^{'}_{N}\} \tag{1}$$

$T^{'}_{vector}$ learns the relationship with N image feature vectors derived through Image Encoder. The I_{vector} derived through Image Encoder can be expressed as follows:

$$I_{vector} = \{I_1,I_2,I_3,\ldots,I_N\} \tag{2}$$

I_{vector} and $T^{'}_{vector}$ use Transformer base's network to learn the relationship. Therefore, the learning of I_{vector} and $T^{'}_{vector}$ is shown as follows:

$$I_{vector} \cdot T^{'}_{vector}=\{I_1 \cdot T^{'}_{1},I_2 \cdot T^{'}_{1},I_3 \cdot T_1{'},\ldots I_N \cdot T^{'}_{N}\} \tag{3}$$

The result C of $I_{vector} \cdot T^{'}_{vector}$ can be derived. It contains information that shows which features of the image have been distinguished.

Zero-shot Learning Zero-shot learning refers to evaluating performance for problems or Data Sets that have never been seen in the learning process. In this paper, both unused and used data were measured to evaluate the performance of Zero-shot learning.

3.4 StyleGAN

As shown in Fig. 6, StyleGAN receives the data C and initial Input Image I received from the CLIP Model as input. StyleGAN uses the C received through CLIP as a Style value. Style mixing is a method of reducing the style correlation between adjacent layers so that styles are well localized on the image. The StyleGAN Model performs Style mixing to reflect the conditions of C. Therefore, put a vector for noise with a Style value on top of the input image I. In StyleGAN, C and image I are combined to return the resulting image, $I^{'}$.

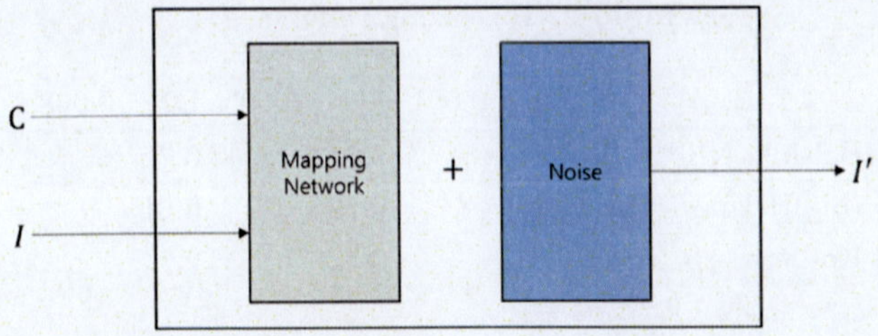

Fig. 6. Process of obtain $I^{'}$ from StyleGAN

4 Performance Evaluation

In this section, we analyze the results of CLIP processing Korean input text. When inputting Korean text, the most influential part was the translator part and the input text itself. Explain the experimental environment and check the differences in the results.

Table 1. Validation Test Environment

Target	Detail
OS	Ubuntu 20.04.5 LTS
CPU	1 CORE
Memory	16 GB
GPU	Tesla T4
Python version	3.8.10

4.1 Environment

The environment in which the experiment was conducted is shown in Table 1. The experiment recently used Google's colab, which is widely used as a cloud server platform. The environment of Table 1 allocated as an account may vary, but it did not significantly affect the results. Google's colab provides cloud server, so you can access source code and server anywhere, so the results were constant even if the location changed. In addition, it can be said that the reliability of the results is also high because it is stable.

Table 2. Tokenizing Difference

	L2 = 0.001	L2 = 0.005	L2 = 0.008
Helsinki-NLP	0.68	0.7	0.72
(big)Helsinki-NLP	0.75	0.78	0.81
Ko-bert	0.67	0.72	0.74

4.2 Tokenization

Several Bert models were used for Tokenizing. Tokenizing was performed using three models: Helsinki-NLP, (big)Helsinki-NLP, and Ko-bert and the differences in the results were confirmed. Helsinki-NLP, (big) Helsinki-NLP, is a model in which larger data is learned in (big) Helsinki-NLP. It was used to understand how much the size of the data affects Tokenizing. As a result, it can be seen that the difference in loss occurred as shown in Table 2. Overall, there was no significant difference in Loss itself. Although the difference between the loss of Helsinki-NLP and (big) Helsinki-NLP was not significant, the loss of (big) Helsinki-NLP was larger and the resulting image showed a larger difference. There was no noticeable difference between Helsinki-NLP and Ko-bert, but the resulting image showed better quality of Ko-bert. The difference can be confirmed through Figs. 7. In addition, the L2 value was adjusted to determine the degree to which the image was distorted, and an appropriate L2 value was selected. The most ideal value of the L2 value was selected as 0.008 and the experiment was conducted. The first picture on the left is the original picture, and the L2 value increases as it goes to the right. As the L2 value decreases, the distortion increases, and as the L2 value increases, the image changes closest to the intention.

Fig. 7. Korean text that requires blue eyes

5 Conclusion

In this paper, we recognized the problem of CLIP, which was not able to handle the existing Korean input, and conducted a study for improvement. Since CLIP was previously a model made only in English, it could not be processed in the Korean input, so it was necessary to translate the Korean input into English. Therefore, using the Tokenizer and translation module of the BERT Model, the input text in Korean was translated into English. The English text that the CLIP Model can recognize was put back into the input value of the CLIP Model and provided an image to be manipulated. The CLIP Model recognized the English text and image and recognized obejct and the conditions to be changed. The CLIP Model returned the object and conditions and entered the StyleGAN Model. The StyleGAN Model successfully reconstructed the image by reflecting the conditions on the image and reflected the input content in Korean. As a result, we found a way to operate the CLIP Model in Korean text. As a result, the results of the CLIP Model and StyleGAN Model that successfully reflected the given conditions in the Korean text could be confirmed.

6 Limitations of the Study

The limitations of the research in this paper are as follows. First, it was natural that the Korean text was not processed because the CLIP Model was initially a model made up of only English. Therefore, the work of translating the Korean text into English became a work of delaying the processing time and increasing the steps. When input is given in Korean, the translation model is passed once more, so the speed is slowed down and the parts that affect performance are increased. If the translation itself varies depending on the performance of the tokenizer for translation, there was a risk that the conditions themselves would change, resulting in different results or not being fully reflected. Second, because there is no module that recognizes various languages, only Hangul could be recognized. In order to improve CLIP, which recognizes only English input, a model that can recognize only Korean was created. In future studies, it is necessary to be able to improve and evaluate the processing speed by recognizing the language itself and reducing the number of models.

Acknowledgement. This work was supported by Institute of Information & communicationsTechnology Planning & Evaluation (IITP) grant funded by the Korea government(MSIT) (No. 2020-0-00107, Development of the technology to automate the recommendations for big data analytic models that define data characteristics and problems).

References

1. Creswell, A., White, T., Dumoulin, V., Arulkumaran, K., Sengupta, B., Bharath, A.A.: Generative adversarial networks: An overview. IEEE Signal Proc. Mag. **35**(1), 53–65 (2018)
2. Deng, J., Dong, W., Socher, R., Li, L.-J., Li, K., Fei-Fei, L.: Imagenet: A large-scale hierarchical image database. In: 2009 IEEE Conference on Computer Vision and Pattern Recognition, pp. 248–255 IEEE (2009)
3. Devlin, J., Chang, M.-W., Lee, K., Toutanova, K.: Bert: Pre-training of deep bidirectional transformers for language understanding. arXiv preprint arXiv:1810.04805 (2018)
4. O'Shea, K., Nash, R.: An introduction to convolutional neural networks. arXiv preprint arXiv:1511.08458 (2015)
5. Radford, A., et al.: Learning transferable visual models from natural language supervision. In: International Conference on Machine Learning, pp. 8748–8763. PMLR (2021)

Internet Olympiad in Computer Science in the Context of Assessing the Scientific Potential of the Student

Nataliya V. Chernousova, Natalia A. Gnezdilova, Tatyana A. Shchuchka(✉), and Lydmila N. Alexandrova

Federal State Budgetary Educational Institution of Higher Education «Bunin Yelets State University», 28, Kommunarov St, Yelets 399770, Russia
tasiaelez@mail.ru

Abstract. Currently, the state of higher education requires solving many tasks, one of which is to work with young people to identify their scientific potential, which is emphasized by the provisions of the Strategy of digital transformation of science and education in the direction of personalized approach. The implementation of this task can be carried out within the framework of the Olympiad movement and can act as a tool for assessing the scientific potential of the student. In this context, digital technologies are also an important factor. Conducting the Olympiad in a digital space provides conditions for the participation of a wide mass of creative young people from different countries, allows to assess the level of development of scientific potential, increases the information and research competence of young scientists, which determines the relevance of this study. The paper states and solves the problem of identifying the capabilities of the Internet Olympiad in Computer Science as a tool for assessing the scientific potential of students. The aim of the study is to identify the level of scientific potential of its participants according to the results of the Internet Olympiad in Computer Science. The objectives of the study determine the necessity of compiling the Olympiad tasks in the framework of the competence approach and the methodology of their assessment as the level of scientific potential; representation of the contingent of participants according to the profile of knowledge; the analysis of the results. The following methods were used in the study: analysis, comparison, comparison, design, testing, systematization of the results of scientific research. Diagnostic materials of assessment of the level of scientific potential were applied. In the present study we proposed the implementation of the process of assessing the level of development of scientific potential of students on the criterion of scientific and creative activity of personality by the tools of Olympiad movement, testing of which showed its viability. The implementation of the process of assessing the level of development of scientific potential of student youth on the criterion of scientific and creative activity of personality with the tools of Olympiad movement allows to accumulate the base of diagnostics in pedagogical science. #CSOC1120.

Keywords: Scientific Potential · Criterion of Scientific and Creative Activity of a Person · Olympiad · Computer Science · Creative Student Youth

R. Silhavy and P. Silhavy (Eds.): CSOC 2023, LNNS 724, pp. 231–239, 2023.
https://doi.org/10.1007/978-3-031-35314-7_22

1 Introduction

In pedagogical science the definition "scientific potential" is presented by researchers as "latent possibilities in mastering of modern educational and scientific environment, stipulating the creative study of its educational program, solution of modern problems of scientific knowledge and manifesting the desire for self-education and creative self-development throughout life" [1].

The state of higher education in the present period requires solving many tasks, one of which is to work with young people to identify their scientific potential, which is emphasized by the provisions of the Strategy of digital transformation of science and education in the direction of personalized approach.

The implementation of this task can be carried out within the framework of the Olympiad movement and can act as a tool for assessing the scientific potential of the student.

In this context, digital technologies are also an important factor. Conducting the Olympiad in a digital space provides conditions for the participation of a wide mass of creative young people from different countries all over the world, allows to assess the level of development of scientific potential, increases the information and research competence of young scientists, which determines the relevance of this study.

The problem of identifying the capabilities of the Internet Olympiad in computer science as a tool for assessing the scientific potential of students is stated and solved in the paper.

The aim of the study is to identify the level of scientific potential of its participants according to the results of the Internet Olympiad in Computer Science.

The objectives of the study determine the necessity of compiling the Olympiad tasks in the framework of the competence approach and the methodology of their assessment as the level of scientific potential; representation of the contingent of participants on the profile of knowledge; analysis of the results.

2 Materials and Methods

In the study the following methods were used: analysis, matching, comparison, design, testing, systematization of the results of scientific research.

Diagnostic materials of assessment of the level of scientific potential in scores B_j depending on the coefficient of the completed task were applied.

$$B_j = \begin{cases} 4, \mathit{if} k_j \leq 0.06 \\ 3, \mathit{if} 0.06 < k_j \leq 0.16 \\ 2, \mathit{if} 0.16 < k_j \leq 0.38 \\ 1, \mathit{if} k_j > 0.38 \end{cases},$$

where k_j is the coefficient of the completed task j, which is defined as the quotient of the number of students who correctly completed the task, to the total number of students who completed the task.

The final score received by i-th student number is:

$$m_i = \sum_{j=1}^{16} B_j \cdot \alpha_{ij};$$

where $\alpha_{ij} = 1$ or $\alpha_{ij} = 0$, provided that the i-th student correctly solved the j-th task and vice versa respectively.

The highest number of scores is determined for the student is:

$$M = \sum_{j=1}^{16} B_j;$$

or as a percentage as:

$$D_i = \frac{m_i}{M} \cdot 100\% = \frac{\sum_{j=1}^{16} B_j \cdot \alpha_{ij}}{\sum_{j=1}^{16} B_j} \cdot 100\%$$

When preparing the tasks for the Internet-Olympiad correspondence to their level of scientific potential on the criterion of scientific and creative activity of the student, which are designated as basic, advanced, high, was carried out.

The requirements, formulated to the designated levels of scientific potential, reveal the ability to define the problem in the subject language of computer science, to analyze the methods of solution, to interpret the results.

The content maps of the Informatics Olympiad tasks for each knowledge profile have been compiled.

3 Results

The competency-based approach became the fundamental approach to compiling the 16 tasks for the Internet Olympiad in Informatics, within the framework of which the practice-oriented tasks were offered, making it possible to reveal the abilities stated in the requirements for problem definition in the subject language of Informatics, analysis of the solution methods, interpretation of results and taking into account the specific features of the knowledge profile:

- Humanities and Law (H&L);
- Economics and Management (E&M);
- Engineering and Technology (E&T);
- Biotechnology and Medicine (B&M);
- Specialized (with advanced study of Informatics discipline) (SP).

The data analysis of the Internet Olympiad in Informatics is considered for all knowledge profiles for all participating universities and is presented in the form of diagrams, coefficient maps, rating lists (posted on the official websites of the corresponding university).

135 educational institutions of higher education from six countries participated in the Internet Olympiad, with a total number of 5262 students (Fig. 1, Table 1).

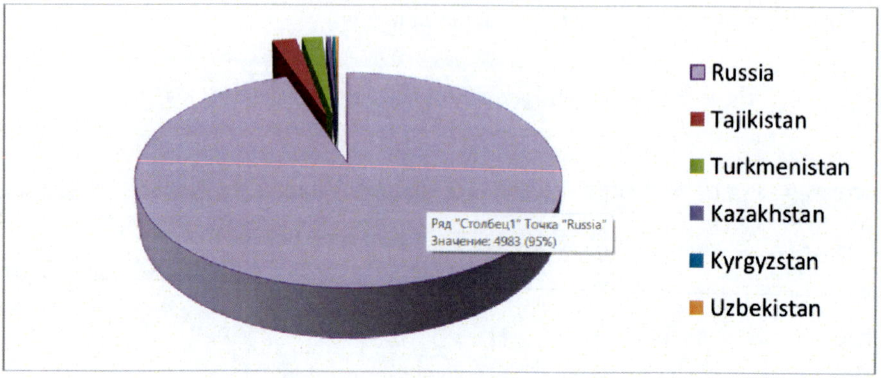

Fig. 1. Number of participants in the Internet Olympiad in Informatics.

Table 1. Number of participants in the Internet Olympiad in Informatics.

№	Country participating in the Olympiad	Number of universities participating in the competition	Number of students
1	Russia	120	4983
2	Tajikistan	4	122
3	Turkmenistan	7	103
4	Kazakhstan	2	26
5	Kyrgyzstan	1	15
6	Uzbekistan	1	13

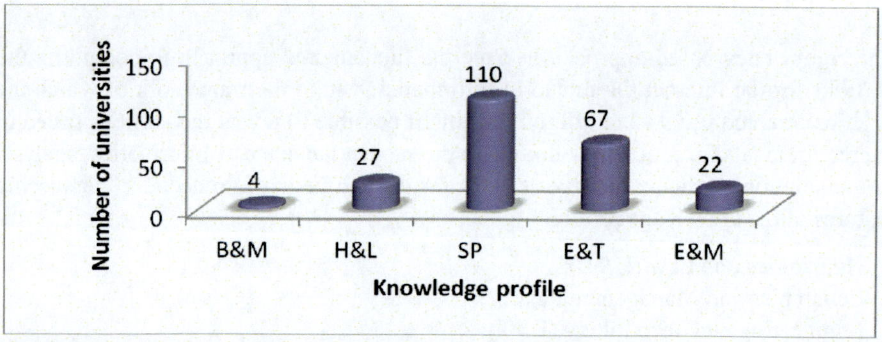

Fig. 2. Number of universities - participants of the Internet Olympiad, by knowledge profile.

Figures 2 and 3 show the number of participants in the Internet Olympiad in Computer Science, respectively universities and students.

In the processing of the results data the correspondence of the coefficients of task completion and evaluation scores of the level of scientific potential for each task has been carried out.

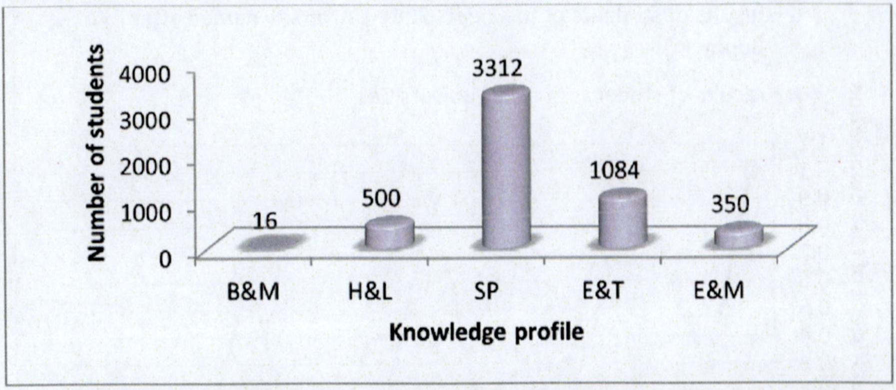

Fig. 3. Number of students participating in the Internet-Olympiad, by knowledge profile.

A comparative analysis of the results of the students of Yelets State University named after I. A. Bunin with the results of the students of other educational organizations - participants of the Internet-Olympiad in the relevant knowledge profiles was carried out.

Let us summarize the results of the Internet-Olympiad on the example of the knowledge profile “Humanities and Law”.

Figure 4 shows the conformity of the results data in percents of 500 students from 27 educational organizations, who participated in the Internet Olympiad.

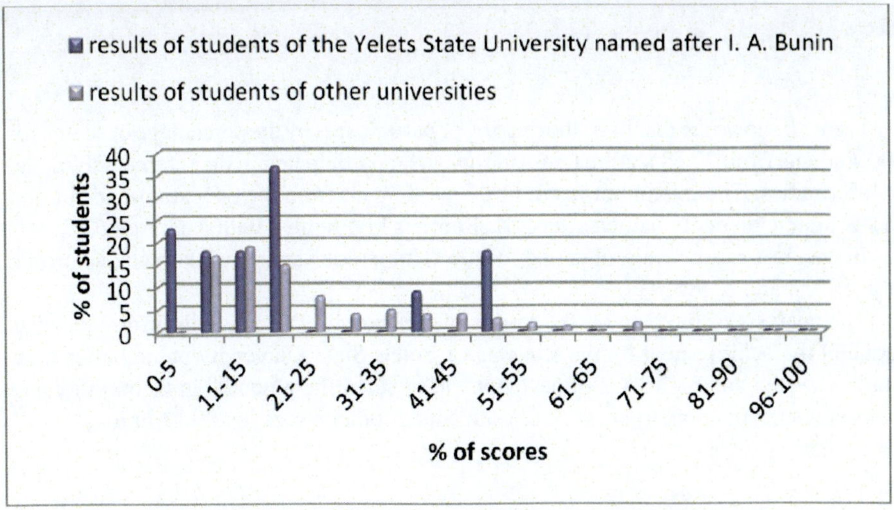

Fig. 4. Final data of students on a profile of knowledge “Humanities and law”.

Figure 5 shows a map of tasks completion ratio, while Table 2 shows the correspondence of tasks to the specified scores.

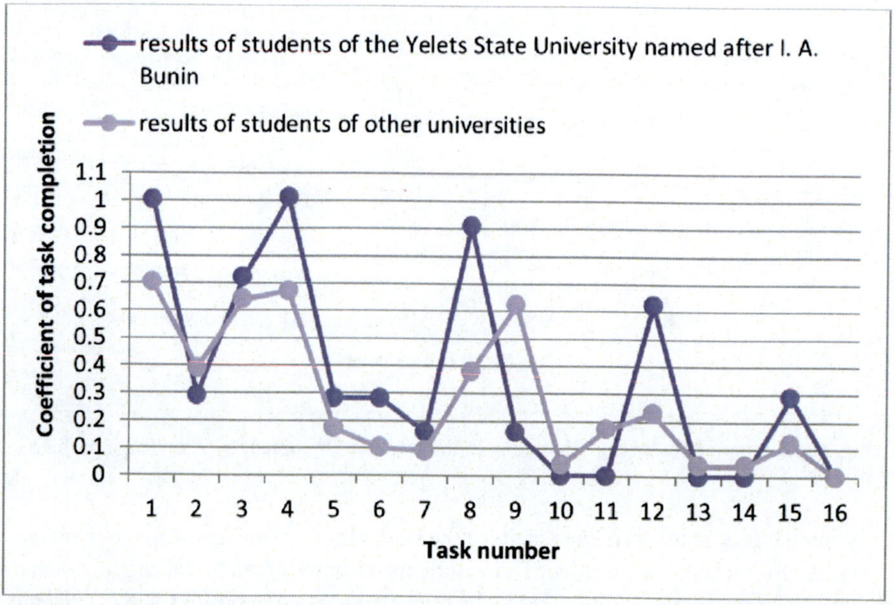

Fig. 5. Map of the coefficients of task completion. The knowledge profile "Humanities and law".

Table 2. Correspondence of tasks to the established points.

Task number	1	2	3	4	5	6	7	8	9	10	11	12	13	14	15	16
Score	1	2	1	1	2	3	3	2	1	4	3	2	4	4	3	4

Figure 6 shows the results of the ranking of participants by the percentage of scores for 500 students from 27 educational institutions, who participated in the Internet Olympiad in Informatics. The maximum result of the participant from the educational institution Yelets State University named after I. A. Bunin is highlighted with a dark tone.

Eleven test results were obtained for the "Humanities and Law" knowledge profile (Tables 3, 4 and 5, Fig. 7).

The analysis of the data on the results of the Internet Olympiad in Informatics has revealed the achievement by the students of Yelets State University named after I. A. Bunin elevated and high level of development of scientific potential in its majority and the best results in comparison with the students of other educational institutions.

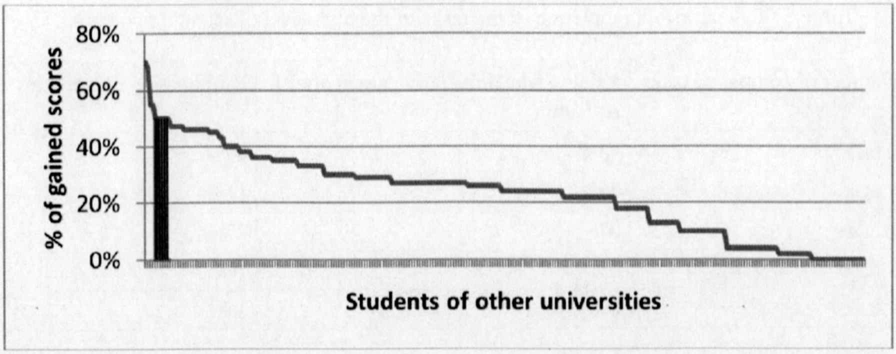

Fig. 6. Ranking of the results of the students of participating educational institutions by the percentage of gained scores. The knowledge profile "Humanities and law".

Table 3. Indicators of completion of the tasks of the basic level of scientific potential.

Number of completed tasks	Yelets State University named after I. A. Bunin	Participating universities
0 task	0%	10%
1 task	0%	17%
2 tasks	28%	24%
3 tasks	45%	31%
4 tasks	27%	18%

Table 4. Indicators of completion of the tasks of the elevated level of scientific potential.

Number of completed tasks	Yelets State University named after I. A. Bunin	Participating universities
0 task	10%	24%
1 task	9%	26%
2 tasks	54%	24%
3 tasks	0%	13%
4 tasks	9%	7%
5 tasks	18%	5%
6 tasks	0%	1%

Table 5. Indicators of completion of the tasks of the high level of scientific potential.

Number of completed tasks	Yelets State University named after I. A. Bunin	Participating universities
0 task	73%	85%
1 task	27%	11%
2 tasks	0%	2%
3 tasks	0%	1%
4 tasks	0%	1%

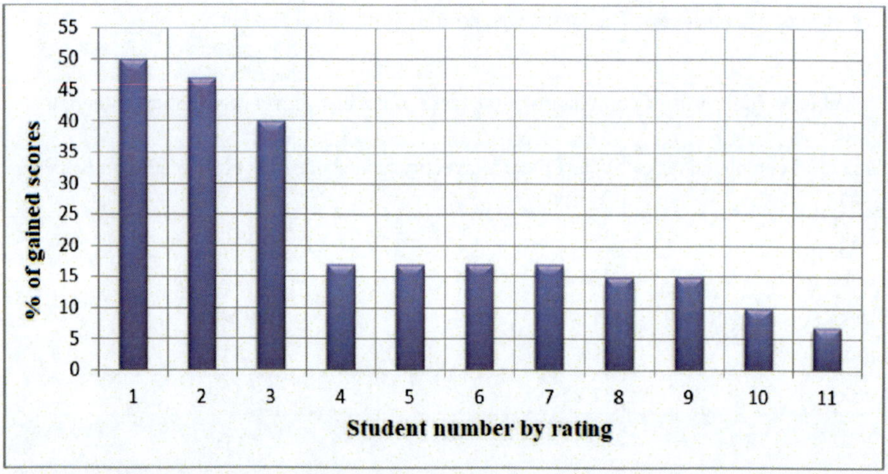

Fig. 7. Ranking of students of FSBEI HE "Yelets State University named after I.A. Bunin" by the percentage of points scored. The knowledge profile "Humanities and law".

4 Discussion

The discussion in the academic environment of the processes of development and assessment of the creative student's potential in the light of digital technology application justifies the relevance in the process of educational practice [2–5]. Researchers have developed many methods for assessing the creative potential of young people [6–8]. Among them are the research program of A.R. Gregos on determining the style of information assimilation, and methodological developments on diagnosing personal creativity of E.E. Tunik [9]), and N.P. Fetiskin's method of assessing the level of creative potential [10], and diagnostic materials of achievement motivation of A. Mehrabian.

In the present study we proposed to carry out the process of assessing the level of development of scientific potential of student youth by the criterion of scientific and creative activity of personality using the tools of the Olympiad movement.

5 Conclusion

Purposeful testing of the identified productive ideas for the implementation of the assessment of the level of development of students' scientific potential on the criterion of scientific and creative activity of personality with the tools of Olympiad movement in Yelets State University named after I.A. Bunin has shown their viability.

Implementation of the process of assessing the level of development of students' scientific potential according to the criterion of scientific and creative activity of a person by the Olympiad movement tools allowed to accumulate the diagnostic base in pedagogical science.

References

1. Isaev, I.F., Isaeva, N.I., Makotrova, G.V.: Development of Personal Scientific Potential: Theory, Diagnostics, Technology: Collective Monograph. Publishing House of NRU "BelGU", Belgorod p. 361 (2011)
2. Andryukhina, L.M., Sadovnikova, N.O., Utkina, S.N., Mirzaahmedov, A.M.: Digitalization of professional education: prospects and invisible barriers. Educ. Sci. **2**(2)(119), 365–368 (2020)
3. Blinov, V.I., Dulinov, M.V., Esenina, E.Y., Sergeev, I.S.: Draft didactic concept of digital professional education and training. Pero Moscow p. 71 (2019)
4. Gerasimova, A.G.: Preparation of students for professional activity in the conditions of digitalization of education. Mod. Sci.-Intens. Technol. **7**, 136–140 (2020)
5. Markov, V.N., Sinyagin, Y.: Potential of personality. World Psychol. **1**(21), 250–261 (2000)
6. Druzhinin, V.N.: Cognitive Abilities: Structure, Diagnosis, Development, p. 224. PER SE, Moscow. Imaton-M, SPb (2001)
7. Martynovskaya, S.N.: About criteria of actualization of creative potential of a future specialist. Mod. Prob. Sci. Educ. **1**, 70 (2006)
8. Shafikov, M.T.: Potential: essence and structure. Soc. Hum. Knowl. **1**, 236–245 (2002)
9. Tunik, E.E.: Modified Creative Williams Tests, p. 96. Rech, St. Petersburg (2003)
10. Fetiskin, N.P., Kozlov, V.V., Manuilov, G.M.: Social-Psychological Diagnostics of the Development of Personality and Small Groups, p. 490. Publishing House of the Institute of Psychotherapy, Moscow (2002)

Evaluation of the Prognostic Significance and Accuracy of Screening Tests for Alcohol Dependence Based on the Results of Building a Multilayer Perceptron

Michael Sabugaa[1](✉), Biswaranjan Senapati[2], Yuriy Kupriyanov[3], Yana Danilova[3], Shokhida Irgasheva[4], and Elena Potekhina[5]

[1] Agusan del Sur State College of Agriculture and Technology, Bunawan, Philippines
msabugaa@asscat.edu.ph
[2] Capitol Technology University, Laurel, USA
[3] National Research Mordovian State University N.P. Ogaryov, Saransk, Russian Federation
[4] Tashkent Institute of Finance, Tashkent, Uzbekistan
[5] Russian State Social University, Moscow, Russia

Abstract. The number of alcohol addicts is steadily increasing year by year all over the world. Screening of abusers of alcoholic beverages in a timely manner will allow early diagnosis of alcohol dependence syndrome, determine the stage of alcoholism and prescribe timely treatment aimed at the elimination of mental and alcohol dependence. There are many screening-tests identifying persons suffering from chronic alcohol consumption, but at the same time it is impossible to assert with certainty which of them can prognostically accurately assess the effectiveness of these tests. To assess the prognostic accuracy of screening tests, a direct propagation neural network based on a multilayer perceptron was used. A multilayer perceptron was built to evaluate popular tests: MAST, Jellinica, and AUDIT. This paper presents data on a suitable model of the multilayer perceptron, at which the least number of training sample errors with a suitable activation function was obtained. A test model for the tests under study is constructed, resulting in the calculation of correlation coefficients reflecting the predictive accuracy of the screening. The results of the study can be used to select the best screening model and assess the significance of testing by building a neural network. #CSOC1120.

Keywords: Multilayer Perceptron · Neural Network · Screening Test · Alcoholism

1 Introduction

In today's realities, alcoholism of different age groups around the world remains a major topic of discussion in modern drug treatment [1]. The global medical community uses the most appropriate term for this problem, "alcohol use disorders". The term reflects and communicates the essence of the problem well, thereby separating alcohol consumption (abuse and dependence) from the safe level.

R. Silhavy and P. Silhavy (Eds.): CSOC 2023, LNNS 724, pp. 240–245, 2023.
https://doi.org/10.1007/978-3-031-35314-7_23

To improve and quickly diagnose alcohol use disorders, many screening tests have been developed with different scales and algorithms for assessing the propensity or presence of alcohol dependence. The problem of assessing the accuracy of these tests and their value in prognostic terms is relevant.

Recently neural networks have become very popular [2–4]. Neural networks are universal computational tools in statistics and programming which are similar in principle to biological networks of brain neurons. Due to their resemblance to biological brain structures neural networks allow to extract important properties from received information, evaluate precisely the value of input parameters and give appropriate forecasts about the relationships between the received parameters. Today neural networks are experiencing a new round of evolution, their application in a variety of fields shows their efficiency [5–10].

There are a number of basic models of neural networks: feedforward artificial neural networks, recurrent neural networks, multilayer perceptron neural networks, radial basis function artificial neural, modular neural networks [11–16]. These neural networks have a number of disadvantages and advantages and are used depending on the task specified. For instance, feedforward neural networks are able to solve quickly and simply the set problem, but only linear classification of separable objects is possible. A multilayer perceptron is capable of generating any input-output function, but the choice of the number of layers, neurons and connections between them is difficult. Recursive neural networks have the largest number of neurons in layers, predicting the result in all directions forming a spatiotemporal structure, but it is impossible to distinguish separate layers in them.

2 Method

In this study a neuron was considered as a formal (abstract) unit. Other qualitative descriptions of neuron's behavior (physiological and phenomenological) and their mathematical interpretations were not taken into account. Abstract neuron was considered in the following mathematical description:

$$y = f(u),$$

where $u = \sum_{i=1}^{n} w_i x_i + w_0 x_0$, x_i and w_i are signals at the neuron inputs and weights of inputs respectively.

To solve the task we have chosen the model of direct propagation neural network with multilayer perceptron, because this neural network is able to classify any objects, divided in the feature space. The basic principle of the multilayer perceptron was to feed and further summarize the input data (input layer) through the activation function to the hidden layers, where the best hidden neuron was determined to output a predictively accurate value (Fig. 1).

For the best neuronal activation the hyperbolic tangent function $f(x) = \tanh(x) = \frac{(e^x - e^{-x})}{(e^x + e^{-x})}$ was applied (Fig. 2).

This activation function showed the best result both at the level of hidden neurons, and at the level of output neurons. The minimum and maximum number of hidden neurons on which the neural network was trained was set to 24 and 30 neurons, respectively.

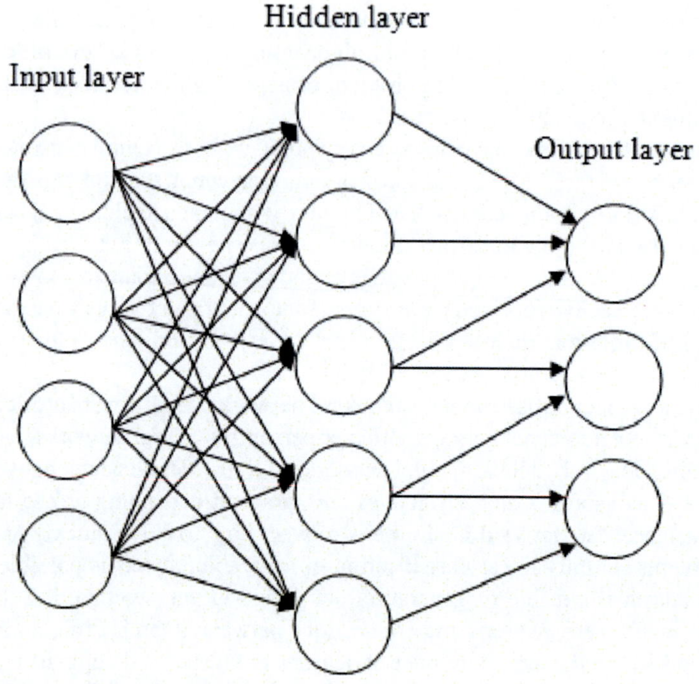

Fig. 1. The basic principle of a multilayer perceptron.

The initial data fed to build the multilayer perceptron were answers of 124 respondents to the verified tests: Michigan Alcoholism Screening Test ("MAST"), "Jellinica" Test, Alcohol Use Disorders Identification Test ("AUDIT"). Respondents' responses to the test were further interpreted as categorized data [17, 18].

The resulting categorized ordinal data were recoded by One-Hot Encoding. The predictive significance and accuracy of the questionnaires were assessed by constructing a multilayer perceptron in the STATISTICA 13 software package. Predictive significance (quality of the constructed test sample models) was assessed by means of Spearman's rank correlation coefficient *r*.

The formula for *r* calculation is as follows:

$$r = 1 - \frac{6 \cdot \sum (d^2)}{N \cdot (N^2 - 1)},$$

where *d* is difference between ranks of two variables, *N* is number of ranked values.

The qualitative assessment of the correlation coefficient is given by the Chaddock scale (Table 1) [19].

The multilayer perceptron constructed in this way made it possible to determine the most appropriate screening test for identifying alcoholics.

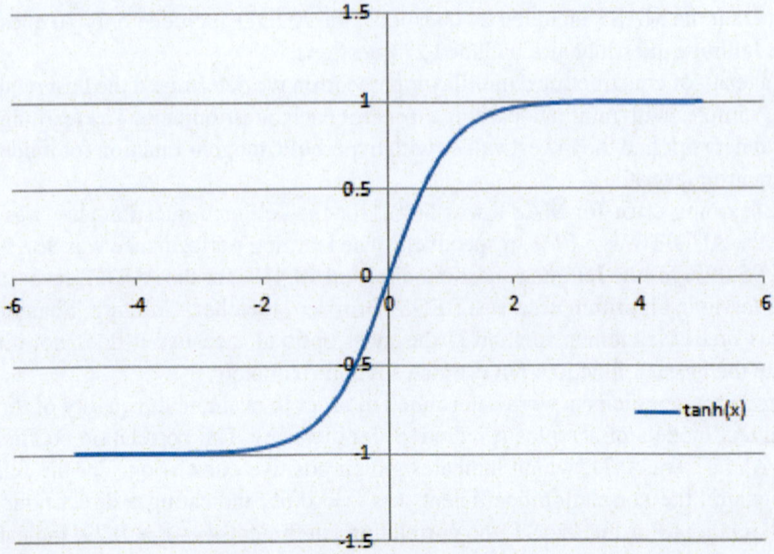

Fig. 2. Hyperbolic tangent function.

Table 1. Chaddock scale.

Absolute Value of Correlation, *r*	Interpretation
0–0.3	Negligible correlation
0.3–0.5	Weak correlation
0.5–0.7	Moderate correlation
07.–0.9	Strong correlation
0.9–1	Very strong correlation

3 Results and Discussion

According to the results of statistical processing of the respondents it was determined that 94 respondents (75.8%) had a low level of problems (less than 8 points), 23 respondents (18.5%) had hazardous and harmful alcohol consumption (8–19 points), 7 respondents (5.6%) had possible alcohol addiction (more than 19 points).

According to the results of the Jellinica questionnaire, 94 participants of the survey (75.8%) had early stage alcoholism (5–8 points). Twenty two respondents (17.7%) had no alcohol dependence (less than 5 points). The remaining 8 interviewees (5%) could be assumed to have an initial intermediate stage of alcohol dependence (9–15 points).

The MAST Alcohol Screening Test revealed that 88 respondents (70.9%) had a low risk of alcohol-related problems (less than 5 points). Suspicion of alcohol abuse or alcohol dependence was detected in 36 respondents (29.1%).

Note that the MAST included 24 questions, the AUDIT included only 10 questions, and the Jellinica questionnaire included 27 questions.

As a result of constructing a multilayer perceptron we determined the best models of training samples with minimal learning errors for each questionnaire. The learning error of neural network had the lowest values with hyperbolic tangent function for hidden and output neuron layers.

The learning error for MAST was 5.7%, for the Jellinica questionnaire was 5.75, and for the AUDIT was 5.01%, respectively. The learning performance was 96.99% for MAST, 95.09% for the Jellinica questionnaire, and 98.41% for the AUDIT, respectively.

The learning algorithm used was BFGS (Broyden, Fletcher, Goldfarb, Shanno) [20, 21]. This quasi-Newtonian method is the most optimal, because it does not directly calculate the hessian function, but estimates it approximately.

Correlation coefficients were determined in order to evaluate the quality of the constructed test models of samples (predictive significance). The correlation coefficient r for the AUDIT was 0.71, which indicates a high positive correlation. For the Jellinica questionnaire, the correlation coefficient was $r = 0.65$, indicating a direct noticeable positive relation. For the MAST, the correlation coefficient was $r = 0.79$, indicating a very high positive relation.

Certainly, the data obtained with screening tests can be considered preliminary, the final assessment is made by clinical data, which are the priority. Therefore, an independent assessment of the accuracy of the tests is also preliminary, requiring clinical verification. However, we find our results interesting in the appropriate context of the sensitivity of the questions asked to the logic of a person with a prior propensity to drink alcohol.

4 Conclusion

As a result of the conducted research, using multilayer perceptron construction, we have found that the best verified test to determine the predictive value and accuracy of the survey is the Michigan Alcoholism Screening Test ("MAST") ($r = 0.79$), which includes the greatest number of sensitive questions. The Jellinica questionnaire had the worst accuracy and predictive value ($r = 0.65$), even though it had the largest number of questions.

References

1. Kelemen, A., Minarcik, E., Steets, C., Liang, Y.: Telehealth interventions for alcohol use disorder: a systematic review. Liver Res. **6**(3), 146–154 (2022). https://doi.org/10.1016/j.livres.2022.08.004
2. Yu, X., Bo, L., Xin, C.: Low light combining multiscale deep learning networks and image enhancement algorithm. Mod. Innov. Syst. Technol. **2**(4), 0214–0232 (2022). https://doi.org/10.47813/2782-2818-2022-2-4-0215-0232
3. Semenova, E.A., Tsepkova, S.M.: Neural networks as a financial instrument. Inform. Econ. Manag. **1**(2), 0168–0175 (2022). https://doi.org/10.47813/2782-5280-2022-1-2-0168-0175

4. Lunev, D., Poletykin, S., Kudryavtsev, D.O.: Brain-computer interfaces: technology overview and modern solutions. Mod. Innov. Syst. Technol. **2**(3), 0117–0126 (2022). https://doi.org/10.47813/2782-2818-2022-2-3-0117-0126
5. Chen, Y., Zhang, N., Yang, J.: A survey of recent advances on stability analysis, state estimation and synchronization control for neural networks. Neurocomputing **515**, 26–36 (2023). https://doi.org/10.1016/j.neucom.2022.10.020
6. Coulibaly, S., Kamsu-Foguem, B., Kamissoko, D., Traore, D.: Deep convolution neural network sharing for the multi-label images classification. Mach. Learn. Appl. **10**, 100422 (2022). https://doi.org/10.1016/j.mlwa.2022.100422
7. Gruzenkin, D.V., et al.: Neural networks to solve modern artificial intelligence tasks. J. Phys: Conf. Ser. **1399**(3), 033058 (2019). https://doi.org/10.1088/1742-6596/1399/3/033058
8. Kesawan, S., Rachmadini, P., Sabesan, S., Janarthanan, B.: Application of neural networks for light gauge steel fire walls. Eng. Struct. **278**, 115445 (2023). https://doi.org/10.1016/j.engstruct.2022.115445
9. Semenenko, M.G., et al.: How to use neural network and web technologies in modeling complex technical systems. IOP Conf. Ser.: Mater. Sci. Eng. **537**(3), 032095 (2019). https://doi.org/10.1088/1757-899X/537/3/032095
10. Wang, P., Bu, H.: Enterprise hierarchical management based on neural network model. Optik **272**, 170326 (2023). https://doi.org/10.1016/j.ijleo.2022.170326
11. Bahtiyar, H., Soydaner, D., Yüksel, E.: Application of multilayer perceptron with data augmentation in nuclear physics. Appl. Soft Comput. **128**, 109470 (2022). https://doi.org/10.1016/j.asoc.2022.109470
12. Banki-Koshki, H., Seyyedsalehi, S.A.: Complexity emerging from simplicity: bifurcation analysis of the weights time series in a feedforward neural network. Commun. Nonlinear Sci. Numer. Simul. **118**, 107044 (2023). https://doi.org/10.1016/j.cnsns.2022.107044
13. Goto, K., et al.: Development of a vertex finding algorithm using recurrent neural network. Nucl. Instrum. Methods Phys. Res. Sect. A **1047**, 167836 (2023). https://doi.org/10.1016/j.nima.2022.167836
14. Rojas, M.G., Olivera, A.C., Vidal, P.J.: Optimising multilayer perceptron weights and biases through a cellular genetic algorithm for medical data classification. Array **14**, 100173 (2022). https://doi.org/10.1016/j.array.2022.100173
15. Zhang, L., Li, H., Kong, X.-G.: Evolving feedforward artificial neural networks using a two-stage approach. Neurocomputing **360**, 25–36 (2019). https://doi.org/10.1016/j.neucom.2019.03.097
16. Zhang, X., Zhong, C., Zhang, J., Wang, T., Ng, W.W.Y.: Robust recurrent neural networks for time series forecasting. Neurocomputing (2023). https://doi.org/10.1016/j.neucom.2023.01.037
17. Briggs, M., Peacock, A.: Screening older adults for alcohol use. J. Nurse Pract. **2022**, 104432 (2022). https://doi.org/10.1016/j.nurpra.2022.08.015
18. Paulus, D.J., Rogers, A.H., Capron, D.W., Zvolensky, M.J.: Maximizing the use of the Alcohol Use Disorders Identification Test (AUDIT) as a two-step screening tool. Addict. Behav. **137**, 107521 (2023). https://doi.org/10.1016/j.addbeh.2022.107521
19. Borovskaya, R., Krivoguz, D., Chernyi, S., Kozhurin, E., Khorosheltseva, V., Zinchenko, E.: Surface water salinity evaluation and identification for using remote sensing data and machine learning approach. J. Mar. Sci. Eng. **10**(2), 257 (2022). https://doi.org/10.3390/jmse10020257
20. Fletcher, R.: Practical Methods of Optimization, 2nd edn. John Wiley & Sons, New York (1987)
21. Irwin, B., Haber, E.: Secant penalized BFGS: a noise robust quasi-Newton method via penalizing the secant condition. Comput. Optim. Appl. (2023). https://doi.org/10.1007/s10589-022-00448-x

Geographic Data Science for Analysis in Rural Areas: A Study Case of Financial Services Accessible in the Peruvian Agricultural Sector

Rosmery Ramos-Sandoval(✉) and Roger Lara

Universidad Tecnológica del Perú, Lima, Peru
rramoss@utp.edu.pe

Abstract. Due to the economic crisis, the agrarian sector could be found in a situation of greater vulnerability. This is due to a reduction in the availability of financial services in the sector, generated by the volatility of the conjuncture, added to the risk that characterizes the activities in the rural agricultural sector. Regarding the family agricultural sector, previous studies indicate that access to financial services is an outstanding challenge, which consequently inhibits the development of more profitable economic and social activities, which in turn generates an increase in the levels of vulnerability in the economies of small producers in the agricultural sector. This study argues that spatial patterns are important because they may provide clues about the causes and potential consequences of the context, which in this study will fall on territory occupation from the Peruvian financial services providers, which are supposed to supply credit to farmers. The proposals explored an emerging open source for geographic analysis that has been applied in Peruvian territory. The expansion and diversity of geographic data tools have the aim to promote digital transformation in cities, therefore, this research proposes to extend the limited diffusion of open-source tools developed applied spatial algorithms that could be modelled under a spatial statistic approach, focusing on the producer location's impact on access to financing.

Keywords: geographic data · open-source tools · data visualization · Peruvian financial services suppliers

1 Introduction

During the past several years, technological and innovation improvements have permitted financial institutions to interact with customers more efficiently and to measure and manage risk more effectively [1]. In this regard, Benni [2] pointed out that digital innovation in financial markets has the potential for alleviating several of the most critical barriers, that currently limit access to financial services for rural and vulnerable actors. Nonetheless, benefits gained from the expansion of customer service channels offered by the financial institution's system might be different in urban than rural areas.

Signs of financial exclusion among rural citizens include the difficulty they have in accessing credit. In this sense, Brevoort et al. [3] describe these customers as "credit

R. Silhavy and P. Silhavy (Eds.): CSOC 2023, LNNS 724, pp. 246–259, 2023.
https://doi.org/10.1007/978-3-031-35314-7_24

invisible", which are those that have no credit history. Furthermore, Jagtianiet al. [4] pointed out that the absence of local banks in a community may lead to a discontinuity in the availability of credit services for local small businesses, which, are often covered by non-bank actors, such as moneylenders. Therefore, bearing in mind that one of the main reasons for rural customers to lack a credit history, could be the distance of rural populations from banking institutions, the geographical information becomes a potential tool for solving the persistent exclusion of rural areas. In that sense, considering that rural areas are predominantly agrarian areas, one population potentially "invisible" to the financial sector, will be the smallholder farmers.

Regarding the importance of distance or more specifically geographic proximity, in the provision, delivery, and use of services provided by the banking sector, Brevoort and Wolken [5] pointed out that contrary to expectations, in recent years the development of U.S. local financial ecosystems, integrating users and providers of financial services, has been uneven. Besides, in developing countries, most of the observable factors are socio-economic characteristics and household capacities, one of the main factors affecting households' credit accessibility is still bank distance [6]. In this regard, considering the importance that the geographical proximity of banking institutions might have on the development of rural areas, this study will explore, using spatial data mining techniques, the current state of the financial ecosystem between institutions and users in the Peruvian context. Therefore, this study focuses on the application of a spatial data mining technique to map the geographical proximity of banking institutions (e.g., Banks, exchange bureaus, ATMs), in order to explore a potential market that strengthens financial inclusion in the Peruvian rural agricultural sector.

The remainder of the study is structured as followed. Section 2 describes the spatial data science theory framework. In Sect. 3, we turn to the data which reflect specific characteristics from the Peruvian financial sector under a geographic data science approach. Section 4 presents the results and develops an estimate of the impact of credit constraints on the agricultural context. Section 5 concludes by pointing to several recent.

2 Geographic Data Science

In the last several years, geographic data science also referred to under different titles as spatial data science or geospatial data, is considered a sub-section within data science focusing on the special characteristics of spatial data, as well as, the different methods and tools [7]. Regarding the geographic information framework, Singleton et al. [8] bring about that data should be accessible within the public domain and available to researchers, pointing out the challenge and practical implication as an open source tool. In addition, spatial indicators under a geographical approach, contribute to local and regional planners to reveal the spatial distribution of infrastructure and resources within cities, facilitating comparisons within cities, highlighting resource distribution and areas needing interventions, empowering communities to advocate for improvement based on their real context [9, 10]. Therefore, based on the previous literature, it is clear the importance of the geographic approach rapidly evolves, becoming transversal between the different areas of study, especially through the open data tools, which are becoming more and more accessible to everyone.

2.1 Open Geographic Information to Solve Societal Issues

Focusing on the openness of the data, information, and knowledge it has been evolving continuously in the last several years. In line with the Open Knowledge Foundation definition [24], regarding openness, "open data and content can be freely used, modified, and shared by anyone for any purpose". Whereas Rey et al. [26], propose that openness in science is a key component of what makes our scientific practices, "science", and pointed out that transparency, accessibility, and inclusiveness are critical for good science. Furthermore, regarding the open-source geospatial software available, most of them include a broad range of utilities including libraries, tools applications, and platforms development and are released under different initiatives [11]. In this sense, this paper focuses on highlighting the benefits of using geographic information from open sources, considering its reproducibility in different contexts and uses.

As previously described, the capacity of these tools to generate information in inaccessible physical locations is one of their potential uses, a feature that has generated great inequalities in contexts predominantly rural. Therefore, consider explore rural areas holds the major potential in order to obtain novel approaches to identifying root causes of issues due to the effects of spatial characteristics.

In this regard, Yuan [12] proposed that geographical information science may become a tool for solving societal challenges as Sustainable Development Goals (SDGs) agenda, mapping problems that conceptualize sustainability principles and processes in spatiotemporal queries, to facilitate search, analysis, and modelling sustainability constructs. Whereas, for country-based study cases, in recent Han's work [13], it has been proposed that spatial distribution of economic and service facilities in classes, could help us to understand how space was structured socio-economically based on principles of socio-spatial inequalities. Furthermore, for the Peruvian context, a recent study from Regal et al. [14] focuses on spatial data to manage public spaces as a resource in order to increase the vulnerable population's accessibility to essential goods and services during the COVID-19 quarantine situation. This previous study, implemented a methodology based on spatial data from open sources in terms of population density, accessibility, and vulnerability, allowing the decision-makers to explore solutions based on spatial local indicators.

2.2 Spatial Considerations and Financial Accessibility in Rural Areas

The Peruvian Agricultural Context. In Peru, according to Trivelli [15], the main activity of the populations living in rural areas is agriculture (about 95%), these populations have a low level of capitalization, in relation to heritage, as well as financial resources. They spend more than 60% of their annual expenses on their food, while the head of the household has a low level of education. In this sense, rural customers could face a series of risks that would imply neglect of their credit demands in the formal financial market [16]. According to the results of the National Agricultural Census-CENAGRO 2012 [17], the demand for credits in the agricultural sector is mainly met by national banks (e.g. *Agrobanco*) and local financial institutions (e.g. Microfinance institutions; Cooperatives); however, since public intervention is the main formal provider, this has not achieved coverage of financial services in the agricultural sector [18]. In this regard, the

Peruvian National Farming Survey (ENA, by its Spanish initials) of the year 2019 [19], provide statistical information to build indicators about the Peruvian agricultural sector, specifically in its subsection: "900. Financial Services"; reports the fact that as low as 20% of farmers (smallholders and big enterprises) across all regions in the country, have access to financial services in the agricultural sector (Fig. 1).

Meanwhile, other non-financial institutions (e.g., lenders; merchants) serve about a third of the sector's credit demand, these being mainly alternative sources to the formal financial sector, companies within the value chain or informal loans [20]. Although this type of financing will be more expensive, consequently reduce the efficiency in the acquisition of credit for farmers. Therefore, insufficient access to finance becoming a critical obstacle that determines the low capitalization of the sector, as well as the low incorporation of technologies and technical models in SMEs in the agricultural sector, leading to the low productivity of the sector.

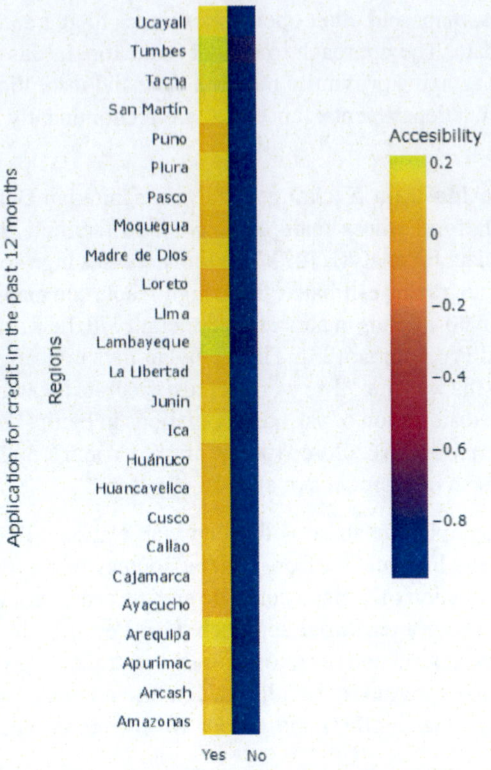

Fig. 1. Peruvian farmers' applications for credit in the past 12 months by regions, 2019. Source: Adapted from the ENA (2019).

3 Material and Method

In this paper, the datasets present were obtained through information extraction techniques. According to Cowie and Lehnert [21], no human can read, understand, and synthesize megabytes of text produced on an everyday basis, therefore, information extraction has been proposed as a method to manage vast amounts of information and thus avoid lost opportunities. Open-source extraction tools such as OpenStreetMap (OSM) and Humanitarian Data Exchange (HDX) were used in this study. OSM extracted the maps, while HDX extracted the population data for Peru and its regions.

3.1 Spatial Data

Peruvian OpenStreetMap Dataset. OpenStreetMap® is an open database that provides data mapping for thousands of websites, mobile applications, and hardware devices [25]. OSM highly appreciates local knowledge, since most of the contributors use aerial images, GPS devices, maps and other open data sources to guarantee that OSM data is accurate and up to date. The approach to spatial indicators in this research was based on measuring the geographic proximity between financial institutions and smallholder farmers in the Peruvian departments, those who are predominantly living in rural areas of Peru.

Peruvian Humanitarian Data Exchange. The Humanitarian Data Exchange® is an open data-sharing platform whose main objective is to facilitate the easy finding and use of humanitarian data for analysis [27]. The tool accesses high-resolution population density maps and demographic estimates, these high-resolution maps calculate the number of people living in 30-m grids in almost every country in the world collected through Meta Data For Good Programme [28]. The Peruvian data collected has the following features according to the website: "The world's most accurate population datasets. Seven maps/datasets for the distribution of various populations in Peru: (1) General population density (2) Women (3) Men (4) Children (0–5 years) (5) Youth (15–24 years) (6) Elderly (60 + years) (7) Women of reproductive age (15–49 years)".

Geo Peru. This geographic database is the Peruvian National Platform of Georeferenced Data building on information from official sources of the Peruvian State. This data allows the identification of social, economic and infrastructure gaps, among others, for decision-making under a territorial approach. Geo Peru [29], contains information grouped into 27 categories, related to cartography, infrastructure, poverty, public investment projects, social programmes, health, education, economy, agriculture, tourism, culture, environment, social conflicts and gender violence in the country.

3.2 Empirical Methodology

This research aim was to explore, using spatial data mining techniques, the current state of the financial ecosystem between institutions and users in the Peruvian context.

Whereas the focus is on both Peruvian urban and rural territories, it was employing an open-source library, UrbanPy.

UrbanPy. This library is an open-source project to automate data extraction, measurement, and visualization of urban accessibility metrics created by Regal et al. [22]. Executing this library will enable us to construct, project, visualize and analyse layers of data in order to construct a Peruvian Region's polygons segmentation, as well as downloading walking, driving and biking networks within these polygons, around financial institutions in local territories. We execute this library within Python [23], along with other python libraries that will help us to process the data such as geopandas, numpy, pandas, shapely, plotly.express and osmnx.

4 Results

4.1 Peruvian Regional Visualization

After installing the *pip3 install urbanpy* command, in addition to the libraries previously mentioned, it selected the 25 political regions officials in Peru, for which we used the function *up.download.nominatim_osm* (Table 1).

Since it was defined each one of the Peruvian regions with the function *up.download.nominatim_osm* and the authentication code OSM, we are able to print the Peruvian region maps. Furthermore, it has been to be able to visualize varied spatial data, previously defined as a unified spatial unit, which means delimiting each one of the Peruvian regions in the study (Fig. 2a). Whereas, through Uber's H3 Hexagonal Hierarchical Geospatial Indexing System in Python (h3-py), Urbanpy [22], divide the city into hexagonal uniforms cells (Fig. 2b). We obtained the maps for the 25 Peruvian regions, although, we only represent one or two regions, aleatoric selected as a sample.

In addition, to get the Peruvian regional population density, it been applied the function *filter_population* (Fig. 3a). Since we had the population density information by region, we proceeded subsequently with the calculation of population density by hexagons (Fig. 3b).

4.2 Spatial Considerations of Local Financial Accessibility by Regions

This study focuses on the application of a spatial data mining technique to map the geographical proximity of banking institutions (e.g., Banks, exchange bureaus, ATMs), to explore a potential market that strengthens financial inclusion in the Peruvian rural agricultural sector. Therefore, the study examined the characteristics of each Peruvian regional territory, to identify the spatial distribution of banking institutions by region.

We gather information on the banks' spatial distribution using the hexagon population density layer and UrbanPy's points of interest functions (POI). These functions allow us to obtain financial institutions' facilities accessibility metrics, estimating travel time by feet, bike, and drive, from each hexagon as a spatial unit, to each banking institution. In order to obtain financial accessibility by Peruvian regions, UrbanPy identified the spatial distribution to the nearest financial services attention points from the previously identified centroid of the hexagons into the 24 Peruvian regions (Table 2).

Table 1. The authentication code for Peruvian regions.

Peruvian region	Code
Amazonas	PE-AMA
Ancash	PE-ANC
Apurimac	PE-APU
Arequipa	PE-ARE
Ayacucho	PE-AYA
Cajamarca	PE-CAJ
Callao	PE-CAL
Cusco	PE-CUS
Huancavelica	PE-HUV
Huánuco	PE-HUC
Ica	PE-ICA
Junio	PE-JUN
La Libertad	PE-LAL
Lambayeque	PE-LAM
Lima	PE-LMA
Loreto	PE-LOR
Madre de Dios	PE-MDD
Moquegua	PE-MOQ
Pasco	PE-PAS
Piura	PE-PIU
Puno	PE-PUN
San Martin	PE-SAM
Tacna	PE-TAC
Tumbes	PE-TUM
Ucayali	PE-UCA

Furthermore, we gather information on the Bank's spatial distributions from the 24 Peruvian regions, which allows us to explore the banking firms that may supply credit lending to Peruvian farmers (smallholders and enterprises). Although, as shown in Table 3, according to the ENA (2019) the Peruvian Farmer's accessibility rate by region is very lowest. nonetheless, the main activity of the Peruvian population living in rural areas is dedicated to agriculture, we expect that the facility location of banking institutions and population density, delivered to different regions becomes a factor for banking institutions considering expanding lending facilities, particularly in areas with a high concentration of agricultural population.

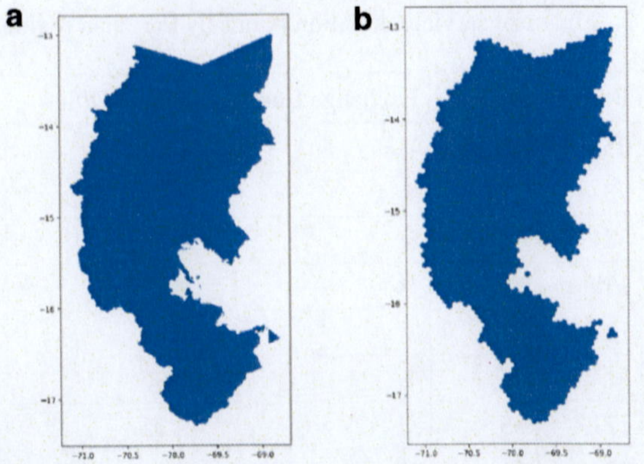

Fig. 2. **a.** Peruvian region, Puno, authentication code 'PE-PUN'. **b.** Peruvian region, Puno, divided into hexagonal cells.

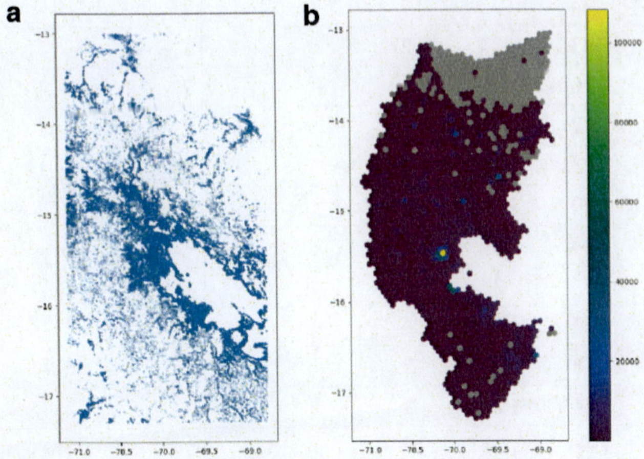

Fig. 3. **a.** Peruvian region, Puno, authentication code 'PE-PUN'. **b.** Peruvian region, Puno, divided into hexagonal cells.

Continuing the analysis on the Puno region, the POI detected by UrbanPy concerning financial services attention points, we obtained: Bank = 71; ATM = 23; Exchange bureaus = 8; and Post office = 3 (Fig. 4a). We identify, diverse financial services attention points, therefore, we must filter the accessibility to only financial organizations which have credit lending as a part of their services, filtering the data unloaded by the type "Bank" (Fig. 4b).

In addition, using the national geo-referenced information viewer, Geo Peru [29], we propose to explore a thematic layered visualization from the Puno region (Fig. 5). The thematic layer selected comprises the national information on the Peruvian agricultural sector, specifically, it was selected a thematic sub-layer which corresponds to farmers who accessed to a credit. The official source of information that provides thematic information about credit accessibility among producers in the Puno region, corresponds

Table 2. Financial services attention points by Peruvian regions (*).

Peruvian region	Bank	ATM	Exchange Bureaus	Post Office	Banking Agent
Amazonas	25	16	-	1	-
Ancash	73	12	5	12	-
Apurimac	38	16	8	5	1
Arequipa	138	56	14	9	1
Ayacucho	127	37	7	8	-
Cajamarca	51	25	4	3	-
Callao	121	35	31	22	3
Cusco	121	61	17	13	2
Huancavelica	81	24	2	3	1
Huánuco	80	13	-	7	1
Ica	54	25	-	7	-
Junio	83	14	2	1	1
La Libertad	100	32	4	7	1
Lambayeque	27	12	-	2	-
Lima	731	215	98	94	12
Loreto	99	34	7	18	2
Madre de Dios	34	19	5	7	-
Moquegua	5	2	-	-	-
Pasco	63	7	-	3	-
Piura	78	32	2	5	-
Puno	74	23	3	7	-
San Martin	28	11	-	8	1
Tacna	29	24	-	-	-
Tumbes	11	3	-	2	-
Ucayali	36	5	-	4	1

(*) = The POI detected by UrbanPy concerning financial services attention points correspond to 2020, which may have changed.

to data provided by the CENAGRO 2012 [17], the scale of data visualization it was building under the "Agricultural Registration Sector" (SEA, in its Spanish acronym) that corresponds to a portion of territory in which are located, on average 100 agricultural units.

Table 3. Financial services attention points by Peruvian regions (*).

Peruvian region	Rate farmers accessing credits (ENA* 2019)	Rate farmers not accessing credits (ENA* 2019)	Bank
Amazonas	0.135	0.881	25
Ancash	0.096	0.912	73
Apurimac	0.142	0.876	38
Arequipa	0.191	0.839	138
Ayacucho	0.066	0.938	127
Cajamarca	0.097	0.911	51
Callao	0.111	0.900	121
Cusco	0.092	0.916	121
Huancavelica	0.095	0.913	81
Huánuco	0.091	0.917	80
Ica	0.181	0.847	54
Junio	0.136	0.881	83
La Libertad	0.081	0.925	100
Lambayeque	0.245	0.803	27
Lima	0.175	0.851	731
Loreto	0.037	0.965	99
Madre de Dios	0.160	0.862	34
Moquegua	0.081	0.925	5
Pasco	0.127	0.887	63
Piura	0.174	0.851	78
Puno	0.093	0.915	74
San Martin	0.192	0.839	28
Tacna	0.217	0.822	29
Tumbes	0.288	0.776	11
Ucayali	0.184	0.845	36

(*) = Adapted from the ENA (2019).

Finally, under an explorative comparative visualization, we identify a potential relationship between population densities (Fig. 6a), the territory occupation from the formal Peruvian financial services providers (Fig. 6b), and the credit accessibility rates among Peruvian farmers in the Puno region (Fig. 6c). According to the datasets representations obtained from geographic open data sources, it has been identified that in those territories where traditional financial institutions have been located, the credit accessibility rate from farmers was higher than in those low habited territories which are furthest away from a banking institution.

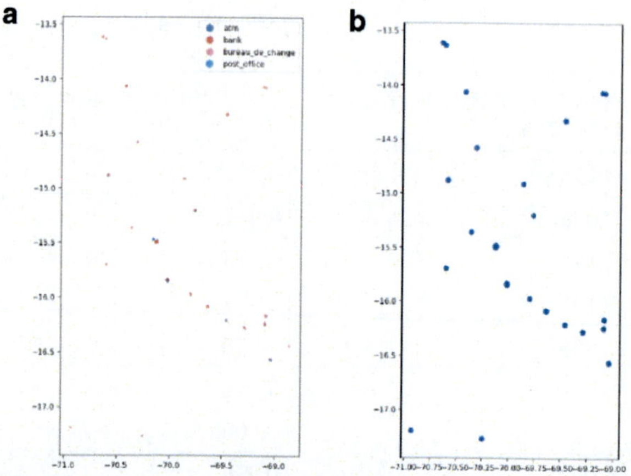

Fig. 4. **a.** POI in Puno region, "finance" category: Bank = 71; ATM = 23; Exchange bureaus = 8; and Post office = 3. **b.** POI in Puno r25egion, filtered by "bank" = 71.

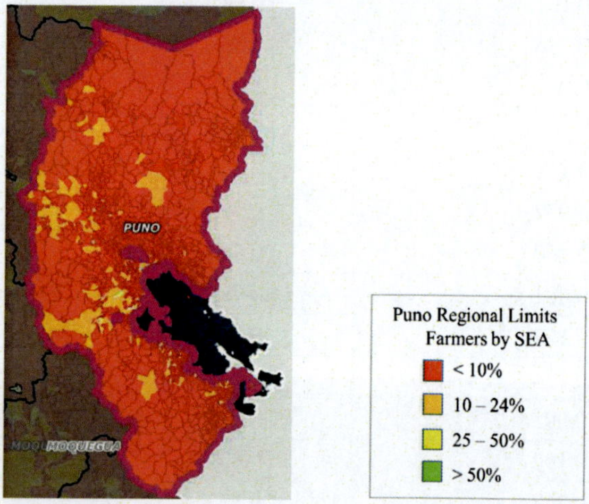

Fig. 5. Farmers from the Puno region who have access a credit. Source: Adapted from Geo Perú (https://visor.geoperu.gob.pe/).

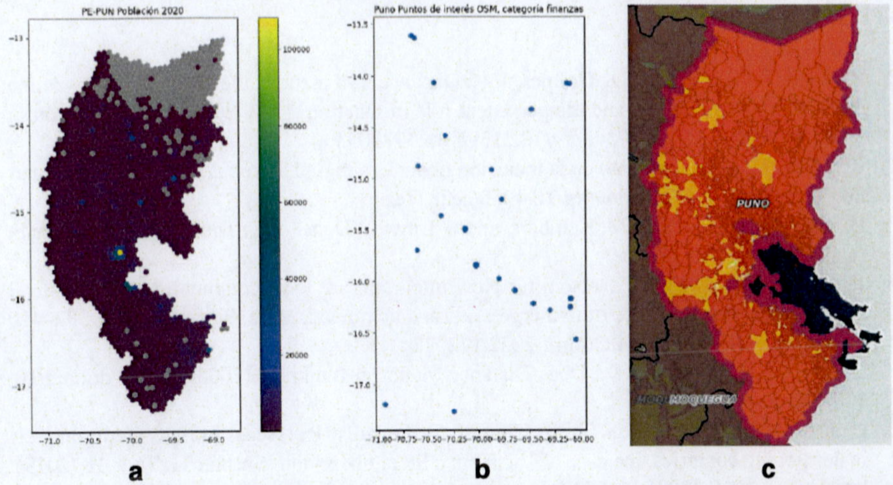

Fig. 6. **a.** Peruvian region, Puno, divided into hexagonal cells, representing population density. **b.** POI in the Puno region, representing the 71 banking institutions. **c.** Farmer's credit accessibility rate in the Puno region.

5 Conclusions

In this study, it has been proposed that geographic data science could become an innovative tool for analysis in rural areas. In line with this, the open-source library, UrbanPy [22], was implemented in order to explore the capacity of these tools to generate information in inaccessible physical locations in the Peruvian urban and rural contexts.

The exploration of the potential uses of geographic data available in open sources allows proposing innovative solutions to persistent gaps arising because of a lack of official data, although, through the generation of user-generated data, this becomes an opportunity in predominantly rural territories.

Since the Peruvian agricultural sector has insufficient access to finance becoming a critical obstacle [15], within these open-source tools, we identify different financial services attention points by Peruvian regions, although, that focused on the banking firms that may supply credit lending to the Peruvian farmers, most of them are agglomerated in urban areas. Therefore, considering the potential uses of population density as information data of non-met demand from financial services as credits, implementing strategies differentiated by regions may expand lending facilities, particularly in areas with a high concentration of agricultural population.

Finally, a limitation in our findings is related to the exploratory nature of the research, which is based on the visualization of the data. We expected that develop a model employing the proximity metrics, also provided by UrbanPy, which will provide better and more accurate information regarding mobility costs in the economic environment in rural areas.

References

1. Brevoort, K.P., Wolken, J.D., Holmes, J.A.: Distance Still matters: the information revolution in small business lending and the persistent role of location, 1993–2003. SSRN Electron. J. 1993–2003 (2012). https://doi.org/10.2139/ssrn.1782053
2. Benni, N.: El fortalecimiento de la inclusión financiera digital en zonas rurales y agropecuarias. Roma (2022). https://doi.org/10.4060/cc2877es
3. K. Brevoort, J. Clarkberg, M. Kambara, and B. Litwin, "Data Point : The Geography of Credit Invisibility," 2018
4. Jagtiani, J., Maingi, R.Q., Dolson, E.: How important are local community banks to small business lending? Evidence from mergers and acquisitions. J. Math. Financ. **12**(02), 382–410 (2022). https://doi.org/10.4236/jmf.2022.122022
5. Brevoort, K.P., Wolken, J.D.: Does Distance Matter in Banking? (2008). https://doi.org/10.1007/978-0-387-98078-2
6. Linh, T.N., Long, H.T., Van Chi, L., Tam, L.T., Lebailly, P.: Access to rural credit markets in developing countries, the case of Vietnam: a literature review. Sustain **11**(5), 1–18 (2019). https://doi.org/10.3390/su11051468
7. Anselin, L., Rey, S.J.: Open source software for spatial data science. Geogr. Anal. **54**(3), 429–438 (2022). https://doi.org/10.1111/gean.12339
8. Singleton, A.D., Spielman, S., Brunsdon, C.: Establishing a framework for open geographic information science. Int. J. Geogr. Inf. Sci. **30**(8), 1507–1521 (2016). https://doi.org/10.1080/13658816.2015.1137579
9. Arundel, J., Lowe, M., Roberts, R., Rozek, J., Higgs, C., Giles-corti, B.: Creating liveable cities in Australia: Mapping urban policy implementation and evidence-based national liveability indicators, no. October. Melbourne: Centre for Urban Research (CUR) RMIT University (2017)
10. Boeing, G., et al.: Using open data and open-source software to develop spatial indicators of urban design and transport features for achieving healthy and sustainable cities. Lancet Glob. Heal. **10**(6), e907–e918 (2022). https://doi.org/10.1016/S2214-109X(22)00072-9
11. Coetzee, S., Ivánová, I., Mitasova, H., Brovelli, M.A.: Open geospatial software and data: A review of the current state and a perspective into the future. ISPRS Int. J. Geo-Inf. **9**(2), 1–30 (2020). https://doi.org/10.3390/ijgi9020090
12. Yuan, M.: Geographical information science for the United Nations' 2030 agenda for sustainable development. Int. J. Geogr. Inf. Sci. **35**(1), 1–8 (2021). https://doi.org/10.1080/13658816.2020.1766244
13. Han, S.: Spatial stratification and socio-spatial inequalities: the case of Seoul and Busan in South Korea. Humanit. Soc. Sci. Commun. **9**(1), 1–14 (2022). https://doi.org/10.1057/s41599-022-01035-5
14. Regal Ludowieg, A., Ortega, C., Bronfman, A., Rodriguez Serra, M., Chong, M.: A methodology for managing public spaces to increase access to essential goods and services by vulnerable populations during the COVID-19 pandemic. J. Humanit. Logist. Supply Chain Manag., **12**(2), 157–181 (2021)
15. Trivelli, C.: Finanzas agropecuarias: Desafío pendiente en la agenda agraria en Perú Washington, DC (2021)
16. Gutiérrez, A.: Microfinanzas rurales: experiencias y lecciones para América Latina. CEPAL (2004)
17. INEI: IV Censo Nacional Agropecuario 2012. Lima (2014). http://iinei.inei.gob.pe/microdatos/
18. Wiener Fresco, H.C.: Perú: un mercado incompleto de servicios financieros al agro. Pensam. Crítico **26**(1), 5–38 (2021). https://doi.org/10.15381/pc.v26i1.20222

19. INEI: Encuesta Nacional Agropecuaria 2017. Lima (2018). http://iinei.inei.gob.pe/microdatos/
20. Demirgüç-Kunt, A., Klapper, L.F., Singer, D., Van Oudheusden, P.: The global findex database 2014: measuring financial inclusion around the world. World Bank Policy Res. Work. Pap. no. 7255 (2015)
21. Cowie, J., Lehnert, W.: Information extraction. Commun. ACM **39**(1), 80–91 (1996). https://doi.org/10.1145/234173.234209
22. Regal, A., Ortega, C., Vazquez Brust, A., Rodriguez, M., Zambrano-Barragan, P.: UrbanPy: a library to download, process and visualize high resolution urban data to support transportation and urban planning decisions BT - production and operations management, pp. 463–473 (2022)
23. VanRossum, G., Drake, F.L.: Python 3 Reference Manual. CreateSpace, Scotts Valley (2009)
24. Open Knowledge Foundation: "The Open Definition". http://opendefinition.org/. Accessed 23 Jan 2023
25. OpenStreetMap: "About us". https://www.openstreetmap.org/about
26. Rey, S.J., Arribas-Bel, D., Wolf, L.J.: Geographic Data Science with Python. CRC Press (2020). https://geographicdata.science/book/intro.html
27. Humanitarian Data Exchange. Datasets. Peru. https://data.humdata.org/dataset?q=Peru
28. Meta. Data for Good. https://dataforgood.facebook.com/
29. Peruano, E.:. Qué es Geo Perú. Geo Perú (2023). https://www.gob.pe/33929-que-es-geo-peru. Accessed 17 Feb 2023

A Sensor Based Hydrogen Volume Assessment Deep Learning Framework – A Pohokura Field Case Study

Klemens Katterbauer(✉), Abdallah Al Shehri, Abdulaziz Qasim, and Ali Yousif

Saudi Aramco, Dhahran 31311, Saudi Arabia
klemens.katterbauer@aramco.com

Abstract. Hydrogen has become a crucial potential energy carrier with significant opportunities to reduce the carbon footprint of power generation as well as an energy alternative for various applications. Hydrogen is abundant as an element on our earth and as a composite widely used in the form of water and other substances. Hydrogen can be readily used in fuel cells and produces as a by-product only water, which makes it a rather clean fuel. Volume assessments have assumed an important parameter for the development of hydrogen storage reservoirs. In the different stages of hydrogen storage utilization, one's cognition for geological conditions and injection and production rules keep on changing, and the precision of parameters acquired for the calculation of the hydrogen volumes improves. Therefore, hydrogen volume calculation and evaluation are meaningful task throughout the entire life cycle of reservoir hydrogen storage. Furthermore, they require the evaluation and cognition for hydrogen volumes to be continually updated, and the methods for hydrogen storage volume calculation to be identified and optimized. We present a new AI driven framework for hydrogen storage volume assessment integrating a random forest approach to accurately estimate expected volume storage fractions. The framework is connected to an uncertainty estimation framework for the determination of the S1 to S3 expected hydrogen storage volumes. The framework was evaluated on the Pohokura field in New Zealand and exhibited strong performance in the determination of the hydrogen storage fractions within the subsurface reservoir. The framework represents an important step towards the assessment of hydrogen storage volumes within subsurface reservoir for long-term hydrogen storage. This will play an important role at determining the productivity potential of hydrogen storage reservoirs and assess the overall storage capacities.

Keywords: 4IR · artificial intelligence · hydrogen storage · hydrogen storage volume assessment · uncertainty · New Zealand

1 Introduction

Hydrogen has become a crucial potential energy carrier with significant opportunities to reduce the carbon footprint of power generation as well as an energy alternative for various applications. Hydrogen is abundant as an element on our earth and as a composite

R. Silhavy and P. Silhavy (Eds.): CSOC 2023, LNNS 724, pp. 260–274, 2023.
https://doi.org/10.1007/978-3-031-35314-7_25

widely used in the form of water and other substances [1, 2]. Hydrogen can be readily used in fuel cells and produces as a by-product only water, which makes it a rather clean fuel.

Hydrogen can be readily derived from a variety of resources, such as natural gas, nuclear power, biomass and renewable power incorporating solar and wind. Given the variety of different sources from which hydrogen can be derived, makes it necessarily a very attractive option as a fuel for both transportation and electricity generation applications, as well as other associated applications [3, 4].

There are several methods in order to produce hydrogens. The most common methods are thermal processes, electrolytic processes, solar-driven processes and biological processes.

Biological processes involve the utilization of microbes in order to produce hydrogen through biological reactions [5, 6]. These microbes include both bacteria and microalgae and depending on whether it is a microbial biomass conversion or photobiological process, the microbes break down the organic matter or use sunlight in order to generate the hydrogen. Organic matter may be in the form of wastewater or any other biomass source. Microbial biomass conversion is amongst the most promising solutions in order to utilize the fermentation process to break down organic matter and produce subsequently hydrogen. The organic matter may be in the form of various materials, such as sugars, raw biomass source and wastewater. For the direct hydrogen fermentation process, the hydrogen is produced directly via the microbes [7]. Challenges arise from the rather slow fermentation process, and the limited yield that is currently derived from these systems. There are several initiatives such as microbial electrolysis cells that can harness the energy produced by microbes in order to produce both hydrogen and electricity. However, these processes are currently rather limited in scope to be able to effectively produce large quantities of hydrogen efficiently [8–10].

Solar driven processes utilize the photobiological, photoelectrochemical and solar thermochemical processes. Photobiological processes utilize the natural photosynthetic activity of bacteria and green already in order to generate the hydrogen from these natural matters. Photoelectrochemical processes use semiconductors in order to separate water into hydrogen and oxygen and extract the hydrogen [11, 12].

Solar-thermochemical hydrogen utilize solar power in order to split water in combination with species such as metal oxides. While being of considerable interest given the ability to utilize thermal energy from solar panels to separate the hydrogen from the oxygen, they are currently rather inefficient and the solar fusion process they drive is expensive to be implemented [13, 14].

Electrolytic processes utilize electrolysis in order to separate water into its molecular components [15]. This takes place in an electrolyzer, which is a reverse modeling of a fuel cell that separates the hydrogen from the water [16]. Electrolyzers contain an anode and cathode that is separated by an electrolyte. Electrolysis may be performed both at the small as well as large scale. At the small scale it may be utilized for small-scale distributed hydrogen production, such as via the production of hydrogen for houses by the utilization of solar energy. Central large scale production facilities, utilizing renewable or other non-greenhouse gas emitting forms for the energy production, may provide an efficient alternative to large scale hydrogen production for a variety of utilizations such as power production, and transportation. There are various forms of electrolyzers such as polymer electrolyte membrane electrolyzers, alkaline electrolyzers and solid oxide electrolyzers. Polymer electrolyte membrane electrolyzers use a polymer electrolyte membrane in order to separate the hydrogen from the oxygen, via an anode cathode configuration with the membrane in the middle [17].

Alkaline electrolyzers use a liquid alkaline solution of sodium or potassium hydroxide for the electrolyte that separates the hydrogen in the hydroxide ions when going from the cathode to the anode. Inefficiency is a major challenge of these electrolyte solutions, however there have been new approaches to utilize solid alkaline exchange membranes for the separation of hydrogen [16].

Solid oxide electrolyzers utilize ceramic materials as the electrolytes that enables a selective conduction of charged oxygen ions at high temperatures in order to produce hydrogen [18]. The steam at the cathode experiences a reaction with the electrons from the circuit forms subsequently a hydrogen gas and oxygen ions that are negatively charged. The resulting oxygen ions pass then through the ceramic membrane, reacting at the anode in order to form oxygen gas, as well as generate the electrons for the circuit. There are several challenges for electrolyzers in that the current cost of producing hydrogen are significant, specifically capital costs of electrolyzers are quite significant, in addition to the challenges related to energy efficiency in the conversion process [19]. Specifically, the stacking of various membranes and performance optimization of the electrolyzer cell is crucial in increase the operational life of the electrolyzer and enhance efficiency.

Finally, thermal processes, such as steam reforming are amongst the most widely utilized processes for the production of hydrogenous from hydrocarbon fuels [20]. There are several hydrocarbon fuels such as natural gas, diesel, renewable liquid fuels, gasified coal and gasified biomass that can be utilized for the production of hydrogen. Natural gas reforming is amongst the most common existing techniques to produce hydrogen from natural gas. Steam-methane reforming is a mature process where high temperature steam is utilized in order to split methane into its hydrogen components. The methane reacts with the high-temperature steam and a catalyst in order to produce hydrogen, carbon monoxide and a minor amount of carbon dioxide. Additionally, the carbon monoxide and the steam react in the presence of a catalyst to produce carbon dioxide and additional hydrogen. This process is typically called the resulting water-gas shift reaction. In the final step, called pressure-swing adsorption, the impurities such as carbon dioxide are removed in order to produce pure hydrogen. While natural gas is the most common fuel type, other fuels such as ethanol, propane or gasoline may be utilized for the production of hydrogen [21]. A different process, called partial oxidation, involves the methane and other hydrocarbons react with a certain amount of oxygen. The resulting process is typically much faster than steam reforming, however, it produces a lesser amount of hydrogen per fuel input as compared to steam reforming. While steam-methane reforming is an endothermic process, requiring heat to be added in order for the reaction to be completed, partial oxidation is an exothermic process implying that heat is removed within the process.

Volume assessments have assumed an important parameter for the development of hydrogen storage reservoirs. In the different stages of hydrogen storage utilization, one's cognition for geological conditions and injection and production rules keep on changing, and the precision of parameters acquired for the calculation of the hydrogen volumes improves. Therefore, hydrogen volume calculation and evaluation are meaningful task throughout the entire life cycle of reservoir hydrogen storage. Furthermore, they require the evaluation and cognition for hydrogen volumes to be continually updated, and the methods for hydrogen storage volume calculation to be identified and optimized [22].

Before the development of a hydrogen storage site, the pre-development stage is to ensure that an initial assessment of the reservoir formation is performed. If the hydrogen storage site is based on an existing reservoir, then existing assessments can be utilized to analyze the storage volume potential. This implies that the volumes have to be re-calculated when the production and injection is conducted. The importance of continuous assessment of the hydrogen storage volumes is a key factor for hydrogen storage volume assessments given that the production and injection is performed in periodical intervals [23].

In this article, we present a new subsurface sensor-based hydrogen storage volume assessment framework incorporating artificial intelligence. The framework incorporates continuous sensor information that determines in real-time maximum storage volume capacity, existing volume occupation and the production potential.

2 Methodology

The volume assessments can be typically divided in S1, S2 and S3. S1 represents the proved storage volumes for hydrogen, while S2 amount to the probable storage volumes. Finally, S3 are the possible storage volumes. In all circumstances, these volumes are considered to be able to be economically storable and the volumes shall be calculated at standard conditions of 15 degree Celsius and 0.101 MPa [24]. The storage volumes can depend on several factors such as legal regimes that limit the amount of hydrogen to be stored in the subsurface reservoir. Additionally, chemical and microbial factors can affect significantly the quality of hydrogen volumes, and hence the volume assessments. S2 applies to volumes that have not been confirmed, which may exist due to uncertainties regarding the full extent of the reservoir or other effects such as microbial that affect the stored hydrogen quality. S3 refers to the potential hydrogen storage volumes inferred from the ideal geological conditions. This implies that S1 shall be at least 90% that it can be really stored within the reservoir, while S2 shall provide a volume assessment that is at least 50% of the expected volume. Finally, S3 may represent a value that shall be at least 10%. The main objective behind the storage volume assessment is determining how much storage volume can be reasonably and reliably maintained, and also to assist in the planning for the possible production capacities in order to satisfy demand for hydrogen. Given the focus on real-time assessment, dynamic volume assessment, also referred to as dynamic geological volumes, represent an important step to connect the volume estimates to the volume injection and production dynamics data.

When focusing on dynamic volume assessments, there are several conventional methods utilized for the estimation. The main traditional methods are the material balance method, elastic two-phase method, unsteady well testing method and a numerical simulation method. There have been several new methods, such as those proposed by Fetkovich and Blasingame, as well as the AG, NPI, FMB and analytical method for evaluating the volumes for hydrogen storage subject to injection and production.

For the assessment of the hydrogen storage volume, a random forest deep ensemble learning framework was developed. The deep learning framework incorporates production and injection figures in addition to microbial measurements, together with salinity, pressure and temperature data retrieved from the wells via well-based sensors. Given the variety of different subsurface processes that affect the hydrogen storage quantity and quality, a deep learning approach is most preferable to take into account the complex dynamics. The framework will be outlined in greater detail below.

3 Results

The Pohokura gas field is located in the north-east of New Plymouth in the Taranaki Basin that is close to the Methanex Motunui site in Waitara (Fig. 1). The Taranaki Basin covers more than 100,000 km^2, primarily beneath the self and the continental slop that is offshore or in central-western of the North Island in New Zealand. The land sections are beneath the Taranaki Peninsula and in the northwest of the South Island. The basin incorporates a late Cretaceous to Quaternary sedimentary fill up that may be up to 8 km thick. The basin comprises an undeformed block and a heavily deformed area [25]. The heavily deformed area contains the Taranaki Fault that leads to a Miocene basement overthrusting into the basin. The sedimentary fill of the Taranaki Basis may have arisen due to an intracontinental manifestation of a transform fault offsetting or a failed-rift (Three-Dimensional Structural and Petrophysical Modeling for Reservoir Characterization of the Mangahewa Formation, Pohokura Gas-Condensate Field, Taranaki Basin, New Zealand). There were several tectonic episodes that influenced the generation of the Taranaki Basin, which included the rifting between the cretaceous – Eocene period, the compression within the Eocene period and an extension that arose from the late Miocene [26]. The Pohokura files is the largest gas-condensate field and is a low-relief N-S trending anticline in the Northern Graben. The boundaries of the field are supported by the Taranaki Fault Zone in the east and the Cape Egmont Fault Zone in the west. The reservoir underwent compressional and extensional stress regimes and most of the drilling was in the northern part of the Pohokura Field. The petroleum play is a transgressive marginal marine sand within an inverted anticline. The coal seams of the Rakopi Formation are the main sources for the hydrocarbons in the pohakura field. The Rakopi Formation represents the deepest stratigraphic unit within the Taranaki Basin, and represents the major source for oil and gas fields within the basin. The Rakopi Formation is heavily dominated by fluvial-to-marginal marine lithofacies. The Mangahewa Formation from the Eocene Epoch is the primary reservoir zone, being the most prolific and thickest. The Mangahewa Formation contains interbedded sandstone, siltstone, mudstone and coal. The Turi Formation represents a top seal for the Pohokura Field and arose from Paleocene to the Eocene. The top seal is composed of non-calcareous, dark colored, micaceous and carbonaceous marine mudstone that is distributed throughout the Taranaki Basin [27].

The water depth is around 35 m in the block PMP 38154. The Taranaki Basin has several reservoirs that range from the Paleocene to Pliocene and is a gas-condensate field in a low-relief anticline, that is north-south elongated. The reservoir is 16 km long, and 5 km wide.

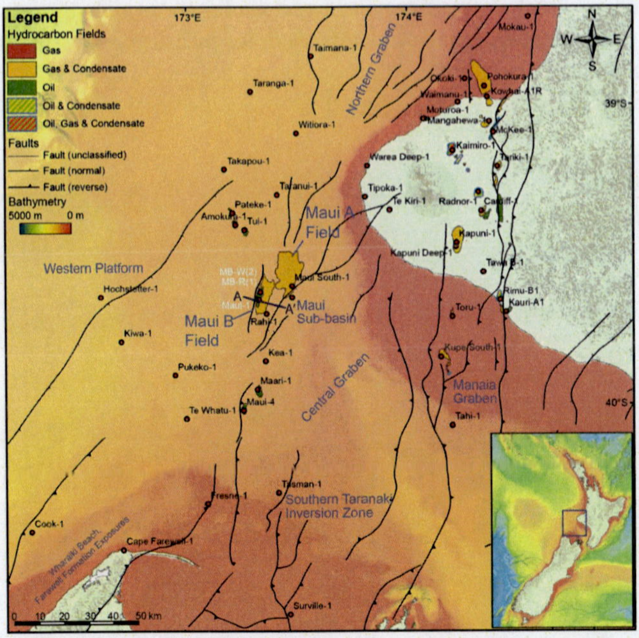

Fig. 1. Map of central and southern Taranaki basin with the Maui and Pohokura fields indicated in orange.

The Pohokura-1 well (Fig. 3) targeted the the Kapuni group Mangahewa Formation sands in the Pohokura structure, where in total 700 m of shallow marine sands were encountered. The overall gas column extended to 130 m in the upper part of the Mangahewa structure. A deepening of the well was performed in order to potentially access the lower Mangahewa formation and reach the TD in the Mid-Eocene shales of the Omata formation. The lower part of the formation was not economically viable, and never produced gas condensates [26].

In order to measure the gas production performance, two drill stem tests were conducted in the upper section within the interval of 3,625 to 3,634 m MD and between 3,553 to 3,570 m MD. The flow measurements for the first interval were 3.5 MMSCF/D, and 16.5 MMSCF/D for the second interval. The well was rather valuable in mapping the reservoir with seismic and have a control point for determine the overall structure of the Pohokura field. The expected gas storage volume for the Pohokura field are assumed to be around 1 trillion standard cubic feet [28].

Given the existing infrastructure and beneficial reservoir properties, the Pohokura reservoir is an attractive option to store efficiently large volumes of hydrogen within the reservoir formation. A key challenge besides the injection and productivity potential is to what extent microbial effects have an impact on longer term hydrogen storage. All the wells contained downhole subsurface sensors for measuring pressure, temperature, microbial composition and salinity. Additionally, the amount of CO2, H2S and corrosion was measured for each of the wells (Fig. 2).

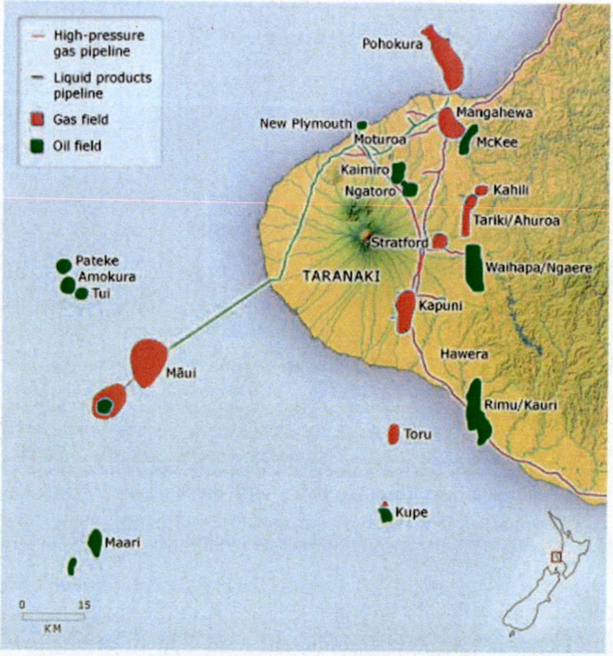

Fig. 2. Map of the oil and gas fields in New Zealand.

We evaluated the impact of microbial effects on hydrogen storage volumes on the Pohokura field as outlined in Fig. 4. For the hydrogen storage assessment, we assumed that only the southern part of the reservoir that is connected to the Pohokura Production Station, and the infrastructure of the Pohokura South station are utilized. Specifically, POH-01 and POH-02 are monitoring wells equipped with sensors that monitor the reservoir, and POS-01B and POW-3 are injector wells. The hydrogen is produced from POW-02 and POW-01 that are connected to the existing Pohokura production station.

The assumption for the reservoir was that the injected hydrogen is sufficiently injected, and based on demand produced from the reservoir via the two producer wells.

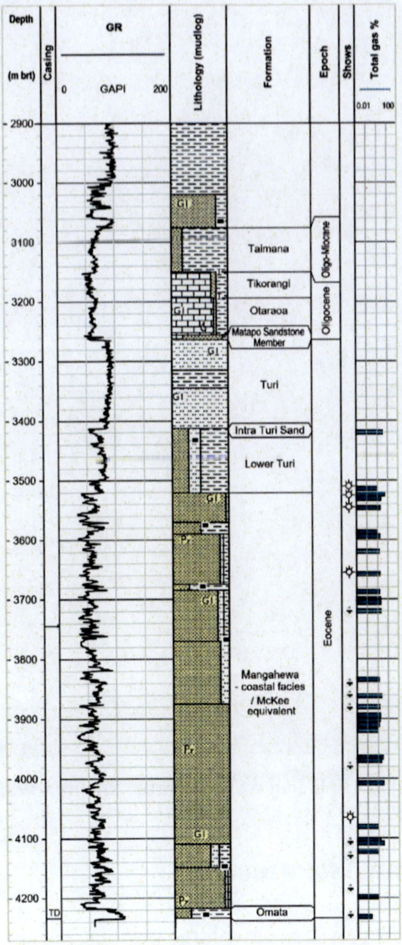

Fig. 3. Pohokura-1 gas well.

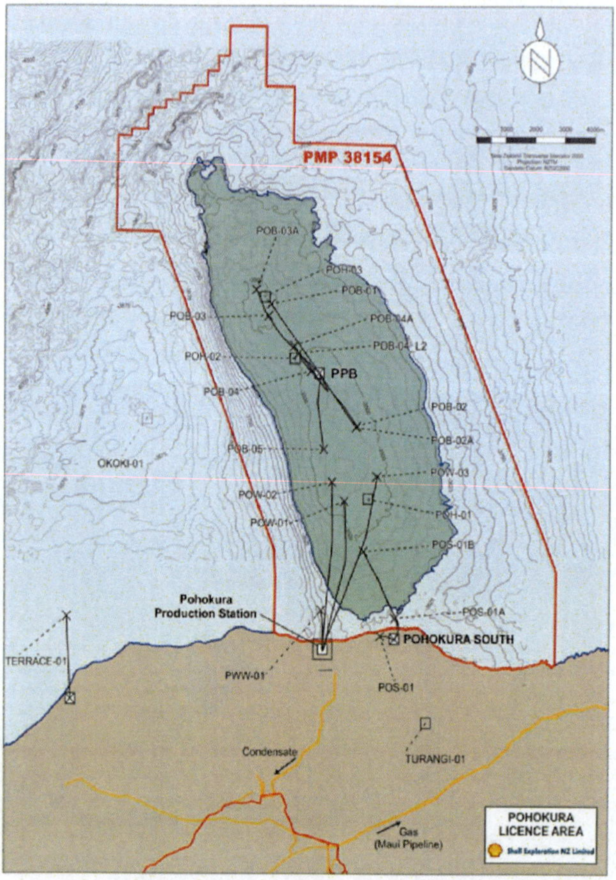

Fig. 4. Pohokura field and wells (Source: Shell Exploration NZ).

4 AI Storage Volume Assessment

In order to analyze the hydrogen storage volume options, we simulated expected hydrogen injection and production in addition to pressure, temperature, salinity and microbial quantities for the reservoir. The production and injection profiles for the Pohokura wells are displayed in Fig. 5 Both the injection and production profiles show fluctuating but relatively stable production and injection patterns. Injection quantities are higher as compared to the production with the intention to gradually fill the reservoir completely with hydrogen and evaluate the effects of longer-term storage on the hydrogen storage. Furthermore, the well pressure rates and the microbial growth rates are displayed in Fig. 6, which exhibit relatively robust pressure levels close to the wells, in addition to an increase in the microbial growth rates. Microbial effects have a significant effect on hydrogen quality that is stored within the subsurface formation and may degrade overall hydrogen storage quality. The major impact arises from the decomposition of the hydrogen into other gases, such as methane, hydrogen sulfide and acids. These factors significantly affect the hydrogen quality stored in the subsurface and also affect recovery rates. Limiting the effects on the hydrogen quality plays a critical role.

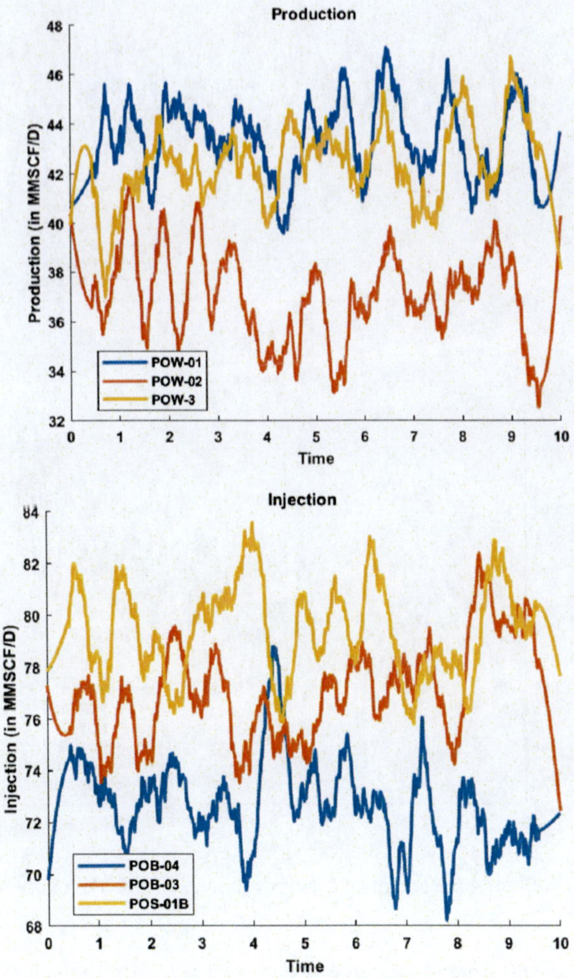

Fig. 5. Hydrogen injection and production profiles for the Pohokura wells.

The cumulative production and injection data, in addition to subsurface measurement data such as pressure, temperature, salinity and microbial quantities were then utilized in order to estimate the storage volume of hydrogen as a fraction of the overall storage volume. Both the training and test data results indicate a significant strong performance for the estimation of the overall hydrogen volume stored within the formation. This implies that within the considered ranges, the deep learning framework may accurately determine the hydrogen storage volume percentages (Fig. 7).

In order to determine the impact of the various input features on the estimation of hydrogen volumes in the reservoir, Shapley values were computed for the deep learning

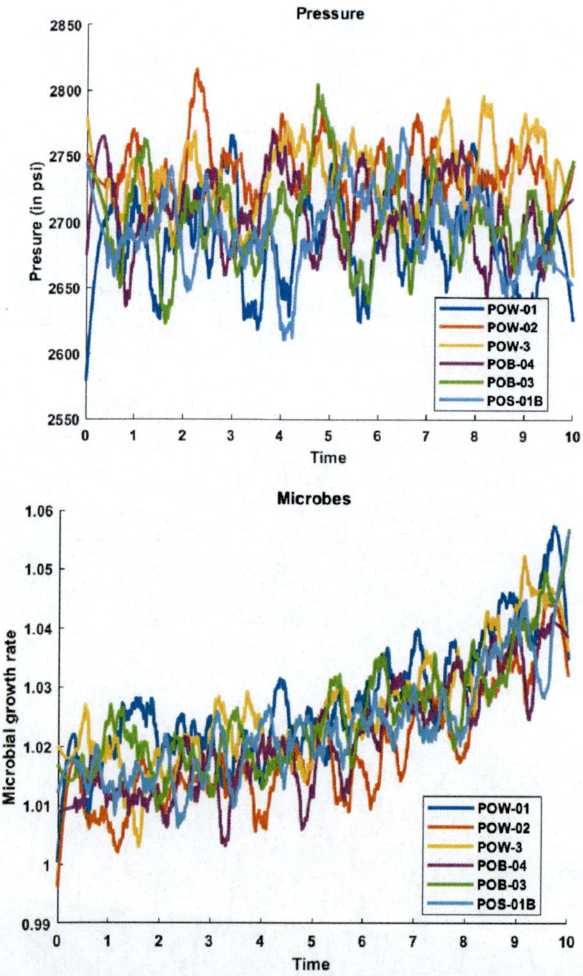

Fig. 6. Well pressure and microbial growth rates for the Pohokura wells.

model. The normalized Shapley parameters are displayed in Table 1. The comparison indicates a strong importance of both production and injection on the volume assessment with microbes and salinity concentration being additional important features.

Finally, the next step is the estimation of the uncertainty of the hydrogen storage volumes quantified by S1 to S3 storage volume curves. After ten years, the anticipated storage volumes range from 23 percent to little more than 65 percent. The variations in the individual estimates arise from a multitude of factors such as microbial and chemically induced changes in the hydrogen composition and overall quantity (Fig. 8).

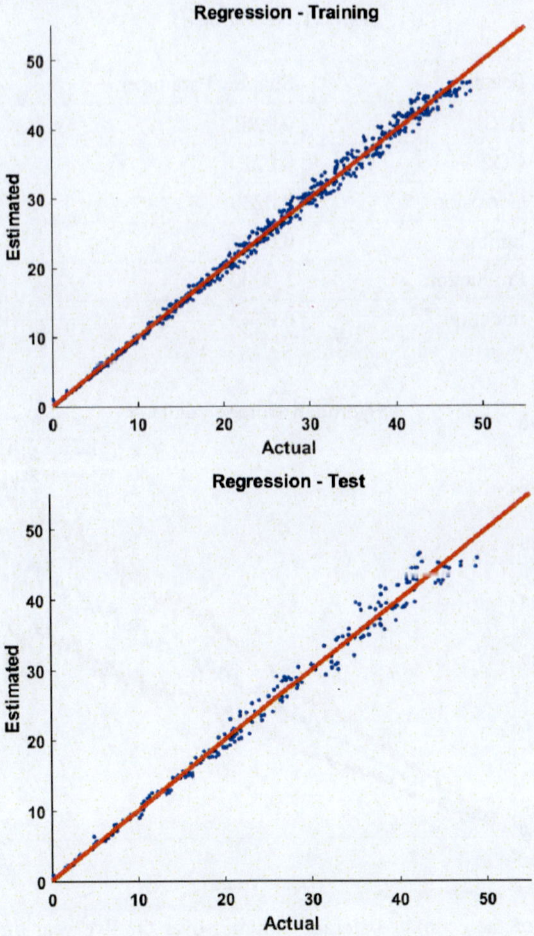

Fig. 7. Estimation comparison of the hydrogen storage percentage for both the training and test data set.

Table 1. Comparison of the scaled Shapley parameters for determining the impact of the various features on the estimation of the deep learning framework.

Feature	Shapley Parameter
Microbes	0.1479
Pressure	0.0022
Salinity	0.0920
Temperature	0.0401

(continued)

Table 1. (*continued*)

Feature	Shapley Parameter
H2S	0.0285
CO2	0.0229
Corrosion	0.0230
Inhibitor	0.0345
Production	1.0000
Injection	0.8742

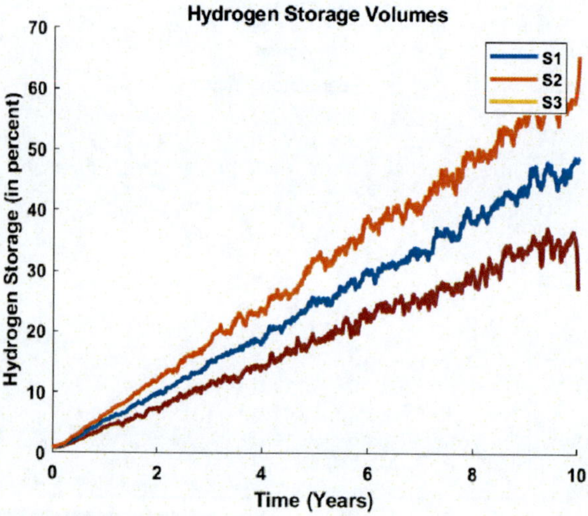

Fig. 8. Comparison of the estimated storage volumes over the five year time horizon for the Pohokura field)

5 Conclusion

In this article, we presented a new AI driven framework for hydrogen storage volume assessment integrating a random forest approach to accurately estimate expected volume storage fractions. The framework is connected to an uncertainty estimation framework for the determination of the S1 to S3 expected hydrogen storage volumes. The framework was evaluated on the Pohokura field in New Zealand and exhibited strong performance in the determination of the hydrogen storage fractions within the subsurface reservoir. The framework represents an important step towards the assessment of hydrogen storage volumes within subsurface reservoir for long-term hydrogen storage. This will play an important role at determining the productivity potential of hydrogen storage reservoirs and assess the overall storage capacities.

References

1. Turner, J.: Sustainable hydrogen production. Science **305**(5686), 972–974 (2004)
2. Dawood, F., Anda, M., Shafiullah, G.M.: Hydrogen production for energy: an overview. Int. J. Hydrogen Energy 3847–3869 (2020)
3. Katterbauer, K., Marsala, A.F., Schoepf, V., Donzier, E.: A novel artificial intelligence automatic detection framework to increase reliability of PLT gas bubble sensing. J. Petrol. Explor. Product. **11**(3), 1263–1273 (2021). https://doi.org/10.1007/s13202-021-01098-1
4. Al Shehri, A., Shewoil, A.: Connectivity analysis of wireless FracBots network in hydraulic fractures environment. In: Offshore Technology Conference Asia, Kuala Lumpur (2020)
5. Sivaramakrishnan, R., et al.: Insights on biological hydrogen production routes and potential microorganisms for high hydrogen yield. Fuel 120136 (2021)
6. Katterbauer, K., Qasim, A., Marsala, A., Yousef, A.: A data driven artificial intelligence framework for hydrogen production optimization in waterflooded hydrocarbon reservoir. In: Abu Dhabi International Petroleum Exhibition & Conference (2021)
7. Balachandar, G., Varanasi, J.L., Singh, V., Singh, H., Das, D.: Biological hydrogen production via dark fermentation: a holistic approach from lab-scale to pilot-scale. Int. J. Hydrogen Energy 5202–5215 (2020)
8. Katterbauer, K., Hoteit, I., Sun, S.: A time domain update method for reservoir history matching of electromagnetic data. In: Offshore Technology Conference, Kuala Lumpur (2014a)
9. Katterbauer, K., Hoteit, I., Sun, S.: EMSE: synergizing EM and seismic data attributes for enhanced forecasts of reservoirs. J. Petrol. Sci. Eng. **122**, 396–410 (2014). https://doi.org/10.1016/j.petrol.2014.07.039
10. Katterbauer, K., Hoteit, I., Sun, S.: History matching of electromagnetically heated reservoirs incorporating full-wavefield seismic and electromagnetic imaging. SPE J. **20**(5), 923–941 (2015)
11. Pourrahmani, H., Moghimi, M.: Exergoeconomic analysis and multi-objective optimization of a novel continuous solar-driven hydrogen production system assisted by phase change material thermal storage system. Energy 116170 (2019)
12. Katterbauer, K., Hoteit, I., Sun, S.: Synergizing crosswell seismic and electromagnetic techniques for enhancing reservoir characterization. SPE J. **21**(03), 909–927 (2016). https://doi.org/10.2118/174559-PA
13. Sadeghi, S., Ghandehariun, S.: Thermodynamic analysis and optimization of an integrated solar thermochemical hydrogen production system. Int. J. Hydrogen Energy 28426–28436 (2020)
14. Katterbauer, K., Al Qasim, A., Al Shehri, A., Yousif, A.: A Novel artificial intelligence framework for the optimal control of wireless temperature sensors for optimizing oxygen injection in subsurface reservoirs. In: Offshore Technology Conference Asia (2022)
15. Ursua, A., Gandia, L.M., Sanchis, P.: Hydrogen production from water electrolysis: current status and future trends. Proc. IEEE **100**(2), 410–426 (2012). https://doi.org/10.1109/JPROC.2011.2156750
16. Chi, J.., Hongmei, Y..: Water electrolysis based on renewable energy for hydrogen production. Chin. J. Catal. **39**(3), 390–394 (2018). https://doi.org/10.1016/S1872-2067(17)62949-8
17. Toghyani, S., Baniasadi, E., Afshari, E.: Numerical simulation and exergoeconomic analysis of a high temperature polymer exchange membrane electrolyzer. Int. J. Hydrogen Energy 31731–31744 (2019)
18. Al Zahrani, A.A., Dincer, I.: Thermodynamic and electrochemical analyses of a solid oxide electrolyzer for hydrogen production. Int. J. Hydrogen Energy 21404–21413 (2017)

19. Khan, M.A., et al.: Seawater electrolysis for hydrogen production: a solution looking for a problem? Energy Environ. Sci. (2021)
20. Chen, S., Pei, C., Gong, J.: Insights into interface engineering in steam reforming reactions for hydrogen production. Energy Environ. Sci. 3473–3495 (2019)
21. Ranjekar, A.M., Yadav, G.D.: Steam reforming of methanol for hydrogen production: a critical analysis of catalysis, processes, and scope. Indust. Eng. Chem. Res. 89–113 (2021)
22. Yu, W., et al.: A systematic method for assessing the operating reliability of the underground gas storage in multiple salt caverns. J. Energy Storage **31**, 101675 (2020)
23. Reitenbach, V., Ganzer, L., Albrecht, D., Hagemann, B.: Influence of added hydrogen on underground gas storage: a review of key issues. Environ. Earth Sci. **73**(11), 6927–6937 (2015). https://doi.org/10.1007/s12665-015-4176-2
24. Wang, X., Pan, L., Lau, H.C., Zhang, M., Li, L., Zhou, Q.: Reservoir volume of gas hydrate stability zones in permafrost regions of China. Appl. Energy **225**, 486–500 (2018)
25. Islam, M.A., Mutiah Yunsi, S.M., Qadri, T., Shalaby, M.R., Eahsanul Haque, A.K.M.: Three-dimensional structural and petrophysical modeling for reservoir characterization of the Mangahewa formation, Pohokura Gas-Condensate Field, Taranaki Basin, New Zealand. Natl. Resources Res. **30**(1), 371–394 (2020). https://doi.org/10.1007/s11053-020-09744-x
26. Knox, D.A., et al.: New Zealand gas development sets new performance and environmental milestones onshore and offshore-the role of drilling fluid selection and performance in the Pohokura development. In: SPE Asia Pacific Oil and Gas Conference and Exhibition (2008)
27. Sim, C.Y., Adam, L.: Are changes in time-lapse seismic data due to fluid substitution or rock dissolution? A CO_2 sequestration feasibility study at the Pohokura Field, New Zealand. Geophys. Prospect. 967–986 (2016)
28. Uruski, C.: New Zealand's deepwater frontier. Mar. Pet. Geol. **27**(9), 2005–2026 (2010)

Complex Network Analysis of the US Marine Intermodal Port Network

Natarajan Meghanathan(✉), Otto Ikome, Opeoluwa Williams, and Carolyne Rutto

Department of Electrical and Computer Engineering and Computer Science, Jackson State University, 1400 Lynch Street, Jackson, MS 39217, USA
natarajan.meghanathan@jsums.edu

Abstract. We consider the 21 intermodal ports through which two or more marine highways go through and are located either in the east coast, gulf coast or the mid western US. We model the US Marine Intermodal Port Network (MIPN) as an undirected graph of nodes represented by the intermodal ports and two nodes are connected by an edge if one or more marine highway goes through the corresponding ports. We present a comprehensive evaluation of the MIPN with respect to node and edge centrality metrics as well as a suite of network-level metrics and identify prospective bottleneck ports and marine highways owing to their topological position. We also conduct cluster analysis and identify the bridge nodes critical for network and cluster connectivity. We conduct principal component analysis on a dataset of these results and observe the MIPN to be topologically unique compared to transportation networks already studied in the literature.

Keywords: Intermodal port · Marine highway · Cluster analysis · Bridge nodes · Complex network analysis · Centrality metrics · Principal component analysis

1 Introduction

In 2007, the US Congress passed public law 110-140 to develop marine highways as waterborne alternatives (with respect to several criteria such as traffic congestion, fuel usage, wear and tear of the roads, public safety and security, etc.) to ease the freight transportation load across the interstate roads connecting the major cities in the country. Accordingly, the US Maritime Administration (MARAD) identified marine highways (navigable waterways) that run parallel or close to the major interstate roads and designated them with numerical identifiers that correspond to the interstate roads for which they are meant to serve as an effective alternative. For example, M-55 refers to the marine highway (on the Mississippi river) that runs along with interstate I-55 from New Orleans, LA to Chicago, IL. We used the marine highway network map posted at the MARAD website [1] and the one-pager description of the marine highways in [2] to build a graph model of the US marine highway network (MHN). A node in the MHN corresponds to a marine highway and an edge connects two nodes if the corresponding marine highways intersect at one or more intermodal ports. An intermodal port is a port that is located on a marine highway and a nearby city supports at least two of the three major forms of transportation: rail, road and air.

R. Silhavy and P. Silhavy (Eds.): CSOC 2023, LNNS 724, pp. 275–284, 2023.
https://doi.org/10.1007/978-3-031-35314-7_26

Among the 22 marine highways reported in MARAD, we did not consider the M-146, M-295, M-495 and M-580 highways as their individual waterway spanning distance is much smaller compared to the rest of the marine highways, and we did not consider the M-5 and M-85 highways in the west coast as they were disconnected from the rest. Among the remaining 16 marine highways spanning the east coast, gulf coast and the mid western portion (EGMW) of the US: M-71 and M-77 had identical waterways (hence used just M-71 and did not consider M-77). We hence focused on the 15 marine highways spanning the east coast, gulf coast and the mid western portion of the US to construct our MHN. We identified a total of 89 intermodal ports located on the 15 marine highways of the MHN. We observed 21 of these 89 intermodal ports to be located on more than one marine highway and chose them as the basis to construct the graph model for a US marine intermodal port network (MIPN) as these ports (referred to as core ports) would serve as crucial hubs (similar to the hub airports in an airport network) for freight transfer across marine highways. The rest of the intermodal ports are more like stub ports located in just a particular marine highway and would not serve as hub for freight transfer. Ours is the first such work in literature to build and present a complex network analysis of the MIPN in the US.

Figure 1 presents an intersection matrix (corresponding to the MHN) of the 15 marine highways (starting with a prefix of M-) and the core intermodal ports (starting with a prefix of IP-) that are present at the intersections of two or more marine highways; the names of the 21 core intermodal ports are shown along with. These 21 core ports form the nodes of the MIPN whose intersection matrix is shown in Fig. 2. There is an edge (represented using an 'x' in the intersection matrix of Fig. 2) between two nodes in the MIPN (modeled as an undirected graph) if the corresponding core intermodal ports are present in one or more of the same marine highways.

		M-10	M-29	M-35	M-40	M-49	M-55	M-64	M-65	M-69	M-70	M-71	M-75	M-87	M-90	M-95
Albany, NY	IP-1													X	X	
Aransas Pass, TX	IP-2	X								X						
Brownsville, TX	IP-3	X								X						
Chicago, IL	IP-4						X								X	
Cleveland, OH	IP-5											X			X	
Corpus Christi, TX	IP-6	X								X						
Detroit, MI	IP-7												X		X	
Galveston, TX	IP-8	X								X						
Houston, TX	IP-9	X								X						
Jacksonville, FL	IP-10	X														X
Kansas City, MO	IP-11		X								X					
Memphis, TN	IP-12						X		X							
Mobile, AL	IP-13	X							X							
Morgan City, LA	IP-14	X				X				X						
New Orleans, LA	IP-15	X					X									
NewYork City, NY	IP-16													X		X
Norfolk, VA	IP-17							X								X
Paducah, KY	IP-18						X		X		X					
Port Arthur, TX	IP-19	X								X						
Rosedale, MS	IP-20				X		X									
St. Louis, MO	IP-21		X	X			X				X					

Fig. 1. Intersection Matrix of the US Marine Highway Network and the Core Intermodal Ports

The rest of the paper will focus on the complex network analysis of the MIPN and is organized as follows: Sect. 2 presents the results of the community detection algorithm (cluster analysis) run on the MIPN and ranks the nodes per the role of "bridge" nodes that connect the clusters. Section 3 presents the ranking of the nodes and edges in the MPIN on the basis of a suite of node and edge centrality metrics. Section 4 presents results of principal component analysis (PCA) conducted on a dataset comprising of various network-wide metrics incurred for the MIPN vis-a-vis other transportation-related networks and demonstrates the uniqueness of the MIPN. Section 5 concludes the paper and presents plans for future work. Throughout the paper, the terms 'node' and 'vertex', 'link' and 'edge', 'network' and 'graph', 'community' and 'cluster' are used interchangeably. They mean the same.

	IP-1	IP-2	IP-3	IP-4	IP-5	IP-6	IP-7	IP-8	IP-9	IP-10	IP-11	IP-12	IP-13	IP-14	IP-15	IP-16	IP-17	IP-18	IP-19	IP-20	IP-21
IP-1				X	X		X									X					
IP-2			X			X		X	X	X			X	X	X				X		
IP-3		X				X		X	X	X			X	X	X				X		
IP-4	X				X		X					X			X			X		X	X
IP-5	X			X			X														
IP-6		X	X					X	X	X			X	X	X				X		
IP-7	X			X	X																
IP-8		X	X			X			X	X			X	X	X				X		
IP-9		X	X			X		X		X			X	X	X				X		
IP-10		X	X			X		X	X				X	X	X	X	X		X		
IP-11																		X			X
IP-12				X									X		X			X		X	X
IP-13		X	X			X		X	X	X		X		X	X			X	X		
IP-14		X	X			X		X	X	X			X		X				X		
IP-15		X	X	X		X		X	X	X		X	X	X				X	X	X	X
IP-16	X									X							X				
IP-17										X						X					
IP-18				X							X	X	X		X					X	X
IP-19		X	X			X		X	X	X			X	X	X						
IP-20				X								X			X			X			X
IP-21				X							X	X			X			X		X	

Fig. 2. Intersection Matrix of Core Ports of the US Marine Intermodal Port Network

2 Cluster Analysis

A cluster or a community in a complex network is a subset of the nodes that are more densely connected among themselves than to the rest of the nodes in the network [3]. We refer to the clusters of a complex network to be more modular [4] if the intra-cluster density is appreciably greater than the inter-cluster density. While most of the nodes in a cluster are likely to be connected "only" to nodes within the same cluster (home cluster), if the underlying graph is connected, there could be one or more "bridge" nodes for each cluster. A bridge node for a cluster has edges to bridge nodes in one or more other (alien) clusters; nevertheless, majority of the edges for a bridge node are expected to be with nodes in its home cluster [5]. Bridge nodes are critical for connectivity among clusters, for freight transportation and for the overall connectivity of the MIPN.

We ran the MIPN through the well-known Louvain community detection algorithm [6] and observed the following three clusters (referred to as the M-10 cluster, M-55 cluster and M-90 cluster) of the core intermodal ports whose individual identifier names coincide with the marine highways in which the majority of the ports in the particular clusters are located. The **M-10 cluster** comprises of the following nine core ports: IP-2: Aransas Pass, TX; IP-3: Brownsville, TX; IP-6: Corpus Christi, TX; IP-8: Galveston,

TX; IP-9: Huston, TX; IP-10: Jacksonville, FL; IP-13: Mobile, AL; IP-14: Morgan City, LA and IP-19: Port Arthur, TX. The **M-55 cluster** comprises of the following seven core ports: IP-4: Chicago, IL; IP-11: Kansas City, MO; IP-12: Memphis, TN; I IP-15: New Orleans, LA; P-18: Paducah, KY; IP-20: Rosedale, MS and IP-21: St. Louis, MO. The **M-90 cluster** comprises of the following five core ports: IP-1: Albany, NY; IP-5: Cleveland, OH; IP-7: Detroit, MI; IP-16: New York City, NY and IP-17: Norfolk, VA. Figure 3 displays the MHN, the 21 core intermodal ports of the MIPN and their clusters (differentiated using the colors of the nodes representing the ports).

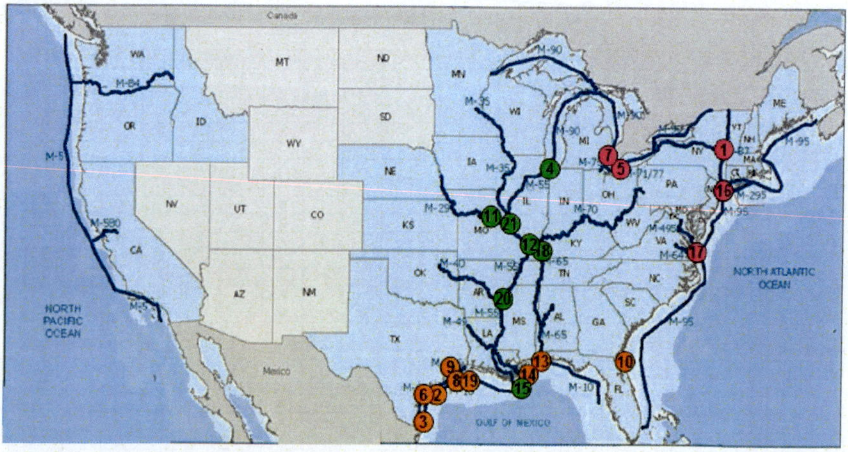

Fig. 3. Map of the MHN and the Clusters of the Core Intermodal Ports of the MIPN

The M-10 cluster is the most densest with the M-10 highway running through all the 9 core ports (in other words, these core ports are directly reachable to each other). Nevertheless, the presence of just one marine highway connecting all the 9 core ports in the Gulf coast also makes these ports vulnerable for closure and disconnection from the rest of the MIPN in the wake of hurricanes approaching the coast. The M-90 cluster is the least densest, comprising of marine highways M-90, M-75, M-71, M-87 and M-95 that connect the five core ports (in other words, any two core ports of this cluster are more likely not directly reachable to each other and are reachable only through multi-hop paths) and the ports are vulnerable for disconnection if a "bridge" port (see more details below) such as the IP-1: Albany, NY port gets closed. The M-55 cluster is also moderately connected with four marine highways involved in connecting seven core ports. The low-moderate intra-cluster density of two of the three clusters is reflected in a low overall modularity score [6] of 0.338 (in a scale of 0.0 to 1.0).

The NBNC (neighborhood-based bridge node centrality) tuple [5] is used to quantify and rank the nodes on the basis of the extent with which they play the role of bridge nodes. The NBNC tuple for a node comprises of three entries and is determined based on the neighborhood graph of the node. The first entry in the NBNC tuple indicates the number of components in the neighborhood graph of the node; the second entry is the ratio of the algebraic connectivity [7] of the neighborhood graph and the degree of the node; the third entry is the degree of the node. Figure 4 presents the NBNC tuple values of the 21 core nodes of the MIPN in the order of their ranking (high to low) with regards to the extent with which they play the role of bridge nodes. The lower the rank number for a node in Fig. 4, the higher is its ranking as a bridge node. If the first two entries in the NBNC tuples of two or more nodes are the same, the tie is broken on the basis of the degree of the nodes (the larger the degree, the higher the ranking). If all the three entries are the same, the nodes are ranked equally. Figure 4 also presents an Yifan Hu Proportional layout [8] of the MIPN (generated using Gephi [9]) that corroborates the NBNC-based ranking of the nodes as bridge nodes. The colors of the nodes in the three clusters of the MIPN layout in Fig. 4 correspond to the colors of the nodes shown in the clusters in Fig. 3.

Node ID	# Components	Algebraic Connectivity ratio	Degree	Ranking
IP-10	2	0	11	1
IP-4	2	0	8	2
IP-1	2	0	4	3
IP-16	2	0	3	4
IP-15	1	0.0332	14	5
IP-13	1	0.0909	11	6
IP-18	1	0.1325	7	7
IP-21	1	0.1667	6	8
IP-12	1	0.3333	6	9
IP-2	1	1	9	10
IP-3	1	1	9	10
IP-6	1	1	9	10
IP-8	1	1	9	10
IP-9	1	1	9	10
IP-14	1	1	9	10
IP-19	1	1	9	10
IP-20	1	1	5	11
IP-5	1	1	3	12
IP-7	1	1	3	12
IP-17	1	1	2	13
IP-11	1	1	2	13

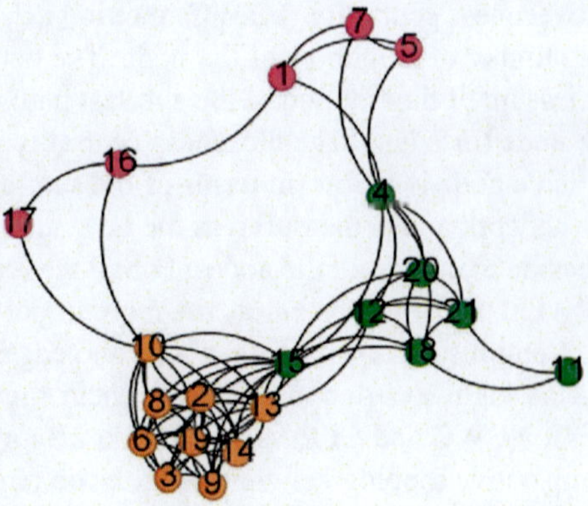

Fig. 4. Neighborhood-based Bridge Node Centrality (NBNC) Tuples of the Nodes (Core Intermodal Ports: IP) in the MIPN and the Topological Layout of the Clusters

We conclude the following nine core ports as the top-ranked bridge nodes (in the order of the ranking shown in Fig. 4) of the MIPN. We observe node IP-10 (Jacksonville, FL) to be critical in connecting the Gulf coast ports to the East coast ports and node IP-4 (Chicago, IL) to be critical in connecting the upper mid-west ports to both the Gulf coast ports and the lower mid-west ports. In the absence of IP-10 (Jacksonville, FL), the shortest paths between the Gulf coast ports and the East coast ports have to go through the mid-west ports in the M-55 cluster. Likewise, in the absence of IP-4 (Chicago, IL), the shortest paths between the Gulf coast ports and the mid-west ports have to go through the entire east coast. IP-1 (Albany, NY) is critical to facilitate connectivity between the East coast ports and the upper mid-west ports in the M-90 cluster. IP-16 (New York City, NY) facilitates connections between the Gulf coast ports and the East coast/upper mid-west ports. IP-15 (New Orleans, LA) is the only core port that provides direct connectivity for the Gulf coast ports to an upper mid-west port IP-4 (Chicago, IL) and is also connected to the majority of the lower mid-west ports in the M-55 cluster. Likewise, node IP-13 (Mobile, AL) connects the Gulf coast ports to the lower mid-west ports, but not to any upper mid-west port. Nodes IP-18 (Paducah, KY), IP-21 (St. Louis, MO) and IP-12

(Memphis, TN) are critical (in this order) for connectivity among the lower mid-west ports as well as their connection to ports in the upper mid-west and Gulf coast.

3 Centrality Metrics

Centrality is a measure of the topological importance of the nodes and edges in a network [3]. Centrality of a node or edge is typically measured as a scalar value [3], but in Sect. 2, we demonstrated the use of the recently proposed NBNC tuple [5] to quantify and rank the extent with which nodes play the role of bridge nodes. In this section, we rank the 21 core ports of the MIPN on the basis of the following major node centrality metrics: degree (DEG) [3], node betweenness centrality (NBWC) [10] and closeness centrality (CLC) [11] as well as rank the 74 edges between the core ports with respect to the edge betweenness centrality (EBWC) metric [12]. The degree centrality of a node is simply the number of neighbors of the node. The betweenness centrality of a node (or edge) is a measure of the fraction of the shortest paths (between any two nodes) that go through the node (or edge). The closeness centrality of a node is a measure of the extent with which a node is closer (in terms of the shortest path lengths) to the rest of the nodes in the network and is measured as the inverse of the sum of the shortest path lengths from the node to the rest of the nodes. For all the four centrality metrics (DEG, NBWC, CLC, EBWC), the larger the value, the more important is a node or edge, as applicable.

Centrality analysis of the nodes and edges in the MIPN identifies the ports and the marine highways that are likely to incur the bulk of the traffic. Figure 5 displays the DEG, NBWC and CLC values for the 21 core ports of the MIPN as well as presents a comparative topological view of the node rankings (on the basis of the node size) with respect to each of these three centrality metrics. Figure 5 also illustrates a comparative view of the edge betweenness centrality (EBWC) values: edges with larger EBWC values are shown thicker and vice-versa. Overall, we observe the ports of IP-15: New Orleans, LA, IP-10: Jacksonville, FL and IP-4: Chicago, IL to incur larger centrality values with respect to all the three node-level metrics as well as the IP-4—IP-15 edge (marine highway M-55) incurs the largest EBWC value of 32.25 that is about 70% greater than the next largest EBWC value of 18.68 incurred for the edge IP-10—IP-16 (marine highway M-95). The IP-4—IP-15 edge could be considered the primary edge for freight transportation between the nine M-10 cluster ports in the Gulf coast to the upper mid-west ports and east coast ports; the IP-10—IP-16 edge could be considered to provide a backup path (a longer shortest path) between the Gulf coast ports and the upper mid-west/east coast ports.

With respect to DEG, we observe the intermodal ports in the M-10 cluster (the ports in the Gulf coast) to incur significantly larger values compared to the ports in the other two clusters. However, a larger DEG centrality value does not guarantee a larger NBWC value for several of these ports. There are nine high-degree ports connected to each other in the M-10 cluster/Gulf coast and closer (through multi-hop minimum hop paths) to the ports in the other two clusters, but only three of them (New Orleans, LA; Jacksonville, FL; and Mobile, AL) are observed to incur high values for the node betweenness centrality and connect the other M-10/Gulf coast ports to the rest of the ports in the MIPN. We observe the ports in the M-90 cluster to incur the smallest centrality values with respect to all the three metrics. However, the EBWC values of the edges incident on these ports (in the upper mid-west/east coast: M-90 cluster) are relatively much larger than the edges in the M-10 cluster. Since the Gulf coast ports are directly connected to each other through M-10, the EBWC values for the edges within the M-10 cluster are much smaller. The edges connecting the ports in the M-90 cluster need to support traffic involving several pairs of ports. With respect to EBWC, we observe six of the top eight edges to involve a port in the M-90 cluster to be at least one of the two end vertices. The ports in the M-10 cluster are closer to the rest of the MIPN ports in terms of the shortest path lengths (CLC metric); whereas, the ports in the M-90 cluster are the farthest from the rest of the ports.

NodeID	DEG	NBWC	CLC
IP-1	4	8.566667	0.022222
IP-2	9	0	0.028571
IP-3	9	0	0.028571
IP-4	8	41.31667	0.030303
IP-5	3	0	0.020833
IP-6	9	0	0.028571
IP-7	3	0	0.020833
IP-8	9	0	0.028571
IP-9	9	0	0.028571
IP-10	11	30.18333	0.03125
IP-11	2	0	0.010868
IP-12	6	1.916667	0.027778
IP-13	11	12.86667	0.03125
IP-14	9	0	0.028571
IP-15	14	63.98333	0.038462
IP-16	3	9.25	0.023256
IP-17	2	0	0.021277
IP-18	7	13.01667	0.028571
IP-19	9	0	0.028571
IP-20	5	0	0.027027
IP-21	6	6.9	0.027778

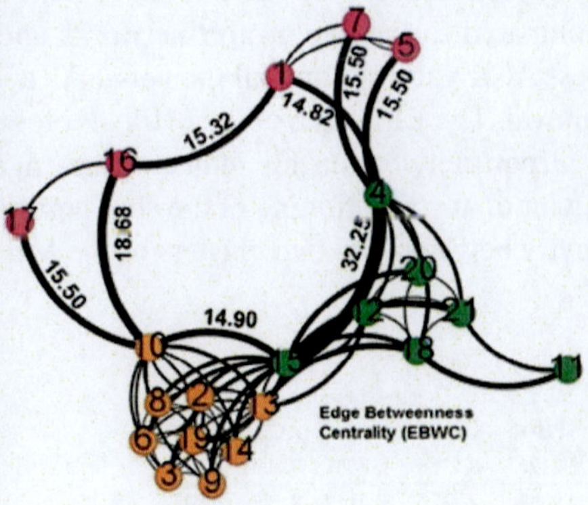

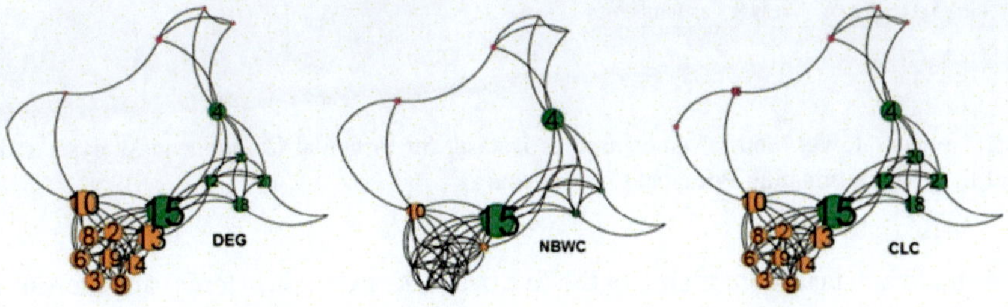

Fig. 5. Node and Edge Centrality Metrics for the MIPN

4 Principal Component Analysis of Network-Level Metrics

We seek to show that the topological characteristics of the MIPN are different from that of the other transportation related networks already studied in the literature. We choose the US Airports Network (USAN: 332 nodes and 2126 edges [13]), London Train Stations Network (LTSN: 381 nodes and 507 edges [14]), the EU Airports Network (EUAN: 418 nodes and 1999 edges [15]) and the EU Road Network (EURN: 1174 nodes an 1417 edges [16]) for our comparative analysis. The topological characteristics of these five networks are captured on the basis of five well-known network-level metrics for complex network analysis: Spectral radius ratio for node degree (SPR-K [17]), Assortativity index (AssI [3]), Randomness index (RanI [18]), Algebraic connectivity ratio (ACR [7]) and Bipartivity index (BPI [19]). All these five metrics are independent of the number of nodes and edges in the network.

Figure 6 presents the values incurred for the above five network-level metrics for the MIPN and the other four transportation networks. We observe the MIPN to incur relatively the lowest value for SPR-K, indicating that its vertices exhibit the least variation in their degree values. The MIPN is also the only one among these five transportation networks to exhibit a positive value for the assortativity index (i.e., the MIPN is assortative with respect to node degree; whereas the other four networks are dissortative). The MIPN (like the LTSN) exhibits more randomness in the associations of nodes as edges compared to the two airport networks and the road network. The MIPN incurs the largest ACR value among all the networks and is thus more resilient to any node or edge removal. The BPI value of the MIPN is close to 0.5 (like that of USAN and EUAN, the two airport networks): this indicates the edge distribution among the core ports is independent of any partitioning of the core ports into two disjoint sets. Figure 4 displays connectivity between any two clusters of the MIPN.

	SPR-K	AssI	RanI	ACR	BPI
USAN	3.2195	-0.1186	-0.8887	0.0004	0.5
LTSN	3.6004	-0.7258	-0.3139	0.0002	0.9819
EUAN	3.8111	-0.2051	-0.8647	0.0011	0.5
EURN	1.6613	-0.5195	-0.5387	0	0.9769
MIPN	1.3124	0.3271	-0.3850	0.0390	0.5029

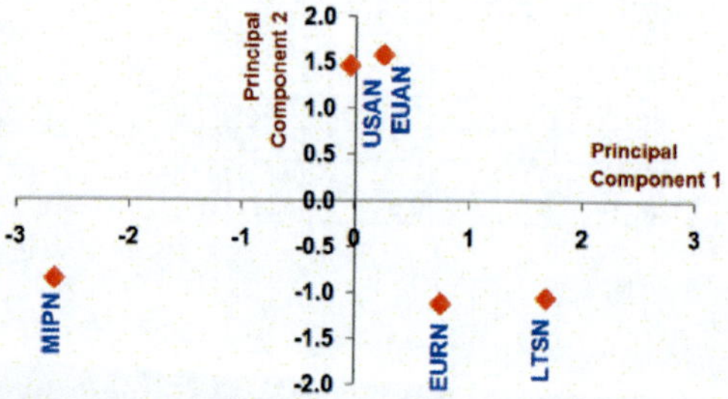

Fig. 6. Network-Level Metric Values used as Dataset for Principal Component Analysis and a Plot of the Two Dominating Principal Components

We build a dataset comprising of the five networks as rows (records) and the values incurred for the above five network-level metrics as columns (features). We then run Principal Component Analysis (PCA [20]) on this dataset and extract the two principal components that make up for more than 80% of the variances among the feature values. We use the entries for the five networks in these two principal components as X and Y coordinates and plot them (see Fig. 6). We observe the MIPN to be isolated from the other four networks: this justifies our claim that the MIPN is indeed topologically different from the other transportation networks already studied. The values incurred for the network-level metrics (lowest SPR-K value, the only network to be assortative, largest ACR value) corroborate our claim that the MIPN is topologically unique among the transportation networks analyzed in this paper.

5 Conclusions and Future Work

The high-level contribution of this paper is a complex network model for the core US marine intermodal ports located on two or more marine highways. To the best of our knowledge, this is the first such work to propose a marine intermodal port network (MIPN) and its analysis. The MIPN is observed to encompass three clusters, each comprising of a majority of ports located in a particular marine highway (and accordingly referred to as M-10, M-55 and M-90 clusters). We used the Neighborhood-based Bridge Node Centrality (NBNC) tuple to identify the critical bridge nodes that are essential for freight transportation and connectivity across as well as within the clusters. We have also identified topologically central ports (with respect to three different node centrality metrics: degree, betweenness and closeness) as well as the marine highways that are to be heavily congested due to point-to-point traffic between these ports. The ports of New Orleans, LA, Jacksonville, FL and Chicago, IL are observed to be topologically critical (with respect to all the three centrality metrics) for the connectivity and transportation between several of the core ports. The New Orleans, LA—Chicago, IL edge (involving M-55) incurs the largest EBWC value. However, we observe the edges involving the ports in the M-90 cluster to be among six of the top eight edges with the largest EBWC values. The average path length for the MIPN is 1.895, which is close to the logarithmic value of 21 (the number of nodes) and the randomness index is −0.3850 (not above 0.0), indicating the network can be expected to exhibit small-world characteristics and edge associations are not completely random in nature. Finally, we have run principal component analysis (PCA) on a dataset comprising values for a suite of five different network-level metrics incurred for MIPN and four other transportation networks that have been already analyzed in the literature. Using PCA, we show that the proposed MIPN is topologically unique compared to the other four transportation networks and is hence a worthy contribution to the literature of complex network analysis. As part of future work, we plan to expand the MIPN to include stub ports (ports that are located on only one marine highway) and reevaluate the topological importance of the core ports with respect to the centrality metrics and the NBNC tuple. We also plan to evaluate the robustness of the MIPN to targeted closures (i.e., the connectivity of the MIPN due to the removal of a core port or a marine highway).

Acknowledgement. This research is supported through a grant received by Jackson State University as part of its partnership with the Maritime Transportation Research and Education Center (MarTREC) at The University of Arkansas. The views and conclusions contained in this document are those of the authors and should not be interpreted as representing those of the funding agency.

References

1. https://www.maritime.dot.gov/grants/marine-highways/marine-highway
2. https://www.maritime.dot.gov/sites/marad.dot.gov/files/2022-08/Route%20Designation%20Descriptions.pdf
3. Newman, M.: Networks: An Introduction. Oxford University Press, 1st edn (2010)
4. Newman, M.: Modularity and community structure in networks. Proc. Natl. Acad. Sci. USA **103**(23), 8577–8696 (2006)

5. Meghanathan, N.: Neighborhood-based bridge node centrality tuple for complex network analysis. Appl. Network Sci. **6**(47), 1–36 (2021)
6. Blondel, V.D., Guillaume, J.-L., Lambiotte, R., Lefebvre, E.: Fast unfolding of communities in large networks. J. Stat. Mech. Theory Exp. **2008**, P10008 (2008)
7. Fiedler, M.: Algebraic connectivity of graphs. Czechoslov. Math. J. **23**(98), 298–305 (1973)
8. https://gephi.org/tutorials/gephi-tutorial-layouts.pdf
9. https://gephi.org/
10. Freeman, L.: A set of measures of centrality based on betweenness. Sociometry **40**(1), 35–41 (1977)
11. Freeman, L.: Centrality in social networks: conceptual classification. Soc. Networks **1**(3), 215–239 (1979)
12. Girvan, M., Newman, M.: Community structure in social and biological networks. Proc. Natl. Acad. Sci. USA **99**(12), 7821–7826 (2002)
13. Batagelj, V., Mrvar A.: Pajek Datasets (2006). http://vlado.fmf.uni-lj.si/pub/networks/data/
14. De Domenico, M., Sole-Ribalta, A., Gomez, S., Areans, A.: Navigability of interconnected networks under random failures. Proc. Natl. Acad. Sci. **111**, 8351–8356 (2014)
15. Cadrillo, A., et al.: Emergence of network features from multiplexity. Scientific Reports 3, # 1344 (2013)
16. Subelj, L., Bajec, M.: Robust network community detection using balanced propagation. Euro. Phys. J. B **81**(3), 353–362 (2011)
17. Meghanathan, N.: Spectral radius as a measure of variation in node degree for complex network graphs. In: Proceedings of the 3rd International Conference on Digital Contents and Applications, pp. 30–33. Hainan, China (2014)
18. Meghanathan, N.: Randomness index for complex network analysis. Soc. Netw. Anal. Min. **7**(25), 1–15 (2017)
19. Ernada, E., Rodriguez-Velazquez, J.A.: Spectral measures of bipartivity in complex networks. Phys. Rev. E **72**(4), part 2, 046105 (2005)
20. Jolliffe, I.T.: Principal Component Analysis. Springer Series in Statistics, Springer-Verlag, New York (2002)

A Maari Field Deep Learning Optimization Study via Efficient Hydrogen Sulphide to Hydrogen Production

Klemens Katterbauer(✉), Abdulaziz Qasim, Abdallah Al Shehri, and Ali Yousif

Saudi Aramco, 31311, Dhahran, Saudi Arabia
Klemens.katterbauer@aramco.com

Abstract. Hydrogen sulfide (H2S) is a very corrosive and toxic by-product of a variety of feedstock including fossil resources, such as natural gas and coal, as well as renewable resources. H2S is also a potential source of an important green energy carrier, namely, hydrogen gas. The recovery of H2 from chemical substances identified as pollutants, as H2S, will be of great advantage to our operation.

Because of the significant amounts of H2S available worldwide, and given the growing importance of hydrogen and its byproducts in the global energy landscape, efforts have been made in recent years to obtain H2 and Sulphur from H2S through different approaches. Hydrogen sulfide can be encountered in a variety of different reservoirs and is commonly encountered in deep gas reservoirs. Economic viability of these gas reservoirs is typically limited due to its presence of limited utilization. Novel techniques for transforming hydrogen sulfide into hydrogen and its residual components have become a gamechanger, allowing to efficiently extract both hydrocarbons and its hydrogen sulfide components.

In this work we present a novel deep learning (DL) framework for the optimization of recovery from the field in time. We examined the framework on the Maari Field. The overall objective is to maximize recovery and attain a certain volume of H2S subject to processing constraints. An uncertainty analysis indicated relatively little variation in potential H2S production levels for the optimal production strategy. This is crucial given the toxicity of the H2S if not being able to be processed.

The developed deep learning framework represents an innovative approach towards enhancing sustainability. The framework can be easily expanded to other types of reservoirs and production environments.

Keywords: H2S to hydrogen · sustainability · deep learning · Maari field · reservoir optimization

1 Introduction

Hydrogen sulphide (H2S) is corrosive and toxic in nature and may be a by-product from fossil sources, such as natural gas, oil and coal, in addition to renewable resources. H2S is also a major waste product of the petrochemical industry in general and may arise

R. Silhavy and P. Silhavy (Eds.): CSOC 2023, LNNS 724, pp. 285–296, 2023.
https://doi.org/10.1007/978-3-031-35314-7_27

from the catalytic hydrodesulfurization processes (HDS) of the hydrocarbon feedstocks. Furthermore, it may arise as a by-product of sour natural gas that is sweetened. The challenge is that being a toxic gas, it may not be easily utilized for other applications. Specifically, for conventional combustion technologies it may lead to acidic precipitation and negative effects on human health [1]. Hydrogen sulphide can be removed with a variety of different technologies that may encounter high costs and limitations in the conversion efficiency. The purification processes for the removal of H2S are based on absorption, adsorption, membrane separation and processes that employ catalytic [2]. Adsorption and catalytic oxidations are rather simple, efficient and low-cost in order to achieve desulfurization. High surface area and large pore volume materials, such as the activated carbons, zeolites, mesoporous silica and other metal-organic frameworks, are frequently utilized for adsorption processes. Catalytic methods are very attractive as they enable the conversion of hazardous hydrogen sulphide into non-toxic marketable product and elementary sulphur. The basic processes are based on the hydrogen sulphide-to-sulphur conversion that experience the direct oxidation of H2S into sulphur. Furthermore, low-temperature reduction of sulphur dioxide is another basic process for the hydrogen sulphide to sulphur conversion. The major existing process for the removal is the Claus process that deals with rich-H2S gas streams [3]. The Claus process is a major technology for the production of sulphur, but the hydrogen may be lost as water. Hence, economic profitability of the process for the production of hydrogen may be limited. In order to generate hydrogen from H2S, thermochemical cycles have attracted significant attention. This may be in the form of electrolysis, photolysis and plasmolysis. There are several other variants of this that may be of interest. Utilizing thermal catalytic decomposition of hydrogen sulphide may be an interesting alternative in order to produce both sulphur and hydrogen at the same time [4]. The challenge is the amount of energy needed to achieve the high temperatures may not compensate for the limited hydrogen yield. The objective is to have a one-reaction step in the presence of an active catalyst in order to have strong selectivity with respect to Sulphur. Hence, the catalyst plays a major role for ensuring high quality H2S removal and have a limited selectivity related to SO2. This requires to have active materials that can be utilized at lower temperatures to improve the H2S abatement, and reduce the operational cost. This helps in optimizing the desulfurization technology [5].

There are several recent advances in the area of new materials for hydrogen sulphate conversion. Chen et al. analyzed a porous carbonaceous material in order to reduce H2S emissions. There were two biochars that are based on highly alkaline and porous material. The surficial biochar treatment enabled to reduce H2S emissions, and be able to extract the hydrogen sulphide [6]. Another research was performed by Bao et al. in terms of utilizing waste solid in order to purify the H2S. The waste solid was utilized as a wet absorbent [7]. This led to the removal of H2S and phosphine utilizing a manganese slag slurry. Specifically, the manganese slag slurry achieved strong removal of both the hydrogen sulphide and the phosphine. Addressing the issue of desulfurization of gas that is sour utilizing a carbon-based nanomaterial was investigated by Duong-Viet et al and delivered some interesting results [8]. The material was an N-doped network coated by a ceramic SiC. The nano-doped carbon phase/SiC-based composite enabled to control both the chemical as well as morphology and achieved effective as well as robust catalysts. This enabled to remove hydrogen sulphide in severe desulfurization environments from the souring gases. Severe desulfurization environments are primarily in the form of high GHSVs and significant amounts of aromatics that can contaminate the sour gas streams. Another application is the oxidative desulfurization of fuel oil. In this instance, dibenzothiophene was removed utilizing imidazole-based polyoxometalate dicationic ionic liquids. The researchers evaluated various catalysts under various conditions, and were evaluated in terms of their catalytic performance. One of the catalysts was especially promising, and had good DBT removal efficiency for optimal operating conditions. Another attempt was by Ahmad et al, that evaluated the removal of hydrogen sulphide and SO2 at low temperatures for eco-friendly sorbents from raw and calcined eggshells [9]. The importance of the study was to investigate the relative humidity and reaction temperatures of the environment. This led to best adsorption capacities to be encountered for H2S and SO2 from high calcination temperatures of the eggshells. Another study on the preparation of zinc acetate for commercial activated carbon was outlined by Zulkefli et al., and the objective was to capture hydrogen sulphide by adsorption. In order to optimize the adsorbent synthesis, RM and Box-Behnken experimental designs were utilized. The authors evaluated several factors and levels, especially the zinc acetate molarity and the soaking period. Furthermore, the soaking temperature in addition to the H2S adsorption capacity was evaluated [10]. Additionally, vanadium-sulfide based catalysts were utilized for both the direct and selective oxidation of the hydrogen sulphide. This enabled to convert it to Sulphur and water at low temperatures. Another screening of catalyst for different vanadium loadings was investigated by Barba et al., that evaluated the catalytic performance for H2S conversion and the selectivity of SO2 [11]. Temperature effects, time of contact and the H2S inlet concentrations were investigated in relation to the catalyst. Furthermore, som new hydrochar absorbent for H2S adsorption was investigated and derived from chitosan and starch. Another different technology based on gas-based phase for the oxidation of hydrogen sulphide into sulphur was evaluated by Kahirulin et al. Direct oxidation reactions are generally preferred for the conversion as they are significantly more efficient.

H2S is also a potential source of an important green energy carrier, namely, hydrogen gas. The recovery of H2 from chemical substances identified as pollutants, as H2S, will be of great advantage to our operation. Hydrogen sulfide can be encountered in a variety of different reservoirs and is commonly encountered in deep gas reservoirs. Economic viability of these gas reservoirs is typically limited due to its presence of limited utilization [12].

Because of the significant amounts of H2S available worldwide, and given the growing importance of hydrogen and its byproducts in the global energy landscape, efforts have been made in recent years to obtain H2 and Sulphur from H2S through different approaches. H2S is both naturally occurring and human-made. In the future, H2S production is expected to increase due to increased heavy oil refining. Currently, H2S is largely converted to sulfur and water using industrial processes such as the Claus process, however, it would be more useful and economical to convert H2S to sulfur and H2 instead [13]. H2 currently comes from the steam reforming of natural gas, which is an energy-intensive process. Because H2 is a valued commodity and global consumption is expected to increase, alternative sources of H2 and hydrogen conservation have become topics of active research. From petroleum-based H2S sources could be reused in petroleum upgrading, as a partial replacement of steam methane reforming [14]. Novel techniques for transforming hydrogen sulfide into hydrogen and its residual components have become a gamechanger, allowing to efficiently extract both hydrocarbons and its hydrogen sulfide components. There are several methods of H2S utilization, such as partial oxidation, reformation and decomposition techniques and approaches that convert H2S to sulfur, water and, more importantly, H2. There are several techniques to use electrolysis to convert H2S into hydrogen and sulfur, as well as utilizing a catalyst together with electrocatalytic decomposition to separate these quantities [15].

Artificial intelligence (AI) practices have allowed to significantly improve optimization of reservoir production in addition to optimizing the processes of conversion of gases into useful substances.

This work utilizes a data-driven physics-inspired AI model for the optimization of recovery while maintaining H2S for an oil and gas reservoir utilizing a deep learning time-series optimization approach [16].

2 Reservoir

The Maari Field was discovered in 1983 by Well Moki-1, drilled by Tricentrol Exploration Overseas Ltd. Hydrocarbon pay was found in the Miocene Sequence 0 ("S0") and Moki Formations as well as the Eocene Mangahewa Formation. The wells Moki-2A, Maari-1, Maari-1A and Maari-2 were developed between the years of 1984 and 2003 in order to assess the economic viability of the field and assess the quantities [17]. The development of the field has led to a total oil production by 2021 of 37.4 MMSTB, and in terms of oil production, the field delivered 4,360 STB/D of oil with a water cut of 38% (Fig. 1). The Maari reservoir delivers currently the majority of the oil production, that is developed by several horizontal wells (MR-3, MR-4, MR-7A, MR-8A and MR-10 and two horizontal water injection wells (MR-1A and MR-5A). Additionally, there is the S0 and Mangahewa reservoir that solely have a single horizontal producer well (MR-9 and MR-6A).

The additional objective is to utilized coiled tubing drilling to drill further laterals to the existing producing wells, where several of these will be in the Moki formation, and one will be in the M1A and S0 formation. Five horizontal wells (MR-1, MR-2 (multi-lateral), MR-3, MR-4 and MR-5) were drilled into the Moki Formation that led to the first oil production in 2009. The additional three sub-vertical water injection wells (MR-6, MR-7 and MR-8) were drilled to increase pressure support whose impact was limited given that they injected primarily in the lower cycle [18].

Horizontal Well MR-9 was drilled into the M2A reservoir within the S0 Formation and went live in 2010. In 2014, Well MR-1 was shut-in and converted to water injection well MR-1A to provide support from the western flank of the field. In 2015, Well MR-6 was shut-in and the slot was reused to drill horizontal producer Well MR-6A into the Mangahewa Formation. Wells MR-7 and MR-8 were also shut-in and the slots re-used to drill horizontal producer Wells MR-7A and MR-8A in the Moki Formation. Further slots were utilize such as to drill the horizontal producer well MR-10. The well MR-5 was shut-in in 2018 and converted to water injection (now Well MR-5A).

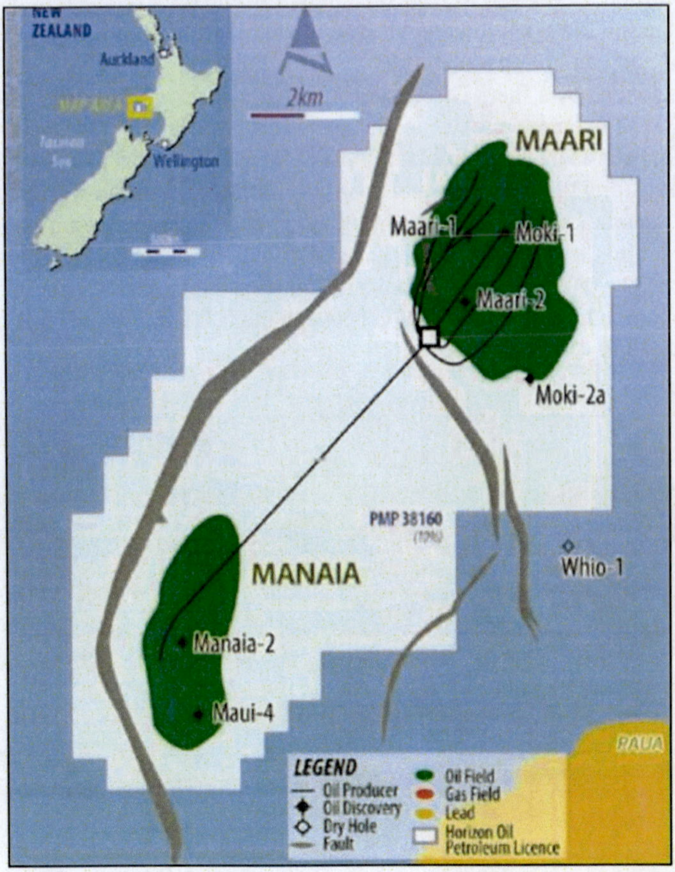

Fig. 1. Location of the Maari and Manaia Fields, PMP 38160, offshore New Zealand

The wells MR-3, MR-4, MR-7A, MR-8A, MR-9 and MR-10 are active oil producers with well MR-6A shut-in due sand screen failure. The remaining wells MR-1A and MR-5A continue to inject water into the formation.

There were two 3D seismic surveys that have been acquired over the Maari area, in 1999 and 2012. Each of these surveys has been re-processed and a multi-azimuth, merged Pre-SDM volume provided the basis for the formation interpretation. There exists a gas cloud over the crest of the Maari field which causes reflection sag, loss of high frequencies and loss of coherency. These challenging effects have corrected partially during the processing, taking account additional structural information, it still represents a challenge at precisely imaging the formation and interpreting its characteristics (Fig. 2).

A stratigraphic column for the Taranaki basin exists and the reservoirs in the Maari fields include the Miocene S0 and Moki Formations and the Eocene Mangahewa Formation. The S0 and Moki reservoirs were deposited in a Miocene turbidite fan system. The Mangahewa reservoirs were deposited in an Eocene fluvio-deltaic system. The S0 reservoir is the shallowest producing interval in the Maari field. It is interpreted as base of slope fans and can be separated into two units, known as the M1A and M2A. Each unit is characterized by a sharp based sand which grades upwards into interbedded silts and shales. These two units have separate free-water levels and the M1A reservoir is of varying quality across the wells in Maari. It has a lower a lower net to gross interval than the M2A formation [19]. The best sand quality is found in the south-western area of the field. There is some correlation between the NTG and the seismic amplitudes but this may be caused by the dimming of the gas cloud, which represents a challenge for the interpretation. The M2A is a thick amalgamated sand which is restricted to the south-western area of the field. The correlation between the NTG and the seismic amplitudes seem to be weaker.

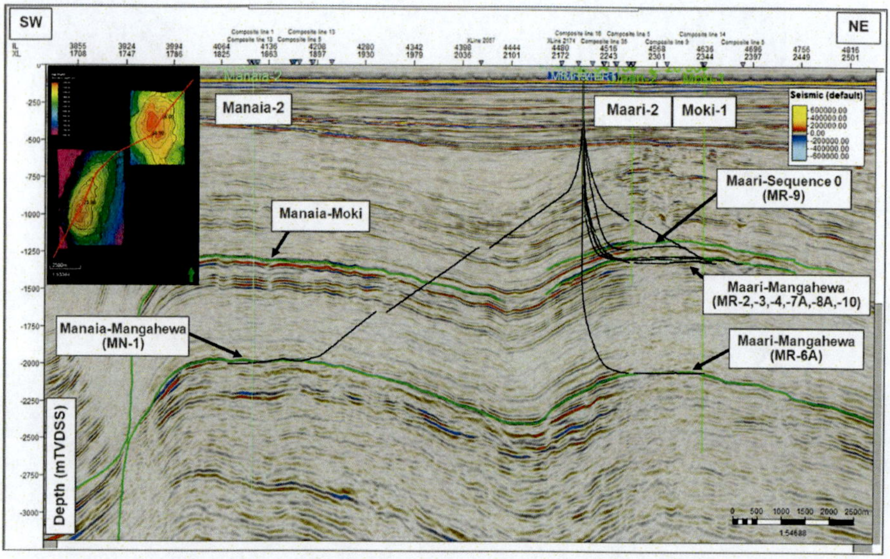

Fig. 2. NE-SW seismic line through Maari and Manaia fields

In the M1A there is a good correlation between amplitude and net thickness, with each well in the dim area in the center of the field having very low net thickness and low NTG. Although the gas cloud covers part of this dim area, the correlation extends to Well Moki-1 which is outside of the gas cloud. The cut-off of 0.25 was utilized on the normalized amplitude map to define areas of god quality reservoir rock and poor quality (Fig. 3). Petrophysical averages used in each area are based on wells within the areas. There is no correlation between net thickness and RMS amplitudes in the M2A unit. As such, the area was treated as one with all wells used.

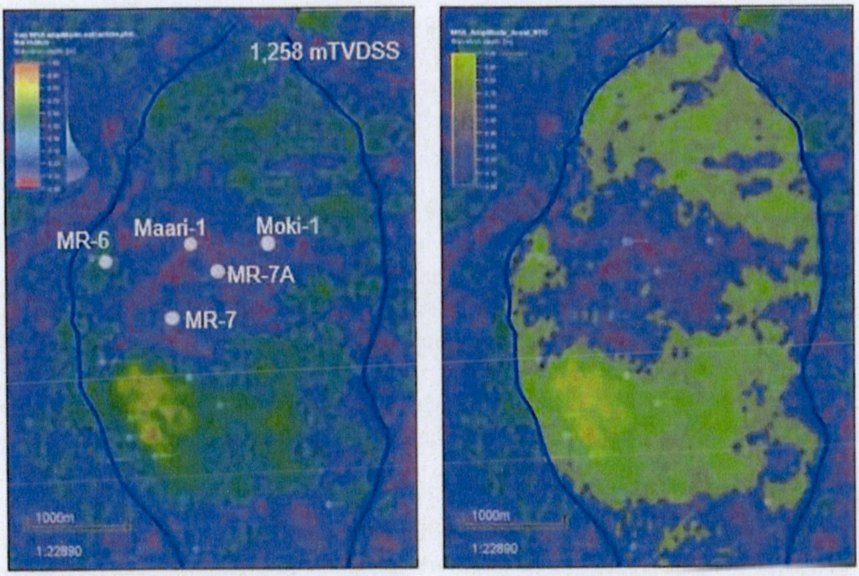

Fig. 3. M1A normalised amplitude extraction (left) with ERCE cut-off applied (right)

3 Optimization Study

We present a novel deep learning (DL) framework for the optimization of recovery of oil, natural gas and hydrogen sulphide. The deep learning methodology utilizes a long short-term memory network that is different from standard feedforward neural networks in terms of the incorporation of feedback connections. This implies that such a recurrent neural network (RNN) can be expanded to process both single data points (e.g. images) but also sequences of data. These sequences of data incorporate speech or video information. This naturally makes LSTM networks adequate for the processing and prediction of data, which may enable it to perform handwriting recognition, speech recognition, machine translation, robot control and time series prediction [14]. The LSTM has both a long term memory, and short term memory. The weights of the connections and the biases in the network, change during each episode of training once. This is analogous to how physiological changes in the synaptic strengths lead to a storing of long-term memories. Additionally, the activation patterns in the network lead to changes for each time-step, which can be considered analogous to how moment-to-moment changes are happening in electric firing patterns. In such a way, the brain stores the short-term memories. The LSTM architecture provides a short-term memory architecture for the RNN that may support the network for a large number of timesteps. This makes it a long short-term memory network [20].

The LSTM is composed of a cell, the input gate and an output gate. Furthermore, it introduces a forget gate. The cell memorizes the values over an arbitrary number of time intervals and the other three gates focus on the regularization of flow of information from and into the cell. The forget gates decide on which of the information is discarded from the previous state. This happens via assigning a previous state a binary value where, the value of 1 implies that the information is kept, while 0 implies that the information is discarded. The input gates decide which of the new information is stored in the current state. The output gate control incorporate the information in the current state to the output [16]. The value may be either 0 or 1 depending on whether the information is discarded or not. The selective output of relevant information from the current state allows the LSTM network to have useful, long-term dependencies and make predictions. This is the case for both current and future time-steps [21].

The methodology was evaluated on the Maari reservoir structure with simulated historical reservoir oil production (Fig. 4) and the gas and H2S production data (Fig. 5). The historical data from 2011 until 2020 indicate a significant increase in natural production that fluctuated with a rising level of H2S. The time series data were then utilized in order to model the three years of production and injection related data that are expected for the efficient natural gas recovery. Furthermore, the H2S conversion was analysed taking into account the uncertainty in the conversion performance.

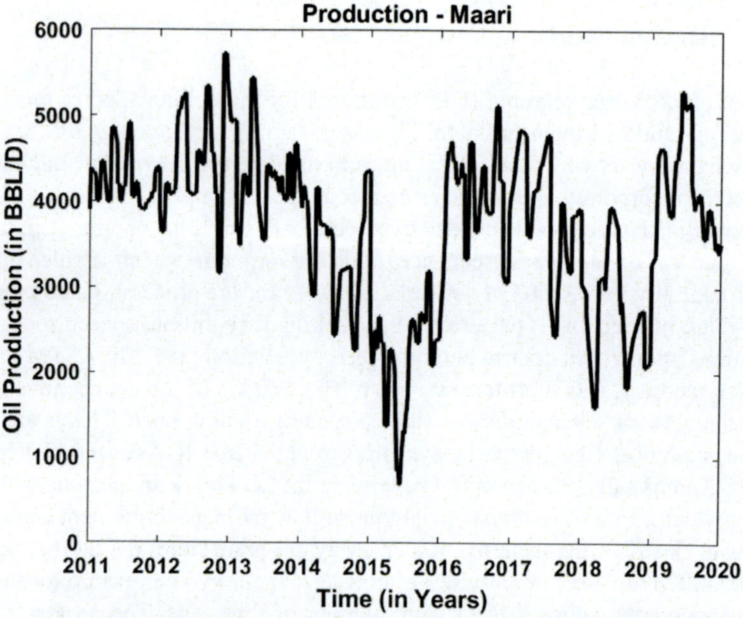

Fig. 4. Historical simulated Maari reservoir oil production (in BBL/D).

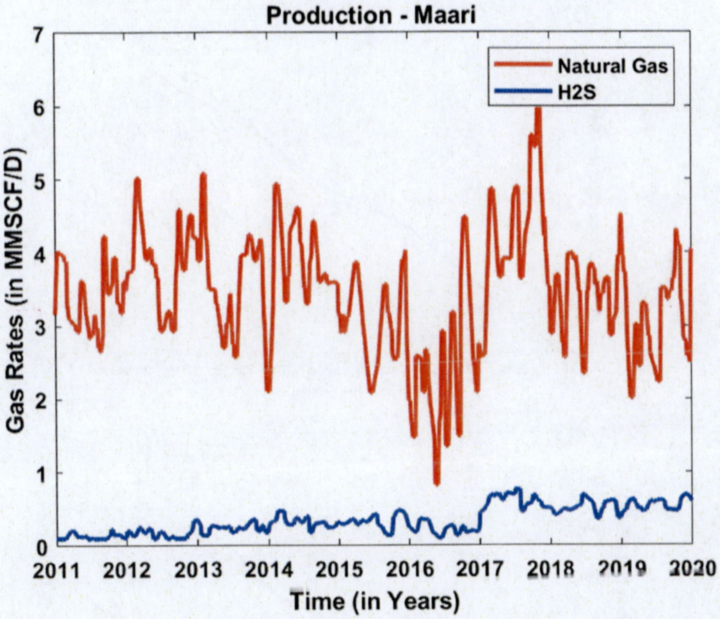

Fig. 5. Historical simulated Maari reservoir gas and H2S production (in MMSCF/D).

The framework was then utilized in order to optimize oil and gas production with respect to H2S constraints and the minimization of H2S for the time period between 2017 and 2020. The results are indicated in Fig. 6 that show the optimized production potential versus the original and indicate a significant improvement in the recovery while maintaining H2S production levels. The optimized results indicate a significant improvement in recovery while maintaining H2S levels within the reservoir. Given H2S being a secondary gas for the production of hydrogen, the optimized production from the field enables to avoid excessive hydrogen sulphide production for which there may not be sufficient capacity to convert into hydrogen and sulphur.

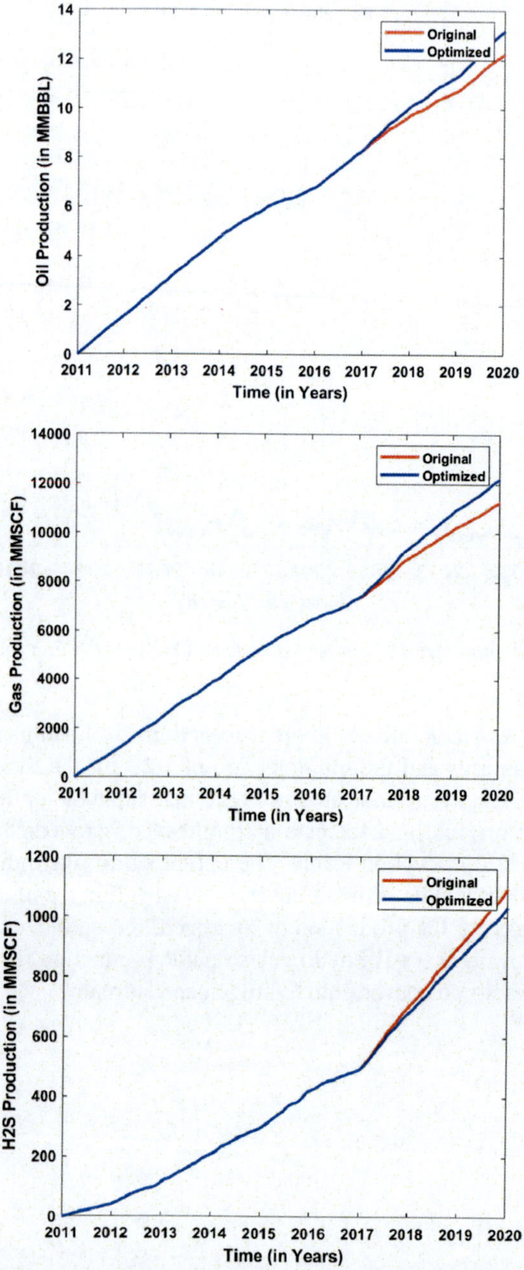

Fig. 6. Original versus optimized Maari reservoir oil, gas and H2S production.

4 Conclusion

Hydrogen sulfide (H2S) is a very corrosive and toxic by-product of a variety of feedstock including fossil resources, such as natural gas and coal, as well as renewable resources. H2S is also a potential source of an important green energy carrier, namely, hydrogen gas. The recovery of H2 from chemical substances identified as pollutants, as H2S, will be of great advantage to our operation. Given the growing importance of hydrogen and its byproducts in the global energy landscape, efforts have been made in recent years to obtain H2 and Sulphur from H2S through different approaches. Hydrogen sulfide can be encountered in a variety of different reservoirs and is commonly encountered in deep gas reservoirs. Economic viability of these gas reservoirs is typically limited due to its presence of limited utilization. Novel techniques for transforming hydrogen sulfide into hydrogen and its residual components have become a gamechanger, allowing to efficiently extract both hydrocarbons and its hydrogen sulfide components. Specifically, catalytic approaches have exhibited significant potential in reducing energy requirements for the conversion, and enhance conversion efficiency.

In this work we present a novel deep learning (DL) framework for the optimization of recovery from the field in time. We examined the framework on the Maari Field. The overall objective is to maximize recovery and attain a certain volume of H2S subject to processing constraints. An uncertainty analysis indicated relatively little variation in potential H2S production levels for the optimal production strategy. This is crucial given the toxicity of the H2S if not being able to be processed.

The developed deep learning framework represents an innovative approach towards enhancing sustainability. The framework can be easily expanded to other types of reservoirs and production environments.

References

1. Dan, M., Yu, S., Li, Y., Wei, S., Xiang, J., Zhou, Y.: Hydrogen sulfide conversion: how to capture hydrogen and sulfur by photocatalysis. J. Photochem. Photobiol. C **42**, 100339 (2020)
2. Li, J., Wang, R., Dou, S.: Electrolytic cell–assisted polyoxometalate based redox mediator for H2S conversion to elemental sulphur and hydrogen. Chem. Eng. J. **404**, 127090 (2021)
3. Reddy, S., Nadgouda, S.G., Tong, A., Fan, L.S.: Metal sulfide-based process analysis for hydrogen generation from hydrogen sulfide conversion. Int. J. Hydrogen Energy **44**(39), 21336–21350 (2019)
4. Spatolisano, E., De Guido, G., Pellegrini, L.A., Calemma, V., de Angelis, A.R., Nali, M.: Hydrogen sulphide to hydrogen via H2S methane reformation: thermodynamics and process scheme assessment. Int. J. Hydrogen Energy **47**(35), 15612–15623 (2022)
5. Iulianelli, A., et al.: Sustainable H2 generation via steam reforming of biogas in membrane reactors: H2S effects on membrane performance and catalytic activity. Int. J. Hydrogen Energy **46**(57), 29183–29197 (2021)
6. Jangam, K., Chen, Y.Y., Qin, L., Fan, L.S.: Mo-doped FeS mediated H2 production from H2S via an in situ cyclic sulfur looping scheme. ACS Sustain. Chem. Eng. **9**(33), 11204–11211 (2021)
7. Bao, J., et al.: Reaction mechanism of simultaneous removal of H2S and PH3 using modified manganese slag slurry. Catalysts **10**(12), 1384 (2020)

8. Duong-Viet, C., et al.: Tailoring properties of metal-free catalysts for the highly efficient desulfurization of sour gases under harsh conditions. Catalysts **11**(2), 226 (2021)
9. Ahmad, W., Sethupathi, S., Munusamy, Y., Kanthasamy, R.: Valorization of raw and calcined chicken eggshell for sulfur dioxide and hydrogen sulfide removal at low temperature. Catalysts **11**(2), 295 (2021)
10. Zulkefli, N.N., Masdar, M.S., Wan Isahak, W.N.R., Md Jahim, J., Md Rejab, S.A., Chien Lye, C.: Removal of hydrogen sulfide from a biogas mimic by using impregnated activated carbon adsorbent. PLoS One **14**(2), 0211713 (2019)
11. Barba, D.: Catalysts and processes for H2S conversion to sulfur. Catalysts **11**(10), 1242 (2021)
12. Turner, J.: Sustainable hydrogen production. Science **305**(5686), 972–974 (2004)
13. De Crisci, A.G., Moniri, A., Xu, Y.: Hydrogen from hydrogen sulfide: towards a more sustainable hydrogen economy. Int. J. Hydrogen Energy, 1299–1327 (2019)
14. Katterbauer, K., Qasim, A., Marsala, A., Yousef, A.: A data driven artificial intelligence framework for hydrogen production optimization in waterflooded hydrocarbon reservoir. In: Abu Dhabi International Petroleum Exhibition & Conference, Abu Dhabi (2021)
15. Ghahraloud, H., Farsi, H.: Process modeling and optimization of an eco-friendly process for acid gas conversion to hydrogen. Int. J. Hydrogen Energy (2022)
16. Katterbauer, K., Al Qasim, A., Al Shehri, A., Yousif, A.: A novel artificial intelligence framework for the optimal control of wireless temperature sensors for optimizing oxygen injection in subsurface reservoirs. In: Offshore Technology Conference Asia, Kuala Lumpur (2022)
17. Mills, C.R., Marron, A., Leeb, W.J.: Maari Field Development–Case Study. In: SPE Asia Pacific Oil and Gas Conference and Exhibition (2011)
18. Singh, D., Kumar, P.C., Sain, K.: Interpretation of gas chimney from seismic data using artificial neural network: a study from Maari 3D prospect in the Taranaki basin, New Zealand. J. Natl. Gas Sci. Eng. **36**, 339–357 (2016). https://doi.org/10.1016/j.jngse.2016.10.039
19. Marron, A.J., et al.: Digital oilfield down under: implementation of an integrated production monitoring and management system for the Maari Field, Taranaki, New Zealand. In: SPE/IATMI Asia Pacific Oil & Gas Conference and Exhibition (2015)

Daeng AMANG: A Novel AIML Based Chatbot for Information Security Training

Irfan Syamsuddin[1](✉) and Mustarum Musaruddin[2]

[1] CAIR Center for Applied ICT Research, Department of Computer and Networking Engineering, State Polytechnic of Ujung Pandang, Makassar, Indonesia
irfans@poliupg.ac.id

[2] Department of Electrical Engineering, Universitas Halu Oleo, Kendari, Indonesia
mustarum@uho.ac.id

Abstract. Security awareness is considered as the main key to maintain institutional cyber security during digital transformation. As digital assets become more diverse and increasingly vulnerable to various cyber security threats, every organization (public and private) require information security training to enhance security awareness among their employees. This study presents the creation of a specific chatbot aimed at delivering fundamental information security training. Program-O as an application framework based on Artificial Intelligence Markup Language (AIML) is used in developing the chatbot called Daeng AMANG. Daeng AMANG chatbot able to respond questions related to information security by providing human like communications in Indonesian language. Finally, Blackbox and System Usability Scale (SUS) evaluations are employed to evaluate Daeng AMANG chatbot which show positive results.

Keywords: Security Awareness · Chatbot · Artificial Intelligence Markup Language · Information Security

1 Introduction

Digital transformation is a necessity for modern institutions, especially the private sector in current industrial 4.0 era [1]. Therefore, we are witnessing growing dependence of all business processes on digital technology as seen in many organizations all around the world. On the other hand, this results in greater security threats, which in turns bring increasingly unintended impacts to organization [2]. Technology alone will not be able to deal with such evolving threats unless the humans behind the technology have proper understanding of the consequences of their actions and carefully follow the rules.

This factor justifies the urgency of information security training to build appropriate information security awareness [3]. Successful information security awareness might be seen in mindset and behavioral changes among the employees of an organization. Eventually, it will build a uniform mindset and pattern of action among employees and management in dealing with any potential digital security threats correctly within an institution [3, 4].

R. Silhavy and P. Silhavy (Eds.): CSOC 2023, LNNS 724, pp. 297–305, 2023.
https://doi.org/10.1007/978-3-031-35314-7_28

There are several ways to deliver information security training such as onsite training as well as online ones. Onsite training is very interactive in nature but costly in terms of renting training facilities, hire trainers, and print training materials [5]. On the other hand online training offers flexible time although sometimes criticized with lack of responsibilities among trainees in comparison to offline training [5]. This is an open gap for further studies in proposing new and sophisticated technologies to enhance online base training in various aspects.

The main objective of this paper is to presents the design of a new chatbot called Daeng AMANG to enhance information security training. It is basically based on Artificial Intelligence Markup Language (AIML) which extracted knowledge from selected sources related to information security and deliver the knowledge through chatbot which mimics human communications.

The rest of paper is organized as follows. Literature review is presented in Sect. 2. Methodology to conduct the research is given in Sect. 3. Results and analysis are discussed in Sect. 4. Finally, Sect. 5 concludes the study.

2 Related Studies

Until recently, various chatbot applications have been reported by experts. Although at first chatbots were implemented more as educational supporting, a number of further studies have shown the application of chatbots in the fields of business, health, government, industry, and many others [6].

In business sector, chatbot has been extensively developed and applied to support business communications with customers [7]. Recent approaches are also seen in SME to show that chatbot also works well for helping small and medium enterprise to extend promotions and communications with their customers [8].

Health care institutions also reported several cases of chatbot applications. Increasing demand for healthcare services has been the driving force to development of various online health services such as chatbot [9]. Efforts to establish chatbot for health enterprise training is increasing which requires proper chatbot platform selection at early stage [10–12].

In terms of industrial services, Castillo, et al [13] reported a novel application of chatbot for training new employees which showed positive results. Similarly, an intuitive dialog based on chatbot introduced in [14]. The chatbot able to give reasonable recommendations to those who have interested in the insurance industry.

From the perspective of chatbot platform as discussed by Syamsuddin and Warastuty [10], several approaches exist to develop a chatbot. Among them is using rule based conversational agent. One of the most popular mechanisms of representing rules is AIML which stands for Artificial Intelligence Markup Language [15]. Several studies have explored effectiveness and usefulness of AIML in creating chatbot for various domains [16–18].

3 Methodology

The study adopts agile methodology to realize the chatbot for information security training purpose. Figure 1 illustrates the methodology which consists four main steps.

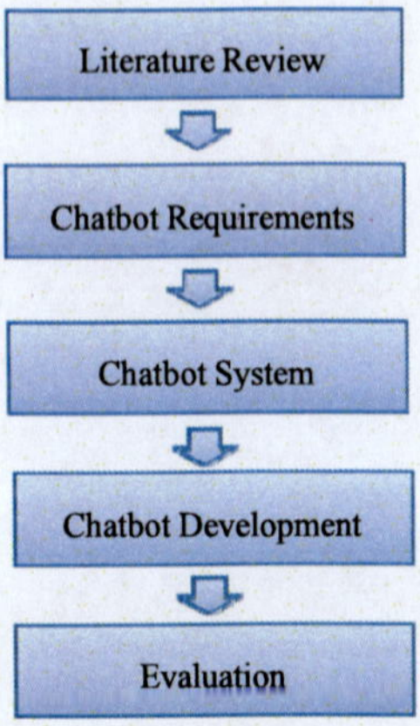

Fig. 1. Research methodology.

After conducting literature review in the first step, the second step consists of fulfilling all requirements to create the chatbot. This step includes preparing all hardware and software requirements to create the intended chatbot. The main part in the second step is selecting the framework based on AIML engine which will be used later when creating the chatbot. Program-O is the selected framework which is a PHP and MySQL based AIML and has many advantages such as community support, well documentations, simplicity and robustness (Fig. 2).

The next step is constructing the chatbot system. In this third step, chatbot system design is established in two perspectives, client and server. Knowledge base is the heart of chatbot ability to respond any given questions. Figure 3 reveals the mechanisms of the chatbot system.

The fourth step is chatbot development which consists of realizing the whole chatbot system both client and server side using Program-O as mentioned in previous step of chatbot systems. Extensive User Interface (UI) and User Experience (UX) are also employed in order to produce excellent chatbot for information security training purposes. Finally, in the last step two evaluations will be applied to assess the chatbot. Firstly, Blackbox testing aims to assess all functions offered by the chatbot [19, 20]. Secondly, usability testing through the use of System Usability Scale (SUS) aimed at measuring user opinions after using the chatbot [21, 22].

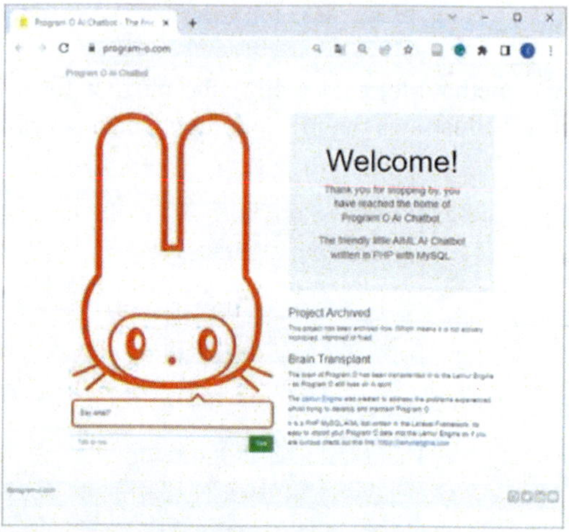


```
<aiml version="1.0">
<topic name="the topic">
<category>
<pattern>PATTERN</pattern>
<that>THAT</that>
<template>Template</template>
</category>
        ..
        ..
</topic>
</aiml>
```

Fig. 2. Program-O as the AIML Engine

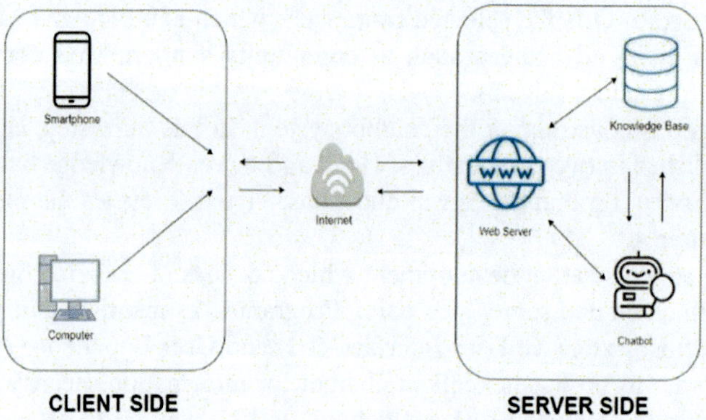

Fig. 3. The chatbot system of both client and server sides.

4 Results and Discussion

Figure 4 shows the final output of the research on mobile phone and tablet. It is a chatbot application called Daeng AMANG. The name is derived from Makassar language, one of many local languages in Indonesia means "*Mr. Secure*" in English.

Fig. 4. Daeng AMANG, the chatbot for information security training

With a target audience of Indonesian users, it can respond to questions in Indonesian language. Users may ask basic security questions such as password combination or email phishing attacks, until giving advice whenever users found unusual files or actions which considered harmful.

The first evaluation is in terms of Daeng AMANG functions as a chatbot for information security training. Main functions of the chatbot are tested by applying Blackbox testing. It is found that all of them are working as expected (see Table 1).

Table 1. Blackbox testing on Daeng AMANG chatbot.

No	Item Tested	Expected Outcome	Validity	
			Y	N
1	Application Icon	Running and displays the interface of Daeng AMANG	√	
2	Category button	Displays the intendedcategory	√	
3	Question box	Responds the givenquestionin the box	√	
4	Answer box	Show related answer according to the question in answer box	√	
5	Suggestion box	Save the suggestedanswer in the database	√	

The second test is user experince evaluation by users in using Daeng AMANG as chatbot for learning information security. The evaluation is assisted with System Usability Scale (SUS) framework. The SUS framework has been used in many information systems domains as a valid measurement for usability analysis [22].

There are 40 respondents participated in assessing the usability aspect of Daeng AMANG chatbot for information security training using SUS survey. There are 68% of them with IT related background, while the rest are with no IT related background at all (Fig. 5).

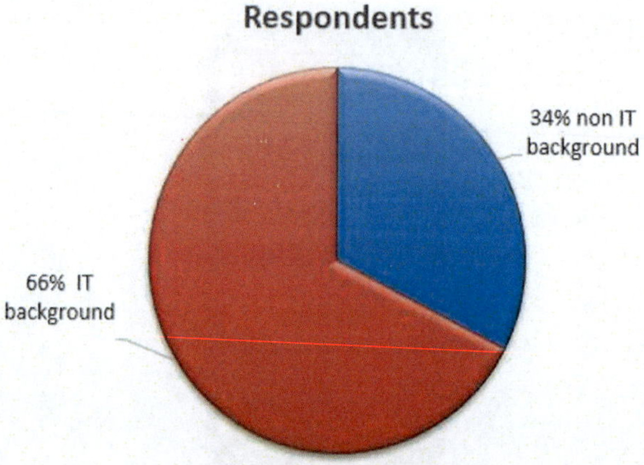

Fig. 5. Respondents of the SUS survey

There are 50 respondents participated in assessing the usability aspect of Daeng AMANG chatbot for information security training using SUS survey. There are 66% of them with IT related background, while the rest are with no IT related background at all (Fig. 5).

Figure 6 displays the individual responses of 50 respondents to each of the 10 SUS survey items. The average score is 70.25, and the SUS framework classifies the Daeng AMANG chatbot as Grade B (Good) (refer to the following Table 2). This clearly indicates that the chatbot is considered effective in distributing information security training materials.

SUS Questions

	q1	q2	q3	q4	q5	q6	q7	q8	q9	q10	Total	x2.5
r1	4	2	4	3	4	2	4	0	4	2	29	72.5
r2	3	3	3	3	3	2	3	3	3	3	29	72.5
r3	3	3	3	3	4	2	4	2	4	3	31	77.5
r4	3	3	3	2	2	4	2	4	3	2	28	70
r5	3	3	2	3	2	4	4	3	3	3	30	75
r6	3	3	4	3	4	3	3	3	4	3	33	82.5
r7	3	3	3	3	3	1	3	1	3	2	25	62.5
r8	4	4	3	4	3	3	3	3	3	3	33	82.5
r9	4	4	2	3	2	4	3	3	4	3	32	80
r11	3	2	3	3	3	3	3	3	3	3	29	72.5
r11	4	3	3	3	3	3	4	2	3	3	31	77.5
r12	3	3	4	3	4	3	3	3	4	3	33	82.5
r13	3	3	3	3	4	3	3	1	4	3	30	75
r14	3	3	3	4	3	3	3	3	3	4	32	80
r15	3	2	3	2	3	2	3	2	3	2	25	62.5
r16	3	3	3	3	3	3	3	3	3	3	30	75
r17	3	3	3	3	3	3	3	3	3	3	30	75
r18	4	3	3	3	3	3	3	3	3	3	31	77.5
r19	4	3	3	3	3	3	3	3	3	3	31	77.5
r20	3	3	3	3	3	3	3	1	3	2	27	67.5
r21	4	3	3	3	3	3	1	3	3	3	29	72.5
r22	3	2	2	2	3	2	3	1	2	1	21	52.5
r23	3	4	3	2	3	3	1	3	3	3	28	70
r24	2	2	3	2	3	3	3	1	3	3	25	62.5
r25	3	3	2	3	2	3	1	3	2	3	25	62.5
r26	3	2	3	2	3	3	4	3	4	3	30	75
r27	**3**	**3**	**3**	**3**	**3**	**1**	**3**	**1**	**3**	**1**	**24**	**60**
r28	3	3	3	3	3	2	3	3	3	3	29	72.5
r29	3	2	3	2	2	1	3	1	3	2	22	55
r30	3	3	3	3	3	3	3	1	3	3	28	70
r31	**3**	[illegible]	[illegible]	[illegible]	[illegible]	**1**	**2**	**2**	**3**	**1**	**22**	**55**
r32	4	2	3	2	2	1	3	0	3	2	22	55
r33	2	3	3	2	4	4	1	1	3	3	26	65
r34	3	2	3	4	3	3	3	3	3	3	28	70
r35	4	3	3	2	4	1	3	1	4	1	26	65
r36	4	3	3	3	3	3	3	1	3	3	29	72.5
r37	4	3	2	2	1	3	3	3	1	3	25	62.5
r38	3	2	2	2	3	2	2	2	3	1	22	55
r39	3	2	3	2	3	2	2	2	3	2	24	60
r40	4	2	2	3	3	3	2	1	3	2	25	62.5
r41	**3**	**2**	**2**	**2**	**3**	**2**	**2**	**2**	**3**	**1**	**22**	**55**
r42	4	2	3	2	2	1	3	0	3	2	22	55
r43	2	3	3	2	4	4	1	1	3	3	26	65
r44	3	2	3	4	3	3	3	3	3	3	28	70
r45	4	3	3	2	4	1	3	1	4	1	26	65
r45	4	3	3	3	3	3	3	1	3	3	29	72.5
r47	4	3	2	2	1	3	3	3	1	3	25	62.5
r48	3	2	2	2	3	2	2	2	3	1	22	55
r49	3	2	3	2	3	2	2	2	3	2	24	60
r50	4	2	2	3	3	3	2	1	3	2	25	62.5
											[illegible]	70.25

Fig. 6. Details and result of SUS survey

Table 2. SUS Survey Interpretation [21, 23].

SUS Score	Grade	Adjective Rating
>80.3	A	Excellent
68–80,3	**B**	**Good**
68	C	Okay
51–68	D	Poor
<51	F	Awful

5 Conclusion

We have described the creation of Daeng AMANG, a brand-new chatbot designed to deliver fundamental information security training. The chatbot's foundation is based on Artificial Intelligence Markup Language (AIML), which is implemented through the free and open-source application called Program-O. It is preferred since the Program-O is a PHP MySQL AIML framework that offers simple to use and effective tools for building chatbots like Daeng AMANG.

Two evaluation techniques were applied to assess Daeng AMANG chatbot, namely Blackbox and System Usability Scale (SUS) evaluation approaches. The Blackbox testing indicates that all chatbot components are correctly functioning as intended. Similarly, positive findings were also obtained through the SUS examination. Daeng Amang chatbot was given a Grade B meaning that the chatbot's usability is considered "Good" by the users in delivering information security training.

References

1. Kleinert, J.: Digital transformation. Empirica **48**(1), 1–3 (2021). https://doi.org/10.1007/s10663-021-09501-0
2. Husák, M., Komárková, J., Bou-Harb, E., Čeleda, P.: Survey of attack projection, prediction, and forecasting in cyber security. IEEE Commun. Surv. Tutor. **21**(1), 640–660 (2018)
3. Wahyudiwan, D.D.H., Sucahyo, Y.G., Gandhi, A.: Information security awareness level measurement for employee: case study at ministry of research, technology, and higher education. In: 2017 3rd International Conference on Science in Information Technology (ICSITech), pp. 654–658. IEEE (2017)
4. Setiadi, F., Sucahyo, Y.G., Hasibuan, Z.A.: An overview of the development Indonesia national cyber security. Int. J. Inform. Comput. Sci. **6**, 108 (2012)
5. Yen, S.C., Lo, Y., Lee, A., Enriquez, J.: Learning Online, Offline, and in-between: comparing student academic outcomes and course satisfaction in face-to-face, online, and blended teaching modalities. Educ. Inf. Technol. **23**, 2141–2153 (2018)
6. Shawar, B.A., Atwell, E.: Chatbots: are they really useful? J. Lang. Technol. Comput. Linguist. **22**(1), 29–49 (2007)
7. Meng-Chieh, K., Lin, Z.H., CardBot: a chatbot for business card management. In: Proceedings of the 23rd International Conference on Intelligent User Interfaces Companion (2007)
8. Selamat, M.A., Windasari, N.A.: Chatbot for SMEs: integrating customer and business owner perspectives. Technol. Soc. **66**, 101685 (2021)
9. Chung, K., Park, R.C.: Chatbot-based heathcare service with a knowledge base for cloud computing. Clust. Comput. **22**(1), 1925–1937 (2018). https://doi.org/10.1007/s10586-018-2334-5
10. Syamsuddin, I., Warastuti, S.W.: Selecting ChatBot platform for health enterprise training: a fuzzy AHP approach. In: 2021 International Conference on Decision Aid Sciences and Application (DASA), pp. 756–760. IEEE (2021)
11. Lommatzsch, A.: A next generation chatbot-framework for the public administration. In: Hodoň, M., Eichler, G., Erfurth, C., Fahrnberger, G. (eds.) Innovations for Community Services. I4CS 2018. CCIS, vol. 863. Springer, Cham (2018). https://doi.org/10.1007/978-3-319-93408-2_10

12. Cantador, I., Viejo-Tardío, J., Cortés-Cediel, M.E., Rodríguez Bolívar, M.P.: A Chatbot for Searching and Exploring Open Data: Implementation and Evaluation in E-Government. In DG. O2021: The 22nd Annual International Conference on Digital Government Research, pp. 168–179 (2021)
13. Casillo, M., Colace, F., Fabbri, L., Lombardi, M., Romano, A., Santaniello, D.: Chatbot in industry 4.0: an approach for training new employees. In 2020 IEEE International Conference on Teaching, Assessment, and Learning for Engineering (TALE), pp. 371–376. IEEE (2020)
14. Nuruzzaman, M., Hussain, O.K.: IntelliBot: a dialogue-based chatbot for the insurance industry. Knowl. Based Syst. **196**, 105810 (2020)
15. Yamaguchi, H., Mozgovoy, M.: Generating AIML rules from twitter conversations. In: vol. Communication Papers of the 2017 Federated Conference on Computer Science and Information Systems (FedCSIS), pp. 59–61 (2017)
16. Nursetyo, A., Subhiyakto, E. R.: Smart chatbot system for E-commerce assistance based on AIML. In: 2018 International Seminar on Research of Information Technology and Intelligent Systems (ISRITI), pp. 641–645. IEEE (2018)
17. Boden, C., Fischer, J., Herbig, K., Spierling, U.: CitizenTalk: application of chatbot infotainment to e-democracy. In: Göbel, S., Malkewitz, R., Iurgel, I. (eds.) Technologies for Interactive Digital Storytelling and Entertainment. TIDSE 2006. LNCS, vol. 4326. Springer, Berlin, Heidelberg (2006). https://doi.org/10.1007/11944577_37
18. Gunawan, D., Putri, F.P., Meidia, H.: Bershca: bringing chatbot into hotel industry in Indonesia. TELKOMNIKA **18**(2), 839–845 (2020)
19. Khan, M.: Different approaches to black box testing technique for finding errors. Int. J. Softw. Eng. Appl. **2**(4) (2011)
20. Sukmandhani, A.A., Rusly, J.K., Pratama, A., Ohliati, J.: The design and implementation of a dashboard web-based service management operations, J. Adv. Inf. Technol. **13**(6) (2022)
21. Brooke, J.: SUS: a 'quick and dirty' usability scale. In: Usability Evaluation in Industry, CRC Press (1996)
22. Syamsuddin, I., Bramantoro, A.: Usability assessment on symmetric cryptography learning module. In: 2020 2nd International Conference on Computer and Information Sciences (ICCIS), pp. 1–6. IEEE (2020)
23. Syamsuddin, I., Yunianta, A.: Design and usability assessment of DocManS: a document management system with security and social media features. Int. J. Adv. Appl. Sci. **9**(1), 48–54 (2022)

Forecasting Oil Production for Matured Fields Using Reinforced RNN-DLSTM Model

Pramod Patil, Klemens Katterbauer(✉), Abdallah Al Shehri, Abdulaziz Qasim, and Ali Yousif

Saudi Aramco, Dhahran 31311, Saudi Arabia
klemens.katterbauer@aramco.com

Abstract. Forecasting of oil production plays a vital role in petroleum engineering and contributes to supporting engineers in the management of petroleum reservoirs. However, reliable production forecasting is difficult to achieve, particularly in view of the increase in digital oil big data. Although a significant amount of work has been reported in the literature in relation to the use of machine learning in the oil and gas domain, traditional forecasting approaches have limited potential in terms of representing the complex features of time series data. The application of a deep supervised machine learning model for oil field production and optimization has been gaining much interest in the research community. Advances in new technologies particularly artificial intelligence (AI) and machine learning (ML) have now made it possible to use AI and ML-based approaches for building reservoir simulation models that do not rely on conventional simulation tools for history matching and production forecasting. These new methods can improve speed, efficiency, and the potential to eventually improve workflow efficiency. The advantages of AI and ML-based simulators are significant time and money savings to achieve statistically accurate and physics-assisted history matching and forecasting results.

AI and ML-based deep learning approaches involve using subsurface reservoir static, dynamic, and well production data to form a fully independent deep learning-based model with capabilities to perform history matching and production forecasting. The results show that supervised deep learning algorithm-based models provide a framework in history matching well patterns. Models are evaluated on the industrial datasets from the North Sea fields. The models are also extended to generate long-term forecasts. The work in this paper will demonstrate the development and use of physics assisted deep-gated recurrent neural network consisting of multiple hidden layers with each layer having multiple nodes.

Keywords: 4IR · artificial intelligence · hydrogen storage · hydrogen storage volume assessment · uncertainty · New Zealand

1 Introduction

Forecasting of oil production has become an increasingly important task in the oil and gas industry and can assist petroleum engineers in the analysis of the characteristics of oil and gas wells. Challenging reservoir environments with multiple intersecting faults

R. Silhavy and P. Silhavy (Eds.): CSOC 2023, LNNS 724, pp. 306–326, 2023.
https://doi.org/10.1007/978-3-031-35314-7_29

and complex fractures lead to waterflooding patterns where sweep efficiency may be uneven. There have been several new technologies that assisted in the enhancement of predictions and support field operations. However, accurate prediction is complex and requires significant resources and time commitment in order to assess the performance of a reservoir (Hammerton, Osborne, Armstrong, & Daelemans 2002; Moreno 2011). The use of numerical simulation for history matching and forecasting of oil, gas, and water (liquids) is the most common approach and enables to simulate of the flow dynamics in detail within the reservoir. Although this approach enables to simulate in detail the dynamics in the reservoir, the simulations are often time-consuming for large reservoirs. As the history-matching process requires an extensive number of simulations in order to achieve an adequate fit of the historical data, the time required for such history-matching operations may be too long for real-time operational decision-making that takes into account the uncertainty. Overfitting in history matching operations is another challenge that requires careful calibration of the simulation model parameters in order to provide adequate forecasts.

In the past, several approaches have been utilized to forecast oil production that can be broadly classified into statistical and numerical computational methods. One of the most popular traditional statistical models is Decline Curve Analysis (DCA). The key limitation of this approach is that it discounts the physics of the process and completely performs analysis based on the hard data collected from the field data and does not provide much information to provide a deep understanding of the factors influencing the oil production and reservoir performance (Fetkovich, Vienot, Bradley, & Kiesow 1987). The most common traditional application is the Autoregressive Integrated Moving Average (ARIMA) model, which has been successfully applied to diverse petroleum-related tasks such as forecasting oil prices, and estimation of crude oil reserves and oil production. Despite the considerable potential of ARIMA, drawbacks related to capturing temporal sequence data represent significant shortcomings of this model. Additionally, ARIMA models are entirely statistical models that do not explicitly take into account the flow dynamics within the reservoir (George, Gwilym, & Gregory 2015).

Recently the use of artificial intelligence has become more prominent in the oil and gas industry. Several machine learning algorithms have been extensively applied such as Fuzzy Logic (FL), Genetic Algorithms (GA), Support Vector Machines (SVM), and intelligent optimization (Braswell 2013; Mohaghegh 2005). A neural network is one of the most popular machine learning models and is capable of handling nonlinear datasets (Zhang, Patuwo, & Hu 1998). Although this approach has shown good prediction outcomes, it has several disadvantages in terms of solving time-series tasks. Since the information transfer in this model is only in the forward direction, it cannot track historical data (Moreno 2011), and hence is incapable of inferring the dependency between past and future events.

Recurrent Neural Networks (RNNs) have been widely applied to perform forecasting activities in various areas of petroleum production. Still, they are inadequate to cope with long-interval time-series data (Hochreiter, Bengio, & Frasconi 2011; Moreno 2011). To overcome this limitation, Long Short-Term Memory (LSTM) was proposed as an extension of RNN. Empirical studies have demonstrated that LSTM can handle dynamic petroleum temporal data efficiently (Hochreiter et al. 2011; Patil et al. 2021).

To make a model physics-assisted, one can bring in the data that represent physics for the pattern-level data. The use of the capacitance resistance model (CRM) as a precursor to the RNN-LSTM model can bring the physics from the pattern level data based on the reservoir heterogeneity and well-to-well connectivity between injectors and producers in the pattern (Albertoni and Lake 2003, Yousef et al. 2006; Liang et al. 2007, and Sayarpor et al. 2009). CRM models have been widely deployed for determining and forecasting oil production from water-injection operations, enabling to provide of an estimate of the transient well connectivity. Nevertheless, potential overfitting and the constraints of the model may pose a challenge in taking into account the large number of data and measurement discrepancies that may arise. The sequential modeling approach where proposed in the current work, where CRM can be used to bring the transient well connectivity and then use that as input to the RNN-LSTM architecture to have a physics-assisted deep ML model for history matching and forecasting the oilfield production data. The objective of this work is to use new AI and ML-based deep learning algorithms for performing history matching and forecasting on a series of industrial datasets from the middle east and the North Sea field.

The results show that a physics-assisted supervised deep ML model based on RNN-LSTM shows a promise to perform history matching and forecasting. Due to the sequential modeling approach, one can perform scenario studies in terms of the impact of changes at the injector (water or gas rate change or any other operating parameter change) on the producers. We demonstrate the performance of the model on the public domain data from the North Sea project.

2 Methodology

The precursor of the LSTM neural network is Recurrent Neural Network (RNN). RNN, the evolution of multi-layer perception, is skilled in settling sequential data. A simple RNN structure whose recurrent layer is unfolded into a full network is shown in Fig. 1. RNN is a power structure because it combines two properties, a distributed hidden state that allows them to store a lot of information about the past, and non-linear dynamics that allow them to update its hidden state in complicated ways. Xt is input, Ot is the hidden state that gives the network memory ability. W_{ho}, b_o, and W_{hx} are the hyperparameters of different layers. The distinctive characteristic lies in that the output of the hidden layer with the present information is transferred to the hidden layer of the next time step as part of the input. This kind of loop can preserve the information from the previous step to keep the data dependency, which improves the ability to learn and abstract from the sequential data. However, the vanishing gradient problem during the calculation of the back-propagation makes the influence of the input cannot be far transferred (Bengio et al. 1994); in other words, the performance of RNN is not satisfying for long -term dependence problems.

Therefore, LSTM neural network is proposed to improve RNN. The inner structure of the hidden layer in LSTM is more complicated compared with RNN (Hochreiter and Schmidhuber 1997). As shown in Fig. 2, a typical LSTM block is configured mainly by memory cell state, forget gate, input gate, and output gate. The crucial element, the memory cell state, runs down through the entire chain to selectively add or remove

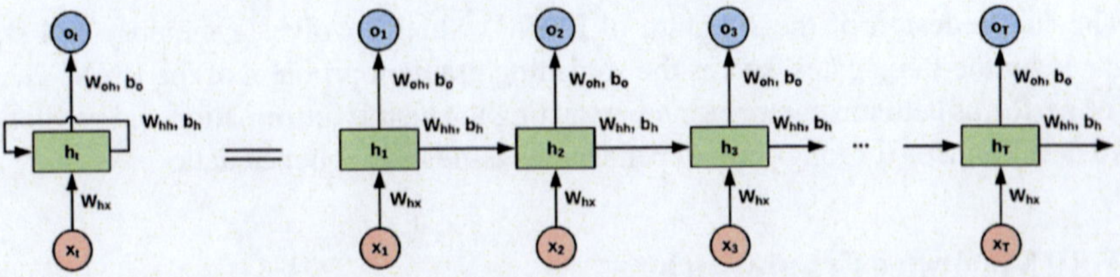

Fig. 1. Typical architecture of RNN

information to the cell state with the help of the three gates. The specific procedures of the LSTM neural network at time step t are as follows:

(a) Decide what information should be discarded from the previous cell state t−1 in the forget gate.
(b) Identify what information of input xt should be stored in the cell state in the input gate, where the input information at t and the candidate cell state Ct should be updated.
(c) Update the cell state of the present time step, which combines the candidate memory at t and the long-term memory at t 1.
(d) Confirm the outcome ht in the output gate where the output information from the output gate and the cell state at t is used.

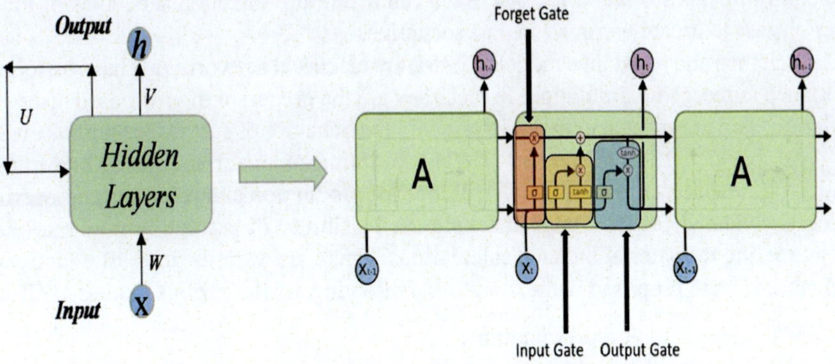

Fig. 2. The typical architecture of RNN-LSTM

Figure 2 shows that W is the input weight, U is the recurrent weight, and b is the bias. The activation function of the three gates is the sigmoidal function which enables the values between 0 and 1. The activation function tanh is the hyperbolic tangent function, which squeezes the value between −1 and 1. Both activation functions are utilized to enhance the nonlinearity of the network, which are expressed as:

$$\sigma(x) = \frac{1}{1 + e^{-x}} \tag{1}$$

$$tanh(x) = \frac{e^{x} - e^{-x}}{e^{x} + e^{-x}} \tag{2}$$

The subtle design of the structure of LSTM, which involves a memory cell state regulated by the three gates, solves the vanishing gradient problem of the RNN. There, LSTM performs better in capturing and extracting the history information and predicting future development for long-term dependencies issues of sequence data.

3 LSRM Model Framework

The productivity of wells changes with time and other influential factors including changes in the injector wells. It is suitable to utilize the LSTM model to explore the production mechanism from the relevant data. The architecture of the proposed LSTM model is depicted in Fig. 3. The dimension of the input data is decided according to the number of influential factors of the oil rate. The LSTM layer is embedded to memorize the extract containing information from input data, where the ADAM optimizer is used to update the weights and bias and the mean square error is set as the fitness function. The final output is given by a fully connected layer after the termination condition is reached.

Figure 3 shows the workflow of the physics-assisted RNN-LSTM model. The critical part of the architecture proposed here is the use of the CRM model to estimate the good connectivity between the injector and producer wells which is being used as an input to the RNN-LSTM model. This approach captures the heterogeneity of the reservoir through the CRM model and hence this model can be utilized for making realistic production forecast for scenarios where we can use future injection schedules of fluids (water or gas) to create various forecast scenarios.

Capacitance and resistance models (CRM) are electrical networks that have historical importance in reservoir simulation. In fact, they are the precursor of grid-based reservoir simulation. The use of such networks to explain the behavior of subsurface porous media flow dates back to 1943 (Bruce 1943). The coupling of material balance and inflow equations has been a simple but powerful tool for production and reservoir engineering for decades. The CRM facilitates checking the feasibility of predicting flow rates and adds a timeline to material balance calculations. There are various times of a family of CRM models were proposed in the literature. Following are the various models of CRM:

(a) CRMT – Single tank representations
(b) CRMP – Produced-based representation
(c) CRMIP – Injector-producer based connectivity

Figure 4 shows the various CRM models that are discussed in a few publications (Yousef et al. 2003; Rafael et al. 2018). In this work, the CRMIP model is being used to estimate the injected fluid allocation during secondary and tertiary recovery. The CRM models have been successfully applied for water flood simulations and tertiary simulations. These models do not intend to capture all of the reservoir physics but are expected to identify the primary drivers and provide a reasonably accurate basis for analysis and forecast.

The CRMIP is the injector-producer pair-based representation (CRMIP) that assigns time constant (τ_{ij}) abd one connectivity (f_{ij}) for each injector(i) and producer (j) pair,

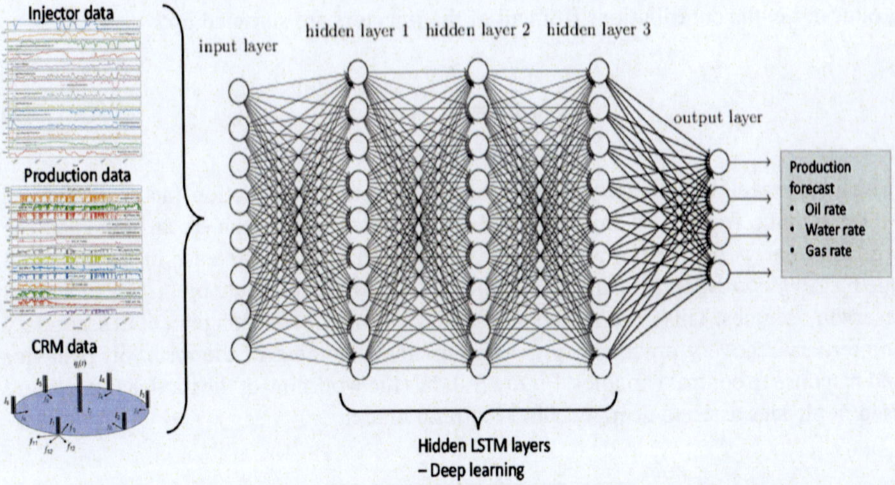

Fig. 3. The complete architecture of the physics-assisted RNN-LSTM model.

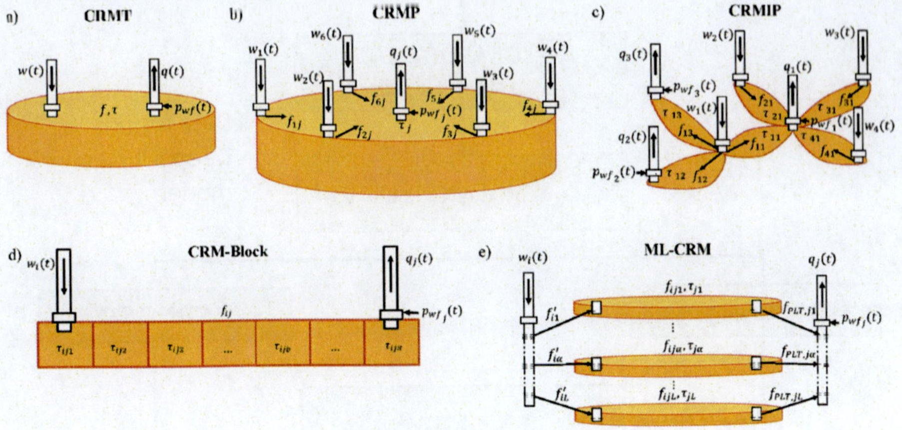

Fig. 4. Various CRM model – (a) single tank (CRMT), (b) producer based (CRMP), (c) Injector-producer pair based (CRMIP), (d) blocks in series (CRM-block) (e) Multi-layer or blocks in parallel (ML-CRM).

shown in Fig. 4. This model was initially proposed by Yousef et al. 2003 and the continuity equation for each control volume is similar to Eq. (3).

$$\tau_{ij}\frac{dq_{ij}}{dt}+q_{ij}(t)=f_{ij}w_i(t)-\tau_{ij}J_{ij}\frac{dp_{wfj}}{dt} \tag{3}$$

Where q_{ij} is the production rate in producer j from the injector(i)-producer(j) pair control volume, as well as J_{ij} is the productivity index associated with such a control volume. Then, supervision in space is applied to obtain the total production rate of each producer, i.e. the contributions from all of the injectors are summed up:

$$q_j(t) = \sum_{i=1}^{N_{inj}} q_{ij}(t) \tag{4}$$

The interwell connectivity, f_{ij} also known as gain or allocation factor, is defined by the volume fraction of injected fluid from injector I that can be attributed to the production at well j. There, at stabilized flow conditions, an increase in the injection rate by Δw_{ij} corresponds to an increase in the total production rate by $\Delta q_{ij} = f_{ij}\Delta w_i$ in reservoir volumes. This information is essential for improved management in secondary and tertiary recovery processes, providing an understanding of the reservoir behavior and response to control variables. Figure 5 shows the workflow of the model architecture used in physics assisted deep machine-learning model.

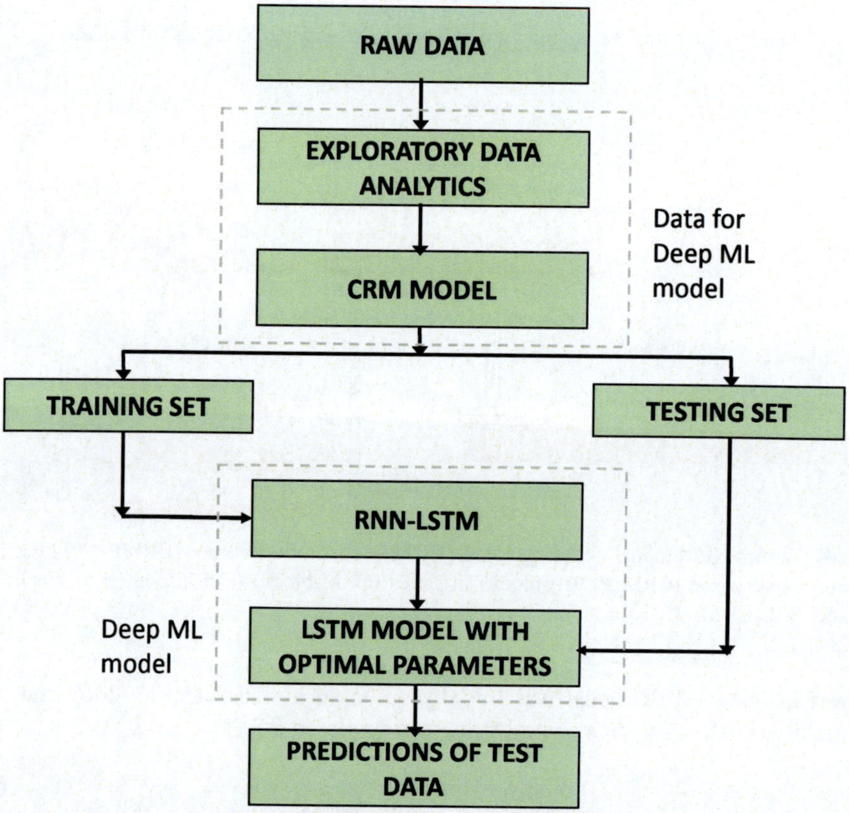

Fig. 5. The flow of the model architecture.

4 Data Set

The datasets were used for the purpose of testing and validating the models. One dataset, which comes from a Volve field (Equinor 2021), was used for model validation. Data processing was one of the key elements of the work, it involved processing raw injection and production data to eliminate outliers and duplicate values. Outlier detection and removal is an intricate process when applied to oil production data as attention must be paid not to remove increased oil/gas or water rates coming from re-opening of wells after shut-in, therefore shut-in detection algorithm must be implemented to avoid removal of useful data points. Table 1 and 2 shows the details of the Volve field producers and injectors information in the scope. Figure 6 shows the aerial map of the Volve field.

Table 1. The data of the producer wellbore.

Well Bore Name (Producer)	Duration
15/9-F-12 H	2/2008–9/2016
15/9-F-1 C	4/2014–4/2016
15/9-F-5 AH(Producing)	4/2016–8/2016
15/9-F-11 H	7/2013–9/2016
15/9-F-14 H	7/2008–7/2016
15/9-F-15 D	1/2014–/2016

Table 2. The data for the injector.

Well Bore Name (Injector)	Duration
15/9-F-5 AH(Injection)	8/2008 —4/2016
15/9-F-4 AH	4/2008–9/2016

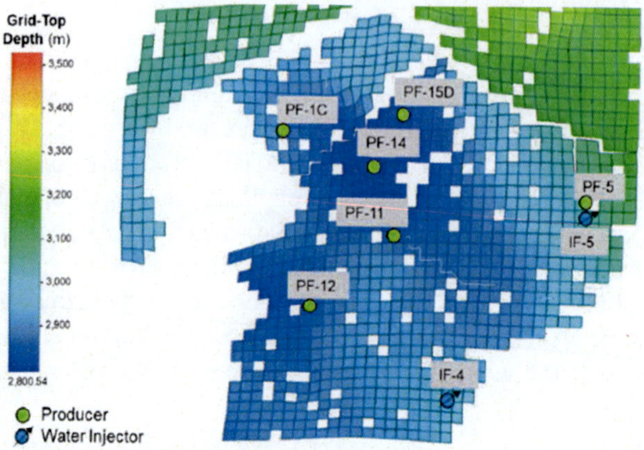

Fig. 6. The map of the Volve field.

For the model development, it was important to find the correlation between different oil, gas, and water production parameters. Three types of correlation coefficients (Spearman, Kendall, Pearson) were taken into consideration. Finally, only the Pearson correlation coefficient was used throughout all this work as it was giving the best results. A sample correlation plot from the analysis is shown in Fig. 7. Pearson correlation coefficient is a measure of linear correlation between two sets of data. It ranges from -1 (negative correlation) to 1 (positive correlation). The further from zero, the stronger is the correlation.

5 RNN-DLSTM Model

Long short-term memory or LSTM, is an artificial recurrent neural network architecture, used for deep learning methods. LSTM is not like typical feedforward neural networks; LSTM has a unique function – feedback connection, see Fig. 3. One of the bigger advantages of LSTM is that it can process not only single point of data but an entire sequence of data, which is the case with time series data from oil and gas production. In this work, LSTM based model was created and used for oil history matching and prediction/forecasting.

Figure 8 shows the LSTM model architecture visualization for the LSTM model used in this work, the LSTM cell has a tanh activation function, but for the work carried out for this paper, ELU (exponential linear unit) activation function was giving much better results, so it was used. This ELU activation function is described as follows:

$ELU = f(x) = \{\alpha(ex - 1x), xx \geq \ < 00$ Below there is an explanation of equations, that are used in the model:

$$ft = \sigma g(Wfxt = Ufht - 1 + bf); \tag{5}$$

$$it = \sigma g(Wixt = Uiht - 1 + bi); \tag{6}$$

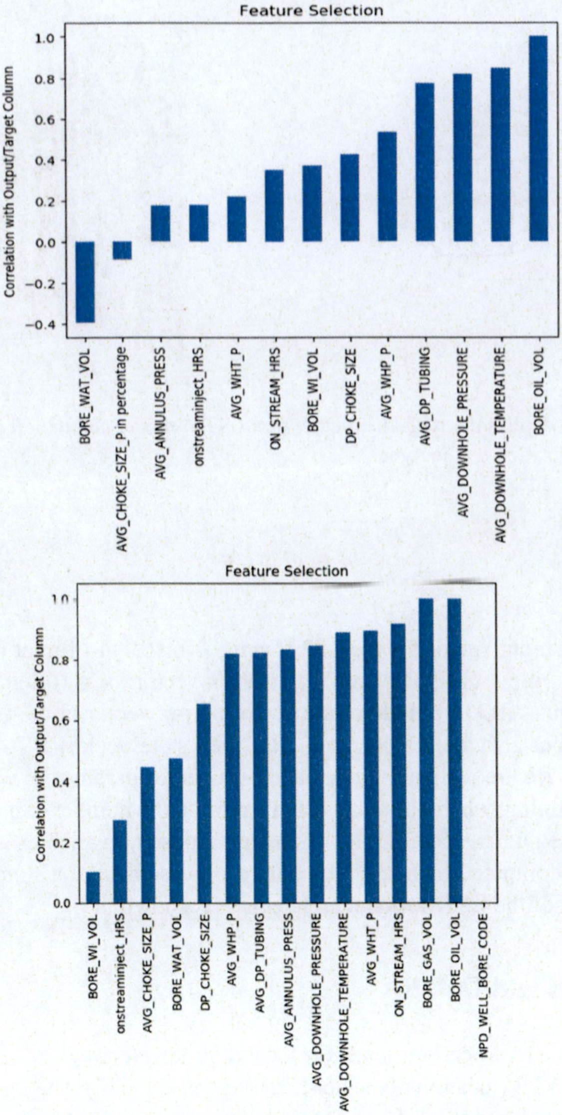

Fig. 7. The Pearson correlation coefficient for various independent variables vs the production rates of oil – (a) Producer #15/9-F-14 & Injector#15/9-F-4 AH and (b) Producer #15/9-F-12 & Injector 15/9-F-4 AH.

$$ot = \sigma g(W0xt = U0ht - 1 + b0); \tag{7}$$

$$\tilde{ct} = \sigma e(Wcxt = Ucht - 1 + bc); \tag{8}$$

$$ct = ft^{\circ}ct - 1 = ot^{\circ}\tilde{ct}; \tag{9}$$

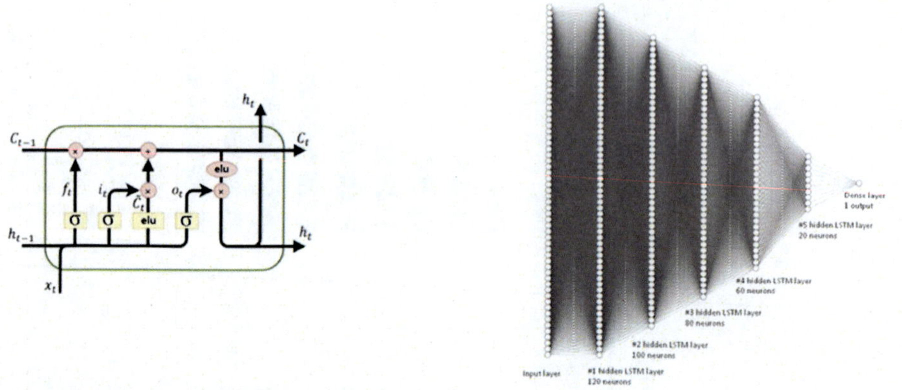

Fig. 8. This figure depicts the typical structure of a LSTM model cell on left and LSTM model architectural visualization on the right.

$$5\ ht = ot^{\circ}\sigma e(ct). \tag{10}$$

Variables:

$xt \in Rd$ is an input vector for the LSTM unit; $ft \in (0,1)h$ - forget gate's activation error; $it \in (0,1)h$ - input or update gate's activation vector; $ot \in (0,1)h$ - outputs gate's activation error; $ht \in (0,1)h$ - hidden state vector/output vector of the LSTM unit; $ct \in (-1,1)h$ - cell input activation error; $ct \in Rh$ - cell state vector; $W \in Rh \times d$, $U \in Rh \times x$, and $b \in Rh$ are weight matrices and bias vector parameters which need to be learned during training, where the small d and h refer to the number of input features and a number of hidden units, accordingly. These equations and variables are in the 'back' and they are not computed manually, it's only a process behind a single LSTM layer. The visualization of the LSTM model architecture is shown in Fig. 8.

6 The Volve Field Details

The Volve field is a hydrocarbon reservoir located in the Norwegian North Sea with an OOIP of 173 MMSTB (location of the field can be seen in Fig. 9). The total production of 63 MMSTB was achieved till 2016. After achieving a recovery of 54% post-waterflood, the field was shut down in late 2016. The water injection in the Volve field started in early 2007. Table 3 shows the details of the available data for the various injectors and producers from the Volve field. We have used the producer 15/9-F-11 H & 15/9-F-12 H and injector 15/9-F-4-H in the current analysis.

Table 3. Table shows the injector and producer data and number of data points available for the analysis.

Wellbore Name	Number of data points	Type
15/9-F-4 AH	3327	Injector
15/9-F-5 H	3306	Injector/Producer
15/9-F-14 H	3056	Producer
15/9-F-12 H	3056	Producer
15/9-F-11 H	1165	Producer
15/9-F-15 D	978	Producer
15/9-F-1 C	746	Producer

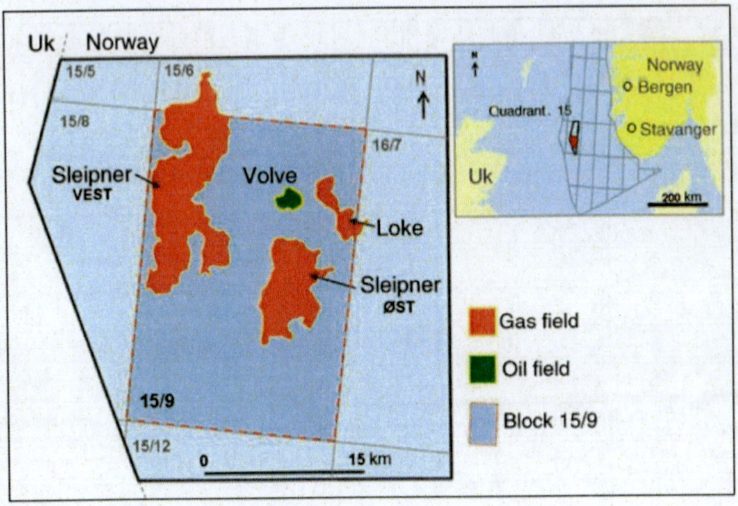

Fig. 9. The location of the Volve field in the North Sea (reference – Plan for the development of Volve, Feb 2005)

7 Results and Discussion

The Volve field data used here involves the water flooding which started in 2007. Figure 11 and 12 show the plots of various variables for the pair injector 15/9-F-4 AH & producer 15/9-F-12 and pair 15/9-F-4 AH & producer 15/9-F-11. There are various variables considered in the exploratory data analytics and later in the RNN-DLSTM (Recurrent Neural Network – Deep long term-short term memory) model.

- Average downhole pressure
- Active production hours (ON_STREAM_HRS)
- Average downhole pressure (AVG_DOWNHOLE_PRESSURE)
- Average downhole temperature (AVG_DOWNHOLE_TEMPERATURE)

- Average pressure drop tubing (AVG_DP_TUBING)
- Average annulus pressure (AVG_ANNULUS_PRESS)
- Average choke size (AVG_CHOKE_SIZE_P)
- Average well head pressure (AVG_WHP_P)
- Average well head temperature (AVG_WHP_P)
- DP choke size (DP_CHOKE_ZISE)
- Bore oil volume (BOR_OIL_VOL)
- Bore gas volume (BOR_GAS_VOL)
- Bore water volume (BOR_WAT_VOL)
- Total injection time in hours (onstreaminject_HRS)
- Wellbore injection volume (BORE_WI_VOL) (Fig. 10)

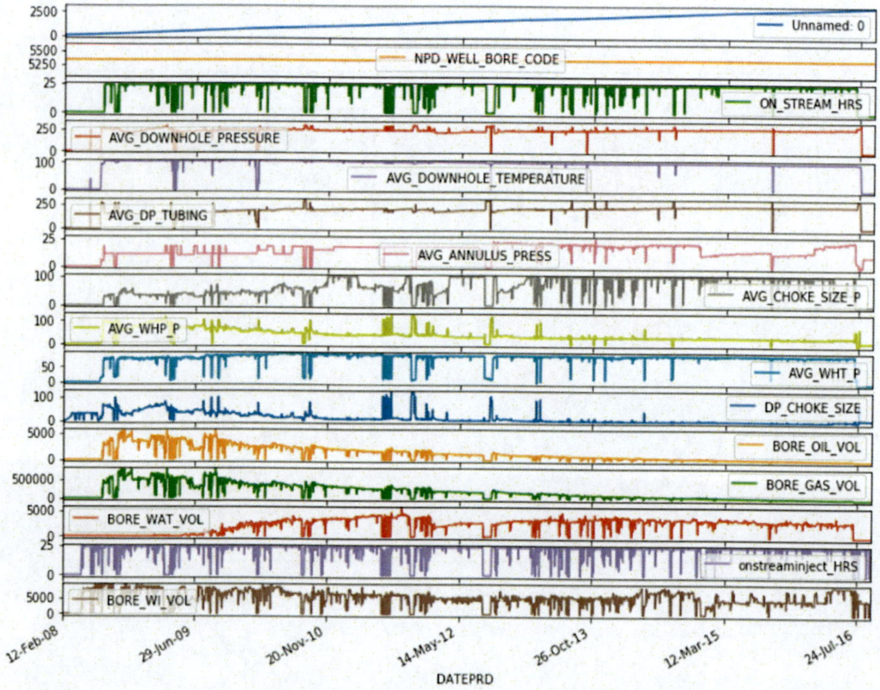

Fig. 10. Data visualization for the injector 15/9-F-4 Ah and Producer 15/9-F-12 H.

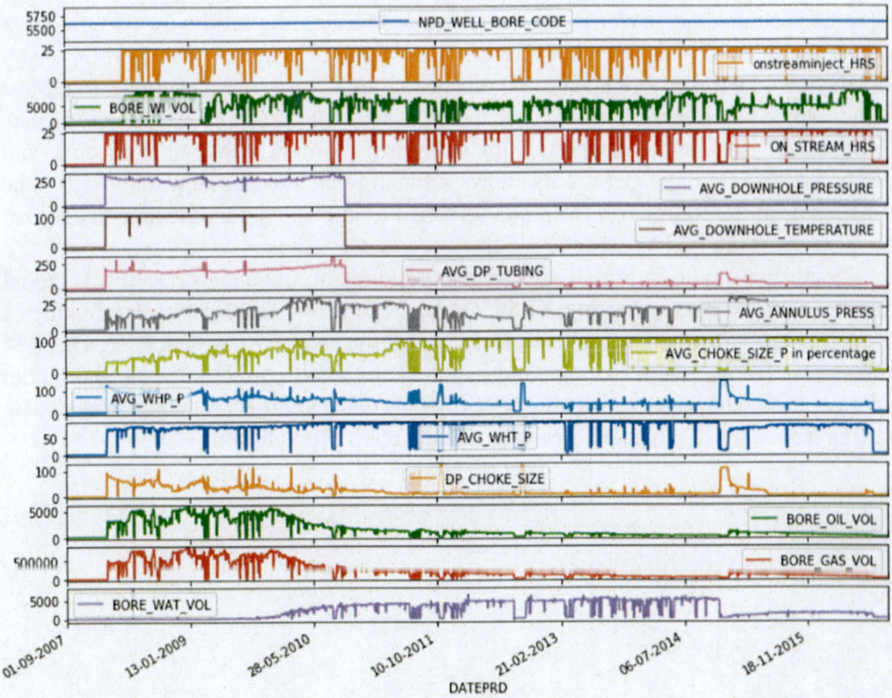

Fig. 11. Data visualization for the injector 15/9-F-4 AH and Producer 15/9-F-11 H.

Figure 12 shows the production rates of oil and water for the producer 15/9-F-12. It also shows the water injection rate of the injector 15/9-F-4 AH. The water injection started in early 2007, and the water breakthrough happened in the producer 15/9-F-12 around early 2009.

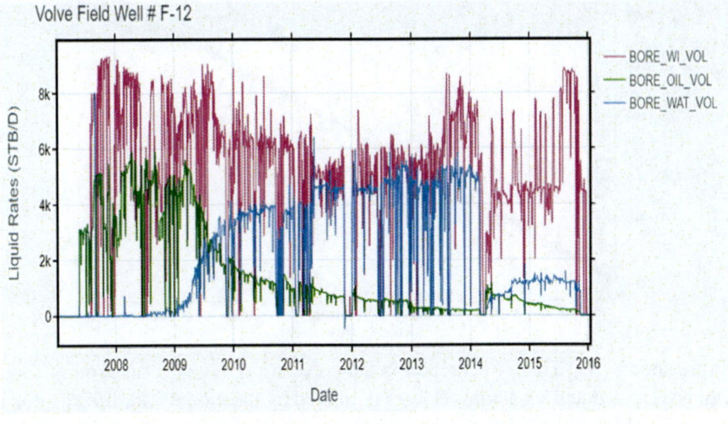

Fig. 12. Figure shows the production rate of oil and water for producer 15/9-F-12 and injection rate of 15/9-F- 4AH.

In this section, we show history matching and forecasting results for two producer wells using the RNN-DLSTM model reinforced by the CRM model. Results show a very good match with the historical data and forecasting trends are shown using the blind-test methodology, where part of the historical data is used for pseudo-prediction to validate the model. Figure 9 shows the correlation analysis performed between the injection and production data for the pattern. This allows estimating the time window that needs to be considered for the RNN-DLSTM model in terms of allocating the past-time history for the memory of the LSTM model.

The results shown in Fig. 13 clearly show the history matching (Deep ML model training) and forecasting (testing) of the model. Figure 14 clearly shows the training of the RNN-DLSTM model using historical data till the end of 2014. The RNN-DLSTM architecture learned from the data and captured the trend correctly during the model training phase. The blind testing was performed by using almost 1 year of historical data starting from early 2016 and it clearly predicted the performance of well.

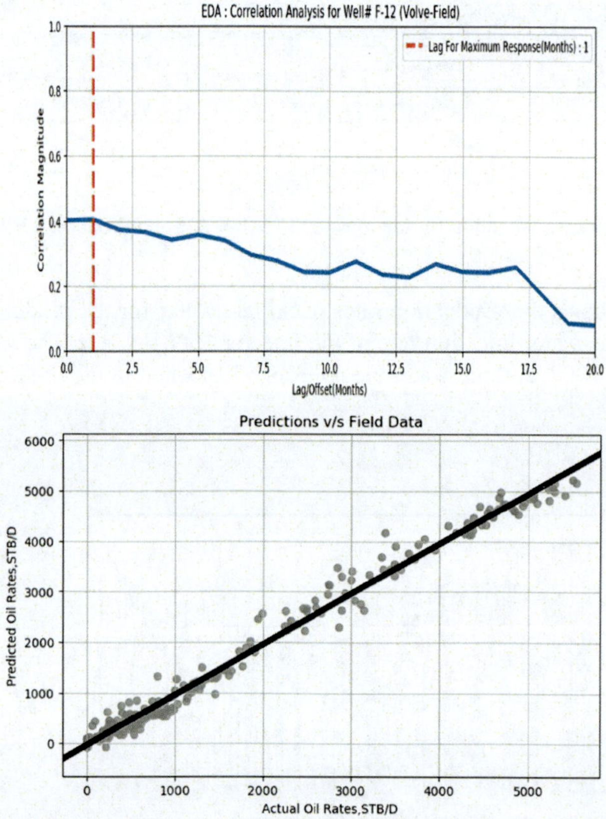

Fig. 13. Figure shows - (a) The left side plot is the correlation plot as a function of time and (b) right side plot is the predicted Vs actual oil rates predicted by the model – for the injector 15/9-F-4 AH and producer 15/9-F-12.

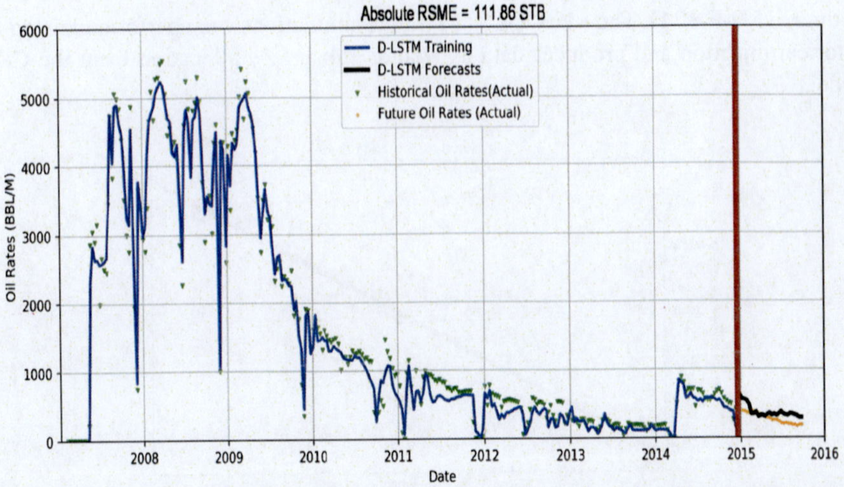

Fig. 14. Figure shows - (a) The left side plot is the correlation plot as a function of time and (b) right side plot is the predicted Vs actual oil rates predicted by the model – for the injector 15/9-F-4 AH and producer 15/9-F-12

Figure 15 shows the oil rate and water rate for the producer 15/9-F-11 H and the water injection rate of injector 15/9-F-4 AH. The producer 15/9-F-11 H was completed in mid of 2013 and it was produced until late 2016. The waterfront had already reached to the 15/9-F11 H well areas, hence the water uptake was immediately seen after putting the well on production.

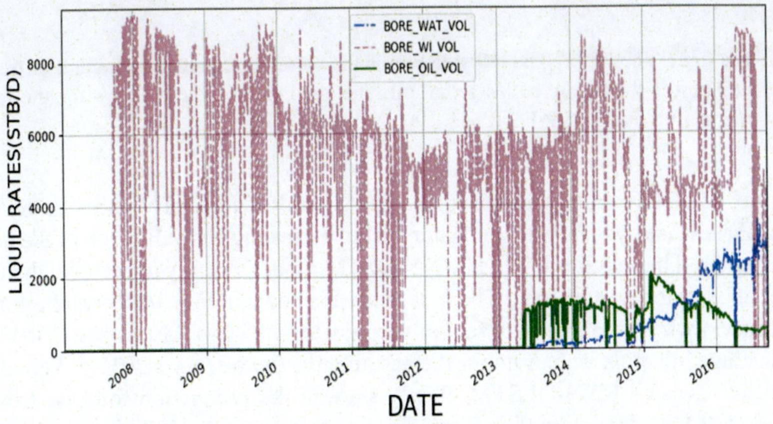

Fig. 15. The figure shows the production rate of oil and water for producers 15/9-F-11 and the injection rate of 15/9-F- 4 AH.

The analysis was performed on the well pair 15/9-F-4 AH and 15/9-F-11 H. The initial exploratory analytics and Pearson correlation coefficient shows that the producer 15/9-F-11 H is weakly connected to the injector 15/9-F- 4H (see Fig. 16) compared the producer 15/9-F-12 H. The RNN-DLSTM model simulations were performed using the historical injection and producer data as well as subsurface allocation from the CRM model.

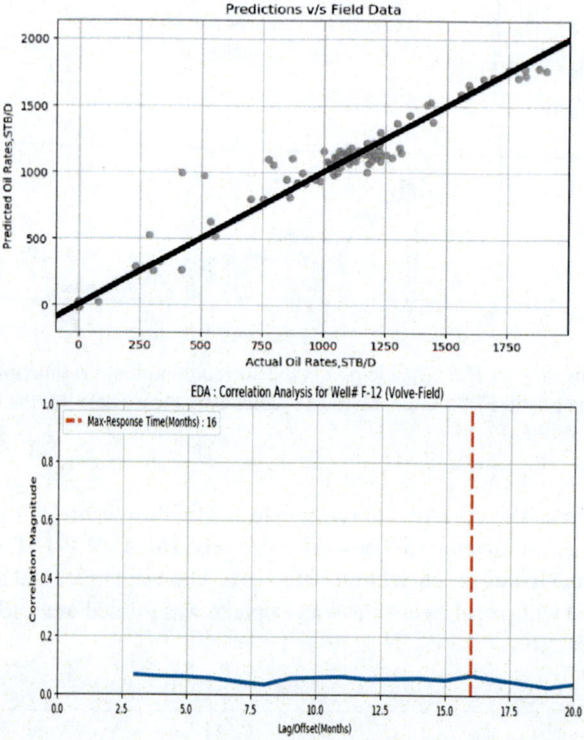

Fig. 16. The figure shows: (a) The left side plot is the correlation plot as a function of time and (b) right side plot is the predicted Vs actual oil rates predicted by the model – for the injector 15/9-F-4 AH and producer 15/9-F-11.

Figure 17 clearly shows the performance of the RNN-DLSTM model during the training phase (history matching) and later during the testing and validation of the model.

We explored the physics-assisted RNN-DSTM model's ability to perform the future forecast under various scenarios. The first scenario shown in Fig. 14 shows the forecast for a scenario where the injection rate of the injector 15/9-F-4 AH was doubled based on the average injection water rate for the past months (between Jan 2015 to Sept 2015). The physics-assisted RNN-DLSTM clearly showed the production forecast from the producer 15/9-F-12 H, where the oil rate jumped from around 230 bbl/d to a peak rate of 1860 bbl/d in few months. If we continue the water flood for 5years, then the standard decline will be observed in the oil rates. If we again increase the water injection rate based on the allowable well bottom hole pressure again in early 2020 then we see another uptake in the oil rate, then oil production declines. The model clearly shows that this well could have rejuvenated by increasing the water injection rate within the operations limit to increase the oil production rate (Fig. 18).

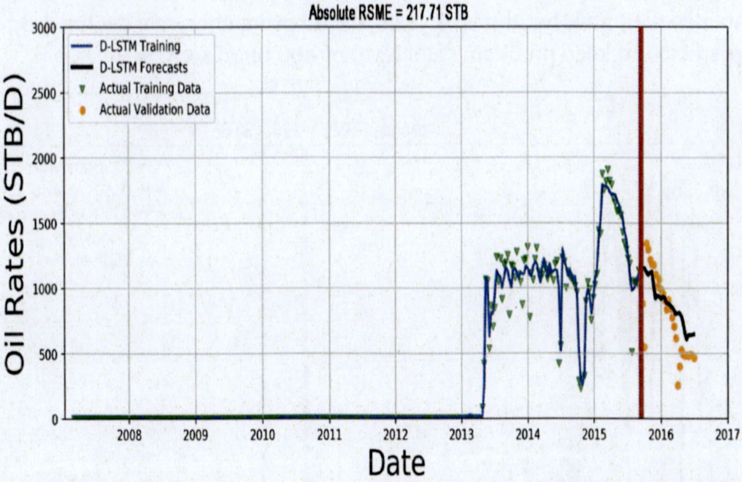

Fig. 17. Figure shows the oil rate Vs date – Physics assisted RNN-LSTM predictions Vs field data for producer 15/9-F-11.

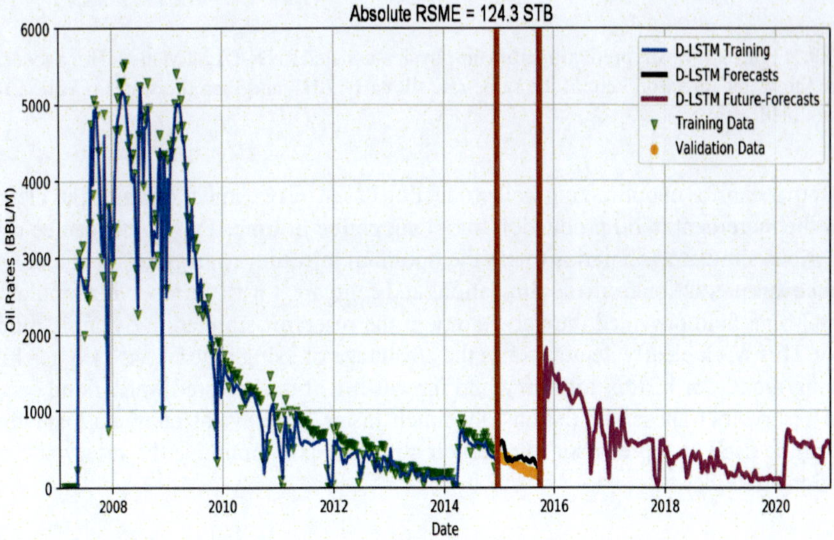

Fig. 18. Figure shows the prediction from the physics assisted RNN-DLSTM model for a scenario where the water injection is doubled from Nov 2015.

Figure 19 shows another scenario where the injection rate of water was kept as it is from Sept 2025 to Dec of 2016. The oil rate was on decline and it continued on the same decline rate until the oil production rate became very low <10 bbl/d. The water injection rate from the injection was doubled in earl7 2017. The oil update was observed in few months and reached to peak value of 860 bbl/d. The higher water injection rate was maintained thereafter for almost 4 years, the oil production rate declined slowly and the well predicted to keep producing at a level of 400 bbl/d for few years.

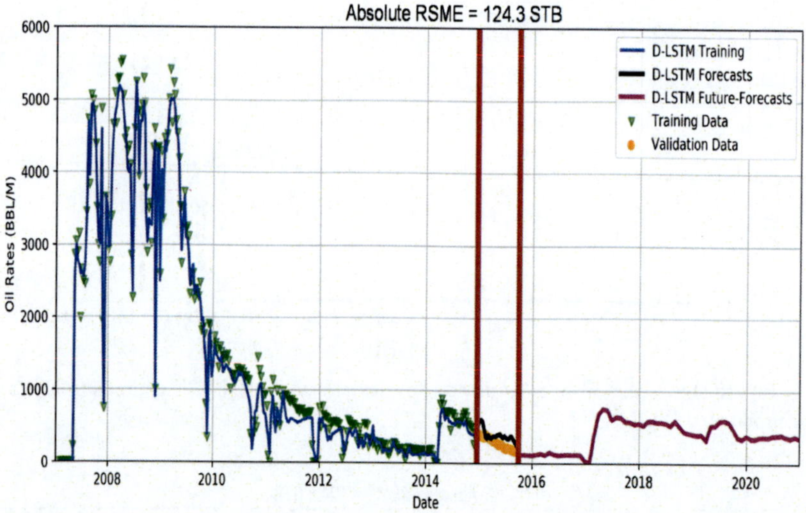

Fig. 19. Figure shows the prediction from the physics-assisted RNN-DLSTM model for a scenario where the water injection kept at the same rate till early 2017 and then the water injection rate was doubled.

Both scenarios of optimizing well production based on two injection strategies clearly show the increment in oil production for a long period of time. This model can be used for various simulated scenarios to see the optimum injection strategy for the field. One can take this prediction from the simulations and compare it with the reservoir simulation of the Volve field provided the uncertainty in the reservoir simulation dynamic model is low. This work clearly demonstrates the advantage of using physics assisted machine learning model for history matching and forecasting of the matured water flood or gas flood reservoirs. These simulations are much faster in terms of computational time and can be used as a precursor to reservoir simulations to optimize the reservoir-level injection strategies.

8 Conclusions and Recommendations

This paper shows that physics-assisted RNN-DLSTM-based models are good alternatives to conventional simulation tools for faster history matching and forecasting. The use of the capacitance and resistance model was investigated to provide the reservoir flow heterogeneity input to the deep supervised machine learning model based on RNN-DLSTM.

Once the model learned from the historical data set, the model's ability in predicting performance was tested extensively. It was shown that the model performed very well for the various injection strategies encountered within the dataset. Due to physics assisted nature of the deep learning model, it was clearly demonstrated that the model performed very well in forecasting similar to history matched reservoir simulations within a shorter time period. We believe that such models can be used as a precursor to reservoir simulation to optimize reservoir-level operational strategies given that there are sufficient data. The model can be then integrated to fine-tune the reservoir simulation models in order to confirm the results and improve the estimation performance. The models may further assist in evaluating various injection strategies and assist in field development to estimate the potential scenarios. Additionally, the models may provide engineers with tools to optimize reservoir operations in real-time and may be a component of a reservoir recommendation system.

References

Albertoni, A., Lake, L.W.: Interfering interwell connectivity only from well-rate fluctuations in waterfloods. SPE Reservoir Evaluation and Engineering (2003)

Arts, R.J., Winthaegen, P.: Monitoring options for CO_2 storage. In: Thomas, D.C., Benson, S. (eds.) Geological Storage of Carbon Dioxide with Monitoring and Verification, pp. 1001–1013. Elsevier, Oxford, UK (2005)

Bengio, Y., Simard, P., Frasconi, P.: Learning long-term dependencies with gradient descent is difficult. IEEE Trans. Neural Networks **5**(2), 157–166 (1994). https://doi.org/10.1109/72.279181

Braswell, G.: Artificial intelligence comes of age in oil and gas. J. Petrol. Technol. **65**(01), 50–57 (2013). https://doi.org/10.2118/0113-0050-JPT

Bruce, W.A.: An electrical device for analyzing oil-reservoir behavior. Pet. Technol. **151**, 112–124 (1943)

Chithra, C.N.: An innovative neural forecast of cumulative oil production from a petroleum reservoir employing higher order neural networks. J. Petrol. Sci. Eng. **106**, 18–33 (2013)

Volve Field dataset Retrieved from, Equinor (2021). https://www.equinor.com/en/what-wedo/digitalisation-in-our-dna/volve-field-data-village-download.html. Accessed 2021

Fetkovich, M.J., Vienot, M.E., Bradley, M.D., Kiesow, U.G.: Decline-curve analysis using type curves—case histories. SPE Formation Eval. **2**(04), 637–656 (1987). https://doi.org/10.2118/13169-PA

Hammerton, J., Osborne, M., Armstrong, S., Daelemans, S.W.: Introduction to special issue on machine learning approaches to shallow parsing. J. Mach. Learn. Res. **2**(3), 551–558 (2002)

Hochreiter, S., Bengio, Y., Frasconi, P.: Gradient flow in recurrent nets: the difficult of learning long-term dependencies (2011)

Hochreiter, S., Schmidhuber, J.: Long short-term memory. Neural Comput. **9**(8), 1735–1780 (1997). https://doi.org/10.1162/neco.1997.9.8.1735

Liang, W., et al.: Optimization of oil production based on a capacitance model of production and injection rates. In: Hydrocarbon Economics and Evaluation Symposium, Dallas, Texas, USA, Dallas, TX (2007)

Mohaghegh, S.D.: Recent developments in application of artificial intelligence in petroleum engineering. J. Petrol. Technol. **57**(4), 86–91 (2005)

Moreno, J.J.M.: Artific. Neural Networks Appl. Forecast. Time Ser. **23**, 322–329 (2011)

Pal, M.: On application of machine learning method for history matching and forecasting of times series data from hydrocarbon recovery process using water flooding. Petrol. Sci. Technol. 1–32 (2021)

Patil, P.: Application of DLSTM-RNN based supervised machine leaning model to forecast and optimize the performance of the water flood and WAG. In: Oman Drilling And Exploration Symposium, Muscat, Oman (2021)

de Holanda, R.W., Gildan, E., Jensen, J.L., Lake, L.W., Kabir, C.S.: A state-of-art literature review on capacitance resistance models for reservoir characterization and performance forecasting. Energies**11**, 3368 (2018)

Sayarpour, M., Zuluaga, E., Kabir, C.S., Lake, L.W.: The use of capacitance–resistance models for rapid estimation of waterflood performance and optimization. J. Petrol. Sci. Eng. **69**(3–4), 227–238 (2009). https://doi.org/10.1016/j.petrol.2009.09.006

Volve-PUD: Plan for development and operation of Volve (2005). https://discovervolve.com/wp-content/uploads/2020/06/Volve-PUD-_upload.pdf

Team, F.D.: (2021). https://research.fb.com/category/data-science/

Terence, C.M.: Time Series Techniques for Economist. Cambridge University Press, Cambridge (1990)

Yousef, A.A., Gentil, P., Jensen, J.L., Lake, L.W.: A capacitance model to infer interwell connectivity from production and injection rate fluctuations. SPE Reserv. Eval. Eng. **9**, 630–646 (2006)

Zhang, G., Patuwo, B., Hu, M.Y.: Int. J. Forecast. **14**(1), 35–62 (1998)

Machine Learning Techniques for Predicting Malaria: Unpacking Emerging Challenges and Opportunities for Tackling Malaria in Sub-saharan Africa

Elliot Mbunge[1,2(✉)], Richard C. Milham[2], Maureen Nokuthula Sibiya[3], and Sam Takavarasha Jr[4]

[1] Department of Computer Science, Faculty of Science and Engineering, University of Eswatini (Formerly Swaziland), Kwaluseni Manzini, Eswatini
mbungeelliot@gmail.com

[2] Department of Information Technology, Faculty of Accounting and Informatics, Durban University of Technology, P. O. Box 1334, Durban 4000, South Africa

[3] Research, Innovation and Engagement, Mangosuthu University of Technology, 511 Griffiths Mxenge Hwy, Umlazi 4031, South Africa

[4] Faculty of Management and Entrepreneurial Sciences, Women's University of Africa, 549 Arcturus Road, Harare, Zimbabwe

Abstract. Malaria resurgence significantly threatens progress made towards malaria elimination in the past years and consequently increases socioeconomic and public health burden, especially in developing countries. This is exacerbated by the lack of intelligent models for predicting, mapping, diagnosing, and detecting malaria to strengthen malaria prevention and control measures. Predicting malaria and understanding risk factors leading to malaria outbreaks can assist policymakers in re-strategizing and re-aligning malaria elimination strategies and optimizing resource allocation by prioritizing malaria-endemic areas. Therefore, this study provides a comprehensive review of machine learning techniques applied to predict malaria using various risk factors. The study revealed that despite the distribution of mosquito nets, indoor spraying of insecticides, community engagement programmes and awareness strategies, socioeconomic factors, climate and environmental conditions significantly contribute towards malaria outbreaks and remain underexploited and poorly understood. Socioeconomic factors such as lower income, living conditions with house type, distance to health facilities, availability, and use of mosquito nets influence malaria outbreaks. Climatic and environmental risk factors including land surface temperature, rainfall, humidity, enhanced vegetation index, normalized difference vegetation index, and normalized difference water index significantly influence malaria incidences. The study further revealed that machine learning models such as support vector machines, decision trees, random forests, Extreme Gradient Boosting, logistic regression, K-Nearest Neighbors, Naïve Bayes, and multilayer perceptron have been greatly used to predict malaria using socioeconomic, climatic and environmental data. Predicting malaria can assist to develop early malaria warning systems, redesign interventions, make informed decision-making and subsequently strengthening malaria prevention and control measures.

R. Silhavy and P. Silhavy (Eds.): CSOC 2023, LNNS 724, pp. 327–344, 2023.
https://doi.org/10.1007/978-3-031-35314-7_30

Keywords: Malaria · machine learning · Prediction · sub-Saharan Africa

1 Introduction

The increase in malaria incidences and deaths threatens the significant progress made towards malaria elimination in the past years. This poses a socioeconomic and public health burden, especially in vulnerable developing countries, especially in sub-Saharan Africa. According to the World Health Organization (WHO)'s recent malaria report, there were an estimated 241 million malaria cases and 627 000 malaria deaths worldwide in 2020 and the African region accounted 95% accounted for 95% of all malaria cases and 96% of deaths [1]. The sudden increase in malaria cases has been exacerbated by the outbreak of coronavirus diseases (COVID-19) which subsequently affects the distribution of malaria commodities [2], interruption of malaria prevention and control measures and also community-based malaria awareness programmes [3]. However, the World Health Organization (WHO) recommended the use of key interventions to intensify malaria prevention and control measures such as increasing universal access to insecticide-treated bed nets (ITNs) or Long-lasting insecticidal bed nets (LLINs) [4, 5], indoor residual spraying (IRS) [6], mass drug administration (MDA) [7], behavioral change and communication strategies [8], cross-border collaborations [9], artemisinin-based combination therapy (ACT) [10], intermittent preventive therapy in pregnancy (IPTp) for pregnant women [11]. These interventions made a significant impact in reducing malaria transmission in the past decades, especially in sub-Saharan Africa. However, factors such as lack of knowledge of ITNs and IRS, insufficient access to ITNs [12], costs of ITNs, lack of malaria awareness [13], misuse of malaria prevention practices [14], social dynamics [7], bad attitudes and perception towards malaria prevention and control strategies [14], insufficient funding, lack of political will and commitment to eliminate malaria, increasing larval breeding sites [9], ecological factors and change of mosquito biting behavior, scarcity of malaria commodities [9], resistant of IRS chemicals and are key barriers affecting to attain global malaria elimination goals. However, to attain complete malaria elimination there is a need for continued investment and diversify malaria prevention and control methods by incorporating and integrating emerging technologies such as computational intelligence techniques to detect, model, map hotspots, diagnosis, and predict malaria.

Predicting malaria outbreaks is increasingly becoming imperative to intensify malaria prevention and preparedness in malaria-endemic areas in order to optimize the distribution of limited malaria commodities [15]. Thus, machine learning can assist to predict malaria with high precision using a dataset that would ordinarily present challenges with manual computations [16]. This can also assist policymakers to expedite the development of early warning systems to monitor the spread of malaria and subsequently strengthen malaria prevention and control measures. Malaria prediction models help to plan, organize, and expeditiously allocate resources [17] and operationalize malaria control and vector management giving priority to high perennial malaria hotspot zones. Despite the substantial evidence in the literature highlighting the importance of machine learning models in predicting malaria in endemic areas, their implementation is still nascent in many sub-Saharan Africa. Predicting malaria can significantly assist in effectively allocating resources, developing early malaria systems, and enhancing the understanding of malaria trends [18], distribution, and hotspots at the ward level. However, the application of machine learning models in malaria surveillance systems has been tremendously affected by scanty, inaccurate, and great lack of quality data due to inconsistencies in malaria data collection methods and data formats. Despite these impediments, machine learning techniques have been applied to solve various malaria-related problems such as malaria detection, diagnosis [16], mosquito identification, predicting malaria outbreaks [19] and forecasting malaria transmission [20]. There is limited literature on the application of machine learning to predict malaria outbreaks. Therefore, this study aimed to provide a comprehensive review of machine learning-based malaria prediction models and associated risk factors/ predictors. The study aimed to achieve the following research objectives:

a. Identify and explore machine learning-based malaria prediction models, performance, and their respective challenges and limitations.
b. Identify various data sources and risk factors associated with malaria outbreaks.
c. Presents identified research gaps and recommendations for future work.

The sections for this paper are structured as follows: Sect. 2 explains the methodology adopted by the study. Section 3 presents machine learning-based malaria prediction models and their respective challenges and limitations. Various predictors/ risk factors associated with malaria outbreaks are also presented in Sect. 3. Recommendations and identified research gaps are presented in Sect. 4. Lastly, Sect. 5 presents the conclusion and future work.

2 Materials and Method

2.1 Search Strategy

This study applied the Preferred reporting items for systematic review and meta-analysis (PRISMA) model [21] to guide literature from reputable and prominent electronic database sources and applied inclusion and exclusion criteria to select relevant papers as illustrated in Fig. 1. The study applied search keywords such as "predicting malaria" OR "machine learning techniques" OR "machine learning-based malaria prediction models" OR "malaria prediction" AND "malaria" OR "malaria forecasting" AND "Africa", OR "sub-Saharan region", OR "African continent". These keywords were searched in the following online electronic databases: Science Direct, Google Scholar, IEEE Xplore Digital Library, PubMed, Springer Link, and ACM Digital Library.

2.2 Study Selection and Eligibility Criteria

All published papers written in English or English translation were included in this study. Commentaries, letters to the editor and preprint were not considered in this study. The articles' publication period was restricted to the period between January 2010 and February 2023. Articles that did not apply machine learning techniques to predict malaria were excluded. All duplicate papers were also not considered. A total of 240 articles were retrieved from electronic databases. Predicting malaria using machine learning techniques is an emerging field hence few articles were retrieved. The selected articles were screened based on the following: titles, abstracts and full text. The selection of relevant studies was guided by the following inclusion criteria: (1) written in English or have English translations; (2) peer-reviewed; (3) applied machine learning techniques or models to predict malaria.

2.3 Data Extraction

The citation chain was done for each retrieved article to ensure that all relevant articles are included in the study. The authors removed 208 articles including duplicates, non-English articles, irrelevant articles, reprints, letters to the editor, and non-peer-reviewed articles. We further checked the abstracts and removed 11 unqualified articles that applied statistical methods such as ARIMA, and time series approaches among others. A total of 21 articles were further checked independently by all authors and 6 unqualified articles were excluded due to missing performance metrics. However, articles that used at least one performance metric were included. After the selection of relevant articles, we further assessed full-text articles for eligibility and removed 6 articles. Only 15 articles

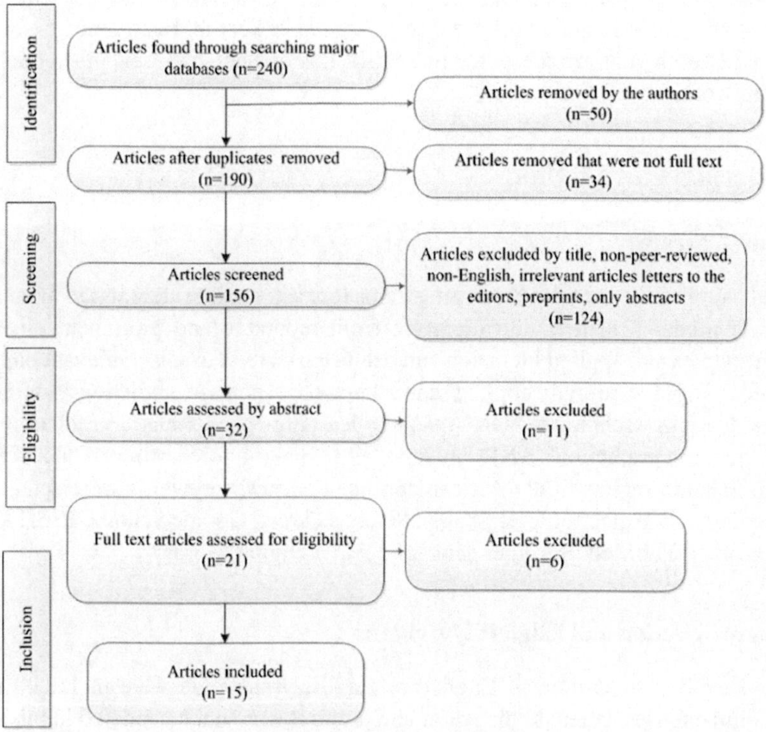

Fig. 1. PRISMA model

were considered in this study. The study designs included were mainly machine learning techniques applied to predict malaria. The steps followed to report the number of published articles are presented in the PRISMA flow diagram in Fig. 1.

The selected articles were examined in full-text, and authors collected and captured references, machine learning prediction models used, data sources or type of data, and performance metrics. The captured performance metrics used to assess malaria prediction models include accuracy, precision, recall and F1-score, as shown in Table 1.

Table 1. Machine learning-based malaria prediction models

Reference	Prediction models	Data sources	Malaria prediction model performance evaluation metrics			
			Accuracy (%)	Precision (%)	Recall (%)	F1-score (%)
[22]	RF, DT, SVM, Naïve Bayes, KNN and XGBoost	Meteorological Centre, India	KNN:86.21, RF:93.94, SVM:92.69, XGBoost:96.26, LR:92.44, Naïve Bayes: 91.69	KNN = 92.9, RF:85.42, SVM:91.14,XGBoost:91.8, LR:90.44, Naïve Bayes: 88.36	KNN = 50.48, RF:92.28, DT:42, SVM:79.42, XGBoost:93.89,LR:79.1, Naïve Bayes: 78.14	Not used to assess the performance
[23]	DT, RF, NB, LR and SVM	Clinical data, Senegal	DT: 97.8, RF: 97.24,	NB:96, LR: 93, DT:92, RF: 92, SVM:93	NB:5, LR: 62, DT:85, RF: 85, SVM:86	NB:10, LR: 75, DT:88, RF: 89, SVM:89
[24]	Decision Tree Multilayer perceptron (MLP)	Malaria cases, rainfall, temperature, humidity,	DT: 89 MLP:83	Not used to assess the performance	Not used to assess the performance	Not used to assess the performance
[25]	SVM, RF, Gradient boosting, CatBoost(CB) and multilayer perceptron, AdaBoost	Parasitic disease dataset using BioPython from the Center for Disease Control and Prevention	SVM: 91.5 RF: 90.3 MLP: 90.9 AdaBoost: 89.4 GB: 89.1 CB: 90.9	SVM:00 RF: 25 MLP: 28.6 AdaBoost: 21.1 GB: 22.7 CB: 25	SVM: 00 RF: 16 MLP: 16 AdaBoost: 16 GB: 20 CB: 12	SVM: 00, RF: 19.5, MLP: 20.5, AdaBoost: 18.2, GB: 21.3, CB: 16.2
[19]	Random Forest	IeDA database	95	83	99	Not used to assess the performance
[20]	XGBoost, SVM, Naïve Bayes and LR	Climate factors and malaria incidences	XGBoost: 86, SVM: 72, Naïve Bayes: 70 and LR: 77	Not used to assess the performance	Not used to assess the performance	Not used to assess the performance
[26]	Decision tree	Clinical malaria data	77	Not used to assess the performance	Not used to assess the performance	Not used to assess the performance
[27]	SVM and RF	Malaria cases, IRS data and climatic factors	SVM: 95 RF: 90	Not used to assess the performance	Not used to assess the performance	Not used to assess the performance
[28]	LR, SVM, DT, and RF	Climate data	LR: 83 SVM: 77 DT: 66 RF: 83	LR: 82 SVM: 78 DT: 79 RF: 87	LR: 100 SVM: 100 DT: 79 RF: 93	LR: 90 SVM: 88 DT: 79 RF: 90
[29]	RF and SVM	Clinical data	RF: 94.2 SVM: 94.9	Not used to assess the performance	Not used to assess the performance	Not used to assess the performance

(continued)

Table 1. *(continued)*

Reference	Prediction models	Data sources	Malaria prediction model performance evaluation metrics			
			Accuracy (%)	Precision (%)	Recall (%)	F1-score (%)
[30]	DT, LR, Naïve Bayes, RF and SVM	Rainfall, temperature and malaria incidences	DT:75.1 LR:66.9 Naïve Bayes:66.9 RF:82.9 SVM:64.8	DT:89.2 LR:66.9 Naïve Bayes:66.9 RF:91.5 SVM:66.2	DT:72.3 LR:100 Naïve Bayes:100 RF:81.1 SVM:96.7	DT:79.5 LR:80.1 Naïve Bayes:80.1 RF:85.9 SVM:78.5
[31]	LR, RF, DT, SVM and KNN	Rainfall and temperature	LR: 83.8 RF: 90.7 DT: 88.5 SVM:56.8 KNN:81.7	LR: 83.8 RF: 90.8 DT: 88.6 SVM:60 KNN:82.1	LR: 83.7 RF: 90.6 DT: 88.5 SVM: 88.5 KNN:81.6	LR: 83.8 RF: 83.8 DT: 88.5 SVM:51.8 KNN:81.6
[32]	SVM. KNN,RF, DT, Naïve Bayes and LR	Temperature and rainfall	SVM: 93.18 KNN: 86.36 RF: 97.72 DT: 95.45 Naïve Bayes: 88.63 LR: 95.45	SVM: 91 KNN: 81 RF:100 DT:100 Naïve Bayes: 84 LR:95	SVM: 95 KNN: 95 RF: 95 DT:91 Naïve Bayes: 95 LR: 95	SVM: 93 KNN: 88 RF: 98 DT: 95 Naïve Bayes: 89 LR: 95
[33]	SVM	Hydrological Drought Indices and Climate data:	99	Not used to assess the performance	Not used to assess the performance	Not used to assess the performance
[34]	Extreme Learning Machine	Clinical data	99	98	Not used to assess the performance	Not used to assess the performance

3 Results Analysis and Discussion

The selected articles shown in Table 1 were extracted and synthesized guided by the PRISMA model to address research objectives. The following subsections present the findings as follows: (i) machine learning-based malaria prediction models, (ii) malaria prediction models performance evaluation metrics, (iii) data sources and risk factors(predictors) associated with malaria outbreaks, (iv) identified research gaps and future opportunities for predicting malaria and (v) recommendations for future work.

3.1 Machine Learning-Based Malaria Prediction Models

Machine learning (ML) is a subset of artificial intelligence that learns from experience and extracts valuable patterns from large, unstructured, and complex datasets to predict future incidences [19]. Machine learning models are generally trained and learn from data using either supervised or unsupervised learning techniques [35]. Unsupervised learning focuses on uncovering structure within a dataset without prior knowledge of how the data are organized, thus expected output is unknown [36]. In contrast, supervised learning relies on prior knowledge and labelled training data to make predictions or classify labelled data [25]. Therefore, supervised machine learning predicts the value of response labels of either categorical or continuous values based on the input dataset [37]. Several machine learning models have been used to predict malaria in various malaria-endemic countries. Such models include support vector machines (SVM) decision tree

(DT), random forest (RF), Extreme Gradient Boosting (XGBoost), logistic regression (LR), K-Nearest Neighbors (KNN), and Naïve Bayes.

Support vector machine is a supervised algorithm that learns by analyzing data used for classification and regression analysis purposes, thus assigning labels to objects [25]. Support vector machine can classify both linear and non-linear data by mapping each input data into an n-dimensional feature space, where n, is the number of features [38]. The classes of data input are then separated by a hyperplane to maximize the marginal distance and minimize classification errors in both classes [39]. In linear SVM, the plane can be split by a line and then the marginal distance for a class is determined by the distance between the nearest input label of the same class and the decision hyperplane [40]. SVM can also solve nonlinear regression problems by transforming support vectors into higher dimensional feature space, which is introduced by kernel functions. In this context, support vectors refer to training points that are nearest to the separating function. The kernel functions map nonlinear regression problems into separable ones by mapping data into better representational space. For instance, [41] applied SVM and artificial neural networks (ANN) to predict malaria outbreaks with environmental data such as temperature, humidity, rainfall, and malaria-positive cases. From their study, SVM recorded a root mean squared error (RMSE) of 0.12 as compared to artificial neural networks which had an RMSE of 0.47. However, SVM requires manual feature extraction and is unsuitable for discovering multifaceted correlations in malaria data though it seems to be the best prediction model. Therefore, SVM cannot be easily integrated alone without other optimization algorithms and process malaria data with unbalanced dimensions because it may affect the prediction results. To solve this problem, [42] applied SVM coupled with the firefly algorithm to determine malaria transmission, and compared it with other models such as SVM, ARMA and ANN. From their study, SVM with the firefly algorithm outperformed other models with the normalized mean square error of 0.21, and 0.32 for SVM, 0.34 for ANN, and 0.39 for ARMA [42].

A decision tree is a tree-like structure that uses several questions with binary answers. Each answer to a new question leads to a new yes or no in the form of nodes until finally, a prediction can be made. Each node represents the splitting rule for a feature to classify the target value. The CART model is based on a computational-statistical algorithm that separates observations in binary and sequential branches of a tree to improve prediction accuracy. The best-known methods for automatically building decision trees are ID3 and the C4.5 algorithms. However, [27] applied decision trees to predict incidences of malaria cases in Mozambique using a combination of IRS datasets, malaria cases and climatic factors. Decision trees suffer from misclassification errors of the multiclass variables since the splitting criterion is based on the normalized information gain (difference in entropy) [43]. Therefore, the attribute with the highest normalized information gain is chosen to make the decision.

Random forest (RF) is a supervised machine learning ensemble classifier that consists of many decision trees [22, 44]. The class that is most frequently predicted across all trees in the forest is chosen as the predicted class by the random forest [45]. It combines decision trees with ensembles of classification or regression trees, which are created by employing bootstrap sampling on the training set and random feature selection during tree induction [27]. It is a machine learning method composed of many decision trees that use randomly picked data attributes as their input. The forest generation process constructs a collection of trees with controlled error variance. The resulting prediction can be decided by a selection scheme, such as averaging or majority voting. In South Korea, [46] applied random forest using climate data while in Mozambique, [27] also applied RF to predict malaria using environmental data. A study by [27] noted that the mean square errors of RF were higher than SVM. This shows the applicability of support vector machines in predicting malaria incidences using environmental data, especially in malaria-affected zones.

Adaptive Boosting (AdaBoost) is a boosting algorithm introduced by Freund and Schapire to solve classification problems [47]. The algorithm is generally integrated with other machine learning classifiers to improve performance. The Adaptive boosting algorithm is trained by training a classifier into input data and weights of miss-classified data are adjusted, and additional copies of the classifier are fitted to the same data [48, 49]. This process is done iteratively till it achieves the best performance. A study by [50] applied GIS, support vector machine, artificial neural networks and AdaBoost to understand malaria spatial variations in Vietnam and further assessed the performance of these techniques using statistical evaluation metrics namely, Mean Absolute Error (MAE) and Root Mean Square Error (RMSE).

Logistic regression (LR) is the most prominent and powerful supervised machine learning classifier used to solve binary classification problems [23]. It is based on the logistic representation of a binary dependent variable and explanatory variables as feedback to calculate the probability of a particular class label [51]. Logistic regression finds the probability that a new instance belongs to a certain class and the outcome lies between 0 and 1 [52]. Therefore, a threshold needs to be assigned to separate the two classes. For example, if a probability score is greater than 0.5, logistic regression classifies it as 1, otherwise 0. The next subsection presents performance evaluation metrics for assessing machine learning-based malaria prediction models.

Multilayer perceptron (MLP) is a feed-forward artificial neural network that uses a supervised training method. Multilayer perceptron consists of an input layer, hidden layer(s) and output layer [25]. The input layer receives input data, while the hidden layer consists of hidden neurons that compute associations in sample data and pass the results to the output layer [24]. The purpose of the output layer is to generate results. The MLP is generally trained using a supervised learning algorithm until the error rate becomes sufficiently small. As shown in Table 1, [25] applied MLP to predict malaria using clinical data and the model achieved an accuracy of 90.9%, a precision of 28.6%, a recall of 16% and an F1-score of 20.5%. Also, a study conducted by [24] applied MLP and decision tree (J48) to predict malaria and achieved a high prediction accuracy of 83% and 89%, respectively in October. This shows that MLP generally performs poorly as compared to other machine learning classifier when predicting malaria with a small dataset and tend to overfit deterring its implementation in real-life [30].

K-Nearest Neighbour (KNN) is the most popular machine learning classification algorithm that has strong predictive power. It is generally used to determine the similarity or associates each new element to the most common class among the elements closest to this one (where k nearest element is used in determining the class). Thus, the value of "k" minimizes the training and validation step's error rates. The description of the

KNN algorithm is fully explained in a study conducted by [53]. However, a study by [22] applied KNN to predict malaria and achieved an accuracy of 86.21%, precision of 92.9% and recall of 50.48%.

Naïve Bayes is a probability-based classifier that is based on the Bayes theorem [54]. Naïve Bayes assumes that all features used for prediction are independent of each other. Based on the training sample data, Naïve Bayes computes the probability of each of the values of the dependent variable corresponding to each combination of the values (or ranges of value) possibly assumable by the independent variables [55]. The naïve Bayes classifier has been greatly used to predict malaria. For instance, a study by [22] applied Naïve Bayes to predict malaria and achieved an accuracy of 91.69%, precision of 88.36% and recall of 78.14%.

Extreme Gradient Boosting is also known as XGBoost. It is used to solve machine learning classification, regression and ranking problems [20]. The XGBoost model is an efficient and scalable algorithm that presents results in a prediction model, mostly in the form of trees. For instance, a study conducted by [22] utilised XGBoost to predict malaria and achieved an accuracy of 96.26%, precision of 91.8% and recall of 93.89%.

3.2 Malaria Prediction Models Performance Evaluation Metrics

Prediction models have generally assessed their performance after the computing model training and testing process. In the past, the performance of machine learning models was centered on accuracy [56]. However, accessing prediction models using accuracy proved insufficient and sometimes misleading due to different factors including the type of the problem being solved [57], and the structure and size of the dataset [22]. In addition, [58] also alluded that there is no general consensus on the best evaluation metrics because values generated in the confusion matrix (especially binary classification) that determine the false positives, false negatives, and true positives as well as true negatives. Thus, the performance of machine learning models to solve binary, multi-label and multiclass classification problems can be assessed using classification metrics such as precision, recall, accuracy, and F1-score [59]. These classification metrics capture important aspects in the model evaluating space. Binary classification ensures that the input is classified into one, and only one, of two non-overlapping classes [56].

Predicting malaria outbreaks is a binary classification problem that can be solved by applying machine learning classification models and assessing their performance by using classification performance evaluation metrics [60]. The study thrives to address a classification problem by predicting malaria (malaria or not) in malaria hotspot areas, hence the use of classification performance evaluation metrics such as precision, recall, accuracy, and F1-score. These performance measurement metrics are calculated using the confusion matrix of each machine learning model. The parameters of confusion matric consist of True Positive (TP), False Positive (FP), True Negative (TN) and False Negative (FN). True positive is the number of positive target variable instances that have been correctly predicted as positive [22]. False-positive is the number of negative target variable instances that have been wrongly predicted as positive [52]. True negative is the number of negative target variable instances that have been correctly predicted as negative [60]. False-negative is the number of positive target variable instances that have been wrongly predicted as negative [23]. From these confusion matrix parameters, we can calculate evaluation standards like recall, precision, accuracy, and F1-score for all machine learning models. From these confusion matrix parameters, we can calculate evaluation metrics such as recall, precision, accuracy, and F1-score for all malaria prediction models. Evaluation metrics are paramount machine learning models' performance indicators. The standard ML performance evaluation metrics are generally used based on accuracy and error rate. These evaluation metrics are explained below.

A recall is also known as sensitivity; it evaluates how many positive targets are lost in the prediction. It is calculated as follows.

$$Recall = \frac{TP}{TP + FN} \tag{1}$$

Precision is also known as positive predictive value; it helps evaluate how many positives are superfluous in the result and thus cannot be trusted. It is calculated as follows.

$$Precision = \frac{TP}{TP + FP} \tag{2}$$

Accuracy is the ratio of correctly predicted observations over the total observations [61]. It is calculated as follows.

$$Accuracy = \frac{TP + TN}{TP + FP + FN + TN} \tag{3}$$

F1-score is a function of precision and recall and it is needed to maintain a balance of precision and recall results. It is calculated as follows.

$$F1 - score = \frac{2 * TP}{2 * TP + FP + FN} \tag{4}$$

3.3 Data Sources and Risk Factors (Predictors) Associated With Malaria Outbreaks

As shown in Table 1, the growth and maturity of parasites mostly depend on climatic factors such as temperature, rainfall, and relative humidity, and thus, any change in climate factors would certainly exert an effect on the mosquito ecology. Therefore, the correlation between climatic and environmental risk factors and malaria incidence has been a predominant research focus [20]. Although many studies indicated that climate and environmental change may influence the distribution of malaria, there is a paucity of information on its trends and association with climatic variables in Zimbabwe, especially at the ward level[62]. For instance, rainfall and temperature influence seasonal malaria transmission. Rainfall creates mosquito breeding sites while temperature regulates the rate of development of the mosquito larvae and influences mosquito survival rates [63]. In addition, [62] alluded that rainfall, temperature and altitude are key factors in determining the habitat suitability of malaria vectors, including Anopheles (common in Zimbabwe), as well as determining malaria incidences. Also, [64] states that enhanced vegetation index (EVI), normalized difference vegetation index (NDVI), and normalized difference water index (NDWI) influence seasonal and spatial variation in malaria transmission in Zimbabwe. Other environmental and socioeconomic factors that perpetuated the increase of malaria include land cover, human activities (especially farming in wetland areas), topographical variations and human migration (imported cases) [24].

These climatic and environmental factors can be used to predict malaria. For instance, [65] applied the autoregressive integrated moving average (ARIMA) model to forecast malaria using clinical predictors and environmental data in Uganda. The environmental data used in the study include day and night land surface temperature (LST) from MODIS (MOD11A2) using 8-day composite images at a 1 km × 1 km resolution, enhanced vegetation index (EVI) from MODIS (MOD13A1) using 16-day composite images at a 0.5 km × 0.5 km resolution, and rainfall estimates with a spatial resolution of 0.25° × 0.25° or 27.8 km × 27.8 km (at the equator). For instance, [23] applied machine learning models such as random forest, support vector machine with Gaussian kernel and artificial neural networks to predict malaria using clinical data such as signs and symptoms in Senegal. Also, [20] utilized machine learning to predict malaria incidences at the national level using data from the National Centre for Atmospheric Research, for sub-Saharan African countries such as Burkina Faso, Cameroon, Mali, Democratic Republic of Congo, Niger Republic and Nigeria. However, their study did not have adequate datasets for confirmed malaria incidences to stratify malaria seasons to enhance real-time prediction. Also, [19] applied time series to predict malaria epidemics in Burkina Faso. In Ethiopia, [66] predicted malaria incidences three months ahead by using an adaptive neuro-fuzzy inference system and support vector regression. These models were trained ad tested using climatic data such as monthly rainfall, average temperature, relative humidity and monthly malaria data.

To improve the performance of machine learning models, [19] utilized the synthetic minority oversampling technique (SMOTE) to address the class imbalance problem and then applied a support vector machine, random forest, multi-layered perceptron, AdaBoost, gradient boosting, and CatBoost to predict malaria using both clinical information and parasitic disease dataset. Class imbalance occurs when instances of one class outnumber the instances of other classes which can affect the performance of machine learning problems [61]. This is because the classifiers, which learn from a minority class with very few instances, tend to be biased towards a high accuracy in the prediction of the majority class [67]. Several techniques such as over-sampling minority class, under-sampling the majority class, and hybrid methods have been utilized to balance class distribution. Undersampling the majority class is a popular most naive method of dealing with class-imbalance problems by eliminating of majority class which consequently leads to discarding potentially useful data that could be important for classifiers [68]. Therefore, to avoid discarding important useful malaria data, some studies including [28] and [20] utilized the synthetic minority oversampling technique to address the class imbalance problem by changing class distributions [69]. The synthetic minority oversampling technique is the most popular over-sampling method due to its simplicity [70], computational efficiency, and superior performance. SMOTE generates a synthetic example rather than a replacement with replication. This technique identifies more specific regions in the feature space for the minority class [71]. However, SMOTE is applicable for binary classification problems with a continuous feature space [43]. For instance, predicting malaria outbreaks is a binary classification problem. A study conducted by [25] applied SMOTE to solve the class imbalance problem and predict malaria using clinical information.

3.4 Identified Research Gaps and Future Opportunities for Predicting Malaria

The literature reflects that several studies applied machine learning models to predict malaria using previously confirmed malaria incidences, climate and environmental data. However, there is a lack of machine learning-based data-driven malaria tools for mapping hotspots, modelling imported cases, optimising resource allocation, and predicting malaria using fine-scale data at the micro-level such as wards and district level. In addition, more studies focused on climatic and environmental data [63] despite the importance of socioeconomic factors such as LLINs use, indoor residual spraying, population growth and income levels [39]. Socioeconomic factors are rarely used to predict malaria yet they significantly contribute to malaria outbreaks [18]. Therefore, future studies can focus on incorporating these important factors into malaria predictive models.

The study also revealed that there is a lack of readily available data that may affect the use and integration of machine learning-based malaria prediction models into healthcare settings. A study conducted by [20] alluded that unavailability of the daily malaria incidences affect machine learning-based malaria prediction models to understand how daily seasonal variations in the climatic factors influence malaria and further predict the likelihood of having malaria outbreak on daily basis without using averaged data. The availability of a huge dataset of malaria cases at the micro-level, especially wards or district level can assist in seasonally stratifying important malaria seasons to enhance real-time prediction, implement proactive intervention strategies [72], and expedite the development of early warning systems.

The lack of policies to integrate emerging technologies such as machine learning-based mobile health applications in healthcare [73] is also a major setback to integrating machine learning-based malaria prediction models. This also contributes to the lack of interpretable machine learning models to process fine-scale determinants leading to malaria transmission [74]. Future work can focus to address these gaps. This could assist health workers and policymakers to understand complex processes leading to malaria outbreaks and re-orient malaria interventions and operational tools for seasonal mapping of malaria hotspots, and early warning systems. In addition, a study conducted by [30] also alluded that there is a lack of machine learning-based malaria prediction models that integrate indigenous knowledge of malaria outbreaks. Also, lack of evaluation ascertaining the scalability of machine learning-based malaria prediction models in other malaria-endemic African countries [30].

4 Recommendations

The literature revealed that there is a significant association between malaria incidences and climatic and environmental factors. Efforts to develop spatial decision support systems have been made progressively in Kenya [24] but the development of malaria prediction systems that utilise risk factors on a seasonal scale is still nascent. Therefore, future works should focus on developing malaria data-driven applications or models for predicting, early warning, and hotspot mapping to adequately support policymakers and health workers, especially in resource-constrained malaria-endemic areas to optimise the distribution of limited resources. Moreso, several studies utilised climate and environmental risk factors as major predictors, however, malaria intensity and distribution are also exacerbated by other factors including clinical data, socioeconomic factors [5], population, migration patterns, behavioural change and communication strategies. Therefore, future application of machine learning models to predict malaria should factor in other important risk factors. The study also revealed that machine learning models shown in Table 1 can predict malaria at either district, provincial or national level using aggregated data on monthly basis. However, we, therefore, recommend future works to predict malaria at the micro/ ward level require finer scale datasets that can provide deep analysis to understand malaria distribution and intensity variations at the ward level within a month on a real-time basis.

5 Conclusion

Malaria remains a significant public health burden despite the progress made towards malaria elimination through intensified malaria prevention and control strategies [66]. The study revealed that predicting malaria using machine learning techniques can tremendously assist to optimise limited resources, re-strategize malaria elimination policies, and understanding seasonal transmission and intensity as well as distribution. Machine learning models such as support vector machines decision trees, random forests, Extreme Gradient Boosting, logistic regression, K-Nearest Neighbors, and Naïve Bayes have been greatly utilised to predict malaria using climatic and environmental risk factors and achieved fairly good performance. However, the application of machine learning models to predict malaria encounters several impediments such as the unavailability of huge datasets (most used datasets are small), class imbalance, limited use of other non-climatic risk factors and lack of policies to integrate and utilise machine learning-based malaria prediction models in healthcare systems. Therefore, there is a need to craft policies to integrate emerging technology such as machine learning into healthcare systems and collect important data including climatic, and environmental data, socioeconomic factors [18], population migration patterns and other important risk factors that can be effectively used to predict malaria. The future work can also focus on developing malaria data-driven applications or models for predicting, early warning, and hotspot mapping to adequately support policymakers and health workers, especially in resource-constrained malaria-endemic areas to optimise the distribution of limited resources.

References

1. Tian, H., et al.: Malaria elimination on Hainan Island despite climate change. Commun. Med. **21**(2), 1–9 (2022) https://doi.org/10.1038/s43856-022-00073-z
2. Mbunge, E., Ndumiso, N., Kavu, T.D., Dandajena, K., Batani, J., Fashoto, S.G:. Towards QR Code Health Systems Amid COVID-19: Lessons Learnt from Other QR Code Digital Technologies, pp. 129–43 (2022). https://doi.org/10.1007/978-3-031-10031-4_7

3. Kebede, Y., et al.: Primary school students' poetic malaria messages from Jimma zone, Oromia, Ethiopia: a qualitative content analysis. BMC Public Health **21**, 1–16 (2021). https://doi.org/10.1186/S12889-021-11641-8/TABLES/2
4. Mbunge, E., Millham, R., Sibiya, M.N., Takavarasha, S.: Impact of COVID-19 on Malaria Elimination: Juxtaposing Indoor Residual Spraying and Mobile Phones in Buhera Rural District, Zimbabwe (2021). https://doi.org/10.21203/rs.3.rs-173130/v2
5. Tarekegn, M., Tekie, H., Dugassa, S., Wolde-Hawariat, Y.: Malaria prevalence and associated risk factors in Dembiya district North-western Ethiopia. Malar J. **20**, 1–11 (2021). https://doi.org/10.1186/S12936-021-03906-9/TABLES/4
6. Mohammed, M.A., Hong, T.: Role of vector control in fighting against malaria: Evidence from Ethiopian health-related indicators. J. Infect. Public Health **14**, 527–532 (2021). https://doi.org/10.1016/J.JIPH.2020.12.002
7. Cheng, B., Htoo, S.N., Mhote, N.P.P., Davison, C.M.: A systematic review of factors influencing participation in two types of malaria prevention intervention in Southeast Asia. Malar J. **20**, 1–9 (2021). https://doi.org/10.1186/S12936-021-03733-Y/FIGURES/1
8. Mbunge, E., Millham, R., Sibiya, N., Takavarasha, S.: Is malaria elimination a distant dream? Reconsidering malaria elimination strategies in Zimbabwe. Public Heal Pract. **2**, 100168 (2021)
9. Li, X.H., et al.: Seven decades towards malaria elimination in Yunnan, China. Malar J. **20**, 1–16 (2021). https://doi.org/10.1186/S12936-021-03672-8/TABLES/2
10. Kamndaya, M., Mfipa, D., Lungu, K.: Household knowledge, perceptions and practices of mosquito larval source management for malaria prevention and control in Mwanza district, Malawi: a cross-sectional study. Malar J. **20**, 1–8 (2021). https://doi.org/10.1186/S12936-021-03683-5/TABLES/3
11. Dong, S., Dong, Y., Simões, M.L., Dimopoulos, G.: Mosquito transgenesis for malaria control. Trends Parasitol **38**, 54–66 (2022). https://doi.org/10.1016/J.PT.2021.08.001
12. Mbunge, E., Millham, R.C., Sibiya, M.N., Takavarasha, S.: Diverging mobile technology's cognitive techniques into tackling malaria in sub-saharan africa: a review. In: Silhavy, R., Silhavy, P., Prokopova, Z. (eds.) CoMeSySo 2021. LNNS, vol. 232, pp. 679–699. Springer, Cham (2021). https://doi.org/10.1007/978-3-030-90318-3_54
13. Mbunge, E., Sibiya, M.N., Millham, R.C., Takavarasha, S.: Micro-spatial modelling of malaria cases and environmental risk factors in Buhera rural district, Zimbabwe. In: 2021 Conference on Information Communications Technology and Society ICTAS 2021 - Proceedings, pp. 2–8 (2021). https://doi.org/10.1109/ICTAS50802.2021.9394987
14. Fikrie, A., Kayamo, M., Bekele, H.: Malaria prevention practices and associated factors among households of Hawassa City Administration, Southern Ethiopia, 2020. PLoS ONE **16**, e0250981 (2021). https://doi.org/10.1371/JOURNAL.PONE.0250981
15. Brown, B.J., et al.: Data-driven malaria prevalence prediction in large densely populated urban holoendemic sub-Saharan West Africa. Sci. Reports **101**(10), 1–17 (2020). https://doi.org/10.1038/s41598-020-72575-6
16. Okagbue, H.I., Oguntunde, P.E., Obasi, E.C.M., Adamu, P.I., Opanuga, A.A.: Diagnosing malaria from some symptoms: a machine learning approach and public health implications. Health Technol. (Berl.) **11**, 23–37 (2021). https://doi.org/10.1007/S12553-020-00488-5/TABLES/9
17. Mwanga, E.P., et al.: Using mid-infrared spectroscopy and supervised machine-learning to identify vertebrate blood meals in the malaria vector, Anopheles arabiensis. Malar J. **18**, 1–9 (2019). https://doi.org/10.1186/S12936-019-2822-Y/FIGURES/7
18. Martineau, P., et al.: Predicting malaria outbreaks from sea surface temperature variability up to 9 months ahead in Limpopo, South Africa, using machine learning. Front. Public Heal. **10**, 962377 (2022). https://doi.org/10.3389/FPUBH.2022.962377/FULL

19. Harvey, D., Valkenburg, W., Amara, A.: Predicting malaria epidemics in Burkina Faso with machine learning. PLoS ONE **16**, e0253302 (2021). https://doi.org/10.1371/JOURNAL.PONE.0253302
20. Nkiruka, O., Prasad, R., Clement, O.: Prediction of malaria incidence using climate variability and machine learning. Inform. Med Unlocked **22**, 100508 (2021)
21. Shamseer, L., et al.: Preferred reporting items for systematic review and meta-analysis protocols (PRISMA-P) 2015: elaboration and explanation. BMJ **349** (2015). https://doi.org/10.1136/BMJ.G7647
22. Kalipe, G., Gautham, V., Behera, R.K.: Predicting malarial outbreak using machine learning and deep learning approach: a review and analysis. In: Proceedings of 2018 International Conference on Information Technologies, ICIT 2018 2018:33–8. https://doi.org/10.1109/ICIT.2018.00019
23. Yadav, S.S., Kadam, V.J., Jadhav, S.M., Jagtap, S., Pathak, P.R.: Machine learning based malaria prediction using clinical findings. In: 2021 International Conference on Emerging Smart Computing and Informatics, ESCI 2021, pp. 216–222 (2021). https://doi.org/10.1109/ESCI50559.2021.9396850
24. Mohapatra, P., Tripathi, N.K., Pal, I., Shrestha, S.: Determining suitable machine learning classifier technique for prediction of malaria incidents attributed to climate of Odisha (2021). https://doi.org/10.1080/09603123.2021.1905782
25. Lee, Y.W., Choi, J.W., Shin, E.H.: Machine learning model for predicting malaria using clinical information. Comput. Biol. Med. **129**, 104151 (2021)
26. Muhammad, B., Varol, A.: A symptom-based machine learning model for malaria diagnosis in Nigeria. In: 9th International Symposium on Digital Forensics and Security, ISDFS 2021 (2021). https://doi.org/10.1109/ISDFS52919.2021.9486315
27. Zacarias, O.P., Bostrom, H.: Comparing support vector regression and random forests for predicting malaria incidence in Mozambique. In: 2013 International Conference on Advances in ICT for Emerging Regions (ICTer 2013) - Conference Proceedings, pp. 217–21 (2013). https://doi.org/10.1109/ICTER.2013.6761181
28. Mbunge, E., Millham, R.C., Sibiya, M.N., Takavarasha, S.: Application of machine learning models to predict malaria using malaria cases and environmental risk factors. In: 2022 Conference on Information Communications Technology and Society, ICTAS 2022 - Proceedings (2022). https://doi.org/10.1109/ICTAS53252.2022.9744657
29. Zafar, A., et al.: Machine learning-based risk factor analysis and prevalence prediction of intestinal parasitic infections using epidemiological survey data. PLoS Negl. Trop. Dis. **16**, e0010517 (2022). https://doi.org/10.1371/JOURNAL.PNTD.0010517
30. Masinde, M.: Africa's Malaria epidemic predictor: application of machine learning on Malaria incidence and climate data. In: ACM International Conference Proceeding Series, pp. 29–37 (2020). https://doi.org/10.1145/3388142.3388158
31. Dukuzumuremyi, A.: Machine learning based prediction of malaria outbreak using environment data in Rwanda (2020)
32. Adamu, Y.A.: Malaria prediction model using machine learning algorithms. Turkish J. Comput. Math. Educ. **12**, 7488–7496 (2021). https://doi.org/10.17762/TURCOMAT.V12I10.5655
33. Phoobane, P., Masinde, M., Botai, J.: Prediction model for malaria: an ensemble of machine learning and hydrological drought indices, vol. 216. LNNS, pp. 569–584 (2022). https://doi.org/10.1007/978-981-16-1781-2_51/COVER
34. Iradukunda, O., et al.: Malaria disease prediction based on machine learning. In: IEEE International Conference on Signal, Information and Data Processing 2019 (2019). https://doi.org/10.1109/ICSIDP47821.2019.9173011

35. Mbunge, E., et al.: predicting student dropout in massive open online courses using deep learning models - a systematic review, vol. 503. LNNS, pp. 212–31 (2022). https://doi.org/10.1007/978-3-031-09073-8_20/COVER
36. Mariki, M., Mkoba, E., Mduma, N.: Combining clinical symptoms and patient features for Malaria diagnosis. Mach. Learn. Appr. (2022). https://doi.org/10.1080/08839514.2022.2031826
37. Mbunge E., et al.: Predicting diarrhoea among children under five years using machine learning techniques, vol. 502. LNNS, pp. 94–109 (2022). https://doi.org/10.1007/978-3-031-09076-9_9/COVER
38. Akinnuwesi, B.A., et al.: Application of support vector machine algorithm for early differential diagnosis of prostate cancer. Data Sci. Manag. (2022). https://doi.org/10.1016/J.DSM.2022.10.001
39. Zinszer, K., et al.: A scoping review of malaria forecasting: past work and future directions. BMJ Open **2**, e001992 (2012). https://doi.org/10.1136/BMJOPEN-2012-001992
40. Golumbeanu, M., et al.: Leveraging mathematical models of disease dynamics and machine learning to improve development of novel malaria interventions. Infect. Dis. Poverty **11**, 1–17 (2022). https://doi.org/10.1186/S40249-022-00981-1/FIGURES/6
41. Sharma, R.K., Thakor, H.G., Saha, K.B., Sonal, G.S., Dhariwal, A.C., Singh, N.: Malaria situation in India with special reference to tribal areas. Indian J. Med. Res. **141**, 537 (2015). https://doi.org/10.4103/0971-5916.159510
42. Sudheer, C., et al.: a support vector machine-firefly algorithm based forecasting model to determine malaria transmission. Neurocomputing **129**, 279–288 (2014). https://doi.org/10.1016/J.NEUCOM.2013.09.030
43. Wang, M., et al.: A novel model for malaria prediction based on ensemble algorithms. PLoS ONE **14**, e0226910 (2019). https://doi.org/10.1371/JOURNAL.PONE.0226910
44. Zhang, H., Guo, J., Li, H., Guan, Y.: Machine learning for artemisinin resistance in malaria treatment across in vivo-in vitro platforms. IScience **25**, 103910 (2022). https://doi.org/10.1016/J.ISCI.2022.103910
45. Pourhomayoun, M., Shakibi, M.: Predicting mortality risk in patients with COVID-19 using machine learning to help medical decision-making. Smart Heal **20**, 100178 (2021). https://doi.org/10.1016/J.SMHL.2020.100178
46. Buczak, A.L., et al.: Fuzzy association rule mining and classification for the prediction of malaria in South Korea Standards, technology, and modeling. BMC Med. Inform. Decis. Mak. **15**, 1–17 (2015). https://doi.org/10.1186/s12911-015-0170-6
47. Sornsuwit, P., Jaiyen, S.: A New Hybrid Machine Learning for Cybersecurity Threat Detection Based on Adaptive Boosting, vol. 33, pp. 462–82 (2019). https://doi.org/10.1080/08839514.2019.1582861
48. Walker, K.W., Jiang, Z.: Application of adaptive boosting (AdaBoost) in demand-driven acquisition (DDA) prediction: A machine-learning approach. J. Acad. Librariansh **45**, 203–212 (2019). https://doi.org/10.1016/J.ACALIB.2019.02.013
49. Feng, D.C., et al.: Machine learning-based compressive strength prediction for concrete: An adaptive boosting approach. Constr. Build. Mater. **230**, 117000 (2020). https://doi.org/10.1016/J.CONBUILDMAT.2019.117000
50. Bui. Q.-T., Nguyen, Q.,-H., Pham, V.M., Pham, M.H., Tran, A.T.: Understanding spatial variations of malaria in Vietnam using remotely sensed data integrated into GIS and machine learning classifiers, vol. 34, pp. 1300–1314 (2018). https://doi.org/10.1080/10106049.2018.1478890
51. Fashoto, S.G., Mbunge, E., Ogunleye, G., den Burg, J.V.: Implementation of machine learning for predicting maize crop yields using multiple linear regression and backward elimination/Stephen Gbenga Fashoto (2021)

52. Uddin. S., Khan, A., Hossain, M.E., Moni, M.A.: Comparing different supervised machine learning algorithms for disease prediction. BMC Med. Informat. Decis. Mak. **191**(19), 1–16 (2019). https://doi.org/10.1186/S12911-019-1004-8
53. Cunningham, P., Delany, S.J.: k-Nearest neighbour classifiers - a tutorial. ACM Comput. Surv. **54** (2021). https://doi.org/10.1145/3459665
54. Grampurohit S, Sagarnal C. Disease prediction using machine learning algorithms. 2020 Int Conf Emerg Technol INCET 2020 2020. https://doi.org/10.1109/INCET49848.2020.9154130
55. Chingombe. I., et al.: Predicting HIV status among men who have sex with men in Bulawayo & Harare, Zimbabwe using bio-behavioural data.In: Recurrent Neural Networks, and Machine Learning Techniques. Trop Med Infect Dis 2022, vol. 7, p. 231 (2022). https://doi.org/10.3390/TROPICALMED7090231
56. Sokolova, M., Lapalme, G.: A systematic analysis of performance measures for classification tasks. Inf. Process. Manag. **45**, 427–437 (2009). https://doi.org/10.1016/J.IPM.2009.03.002
57. Lahmiri, S., Dawson, D.A., Shmuel, A.: Performance of machine learning methods in diagnosing Parkinson's disease based on dysphonia measures. Biomed. Eng. Lett. **8**, 29–39 (2018). https://doi.org/10.1007/S13534-017-0051-2/FIGURES/2
58. Seliya, N., Khoshgoftaar, T.M., Van Hulse, J.: A study on the relationships of classifier performance metrics. In: Proceedings of International Conference on Tools with Artificial Intelligence (ICTAI 2009), pp. 59–66 (2009). https://doi.org/10.1109/ICTAI.2009.25
59. Erickson, B.J., Kitamura, F.: Magician's corner: 9. performance metrics for machine learning models. Radiol. Artif. Intell. **3** (2021). https://doi.org/10.1148/RYAI.2021200126
60. Alaa Khaleel, F., Al-Bakry, A.M.: Diagnosis of diabetes using machine learning algorithms. Mater Today Proc. (2021). https://doi.org/10.1016/J.MATPR.2021.07.196
61. Gonzalez-Cuautle, D., et al.: Synthetic minority oversampling technique for optimizing classification tasks in Botnet and intrusion-detection-system datasets. Appl. Sci. **10**, 794 (2020). https://doi.org/10.3390/APP10030794
62. Gunda, R., Chimbari, M.J., Shamu, S., Sartorius, B., Mukaratirwa, S.: Malaria incidence trends and their association with climatic variables in rural Gwanda, Zimbabwe, 2005–2015. Malar. J. **161**, 1–13 (2017). https://doi.org/10.1186/S12936-017-2036-0
63. Gong, Y.F., Zhu, L.Q., Li, Y.L., Zhang, L.J., Xue, J.B., Xia, S., et al.: Identification of the high-risk area for schistosomiasis transmission in China based on information value and machine learning: a newly data-driven modeling attempt. Infect Dis Poverty **10**, 1–11 (2021). https://doi.org/10.1186/S40249-021-00874-9/FIGURES/3
64. Manyangadze, T., Mavhura, E., Mudavanhu, C., Pedzisai, E.: An exploratory analysis of the spatial variation of malaria cases and associated household socio-economic factors in flood-prone areas of Mbire district, Zimbabwe. GeoJournal, 1–16 (2021). https://doi.org/10.1007/S10708-021-10505-3/FIGURES/3
65. Zinszer, K., et al.: Forecasting malaria in a highly endemic country using environmental and clinical predictors. Malar J. **14**, 1–9 (2015). https://doi.org/10.1186/S12936-015-0758-4/FIGURES/4
66. Chekol, B.E., Hagras, H.: Employing machine learning techniques for the Malaria epidemic prediction in Ethiopia. In: 2018 10th Computer Science and Electronic Engineering Conference (CEEC): Conference Proceedings, pp. 89–94 (2019). https://doi.org/10.1109/CEEC.2018.8674210
67. Seo, J.H., Kim, Y.H.: Machine-learning approach to optimize smote ratio in class imbalance dataset for intrusion detection. Comput. Intell. Neurosci. **2018** (2018). https://doi.org/10.1155/2018/9704672
68. Liu, X.Y., Wu, J., Zhou, Z.H.: Exploratory undersampling for class-imbalance learning. IEEE Trans. Syst. Man Cybern. Part B Cybern. **39**, 539–550 (2009). https://doi.org/10.1109/TSMCB.2008.2007853

69. Guo, X., Yin, Y., Dong, C., Yang, G., Zhou. G.: On the class imbalance problem. In: Proceedings - 4th International Conference on Natural Computation, ICNC 2008, vol. 4, pp. 192–201 (2008). https://doi.org/10.1109/ICNC.2008.871
70. Zhu, T., Lin, Y., Liu, Y.: Synthetic minority oversampling technique for multiclass imbalance problems. Pattern Recognit. **72**, 327–340 (2017). https://doi.org/10.1016/J.PATCOG.2017.07.024
71. Elreedy, D., Atiya, A.F.: A comprehensive analysis of synthetic minority oversampling technique (smote) for handling class imbalance. Inf. Sci. (Ny) **505**, 32–64 (2019). https://doi.org/10.1016/J.INS.2019.07.070
72. Mfisimana, L.D., Nibayisabe, E., Badu, K., Niyukuri, D.: Exploring predictive frameworks for malaria in Burundi. Infect. Dis. Model **7**, 33–44 (2022). https://doi.org/10.1016/J.IDM.2022.03.003
73. Sow, B., Mukhtar, H., Ahmad, H.F., Suguri, H.: Assessing the relative importance of social determinants of health in malaria and anemia classification based on machine learning techniques, vol. 45, pp. 229–41 (2019). https://doi.org/10.1080/17538157.2019.1582056
74. Chingombe, I., et al.: Predicting HIV Status using machine learning techniques and bio-behavioural data from the zimbabwe population-based hiv impact assessment (ZIMPHIA15–16). In: Silhavy, R. (eds.) Artificial Intelligence Trends in Systems. CSOC 2022. LNNS, vol 502. Springer, Cham (2022). https://doi.org/10.1007/978-3-031-09076-9_24

An Ontological Model for Ensuring the Functioning of a Distributed Monitoring System with Mobile Components Based on a Distributed Ledger

Eduard V. Melnik and Irina B. Safronenkova(✉)

Federal Research Center, The Southern Scientific Center of the Russian Academy of Sciences, 41, Chekhov Street, 344006 Rostov-on-Don, Russian Federation
safronenkova050788@yandex.ru

Abstract. The object of this study is distributed monitoring systems with mobile components based on a distributed ledger. The use of such systems is associated with difficulties in the organization of the computing process, including due to the limited energy resources of mobile components. At the same time, one of the problems lies in the effective workload relocation problem solving when solving a complex functional problem by a distributed system. The analysis of existing methods and approaches to workload relocation problem solving has been carried out, which revealed that the ontological approach is the closest to solving the problem, the effectiveness of which has been proven in earlier works of the authors. However, the developed ontological model, on the basis of which candidate nodes are selected, does not take into account the presence of mobile components, distributed ledger nodes and related features of the organization of the computing process in the class of systems under consideration. In this paper, an ontological model of a distributed monitoring system with mobile components based on a distributed ledger is proposed in terms of the organization of the computing process. This model, combined with the method of forming constraints in the workload relocation problem solving in a distributed system based on the ontological analysis procedure, allows you to reduce the search space when organizing calculations in systems of this class and thereby reduce the decision-making time.

Keywords: Ontological Model · Distributed Monitoring System · Distributed Registry · Mobile Components · Fog-Computing

1 Introduction

Modern monitoring systems for various applications operate on the basis of digital economy technology, including cloud computing technologies, distributed registry, Internet of Things and data mining methods [1, 2]. These technologies make it possible to effectively implement new methods of data processing, transmission and storage.

The presence of mobile components as elements of distributed monitoring systems allows us to bring such systems to a fundamentally new level of their use, significantly

R. Silhavy and P. Silhavy (Eds.): CSOC 2023, LNNS 724, pp. 345–355, 2023.
https://doi.org/10.1007/978-3-031-35314-7_31

expanding the scope of their application. However, at the same time, the task of limiting the time of organizing the computing process is urgent, which is a consequence of increasing the dynamism of the system due to the presence of mobile components in it.

The organization of the computing process is directly related to the process of reconfiguration of the system, which consists in the redistribution of subtasks of graph *G(Q,X)* over the operational computing nodes of the system after a failure in the system. The task of reconfiguration is reduced to the task of building schedules, the classical formulation and solution of which is widely covered in the literature [3]. However, the dynamism of the foggy environment, along with the characteristics of a distributed system including mobile components, do not allow the proposed classical methods to be used effectively. In this regard, the task of reconfiguration has not lost its relevance and has attracted the attention of the scientific community in order to find effective methods to solve it in the new field of nebular computing.

In the next section of this paper, the analysis of known methods and algorithms for solving the reconfiguration problem in a foggy environment will be carried out.

2 Analysis of Methods and Algorithms for Workload Relocation Problem Solving in the Fog Computing Environment

Analysis of open sources has shown that there is a wide variety of methods and algorithms for workload relocation problem solving in the fog computing environment [4]. The table provides a list of methods for workload relocation problem solving in the fog computing environment in terms of latency reducing, and lists limitations in terms of their implementation (Table 1).

Table 1. Evaluation of methods and algorithms for scheduling problem

Method/Algorithm	Disadvantages
A two-level resource scheduling technique in fog computing [5]	Not focused on complex topologies
Reallocation mechanism to reduce task delays in accordance with the characteristics of the containers [6]	Does not take into account the influence of the time of workload relocation problem solving on the effectiveness of solving a user task
A task scheduling strategy based on a Hybrid Heuristic algorithm for different fog nodes to solve the problem of terminal devices with limited computing resources and high energy consumption [7]	High computational complexity in the case of a large dimensional computing environment
A four tiers architecture for delay aware scheduling and Load Balancing in the fog environment [8]	Does not take into account the time costs associated with data transmission

(*continued*)

Table 1. (*continued*)

Method/Algorithm	Disadvantages
A task offloading algorithm to complete tasks when their nodes leave [9]	Applicable only in a homogeneous environment
A resource management approach based on the dynamic scheduling of multi-user devices using classification methods for cloud servers and heterogeneous devices [10]	It is used for tasks of small dimension, focused on interaction with cloud resources
An optimized job scheduling algorithm called SJF to minimize the delays for latency-critical applications [11]	Does not take into account the influence of the time of workload relocation problem solving on the effectiveness of solving a user task
Local device group based method [12]	The time to receive a decision can be unacceptably long due to multiple information exchanges
The method of constraints forming in the workload relocation problem solving in a distributed system [13]	Does not take into account the presence of mobile components and distributed registry nodes

It can be concluded that the method of constraints forming in the workload relocation problem solving in a distributed system is the closest to the realization of the goal of this study, but does not take into account the presence of mobile components and distributed registry nodes. Since this method is based on an ontological model, in order to be able to apply it in another subject area, it is necessary to develop a new ontological model reflecting the features of the subject area under consideration.

3 Ontology-Based Method of Workload Relocation

This section provides a brief description of an ontology-based method for workload relocation problem solving in a distributed system operating in the fog-computing environment. More detailed information about this method everyone can learn in the following papers [13, 14]. The key idea of the presented method is to use an ontological model to reduce the search space (P') when solving the workload relocation problem. As a result, the time of solving the distribution task is reduced by reducing the search space of the P' (Fig. 1).

An ontology-based method of workload relocation consists of the following steps that are presented below.

- Step 1. Receiving of input data containing subtask initial data and candidate nodes resource data.
- Step 2. Carrying out an ontological analysis procedure.
- Step 3. Formation of a limited space of candidate nodes.
- Step 4. Procedure for modeling the workload placement on a limited set.
- Step 5. Workload relocation problem solving on a limited set.

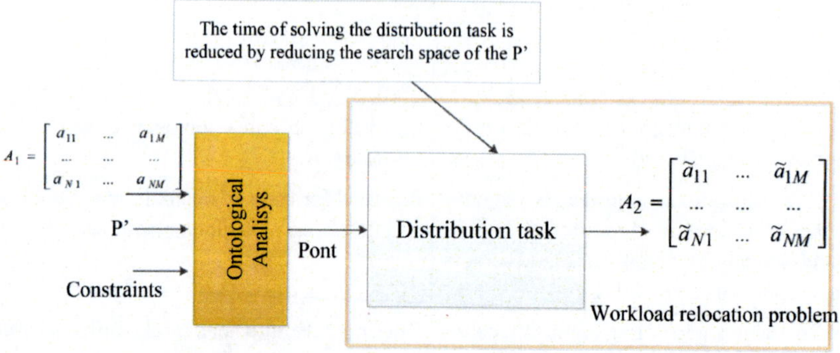

Fig. 1. Search space reduction scheme.

- Step 6. Getting the result of a workload placement.

An implementation scheme of the ontology-based method of workload relocation is presented in Fig. 2.

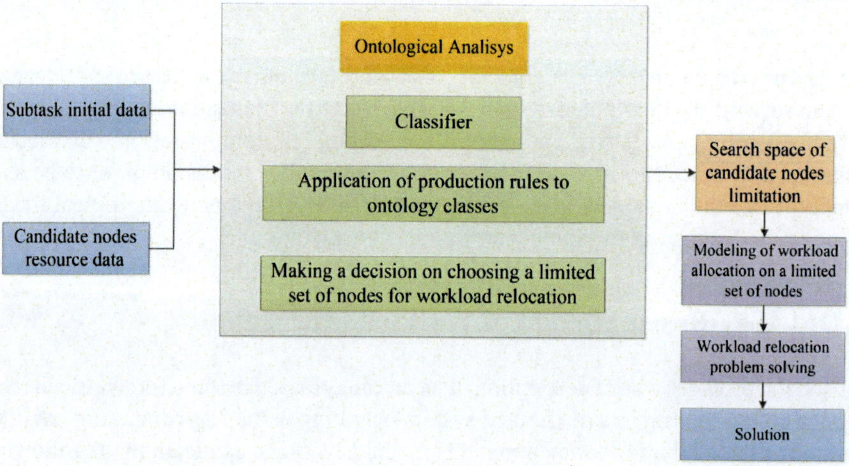

Fig. 2. Implementation scheme of the ontology-based method of workload relocation.

However, this method was developed for the field of distributed systems that do not involve the use of mobile components and a distributed registry. Therefore, to be able to use this method in relation to the class of systems under consideration, it is necessary to develop a new ontological model that takes into account the presence of mobile components and a distributed registry.

4 An Ontological Model for Ensuring the Functioning of a Distributed Monitoring System with Mobile Components Based on a Distributed Registry

The development of the ontological model is based on the methodology described in [15] and includes the following steps:

Step 1. Definition of the scope and scale of ontology.

Step 2. Consideration of options for reuse of existing ontologies.

Step 3. Listing important terms in the ontology.

Step 4. Defining Classes and Class Hierarchy.

Step 5. Defining the properties of slot classes.

Step 6. Defining facets of slots.

Step 7. Creating instances.

We will limit the scope of the study by choosing distributed monitoring systems with mobile components based on a distributed ledger for consideration.

To complete Steps 2–3, it is necessary to conduct a comprehensive analysis of the subject area under study.

The scheme of processes displaying the functioning of a distributed monitoring system with mobile components based on a distributed registry has the form shown in the Fig. 3. In general, to solve some complex problem, a distributed system uses data received from sensors and data read from a distributed ledger. In turn, the data that is the result of solving a complex problem must also be recorded in a distributed ledger. It is worth noting that the number of complex tasks solved by the system at the same time is large. This is related to the problem of multiple updating of data, which are the output data of a complex task. For these purposes, it was proposed to use distributed ledger technologies.

Mobile components which have a limited energy resource are the elements of a distributed system functioning in the fog and edge layers. This imposes some limitations associated with the effective functioning of the system. In this case, we are dealing with a dynamic environment in which there is a multiple workload redistribution. For the efficient organization of the computing process in such a computing environment, it is necessary to optimize the energy consumption of mobile components along with the implementation of time constraints for the execution of a functional task. It is possible to optimize power consumption by reducing the decision-making time when selecting candidate nodes for redistributing the computing load. For this purpose, it is proposed to use the ontology-based method of constraint formation for the workload relocation problem solving, which was described earlier.

Thus, to describe the subject area of a distributed monitoring and forecasting system with mobile components in terms of computing implementation processes when solving a complex functional problem, the following concepts will be used:

- algorithmic task structure;
- information flows;
- subtask;
- subtasks characteristics: computational complexity, input data volume, output data volume;

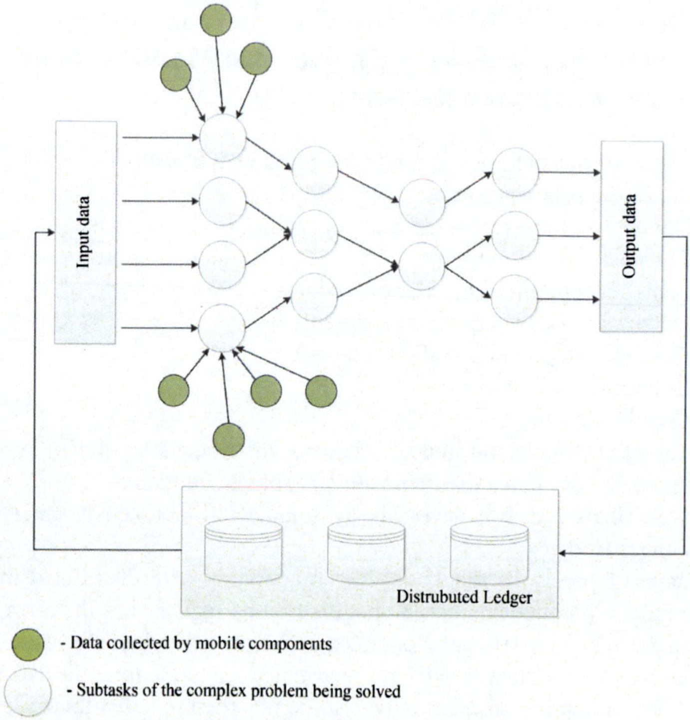

Fig. 3. The scheme of processes.

- flow characteristics: acceptable latency value; data volume; transmitted data frequency;
- graph partition;
- candidate node: cloud, fog, mobile;
- candidate nodes characteristics: efficiency, energy resource, mobility, remoteness, type, workload;
- communication channels characteristics;
- placement method: conveyor, parallel;
- DL node;
- DL nodes characteristics: role, synchronization latency.

Based on the analysis of processes reflecting the functioning of a distributed system with mobile components based on a distributed ledger, the following hierarchy of classes describing the area under consideration was formed (Fig. 4).

Through the properties of objects, various classes can be linked. For example, let's link the class "DL_Node" and the class "DL_Node_Characteristics" through the property "_has", the class "Algorithmic_Task_Structure" and the Class "Subtask" through the property "_includes" (Figs. 5 and 6).

Example of filling classes with instances. Class "direction" has the following instances: "from_the_source", "towards_the_source", which is reflected in the ontological model as it is shown in Fig. 7.

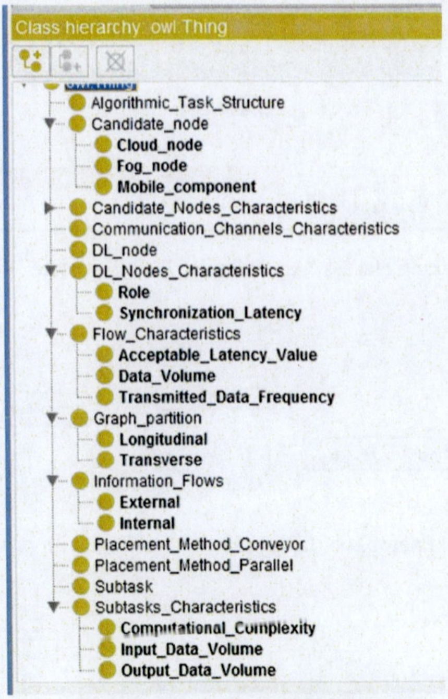

Fig. 4. The hierarchy of classes.

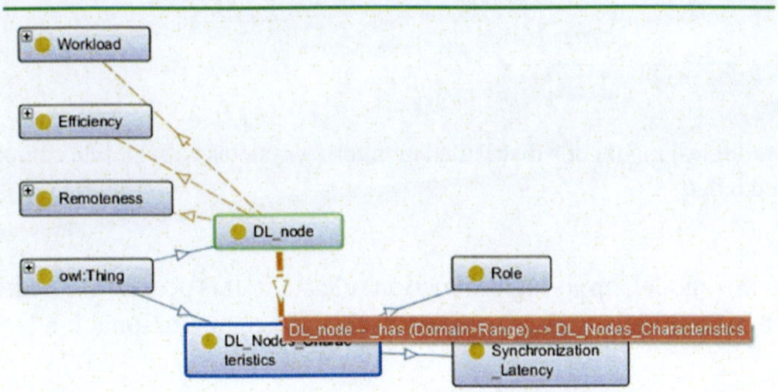

Fig. 5. An example of linking classes "DL_Node" and "DL_Node_Characteristics" through the property "_has".

The result of the development is an ontological model of the subject area of a distributed monitoring systems with mobile components based on a distributed ledge presented in Fig. 8.

This ontological model reflects the knowledge that allows us to describe the processes occurring in a distributed monitoring systems with mobile components based on a distributed ledge during the implementation of calculations.

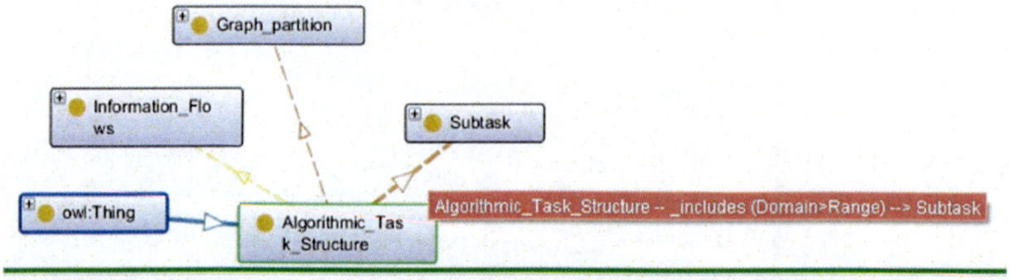

Fig. 6. An example of linking classes "Algorithmic_Task_Structure" and "Subtask" through the property "_includes".

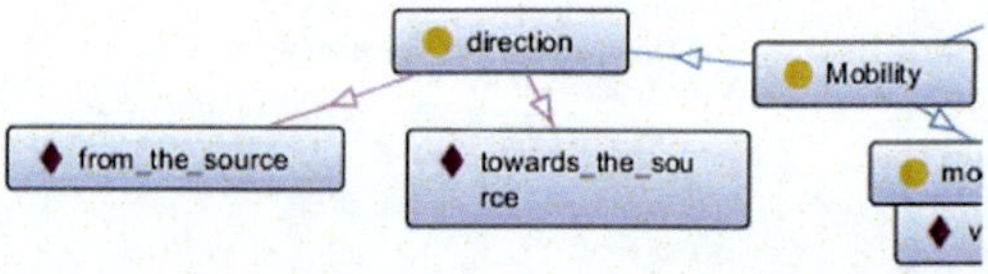

Fig. 7. Example of filling class "Mobility" with instances.

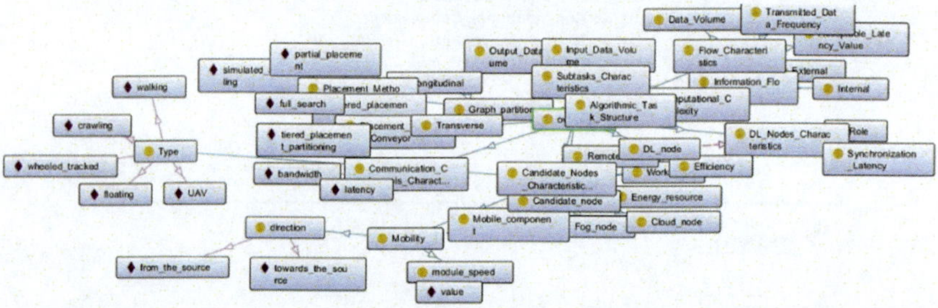

Fig. 8. An ontological model of a distributed monitoring systems with mobile components based on a distributed ledge.

Based on this model, applying production rules to it, it is possible to make a decision on the choice of candidate nodes to accommodate the computational load.

5 Production Rules Development

Based on the conducted research in the field of distributed monitoring systems with mobile components based on distributed ledger, the following general rule of the organization of the computing process can be formulated:

$$T_{cal} + T_{reloc} = T_{solving} \leq T_{solving\_max} \quad (1)$$

where T_{cal} – calculation time,

T_{reloc} – workload relocation time,

$T_{solving}$ – time to solve the problem,

$T_{solving\_max}$ – maximum time to solve the problem, then we choose a solution that assumes the following conditions:

1) minimal involvement of distributed ledger nodes in the computing process;
2) minimum data transmission time.

Based on this general rule, taking into account the analysis of the subject area of a distributed monitoring system with mobile components based on distributed ledger, the following production rules have been developed that are applied to the ontological model to make a decision on limiting the set of candidate nodes for workload relocation.

If Subtask $\{D_i, W_{ij}, W_{jk}\}$ **AND** Information_Flows $\{L_{ij}, V_{ij}\}$ **AND** Candidate_Node $\{\hat{W}_i, M_i, P_i, T_i, D_i, E_i\}$ **AND** Node_DL $\{\hat{W}_k, M_k, P_k, T_k, D_k, E_k, \Psi_{edge}, \Psi_{fog}\}$ **AND** Communication_Channel $\{\Psi_{ij}, \Omega_{ij}\}$ **AND** $T_{cal} \leq T_{solving\_max}$ **AND** Graph_Partition = *longitudinal* **THEN** *discard those nodes that cannot solve the shortest branch of the sub-graph.*

If Subtask $\{D_i, W_{ij}, W_{jk}\}$ **AND** Information_Flows $\{L_{ij}, V_{ij}\}$ **AND** Candidate_Node $\{\hat{W}_i, M_i, P_i, T_i, D_i, E_i\}$ **AND** Node_DL $\{\hat{W}_k, M_k, P_k, T_k, D_k, E_k, \Psi_{edge}, \Psi_{fog}\}$ **AND** Communication_Channel $\{\Psi_{ij}, \Omega_{ij}\}$ **AND** $T_{cal} \leq T_{solving\_max}$ **AND** Graph_Partition = *transverse* **THEN** *for each tier, discard those nodes that cannot participate in the solution of any of the tiers.*

where D_i – computational complexity of the subtask *i*;
W_{ij} – the amount of input data for the subtask *i*;
W_{jk} – the amount of output data for the subtask *i*;
L_{ij} – the amount of data transferred from the subtask *i* to the subtask *j*;
V_{ij} – data transmission frequency from the subtask *i* to the subtask *j*;
$\hat{W}_i$ –workload of the node *i*;
M_i – mobility of the node *i*;
P_i – efficiency of the node *i*;
T_i – type of the node *i*;
D_i – distance of the node *i* from the data source/receiver;
E_i – energy resource of the node *i*;
Ψ_{edge} – latency of the "edge" level;
Ψ_{fog} – latency of the "fog" level;
Ψ_{ij} – communication channel latency;
Ω_{ij} – bandwidth of the communication channel.

The application of these production rules to ontology classes makes it possible to cut off from consideration those candidate nodes that obviously cannot take part in solving a problem. Thus, the search space is reduced and, accordingly, the time for solving the reconfiguration problem is reduced. This makes it possible to increase the efficiency of distributed monitoring systems with mobile components based on a distributed ledger.

6 Conclusions

In this paper, the problem of organizing the computing process in distributed monitoring systems with mobile components based on distributed ledger technologies was considered. The analysis of methods and approaches to solving the problem of reconfiguration revealed that the method based on ontological analysis is the most effective in workload problem solving in the "fog"- computing environment. However, changes in the architecture of the distributed system required the development of a new ontological model that takes into account the presence of mobile components and distributed ledger nodes. The paper analyzes the processes reflecting the functioning of a distributed monitoring system with mobile components based on a distributed ledger, and develops an ontological model. Examples of production rules that can be used in the decision-making process for the selection of candidate nodes for the workload relocation are given.

Acknowledgements. This study is supported by the GZ SSC RAS N GR project 122020100270-3.

References

1. Chakraborty, M., Kundan, A.P.: Architecture of a modern monitoring system. In: Monitoring Cloud-Native Applications. Apress, Berkeley, CA (2021)
2. Orda-Zhigulina, M.V., Melnik, E.V., Ivanov, DYa., Rodina, A.A., Orda-Zhigulina, D.V.: Combined method of monitoring and predicting of hazardous phenomena. In: Silhavy, R. (ed.) CSOC 2019. AISC, vol. 984, pp. 55–61. Springer, Cham (2019). https://doi.org/10.1007/978-3-030-19807-7_6
3. Pinedo, M., Hadavi, K.: Scheduling: theory, algorithms and systems development. In: Gaul, W., Bachem, A., Habenicht, W., Runge, W., Stahl, W.W. (eds.) Operations Research Proceedings 1991, vol. 1991. Springer, Heidelberg (1992). https://doi.org/10.1007/978-3-642-46773-8_5
4. Matrouk, K., Alatoun, K.: Scheduling algorithms in fog computing: a survey. Int. J. Neworked Distrib. Comput. **9**, 59–74 (2021)
5. Sun, Y., Lin, F., Xu, H.: Multi-objective optimization of resource scheduling in fog computing using an improved NSGA-II. Wirel. Pers. Commun. **102**, 1369–1385 (2018)
6. Yin, L., Luo, J., Luo, H.: Tasks scheduling and resource allocation in fog computing based on containers for smart manufacturing. IEEE Trans. Ind. Inform. **14**(10), 4712–4721 (2018)
7. Wang, J., Li, D.: Task scheduling based on a hybrid heuristic algorithm for smart production line with fog computing. Sensors (2019)
8. Wu, C., Li, W., Wang, L., Zomaya, A.: Hybrid evolutionary scheduling for energy-efficient fog-enhanced internet of things. IEEE Trans. Cloud Comput. **9**, 641–653 (2018)
9. Li, G., Yan, J., Chen, L., Wu, J., Lin, Q., Zhang, Y.: Energy consumption optimization with a delay threshold in cloud-fog cooperation computing. IEEE Access **7**, 159688–159697 (2021)
10. Ghenai, A., Kabouche, Y., Dahmani, W.: Multi-user dynamic scheduling-based resource management for Internet of Things applications. In: 2018 International Conference on Internet of Things, Embedded Systems and Communications (IINTEC), Hamammet, Tunisia, pp. 126–131 (2018)
11. Jamil, B., Shojafar, M., Ahmed, I., Ullah, A., Munir, K., Ijaz, H.: A job scheduling algorithm for delay and performance optimization in fog computing. Concurr. Comput. Pract. Exp. **32**(7) (2020)
12. Melnik, E.V., Klimenko, A.B., Ivanov D.Y.: The model of the problem of forming communities of information and control system devices in fog computing environments. In: XIII All-Russian Meeting on Management Problems-2019, pp. 2979–2984. Tula State University, Tula (2019)
13. Klimenko, A., Safronenkova, I.: An ontology-based approach to the workload distribution problem solving in fog-computing environment. Adv. Intell. Syst. Comput. **985**, 62–72 (2019)

14. Klimenko, A.B., Safronenkova, I.B.: A technique of workload distribution based on parallel algorithm structure ontology. In: Silhavy, R., Silhavy, P., Prokopova, Z. (eds.) CoMeSySo 2019 2019. AISC, vol. 1046, pp. 37–48. Springer, Cham (2019). https://doi.org/10.1007/978-3-030-30329-7_4
15. Noy, N., McGuinness, D.: Ontology development 101: a guide to creating your first ontology. Stanford knowledge systems laboratory Technical report KSL-01–05 and Stanford Medical Informatics Technical report SMI-2001-0880 (2001)

MIREAHMATY—Three-Dimensional Chess in the System of Virtual Reality VR

Roman I. Dzerjinsky(✉), Daniil A. Boldin, and Sofia G. Daeva

MIREA – Russian Technological University, 78, Vernadsky pros., Moscow 119454, Russia
9015111295@mail.ru

Abstract. The rules of three-dimensional chess are being developed. Some variants of the rule of moves and capture for each piece are proposed. A unique description of the game situation is analyzed. Various options for placing check, checkmate and stalemate are considered.

1 Introduction

There are a large number of games on the game board and even more on a specially prepared mapping plane, aimed at fighting and competitiveness. These games bring the principle of fighting to the intellectual realm, which allows players to improve their logical thinking and mental abilities. These games of the historical past are a universal heritage.

One of the oldest games on an 8×8 board is chess. According to some sources, its age is estimated at one and a half thousand years. Recall that at that time it was completely different from the version of chess that we are used to seeing. It was only by the XV century that the rules of the so-called "classical" chess were formed [1].

Nowadays, chess is no longer a symbolic object for modeling military operations, as it was when it was created. It is a way to develop your strong gaming and psychological sides. It allows you to develop a sense of beauty, revealing beauty in the logic of thought, including in a complex psychological game with cold calculation or in a game of passions, which is more like an adventure in which a person learns himself [2]. It is thanks to this that the so-called chess culture has emerged, in which not only professional chess players are involved but also ordinary people, who bring in their ideas and visions of this game. Many of these ideas move away from the generally accepted form of "classical" chess and give a completely new look at this game while maintaining some of the fundamental rules inherent in it.

For many years, just as once the simulation of military operations was schematically reduced to chess battles, modern military operations have actually turned into a hybrid war connected with many unrelated influences, with many participants, political blocs and neutral factions, suggests to us the need for a cardinal complication of ordinary chess. Participation in such a game, as well as in modern combat operations, is impossible without a computer. Attempts have been made to come up with rules for three-dimensional chess, wherein a three-dimensional surface in the form of a cube or a parallelepiped consisting of ordinary two-dimensional game boards superimposed on

R. Silhavy and P. Silhavy (Eds.): CSOC 2023, LNNS 724, pp. 356–366, 2023.
https://doi.org/10.1007/978-3-031-35314-7_32

each other would be used as a playing field. However, many of the invented games had significant drawbacks that did not allow players to enjoy some important features of "classic" chess. Therefore, many of these games rely mainly on traditional rules, only transferring them to three-dimensional space.

A good example of such a game is "Millennium Three Dimensional Chess", where three ordinary game boards are used, superimposed on each other and forming three levels, and opposing teams are located on two extreme and opposite levels. The figures themselves are able to walk not only in their own plane, but also up and down, respectively. Thus, this version of three-dimensional chess took the rules of two-dimensional chess and adapted them to three-dimensional space [3].

Another example is "Raumschach", which was one of the first types of three-dimensional chess where a $5 \times 5 \times 5$ cube acts as a playing field. And the main feature is the presence of a new "The Unicorn" figure, which moves along the diagonals of the cube. The rest of the pieces retain their functionality, adding to it only the possibility of moving up and down [4].

Another "classic" example of three-dimensional chess is "Cubic chess", otherwise known as "Kubikschack", where the pieces are located in an $8 \times 8 \times 8$ cube and the opposing teams are on the opposite sides of the cube, but not inside of it, as in "Raumschach", but outside, and the figures move on its surface [5].

The examples of three-dimensional chess discussed above lead us to the fact that to display several boards, that are not superimposed on each other, we can use a cube similar to "Raumschach", where the pieces will be inside of it and each cell of each board will be a small cube, leading us to the possibility of turning a piece on this small cube like in "Cubic chess" [6]. Thus, we have a new version of the move—a turn on the cell, while the rest of the functionality of the figures seems similar to "Millennium Three Dimensional Chess" [7]. And then in our game we get the opportunity to use three-dimensional space in all its coordinate planes with the help of a rotation, and not as in other implementations where in fact the figures have only one orientation.

2 Basic Axioms of the Game

Understanding how to play ordinary one-dimensional chess makes the three-dimensional standard of the game potentially very attractive to use in a set of standard VR, which in this case allows you to change the angle of your view on the playing space and shows the areas of potential damage to the pieces. It opens up the possibility of simultaneously 12 players, sorted by colors of a pack of markers at the same time, whom it is possible to conclude block and neutral agreements that can change during the game, as in a modern military-political battle. Also a significant difference will be this: in the case of a victory of one block of participants over (another in parentheses or a separate color), the use of the remaining figures of this color by the winning blocks though assigning them a new color will be possible. Assignment operation ":=".

Naturally, the game position can be written by analogy with ordinary chess in the form of a table of states. The heading of the table of states is the presence of player blocks, i.e., the indication of groups fighting a chess battle against each other. A trivial example is 12 players, all against all, written as a Roman numeral (I), color; etc. Player blocks will be written as Roman numerals; a list of colors inside blocks allied to each other.

Obviously, the recording of the game position implies the presence of a color printer, but the color rendering often leaves much to be desired and we duplicate the colors with two letters of their English abbreviation (1. WH-white, 2. BL-black, 3. BR-brown, 4. RE-red, 5. DG-dark green, 6. LG-light green, 7. OR-orange, 8. LB-light blue, 9. PU-purple, 10. PI-pink, 11. DB-dark blue, 12. YE-yellow). An example of more complex relationships within blocks is in Tables 1, 2 and 3.

Table 1. Game table with two blocks of players.

1 block	2 block
WH	BL
BR	RE
DG	LG
OR	YE
PU	PI
DB	LB

Table 2. Game table with four blocks of players.

1 block	2 block	3 block	4 block
WH	OR	BL	YE
BR	PU	RE	PI
DG	DB	LG	LB

Table 3. Game table with six blocks of players.

1 block	2 block	3 block	4 block	5 block	6 block
WH	OR	BL	YE	BR	PU
RE	PI	DG	DB	LG	LB

Neutrality can be established between the blocks, we will designate it as "¿". Neutrality means an inability to capture or impose a checkmate or stalemate on a player of a block neutral to your own. Also, if a figure of a player of one block stands on a space occupied by a figure of a player of a neutral block, then the figure of a neutral player becomes interned. Respectively, if the player loses, the interned figure returns to its space. If the space is occupied, the figure will not return to the space until it is free. A player has the right to create a vote before the move to break neutrality with another block, if the majority of players in the block vote for it (more than half), then neutrality disappears and these blocks of players can compete again. To create neutrality with another block, a player creates a vote, if the majority of players in both blocks are in favor (more than half), then a neutrality is formed between these blocks of players. It is impossible to create neutrality if this leads to the situation where there will no longer be a block of players against whom you can continue to play. Also, the neutrality is broken automatically if a block of players has no rivals left after a checkmate is delivered to the enemy block.

The following convention for recording a game situation assumes the standard fixing of Cartesian coordinate axes on an 8 by 8 cube, namely, this rule is mnemonically reduced to "white figures are located on the YoX plane from above, black on YoZ from the reverse side".

A cube with $8 \times 8 \times 8$ sides implies that three pairs of mutually perpendicular planes are significant for each elementary cell.

The player's goal is a tactical game within the group, and in the case of all against all - to deliver as many checkmates as possible and thereby conquer the site, winning the remaining resources of the defeated. The defeated side loses the King and Queen and the rest of the pieces are assigned the color of the winner.

The number of players is from 2 to 12. Movement is the change of a figure from one space of a cell to another. The main innovation of the game is the orientation of the figure, which is possible thanks to a new move—the turn of the figure. The turn of the figure is available to all figures and is considered a full-fledged move. A turn is a change in the position of a figure inside the cell relative to the axes i, j or k (Fig. 1).

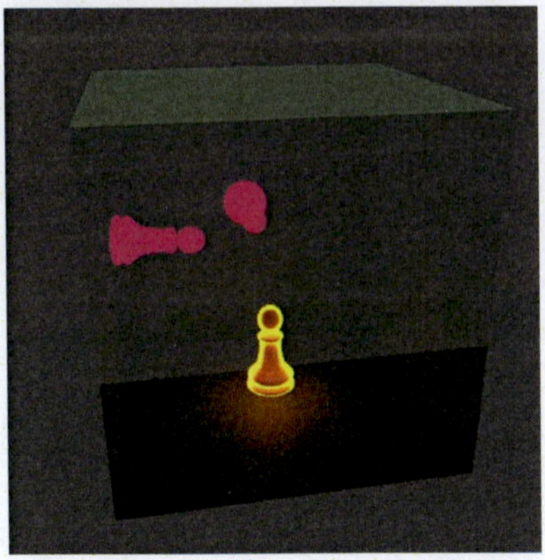

Fig. 1. Visualization of the turn progress.

The rules for declaring a checkmate are 2 or more checks (which you cannot escape) and a classic checkmate (clarification is possible as the VR is implemented). The stalemate does not change in its rules, except that it now operates in three-dimensional space.

After the announcement of a checkmate, the opponent's pieces come into possession of the player who has delivered the checkmate. If there are several checkmates or several player's figures create a checkmate then the losing player has the right to choose which of the opponents his figures will pass.

The position of the figure is encoded by 4 coordinates. Coordinate designations: 1(z) – sets the cell coordinate by z (0, 1, 2, 3, 4, 5, 6, 7, 8); 2(y) – sets the y coordinate (0, 1, 2, 3,

4, 5, 6, 7, 8); 3(x) – sets the cell coordinate by x (0, 1, 2, 3, 4, 5, 6, 7, 8); 4 – the orientation of the figure (i, j, k).

The recording of moves is encoded by the color, the name of the figure, the initial coordinate of the figure and the final coordinate of the figure. Thus, at the beginning of the game, the pawn of the owner of the red pieces (RE) from its initial position (1;1;1;k) makes a turn relative to the i axis, then the recording of its move will be as follows: (RE, pawn, 1;1;1;k, 1;1;1;i). Then the pawn of the owner of the black pieces (BL) makes a move from the initial position 7;7;7;k on the space 6;7;7;k, then the recording of the move will look like this: (BL, pawn, 7;7;7;k, 7;8;7;k). This segment of the game will be displayed in the history of moves as (RE, pawn, 1;1;1;k, 1;1;1;i), (BL, pawn, 7;7;7;k, 6;7;7;k).

The logo of the game "MIREAHMATY" was created (Fig. 2).

МИРЭАХМАТЫ

Fig. 2. Logo of "MIREAHMATY".

3 Moves of the Figures

1. Knight– all types of the letter "L" parallel to the faces of the cube. The rule is as close as possible to the rules of the previously considered variants of three-dimensional chess (Figs. 3 and 4).
2. Pawn – moves though a plane according to the rules of classical chess. Taking through the aisle remains in the plane, with the aim of focusing on classical chess. Similarly, a pawn becomes a queen when it reaches the opposite starting edge of the cube. The pawn has a basic direction towards the opposite edge, similar to classical chess (if the pawn needs to move to the corner of the field, the player can move in both directions adjacent to this corner).
3. Bishop – moves through a plane according to the rules of classical chess. It also captures across 2 space diagonally, in order to balance its value among other figures (Fig. 5).
4. Rook – moves though a plane according to the rules of classical chess and across 2 spaces up and down (Fig. 6).
5. Queen – moves like a bishop + rook + as she moves in the plane of classical chess.
6. King – up to 12 variations of the move (8 moves in the plane, 1 move up, 1 down, 2 turns) (Figs. 7, 8, 9, 10, 11 and 12).

(a) (b) (c)

Fig. 3. Visualization of possible knight moves in different planes: (a) Possible knight moves in plane i; (b) possible knight moves in plane j; (c) possible knight moves in plane k.

Fig. 4. The final visualization of all possible moves of the knight in space.

Fig. 5. Visualization of possible moves of the bishop.

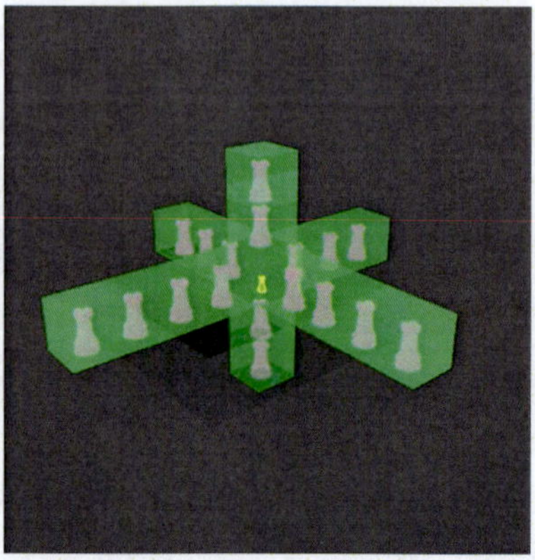

Fig. 6. Visualization of possible moves of the rook.

4 Arrangements

Fig. 7. Arrangement for two players.

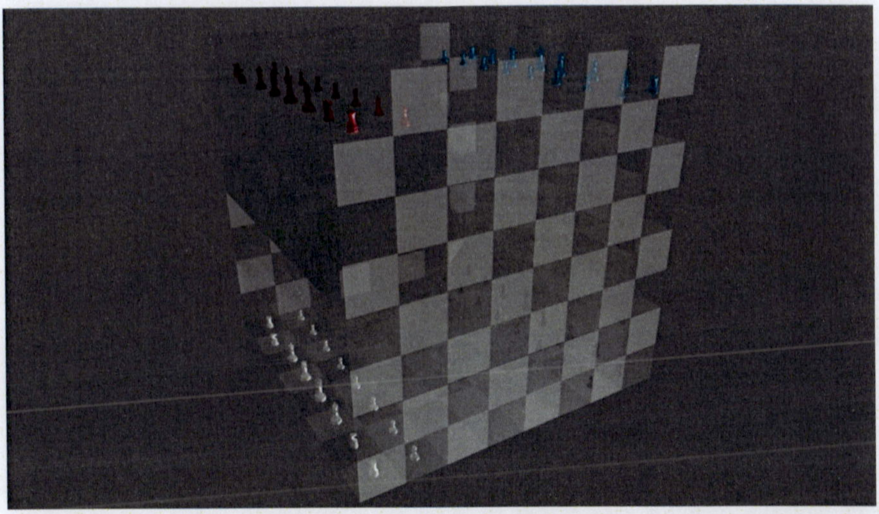

Fig. 8. Arrangement for four players.

Fig. 9. Arrangement for six players.

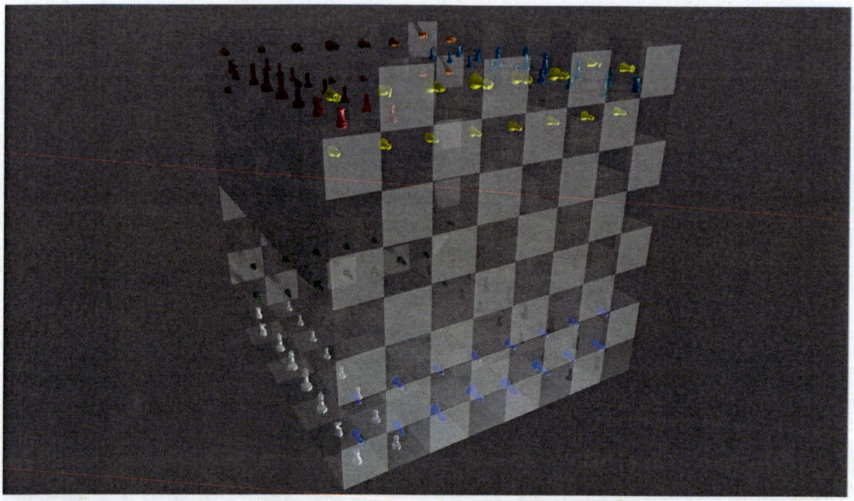

Fig. 10. Arrangement for eight players.

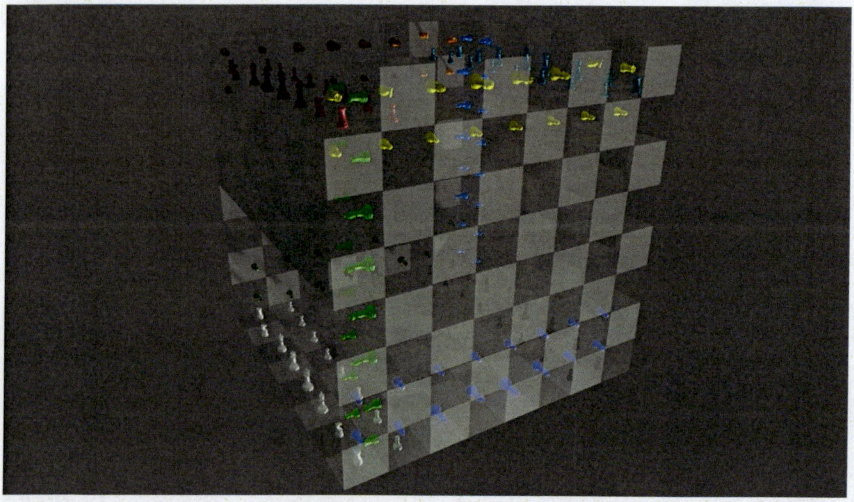

Fig. 11. Arrangement for ten players.

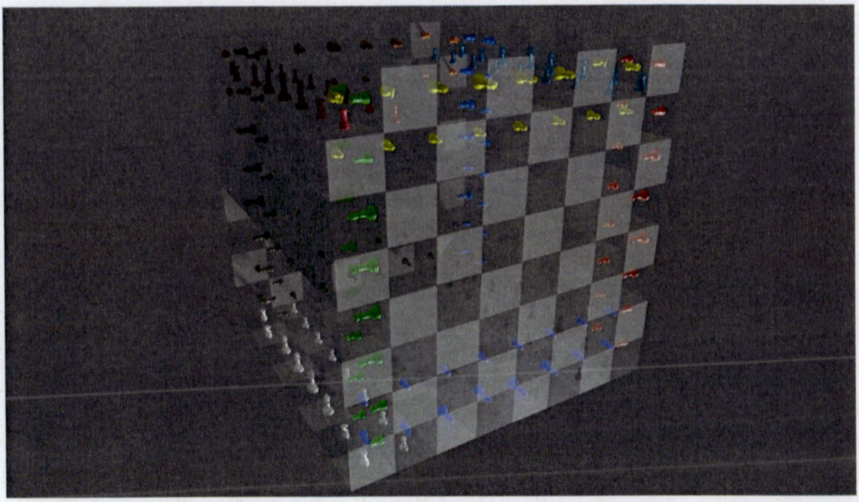

Fig. 12. Arrangement for twelve players.

5 Conclusion

In the standard of the game created by us there was a move towards a number of players other than two, it became possible to form blocks and neutral blocks, taking into account the change in blocking during the game. A recording form has been developed, which is one of the advantages of the proposed design and the creative approach to the development of chess.

In addition, the player who defeats another player gets all the surviving pieces of this player at his disposal, and they change color accordingly when moving to this player.

The rules we have developed described the basic axioms of the game, such as the type of playing field, the rules for declaring a checkmate, the statemate remained the same as in "classical" chess. It was described how the coordinates of the figures are set using a four-dimensional vector, as well as how the moves are recorded, in which the coordinates of the initial position of the figure and the coordinates of the figure after the move are used.

We have described the moves of all the pieces used. Some pieces have retained the functionality of "classic" chess, while others have added new functionality, such as moving up, down and along the diagonals of the cube, and the main innovation is the rotation of the figure relative to the coordinate axes.

As a result, developed at the Department of Applied Mathematics, the game "MIREAHMATY" is an example of a strategy game in which dynamic change and aggregation of participating blocks is possible, internment, formation of permanent and temporary coalitions. This brings this game potentially into the category of the most interesting network games with the maximum number of participants being 12. The system of Recording moves allows you to analyze offline the correctness of the participants' moves and actually form optimal game strategies based on the "MIREAHMATY" code.

References

1. Schmittberger, R.W.: New Rules for Classic Games. Wiley, Hoboken (1992)
2. Gritsay, A.A.: Mathesis. Odessa, USSR (1924)
3. Evans, L.: Space Chess. Popular Science (1969)
4. Scherbin, M.N., Martish, V.S.: Chess as a way of self-knowing. Bull. ChelGU **13**, 409 (2017)
5. Pritchard, D.B.: The Encyclopedia of Chess Variants Games and Puzzles Publications. Surrey (1994)
6. Dickins, A.: A Guide to Fairy Chess. Dover Publications Inc., New York (1971)
7. Pritchard, D.B.: Millennium 3D Chess. John Beasley (2007)

Overview of Machine Learning Processes Used in Improving Security in API-Based Web Applications

Emil Marian Pașca(✉), Rudolf Erdei, Daniela Delinschi, and Oliviu Matei

Technical University of Cluj Napoca, North University Centre of Baia Mare, Baia Mare, Romania

emil.pasca@campus.utcluj.ro

Abstract. Due to being scalable, easily cacheable, and well-suited for high-traffic web applications, RESTful APIs are highly used today, hence in this paper, we present a selective review of how the security evolved around APIs and how Machine Learning (ML) and Artificial Intelligence (AI) techniques have been used in developing new solutions for testing, monitoring, or detecting/classifying malicious attacks, and preventing them from tainting devices that use APIs. The study also highlights the limitations of the current automated API testing tools which are security and non-functional testing and suggests a few possible research directions based on ML procedures that could be integrated and used to address these deficiencies.

Keywords: software testing · cybersecurity · restful api · automation testing tools · api security

1 Introduction

The popularity of APIs has significantly increased in the past years due to the migration of applications from monoliths to distributed systems [16]. With the modern software architecture known as Service Oriented Architecture (SOA), the computational backend (server-side) can be executed separately from the front-end or GUI and accessed through an Application Programming Interface (API). Our concern with this trend is that bad actors could try to exploit vulnerabilities. Therefore, we analyze API elements, usage patterns, and various other cybersecurity threats that can be prevented with the benefit of novel approaches such as artificial intelligence and ML solutions [50].

1.1 How this Research is Structured

After providing a short introduction and presenting the research questions that form the foundation of this study in Sect. 1, we gather accurate definitions for the terms used in our study and describe the most common API software development techniques and standards used in Sect. 2.

R. Silhavy and P. Silhavy (Eds.): CSOC 2023, LNNS 724, pp. 367–381, 2023.
https://doi.org/10.1007/978-3-031-35314-7_33

In Sect. 3, we describe the most popular cybersecurity threats in the world of APIs, the common vulnerabilities found, and the challenges in securing APIs. In Sect. 4, we present examples of automated test solutions, including novel AI/ML cybersecurity approaches used in RESTful APIs. In addition, in Sect. 5, we present ideas related to our subject, gaps found in the researched tools, and topics that may be of interest in future studies.

Finally, we end our study with an examination of potential future research directions in Sect. 6. We conclude our study with a comprehensive summary of our findings and provide insights on the implications of our research in Sect. 7.

1.2 Research Questions

While reviewing articles, journals, and conference papers, we will also try to answer a few research questions that stand at the base of this study.

Our questions are targeted to RESTful API automated testing solutions, preferably tools or testing techniques targeting Web API applications that may also have API Specification documents.

RQ	Research Question	Motivation
RQ1	What kind of ML approach has been used in the solution?	There are many types of ML, such as Natural Language Processing (NLP), Neural Networks (NN), Deep Learning (DL), and others that might have been used in cyber security solutions, hence, we want to find out
RQ2	What types of security vulnerabilities had ML testing solutions managed to uncover?	There are numerous security threats in the world of web applications, however, not all of them are practically suitable to be detected using AI or ML, in this research, we try to identify which ones were and could address
RQ3	How effective were the previously used solutions?	The results of a used solution, are probably, the one which is valued the most. We should critically evaluate the solutions, in terms of the number of tested applications, the algorithm used, accuracy in the predictions, and other metrics that can consider relevant in evaluating the solution

2 RESTful API Analysis

Web 3.0 is highly dependent on RESTful services, defined in [39] as representing services based on the architectural Representation of State Transfer (REST), which are intended to solve enterprise application integration problems implementing service-oriented-architecture, which its objective is to have loosely coupled services, independent, scalable and distributed systems. Also, in other words, as mentioned in [43], an API serves as a foundation for creating applications by saving programmers the time necessary to code functionality from scratch.

In general, a RESTful API is defined by the following characteristics:

- It is based on the HTTP protocol, which is the standard protocol for communication on the web;
- Is usually moldered on one or more create, read, update or delete (CRUD) operations upon a specific resource;
- It is stateless, meaning that the server does not store any state about the client session on the server side. This allows the API to be scalable, easily cacheable, and decoupled [49];
- It follows a uniform interface, which means that all resources are accessed and manipulated in the same way, using a standardized set of URL structures and HTTP methods;
- It is self-describing, meaning that the API usually includes all the necessary information for clients on how it can be used. It includes documentation, data types, and structure.

These characteristics make RESTful APIs easy to use and understand, and they also make it possible to build scalable, distributed systems that can be accessed from any device with an internet connection.

2.1 OpenAPI

Also, commonly known as *Swagger* [51], Open API Specification (OAS) [52], is a standard commonly used in RESTful services. These are not only generated from API-based applications, but, used in describing the functional characteristics of applications which are defined in a JSON, YAML [48], or RAML [1] file formats, from which application modules can be automatically produced [33].

Although OAS defines functional features and requirements of an application, and even security properties such as authentication, there are many non-functional aspects and inter-parameters dependencies that cannot be properly described in this OAS file.

2.2 Inter-parameter Dependencies in Web APIs

Modeling, testing, and validating input parameters, are essential activities in testing web APIs, and one of the major challenges when it comes to developing a tool for testing, is the **inter-parameter dependencies (IPD)**. The IPD concept is fairly brought up in research like [25,30], this subject has been deeply covered also by Alberto et al. [27], where they present a catalog of seven types of IPD consistently found in real-world industry web APIs.

Without an understanding of IPD, any AI-based framework that could be developed to predict vulnerabilities in the API might be subject to a more "brute-force" testing mode, since the tool would try various combinations of parameters until the request is validated, **it would likely result in a high number of invalid requests**, which will result in an untested API. Also, the number of steps in performing a valid request to the API could exponentially grow which may lead to unnecessary server load, resources and time wasted.

An example of an invalid request could be when the client performs a PUT request on a non-existing resource ID (e.g. /product/id=123), if the id doesn't exist it will return an unsuccessful request ("400 bad request" response) what the client should have done instead is to first create the resource and then use a PUT request to update it, this shows that we should pay attention to the IPD such as the resource id, or other property before performing any request or action on a resource.

3 Cybersecurity Threats for APIs

There are common cybersecurity threats that have been exploited in Web Applications, which are tracked in online databases such as CVE [10], the Common Vulnerability Scoring System (CVSS) [35], and the Common Weakness Enumeration CWE [11]. In particular, the OWASP TOP 10 API Security vulnerabilities [38] study is a reliable starting point for considering vulnerabilities when testing APIs due to the fact that they take into account information based on CWE and other trustworthy sources.

3.1 API Security Preventing Challenges

As also described by the OWASP community in [38], we can enumerate the following challenges in securing an API:

- Authentication and Authorization: Ensuring that only authenticated and authorized users can access the API;
- Validating user input to ensure that malicious code is not injected into the system;
- Encrypting data in transit and at rest to protect it from unauthorized access;
- Limiting the number of requests that can be made to the API in a given time period to prevent malicious attacks;
- Monitoring API usage and logging all activities to detect and respond to security incidents;
- Regularly testing the API for security vulnerabilities and patching any discovered flaws.

3.2 API Software Security Focused Metrics

In Serme et al. [44] we found that the security of RESTful services relies on ad-hoc security mechanisms or on the transport layer security, both of which are error-prone.

Yarygina et al. [54] state that the REST style was not designed with security properties in mind and does not fit the security requirements of modern web applications.

As mentioned in works like Kornienko et al. [21] and Brachmann et al. [6], the most common recommendations in terms of security for APIs are related to authentication scheme and authorization protocols which should ensure that the web application, database, and web server are properly configured and are using trusted environments. As such, several metrics can be used to determine if a Restful API is secure. One metric is the authentication mechanism used by the API. If, for example, the API uses a strong authentication mechanism, such as OAuth 2.0 [37], this can be considered a good sign that the API is secure.

Another metric is the use of HTTPS for all API requests which ensures that all data transmitted between the API and clients is encrypted, making it more difficult for attackers to intercept and steal sensitive information.

Additionally, the API should have appropriate access controls in place to ensure that each user has access only to the required information and also the privileges for that user don't exceed his role (Guest, Registered User, Admin, etc.).

3.3 AI/ML Usage in Cybersecurity Solutions

Intelligent approaches in attack detection, prediction, and analysis that use AI, ML, and DL algorithms have been fairly discussed before [42]. Although, as mentioned by Morel et al. [32] with the help of AI, there has been made many achievements, and there is great potential in cybersecurity as well, many new cybersecurity-appropriate technologies could be inspired by existing AI technologies.

Other solutions which imply ML techniques have been proposed for testing the security and reliability of network protocols [47], ML solution for the black-box detection of cross-site request forgery vulnerabilities [7], and malicious URL detection [41].

4 Automated Tools and ML Techniques in Testing APIs Solutions

Usually manual and automated testing is required for testing APIs, however, considering the rigidity of an API in terms of changeability in its lifetime, it is considered that having automated tests which can be run periodically is the preferred approach used in the software development industry.

The application or API can be put to various tests, depending on the level of readiness, this usually is in accordance with the environment where it is used

(Development, Staging, Testing, Production) [19]. Each level can contain its own types of tests and target audience.

4.1 Related Work

In a recent survey from 2022 by Golmohammadi et al. [14], we found out that the number of papers that address various challenges and issues regarding Restful APIs, has been growing in the last 12 years, with a number of research questions addressed and new innovative ways to improve the automated testing of the APIs, in regards to security, it shows that a limited number of papers had managed to successfully assess security characteristics of RESTFul web services, like Access Control by Atlidakis et al. [2] and Zha et al. [55], JSON Hijacking in Masood et al. [29], Broken Object Level Authorization(BOLA) by Barabanov et al. [3] and Barlas et al. [4]. A comparison between these studies and tools could be further asses, however, these aren't using an AI or ML approach for security testing.

Comparison between various API Restful tools, has been achieved in older studies by Martin et al. [28], Corradin et al. [9] as well as a comparison between various techniques has been fairly mentioned in Golmohammadi et al. [14].

The use of AI and ML techniques in Restful API testing seems to be rather limited, not only for security testing but also for another functional testing as well. A few solutions and tools have been proposed and developed as covered in Subsects. 4.2, 4.3, 4.4, 4.5, 4.6, 4.7 and 4.8.

4.2 API Prober 2.0

Feasible collaborative designed approaches were proposed with the help of API Prober 2.0 [24], a tool initially designed to be capable of transforming OAS documents into graph structures, providing analysis on service design practices and REST characteristics, and retrieving practical examples from code sources like GitHub, later in mentioned in Hsiao et al. [8] various security annotations and testing mechanism were added which showed that the tool could be further improved to help developer asses API security issues like the authentication mechanism. Although based on OAS, and facilitates the development effort, it is still a semi-automated process without the use of ML or other AI technique.

4.3 IDLReasoner

Interesting tools which have successfully implemented IDL [17], is IDLReasoner [18], which could provide a feasible solution to interpreting, monitoring, and testing APIs based on IDL, as described by Alberto et al. [26].

The functionality of the IDLReasoner is using the IDL and translating them into a Constraint Satisfaction Problem (CSP), which is later used in analysis operations, an illustration of flow can be seen in Fig. 1.

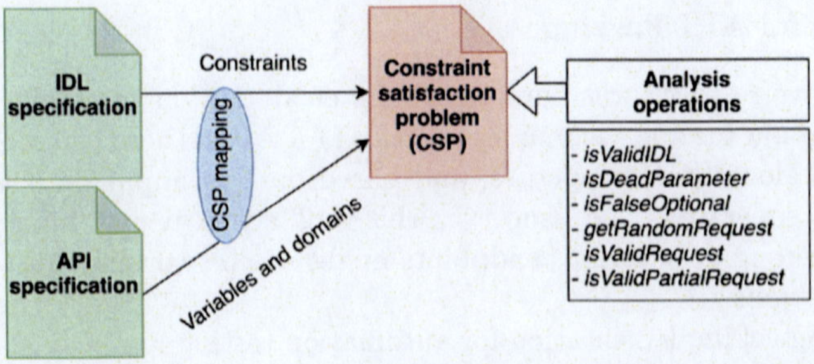

Fig. 1. IDLReasoner - functionality diagram from Alberto et al. Medium Article [23].

4.4 RESTest

As discussed by Martin et al. in [25] the potential of AI in web API testing has been questioned and later managed to develop a solution. RESTest [28], a black-box solution for testing RESTful APIs, which from an OAS specification file and a configuration file can automatically generate and execute tests, and finally produce reports.

In previous work as described in [13] we saw that in terms of API, there are several approaches for identifying vulnerabilities in APIs that had been used, such as:

- Template-based rule extraction
- Association rule
- Frequently closed
- Frequent partial-order
- Imbalanced frequent
- Frequently closed
- k-Nearest Neighbour

There are several approaches to detecting anomalies in a system, some of which are commonly used with the "API Usage Pattern" [12,46]. However, these have their limits, namely, they should be used only for mature software, where conditional checks and API Usage Checks [34] can be neglected.

4.5 API Traffic Classification Techniques and Prediction

In Baye et al. [5], we've discovered how an ML-based technique was used to detect and classify API traffic based on specific features like bandwidth and the number of requests per token. We found how a DL-based approach for automatically predicting the validity of an API request in Mirabella et al. [30] paper, where the validity of an API request was predicted with accuracy thanks to DL.

4.6 RESTful API Fuzzing

Comprehensive research was done by Zhang et al. [56] on the tools used in the sector that apply fuzzing techniques on RESTful APIs, these commonly require an OpenAPI document to operate, and based on this, input data and various payloads are generated. The issue with this tool is that it may interpret strings in the URI, wrongly, or various endpoints might not be working due to outdated API specifications.

Another novel implementation for automation testing has been also provided by Zhang et al. in another study [57], however, here it was targeted to RPC protocol rather than HTTP.

4.7 Test Generation Tools Based on OpenAPI Documents

Automatic test generation tools, based on OpenAPI documents, is not uncommon, such tool has been used in RESTful APIs testing tools as described in [20], which has been effective since the document is directly dependent on the API and is kept up to date as the API is updated.

4.8 ML Testing Tools with the Scope of Security Vulnerability Exposure

Mohamed et al. [31] identified various methods used in detecting malware in android apps by analyzing malicious applications and Maldroid data-set, getting the most common API calls and permissions from them in the pre-processing phase, afterward applying ML to differentiate between malicious applications and good ones.

5 Results and Discussions

5.1 Challenges in Using ML Techniques in Cybersecurity

As mentioned in the work of Shaukat et al. [45], and other related works, ML is widely spread in various software solutions, along with cybersecurity solutions, here, we see that using ML models in cybersecurity we can elaborate solutions that can potentially prevent malware, adware, or other malicious attacks.

Among the ML techniques used in cybersecurity, in the study [36] it shows are gaps in:

- Security training - management people, engineering and other members aren't yet fully aware of the importance of security, when and how it could be addressed in each stage of the Software Development Life Cycle (SDLC);
- Traceability in the SDLC Methodology - From development design requirements to the release product there are un-tracked changes that can lead to security flaws;

- Weaknesses and threat to validity - Weaknesses among developed systems which proven to be valid in certain circumstances, and could behave in another way under edge cases that could also lead to security issues.

Depending on the business area in which we are treating security testing, there are many challenges in industries like self-driving cars, financial applications and healthcare, where there is a high need for correctness in predictions, and trustworthiness from ML techniques, in these areas, if Web APIs are used, high attention to detail and testing is imposed.

Two of the most discussed challenges which are frequently met have been covered in Subsects. 5.2 and 5.3.

5.2 Data Collection

As smart as the new existing solutions for preventing various malicious threats are, coming up with a dataset that could be used to train the ML in uncovering new types of attacks, is a challenging objective. The evaluated ML models, showed, that the data models used are outdated, which can present a serious threat in terms of having trust in AI predictions in critical areas.

One of the most time-wasting activities in ML applications are data collection, exploration, and preparation, in order to develop a promising solution, it is vital that we manage to filter the available data and prepare it correspondingly. We should explore the cybersecurity data available and see how we can use it in a future solution. There are a few sources of data mentioned in the reviewed literature, however, we observed, that most data, comes from experimentation with Open Source APIs, we consider this, a convenient approach since we cannot expect sensitive data such as from applications to be widely available for access and research. However, setting up, deploying, and running test data, is time-consuming.

5.3 Security Threats Towards ML

As the interest for ML solutions in regards to discovering new solutions for various issues, there are threats against ML training and testing, defensive techniques of ML have been proposed by Liu et al. [22] and future research in this area is considered.

6 Future Work and Research Paths

Is it necessary to use AI and ML in automated testing? Is there something we're overlooking? In the study by Igor Kirilenko et al. [15], they considered that AI may be used to ease the automation process and reduce the quantity of monotonous and time-consuming jobs in development and testing, making the SDLC simpler, quicker, and ideally less expensive.

As mentioned above there is a well-known need for testing tools focused on non-functional aspects of API Web Applications, such as security characteristics or performance. Using state-of-the-art API tools, and ML algorithms, we can experiment and develop an ML solution that could possibly fill this need.

However, due to the broad variety of possible ML solutions that can be used, and vulnerabilities that can be addressed at a time, we could take into account a number of research directions that are mentioned in the following subsections.

6.1 Test Case Generation and Report Analysis Using NLP

In order to perform different testing types, using NLP features, we could leverage this tool to extract information from the OAS of the application, to dynamically create test cases with test data, that can be used to uncover functional and non-functional issues including security flaws. The reports and test summaries could be future analyzed and based on a trained data set of previous security testing reports, alerts of possible security vulnerabilities could be generated.

6.2 Anomaly Detection that Uses Unsupervised Learning

Trained on a dataset of normal API traffic, which would be used to learn the typical patterns of activity of the application, when deployed on a testing/production environment, the traffic could be monitored and the API calls could be filtered, in order to be compared with the learned patterns of normal traffic. Whenever a request deviates from a normal pattern, a security action could be triggered in order to prevent/alarm the system of possible malicious attacks, security misconfiguration, or another vulnerability flaw.

The solution could also be updated and re-trained periodically to adapt to changes in the traffic patterns of the application, to ensure that it remains effective over time.

6.3 Access Control Using Reinforcement Learning

Learning from the user access patterns along with features from the OAS of the application, a solution that uses a Reinforcement Learning (RL) system could be trained to test access control policies of the application and uncover flaws, that could be later addressed in order to maximize the security and efficiency of the system.

6.4 Requirements and Compliance Checking Using ML

By having a ML system that could be trained on the dataset of API requests, responses, OAS, and other requirements compliance data sources, the system could ensure that the application meets the required criteria.

6.5 Vulnerability Detection Using ML

Trained on a dataset of known vulnerabilities in RESTful APIs, the ML tool could scan the API specification and look for patterns that match those of known vulnerabilities, when such findings are encountered it can alert the developers and also suggest a potential fix if exists.

7 Conclusions

We have divided this Section into three main subsections that correspond to the research questions which we aimed to answer. The first Section, recalls the highlights of AI/ML techniques used in various cybersecurity solutions. In the second subsection, we presented the types of security flaws that have been successfully fixed with the aid of AI/ML approaches. In the third Section, we describe the current solutions' effectiveness, results, and issues. Finally, in the last Section, we concluded with some final toughs on our findings and future research.

7.1 RQ1: ML Approaches Used

In the scientific literature, we had read multiple proposals of test case generation tools that use NLP techniques, however, few of them seems to cover non-functional characteristics of Web APIs, such as security or performance. We've also reviewed a DL technique used by Mohamed et al. [31], which has been used in analyzing and classifying the traffic that an Android Application is making through APIs, this seems to be an interesting procedure, that maybe could be extended to Web APIs as well.

Automated test case selection using AI that simulates various user inputs has been proposed by Mysiuk et al. [40], although this particular approach uses AI, we believe that the features of AI, in this case, are rather limited and not used to their full potential.

7.2 RQ2: Types of Security Vulnerabilities Addressed

Although this is not explicitly mentioned, when using Test Generation Tools that leverage NLP techniques, we believe that vulnerabilities such as broken access control, security miss-configuration or insecure design flaws could be detected if the generated test scenarios have generated tests for a combination of paths and users with various access levels. Security logging and monitoring failure has been covered by a solution proposed in the work of Mohamed et al. [31].

Finally, other than the results above, we didn't manage to see that any of the reviewed tools had successfully used AI in the scope of unveiling security issues at the API level, **any of the security flaws that could have been identified, could be discovered rather by chance, and not explicitly targeted by the testing tools**.

7.3 RQ3: Solution Effectiveness

The effectiveness of these tools has been measured, usually, by the path coverage metrics [53], and, also there were papers where there were mentioned reports of issues found in the tested APIs with the help of the tested solutions, although is not considered a metric, we believe that this also could indicate that the solution is pretty effective since it managed to uncover issues.

Classification and statistics regarding different API testing tools are covered in works like [9,14,56] but, again, we saw a classification based on effectiveness in covering functional characteristics of the API. There wasn't any detailed report on how many non-functional characteristics of the API have been tested, and, as we reviewed these studies, we observed that in terms of security, there are limited results.

We can agree that the above tools are effective in providing a good solution in automating particular activities from the range of testing activities that a test engineer does (Generating Test Cases, Functional Testing), and there are solutions that are using AI techniques for such purposes, but, all the reviewed solution seem rather focused on functional characteristics, not used in the scope of testing Web API for security vulnerabilities and generating test scenarios which could predict security flaws based on API Specification, as such we had identified a particular gap in Restful API test automation tools, that hasn't yet been covered, and also a very important demand which today's IT industry requires. It would be beneficial if such a tool, could exist, so that along with the functional issues that can be discovered, it could also reveal security flaws or security testing scenarios with the help of the API Specification docs.

Although, widely used, standardized, and available automated tools for testing APIs, there isn't yet a cost-effective and general-use application that can help us prevent cybersecurity threats for RESTful Web API applications with the help of AI, ML, or other ML derivatives.

7.4 Final Thoughts

Treated as a non-functional characteristic, security will still be considered a lower priority in the SDLC. As such, investment in it will be left until the later stages of development. What would serve the industry is a tool that could ease the efforts and costs, though as proven in previous work, developing such tools with high efficiency requires extensive labor and resources.

The OAS standard for defining API functional features alone is not sufficient for providing accurate data as input for an ML solution. Additional input, such as configuration files, IDL specifications, and other data is required for developing a possible solution for testing API security issues.

As previously mentioned, there are several future work and research paths that we plan to explore and experiment with, to fill this gap in the industry and find a dedicated solution that can use ML techniques to discover security flaws in RESTful APIs. We plan to exercise the multitude of AI/ML techniques and available options on Open Source APIs, observe the results, and compare them to the latest state-of-the-art tools.

References

1. RAML, December 2020. Accessed 19 Nov 2022

2. Atlidakis, V., Godefroid, P., Polishchuk, M.: Checking security properties of cloud service rest APIs. In: 2020 IEEE 13th International Conference on Software Testing, Validation and Verification (ICST), pp. 387–397, October 2020
3. Barabanov, A., Dergunov, D., Makrushin, D., Teplov, A.: Automatic detection of access control vulnerabilities via API specification processing. CoRR, abs/2201.10833 (2022)
4. Barlas, E., Du, X., Davis, J.C.: Exploiting input sanitization for regex denial of service. In: 2022 IEEE/ACM 44th International Conference on Software Engineering (ICSE), pp. 883–895 (2022)
5. Baye, G., Hussain, F., Oracevic, A., Hussain, R., Ahsan Kazmi, S.M.: API security in large enterprises: leveraging machine learning for anomaly detection. In: 2021 International Symposium on Networks, Computers and Communications (ISNCC), pp. 1–6 (2021)
6. Brachmann, E., Dittmann, G., Schubert, K.-D.: Simplified authentication and authorization for RESTful services in trusted environments. In: De Paoli, F., Pimentel, E., Zavattaro, G. (eds.) ESOCC 2012. LNCS, vol. 7592, pp. 244–258. Springer, Heidelberg (2012). https://doi.org/10.1007/978-3-642-33427-6_21
7. Calzavara, S., Conti, M., Focardi, R., Rabitti, A., Tolomei, G.: Machine learning for web vulnerability detection: the case of cross-site request. IEEE Secur. Priv. **18**(3), 8–16 (2020)
8. Chen, H.-J., Ma, S.-P., Lu, H.-C.: Collaborative security annotation and online testing for web APIs. In: 2021 IEEE International Conference on e-Business Engineering (ICEBE), pp. 9–15 (2021)
9. Corradini, D., Pasqua, M., Ceccato, M.: Automated black-box testing of mass assignment vulnerabilities in restful APIs (2023)
10. Home | CVE, November 2022. Accessed 13 Nov 2022
11. CWE - Common Weakness Enumeration, October 2022. Accessed 13 Nov 2022
12. Ding, X., Niu, J., Wang, J.: API usage pattern search based on model checking. In: Qin, S., Woodcock, J., Zhang, W. (eds.) SETTA 2021. LNCS, vol. 13071, pp. 299–319. Springer, Cham (2021). https://doi.org/10.1007/978-3-030-91265-9_16
13. Ghaffarian, S.M., Shahriari, H.R.: Software vulnerability analysis and discovery using machine-learning and data-mining techniques: a survey. ACM Comput. Surv. **50**(4) (2017)
14. Golmohammadi, A., Zhang, M., Arcuri, A.: Testing RESTful APIs: a survey (2022)
15. Hamilton, D.: Artificial Intelligence in Software Testing & API Automation. Parasoft, September 2022
16. Hora, A., Valente, M.T.: Apiwave: keeping track of API popularity and migration. In: 2015 IEEE International Conference on Software Maintenance and Evolution (ICSME), pp. 321–323 (2015)
17. ISA group. IDL, November 2022. Accessed 22 Nov 2022
18. ISA group. IDLReasoner, November 2022. Accessed 22 Nov 2022
19. ISTQB Glossary, November 2022. Accessed 13 Nov 2022
20. Karlsson, S., Karlsson, S., Karlsson, S., Karlsson, S., Causevic, A., Sundmark, D.: QuickREST: property-based test generation of OpenAPI-described restful APIs. In: 2020 IEEE 13th International Conference on Software Testing, Validation and Verification (ICST) (2020)
21. Kornienko, D.V., Mishina, S.V., Shcherbatykh, S.V., Melnikov, M.O.: Principles of securing restful API web services developed with python frameworks. J. Phys: Conf. Ser. **2094**(3), 032016 (2021)

22. Liu, Q., Li, P., Zhao, W., Cai, W., Yu, S., Leung, V.C.M.: A survey on security threats and defensive techniques of machine learning: a data driven view. IEEE Access **6**, 12103–12117 (2018)
23. López, A.M.: Handling inter-parameter dependencies in REST APIs with IDL4OAS. Medium, March 2022
24. Ma, S.-P., Hsu, M.-J., Chen, H.-J., Su, Y.-S.: API prober – a tool for analyzing web API features and clustering web APIs. In: Chao, K.-M., Jiang, L., Hussain, O.K., Ma, S.-P., Fei, X. (eds.) ICEBE 2019. LNDECT, vol. 41, pp. 81–96. Springer, Cham (2020). https://doi.org/10.1007/978-3-030-34986-8_6
25. Martin-Lopez, A.: AI-driven web API testing. In: Proceedings of the ACM/IEEE 42nd International Conference on Software Engineering: Companion Proceedings, ICSE 2020, pp. 202–205. Association for Computing Machinery, New York (2020)
26. Martin-Lopez, A., Segura, S., Müller, C., Ruiz-Cortés, A.: Specification and automated analysis of inter-parameter dependencies in web APIs. IEEE Trans. Serv. Comput. **15**(4), 2342–2355 (2022)
27. Martin-Lopez, A., Segura, S., Ruiz-Cortés, A.: A catalogue of inter-parameter dependencies in RESTful web APIs. In: Yangui, S., Bouassida Rodriguez, I., Drira, K., Tari, Z. (eds.) ICSOC 2019. LNCS, vol. 11895, pp. 399–414. Springer, Cham (2019). https://doi.org/10.1007/978-3-030-33702-5_31
28. Martin-Lopez, A., Segura, S., Ruiz-Cortés, A.: RESTest: automated black-box testing of RESTful web APIs. In: Proceedings of the 30th ACM SIGSOFT International Symposium on Software Testing and Analysis, ISSTA 2021. Association for Computing Machinery (2021)
29. Masood, A., Java, J.: Static analysis for web service security - tools & techniques for a secure development life cycle. In: 2015 IEEE International Symposium on Technologies for Homeland Security (HST), pp. 1–6 (2015)
30. Giuliano Mirabella, A., Martin-Lopez, A., Segura, S., Valencia-Cabrera, L., Ruiz-Cortés, A.: Deep learning-based prediction of test input validity for restful APIs. In: 2021 IEEE/ACM Third International Workshop on Deep Learning for Testing and Testing for Deep Learning (DeepTest), pp. 9–16 (2021)
31. Mohamed, S.E., Ashaf, M., Ehab, A., Shereef, O., Metwaie, H., Amer, E.: Detecting malicious android applications based on API calls and permissions using machine learning algorithms. In: 2021 International Mobile, Intelligent, and Ubiquitous Computing Conference (MIUCC), pp. 1–6 (2021)
32. Morel, B.: Artificial intelligence and the future of cybersecurity. In: Proceedings of the ACM Conference on Computer and Communications Security (2011)
33. REST Connect Connector Generator | MuleSoft Documentation, November 2022. Accessed 19 Nov 2022
34. Niu, H., Keivanloo, I., Zou, Y.: API usage pattern recommendation for software development. J. Syst. Softw. **129**, 127–139 (2017)
35. NVD - Vulnerability Metrics, November 2022. Accessed 13 Nov 2022
36. Núñez, J.C.S., Lindo, A.C., Rodríguez, P.G.: A preventive secure software development model for a software factory: a case study. IEEE Access **8**, 77653–77665 (2020)
37. OAuth 2.0 — OAuth, December 2022. Accessed 10 Dec 2022
38. OWASP API Security Project | OWASP Foundation, October 2022. Accessed 13 Nov 2022
39. Pautasso, C., Wilde, E.: RESTful web services: principles, patterns, emerging technologies. In: Proceedings of the 19th International Conference on World Wide Web, WWW 2010, pp. 1359–1360. Association for Computing Machinery, New York (2010)

40. Mysiuk, R., Yuzevych, V., Mysiuk, I.: API test automation of search functionality with artificial intelligence. Artif. Intell. **27**(1), 69–274 (2022)
41. Sahoo, D., Liu, C., Hoi, S.C.H.: Malicious URL detection using machine learning: a survey. ArXiv, abs/1701.07179 (2017)
42. Salih, A., Zeebaree, S.T., Ameen, S., Alkhyyat, A., Shukur, H.M.: A survey on the role of artificial intelligence, machine learning and deep learning for cybersecurity attack detection. In: 2021 7th International Engineering Conference "Research and Innovation amid Global Pandemic" (IEC), pp. 61–66 (2021)
43. Scaffidi, C.: Why are APIs difficult to learn and use? XRDS **12**(4), 4 (2006)
44. Serme, G., de Oliveira, A.S., Massiera, J., Roudier, Y.: Enabling message security for restful services. In: 2012 IEEE 19th International Conference on Web Services, pp. 114–121 (2012)
45. Shaukat, K.: Performance comparison and current challenges of using machine learning techniques in cybersecurity. Energies **13**(10) (2020)
46. Shen, Q., Wu, S., Zou, Y., Xie, B.: Comprehensive integration of API usage patterns (2021)
47. Shu, G., Lee, D.: Testing security properties of protocol implementations - a machine learning based approach. In: 27th International Conference on Distributed Computing Systems (ICDCS 2007), p. 25 (2007)
48. The Official YAML Web Site. The Official YAML Web Site, November 2021. Accessed 31 Jan 2023
49. Building a Hypermedia-Driven RESTful Web Service, December 2022. Accessed 10 Dec 2022
50. Storozhuk, D., Ryabchykov, O., Popp, J., Bocklitz, T.: RAMANMETRIX: a delightful way to analyze Raman spectra (2022)
51. Swagger API Resources, January 2023. Accessed 19 Jan 2023
52. Swagger Petstore - OpenAPI 3.0, November 2022. Accessed 18 Nov 2022
53. SW testing concepts. Statement, Branch and Path coverage - SW Testing concepts, January 2023. Accessed 22 Jan 2023
54. Yarygina, T.: RESTful Is Not Secure. In: Batten, L., Kim, D.S., Zhang, X., Li, G. (eds.) ATIS 2017. CCIS, vol. 719, pp. 141–153. Springer, Singapore (2017). https://doi.org/10.1007/978-981-10-5421-1_12
55. Zha, M., Wang, J., Nan, Y., Wang, X., Zhang, Y., Yang, Z.: Hazard integrated: Understanding security risks in app extensions to team chat systems. In: Hazard Integrated: Understanding Security Risks in App Extensions to Team Chat Systems (2022)
56. Zhang, M., Arcuri, A.: Open problems in fuzzing restful APIs: a comparison of tools (2022)
57. Zhang, M., Arcuri, A., Li, Y., Liu, Y., Xue, K.: White-box fuzzing RPC-based APIs with EvoMaster: an industrial case study (2022)

The Way of Application of Mobile Components for Monitoring Systems of Coastal Zones

E. V. Melnik and Marina V. Orda-Zhigulina(✉)

Southern Scientific Center of the Russian Academy of Sciences, st. Chehova, 41, Rostov-on-Don 344006, Russia
jigulina@mail.ru

Abstract. A method is proposed for ensuring the functioning of mobile components for monitoring systems for coastal zones in the current paper. This method is based on the searching of the optimal composition of a group of mobile robots and the restriction of search space when it is redistributing the computational load in monitoring systems and the ontological model describing the subject area at the same time.

The proposed method implements the interaction of mobile components of the system due to horizontal connections between all participants of the system without any centralizing. Also, the proposed method provides an efficient organization of calculations in a distributed monitoring system. At the same time, the decision-making time is reduced by optimizing the distribution of any computing load associated with the transfer, processing and storage of data within this method. Moreover, the decision-making time is reduced as there are individual copies of the distributed ledger can be transferred "closer" to consumers on the fog layer as much as possible.

Keywords: Data Collection · Data Mining · Mobile Robots · Decentralized Decision Support System · Blockchain · Robotics · Hazardous Phenomena

1 Introduction

The task of connecting a large number of mobile components to a distributed storage is an important task in building monitoring systems for coastal zones recently. Such mobile components may include unmanned aerial vehicles (UAVs), unmanned surface vehicles (UAVs), and smart sensors. At the same time, the monitoring system solves the problems of collecting, processing and storing data of various types from a large number of heterogeneous devices. After that, the processed data should be placed in a distributed storage and be available to the system participants as users and experts.

The general principles of organization and structure of a distributed highly reliable reconfigurable monitoring system with mobile components based on a distributed registry, the Internet of things and data mining methods were previously developed by the authors and are given in [1]. In this work, the method of organizing a distributed data storage based on a distributed ledger has been finalized for a large number of system

R. Silhavy and P. Silhavy (Eds.): CSOC 2023, LNNS 724, pp. 382–389, 2023.
https://doi.org/10.1007/978-3-031-35314-7_34

participants. Mobile components can be such participants. The peculiarity of the method is that the system data storage has a decentralized structure. Also, the data storage is distributed. This data storage allows to store heterogeneous data and provides the maximum speed of access to stored data when a data processing.

2 Description of Application of Mobile Components for Monitoring Systems of Coastal Zones

The method for ensuring the functioning of mobile components for monitoring systems based on a distributed ledger was developed to solve the problem of efficient data processing when integrating mobile components. This method is based on a combination of the previously developed method for forming search space constraints when redistributing the computational load in monitoring systems based on a distributed ledger [2–4], a new ontological model [5] and a modified method for searching the optimal composition of a group of mobile components or mobile robots for the monitoring system [6–11].

External developers can create their own functional applications based on the developed method. These functional applications provide interfacing of connected external systems with the developed components of the monitoring and diagnostic system. Such functional applications provide the necessary data processing in addition to reading and writing data to the registry. The interaction of functional applications in a distributed environment is carried out due to the algorithm of the local dispatcher. This dispatcher ensures the distribution of computing tasks between all computing nodes of the system, including mobile components. The distributed ledger is a backbone element in this case. The entire system is built around this distributed ledger. That is, all other components of the system are connected to the distributed ledger, including groups of mobile robots and smart sensors. This approach reduces decision time in a distributed system. This happens by optimizing the distribution of the computational load associated with the transmission, processing and storage of data. Also, separate copies of the distributed ledger can be "moved" as close as possible to consumers due to the foggy layer. Therefore, it makes sense to associate functional applications that process data with each copy of the distributed ledger.

The scheme of the computing environment, for example, is shown for several users within the SSC RAS as shown in Fig. 1.

These users own heterogeneous computing nodes of various types and smart sensors. These nodes and sensors can take on the roles of computing nodes of the foggy and cloudy layers L_i, W_i, respectively. This is true for the developed method and for solving the functional problem T.

The main elements of each computing node are:

- distributed ledger software providing smart contracts for job boards and transaction data storage,

– a multi-agent dispatcher that monitors the job announcement board and controls the progress of work and data flows.

Additional elements of each computing node are:

– distributed storage for storing large static data such as tables, images or videos. The distributed storage also stores data on all transactions of the distributed ledger.
– functional applications. That is, it is a set of functional programs that perform certain functions of the system itself, such as collecting information from sensors, processing and classifying this data. This set of functional programs can be expanded by developers and further complicate the task graph, the progress of which is monitored by a multi-agent dispatcher. Also, functional applications may include any third-party applications that any of the participants in the system considers necessary to perform within the system. For example, applications of the cognitive modeling module.

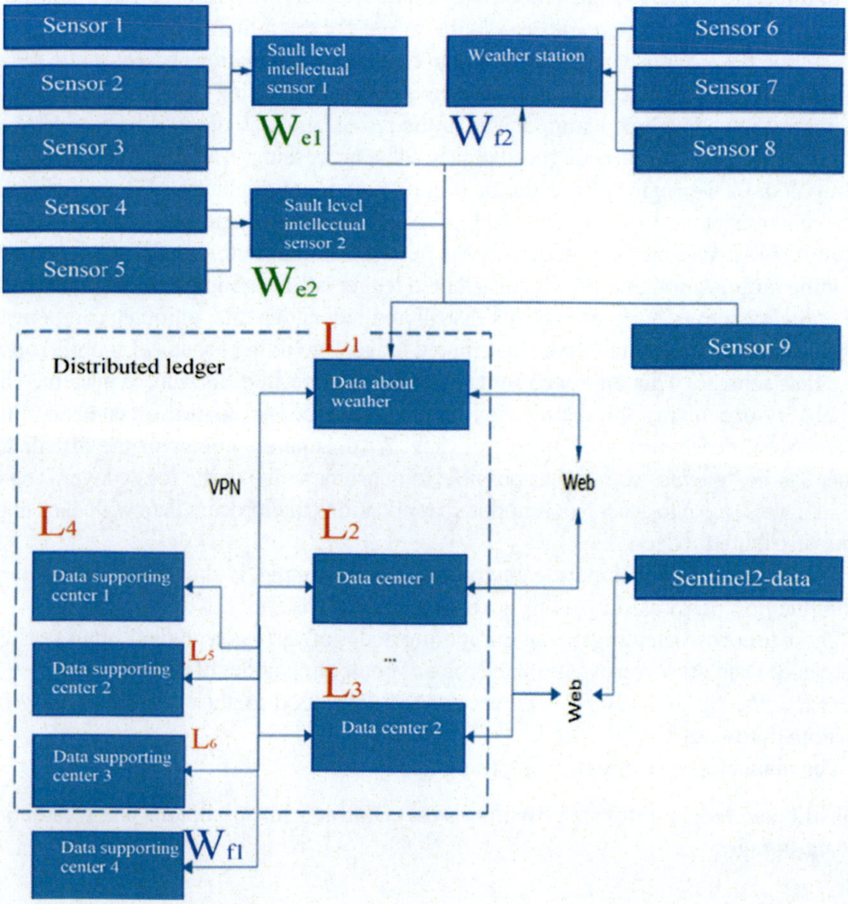

Fig. 1. An example of building a computing environment which is based on distributed ledger for solving monitoring problems.

Each of the software components has its own limitations on the configuration and capacity of resources.

For a distributed registry, for the nodes involved in the calculation of hashes, a certain amount of RAM is required. A distributed ledger can operate in two modes. The first mode is full. In this case, the entire chain of transactions is stored in the ledger. The second mode is lightweight. In this case, only the hash of the set of transactions is stored in the ledger. The user can integrate into the system, both in one and in the other case.

For nodes with distributed storage, intensive work with data carriers is implied, therefore, the nodes participating in the formation of distributed storage must be provided with data storage devices, since in the process of storing data they are replicated to all nodes of the distributed storage.

For the correct operation of functional programs, there may be various conditions, and they are imposed specifically by the functional program itself. This can be high performance for programs involved in data processing. This may be accessibility to a certain type of sensors for programs that are data sources.

Thus, for the system to work on each specific computing node, the presence of any of the four software components is mandatory. Only a distributed ledger is required. This corresponds to the presence of only software on the node to ensure the operation of a distributed ledger and a multi agent dispatcher for "light" wallets W_{ei}, W_{fi} in a lightweight form. The distributed storage of static data contains processed data from sensors, forecasting data, transaction data. In this case, functional applications can be placed on any W_{ei}, W_{fi} nodes with sufficient computing power, if necessary. But functional applications will be placed on the nodes of the core of the L_i system closest to the place of information collection.

A system on several computing nodes can schematically look like this, as shown in Fig. 2.

Here, the computing nodes L_1, L_m, for example, are full-fledged computing nodes and can provide the operation of a distributed registry, data storage, and can also participate in ensuring the operation of functional programs.

Computing node L2 does not have the necessary amount of data storage and computing power is involved only in ensuring the operation of the distributed registry.

Computing node W_{f1} has the necessary computing power and can be used for data processing by a functional program.

The W_{fi} computing node has the necessary computing power and can be used for data storage.

Computing nodes We1 and Wei have access to data from certain types of sensors, which means they can act as data sources.

At the same time, the computing node that should carry out the analysis and processing of the initial data is the most demanding on computing resources. Data analysis and processing can be carried out by any node closest to the data source. It could be L_i W_{ei}, W_{fi}. The general scheme of such a computing node assumes the presence of all components, except for a distributed data storage.

For example, the RStudio functional application can be used to analyze data from sensors. It is a free and open source software development environment for the R programming language, which is designed for statistical data processing and graphics work.

The official Sentinel-2 Toolbox or Sen2Cor can be used to analyze data from the Sentinel 2 satellite.

Also, all of the above computing nodes must be combined into a common distributed data storage system and a distributed task execution system based on a distributed ledger and a local dispatcher. All computing nodes are united in a VPN.

Computing nodes for supporting the computing environment, as well as computing nodes for storing data, are used to distribute the computing load between system participants and support data replication. Mining can also take place on such nodes, that is, the creation and verification of distributed ledger transactions. Such nodes, from the point of view of the software hosted on them, can refer not only to the core of the L_i system, but to the «near» periphery of the W_{fi} system.

At the same time, computing nodes supporting the computing environment are divided into different types according to the power of their computing resources. By computing resource, we mean processor power, the amount of RAM and permanent memory, and network bandwidth.

Nodes of the first type can mine transactions. The main requirement for such nodes is the presence of a minimum of RAM of at least 4 Gb. Nodes of the second type allow you to redistribute part of the computational tasks. Consequently, the requirements for such nodes are determined by the functional programs launched on them. Distributed data storage support nodes require a resource to store data.

The integration of mobile components into such a system should be carried out taking into account the types, power of computing resources of other system participants, characteristics, methods of interfacing and transmission speed of communication channels. This is especially critical for mobile devices and components. Mobile components are groups of UAVs and UAVs as terminal devices for collecting information.

Therefore, to integrate a distributed data storage with mobile components, the previously developed algorithm for the operation of the main loop of the local dispatcher [12] was finalized, which provides decentralized interaction of functional applications in a distributed environment without a leader. At the same time, each executing functional application is provided with a software library for accessing the functionality of the local dispatcher prototype. Algorithms for organizing access to distributed ledger data for mobile components were also developed.

Thus, as part of the development of an approach to ensure the functioning of mobile components for monitoring and diagnostic systems based on a distributed ledger for various areas of application, a method for organizing a distributed library based on a distributed ledger was finalized to ensure the functioning of mobile components based on distributed ledger technology. This approach allows you to combine various mobile components into a decentralized structure, which is the most effective approach in building modern monitoring and diagnostic systems [1]. At the same time, the distributed ledger technology allows you to effectively combine disparate data sources, heterogeneous computing resources and communication channels with different bandwidth and reliability. Mobile components are combined into a system based on the integrated use of such digital economy technologies as a distributed registry, data mining, wireless communication technologies, "fog and cloud" computing in monitoring and diagnostic systems for various fields of application. A set of distributed ledger programs, a set of scripts for organizing access and storing data in a distributed registry, and a local dispatcher for organizing distributed computing and launching and scheduling services of external subsystems in a decentralized heterogeneous reconfigurable monitoring and diagnostic system when connected to such a system of mobile components were finalized.

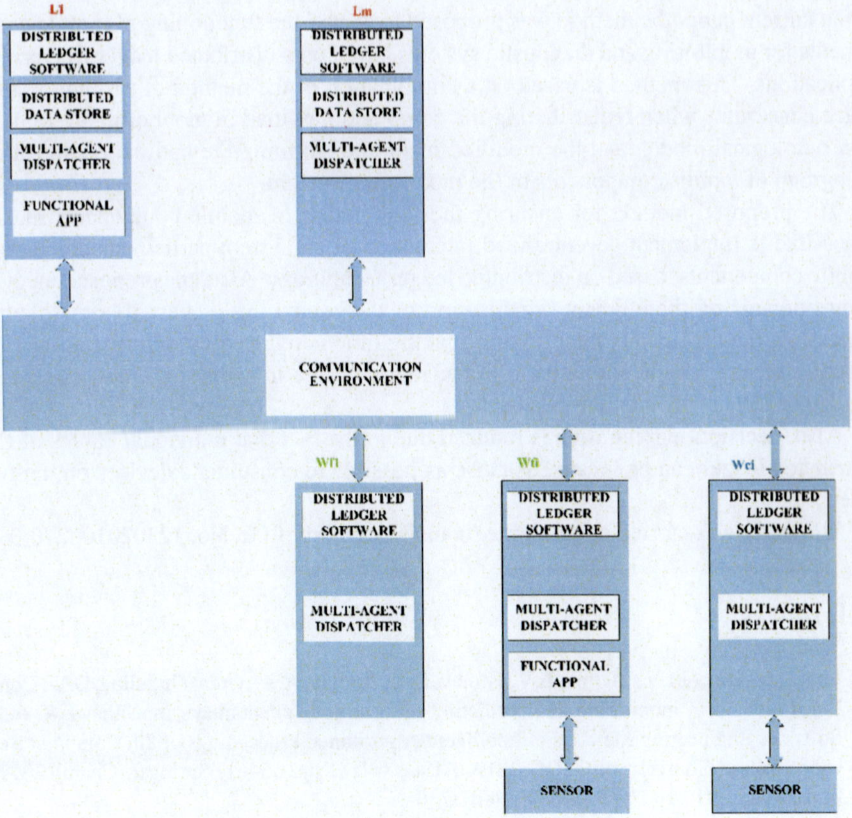

Fig. 2. System on multiple computing nodes.

In addition to the improved method of organizing a distributed data storage based on a distributed registry, a method was developed to ensure the functioning of mobile components for systems, which is based on a combination of a previously developed method for forming search space restrictions when redistributing the computing load in monitoring systems based on a distributed registry, a new ontological model, describing the subject area under consideration, and a modified method for forming the optimal composition of a group of mobile components. An analysis of the use of ontological models for solving optimization problems operating in a foggy environment showed that the time for solving the problem of redistributing the computational load in the system is on average reduced by 2.5 times compared to the analogue method. The modified method of forming the optimal composition of a group of mobile components makes it possible to ensure the value of such an efficiency indicator as the robot load factor of at least 75%, which is up to 1.5 times higher than when using other similar methods, which is confirmed by experimental studies using simulation modeling.

3 Conclusion

In the current paper the method was proposed to ensure the functioning of mobile components for monitoring and diagnostic systems based on a distributed ledger for various applications. This method is based on a combination of the method of forming search space constraints when redistributing the computational load in monitoring systems, a new ontological model, and the modified method of forming the optimal composition of a group of mobile components of the monitoring system.

The proposed method for ensuring the functioning of mobile components makes it possible to implement decentralized interaction of all system participants, including mobile components, based on distributed ledger technology. Also, the proposed method allows organizing the efficient organization of the computing process in a distributed system. At the same time, the decision-making time is reduced by optimizing the distribution of any computational load associated with the transmission, processing and storage of data.

Also, decision-making time is reduced due to the fact that individual copies of the distributed ledger can be located as close as possible to consumer's devices on the fog layer.

The study was carried out with the financial support of GZ No. 122020100270-3.

References

1. Orda-Zhigulina, M.V., Melnik, E.V., Ivanov, D.Y., Rodina, A.A., Orda-Zhigulina, D.V.: Combined method of monitoring and predicting of hazardous phenomena. In: Silhavy, R. (ed.) Software Engineering Methods in Intelligent Algorithms: Proceedings of 8th Computer Science On-line Conference, CSOC 2019. AISC, vol. 1, pp. 55–61. Springer, Cham (2019). https://doi.org/10.1007/978-3-030-19807-7_6
2. Lücking, M., et al.: The merits of a decentralized pollution-monitoring system based on distributed ledger technology. IEEE Access **8**, 189365–189381 (2020)
3. Ali, M.S., Dolui, K., Antonelli, F.: IoT data privacy via blockchains and IPFS. In: Proceedings of the Seventh International Conference on the Internet of Things, p. 14. ACM (2017)
4. Zheng, Z., Xie, S., Dai, H.N., Wang, H.: Blockchain challenges and opportunities: a survey. Work Pap. - 2016 (2016)
5. Melnik, E.V., Safronenkova, I.B., Klimenko, A.B.: A modified ontology-based method of workload relocation problem solving for monitoring and forecasting systems. In: Silhavy, R. (eds.) Artificial Intelligence in Intelligent Systems: Proceedings of 10th Computer Science On-line Conference, CSOC 2021. LNNS, vol. 2, pp. 145–152. Springer, Cham (2021). https://doi.org/10.1007/978-3-030-77445-5_13
6. Davids, A.: Urban search and rescue robots: from tragedy to technology. IEEE Intell. Syst. **17**(2), 81–83 (2002). IEEE, Varna
7. Gajduk, A.R., Kapustjan, S.G.: Gruppovoe upravlenie BLA s uchetom vremennogo zapazdyvanija v kanalah svjazi mezhdu nimi. In: Matematicheskie metody v tehnike i tehnologijah, vol. 2, pp. 52–55. MMTT, Moscow (2020)

8. Kapustjan, S.G., Gajduk, A.R.: Informacionnoe obespechenie gruppovyh dejstvij BLA pri reshenii zadach monitoringa bol'shih territorij. In: Zakonomernosti formirovanija i vozdejstvija morskih, atmosfernyh opasnyh javlenij i katastrof na pribrezhnuju zonu RF v uslovijah globalnyh klimaticheskih i industrialnyh vyzovov, Opasnye javlenija-II, vol. 1, pp. 332–335. Izdatelstvo SSC RAS, Rostov-on-Don (2020)
9. Dorigo, M.: Blockchain technology for robot swarms: a shared knowledge and reputation management system for collective estimation. In: Swarm Intelligence: 11th International Conference, ANTS 2018, vol. 11172, p. 42. Springer, Rome (2018)
10. Valentini, G., Brambilla, D., Hamann, H., Dorigo, M.: Collective perception of environmental features in a robot swarm. In: International Conference on Swarm Intelligence, ANTS 2016, vol. 9882, pp. 65–76. Springer, Cham (2016). https://doi.org/10.1007/978-3-319-44427-7_6
11. Ferrante, E., Turgut, A.E., Huepe, C., Stranieri, A., Pinciroli, C., Dorigo, M.: Self-organized flocking with a mobile robot swarm: a novel motion control method. Adapt. Behav. **20**(6), 460–477 (2012). Sage Publications Sage UK, London
12. Melnik, E.V., Orda-Zhigulina, M.V., Orda-Zhigulina, D.V.: Distributed library model based on distributed ledger technology for monitoring and diagnostics system. In: Silhavy, R. (eds.) Software Engineering and Algorithms: Proceedings of 10th Computer Science On-line Conference 2021, CSOC 2021, vol. 230, pp. 501–509. Springer, Cham (2021). https://doi.org/10.1007/978-3-030-77442-4_43

Network Traffic Analysis and Control by Application of Machine Learning

Zlate Bogoevski(✉), Ivan Jovanovski, Bojana Velichkovska, and Danijela Efnusheva

Computer Science and Engineering Department, Faculty of Electrical Engineering and Information Technologies, Ss. Cyril and Methodius University, Skopje, North Macedonia
{zlateb,bojanav,danijela}@feit.ukim.edu.mk

Abstract. The purpose of this paper is to analyze how networks work, how data is transmitted, and what information we get from each router during data transmission, as well as getting to know the basics of machine learning and how to create models that will understand and learn how networks work. By applying machine learning methods, the obtained results show us the shape of a network. With different methods and the resulting analysis we can get information about how we can plan the network, in terms of expanding the network if the capacity of the links is almost full or when one of the links has predispositions to go from an active state to an inactive one. The results show satisfactory outcomes through the use of three different machine learning models that were capable of accurately detecting the functionality of a port, calculating its utilization and learning when the utilization hits a threshold of above 75%.

Keywords: machine learning · analysis · network expansion plan · control · network traffic

1 Introduction

In the past couple of years, telecommunication networks have been in a constant state of change and transition. The primary reason for this stems from the fact that many new services and applications require increasingly higher data transfer speeds, surpassing the capabilities of traditional global telephone networks.

Because of this, the capacities of the links of existing networks developed in the past need to be constantly renewed, supplemented and expanded. A computer network is a set of two or more computers that are connected through a communication medium and that can communicate with each other and share resources. The links serve to connect the nodes and they themselves constitute the basic transmission network that physically carries the messages through the links. With constant analysis and control of the traffic that passes through a network, many anomalies that exist in it can be discovered and detected in real time.

Therefore, by applying machine learning methods, results are obtained that show us the shape of a network. With different methods we can get information crucial in understanding how to plan a network, in particular when it comes to network expansion

R. Silhavy and P. Silhavy (Eds.): CSOC 2023, LNNS 724, pp. 390–399, 2023.
https://doi.org/10.1007/978-3-031-35314-7_35

or in cases when one of the links has predispositions to go from an active state to an inactive state.

Analysis and control of network traffic can serve as an approach to predict the upgrade of network elements connected in computer networks, as well as on the Internet. With the increase in the load on network elements, there comes a point when it is necessary to expand the capacities of certain links. With the exponential increase in the number of computers and devices connected to the Internet, the efficient utilization of network devices becomes an important topic to consider when analyzing network traffic, as well as increasing the capacity of network links terminating through a given network.

The aim of this paper is to address the discussed issues and propose a machine learning solution for the purpose of analyzing and controlling network traffic in real time. The rest of the paper is organized as follows: Section 2 analyzes the state of the art. Section 3 gives details about the methodology used in the design of the proposed solution. Section 4 discusses the practical implementation and results. The paper concludes with Sect. 5.

2 State of the Art

Network traffic analysis and control systems can be applied during planning and capacity development of network devices in order to enable control over network traffic, using mechanisms and rules for filtering known traffic circulating through a network [1].

With this approach, it is possible to discover the capacity of the links that connect each following element in a network, and thus contribute to increasing the performance and ensuring the best data flow, as well as improving the quality of the network offered by a certain service provider to its end users. Today, networks are of particular significance and can play key roles within any institution or legal entity, as nowadays almost every company has its own communications network, typically connected to the global Internet.

To solve problems related to network traffic, this research proposes the implementation of machine learning, where through the use of certain methods for analysis and control of network traffic, desired solutions can be reached. The applied methodology uses a database which was created in a real-world network environment. Using this database, it is planned to achieve a detailed analysis of its data, namely by examining the columns that show information about actual traffic passing through given ports of the network elements, comparing them to the columns that show the maximum capacity that those same ports are capable of supporting [2]. In addition, by controlling and detecting the network capacity of the ports and network elements included in the network, it is possible to determine when to expand the network and add new links and even new network elements, which would all in all result in an improved performance.

This paper will consider a database created using a real network environment, which includes information from interfaces of network elements. Machine learning methods will be applied to analyze this database. Testing will be done for the accuracy and precision in detection of the capacity of each network element, thus determining how much data can actually pass through it in a given time period.

This will produce results that will help network administrators who work as network planners and developers. Additional investigations will also be carried out such as finding reasons for why a port or an interface is predicted to fail in a given time frame.

The goal of using machine learning is to interpret past network traffic information in order to develop better routing configurations for future conditions. It is reasonable to assume that traffic history also contains information about future traffic such as, for example, the distribution of traffic at different times of the day.

Many researchers have proposed the use of machine learning techniques to solve network problems including traffic classification, latency prediction, and network traffic optimization. To our knowledge, this is the first attempt to use network traffic prediction for the purpose of planning and expanding links with higher capacities [3].

The purpose of this paper is a detailed consideration of the reasons for the implementation of such a solution and its benefits. The reasons for applying machine learning in predicting how many of the network links and network elements are actually used, in terms of their predicted capacity, will also be put forward. Furthermore, several machine learning methods will be compared and analyzed. Existing datasets for training network-based models using machine learning algorithms and methods will be reviewed and explained in detail.

With these results, the most appropriate machine learning model that can be realized and used will be selected. This model will be tested for accuracy and precision as a result of previous training [4]. The utilization of the links (in terms of network traffic) will also be set as a threshold. If the utilization of a link is over 75% of its total capacity, the model should react and let the network know that that location needs to be expanded, i.e., increased.

Finally, a conclusion will be drawn from the entire research, indicating where the proposed solution can be applied for the analysis and control of network traffic using machine learning methods during capacity planning in network devices.

3 Methodology

The application that does the analytics and predictions on the database achieves this using three separate scripts as it's concurrently doing three separate predictions at the same time. The analytics portion is done by all three of the scripts and is dependent on the type of the prediction model used (classification or regression).

The data used in this application is collected by sampling network traffic every two minutes from multiple different links, both fiber and non-fiber, with different capacities, bandwidths and utilization. Analysis starts by converting all of this data into a numeric format and scaling it to a uniform unit of measurement (in our case this unit is kilobytes).

A further step included formatting of the data by dividing it into groups so that the relation of every group to the target variable can be tracked in the form of a correlation matrix [5].

A correlation matrix is a metric that allows us to quantify and visualize the relationship between data. This relationship is expressed using the correlation coefficients of variable pairs which can range between the values of -1 and 1. In a correlation matrix, positive values indicate positive linear correlation between two variables, and the same can be said for negative values and negative linear correlation. Accepted threshold ranges that determine the level of interrelation between different indicators, depending on the correlation coefficient, are as follows:

0–0.3 - no relation;
0.3–0.5 - the relationship is weak;
0.5–0.7 - medium connection;
0.7–0.9 - high;
0.9–1 - very strong.

If the correlation coefficient is negative, it means that the relationship of the parameters is inverse.

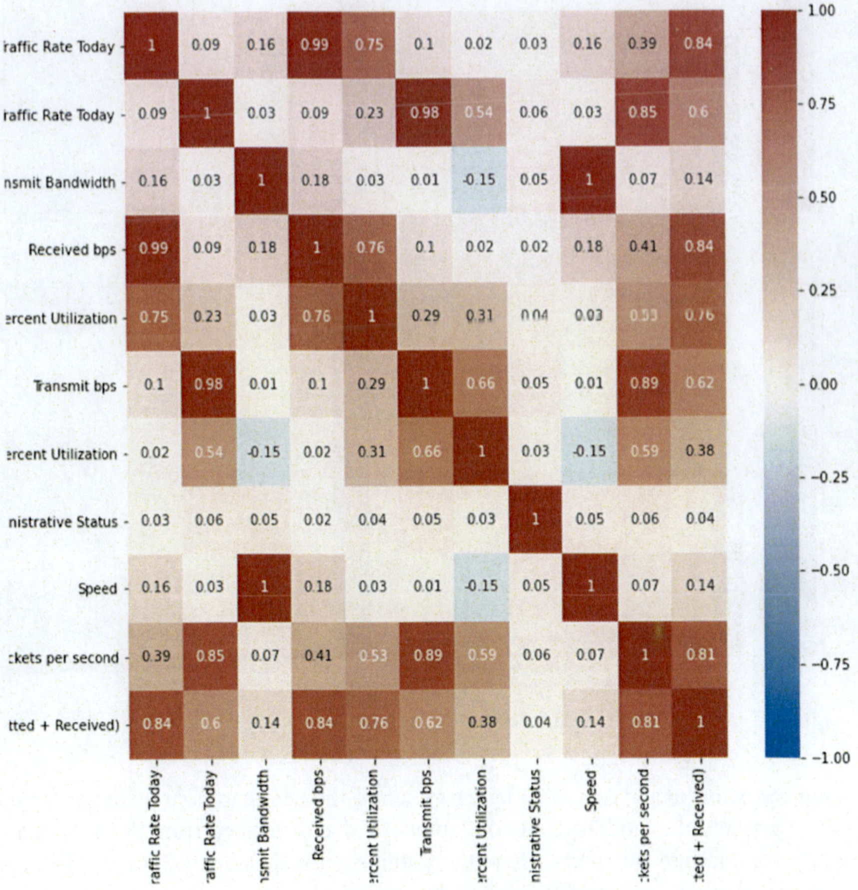

Fig. 1. Filtered correlation matrix

Figure 1 represents our data correlation matrix where for this research, values with a higher correlation than 0.7 (this value may be different in other research) are considered highly correlated.

Here we can see a high positive linear correlation between variables such as the transmit bits per second with the daily traffic rate of the network as well as a perfect positive linear correlation between the network's speed and its transit bandwidth.

Other steps in the analytics portion include representing data in graphs and visualizing it, and also singling out and representing numeric data only (see Fig. 2).

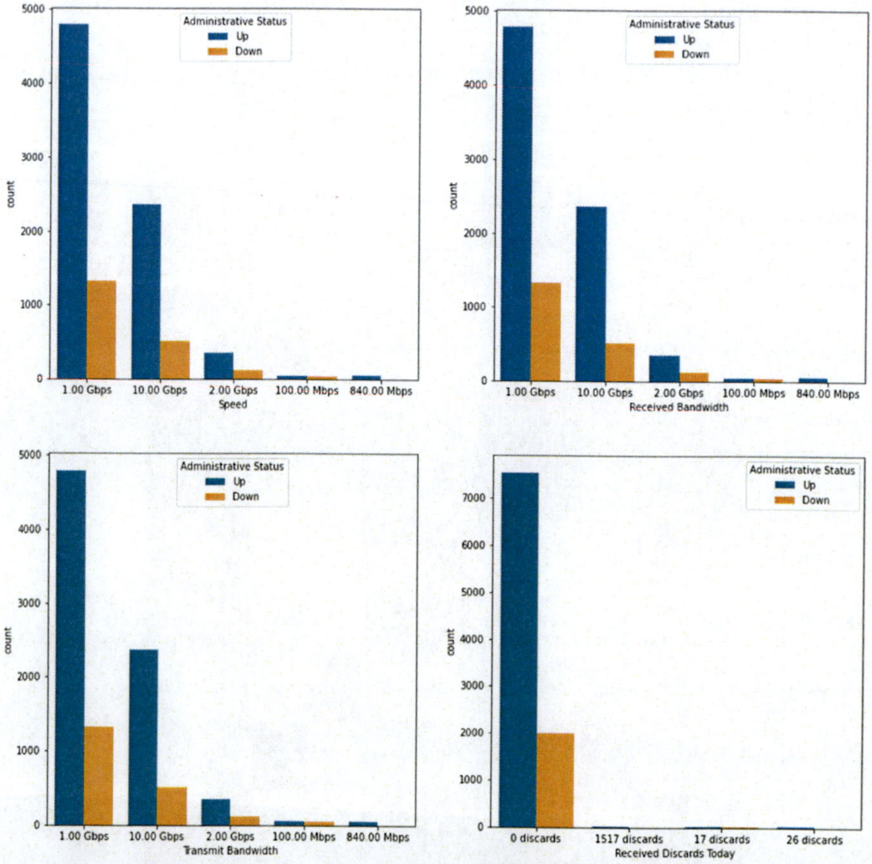

Fig. 2. Ratio of numeric unrounded data to the target column

After this short analysis is done by every script, the next part is preparation for the data to be given to the models. Here, only numeric data is filtered from the dataset and randomly divided into two parts where the training set makes up 30% of the data and the test set makes up the other 70% of the data.

In our case we are making three different prediction models, two classification models and one regression model. All three of the models are made in a similar fashion. The algorithms used for the classification models are: Decision Tree, K nearest neighbors, Regression classifier and AdaBoost classifier. For the regression model we also used a Linear Regression algorithm.

All three of the models were trained and tested on the dataset, receiving three different tasks with their goal being to determine the variation of a target variable. The first task involved detecting the functionality of a port, the second task attempted to calculate the utilization of the port/link and the third task was responsible for learning when the utilization hit a threshold of above 75%. Different target variables were used for individual tasks.

Furthermore, some of these tasks required modification of the data set (addition of multiple new columns) for helping the models during the training process [6]. For detecting the functionality of the port, the script uses classification models in order to determine when the port/link is in operation. Here, for the target variable, the "Administrative Status" column was used from the database. At the start of the script, all of the models for classification are created and a confusion matrix is returned for all of them as a method for comparing their accuracy. Afterwards, the best three of the models are further trained and compared. The next step in training the models is using Grid Search. For every model, a grid map is created using different arguments and values. The results of the grid search are again an indicator for the accuracy of the model (Fig. 3).

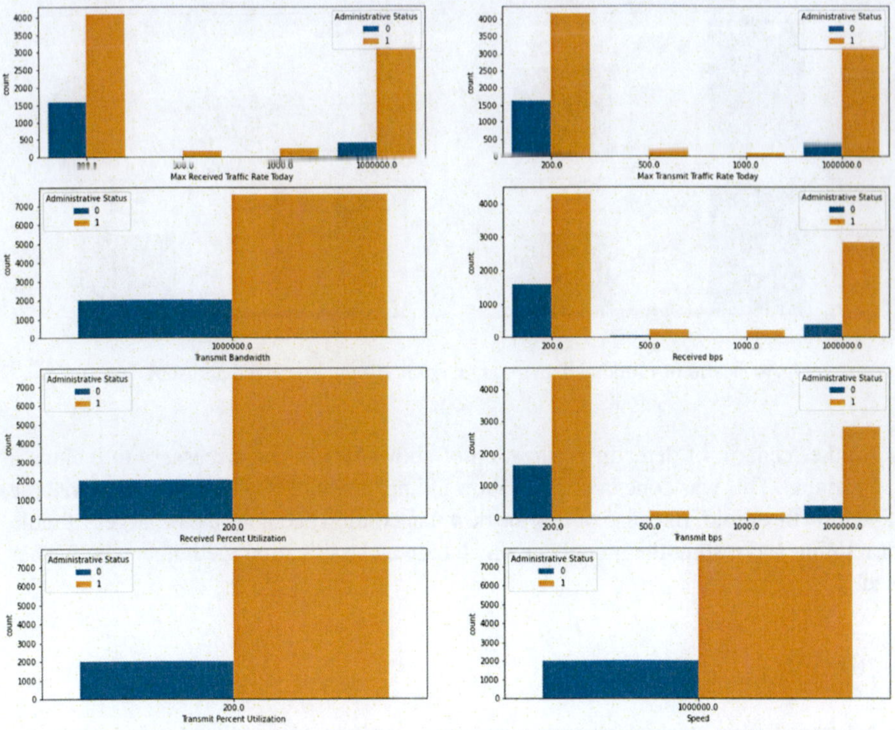

Fig. 3. Ratio of rounded data to target column - total received and sent data

For the scenario of detection of utilization of a port/link we used regression models for determining its usage. This is done by calculating the total amount of bytes that are sent and received through each individual port/link. Training and the testing of these models was done in the same fashion as before.

After these steps, the models are compared using the "Mean Absolute Error" and the "Mean Squared Absolute Error". These metrics show the difference between the real and the predicted values [7, 8]. After training and testing, our models return the predicted values for every port where the metric finds the mean value of every model (Fig. 4).

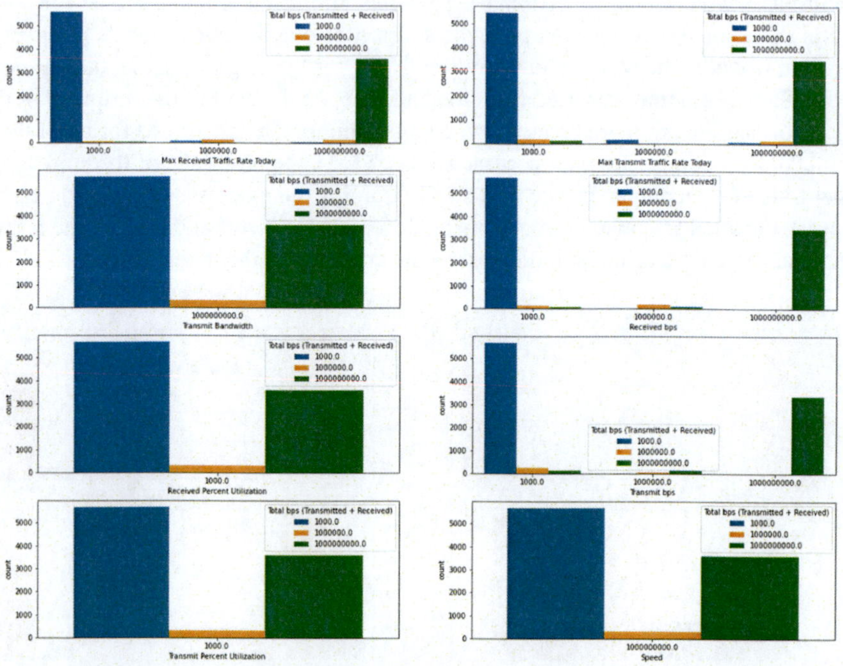

Fig. 4. Ratio of rounded data to target column - total received and sent data

In the scenario of detecting utilization of above 75% it was necessary to include a flag variable. This was done in order to show us and the models which ports achieved the utilization threshold. Because of this the new flag column becomes a new target variable for training and testing the models in which case a classification method can be further used.

4 Results

In this case, the rate at which all three scripts were executed was reasonably fast, although this speed is largely dependent on the size and type of the dataset given for analysis.

After comparing the three models, all of them were capable of learning from the training set and later returned satisfactory results using the test set. To begin with, the classification model exhibited the highest accuracy where only a small number of values were incorrectly classified or outright missed.

In Fig. 5, we can clearly inspect the confusion matrix for every model. In almost every case the accuracy of the model is nearly a 100%. The predictive capabilities of SVM are subpar, in that the performance of the model is a baseline case scenario, which carries no significant research-wise or application-wise contribution. However, the performance

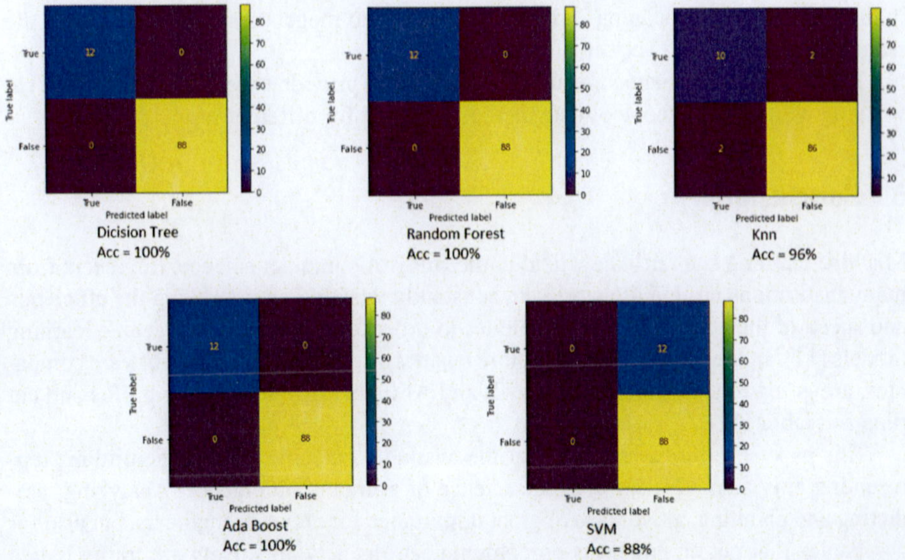

Fig. 5. Results for classification models

provided by the Decision Tree, Random Forest and Ada Boost algorithms shows that these three classifiers can tackle this problem exceptionally well.

On the other hand, the regression models had only a small mean difference, meaning that they were capable of predicting the utilization of a port/link almost perfectly [10]. Still, these results are indicative of the functionality of our models, that are only using one specific dataset, and therefore might show worse results under different conditions.

a)

	DecisionTree	RandomForest	Knn	AdaBoost	LinearRegression
TenG	554582.577163	565513.373539	617480.013969	554614.647594	1860.882552
Gigabits	6702.396061	5572.994915	10555.18258	5137.980617	4.57536
Vlan	0.026471	0.031212	0.118062	0.054581	0.000168

б)

	DecisionTree	RandomForest	Knn	AdaBoost	LinearRegression
TenG	307561834892.260132	319805375651.447327	381281567651.443237	307597407325.520264	3462883.871269
Gigabits	44922112.963324	31058272.319114	111411879.291028	26398844.817607	20.933917
Vlan	0.000701	0.000974	0.013939	0.002979	0.0

Fig. 6. a) Table for the mean square error for regression models, b) Table for the root of the mean square error for regression models

Also, Fig. 6 gives a representation of metric values obtained for the regression models, that provides the comparison between them. In our case, the calculated metrics were Mean Squared Absolute Error and Root Mean Squared Absolute Error. Representation of the results in the table is in kilobytes. Therefore, if we correct for the size of the data in the first two rows, meaning if they are converted to megabytes or terabytes, then the resulting error will be a very small number.

To conclude, the obtained results for this dataset provide excellent results, however working with a wider scope of data there is potential for different results.

5 Conclusions

Machine learning is a fairly new field in the study of computer science. Engineers from many institutions around the world are constantly working on improving the efficiency and speed of their methods and techniques in order to develop better machine learning models [11]. In the future, more and more businesses, primarily led by software companies, are going to start incorporating ML and AI in an effort to increase profits and cut time and labor [12].

This paper examined just a few of the available machine learning algorithms representing only a small fraction of the science of artificial intelligence. Analyzing, predicting and planning networks is of great importance for network engineers, but also for companies that put an emphasis on communications networks. Network traffic is in a way a "living matter" that is constantly changing and growing day by day, therefore, an expansion plan should be constantly monitored and implemented, where necessary.

As part of the idea to improve our work in the future we intend to implement the use of neural networks. Because the data that we were working with included sampling network traffic only every two minutes, when represented on a time axis graph, it behaved like a discrete signal. If the sampling rate of our dataset was increased (sampling every two seconds as opposed to every two minutes for example), using the help of neural networks could possibly allow for more accurate detection of connection drops in individual elements of a certain network.

References

1. Forrest, S., et al.: A sense of self for unix processes. In: Proceedings of 1996 IEEE Symposium on Computer Security and Privacy, Los Alamos, CA, pp. 120–128 (1996)
2. Caswell, B., Beale, J., Baker, A.: Snort Intrusion Detection and Prevention Toolkit. Elsevier (2007)
3. Networking Fundamentals, Crystal Panek, pp. 300–480 (2020)
4. Ajami, R., Dinh, A.: Embedded network firewall on FPGA. In: 8th International Conference on Information Technology: New Generations (2011)
5. Sourdis, I., Dimopoulos, V., Pnevmatikatos, D., Vassiliadis, S.: Packet pre-filtering for network intrusion detection. In: ACM/IEEE Symposium on Architecture for Networking and Communications Systems (2006)
6. Cover, T.M.: Geometrical and statistical properties of systems of linear inequalities with applications on pattern recognition. IEEE Trans. Electron. Comput. **EC-14**, 326–34 (1965)
7. Munkres, J.: Algorithms for the assignment and transportation problems. J. SIAM **5**(1), 32–38 (1957)
8. Bousquet, O., Boucheron, S., Lugosi, G.: Theory of classification: a survey of recent advances, ESAIM: Probab. Stat. **9**, 323–375 (2005)

9. Comer, D.: Fundamentals of Computer Networking and Internetworking, Purdue University, pp. 50–250 (2014)
10. Shi, Q., Petterson, J., Dror, G., Langford, J., Smola, A.J., Vishwanathan, S.V.N.: Hash kernels for structured data. J. Mach. Learn. Res., 2615–2637 (2009)
11. Zadrozny, B., Elkan, C.: Obtaining calibrated probability estimates from decision trees and Naive Bayesian classifiers. In: Proceedings of the International Conference on Machine Learning (2001)
12. Tanenbaum, A.S., Wetherall, D.J.: Computer Networks, pp. 111–355 (2011)

Genetic Algorithm for a Non-standard Complex Problem of Partitioning and Vehicle Routing

Rudolf Erdei, Daniela Delinschi(✉), and Oliviu Matei

Technical University of Cluj Napoca, North University Centre of Baia Mare, Baia Mare, Romania
{rudolf.erdei,daniela.delinschi,oliviu.matei}@cunbm.utcluj.ro

Abstract. This article proposes a non-standard complex problem of Partitioning and Vehicle Routing Problem (PVRP), that assumes an infinite capacity for vehicles and demands that routes will have the least possible time discrepancy. The problem was presented in a real-life scenario, with real constraints and motivations for requirements. Current established VRP-class problems failed to address this, so we formalized it into a new problem and provided a genetic algorithm for solving it as close to optimal as possible. Some Local Minima Detection and Avoidance strategies are presented, and also Self-Adaptation of the algorithm, in order to improve and optimize execution and the resulting solution.

Keywords: Vehicle Routing Problem · Partitioning Problem · Genetic Algorithm · Combinatorial Optimization · Local Minima Detection · Local Minima Avoidance · Genetic Algorithm Self-Adaptation

1 Introduction

The Vehicle Routing Problem (VRP) is currently a subject of extensive interest and study. Route optimization is an essential component of the improvements that can be implemented with relative ease, even on a global scale, bringing enormous benefits not only in terms of revenue/expense ratio for operators but also in terms of greenhouse gas emissions (pollution), which are determined primarily by the vehicles operating these transports, and also supporting machinery.

With the goal of making Europe the first continent to achieve emission-free status by 2050, the European Union has created the **Green Deal** [1] program. This noble, urgent, and vital ambition poses many issues for which scientists must provide sophisticated, all-encompassing, and most importantly sustainable solutions.

Through the integration of IoT (Internet of Things) systems [10,19], the collection and analysis of vehicle parameters, and the isolation and mitigation of problems that may arise, inevitable or not, the topic is currently bringing together researchers from various fields, from pure theorists to experts in Data

R. Silhavy and P. Silhavy (Eds.): CSOC 2023, LNNS 724, pp. 400–412, 2023.
https://doi.org/10.1007/978-3-031-35314-7_36

Mining and even electronics. Clearly, simple routing optimization is insufficient; however, when viewed in a global view, such as vehicle operation optimization, multimodal transport analysis, load optimization, dead time optimization, and predicting the company's future transportation needs, routing plays an important, if not critical, role.

2 Case Study

This paper presents an approach to solving a problem that encompasses both the *Vehicle Routing Problem (VRP)* [7,18,34] and the *Partition Problem* [3,8,13]. The problem is real, inspired by a company with a given set of trucks picking up medical waste from multiple sources.

The trucks are assigned daily destinations they must visit, but there is no information available regarding the volume or weight of waste to be picked up from them. In some situations, regressions could be used to estimate such information to some extent based on historical data (*VRP with stochastic demands* [5,20,24]), but this approach will be tested when such data will be available for this case study. Medical waste itself does not raise capacity concerns, as it can be compacted to a large extent within the truck and the daily production of waste is low in volume. At least in theory, one truck could collect all the waste, but, for time constraints, a number of six trucks (at the time of implementation) were used by the company to visit their clients.

Until now, the company used a manual method of grouping destinations (elementary clustering), and the order of travel was determined by each driver individually through navigation applications available on mobile devices. This led to routes with different travel times, where the time discrepancies were very large (sometimes even exceeding three times the allotted time), so they found the need to find an algorithmic optimization for this problem. Also, new clients are added to the portfolio, exacerbating the problem. Optimizations would allow them better management of the fleet and delay the acquisition of new vehicles.

Figure 1 shows the classical way of defining a VRP-type problem, where the distances between points are calculated in the Euclidean way (physical distances). But, as the mathematical modelling and solution of this problem will show (see Sect. 4), a better approach to this problem is to consider a time-based distance, as shown in Fig. 2.

Due to the nature of the business, when a truck is called for collection, both the weight and volume of waste to be picked up are completely unknown and can vary quite a bit from day to day. Because of this largely unknown, the problem cannot be classified as a simple *VRP*, since, in general, the *VRP* class requires information about weight or volume, and truck division is done based on their capabilities (the capacity of each truck).

Another important constraint concerns the need for the time spent on the route for each truck to be equal or as close as possible to the rest of the trucks. This constraint is specific to Number Partitioning Problems (in this case being the *Multi Way Number Partitioning Problem, MWNPP* [21,28,32]), where the

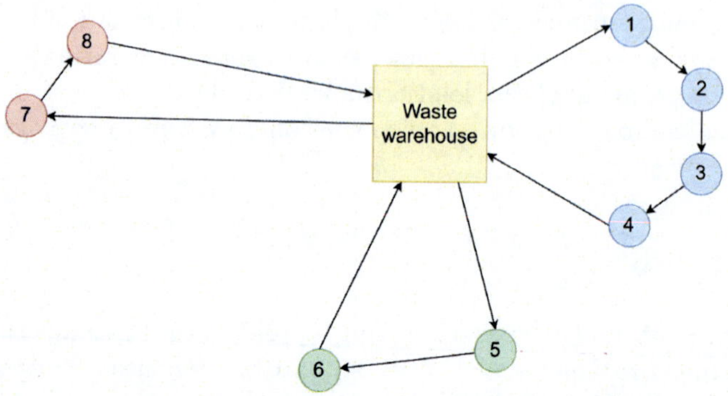

Fig. 1. Example of a three-lane vehicle routing and partitioning system

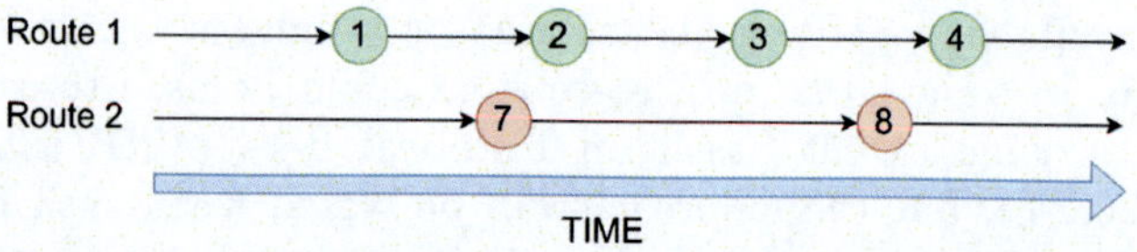

Fig. 2. Example routes (Route1, Route 2) of vehicles in the same time interval

requirement is to partition certain elements so that the final sums of the partitions are equal. The difference between the minimum and maximum partition values is called the **discrepancy** and will be an important element in the fitness function developed within the algorithm presented in this paper. Therefore, the problem is a complex one including clustering/partitioning and vehicle routing. As constraints, the company would like to keep all routes under 8 h (including time spent at stops), and a discrepancy of no more than 30 min.

3 Literature Review of Existing Problem Formalization

3.1 Vehicle Routing Problem

Currently, although studied for a long time, a considerable research effort is devoted to solving classical routing problems (VRP class of problems). Over time, heuristic algorithms have been developed, such as Clarke-Wright [9], Gillett-Miller [17], Foster-Ryan [15] or Fisher-Jaikumar [14]. Each of these algorithms seeks to generate a list of routes (a feasible and qualitative solution to the problem) for a given number of customers, who have a certain quantitative need (a *demand*).

Being an intensively studied problem, due to its practicality but also its complexity, the class of VRP problems also has many variations, as described in the works published by Toth and Vigo [2,33]. Each of these has its own approaches, complexities, and real-world applications that make it worthy of study.

The generalized variant (or **GVRP**) [16] deals with the situation of courier companies, where the vehicles do not visit each individual customer but only a single point in a cluster of customers. The problem, therefore, boils down to the optimal choice of the visited client within each cluster, the order in which the clusters are visited, and the available capacity correlated with the demand. Several genetic-based approahces have been proposed by Pop et al. in [27,29].

A VRP variant similar to the problem in the current study is the VRP with stochastic demands (*VRP with stochastic demands*) [5] variant, where probabilistic estimates of customer demands are used. Through these estimates, for which even machine learning mechanisms such as regression can be used, the aim is actually to simplify the problem and transform it into a classical VRP and solve it by known means.

The major constraint discussed and addressed in these algorithms concerns the *capacity* [30] of the vehicle doing the hauling while analyzing the quality of a solution is generally done by the total distance travelled by the suite of trucks trained in this solution.

The choice of total distance as the quality factor of a candidate route is not random, in fact, it is a simplification of the complex way of operating a company [23] from the point of view of transportation. Each vehicle has a certain fuel consumption relative to the distance travelled, service intervals, and average travel speed. Also, hired staff (drivers) are required by law to take breaks at certain intervals and obviously have to be paid. All these parameters can be reduced to the distance travelled by the vehicle and the average travel speed. The shorter this distance, the lower the total expenses will be.

Another intensively studied approach for solving problems of the two classes (VRP and partitioning) is meta-heuristic algorithms, such as genetic algorithms [22,25], ant-colony optimizations [6,11], simulated annealing [4,12,31], and others. These approaches allow for solving complex problems with a variable number of constraints. They allow introducing new constraints, as in the case of the present problem, and even modifying them at runtime, with potential solutions having in this case a deviated path towards the new global optimal solution.

So, a meta-heuristic approach is a natural choice in this context, a choice that allows designing a flexible algorithm and solving the problem in a relatively short time. The only drawback, introduced by the random nature of some functions, is that the results will not be identical when repeatedly run on the same data set, they will vary more or less depending on the generation parameters and the complexity of the problem.

3.2 Number Partitioning Problem

Another problem in which the topic of this topic can fit is the *partition problem*, known in the literature as *Number Partitioning Problem* or NPP [21]. Depending on its complexity, there are several generalizations, of which the most complex variant is the *MWMDNPP* (Multi-Way Multi-Dimensional Number Partitioning Problem), in which the partitioning criteria are assimilated to the *dimensions*, and the set initial is partitioned into several *sets*.

In the context of the theme, the concepts in the NPP can be defined as follows:

- **The sets** are the trucks to which the destinations must belong;
- **The dimensions** can be the distances travelled by trucks, the number of customers served, the times spent on the route, etc.

Special attention will be paid to the clear, correct and synthesized approach to the problems, so as not to introduce unnecessary complexities that would lead to a very high calculation time. Thus, through certain procedures, the specific dimensions of the NPP can be reduced; strategies can be introduced to optimize the calculation of truck grouping, so that the convergence typical of genetic algorithms is more direct, but without decreasing the quality of the solution.

4 Mathematical Modelling of Proposed Problem

In order to ensure a correct understanding of the given problem, it is necessary to abstract it by formalizing it in a mathematical model, easy to implement in the code and of minimum complexity.

Particular attention should be paid to the *fitness* function, which evaluates the quality of a potential solution within the algorithm, due to the fact that it must combine several concepts:

- **total route quality**, which is traditionally evaluated by its length;
- **truck route discrepancy**, which is the difference between the longest route and the shortest;
- taking into account, however, at any moment, also the **time spent at the medical operators** to pick up their waste and sign the various necessary documents.

Thus, for the present problem we can define the following concepts and calculation formulas:

Regarding a Specific Truck:

$$D_i = \sum_{n=1}^{c} d(s_n, s_{n-1}) \tag{1}$$

$$T_i = D_i \cdot V_m \tag{2}$$

where:

- D_i - the total distance travelled by the truck. This distance also includes the distances from and to the central depot;
- V_m - the average speed of the truck;
- T_i - the total time spent on the route by a truck;
- $d(s_n, s_{n-1})$ - the distance between the destination s_n and s_{n-1} calculated as the straight-line Euclidean distance between the two geographic points;

Thus, the relevant metrics for a truck are: **total distance traveled**, defined as the sum of the distances between points, but also the distances related to the movement from the central warehouse to the first destination and the return from the last destination to the warehouse; **total travel time**, defined as the total travel time added to the total time spent at the destinations.

The total time spent at destinations is important because, for close destinations, a truck may spend more time picking up waste (if it has many destinations to serve) than moving between them. This situation is not reflected in the total distance travelled by the truck, so distance cannot be considered an absolute metric in fitness calculation.

Regarding a Potential Solution:

$$T_t = \sum_{i=1}^{t}(T_i + t_{i_{client}}) \tag{3}$$

$$D_s = max(T_i) - min(T_i) \tag{4}$$

where:

- T_t - the total time spent by all trucks on the route;
- $t_{i_{client}}$ - the average time spent by the truck at the destination i;
- D_s - solution discrepancy, defined as the difference between the longest and shortest time a truck spends on the route in a day.

In the case of a potential solution, we have two important metrics: **total time**, defined as the sum of the time of all trucks on the route; **discrepancy**, defined as the difference between the maximum and minimum time within a potential solution.

These two metrics are important because we want to minimize both of them, but this most likely won't be possible, since a minimum total time path will definitely have a large discrepancy.

The Way to Calculate the Fitness of a Potential Solution:

$$F = T_t + D_s \cdot I_D \tag{5}$$

where:

- I_D - the importance of discrepancy in the calculation of solution fitness;
- F - the calculated solution fitness, defined as the sum of the total solution time and the product of the solution discrepancy and the discrepancy importance;

As can be understood from these formulations, especially from the 5 relation, the discrepancy is defined as a **penalty** in the fitness calculation, which has some importance. By increasing the value of the I_D parameter, the penalty for a large discrepancy can be set numerically. Thus, with a large value for I_D the algorithm will favour small discrepancies but will allow longer time for the solution, and at small values for I_D the algorithm will favour the total time at the expense of the discrepancy (i.e. it will allow larger discrepancies if the total route will be shorter in duration).

When testing and settling on a suitable value, several runs with different values for I_D will need to be performed so that the right balance between discrepancy and total solution time can be established. The trucking company accepts a maximum discrepancy of 30 min and a maximum total time per truck of 8 h.

Distance Calculation Between Two GPS Coordinates: In the design phase, to demonstrate the correct operation of the algorithm, the actual distances (on the road) between the points are not used, since the accuracy of the solution is not critical at this point. This distance is calculated by a simple mathematical method, a method proposed by Veness [35]:

$$a = \sin^2(\frac{\Delta\varphi}{2}) + \cos(\varphi_1) \cdot \cos(\varphi_2) \cdot \sin^2(\frac{\Delta\lambda}{2}) \tag{6}$$

$$c = 2 \cdot \arctan 2(\sqrt{a}, \sqrt{(1-a)}) \tag{7}$$

$$d = R \cdot c \tag{8}$$

where:

- λ - longitude;
- φ - latitude;
- R - average radius of the Earth (6371 km);
- d - the distance calculated between the two points.

This method calculates the distance in a straight line (but taking into account the curvature of the Earth) between two GPS coordinates, so this distance cannot be considered final. For these distances to be correct it is necessary to use external services (with API access) to find the actual physical distance between two such locations and save them in a distance matrix and in a permanent, locally usable database.

5 Design and Implementation

The GA was implemented using the following parameters:

- **Individual representation:** the classical VRP representation, an array of integers (clients) where routes are marked with zeros at the ends;

- **Selection:** multiple types of selection mechanisms, namely `randomSelection`, `rankSelection`, `tournamentSelection`, `rouletteSelection`, were implemented. This is subject to **self-adaptation**;
- **Crossover:** multiple point crossover (random number of cut points);
- **Mutation:** a swap between two positions within the chromosome.

Within the implementation, several hybridizations were tested and implemented within the proposed algorithm, from which we enumerate:

- **Adaptive Selection**: each individual will keep a pointer to the selection method used for his own birth, enabling the estimation of the best selection methods for each generation;
- **Survival**: each individual will be able to participate in more than one generation, but its life will be limited to a set number of generations;
- **Adaptive Mutation**: the absolute amount of mutations a single individual will receive, is dependent on its own position within the population (social position);
- **Consanguinity Check (Genealogy)**: enables the limitation of *inbreeding*. This mechanism implements genetic compatibility restrictions between two individuals [26]. Each individual is assigned a unique identifier and the history of the last six generations (so a total of 62 ancestors) is kept for each individual;
- **Diploidy**: genetic information from both parents is kept on the individual. Generating a new individual involves using genetic information from all four chromosomes. This method generates 8 offspring for each couple of parents.

Mutation is implemented as an exchange between two different, randomly chosen positions within the chromosome. Depending on the individual's relative fitness and the scaling factor calculated by the `Autoadapt` class, a specific number of swaps is applied to each individual.

Two adaptive mechanisms for mutation are present in the algorithm, namely *passive* and *active* self-adaptive mutation. The passive one is based on a simple calculation that returns a certain number of mutations depending on the relative position of the individual within the population (Fig. 3), namely:

- if the individual has a high fitness (i.e. is in the top 10%), a small number of mutations will be applied to it, in an effort to preserve its quality as best as possible;
- If the individual has low fitness (i.e. between 90% and 100%), their utility is limited, so they can be subjected to a greater number of mutations.

The active self-adaptive mutation is based on a scaling factor, which is applied to the result obtained in the previous step, a factor that increases when the algorithm gets stuck in a local minimum and decreases when it gets out of it.

Changing the scaling factor is done gradually, with a slight delay (called *growth rate*), to allow for better exploration and diversification of the resulting

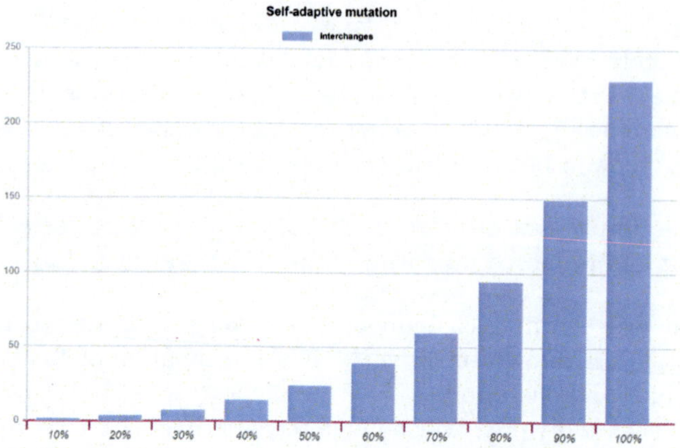

Fig. 3. Passive self-adaptive mutation used within the algorithm

individuals. The gradual change of parameters is critical for the accelerated increase in individual diversity. Because of the limited initial level, a sudden change would result in lower diversity, making diversification difficult to achieve.

6 Validation and Results

The proposed algorithm was implemented in Java 11, as a standalone console application, providing a small and quick framework to test and validate its execution. A complete dataset from the company was used as validation, benchmarking it and comparing it to existing manual (suboptimal) solutions. Practical testing of the resulting application was done on a machine with Ryzen 5 4600H laptop, with 16 GB of RAM.

During testing and validation, the presented algorithm consistently reached good results, depending on the options passed to it. The relationship between the discrepancy and the total time of the paths in a given instance was tested, resulting in an inversely proportional relationship between them.

The algorithm was run, in the parameter testing phase, 240 times on the same instance, in order to be able to estimate the best values. The Table 2 summarizes the ten best results obtained after running the algorithm. Table 1 shows the statistics for all 240 runs.

From analysing Tables 1 and 2, it is visible that the algorithm does reach zero discrepancy, but these instances did not reach the top 10, because, for small discrepancies, the overall fitness decreases qualitatively.

Although a dynamic population was used in terms of numbers, the base population is the one expressed in the results table. It has been observed that the number of individuals within the populations, although it affects the quality of the result to a certain extent, has a much greater direct effect on the running time.

Table 1. Algorithm Execution Statistics

	Execution Time	Individuals	Fitness	Total time	Discrepancy
Min	4540	100	1699	1642,00	0,00
Max	24101	1600	2172	2154,00	83,00
Med	10913	326	1790	1731,81	29,50
Std Dev	2307	405	57	71,06	12,42

Table 2. Summary of the 10 best runs of the algorithm

#	Individuals (no)	Generations (no)	Cutting points (no)	Imp. discr. *	Exec. time (ms)	Discr. ** (min)	Time total route (min)	Fitness
1	**200**	**2000**	**10**	**2**	**9930**	**12**	**1746**	**1770**
2	200	2000	14	2	11132	7	1756	1770
3	200	2000	4	2	9754	7	1759	1773
4	200	2000	20	2	11253	10	1755	1775
5	200	2000	8	2	9750	14	1758	1786
6	200	2000	6	2	10272	13	1769	1795
7	200	2000	20	2	12392	14	1767	1795
8	200	2000	18	2	11139	6	1799	1811
9	200	2000	18	2	12296	14	1817	1845
10	200	2000	6	2	9755	7	1835	1849

* **Imp. discr.** - the importance of the discrepancy
** **Discr.** - discrepancy

Through successive runs, with different parameters, we concluded that a relatively small number of individuals, such as 200 (the value in the table) can produce high-quality solutions, in a relatively short time (the average being 10 s with a standard deviation of 2.3 s).

Also, also from Table 1 the very small standard deviation, of only 3.18%, is very apparent, compared to the average value, and also the very small difference between the average value and the minimum fitness, of only 5.35%. These values confirm that this algorithm, through self-adaptation mechanisms, manages to consistently reach very good results.

For the case study problem discussed in this paper, the genetic algorithm approach, therefore, provides an efficient solution because, in complete opposition to the concept of *Divide et Impera*, it solves in one step both the generation of optimized routes as well as limiting the discrepancy between trucks.

7 Conclusions

This paper presented an implementation for a metaheuristic algorithm from the class of genetic algorithms, for solving a complex vehicle partitioning and routing problem. The unique requirements of the presented problem made it impossible to apply a classical solution for the VRP class, so it was necessary to design a new, combined approach.

To solve the problem in an efficient way, different hybridizations of the algorithm have been researched and implemented, approaches inspired by how evolution works in the natural world, such as inbreeding, but also heuristic methods of adaptive control of the algorithm.

By using these hybridization methods, our internal tests revealed an improvement of about 20% in runtime and 5% in the quality of the result, compared to a standard GA.

As further developments, work will be done on a variant that uses the capacity of the trucks and an estimated amount of waste, obtained from the history of the medical operator through certain procedures such as regressions. Also, the research will focus on limiting the memory used, simplifying the algorithms used, optimizing the way of self-adaptation, and integrating other hybridizations, in an effort to obtain a result as close to optimal as possible, in the shortest possible time.

This paper presents and proposes a very interesting problem, of combining VRP with NPP, resulting in a complex problem that can disregard vehicle capacity. As the problem is rarely found in real-life scenarios, research on it is rare (we haven't found a similar problem being proposed), but can be of interest for certain scenarios, like the one presented in the current paper.

Finally, research in the field of GA hybridizations is not complete, there are still many aspects that can be better explored, such as the migration of individuals between populations, aspects that could bring more quality and robustness to the algorithms that use them. But today, optimization efforts are increasing, even in the context of the drastic climate changes we are witnessing, so research in this field will certainly gain momentum even in the near future.

References

1. Eu green deal. https://ec.europa.eu/info/strategy/priorities-2019-2024/european-green-deal. Accessed 13 Mar 2021
2. Toth, P., Vigo, D.: Vehicle Routing: Problems, Methods, and Applications. SIAM (2014)
3. Babel, L., Kellerer, H., Kotov, V.: The k-partitioning problem. Math. Methods Oper. Res. **47**(1), 59–82 (1998)
4. Bertsimas, D., Tsitsiklis, J.: Simulated annealing. Stat. Sci. **8**(1), 10–15 (1993)
5. Bertsimas, D.J.: A vehicle routing problem with stochastic demand. Oper. Res. **40**(3), 574–585 (1992)
6. Blum, C.: Ant colony optimization: introduction and recent trends. Phys. Life Rev. **2**(4), 353–373 (2005)
7. Braekers, K., Ramaekers, K., Van Nieuwenhuyse, I.: The vehicle routing problem: state of the art classification and review. Comput. Ind. Eng. **99**, 300–313 (2016)
8. Chopra, S., Rao, M.R.: The partition problem. Math. Programm. **59**(1–3), 87–115 (1993)

9. Clarke, G., Wright, J.W.: Scheduling of vehicles from a central depot to a number of delivery points. Oper. Res. **12**(4), 568–581 (1964)
10. Desai, M., Phadke, A.: Internet of things based vehicle monitoring system. In: 2017 Fourteenth International Conference on Wireless and Optical Communications Networks (WOCN), pp. 1–3. IEEE (2017)
11. Dorigo, M., Blum, C.: Ant colony optimization theory: a survey. Theoret. Comput. Sci. **344**(2–3), 243–278 (2005)
12. Dowsland, K.A., Thompson, J.: Simulated annealing. In: Handbook of Natural Computing, pp. 1623–1655 (2012)
13. Feder, T., Hell, P., Klein, S., Motwani, R.: Complexity of graph partition problems. In: Proceedings of the Thirty-First Annual ACM Symposium on Theory of Computing, pp. 464–472 (1999)
14. Fisher, M.L., Jaikumar, R.: A generalized assignment heuristic for vehicle routing. Networks **11**(2), 109–124 (1981)
15. Foster, B.A., Ryan, D.M.: An integer programming approach to the vehicle scheduling problem. J. Oper. Res. Soc. **27**(2), 367–384 (1976)
16. Ghiani, G., Improta, G.: An efficient transformation of the generalized vehicle routing problem. Eur. J. Oper. Res. **122**(1), 11–17 (2000)
17. Gillett, B.E., Miller, L.R.: A heuristic algorithm for the vehicle-dispatch problem. Oper. Res. **22**(2), 340–349 (1974)
18. Golden, B.L., Raghavan, S., Wasil, E.A., et al.: The Vehicle Routing Problem: Latest Advances and New Challenges, vol. 43. Springer, New York (2008). https://doi.org/10.1007/978-0-387-77778-8
19. Guerrero-Ibanez, J.A., Zeadally, S., Contreras-Castillo, J.: Integration challenges of intelligent transportation systems with connected vehicle, cloud computing, and internet of things technologies. IEEE Wirel. Commun. **22**(6), 122–128 (2015)
20. Gupta, A., Nagarajan, V., Ravi, R.: Approximation algorithms for VRP with stochastic demands. Oper. Res. **60**(1), 123–127 (2012)
21. Korf, R.E.: Multi-way number partitioning. In: IJCAI, vol. 9, pp. 538–543 (2009)
22. Kramer, O.: Genetic Algorithms. Springer, Cham (2017). https://doi.org/10.1007/978-3-319-52156-5
23. Lee, S.M., Moore, L.J.: Optimizing transportation problems with multiple objectives. AIIE Trans. **5**(4), 333–338 (1973)
24. Mendoza, J.E., Castanier, B., Guéret, C., Medaglia, A.L., Velasco, N.: A memetic algorithm for the multi-compartment vehicle routing problem with stochastic demands. Comput. Oper. Res. **37**(11), 1886–1898 (2010)
25. Mitchell, M.: An Introduction to Genetic Algorithms. MIT Press (1998)
26. Petrovan, A., Erdei, R., Pop-Sitar, P., Matei, O.: A self-adapting immigrational genetic algorithm for solving a real-life application of vehicle routing problem. In: Silhavy, R., Silhavy, P., Prokopova, Z. (eds.) CoMeSySo 2019 2019. AISC, vol. 1047, pp. 144–156. Springer, Cham (2019). https://doi.org/10.1007/978-3-030-31362-3_15
27. Pop, P., Matei, O., Valean, H.: An efficient hybrid soft computing approach to the generalized vehicle routing problem. In: Corchado, E., Snášel, V., Sedano, J., Hassanien, A.E., Calvo, J.L., Ślęzak, D. (eds.) Soft Computing Models in Industrial and Environmental Applications, 6th International Conference SOCO 2011, pp. 281–289. Springer, Heidelberg (2011). https://doi.org/10.1007/978-3-642-19644-7_30
28. Pop, P.C., Matei, O.: A memetic algorithm approach for solving the multidimensional multi-way number partitioning problem. Appl. Math. Model. **37**(22), 9191–9202 (2013)

29. Pop, P.C., Matei, O., Sitar, C.P., Chira, C.: A genetic algorithm for solving the generalized vehicle routing problem. In: Corchado, E., Graña Romay, M., Manhaes Savio, A. (eds.) HAIS 2010. LNCS (LNAI), vol. 6077, pp. 119–126. Springer, Heidelberg (2010). https://doi.org/10.1007/978-3-642-13803-4_15
30. Ralphs, T.K., Kopman, L., Pulleyblank, W.R., Trotter, L.E.: On the capacitated vehicle routing problem. Math. Program. **94**(2), 343–359 (2003)
31. Rutenbar, R.A.: Simulated annealing algorithms: an overview. IEEE Circuits Devices Mag. **5**(1), 19–26 (1989)
32. Schreiber, E.L., Korf, R.E., Moffitt, M.D.: Optimal multi-way number partitioning. J. ACM (JACM) **65**(4), 1–61 (2018)
33. Toth, P., Vigo, D.: Models, relaxations and exact approaches for the capacitated vehicle routing problem. Discret. Appl. Math. **123**(1–3), 487–512 (2002)
34. Toth, P., Vigo, D.: The Vehicle Routing Problem. SIAM (2002)
35. Veness, C.: Calculate distance and bearing between two latitude/longitude points using haversine formula. Movable Type Script (2011)

A Simple yet Smart Welding Path Generator

Mihail Z. Avramov(✉)

Sofia University "St. Kliment Ohridski", Bourchier Str 1, Sofia 1164, Bulgaria
mavramov@chem.uni-sofia.bg

Abstract. In this paper, we propose a novel approach for automated path planning in robotic welding using point cloud data obtained from laser seam tracking. Our approach involves hierarchical segmentation of the point cloud using 3D and 2D RANSAC [12] techniques, combined with Alpha Shapes [10] and appropriate distance measures to deliver different pathfinding results. We conduct experiments using point cloud data from a real welding scenario to evaluate the effectiveness of our proposed approach. The results show that our approach can generate accurate welding paths with minimal manual intervention and in a fraction of the time required for manual path generation. Future work could explore the use of advanced path planning algorithms to generate more complex paths that take into account welding parameters.

Keywords: robotic welding · path optimization · hyperautomation · point cloud data · laser seam tracking · hierarchical segmentation · RANSAC · Alpha Shapes

1 Introduction

The process of robotized welding requires extensive preparation and testing of a robot program. While this procedure is effective for large series of welded products, it becomes a significant cost for small series or unique workpieces. Thus, automating or simplifying welding path programming is important.

Efforts have been made in recent years to find and optimize welding paths. To simplify the programming process, it is useful to have the application operational on the controller of the camera system or the robot itself, without requiring additional hardware or an offline workspace.

This paper focuses on the path extraction process and presents a lightweight implementation that can be integrated into the controller of a seam tracker, providing an easy way to program.

Various approaches to pathfinding have been introduced in recent decades. According to the evaluation in the Welding Handbook, the use of a MIG/MAG welding robot is efficient for production series between 1,000 and 60,000 [1, 7] pieces per year. Even for shorter series, the high quality of the obtained welds makes the use of a robot desirable despite the significant cost for programming.

To meet the need for more flexible operation, particularly in shorter or small but complicated series, significant efforts have been made in the development of hyperautomation tools [8]. This recent trend in customer needs has led to our interest in

R. Silhavy and P. Silhavy (Eds.): CSOC 2023, LNNS 724, pp. 413–425, 2023.
https://doi.org/10.1007/978-3-031-35314-7_37

developing an automation path generation tool that can simplify programming and can be run on the field using the installed laser seam tracking hardware and software.

2 Related Works

The use of various sensors and technologies for seam tracking in robotic welding applications has been explored in several studies, with laser sensors found to be particularly effective due to their high accuracy and fast response time. Additionally, there is a growing body of research focused on the automation of robotic welding processes, including path planning, trajectory optimization, and control, with a focus on the development of algorithms and tools to optimize welding parameters and ensure high-quality welds [1, 3].

Research has also investigated the use of 3D point cloud data obtained from laser sensors or structured light sensors for stereo reconstruction in robotic welding applications. These studies have shown the feasibility and effectiveness of using point cloud data to improve the accuracy and efficiency of welding processes. Some studies have specifically [2] focused on the use of point cloud data for path planning in robotic welding applications, developing algorithms and techniques to extract features and generate optimized welding paths.

Efforts to develop intelligent path optimization algorithms for welding robots have been extensive, with traditional path searching methods such as the artificial potential field algorithm, A*, and RRT widely used for robot path searching. Meanwhile, intelligent path searching methods including artificial neural network, ant colony algorithm [18], and tabu search [19] have been applied to robot path optimization.

To solve the welding robot path planning problem, environment modeling, obstacle avoidance, and multi-objective path optimization have been studied. The grid method [20] is often used for environmental modeling, and the ant colony algorithm is applied for path searching [4]. For robot joints obstacle avoidance, the two-level collision detection and geometrical obstacle avoidance strategy [21] are often introduced due to the complex obstacle avoidance of welding processes. Multi-objective path optimization [24] has also been widely studied for arc welding robot path planning, with optimization objectives including welding deformation, obstacle avoidance, path length, and energy consumption [5]. Surrogate models [23] have been applied to enhance computational efficiency, and the relationship between welding parameters and welding sequence has been studied to establish proxy models for multi-objective particle swarm optimization.

Existing methods for welding robot path optimization can be broadly classified into traditional and intelligent methods. Traditional methods, such as the artificial potential field algorithm [25], A*, and Dijkstra's algorithm [26], are deterministic and often have good performance for 2-dimensional path planning but may suffer from the problem of local optima, deadlock, and unsteady movements near obstacles. Intelligent methods, including artificial neural network, ant colony algorithm, genetic algorithm, and particle swarm optimization [22], are more flexible and can handle complex optimization problems with multiple objectives.

3 Methodology

In this paper, a novel approach for automated path planning in robotic welding using point cloud data obtained from laser seam tracking is proposed. Our method involves hierarchical segmentation of the point cloud to obtain using 3D and 2D RANSAC [12] techniques with the application of Alpha Shapes [10] and appropriate distance to deliver different variants of path finding results.

3.1 A Brief of the Algorithm

We start here with a brief description of the main cascade of processing the data (Fig. 1).

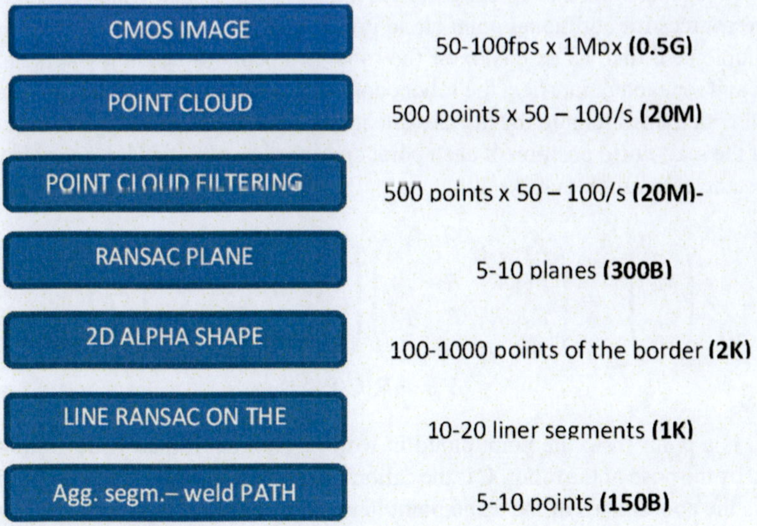

Fig. 1. A cascading reduction of the information generated in the system reduces an example scan from 0.5 G to 150 bytes

The significant reduction of the data through the flow of the algorithm simplify the decision for path selection.

3.2 Data Collection (CMOS IMAGE-POINT CLOUD)

The data collection in this work is performed using a real-time laser seam tracking camera, the SKYBLUE - SB0120, designed and developed by the Bulgarian company, Metrology-lab [14]. The seam tracker is a triangulation device that obtains the reflection of the laser beam on a workpiece in a CMOS sensor. The system is constructed based on the Scheimpflug principle [13], which provides an in-focus projection of the beam from each distance where the camera operates. The collected grayscale image contains the projection of the observed profile. As the system's optics' geometry is known, the

CMOS image can be converted into world coordinates, which are the coordinates of the laser beam's cross-section with the observed profile.

To process the image, the camera uses a line scanning algorithm that finds the integral intensity of the projection and estimates the center of the projection distribution. This averaging allows sub-pixel precision in estimating the vertical coordinate of the camera measurement. The camera can deliver up to 100 "slices" with 640 or 1280 points per second. To obtain a 3D point cloud from the 2D coordinates measured in the laser plane of the seam tracker, an additional step is required. The seam tracker is mounted on the last flange of the welding robot, which is the default mounting of a seam tracker sensor to the robot. An important function of the sensor is to deliver real-time corrections to the robot's position so that the welding torch remains properly aligned to the welded seam. To support this real-time correction delivery, the processing of the images obtained by the sensor is executed on a small but powerful industrial controller, which has significant unused resources for additional point cloud processing.

To implement this, an extension of the robot protocols for seam trackers is implemented, and for each point slice, the robot coordinates are retrieved. The seam tracker is mounted with a fixed joint to the last flexible joint of the robot, allowing for the computation of the real-world position of each point on each slice obtained by the seam tracker using a static calibration matrix.

$$\begin{pmatrix} x_w \\ y_w \\ z_w \\ 1 \end{pmatrix}_w = \begin{pmatrix} R_{11} & R_{12} & R_{13} & T_x \\ R_{21} & R_{22} & R_{23} & T_y \\ R_{31} & R_{32} & R_{33} & T_z \\ 0 & 0 & 0 & 1 \end{pmatrix} \begin{pmatrix} C_{11} & C_{12} & C_{13} & C_x^t \\ C_{21} & C_{22} & C_{23} & C_y^t \\ C_{31} & C_{32} & C_{33} & C_z^t \\ 0 & 0 & 0 & 1 \end{pmatrix} \begin{pmatrix} 0 \\ y_c \\ z_c \\ 1 \end{pmatrix}_c \quad (1)$$

$$\mathbf{P}_w = \mathbf{R}.\mathbf{C}.\mathbf{P}_c$$

Here $\mathbf{P}_w$ is a point from the point cloud in world (robot coordinates), $\mathbf{R}$ is the matrix obtained by the pose of the robot, $\mathbf{C}$ is the calibration matrix in homogeneous coordinates and $\mathbf{P}_c$ is the coordinates of the same point in camera coordinates, R_{ij}- the angular part of robot pose[1], $T_{x|y|z}$ is the position of the robot tool coordinates, C_{ij} is the rotation part of the calibration matrix and $C^t_{x|y|z}$ is the translation part. Indexes $_w$ denote a point in world (robot) coordinates and $_c$ a point in camera coordinates.

3.3 Point Cloud Filtering

The data filtering process in this work involves two procedures: voxel down-sampling and statistical outlier removal. Voxel down-sampling reduces the size of a point cloud while preserving its overall structure. The process involves dividing the 3D space into small cubic voxels and retaining only one point from each voxel. This results in a more equalized point density and can be done quickly and efficiently.

Statistical outlier removal is a method for removing outliers from a point cloud. The algorithm computes the mean and standard deviation of the distances between each point

[1] The rotation (angular) matrix parts can be obtained using Euler or Tait-Brian definitions [wiki] for rotation matrixes which are dependent of the robot type. E.g. for Yaskawa and Fanuc – XYZ transformation, Stäubli uses a XY'Z, for KUKA - ZY'X" etc.

in the cloud and its k nearest neighbors [16].

$$\text{keep}\,\mathbf{p}_i \text{ if } \overline{\left|\mathbf{p}_i - \mathbf{p}_j\right|\big|_{j \text{ in } k \text{ nearest neighbors of } i}} < L \tag{2}$$

Here $\mathbf{p}_i$ is each point of the point cloud $\mathbf{p}_j$ is any point in k nearest neighbors to $\mathbf{p}_i$ and L is the allowed threshold filter level the dash on top of the equation designates average value.

Points that have a distance greater than a specified number of standard deviations from the mean are considered outliers and are removed from the cloud. This process can help mitigate the effects of optical anomalies in the data like reflections or speckles from beam. The approach is cluster oriented i.e. the filter removes independent points located far from other points.

3.4 Plane RANSAC Segmentation

To obtain plane segmentation, an iterative RANSAC process is applied to the filtered point cloud data. The Random Sample Consensus (RANSAC) algorithm is a widely used method for extracting planar regions from noisy and incomplete data [11, 15]

$$ax + by + cz + d = 0 \tag{3}$$

Here *a, b, c* are the components of the normal to the plane vector, *d* is the distance of the plane to the origin of the coordinate system and *x, y, z* are the coordinates. As a total Eq. (3) is an equation of plane in homogeneous form.

RANSAC segmentation with arbitrary models works by randomly selecting a subset of points from the input data and fitting a model to these points. The algorithm then computes the distance between each point in the input data and the model. Points that fall within a certain distance threshold are considered inliers and are added to the consensus set. The model with the most inlier points is the optimal plane model.

For plane segmentation, the RANSAC algorithm is used to fit a plane to a subset of points in the input data. The plane is defined by its normal vector and distance from the origin (3). The algorithm randomly selects three points from the input data and computes the normal vector and distance of the plane passing through these points. It then computes the distance between each point in the input data and the plane. Points that fall within a certain distance threshold are considered inliers and are added to the consensus set (Fig. 2).

The algorithm repeats this process for a specified number of iterations, each time selecting a random set of three points, and returns the plane with the largest consensus set. In this study, the algorithm is applied to extract a plane with some allowed tolerances and appropriate probability.

After the first plan is found the inlier points are removed from the set and the remaining set is submitted to the same procedure to extract the second plane. The iterations continue until the remaining points are small in comparison to the initial point cloud size. The result is a more accurate and robust point cloud segmentation that can be used for further analysis and processing.

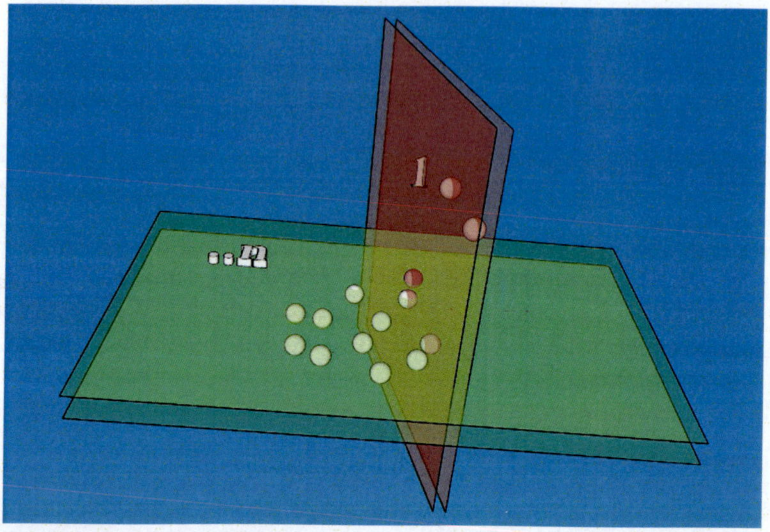

Fig. 2. When RANSAC procedure is started a random 3 points are selected and a plane with thresholds is build on them (1 on the figure). Then the points entering in the threshold range are counted. The procedure repeat with another 3 points until a reasonable amount of points fits in the threshold (..n)

3.5 Finding Cloud Borders with Alpha Shape

Having obtained list of planar segments with respective points one can apply a 2D Alpha Shape algorithm [9, 10] (Fig. 3).

Fig. 3. All planes (left), 2D processed alpha shape on first plane in red (right) (Please Note that outer bounds of the processing area are edges of camera viewing range and are labeled so to be excluded from the final results (fake edges))

Alpha Shape is a geometric algorithm that calculates a shape around a set of points in a way that captures the spatial structure of the points. Specifically, the algorithm constructs a network of Delaunay triangles from the points [17], and then removes triangles whose circumcircles have a radius larger than a certain threshold called the alpha value. This creates a shape that fits tightly around the points while still maintaining their spatial structure.

In practice, the Alpha Shape algorithm can be used to find the borders of a point cloud by setting an appropriate alpha value. Points that fall outside of the shape defined by the alpha value are considered to be outside the borders of the cloud, while points inside the shape are considered to be part of the cloud.

The Alpha Shape algorithm has several advantages over other methods for finding cloud borders. First, it is computationally efficient and can handle large point clouds. Second, it is robust to noise and outliers, as it relies on the spatial structure of the points rather than their exact positions. Finally, it can capture the shape of irregularly shaped point clouds, such as those generated by scanning complex surfaces or objects.

3.6 Linear RANSAC Segmentation of the Borders

The next step in our pipeline is a linear RANSAC approximation of shape to smooth the nonimportant deviations obtained by the alpha shape algorithm. It is convenient to apply (or not) the linear aggregation which is actually controlled with defined tolerance in metric values. So we can define precision of aggregation e.g. 200 microns. In case we target a very precise path motion this step can be omitted.

3.7 Aggregation of the Segments to Welding Path

Following this pipeline idea defined up to now the last step of the process would need a identification of the position of the plate borders which needs to be weld. This is done by doing some detailed analyzes steps on the already significantly reduced volume of information. On first step an aggregation of segments which have a end segments which are near to each other are aggregated or broken. After that step a validation for segments proximity is done. If two segments are near to each other the combine segment is considered as welding path (Fig. 4).

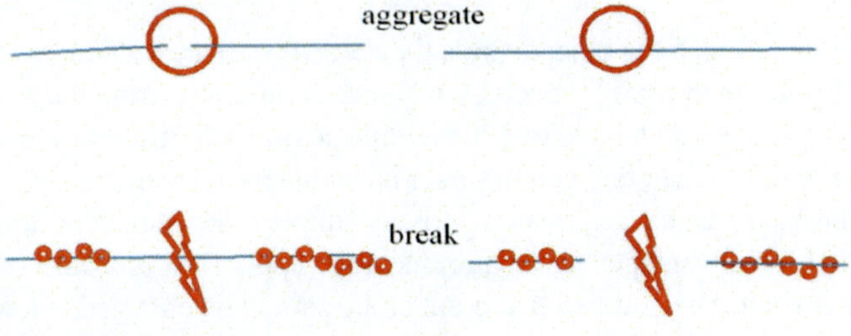

Fig. 4. Illustration of aggregation and breaking of linear segments based on the distance between points of the respective segments.

4 Results

The example point cloud shown here are obtained with Yaskawa GP24 robot with YRC 1000 controller, SkyBlue SB0120 system on custom test plate and some real life objects.

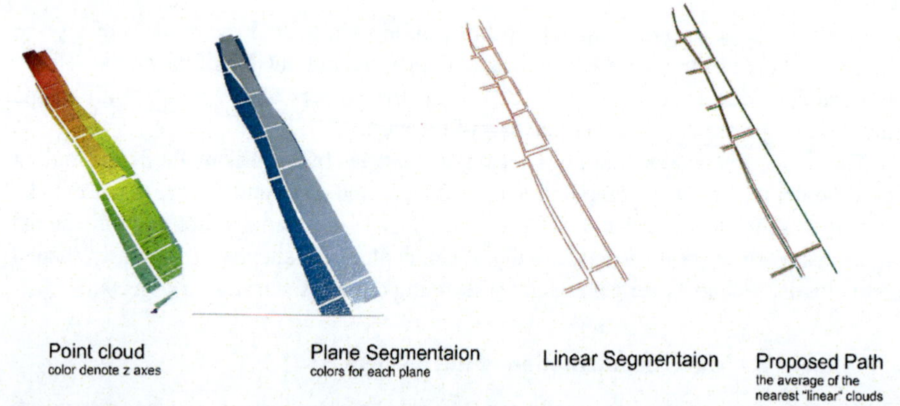

Fig. 5. The steps of the algorithm are illustrated below – Real world work piece with curved shape for welding

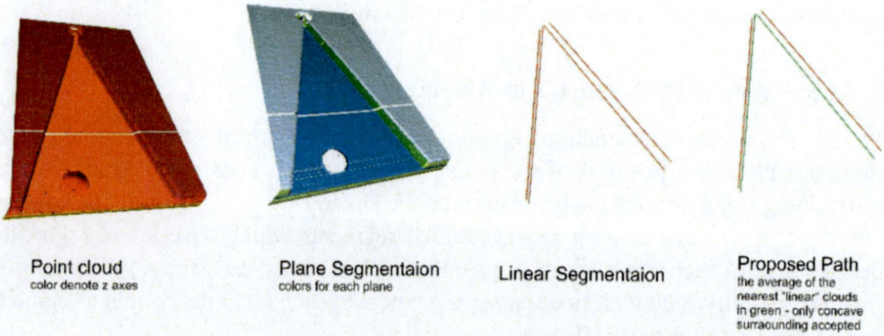

Fig. 6. A calibration plate scanned for welding path finding

As it is visible in Fig. 5 the object where the vertical wall are seen the welding paths are found well (the green lines) especially when the concave surrounding criteria was enforced. This problem can be avoided if multiple scans with different scanning angle are combined together. Than the welding path is visible even when the sight view does not show It. The used here examples are relatively simple cases of path finding, but even on them some of the advantages and requirements to apply the procedure can be seen. As a first point it should be outlined that point cloud should be obtained in a way that all joints should be well scanned. In general the proposed algorithm should be considered as a tool helping the operator not as a tool generating without intervention a welding program. This observation can be considered true for most of the existing algorithms as well (Fig. 6).

We propose a new approach for automated path planning in robotic welding using point cloud data obtained from laser seam tracking. Compared to traditional methods, our approach is lightweight and can be integrated into the controller of a seam tracker without requiring additional hardware or an offline workspace. It allows for fast generation of welding paths from point cloud data, which can be processed on the spot with the live system, providing immediate access to a programmed path. Our approach also utilizes hierarchical segmentation techniques, such as 3D and 2D RANSAC, Alpha Shapes, and linear RANSAC segmentation, which helps to filter out noise and outliers and gradually simplify the problem, resulting in more accurate and robust point cloud segmentation that can be used for further analysis and processing. The accuracy and robustness of our approach make it a valuable addition to the existing welding automation tools, particularly for small series or unique workpieces.

5 Discussion

In this work the main motivation of the authors is to provide a simple approach for path planning which can simplify the work of the operator. With the rise of hyperautomation and Industry 4.0, there is a growing demand for more flexible and cost-effective solutions in manufacturing, particularly for small series or unique workpieces [6]. The proposed approach for automated path planning in robotic welding using point cloud data obtained from laser seam tracking can be a valuable addition to the existing welding automation tools.

An important advantage of this approach is its lightweight implementation, which can be integrated into the controller of a seam tracker without requiring additional hardware or an offline workspace. This can reduce the cost of programming and improve the efficiency of the welding process, particularly for small series or unique workpieces. The proposed algorithm allows for fast generation of welding paths from point cloud data, which can be processed on the spot with the live system, providing immediate access to a programmed path. It represent a good example of an enhance collaboration between humans and robots in the welding process. It simplify operator tasks and increase the efficiency of the operator but do not replaces him and is compliant in that sense to the new paradigms of Industry 5.

Another advantage of this approach is its accuracy and robustness. By using laser seam tracking to collect point cloud data, the proposed method can generate welding paths that are more precise and efficient compared to traditional methods. The use of hierarchical segmentation techniques, such as 3D and 2D RANSAC, Alpha Shapes, and linear RANSAC segmentation, helps to filter out noise and outliers, and gradually simplify the problem, resulting in more accurate and robust point cloud segmentation that can be used for further analysis and processing.

However, there are limitations to this approach that need to be addressed. For example, the proposed method relies on the accuracy of the laser seam tracking system to collect point cloud data, which can be affected by various factors such as environmental conditions, surface quality, and calibration errors.

The accuracy of the laser seam tracking sensor is critical for the success of the algorithm. Any noise or errors in the sensor readings which remains after the filtering and the plane segmentation can result in inaccurate border finding, which will in turn result in incorrect path planning.

The algorithm is specifically designed for robotic welding applications and may not be directly applicable to other domains or tasks.

The algorithm relies on several parameters such as the alpha value in the Alpha Shape algorithm, the distance threshold in the plane RANSAC segmentation, and the precision of linear aggregation. These parameters must be carefully tuned to achieve optimal results, and improper settings can lead to inaccurate path planning. For the time being an analyses of the robustness of the algorithm and parameter sensibility is not completely done.

The plane RANSAC segmentation iterates until a defined percentage of the initial cloud is transformed into planes. This gives on one site a more or less clear definition of the target of the segmentation – a given part of the cloud to be approximated with linear (plane) segments. On the other side it could be confusing for complex objects being scanned. It is important in that aspect to find some more reliable criteria like e.g. minimum plane size to be considered and/or eventually to consider the density of the points for decision of what to include in plane segmentation.

Depending on the shape and size of the workpiece, there may be areas that are not adequately covered by the laser seam tracking sensor, leading to incomplete point clouds and incomplete path planning. This defect is not detected and may result in unexpected behavior. The detection of such artifacts can be addressed by analyzing the point density and react on areas with degrading cases (Fig. 7).

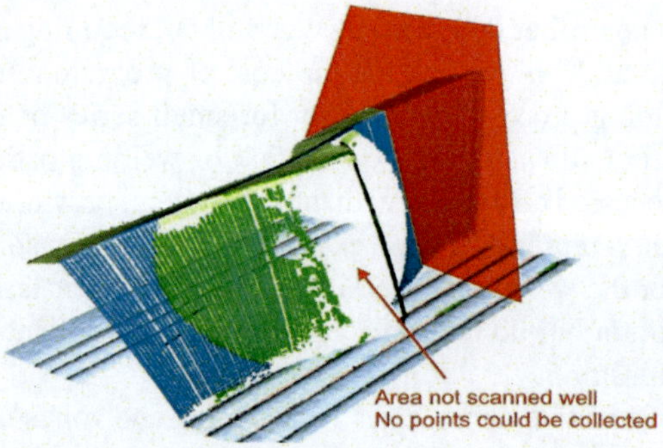

Fig. 7. An example of incorrect scanning angle. Due to not proper selection of scanning angle part of the workpiece which is important for the potential welding path.

Although the algorithm is designed to be lightweight, processing large point clouds can still be computationally expensive and may require significant processing power.

Robustness to environmental changes: The algorithm may be sensitive to changes in the environment such as changes in lighting or the presence of obstacles. These changes can affect the accuracy of the laser seam tracking sensor and lead to incorrect path planning.

While we found that Alpha Shapes and appropriate distance provided effective results for our path finding, we acknowledge that there may be limitations in certain scenarios.

For instance, complex geometries or surfaces with irregularities may pose challenges in accurately segmenting the point cloud data. To address these limitations, future research could explore the use of alternative segmentation techniques or adapt the current methods to better handle these scenarios. Additionally, optimizing the parameter settings for Alpha Shapes and distance thresholds could further improve the accuracy and robustness of the path planning results.

The algorithm focuses primarily on path extraction and segmentation and does not optimize the path for other factors such as welding deformation, obstacle avoidance, path length, and energy consumption. Additional optimization steps may be required to achieve optimal welding performance.

6 Conclusion

Collecting point cloud data using an existing hardware - Real time Laser seam tracking sensor add a significant value to the camera system. Having the processing executed on the spot with the live system provide fast and immediate access the a programmed path which makes.

To ensure the quality and accuracy of the trained models, one could implement a system for validating and verifying the data and models uploaded by users. This could include automated quality checks, manual review by experts, and benchmarking against existing datasets or models.

The proposed method provide a stable starting point for a simplified solution of automated human supervised path generation. Using this a starting point an extension with proper centralized convolutional neural network used by the customers to train the model could be developed.

A customer-centered community for training CNNs for point cloud data could provide a powerful resource for advancing the state of the art in robotic welding and other applications that rely on point cloud data. By harnessing the collective knowledge and expertise of a diverse group of users, the platform could facilitate rapid innovation and progress in this field.

In conclusion, the proposed approach for automated path planning in robotic welding using point cloud data obtained from laser seam tracking can be a valuable addition to the existing welding automation tools, particularly for small series or unique workpieces. The lightweight implementation, accuracy, and robustness of the proposed method can improve the efficiency and quality of the welding process while enhancing collaboration between humans and robots. Further research and development are needed to address the limitations of this approach and to explore its potential applications in the welding industry.

It is important to note that these potential problems are not unique to this algorithm and are common issues that arise in many robotic path planning and optimization algorithms. Therefore, it is essential to carefully evaluate the algorithm's performance and limitations before applying it in a practical setting.

Acknowledgment. *Author is expressing his special thanks to the team of Eiron LTD (Hungary) and Metrology LAB (Bulgaria) for the fruitful discussions and involvement on the subject of the proposed work.*

This Focused Technology Transfer Experiment has received funding from the European Union's Horizon 2020 research and innovation programme under the SMART4ALL grant agreement no 872614. .

References

1. Wang, X., Zhou, X., Xia, Z., Xingsheng, G.: A survey of welding robot intelligent path optimization. J. Manuf. Process. **63**, 14–23 (2021). https://doi.org/10.1016/j.jmapro.2020.04.085
2. Xu, C., Wang, J., Zhang, J., Lu, C.: A new welding path planning method based on point cloud and deep learning. In: 2020 IEEE 16th International Conference on Automation Science and Engineering (CASE), Hong Kong, China, pp. 786–791 (2020). https://doi.org/10.1109/CASE48305.2020.9216866
3. Hao, G., Deng, Z.: Path planning of scanning robot for tubular joint. In: 2006 6th World Congress on Intelligent Control and Automation, Dalian, China, pp. 8178–8182 (2006). https://doi.org/10.1109/WCICA.2006.1713568
4. Li, D., Xiao, P., Zhai, R., Sun, Y., Wenbin, H., Ji, W.: Path planning of welding robots based on developed whale optimization algorithm. In: 2021 6th International Conference on Control, Robotics and Cybernetics (CRC), pp. 101–105, Shanghai, China (2021). https://doi.org/10.1109/CRC52766.2021.9620117
5. Geng, Y., Zhang, Y., Tian, X., Shi, X., Wang, X., Cui, Y.: A method of welding path planning of steel mesh based on point cloud for welding robot. Int. J. Adv. Manuf. Technol. **116**(9–10), 2943–2957 (2021). https://doi.org/10.1007/s00170-021-07601-6
6. Gartner Forecasts Worldwide Hyperautomation-Enabling Software Market to Reach Nearly $600 Billion by 2022. https://www.gartner.com/en/newsroom/press-releases/2021-04-28-gartner-forecasts-worldwide-hyperautomation-enabling-software-market-to-reach-nearly-600-billion-by-2022. https://intellias.com/path-planning-for-autonomous-vehicles-with-hyperloop-option/. Accessed 9 Mar 2023
7. Introduction to Industrial Robot Path Planning. https://www.automate.org/blogs/introduction-to-industrial-robot-path-planning,last. Accessed 6 Mar 2023
8. Automated guided vehicle systems. https://www.kuka.com/en-us/products/mobility/automated-guided-vehicle-systems. Accessed 6 Mar 2023
9. Still, E.K., Schreiber, D.K., Wang, J., Hosemann, P.: Alpha shape analysis (ASA) framework for post- clustering property determination in atom probe tomographic data. Microscopy Microanal. **27**(2), 297–317 (2021)
10. Fischler, M.A., Bolles, R.C.: SRI international random sample consensus: a paradigm for model fitting with Apphcatlons to image analysis and automated cartography. Commun. ACM **24**(6) 381–395 (1981). https://doi.org/10.1145/358669.358692
11. Poux, F., Billen, R.: Voxel-based 3D point cloud semantic segmentation: unsupervised geometric and relationship featuring vs deep learning methods. ISPRS Int. J. Geo-Inf. **8**(5), 213 (2019). https://doi.org/10.3390/ijgi8050213
12. OptoWiki Knowledge Base The Knowledge Base for Technical Optics. https://www.optowiki.info/glossary/scheimpflug-principle/. Accessed 28 Feb 2023
13. Metrology-LAB SkyBlue system. https://sites.google.com/metrology-lab.com/metrology-lab/the-blue-series/skyblue/. Accessed 6 Mar 2023
14. Fischler, M.A., Bolles, R.C.: Random sample consensus: a paradigm for model fitting with applications to image analysis and automated cartography. Commun. ACM **24**(6), 381–395 (1981). https://doi.org/10.1145/358669.358692

15. Nurunnabi, A., West, G., Belton, D.: Outlier detection and robust normal-curvature estimation in mobile laser scanning 3D point cloud data. Pattern Recogn. **48**(4), 1404–1419 (2015). https://doi.org/10.1016/j.patcog.2014.10.014
16. Shewchuk, J.R.: Triangle: Engineering a 2D quality mesh generator and delaunay triangulator (1996). https://doi.org/10.1007/bfb0014497
17. Liu, G., Li, T., Peng, Y., Hou, X.: The ant algorithm for solving robot path planning problem. Paper presented at the Proceedings - 3rd International Conference on Information Technology and Applications, ICITA 2005, II, pp. 25–27 (2005)
18. Balan, K., Luo, C.: Optimal trajectory planning for multiple waypoint path planning using tabu search. Paper presented at the 2018 9th IEEE Annual Ubiquitous Computing, Electronics and Mobile Communication Conference, UEMCON 2018, pp. 497–501 (2018). https://doi.org/10.1109/UEMCON.2018.8796810
19. Yap, P.: Grid-based path-finding. In: Cohen, R., Spencer, B. (eds.) AI 2002. LNCS (LNAI), vol. 2338, pp. 44–55. Springer, Heidelberg (2002). https://doi.org/10.1007/3-540-47922-8_4
20. Maček, K., Becker, M., Siegwart, R.: Motion planning for car-like vehicles in dynamic urban scenarios. Paper presented at the IEEE International Conference on Intelligent Robots and Systems, pp. 4375–4380 (2006). https://doi.org/10.1109/IROS.2006
21. Zeng, N., Zhang, H., Chen, Y., Chen, B., Liu, Y.: Path planning for intelligent robot based on switching local evolutionary PSO algorithm. Assem. Autom. **36**(2), 120–126 (2016). https://doi.org/10.1108/AA-10-2015-079
22. Kennedy, J., Eberhart, R.: Particle swarm optimization. Paper presented at the IEEE International Conference on Neural Networks - Conference Proceedings, vol. 4, pp. 1942–1948. IEEE (1995)
23. Davoodi, M., Panahi, F., Mohades, A., Hashemi, S.N.: Multi-objective path planning in discrete space. Appl. Soft Comput. J. **13**(1), 709–720 (2013). https://doi.org/10.1016/j.asoc.2012.07.023
24. Itamura, Y., Tanaka, T., Kishino, F., Yachida, M.: 3-D path planning in a dynamic environment using an octree and an artificial potential field. Paper presented at the IEEE International Conference on Intelligent Robots and Systems, vol. 2, pp. 474–481 (1995)
25. Hornung, A., Dornbush, A., Likhachev, M., Bennewitz, M.: Anytime search-based footstep planning with suboptimality bounds. Paper presented at the IEEE-RAS International Conference on Humanoid Robots, pp. 674–679 (2012). https://doi.org/10.1109/HUMANOIDS.2012.6651592

Moving Object 3D Detection and Segmentation Using Optical Flow Clustering

Dmitriy Zhuravlev(✉)

T. Shevchenko National University, Kiev 01601, Ukraine
dzhuravlev@ukr.net
https://www.linkedin.com/in/dmitriy-viktorovich-zhuravlev/

Abstract. Deep learning methods have recently improved the performance and accuracy of prediction using big data and abundant computing resources. However, many complex problems in Computer Vision, including 3D detection, motion estimation, and image segmentation, cannot be easily solved. All the same, it is still possible to benefit from using 'traditional' methods.

This article analyzes the relationship between optical flow and object localization. The proposed algorithm works with moving objects and performs not only instance segmentation but also determines the type of objects and estimates 3D bounding boxes using geometric constraints. Since the method is based only on an algorithmic understanding of three-dimensional data, it does not require data and time for the training process and can be deployed on constrained devices.

Keywords: 3D object detection · instance segmentation

1 Introduction

Moving object detection and segmentation is an important task in many *Computer Vision* (CV) applications such as surveillance, autonomous driving, and robotics. The ability to accurately detect and segment moving objects is crucial for tasks such as tracking, path planning, obstacle avoidance, and behavior analysis. Traditional 2D object detection methods have limitations when dealing with occlusions, dynamic backgrounds, and changes in lighting and appearance. 3D object localization can provide more accurate and robust results, but it is a challenging task that often requires the integration of multiple sensing modalities. Developing accurate and efficient methods for moving object 3D detection and segmentation can have a significant impact on various domains including transportation, security, and entertainment.

Especially important is the ability to determine the location without the use of additional equipment [1,2] but only with the help of a monocular camera due to its cheapness. While there are limitations to using such cameras, including limited depth perception, researchers are actively developing new algorithms and

R. Silhavy and P. Silhavy (Eds.): CSOC 2023, LNNS 724, pp. 426–443, 2023.
https://doi.org/10.1007/978-3-031-35314-7_38

techniques to address these challenges and improve the accuracy and robustness of 3D object detection using monocular cameras.

Nowadays, deep learning methods have shown great promise in recent years for 3D object detection, particularly with the advent of new 3D sensing technologies and datasets. These methods [3,4] typically rely on convolutional neural networks (CNNs) and other deep learning architectures to automatically learn features and representations from the data. However, deep learning methods may require a large amount of labeled data to train, and can be computationally intensive and difficult to optimize. Also, in the end, it is hard to guarantee that the trained model will perform equally well with any data.

On the contrary, methods [5,6] based on classical CV use mathematical operations on extracted features and geometrical constraints to minimize prediction errors. These methods can be effective in certain scenarios, such as when the objects have well-defined geometric shapes and the scene is well-structured. This gives confidence that the system will behave the same with any data, and does not require training time and data. Also, it remains available for deployment on low-cost microcontrollers.

We will describe the relationship between object localization and movement to provide a joint method for 3D detection and instance segmentation. The main contribution of this paper can be summarized as follows:

- a method for the estimation of the object orientation by using Optical Flow clustering;
- an approach for the determination of an object type from the corresponding instance segmentation;
- a technique for the construction of a 3D box based on the object orientation and type;
- an algorithm for instance segmentation based on 3D localization;
- a design of the end-to-end baseline architecture that provides good performance with high accuracy;
- open-sourced PoC implementation, available for download at https://github.com/DmitriyZhuravlev/3dim-optical-flow

The rest of this paper is organized as follows. In the next section, a review of popular methods of segmentation and object detection is given. Section 3 introduces a set of necessary concepts that will be used later in our method from Sect. 4. The results of real-world testing with a concluding summary are presented in Sect. 5.

2 Related Studies

In this section, we discuss the most relevant work on object detection and instance segmentation. Due to the fact that the proposed method uses both of them inseparably, we first address previous work in the area of motion segmentation. Subsequently, we attempt to summarize the most relevant work on existing methods for 3D localization that intend to recreate the depth information from an RGB image.

However, in our work, these methods complement each other, so the goals and methods of the previous studies are principally different from the scope of this paper. This paper is assumed the natural continuation of the previous [6] with removing 2D detector dependency.

2.1 Video Segmentation

Semantic object segmentation is a crucial first step in video understanding applications for video search and retrieval purposes. Therefore, finding reliable solutions has become an active research problem. While segmentation methods have made significant progress in recent years, there are still several limitations and challenges that need to be addressed to improve the accuracy, efficiency, and generalizability of segmentation models.

Methods for segmenting video objects are aimed at automatic object extraction from video. These methods use features such as clustering of point trajectories [7,8], motion characteristics [12], appearance [9,10] to achieve object segmentation. Recently, in [14], the main object was separated from the background in the video based on the scheme of variable convex optimization. Also, in the article [15] end-to-end learning structure combining motion and appearance information is proposed in order to create binary segmentation by pixels for each frame.

Standard approaches to the problem include trajectory analysis and appearance-based methods. Several studies [11,13] have shown a significant improvement in the use of a stream of visual features for object segmentation. Previous research has shown that a simple clustering technique can be used for semantic segmentation. However, this requires human intervention to estimate the amount of clusters by watching a short video clip.

Since semantic segmentation and motion segmentation do not separate instances of objects, a segment can consist of several objects. In order to extract individual objects, methods use a pre-trained object detection model [23] as its backbone.

The above methods either cannot separate one object from another or use object detection for this, whereas the proposed method uses its own 3D estimation.

2.2 Monocular 3D Object Detection

This part of the article discusses the existing methods of 3D localization that intend to recreate the depth information from a monocular image.

Like most of the works related to 3D object detection from a single image, [5] proposed a method for regression of an observation angle and 3D object sizes from the cropped image patch enclosed by 2D detection box. Using the regressed dimensions and orientations of the 3D box and 2D detection box, they try to minimize the re-projection error related to the initial 2D detection box constraints. Since each side of the 2D detection box can correspond to any of the eight corners of the 3D box, there are 8^4 configurations.

In the [16] the number of plausible configurations is reduced to 64. These configurations go through equations to find the vertices of a cuboid and the resulting solutions are ranked by approximation error or intersection over union (IoU) score between the tightest bounding box for the projected vertices of the fitted cuboid and the 2D detection rectangle.

To simplify object distance measurement, the Bird's Eye View (BEV) based architecture is proposed in [17] for vehicle detection with uncalibrated traffic cameras. Since all traffic participants are in contact with the ground, the distance can be estimated as the detection of points of contact of the vehicle with the road. The proposed method involves two stages: motion cancellation and the detection of position, size, and orientation using convolutional neural networks. In this work, two types of detectors for rotated box prediction on BEV are proposed. The one-stage detectors have the advantage of fast operating speed because of the direct prediction of the bounding box from the feature map grid without refinement. On the other hand, two-stage detectors predict the position and size of the object more precisely since the network predicts the box proposal and performs refinement in two steps. However, such defectors operate slower.

[18] extends the previous approach for uncalibrated traffic cameras using dual-view architecture. In this architecture, both the original view and BEV images are taken as input. The original view feature maps are warped to BEV and concatenated with BEV feature maps. The dual-view structure enables feature learning before warping, where the object shapes are regular and the knowledge propagation is easier, but it still has the two-stage detectors issues.

In [19] the idea of learning moving object detection directly in BEV space is explored. The architecture based on two-stream RGB and optical flow fusion is proposed for object localization. However, the lengths of some objects are not captured perfectly and some of the objects are misclassified as false static objects, which shows that there is still room for improvement.

All the mentioned methods study 3D object localization using different neural networks which not only get a 2D box but also extract additional information from the image to predict the distance and orientation. However, the geometric constraints of moving objects are not explored thoroughly enough. In this work, we identify real-world scenarios where high reconstruction accuracy is guaranteed via geometric analysis and apply our algorithm to localize 3D objects in real-time.

3 Background

3.1 Camera Model

The camera model is a basis for developing a computer vision application. We use the Pinhole model with no distortion factor to describe the projection of 3D points onto an image plane.

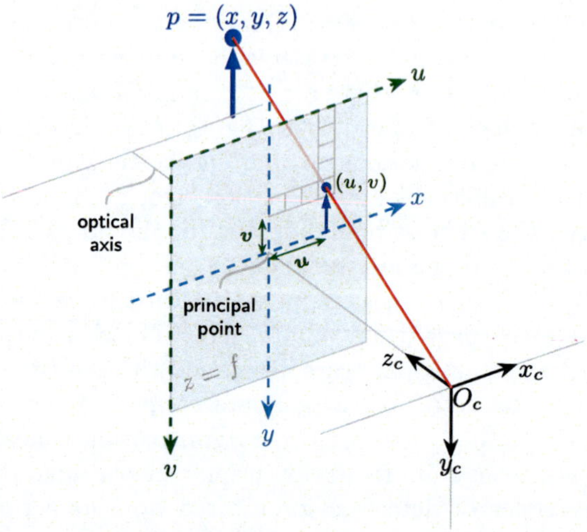

Fig. 1. Pinhole camera model

In this model, the image plane is positioned a fixed distance behind the camera's optical center as shown in Fig. 1. The relationship between a point in 3D space and its projection onto the image plane is defined by a set of equations, which take into account the focal length of the camera lens, the position of the point relative to the camera's optical center, and the orientation of the camera relative to the scene.

Let us describe the correspondences between the camera, world, and picture coordinate systems. Let $p = (x_1^w, x_2^w, x_3^w)$ be a 3D point in the world coordinate system. It can be transformed into camera coordinates using the *extrinsic matrix* which consists of the rotation $R \in \mathbb{SO}(3)$ and translation $t \in \mathbb{R}^3$ between the two coordinate systems:

$$\begin{bmatrix} x_1^c \\ x_2^c \\ x_3^c \end{bmatrix} = R \begin{bmatrix} x_1^w \\ x_2^w \\ x_3^w \end{bmatrix} + t.$$

The new 3D point in the camera coordinate system is projected onto the image plane using the *intrinsic matrix* K which consists of internal camera parameters like the focal length, optical center, etc. :

$$\begin{bmatrix} u' \\ v' \\ w' \end{bmatrix} = K \begin{bmatrix} x_1^c \\ x_2^c \\ x_3^c \end{bmatrix},$$

where $u = \frac{u'}{w'}$, $v = \frac{v'}{w'}$.

The overall projection in homogeneous coordinates can be expressed by:

$$\begin{bmatrix} u' \\ v' \\ w' \end{bmatrix} = K \left(R \begin{bmatrix} x_1^w \\ x_2^w \\ x_3^w \end{bmatrix} + t \right) .$$

3.2 Optical Flow

Optical Flow (OF) is a technique used to estimate the motion of objects in a video sequence. It is based on the assumption that pixels in an image sequence move smoothly over time and that their motion can be represented as a 2D vector field.

The OF algorithm works by analyzing the brightness pattern of pixels in successive frames of a video sequence. The algorithm assumes that the intensity of a pixel should remain constant over time, so any changes in intensity must be due to motion: $I(p,t) = I(p+u,t+1)$ where $I(p,t)$ is image intensity as a function of space $x = (x_1, x_2)$ and time t, and $u = (u_1, u_2)$ is the 2D velocity.

Several different approaches to motion estimation have been proposed. For example, the Horn-Schunk method [20]. It assumes that the optical flow is smooth over the entire image and computes an estimate of the velocity field by an equation minimization:

$$E = \iint \left(I_{x_1}u_1+I_{x_2}u_2+I_t\right)^2 dx_1dx_2+\alpha \iint \left(\frac{\partial u_1}{\partial x_1}^2+\frac{\partial u_1}{\partial x_2}^2+\frac{\partial u_2}{\partial x_1}^2+\frac{\partial u_2}{\partial x_2}^2\right) dx_1dx_2$$

where α is the smoothness term of the velocity field and $\frac{\partial u_1}{\partial x_1}$ and $\frac{\partial u_2}{\partial x_1}$ are the partial derivatives of the optical velocity component u. The α regularization parameter controls the rate of the smoothness constraint.

3.3 Segmentation

Image segmentation is the process of dividing an image into its constituent areas or objects. Subdivision-level images depend on the problem being solved. The segmentation process can stop when the objects of interest are isolated for further processing.

The result of image segmentation is a set of segments that collectively cover the entire image or a set of edges extracted from an image. Each of the pixels in an area has some characteristic or computed property such as color, intensity, or texture.

Let the function $g(x,y) \in \mathbb{R}^3$ represent an RGB color at any point (x,y) of domain $\Omega \in \mathbb{R}^2$ where an image is captured. The aim is to find regions $\Omega_1, \Omega_2, \ldots, \Omega_n$ for $n \in N$, of a domain $\Omega \in \mathbb{R}^2$, corresponding to different n objects or their parts or shadows in the image.

Segmentation can be achieved by several methods, which include thresholding-based, edge-based, region-based, and clustering techniques.

3.4 Clustering

Clustering is a data analysis technique for separating a data set into subgroups of related data. K-means [21] is one of the most widely used unsupervised clustering methods.

In the clustering problem, it is required to group a set $\{x_1, \ldots, x_m\}$ into several connected *clusters*. Here are the feature vectors for each data point $x_i \in \mathbb{R}^n$ but without the labels (making this an unsupervised learning problem). The goal of the algorithm is to predict k centroids and label c_i for each data point.

The classical algorithm can be expressed as an optimization problem:

$$\mathcal{U}^* = \underset{U \in \mathcal{P}}{\operatorname{argmin}} \max_{c_k \in \mathcal{X}} \sum_{k=1}^{C} \sum_{i=1}^{n} U_{k,i} ||c_k - x_i||_d^2$$

for some distance $|| \cdot ||_d$, membership matrix U (which assigns clusters to each x, c_k is the k-th cluster center, and x_j is the j-th data point. Principally, $U_{i,k}$ is 1 if x_i is in the k-th cluster and 0 otherwise. The summations compute the total distance of each data point to its cluster center.

Let us denote this algorithm by KMEANS for further use in the proposed algorithm.

3.5 Inverse Perspective Mapping

Inverse Perspective Mapping (IPM) is a technique used to convert an image captured from a perspective camera to a *Bird's Eye View* (BEV) representation. This is achieved by estimating the homography matrix that maps points in the original image to their corresponding locations in the BEV image. The example of the IPM is shown in Fig. 2.

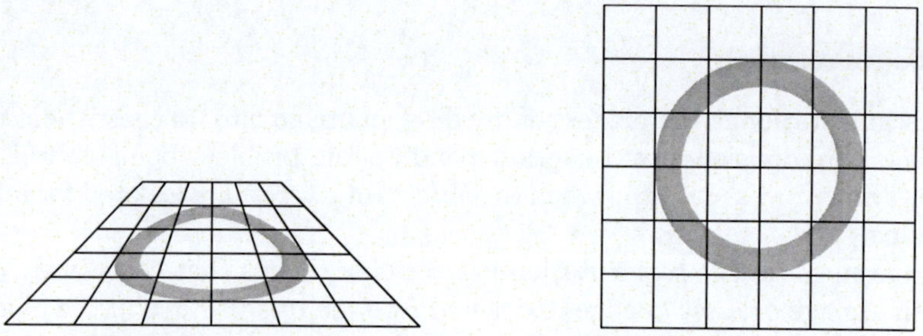

(a) Perspective view. (b) Top-down view.

Fig. 2. BEV image generated by IPM.

In general, the perspective transformation can be expressed as

$$p' = Gp, \tag{1}$$

where G can be expressed in terms of the intrinsic matrix K and the Euclidean *homography* matrix $H \in \mathbb{SL}(3)$ related to a planar viewed object. It should be noted that H depends on the rotation matrix R and the translation vector t corresponds to the rigid transformation between the real camera pose image and the generated one.

IPM can be conveniently computed automatically [25,26] or constructed using scene measurements and satellite images from publicly available map services (e.g. *Google Maps*).

3.6 Object Detection

2D object detection is a common CV technique for identifying and locating objects of certain classes in the image. A 2D bounding box (rectangle) is used to describe the location of an object. It is determined by the rectangle center $c = (c_1, c_2)$, dimension $d = (d_1, d_2)$ with the width and height of the rectangle and type of the object $t \in \{T\}$.

(a) 2D detection. (b) 3D detection.

Fig. 3. Vehicle detection variants.

In the case of *3D detection*, the bounding cuboid (or rectangular parallelepiped) is described by its center $c = (c_1, c_2, c_3)$, dimensions $d = (d_1, d_2, d_3)$, and orientation $\varphi = (\varphi_1, \varphi_2, \varphi_3)$ parameterized by elevation, azimuth and roll angles.

4 Joint Method for 3D Detection and Instance Segmentation

Instance segmentation and 3D object detection are related because they both involve identifying and localizing objects in a scene. By combining instance segmentation with 3D object detection, it is possible to obtain a more comprehensive understanding of the scene, including the shape and orientation of each individual object instance in 3D space. The basic idea of the proposed method is to find the best possible segmentation and related 3D bounding box, using the moving direction of the segment and the overage dimensions of the object.

Building a 3D box not only allows you to localize an object but, based on its dimensions, determine the most appropriate object type from the list of possibilities. We will use this observation after we construct a 2D bounding box around the segment.

4.1 Model Assumptions

Our main task is to detect vehicles on the road and in parking lots, but it is not limited to this application. Several assumptions have been made in order to build a mathematical model. We are most interested in the planar position of the objects and assume a vehicle is always on flat ground. It allows us to assume that only the azimuth (noted φ_2) matters and thus we do not predict the orientation φ characterized by elevation and roll, and we set $\varphi_1 = 0$ and $\varphi_3 = 0$.

The camera position must be stationary. In another case, object moving vectors obtained from OF have to be recalculated. The vehicle being detected must be attached to the plane of the ground because only the points on the ground have the distance information.

In the context of our model, we consider that the shape of the objects is close to a cuboid and also that the component parts of the object move as a whole. We use a cuboid approximation of a 3D object which coincides with the definition of 3D detection as shown in Fig. 3. We assume that 3D detection perfectly fits into 2D detection. Therefore, at least one vertex of the bottom box lies on the lower side of the 2D detection rectangle. Accordingly, we can see a similar situation in BEV as shown in Fig. 5.

Further, we will show how to find the projection of the lower face of the cuboid in BEV. This makes it possible to determine the center of the cuboid with known dimensions. It should be noted that the orientation of the object is determined according to the moving direction of the object and is set up to the opposite since we do not geometrically distinguish between the front and back of a vehicle, which is appropriate with cars as they can go both forwards and backwards but not sidewards.

4.2 HSV Color Quantization

The color and shape features of an image are the most widely used features to analyze it. The proposed method integrates the HSV (hue, saturation, value) color representation of the dense OF.

For this, we visualize the angle (direction) of flow by hue h and the distance (magnitude) of flow by value v with fixed saturation s and then group similar colors together as shown in Fig. 4. After applying KMEANS to the image Ω in HSV color space, the image will be divided into regions with fixed color values $\overline{\Omega} = \{\Omega_1 = (h_1, s, v_1), \ldots, \Omega_k = (h_k, s, v_k)\}$.

(a) Original video frame. (b) Dense OF visualization in HSV space.

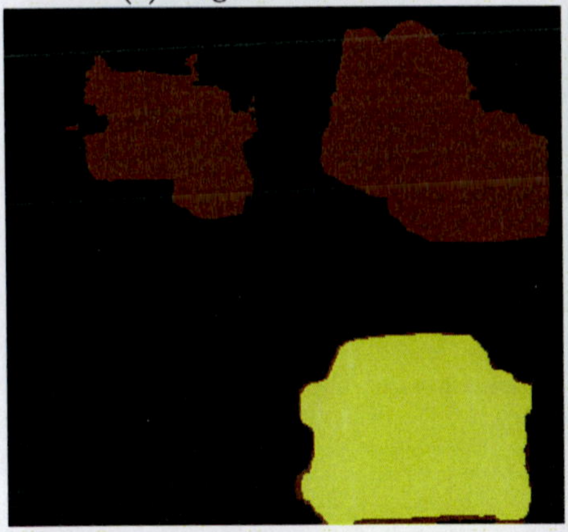

(c) $k = 3$ - the picture is under-segmented. (d) $k = 32$ - the picture is over-segmented.

Fig. 4. Example of applying k-means clustering to dense OF.

After applying KMEANS, each pixel in the image Ω is replaced with the color value of the nearest cluster centroid $\overline{\Omega} = \{\Omega_1 = (h_1, s, v_1), \ldots, \Omega_k = (h_k, s, v_k)\}$.

The parameter k is set so that different objects are separated (if excessive segmentation is allowed, such a value can always be chosen a-priori) as illustrated in Fig. 4. This is rarely sufficient for image segmentation, because there is no control over the shape of the segments. Further, we will build an iterative algorithm and a sufficiency measure that will merge the segments into one until the goal is reached.

4.3 Lifting 2D Object Detection to 3D

All calculations that are given in this section will refer to calculations in BEV unless it is stated otherwise. The domain of an IPM is an image with a 2D bounding box.

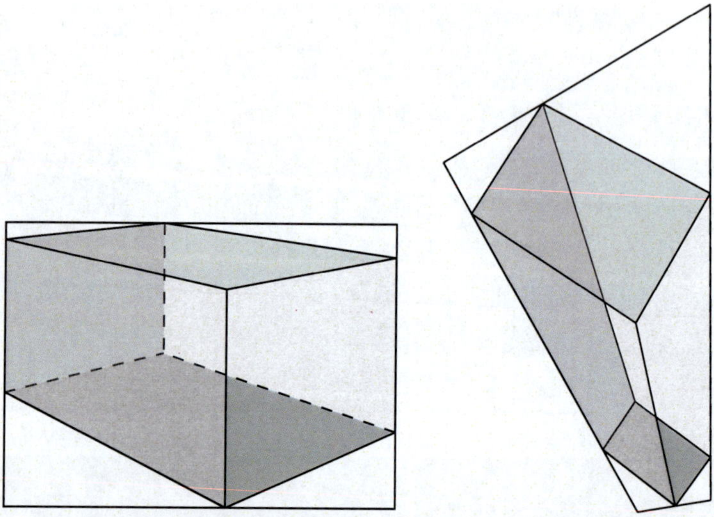

(a) 2D detection of a cuboid. (b) IPM transformation.

Fig. 5. IPM of the 2D detection.

Let the IPM image of a bounding rectangle be a non-degenerate quadrangle A, R, T, B with an inner rectangle where three vertices lie on adjacent sides. Moreover, one of the vertices of the rectangle lies on the bottom side of the outer quadrangle. Our goal is to find touchpoints K, C, D. Further, we will refer to such quadrangles as *bounding* ones.

Let the rectangle be built on points K, C, and D with fixed sizes $KC = a$ and $CD = b$ as shown in Fig. 6. Let us draw a ray from B with the angle equal to the orientation. We denote E the intersection point of this ray with segment AR, and $\angle ABE = \varphi$, and let $EB = l$. According to the construction $\triangle KAC \backsim \triangle EAB$ and from the similarity we have $\dfrac{AC}{AB} = \dfrac{KC}{EB} = \dfrac{a}{l}$.

Hereof $AC = \dfrac{a}{l}(AC + CB)$ and $AC = \dfrac{a}{l-a}CB$.

Consequently $C = (C_x, C_y) = \left(\dfrac{(l-a)A_x + aB_x}{l}, \dfrac{(l-a)A_y + aB_y}{l}\right)$, which makes it possible to recover the rectangle.

A rectangle with one dimension and orientation known can be reconstructed from a bounding rectangle.

Since we only used one of the sides to calculate the taught-point C, the second side of the rectangle is calculated from Fig. 6 as the length of segment CD. Therefore, the expected value b may differ from the calculated one, which will be used further to determine the fitting error.

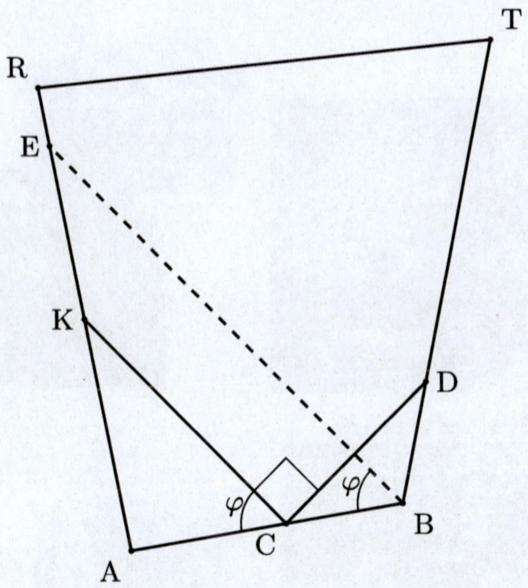

Fig. 6. Rectangle reconstruction by angle and one side.

Returning from the bottom projection in BEV to 3D, we calculate the height of the cuboid so that the top face touches the bounding quadrilateral. The detailed analysis of the lifting procedure (we will refer to it as LIFT) can be found in [6].

4.4 Putting It All Together

Now we are ready to describe the joint algorithm for 3D detection and instance segmentation.

First of all, let's calculate an optical flow $F = \{u = (u_1, u_2)\}$ for each pixel (x_1, x_2) in the video frame. We apply the KMEANS clustering algorithm for the initial segmentation $\overline{\Omega} = \{\Omega_1, \ldots, \Omega_n\}$ of the image Ω, where $\Omega_i \cap \Omega_j = \varnothing$ are non-intersecting segments. The essence of the Ω_i segment is the cluster with the closest angle and magnitude of pixels movement, as described in Fig. 8. Without loss of generality, we can assume that the set $\overline{\Omega}$ is sorted by area $\mathrm{S}(\Omega_{i+1}) \leq \mathrm{S}(\Omega_i)$.

Next, we will merge some segments so that they correspond to the best 3D localization – this will be the desired instance segmentation. To do this, we choose the largest segment $\hat{\Omega}$ and build a bounding rectangle for it. Using IPM we find its correspondence on BEV. Knowing the average size $d_t = (d_1, d_2, d_3)$ of the object and its orientation and type t, it is possible to determine the error e of fitting the cuboid with estimated center $\hat{c}$, sizes $\hat{d}$, and object type $\hat{t}$ into the bounding rectangle of the segment $\hat{\Omega}$ by error minimization:

$$\hat{t} = \underset{t \in T}{\operatorname{argmin}} \left(\left\| d_t - \hat{d} \right\| \right),$$

(a) Oversized 3D box.

(b) Incorrect segmentation.

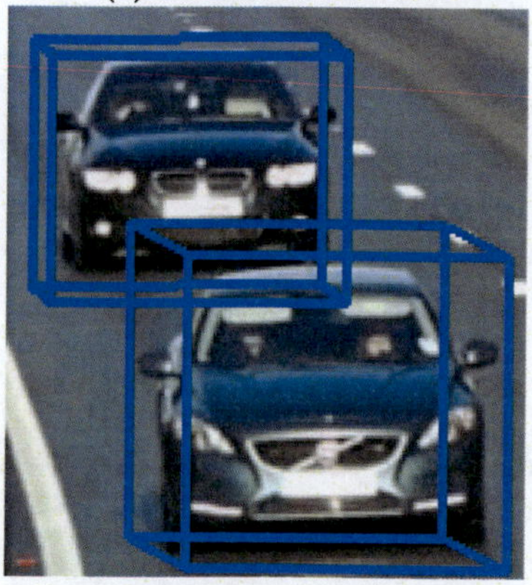

(c) Tight 3D box.

(d) Accurate segmentation.

Fig. 7. Relation between 3D estimation and segmentation.

where T is a set of object types, and $\hat{d}$ is computed dimension. It should be noted that in some cases it is impossible to fit the cuboid and therefore we put $e = \infty$.

Further, if the error e satisfies some small threshold $\mathcal{E}$, we consider that the object of type t is localized by the 3D box and segmented by $\hat{\Omega}$. Otherwise, we merge the next segment Ω_k that is the largest in size and closest to the cluster orientation neighboring segment of cluster $\hat{\Omega}$, and repeat the LIFT procedure for lifting the 2D bounding box to 3D. An example of the correspondence between 3D localization and instance segmentation can be found in Fig. 7.

In fact, we can even process objects of different types if their dimensions are different. To do this, we only choose the best segmentation and localization with the smallest embedding error as shown in Alg. 1.

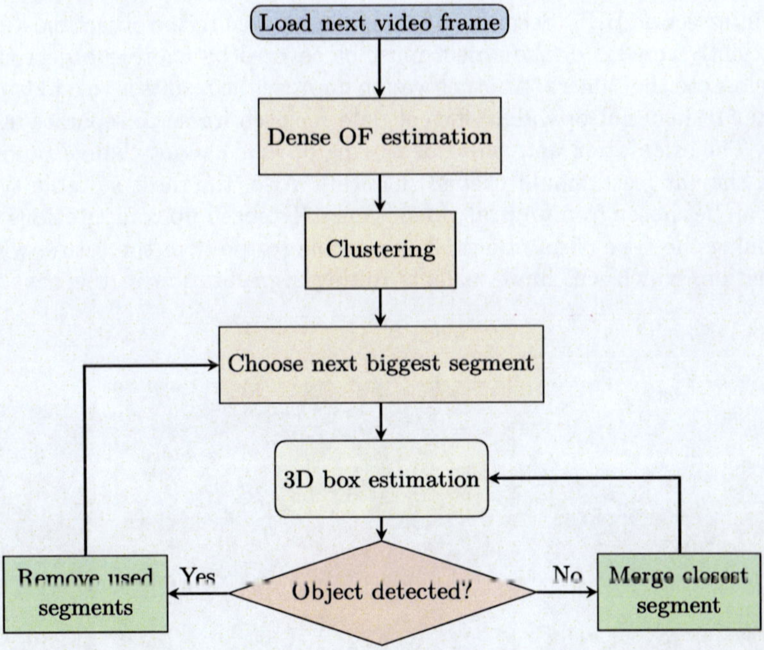

Fig. 8. Block diagram of the moving instance detection algorithm.

Algorithm 1. Moving instance detection.

Require: $\mathcal{V}$ ▷ video, OF, IPM

Initialize $\mathcal{V} \leftarrow \{f \mid \text{ordered set of video frames}\}$
for frame $f \in \mathcal{V}$ **do**
    $\Omega \leftarrow \{(u, v)\}$ ▷ compute flow field updates for each pixel
    $\overline{\Omega} \leftarrow \text{KMEANS}(\Omega) = \{\Omega_1, \ldots, \Omega_n\}$
    while $\overline{\Omega} \neq \varnothing$ **do**
        $\hat{\Omega} \leftarrow \varnothing$
        repeat
            $\Omega^{\star} \leftarrow \underset{\omega \in \overline{\Omega}}{\text{argmax}}\ \text{S}(\omega)$
            $\hat{\Omega} \leftarrow \hat{\Omega} \cup \Omega^{\star}$
            $(\hat{c}, \hat{d}, \hat{\varphi}) \leftarrow \text{LIFT}(\hat{\Omega})$
            $\hat{t} \leftarrow \underset{t \in T}{\text{argmin}} \left\| d_t - \hat{d} \right\|$
        until $\left\| d_{\hat{t}} - \hat{d} \right\| > \mathcal{E}$ (a small positive number)
        $\overline{\Omega} \leftarrow \overline{\Omega} \setminus \hat{\Omega}$
    end while
end for
return 3D boxes and related segments

An additional opportunity to avoid over-segmentation is the analysis of received bottom box projections on BEV. Firstly, the projections of the objects cannot intersect on BEV. Secondly, the segment related to the object must either be sufficiently convex, or the object must be covered by another object that is placed closer to the camera. Another way to improve the results is to add tracking since the OF may not provide sufficient data for each frame to separate moving objects. The orientation and speed of moving objects cannot change randomly, because they at least should change smoothly. Also, the most suitable type of object can be chosen from several video frames. To speed up computations, after determining the type of an object, it is possible to use it in the future without recalculation, which will limit variants during segmentation in the next video frames.

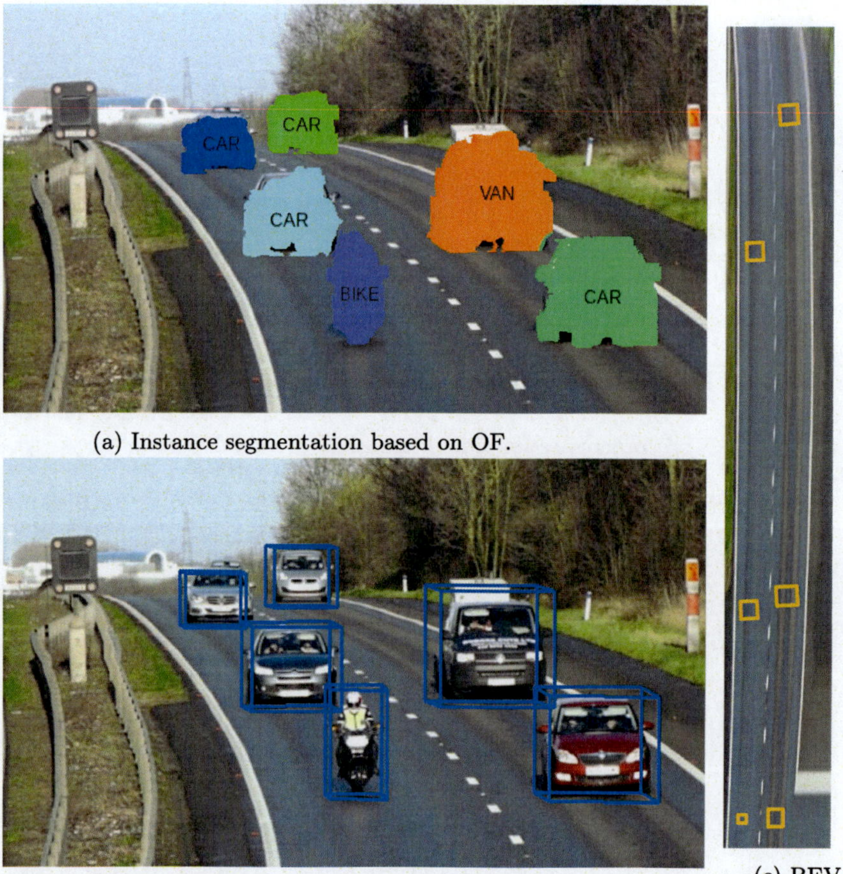

(a) Instance segmentation based on OF.

(b) 3D box estimation based on segmentation.

(c) BEV.

Fig. 9. The relation between segmentation (a), 3D detection (b), and corresponding projection on BEV (c).

5 Conclusion and Future Work

In this paper, a novel approach for joint monocular 3D object detection and instance segmentation is introduced. Despite the fact that it is based exclusively on classical methods of computer vision, it shows promising performance results. Summing up the description of the algorithm working sequence, the initial segmentation of the frame is obtained by using the clustering of the optical flow. Then, with the help of the reconstructed 3D information from the 2D bounding box of a segment, the proposed method merges the original segments into clusters that best fit the observed objects. With post-processing steps, we can eliminate most of the false alarms and false positives from segmentation and 3D localization.

Experiments have shown that the described method not only requires less time and computational resources than the deep-learning-based algorithms, which makes this method more applicable, but can also successfully handle overlapping objects in the image plane. It is also possible to determine the type of objects based on their average sizes as shown in Fig. 9.

It shows that traditional methods can not only be used to solve complex modern problems but can also be better suited to perform tasks under conditions of limited time and resources. In particular, the proposed method does not require training data and time. Its prediction error of 3D localization is fully described mathematically through the errors of the optical flow and deviations of the sizes of objects from average values. The method is general and can be used with any optical flow and clustering algorithms. The prospects for future studies of the non-static case with a moving camera will be considered in order to provide a 3D view for a moving vehicle.

References

1. Shi, S., Wang, X., Li, H.: PointRCNN: 3D object proposal generation and detection from point cloud. IEEE/CVF Conf. Comput. Vis. Pattern Recogn. (CVPR) **2019**, 770–779 (2019). https://doi.org/10.1109/CVPR.2019.00086
2. Wang Y., Zell A.: Yolo+FPN: 2D and 3D Fused Object Detection With an RGB-D Camera. In: 2020 25th International Conference on Pattern Recognition (ICPR), 2021, pp. 4657–4664. https://doi.org/10.1109/ICPR48806.2021.9413066
3. Redmon J., Farhadi A.: Yolov3: An incremental improvement. arXiv preprint arXiv:1804.02767, 2018
4. Fedorov, A., Nikolskaia, K., Ivanov, S., Shepelev, V., Minbaleev, A.: Traffic flow estimation with data from a video surveillance camera. J. Big Data **6**(1), 1–15 (2019)
5. Mousavian, A., Anguelov, D., Flynn, J., Košecká, J.: 3D bounding box estimation using deep learning and geometry. IEEE Conf. Comput. Vis. Patt. Recogn. (CVPR) **2017**, 5632–5640 (2017). https://doi.org/10.1109/CVPR.2017.597
6. Zhuravlev, D.: Lifting 2D Object Detection to 3D: Geometric Approach in Bird-Eye-View. In: Silhavy, R. (eds) Artificial Intelligence Trends in Systems. CSOC 2022. Lecture Notes in Networks and Systems, vol 502. Springer. https://doi.org/10.1007/978-3-031-09076-9_21

7. Brox, T., Malik, J.: Object segmentation by long term analysis of point trajectories. In: Daniilidis, K., Maragos, P., Paragios, N. (eds.) ECCV 2010. LNCS, vol. 6315, pp. 282–295. Springer, Heidelberg (2010). https://doi.org/10.1007/978-3-642-15555-0_21
8. Fragkiadaki, K., Zhang, G., Shi, J.: Video segmentation by tracing discontinuities in a trajectory embedding. In: Proceedings of IEEE Conference on Computer Vision Pattern Recognition, pp. 1846–1853 (2012)
9. Lee, Y., Kim, J., Grauman, K.: Key-segments for video object segmentation. In: Proceedings of IEEE International Conference Computer Vision, pp. 1995–2002 (2011)
10. Khoreva, A., Galasso, F., Hein, M., Schiele, B.: Classifier based graph construction for video segmentation. In: Procedings of IEEE Conference Computer Vision Pattern Recognition, pp. 951–960 (2015)
11. Banica, D., Agape, A., Ion, A., Sminchisescu, C.: Video object segmentation by salient segment chain composition. In: 2013 IEEE International Conference on Computer Vision Workshops (ICCVW), pp. 283–290 (2013)
12. Papazoglou A., Ferrari V.: Fast object segmentation in unconstrained video. In: 2013 IEEE International Conference on Computer Vision (ICCV), pp. 1777–1784 (2013)
13. Bandara A., Ranathunga L, Abdullah N. : Visual Feature Clustering Using Temporal, Color and Spatial Information. In: Electronics, Communications and Networks IV (pp. 677–681)
14. Jang, W., Lee, C., Kim, C.: Primary object segmentation in videos via alternate convex optimization of foreground and background distributions. In: Proceedings of IEEE Conference on Computer Vision Pattern Recognition, pp. 696–704 (2016)
15. Jain, S., Xiong, B., Grauman, K.: Fusionseg: Learning to combine motion and appearance for fully automatic segmention of generic objects in videos. In: Proceedings of the IEEE Conference on Computer Vision Pattern Recognition, pp. 3664–3673 (2017)
16. Fang, J., Zhou, L., Liu, G.: 3d bounding box estimation for autonomous vehicles by cascaded geometric constraints and depurated 2d detections using 3d results (2019). arXiv preprint arXiv:1909.01867
17. Zhu, M., Zhang, S., Zhong, Y., Lu, P., Peng, H., Lenneman, J.: Monocular 3D vehicle detection using uncalibrated traffic cameras through homography. IEEE/RSJ Int. Conf. Intell. Robots Syst.(IROS) **2021**, 3814–3821 (2021). https://doi.org/10.1109/IROS51168.2021.9636384
18. Kim, Y., Kum, D.: Deep learning based vehicle position and orientation estimation via inverse perspective mapping image. IEEE Intell. Veh. Symp. (IV) **2019**, 317–323 (2019). https://doi.org/10.1109/IVS.2019.8814050
19. Rashed H., Essam M., Mohamed, Ei Sallab A., Yogamani S.: BEV-MODNet: Monocular Camera based Bird's Eye View Moving Object Detection for Autonomous Driving. In: 2021 IEEE International Intelligent Transportation Systems Conference (ITSC), 2021, pp. 1503–1508. https://doi.org/10.1109/ITSC48978.2021.9564667
20. Kanawathi, J., Mokri, S., Ibrahim, N., Hussain, A., Mustafa M.: Motion detection using Horn Schunck algorithm and implementation. In: 2009 International Conference on Electrical Engineering and Informatics, Bangi, Malaysia, pp. 83–87 (2009). https://doi.org/10.1109/ICEEI.2009.5254812
21. Lloyd, S.P.: Least squares quantization in PCM. IEEE Trans. Inform. Theor. **28**(2), 129–137 (1982)

22. Hafiz, A.M., Bhat, G.M.: A survey on instance segmentation: state of the art. Int. J. Multimed. Info. Retr. **9**, 171–189 (2020). https://doi.org/10.1007/s13735-020-00195-x
23. Jain, S., Xiong, B., Grauman, K: Fusionseg: Learning to combine motion and appearance for fully automatic segmentation of generic objects in videos. In: IEEE Conference on Computer Vision and Pattern Recognition, pp. 2117–2126 (2017)
24. Hartley, R., Zisserman, A.: Multiple view geometry in computer vision. Cambridge University Press (2003)
25. Zhou J., Li, B.: Homography-based ground detection for a mobile robot platform using a single camera. In: Proceedings 2006 IEEE International Conference on Robotics and Automation, 2006. ICRA 2006, pp. 4100–4105 (2006.) https://doi.org/10.1109/ROBOT.2006.1642332
26. Jeong, J., Kim, A.: Adaptive Inverse Perspective Mapping for lane map generation with SLAM. In: 2016 13th International Conference on Ubiquitous Robots and Ambient Intelligence (URAI), 2016, pp. 38–41 (2016). https://doi.org/10.1109/URAI.2016.7734016

Fuzzy Inference Algorithm Using Databases

Mikhail Golosovskiy[1] and Alexey Bogomolov[2(✉)]

[1] St. Petersburg Federal Research Center of the Russian Academy of Sciences, 39, 14th Line Vasilyevsky Island, St. Petersburg 199178, Russia

[2] Federal Research Center "Computer Science and Control" of the Russian Academy of Sciences, 44 Building 2 Vavylova Str., Moscow 119333, Russia

a.v.bogomolov@gmail.com

Abstract. An algorithm for fuzzy logical inference is proposed for systems consisting of a large number of inference rules occupying a significant amount of random access memory. A distinctive feature of the algorithm is the organization of storage of a set of inference rules in a relational database. In order to reduce the number of rules involved in fuzzy inference, restrictive rules for the formation of fuzzy inference systems are proposed. At the same time, the classical fuzzy inference procedure is supplemented with mechanisms that determine the involvement of a relational database in the fuzzy inference procedure. The effectiveness of the proposed algorithm was tested using a computational experiment for systems of fuzzy inference of the Sugeno type of zero order. It is shown that the execution time of fuzzy inference using the developed algorithm practically does not depend on the number of inference rules used, since due to the introduced restrictions for each input variable, a set of one or two inference rules is always selected for calculation. Compared to the classical fuzzy inference algorithm, the calculation speed using the developed algorithm increases significantly, while there is no need to permanently store the entire set of inference rules in random access memory. This determines the practical value of the developed algorithm.

Keywords: fuzzy logic · Sugeno type systems · fuzzy inference · relational database · fuzzy inference algorithm

1 Introduction

Fuzzy inference systems (FIS) are widely used in various expert systems, including medical systems [1–3]. The use of such systems is due to the relative simplicity of the derivation and the interpretability of the result obtained.

Fuzzy inference systems can become large due to the exponential growth in the number of inference rules used. In the limiting case, the number of such rules is limited by N^M, where N is the number of input variables, M is the maximum number of membership functions in the input variable. The amount of memory occupied by the inference rules of such systems can reach tens and even hundreds of gigabytes. In this regard, it becomes economically inexpedient to store the rules for such systems in random access memory (RAM) [4–6].

R. Silhavy and P. Silhavy (Eds.): CSOC 2023, LNNS 724, pp. 444–451, 2023.
https://doi.org/10.1007/978-3-031-35314-7_39

In this case, it is proposed to use relational databases to store the inference rules of FIS.

2 Methods

The article discusses the fuzzy inference algorithm using rules storage in the database, which consists in the fact that in the table each column corresponds to one input or output variable. Each line in this case corresponds to a separate inference rule of the FIS. An example of the correspondence of rules to the rows of a database table is given in Table 1.

Table 1. An example of a set of rules and the corresponding table in the database for the Sugeno system (x_1–x_2 are the values of the input variable, *A* is the term set of the membership function)

Rule		*In*1	*In*2	*In*3	Out
If (x_1 *this* A_{11}) *and* (x_2 *this* A_{21}) *and* (x_3 *this* A_{31}) *then y this* 2.4	→	1	1	1	2.4
If (x_1 *this* A_{14}) *and* (x_2 *this* A_{21}) *and* (x_3 *this* A_{33}) *then y this* 1.2	→	4	1	3	1.2
If (x_1 *this* A_{12}) *and* (x_2 *this* A_{21}) *and* (x_3 *this* A_{31}) *then y this* 5.4	→	2	1	1	5.4
If (x_1 *this* A_{16}) *and* (x_2 *this* A_{25}) *and* (x_3 *this* A_{32}) *then y this* 6.7	→	6	5	2	6.7

The restrictions and assumptions imposed on the fuzzy inference system to ensure the possibility of using the algorithm are as follows:

1. For input variables, the cells store the number of the membership function used in the rule corresponding to the row of the table; for output variables, the cell stores the conclusion of the fuzzy inference rule. If the input variable is not involved in the inference rule, then the value 0 is stored in the corresponding cell. The numbering of membership functions for each input variable starts from 1. The output variable column is called out. It stores either the number of the membership function for a Mamdani-type system or the value of the conclusion (a real number) for a zero-order Sugeno-type system.

The classical fuzzy inference algorithm implies the use of all inference rules in the inference process [7–9]. In this case, in order to carry out the calculation from the database, it will be necessary to select all inference rules into RAM, the size of which can significantly exceed the size of the available RAM [10–13]. This necessitates the development of an algorithm that can reduce the number of rules required for the calculation and reduce the amount of rules retrieved from the database [14, 15].

2. The fuzzy inference system must be complete - for each point of the domain of definition of each input variable, there must be at least one membership function that is part of the inference rule.

3. Membership functions must be of limited scope, such as triangular or trapezoidal membership functions. For membership functions with an infinite domain of definition, for example, Gaussian ones, the domain of definition can be artificially limited by the domain of definition by setting restrictive rules. For example, there are two possibilities.

Rule 1: the domain of the membership function is the domain in which the value of the membership function becomes greater than the given threshold value e:

$$\mu(x) > e,$$

where $\mu(x)$ is the membership function. Rule 2: the domain of the membership function with initially infinite support ends at the point of the core of the membership function of the neighboring membership function. Illustrations of setting the domain of definition using the proposed rules are shown in Fig. 1.

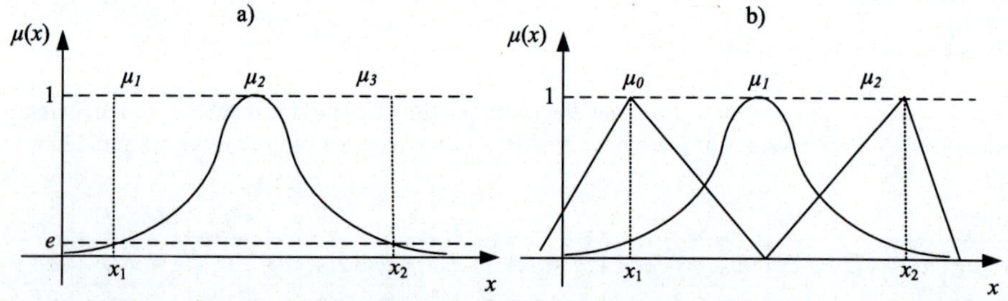

Fig. 1. Examples of limiting the domain of definition by points x_1 and x_2 for the Gaussian membership function μ_1 according to rule 1 – a) and rule 2 – b)

It is important that each membership function has clearly defined left and right boundaries of the domain of definition [7].

4. Each point of the domain of definition of the input variable can belong to no more than two domains of definition of membership functions. An example of partitioning is shown in Fig. 2

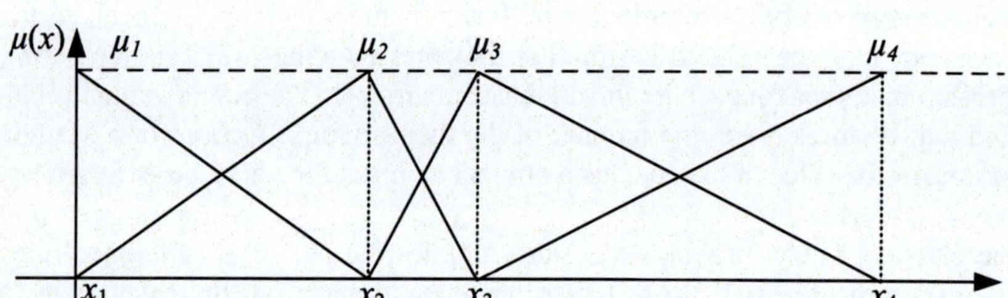

Fig. 2. An example of splitting the domain of definition of an input variable by membership functions μ_1–μ_4

5. In each inference rule, only one membership function can be used from each input variable.

3 Results and Discussion

The scheme of the proposed algorithm in UML notation is shown in Fig. 3. The essence of the description of the fuzzy inference algorithm is as follows:

Step 1. For input values, for each input variable, select membership functions whose values are not equal to zero. The selected values are stored in a special list, in which for each element the number of the input variable, the number of the membership function, and the value of the membership function are stored. A more sequence of selection steps is shown in the algorithm diagram (Fig. 3).

Step 2. Initialize the parameters of the sum of the rule weight (firing strength) triggering Ws and the sum of the products of the degree of rule triggering by the conclusions of the rules WBs equal to zero.

Step 3. Select Nb rows from the list of membership functions formed at step 1. The selected rows are removed from the list. It is done to limit the used RAM. If a programming language is used to implement a fuzzy inference system, in which it is possible to force free occupied RAM, then the memory allocated for storing data for processing rules in subsequent steps must be cleared after returning to step 3. The Nb parameter is set based on the size of the fuzzy system. Logical inference (number of input variables) and the size of the device on which the processing is performed. If there is no need to introduce a restriction, then it is possible to immediately select all the rules at step 4 in accordance with the list formed at step 1; for this, set Nb equal to the number of elements of this list. If the number of rows in the list is less than Nb, then all remaining rows are selected.

Step 4. From the table in the database, select rows in which each input variable has membership functions, with numbers in accordance with the list selected in step 1. In this case, a database query will be generated in SQL. An example query for a table with columns named In1, In2, In3, Out and table name RullesTable:

```
SELECT In1, In2, In3, Out
FROM RulesTable
WHERE In1 in (1, 2) AND In2 in (4, 5) AND In2 in (3,4);
```

Due to constraint 3, no more than 2 membership functions will be selected for each input variable at step 1, and the total number of selected rules will not exceed N^2, where N is the number of input variables, which gives a runtime gain compared to the classical algorithm, in which it is required to calculate the degree of operation for all inference rules.

Step 6. For the resulting string, form a inference rule - replace the numbers of membership functions with the values of the membership functions calculated in step 1.

Step 7. Calculate the rule firing strength w using formula 1.

$$w_k = t(\mu_{l_1^k}(x_1), \mu_{l_2^k}(x_2), .., \mu_{l_n^k}(x_n)), \tag{1}$$

where k is the number of the selected rule, n is the number of the input variable, l is the membership function number, t is the t-norm operator. The min operator is used as the t-norm further in the article. Add the calculated value to the sum of the levels of the rules Ws.

Step 8. Calculate the product of the degree of operation of the rule w calculated in step 6 and the conclusion of the rule b selected from the database. Add the result to the sum of WBs.

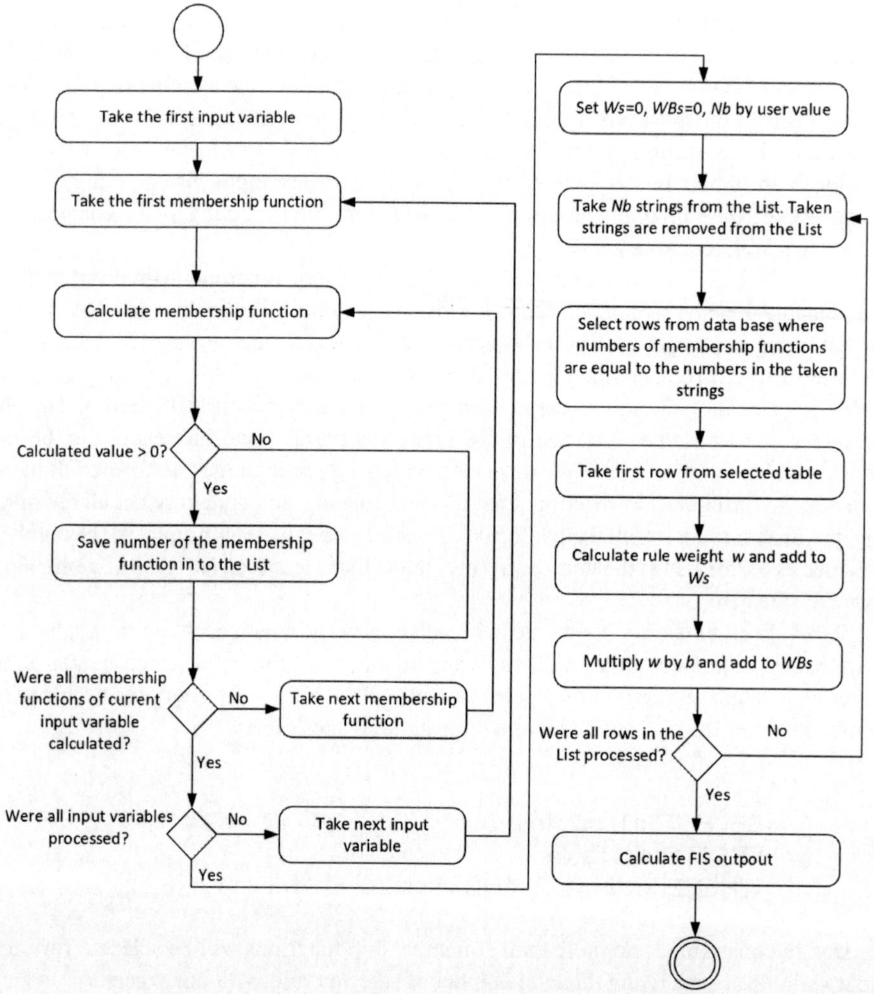

Fig. 3. Scheme of the proposed algorithm in UML notation

Step 9. Check if all lines selected in step 4 have been processed. If not, then take the next line and go to step 6. If all lines are processed, then go to step 10.

Step 10. Check if all lines from the list formed in step 1 have been processed. If no, then go to step 3, otherwise go to step 11.

Step 11. Calculate the resulting value of y using formula (2) for Sugeno-type systems:

$$y = \frac{WBs}{Ws}. \tag{2}$$

The effectiveness of the proposed algorithm was verified for systems of the Sugeno type of zero order.

We will compare the proposed algorithm with the classical Sugeno-type fuzzy inference algorithm. The comparison criterion will be the fuzzy inference time. Both algorithms are implemented in the Python 3 programming language, SQLite is used as a database for storage.

Execution time measurements were carried out for fuzzy inference systems with 8 input variables and the number of membership functions from 2 to 8.

In the system, the domain of each input variable is set from 0 to 10. To calculate the value, a random value is generated that lies in the domain of the input variable, and based on this value, the resulting value is calculated using both algorithms.

Dependence graphs are shown in Fig. 4 and the calculation results in Table 2.

Table 2. Comparison of the classical and proposed algorithms

Number of membership functions	Classical algorithm	Proposed algorithm
2	0,0017	0,00085
3	0,0886	0,00127
4	0,4035	0,00136
5	2,256	0,00148
6	3,202	0,00131
7	16,624	0,00178
8	48,483	0,00187

The performed calculations show that the execution time of the classical algorithm strongly depends on the number of inference rules. The execution time of the proposed algorithm practically does not depend on the number of inference rules, since due to restriction (3) for each input variable, no more than two inference rules are always selected for calculation.

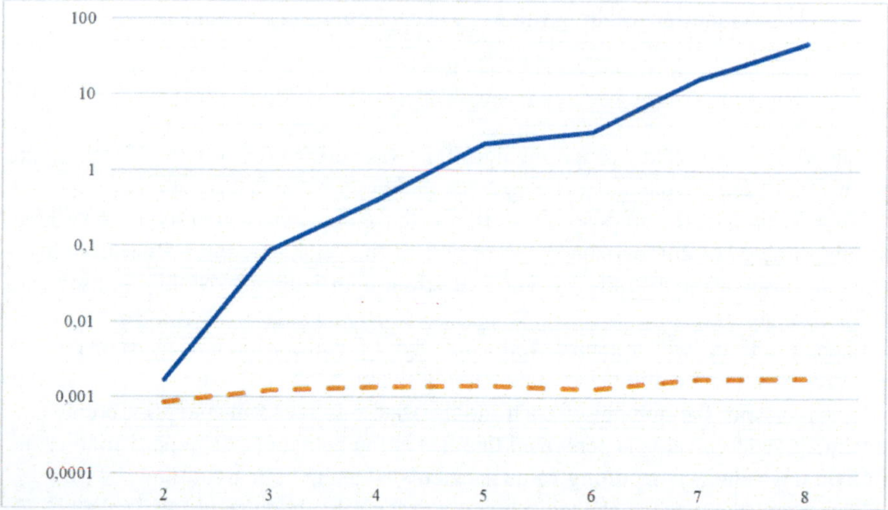

Fig. 4. Graphs of the dependence of the calculation time (ordinate axis) on the number of membership functions (abscissa axis) for the classical algorithm (solid line) and the proposed algorithm (dashed line)

4 Conclusion

The above study showed the effectiveness of the proposed algorithm. Compared to the classical algorithm, the computational speed can be four orders of magnitude faster. In this case, there is no need to constantly store the entire fuzzy inference system in RAM.

Acknowledgements. This research was supported by the grant of the President of the Russian Federation for leading scientific schools of the Russian Federation NSh-122.2022.1.6.

References

1. Bogomolov, A.: Information technologies of digital adaptive medicine. Info. Autom. **20**(5), 1153–1181 (2021)
2. Buldakova, T.I., Krivosheeva, D.A.: Application of biosignals in the end-to-end encryption protocol for telemedicine systems. In: Kravets, A.G., Bolshakov, A.A., Shcherbakov, M. (eds.) Society 5.0: Human-Centered Society Challenges and Solutions. SSDC, vol. 416, pp. 29–39. Springer, Cham (2022). https://doi.org/10.1007/978-3-030-95112-2_3
3. Zarubina, T.V., Kobrinsky, B.A., Kudrina, V.G.: The medical informatics in health care of Russia. Problemy sotsial'noi gigieny, zdravookhraneniia i istorii meditsiny **26**(6), 447–451 (2018)
4. Golosovskiy, M.S., Bogomolov, A.V., Balandov, M.E.: optimized fuzzy inference for sugeno-type systems. Autom. Doc. Math. Linguist. **56**(5), 237–244 (2022)
5. Golosovskiy, M.S., Bogomolov, A.V., Evtushenko, E.V.: An algorithm for setting Sugeno-type fuzzy inference systems. Autom. Doc. Math. Linguist. **55**(3), 79–88 (2021)

6. Golosovskiy, M.S., Bogomolov, A.V., Terebov, D.S., Evtushenko, E.V.: Algorithm toadjust fuzzy inference system of Mamdani type. Vestn. Yuzhno-Ural. Gos. Univ. Ser. Mat. Mekh. Fiz. **10**(3), 19–29 (2018)
7. Kosko, B.: Global stability of generalized additive fuzzy systems. IEEE Trans. Syst. Man Cybern. Part C: Appl. Rev. **3**(28), 441–452 (1998)
8. Buldakova, T.I., Suyatinov, S.I.: Biological principles of integration information at big data processing. In: Proceedings of the 2019 International Russian Automation Conference (2019). https://doi.org/10.1109/RUSAUTOCON.2019.8867710
9. Golosovskiy, M., Bogomolov, A., Balandov, M.: Algorithm for configuring Sugeno-type fuzzy inference systems based on the nearest neighbor method for use in cyber-physical systems. In: Kravets, A.G., Bolshakov, A.A., Shcherbakov, M. (eds.) Cyber-Physical Systems: Intelligent Models and Algorithms. SSDC, vol. 417, pp. 83–97. Springer, Cham (2022). https://doi.org/10.1007/978-3-030-95116-0_7
10. Kosko, B.: Fuzzy systems as universal aproximators. IEEE Trans. Comput. **11**(43), 1329–1333 (1994)
11. Maistrou, A.I., Bogomolov, A.V.: Technology of automated medical diagnostics using fuzzy linguistic variables and consensus ranking methods. IFMBE Proc. **25**(7), 38–41 (2009)
12. Sokolova, A.V., Buldakova, T.I.: Network architecture of telemedicine system for monitoring the Person's condition. In: Proceedings 3rd International Conference on Control Systems, Mathematical Modeling, Automation and Energy Efficiency, pp. 361–365 (2021)
13. Kukushkin, Y., Vorona, A., Bogomolov, A., Chistov, S.: Risk-metriyc staff health facilities for the disposal of chemical weapons. Health Risk Anal. **3**, 26–34 (2014)
14. Tobin, D., Bogomolov, A., Golosovskiy, M.: Model of organization of software testing for cyber-physical systems. In: Kravets, A.G., Bolshakov, A.A., Shcherbakov, M. (eds.) Cyber-Physical Systems: Modelling and Industrial Application. SSDC, vol. 418, pp. 51–60. Springer, Cham (2022). https://doi.org/10.1007/978-3-030-95120-7_5
15. Shamaev, D.: Synthetic datasets and medical artificial intelligence specifics. In: Silhavy, R., Silhavy, P., Prokopova, Z. (eds.) Data Science and Algorithms in Systems: Proceedings of 6th Computational Methods in Systems and Software 2022, Vol. 2, vol. 597, pp. 519–528. Springer International Publishing, Cham (2023). https://doi.org/10.1007/978-3-031-21438-7_41

An Algorithm for Constructing a Dietary Survey Using a 24-h Recall Method

R. S. Khlopotov(✉)

St. Petersburg Federal Research Center of the Russian Academy of Sciences, Saint Petersburg, Russia

gniiivm-h@yandex.ru

Abstract. The article provides a critical analysis of the advantages and disadvantages of existing methods of analyzing the actual nutrition of a socio-hygienic group and a comparative analysis between the particular features of research conducted using each method of nutrition analysis and the general requirements for studying actual nutrition stated by nutritionists (accuracy, representativeness, simplicity, cost-effectiveness, and others). Based on the analysis results, the 24-h recall method should be chosen as the basic method for studying the actual nutrition of an individual by a nutritionist. The main limitation for effective practical application of this method is the strict standardization of the survey's process and of the evaluation of the results obtained. One of the solutions to this problem is the development of a software based on the algorithm we propose.

Keywords: Nutritiology · 24-h recall method · actual nutrition · социально-гигиенические методы · socio-hygienic methods · algorithm for constructing a survey · software

1 Introduction

A person's diet greatly influences their health status, more specifically, the way it corresponds to the needs of their body. It is nutrition that determines body's proper growth and adaptation to environmental influences, helps to develop immunity, and maintain mental and physical efficiency [1–3].

Nutritional status is an integral indicator in the analysis of the quality of a person's diet. It reflects the correlation between the state of health and actual nutrition, taking into account the accumulated impact of human environmental factors [4, 5].

The study and analysis of the nutritional status is carried out with a consistent assessment of the following indicators:

- Actual nutrition, by analysing the list of nutrients consumed and their quantitative characteristics;
- Health status, by analysing nutritional status and structure of nutrition-related morbidity;

R. Silhavy and P. Silhavy (Eds.): CSOC 2023, LNNS 724, pp. 452–462, 2023.
https://doi.org/10.1007/978-3-031-35314-7_40

- Environmental status, by analysing the data on the sources of environmental hazards and ways and mechanisms of foreign influence on the body [6–10].

The purpose of the study is to conduct a comprehensive analysis and evaluation of methods for studying the actual nutrition of a single person, which will allow you to choose the most appropriate information requirements of a nutritionist. This will make it possible to develop an algorithm for the formation of nutrition survey questionnaires and their processing, which will be the basis for the creation of software for practicing nutritionists.

2 Materials and Methods

The theoretical and methodological basis for the study is the scientific achievements of foreign and domestic scientists in the research of actual nutrition [11–13]. When setting and solving the described problem, such general scientific research methods were used as: synthesis, comparative analysis – when studying modern methods of studying actual nutrition; structural modeling – when developing an algorithm for forming a nutrition analysis survey and processing its results.

3 Results and Discussion

All methods of studying actual nutrition are divided into socio-economic (used when the general dietary patterns of a specific territorial district residents or the levels of consumption of specific food stuffs are studied) and socio-hygienic (used when individual nutrition or nutrition of a family or a group is studied) [2, 14] (see Fig. 1).

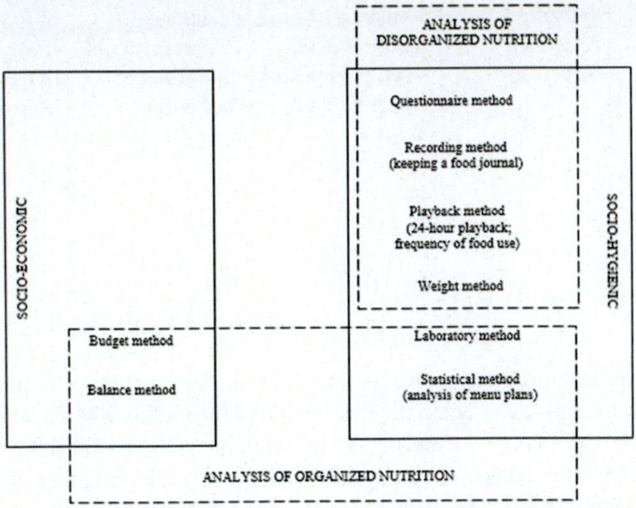

Fig. 1. Classification of the methods for studying actual nutrition.

However, it should be noted, that the classification of the methods is purely customary. So, while using a socio-hygienic method, the researcher, along with the data he needs, receives information of a more socio-economic kind. In some cases, a set of methods is used to analyze the nutrition of both organized and unorganized groups of people. For example, at some workspaces there are common regulations for everyone and people eat at work in accordance with them, so the nutrition is organized.

Outside of work, the same workers belong to unorganized groups of people. In this instance, while studying these population groups, different methods should be used at periods of organized and unorganized nutrition. The choice of the method depends on the organization of nutrition of the observed [15–19].

The choice of the method should be based on a critical assessment of its advantages and disadvantages, since there is no universal method that is suitable for solving all problems (Table 1, 2).

Table 1. Comparative analysis of socio-hygienic nutrition study methods (combined methods).

Title	Peculiarities of use	Advantages	Disadvantages	Research results
Survey-weight method	Daily records of food consumption by means of interviewing the subjects	- large sample size; - provides with an opportunity to do a medical examination	- is used exclusively for studying nutrition of a family	The data on the family budget, the amount of food purchased and other details are collected
Questionnaire -survey method	An interview is conducted with the help of specially composed surveys	- simplicity of implementation; - relatively low level of laboriousness; - fast examination - high level of accuracy concerning the qualitative characteristics of the consumed food stuffs	- low level of accuracy concerning the quantitative characteristics of the consumed food stuffs	The data allow us to assess a separate problem of food hygiene (for which the list of questions has been developed)

To make a more appropriate choice of a method of studying the actual nutrition of an individual, a comparative characteristic description of the methods of the socio-hygienic group should be carried out considering the needs of an analysis (Table 3).

Thus, based on the results of the analysis presented in the Table 2 given above, and the fact that nutritiology is a more comprehensive approach to studying nutrition problems than dietology (from the study of the motives of a person's choice of food to the way it

Table 2. Comparative analysis of socio-hygienic nutrition study methods (traditional methods).

Title	Peculiarities of use	Advantages	Disadvantages	Research results
1	2	3	4	5
Questionnaire method	Limited number of questions	- low level of laboriousness; - fast examination; - large sample size	- low level of accuracy concerning the quantitative and qualitative characteristics of the consumed food stuffs; - lack of representativeness in the data obtained; - lack of opportunity to do a medical examination	The data allow us to assess a separate problem of food hygiene (for which the list of questions has been developed)
Recording method	By means of a survey obtaining information about the time of eating, the place of cooking and eating, a description of the nature of the dish and the food stuff, the methods of its preparation and the quantity of food	- information about a nutrition over long time period; - low level of laboriousness; - fast examination	- low level of accuracy concerning the quantitative and qualitative characteristics of the consumed food stuffs; - lack of representativeness in the data obtained; - hard to determine the time period when the information was obtained	The data allow us to assess the role of nutrition as a risk factor for the development of diseases

(*continued*)

Table 2. (*continued*)

Title	Peculiarities of use	Advantages	Disadvantages	Research results
1	2	3	4	5
24-h playback	By means of a survey obtaining information about the time of eating, the place of cooking and eating, a description of the nature of the dish and the food stuff, the methods of its preparation and the quantity of food	- low level of laboriousness; - fast examination; - large sample size; - the highest representativeness rate in the data obtained	- it is necessary to comply with certain requirements for its participants and also while planning the survey in order to obtain reliable results	The data allows us to estimate the consumption of energy and nutrients
Frequency of food use	A special questionnaire is used, that allows us to estimate the frequency of food stuff consumption over a certain period of time	- fast examination; - large sample size	- low level of accuracy concerning the quantitative characteristics of the consumed food stuffs; - lack of representativeness in the data obtained	The data allow us to assess the role of nutrition as a risk factor for the development of diseases, to divide people into categories depending on the level of consumption, to identify the correlation between morbidity and the patterns of nutrition

(*continued*)

Table 2. (*continued*)

Title	Peculiarities of use	Advantages	Disadvantages	Research results
1	2	3	4	5
Weight method	An executor is attached to each family under study for its implementation	- high level of accuracy concerning the quantitative characteristics of the consumed food stuffs; - high level of representativeness in the data obtained	- high level of laboriousness; - small sample size; - burdensome for the subjects; - inadmissible for large epidemiological surveys	The data allows us to estimate the amount of food consumed by each family member per day. Then the chemical composition of the diet and its energy value are calculated using tables of the food stuffs chemical composition
Laboratory method	Laboratory determination of chemical composition (content of proteins, fats, carbohydrates, solid matter, ash) and energy value of cooked food	- the highest level of accuracy concerning the quantitative and qualitative characteristics of the consumed food stuffs; - allows to evaluate the diet with the greatest accuracy	- expensive; - high level of laboriousness; - takes a long time; - is used to study the nutrition of exclusively organized groups	The data allow us to identify the violations in the technology of cooking and storaging of food stuffs, as well as seasonal fluctuations in their chemical composition
Statistical method	It is carried out in order to establish the correctness and rationality of the menu, the content and ratio of the main ingredients of food and the calculation of energy value	- fast examination; - large sample size	- low level of accuracy concerning the quantitative and qualitative characteristics of the consumed food stuffs; - lack of representativeness in the data obtained	The data are compared with «Norms of physiological energy and nutritional needs for various groups of the population of the Russian Federation»

Table 3. Comparison between capabilities of socio-hygienic methods and the needs of studying actual nutrition.

Research requirements	Methods of the socio-hygienic group								
	Questionnaire method	Recording method	**24-h playback**	Frequency of food use	Weight method	Laboratory method	Statistical method	Survey-weight method	Questionnaire -survey method
Low level of laboriousness	+	+	+	–	–	–	–	–	+
High level of accuracy	–	+	+	–	+	+	–	+	+
Fast examination	+	+	+	+	–	–	+	–	+
Large sample size	+	+	+	+	–	–	+	–	+
Simplicity of implementation	+	–	+	–		–	+	–	+
Representativeness in the data obtained	–	–	+	–	+	+	–	+	–
Identifying the correlation between morbidity and the patterns of nutrition	–	+	-	+	–	–	–	–	–
Economical	+	+	+	+	+	–	+	+	+
Opportunity to do a medical examination	–	–	+	+	–	–	–	+	–
Comprehensiveness	–	–	+	–	–	+	–	+	–
Not burdensome for the subjects	+	–	+	–	–	–	+	–	+

influences the health of their entire body), a 24-h playback method should be chosen as a basic method for studying the actual nutrition of an individual by a nutritionist.

The 24-h playback method allows you to collect, process, and also thoroughly analyze data on an individual's food consumption, taking into account the anthropometric data, information about their physical activity, work and rest. This method evaluates a person's nutrition balance, their energy needs on weekdays and weekends and reveals a deficiency of vitamins, as well as macro and micro elements. All this information is necessary for the selection of the. Necessary individual medical-preventive diet, which would not only stabilize the weight of a person surveyed, but also prevent the development of chronic diseases.

The 24-h playback method is used as a reference method when validating other nutrition assessment tools. Due to the high speed of data collection and the opportunity to obtain quantitative characteristics of the diet (energy and nutritional value), this method is widely used in scientific research. The data that can be collected by means of the 24-h playback method provides great opportunities for analysis and interpretation of the results.

However, it should be noted that the 24-h playback method has some limitations (the data covers the one-day diet and does not give a picture of the nutrition over a longer period). To offset this disadvantage, data should be collected repeatedly, taking into account differences in the diet on weekdays and weekends, as well as seasonal changes [20–23].

The 24-h playback method needs meticulous standardization of the interview procedure, the evaluation of the quantity and the quality of the food consumed, the complete description of the dishes eaten and requires filling out forms. This is one of its most difficult stages.

At the moment there is no generally accepted food coding system available for scientific use. In this regard, the results of research concerning certain food stuff and groups they belong to are often not comparable. That is also caused by the lack of a standard food stuff classification system.

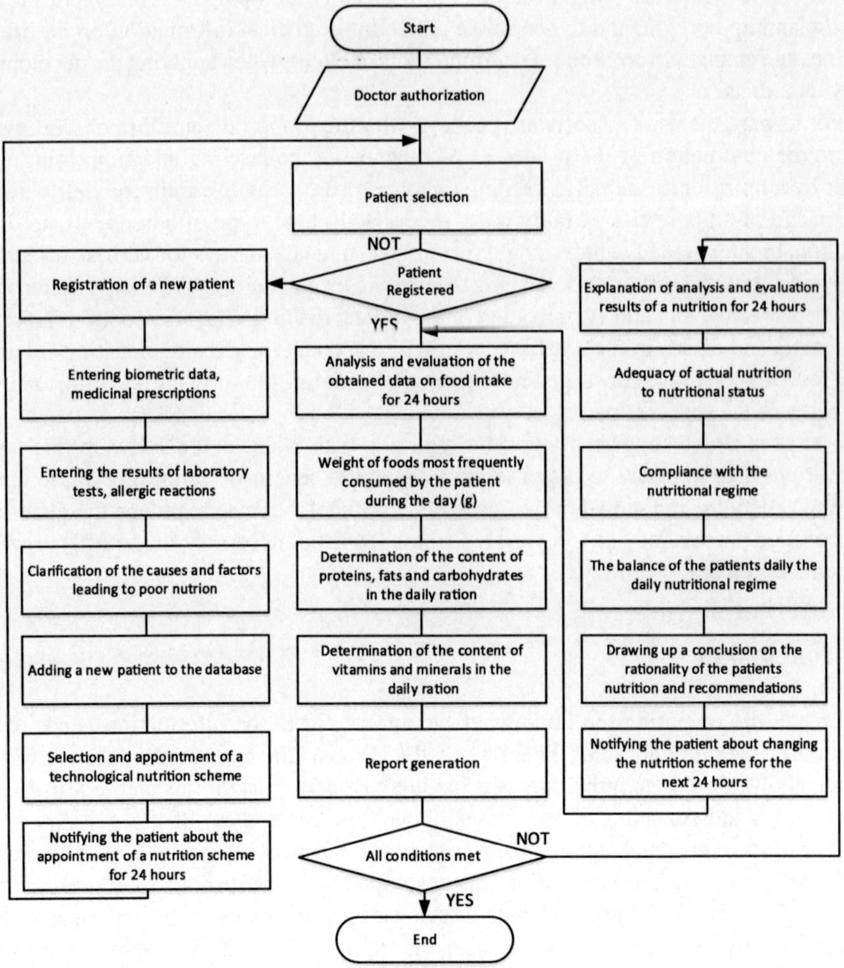

Fig. 2. Algorithm for forming (constructing) a dietary survey using 24-h recall method and their processing.

There is an urgent need to standardize reference data on the chemical composition of food stuffs. Currently, the European Union has made significant investments to support cooperation between European States in creating a single pan-European food coding system and a database with the nutritional profiles. This is the most important step in the development of a standardized 24-h playback method assessment tool for all European countries when conducting any type of research [4, 5, 11, 24].

Thus, for the effective practical application of the 24-h playback method by a nutritionist, it is necessary to develop appropriate software. To do this, first of all, it is necessary to create an algorithm for forming (constructing) of dietary survey on the 24-h playback method (Fig. 2).

The suggested algorithm is designed to automate the work of nutritionists, as well as their clients while assessing the actual nutrition and developing diets and menus for a given planning horizon, and to centralize collecting statistical information on nutrition regimes and changes in anthropometric indicators of clients when applying the developed individual diets.

For its part, the use of a software package with the proposed algorithm of work will ensure the automation of the processes of diagnosing, counseling and supporting the client by a nutritionist, namely: keeping client records, assessing anthropometric data, percentage of adipose tissue, body mass index, body type, type of adipose tissue distribution, recommended weight range; forming 24-h recall surveys for various research tasks and population groups; identifying the chemical composition and energy value of a consumer's actual diet and visualization of these data; making a list of exceptions among food stuffs and dishes in the database, taking into account the existing nutrition-related diseases and food preferences; generating reports; updating the software and information databases of the software package.

Thus, the above suggested algorithm will allow us to develop a software that will make it possible not only to increase the therapeutic benefit by proposing an optimal diet and satisfy the individual preferences of each client, but also to reduce the material costs of food purchasing due to a clearly formulated list of necessary food stuffs, as well as their amount and weight.

4 Conclusion

As a result of comparing the findings of the analysis with the information needs of a nutritionist, the rational use of the 24-h recall method as the main method for studying the actual nutrition of an individual was justified. Effective use of this method involves meticulous standardization of the survey procedure, assessment of the quantity and quality of food consumed, complete description of dishes eaten, and requires filling out forms and reports. In this connection, the development of software that will ensure the offset of all restrictions concerned with the standardization process becomes relevant. To accomplish this task, we have suggested an algorithm for constructing surveys using the 24-h recall method in an information and computer area, which will allow us to develop a software that can help a nutritionist to provide high-quality services.

Acknowledgements. This research was supported by the grant of the President of the Russian Federation for leading scientific schools of the Russian Federation NSh-122.2022.1.6.

References

1. Shvabskaya, O.B., Karamnova, N.S., Izmailova, O.V.: Healthy nutrition: new diets for individual use. Rational Pharmacotherapy Cardiol. **16**(6), 958–965 (2020). https://doi.org/10.20996/1819-6446-2020-12-12
2. Tutelyan, V.A.: Healthy eating for public health. Public Health **1**(1), 56–64 (2021). https://doi.org/10.21045/2782-1676-2021-1-1-56-64
3. Khlopotov, R.S.: Analysis of trends in the development of automated systems for solving problems of food hygiene. Models Syst. Netw. Econom.Technol. Nat. Soc. **3**, 140–157 (2022). https://doi.org/10.21685/2227-8486-2022-3-9
4. Karamnova, N.S., Izmailova, O.V., Shvabskaya, O.B.: Methods for studying nutrition: use cases, opportunities and limitations. Prev. Med. **24**(8), 109–116 (2021). https://doi.org/10.17116/profmed202124081109
5. Maksimov, S., Karamnova, N., Shalnova, S., Drapkina, O.: Sociodemographic and regional determinants of dietary patterns in Russia. Int. J. Environ. Res. Public Health **17**(1), 328 (2020). https://doi.org/10.3390/ijerph17010328
6. Ushakov, I.B., Bogomolov, A.V.: Informatization of personalized adaptive medicine programs. Bull. Russian Acad. Med. Sci. **69**(5–6), 124–128 (2014). https://doi.org/10.15690/vramn.v69i5-6.1056
7. Kukushkin, Y., Vorona, A., Bogomolov, A., Chistov, S.: Risk-metriyc staff health facilities for the disposal of chemical weapons. Health Risk Anal. **3**, 26–34 (2014)
8. Ushakov, I.B., Bogomolov, A.V., Dragan, S.P., Soldatov, S.K.: Methodological bases of personalized hygienic monitoring. Aerospace Environm. Med. **51**(6), 53–56 (2017). https://doi.org/10.21687/0233-528X-2017-51-6-53-56
9. Bogomolov, A.V.: Information technologies of digital adaptive medicine. Informat. Automat. **20**(5), 1154–1182 (2021). https://doi.org/10.15622/20.5.6
10. Maksimov, S.A., Karamnova, N.S., Shalnova, S.A., Drapkina, O.M.: Empirical patterns of nutrition and their impact on health status in epidemiological studies. Nutrition Issues **89**(1), 6–18 (2020). https://doi.org/10.24411/0042-8833-2020-10001
11. Popova, A., Tutelyan, V.A., Nikityuk, D.B.: On new norms of physiological needs for energy and nutrients for various groups of the population of the Russian Federation. Nutrition Issues **90**(4), 6–19 (2021). https://doi.org/10.33029/0042-8833-2021-90-4-6-19
12. Tutelyan, V.A., Nikityuk, D.B.: The global challenge of the 21st century - COVID-19: the answer to nutrition. Nutrition Issues **90**(5), 6–14 (2021). https://doi.org/10.33029/0042-8833-2021-90-5-6-14
13. Osmolovsky, I.S., Zarubina, T.V., Shostak, N.A., Kondrashov, A.A., Klimenko, A.A.: Development of medical nomenclature and algorithms for diagnosis and treatment of gout in outpatient settings. Bull. Russian State Med. Universitythis , 51–57 (2021). DOI: https://doi.org/10.24075/brsmu.2021.014
14. Budanova, E.I., Bogomolov, A.V.: Characteristics of the quality of life and health of contract servicemen. Hygiene Sanitation **95**(7), 627–632 (2016). https://doi.org/10.18821/0016-9900-2016-95-7-627-632
15. Khlopotov, R.S.: Analysis of medical informatics trends. Fundam. Appli. Problems Eng. Technol. **3**, 135–147 (2022). https://doi.org/10.33979/2073-7408-2022-353-3-135-147
16. Ushakov, I.B., Bogomolov, A.V.: Diagnosis of human functional states in priority studies of domestic physiological schools. Medico-biological and socio-psychological problems of safety in emergency situations. **3**, 91–100 (2021). https://doi.org/10.25016/2541-7487-2021-0-3-91-100
17. Zarubina, T.V., Kobrinsky, B.A., Kudrina, V.G.: The medical informatics in health care of Russia. Problems Social Hygiene Public Health History Med. **26**(6), 447–451 (2018). https://doi.org/10.32687/0869-866X-2018-26-6-447-451

18. Bogomolov, A.V., Ushakov, I.B.: Information technologies for the development and implementation of restorative medicine programs. Issues Balneology Physio. Therapeutic Phys. Culture **99**(5–2), 17 (2022)
19. Bogomolov, A.V., Chikova, S.S., Zueva, T.V.: Information technologies for data collection and processing when establishing determinants of epidemic processes. Health Risk Anal. **3**, 144–153 (2019). https://doi.org/10.21668/health.risk/2019.3.17.eng
20. Golosovskiy, M., Tobin, D., Balandov, M., Khlopotov, R.: Architecture of software platform for testing software of cyber-physical systems. In: Silhavy, R., Silhavy, P., Prokopova, Z. (eds) Data Science and Algorithms in Systems. LNNS. Springer, Cham. 597, 488–494 (2023). DOI: https://doi.org/10.1007/978-3-031-21438-7_38
21. Blinov, V.V., Bogomolov, A.V., Dlusskaya, I.G.: Prediction of disadaptation disorders in terms of the immune status. J. Stress Physiol. Biochem. **12**(2), 17–26 (2016)
22. Bubeev, Yu.A., Vladimirskiy, B.M., Ushakov, I.B., Usov, V.M., Bogomolov, A.V.: Mathematical modelling of spread covid-19 epidemic for preventive measures to protect life and health of elderly. Bull. South Ural State Univ. Ser.: Mathem. Model. Program. Comput. Softw. **14**(3), 92–98 (2021). DOI: https://doi.org/10.14529/mmp210307
23. Shamaev, D.M.: Synthetic Datasets and Medical Artificial Intelligence Specifics. In: Silhavy, R., Silhavy, P., Prokopova, Z. (eds.) Data Science and Algorithms in Systems. CoMeSySo 2022. LNNS, vol. 597. Springer, Cham (2023). https://doi.org/10.1007/978-3-031-21438-7_41
24. Ushakov, I.B., Bogomolov, A.V.: Informatization of programs of personalized adaptation medicine. Bull. Russ. Acad. Med. Sci. **69**(5–6), 124–128 (2014)

Machine Learning Methods and Words Embeddings in the Problem of Identification of Informative Content of a Media Text

Klyachin Vladimir(✉) and Khizhnyakova Ekaterina

Volgograd State University, Volgograd University Av., 100, 400062 Volgograd, Russia
klyachin.va@volsu.ru

Abstract. The central problem solved in the article is the task of automating the process of revealing the information content of the media text. This problem is relevant, since its solution allows you to identify texts that do not carry useful information, but try to influence the reader in order to encourage him to take various actions. First, we present a modification of the CBOW algorithm and describe the process of training a neural network to build word embedding. Next, we describe the Russian word embeddings available on the site https://rusvectores.org/ru/models/. Here we also show how ready-made word embeddings can be used to solve problems of thematic classification of texts. To do this, we use a set of news data from the resource https://lenta.ru. Finally, to solve the problem of information content of texts, we train the embedding of words based on the given set of texts and conduct testing. The quality check of the model is carried out on the basis of a dataset prepared by the employees of the Volgograd State University.

Keywords: nature language model · machine learning · text classification · media text · information content

1 Nature Language Models

1.1 Embeddings of Words

Nature language processing by machine learning methods is connected with very important problem. Using machine learning methods demand transformation a text into numerical vector type. This problem is known as a problem of vectorization text. Now it is known some methods of solving the problem. For example, we consider the methods, which corresponds any word from the constructed vocabulary into numerical vector. These methods are known as embeddings of words. We mention the article [1], where the models CBOWS (Continuous Bag-of-Words) and CSG (Continuous Skip-gram) model are considered. It is shown how to learn these models. Also, it is given comparison predictive properties of the models for some semantic relationships between pairs of words. We remember that CBOW model predicts the current word by its context whereas CSG model predicts surrounding words given the current word.

R. Silhavy and P. Silhavy (Eds.): CSOC 2023, LNNS 724, pp. 463–471, 2023.
https://doi.org/10.1007/978-3-031-35314-7_41

Usually, in neural networks there is the first layer as an embedding layer for vectorization inputs sequences of words from texts. The learning of this layer depends on the solving problem. After learning we obtain transformation which transform word into vector. This transformation can be used as a pretrained model to solve some problems of natural language processing.

We introduce some modification of the well known algorithm CBOW. Our idea is to use sentences of the texts instead the n-grams as a context for prediction words. We consider our algorithm in more detail. We name our algorithm Continuous Bag of Sentences (CBOS).

- On the first step we collect text from txt files which are located in the given file system path.
- Next we delete some characters such as braces ([]), comma (,), minus (-), percent (%) etc. as well as stop words using NLTK Python module and build the vocabulary of the extracted words. Also we build indexes for this vocabulary.
- Further we split the text to array of sentences. We use some integer $m > 0$ as a size of sliding window for processing sequence of m sentences.
- We transform this sequence into sequence of the words $w_1, w_2, \ldots, w_k$.
- Transforming words to theirs indexes we construct the sequence of integers $i_1, i_2, \ldots, i_k$.
- We use some integer $n > 0$ as a size of input sequence for neural network. For example, we can calculate n as a length of the longest sentence. In any case if $n < k$ we cut the sequence and obtain sequence $i_1, i_2, \ldots, i_n, i_{n+1}$. In case n more or equal k we complement the sequence to the required length with its first elements.
- For each $j = 1, 2, \ldots, n + 1$ we put in data set array X the sequence $i_1, i_2, \ldots, i_n, i_{n+1}$ without i_j which we transform in unitary vector by its word and put in array y. Unlike the mentioned algorithms and methods for topic modeling (see, for example [2, 3]) we save an order of words in sentences. In fact, we take in attention that words which meet in a sentence are semantic close. This approach is described in [4–6]. Moreover, the skipped word does not have to be symmetrical to the context.
- We construct neural network which we learn for predict the skipped word.

Below the structure of the neural network is given. Here dim is a dimension of the embedding space and vocab_size is a size of vocabulary. We use stochastic gradient descent (SGD) optimizer and categorical cross-entropy loss function (Fig. 1).

After learning we extract from the first layer values of the weights as a matrix W of size ($vocab\_size \times dim$). To obtain the vector representation $v(w)$ of the word w it needs to multiply unitary vector $U(w)$ and matrix W:

$$v(w) = U(w) \cdot W$$

We process some experiments. The results are presented in the following Table 1.

Result is calculated as a percents correctly reconstructed skipped word from sentences in the text which contains number of unique words (after normalization) pointed in the second column of the table.

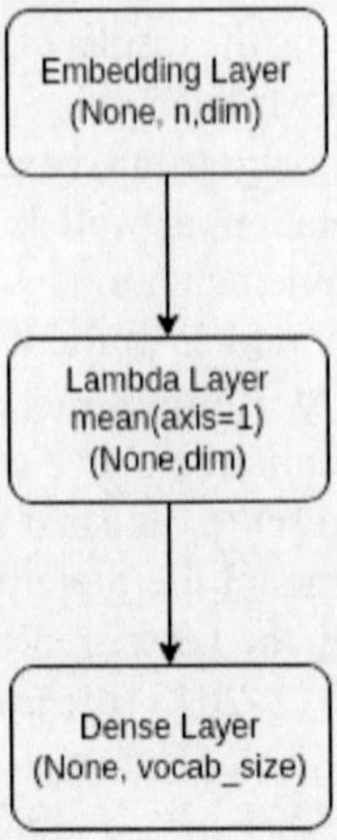

Fig. 1. A structure of the neural network for build the model of nature language.

Table 1. The results of the experiments.

Number words in texts	Number of unique normalized words	Dimension of embedding space, *dim*	Word window size, dimension of input layer, *n*	Number of sentence for sliding window, *m*	Number learning epochs	Result, %
487	421	47	6	1	15000	81%
1171	725	47	6	1	15000	69%
1171	725	97	6	1	15000	82%
1381	1100	97	6	1	15000	80%
1381	1100	97	6	1	21000	97%
3324	2015	97	6	1	21000	98%
23545	4183	97	6	1	21000	99%
23545	4183	64	6	2	21000	99%
119983	12492	32	4	2	16000	88%
119983	12492	48	6	2	16000	97%

1.2 Russian Language Models

One of the most important problem in the nature language processing is a problem of modeling of nature language. For example, work [7] describes the process of automatic the work of the text corpus, assembled from the news feed of a number of Internet sites, creation of a probabilistic n-gram model of the colloquial Russian language. Drive statistical analysis of this corpus, giving the results of the frequency calculation The laziness of various n-grams of words. In work [8], the algorithms definition of the semantic proximity of keywords: the Ginzburg algorithm, based on the frequency characteristics of words and its software implementation, as well as taking into account the frequency of speech and the problems of its implementation.

There are currently over a dozen models of the Russian Language. They are free to download from URL: https://rusvectores.org/ru/models/. These models are built using or CBOW either CSG algorithm mentioned above on the various corpora of Russian-language texts. Most models are represented as a text file in which each line corresponds alone word. In begin of line in the model file there is the word and label of its part of speech. Further in the line we can see the vector representation this word. For example, the model taiga_upos_skipgram_300_2_2018 (file is named by tayga_1_2.vec) has 300 dimensions and line of word 'книга' is looked as.

книга_NOUN 0.043357264 -0.020433554 -0.069780454 0.0016495131 0.022128297 0.02492341 0.08925131 0.029605178 0.068326026 -0.05487236 0.005218216 0.011833937 -0.06762027 -0.01797972 -0.019424992 0.009864659 -0.10268338 0.07266949 -0.065126285 0.009963145 -0.033227425 -0.0052293823 0.068378925 0.008288159 -0.023799416 0.07071366 0.14623627 0.027164182 -0.057159528 -0.036483 0.011825919 -0.0046275635 0.047086667 -0.008079778 -0.044185657 0.10579075 -0.029857717 0.021735663 0.0757409 0.027778205 0.037488084 0.06375316 0.04570305 0.008514274 -0.06026029 -0.03606508 0.0027726388…

These vector models can be used for solving various problems of the text processing. For example, we consider topic classification problem of Russian texts. Let we have some corpus of texts, each of which is labeled some topic marker. We construct machine learning model for predict topic of text. The first step is preprocessing texts and transform its into set of vectors. We consider this in more detail.

1. At the first, we read the text files and build indexed vocabulary, also as two mappings: a map from set of words to index and inverse.
2. The following step is a deleting of stop-words using Python NLTK module for Russian language.
3. Calculate frequency of the words for each text. Then we sort the frequences in descending order and choose the first *dim_words* words with maximal frequency.
4. Further, we transform the sequences of words into matrix D of dimension *dim_words* $\times$ n, where n is dimension of the embedding space of the Russian language model.
5. Next we transform the calculated matrix into n dimensional vector multiplying some *dim_words* dimensional vector V on this matrix: $v = V.D$. We collect vectors v as a data set for machine learning models.

We consider the following ML models:

- Naive Bayesian Classifier;
- K nearest neighbors model;
- Tree Decision model;
- Random Forest model.

Choosing

V = (1/dim_words, 1/dim_words,..., 1/dim_words).
we obtained the results which is collected in the following Table 2.

Table 2. The results of the experiments.

ML model	Vectorizer model	Number of the extracted words, *dim_words*	Number of topics	Size of dataset, number news messages	Result, %
Naive Bayesian model	taiga_upos_skipgram_300_2_2018	20	22	48000	55%
K nearest neighbors model	taiga_upos_skipgram_300_2_2018	20	22	48000	70%
Tree Decision model	taiga_upos_skipgram_300_2_2018	20	22	48000	40%
Random Forest model	taiga_upos_skipgram_300_2_2018	20	22	48000	68%
Naive Bayesian model	taiga_upos_skipgram_300_2_2018	20	25	64000	56%
K nearest neighbors model	taiga_upos_skipgram_300_2_2018	20	25	64000	70%
Random Forest model	taiga_upos_skipgram_300_2_2018	20	25	64000	67%
Tree Decision model	taiga_upos_skipgram_300_2_2018	20	25	64000	42%
Naive Bayesian model	taiga_upos_skipgram_300_2_2018	35	7	15065	74%
K nearest neighbors model	taiga_upos_skipgram_300_2_2018	35	7	15065	83%
Random Forest model	taiga_upos_skipgram_300_2_2018	35	7	15065	80%

(*continued*)

Table 2. (*continued*)

ML model	Vectorizer model	Number of the extracted words, *dim_words*	Number of topics	Size of dataset, number news messages	Result, %
Tree Decision model	taiga_upos_skipgram_300_2_2018	35	7	15065	67%
Naive Bayesian model	taiga_upos_skipgram_300_2_2018	45	7	89608	74%
K nearest neighbors model	taiga_upos_skipgram_300_2_2018	45	7	89608	85%
Random Forest model	taiga_upos_skipgram_300_2_2018	45	7	89608	81%
Tree Decision model	taiga_upos_skipgram_300_2_2018	45	7	89608	70%
Naive Bayesian model	ruwikiruscorpora_upos_cbow_300_10_2021	45	7	61248	73%
K nearest neighbors model	ruwikiruscorpora_upos_cbow_300_10_2021	45	7	61248	83%
Random Forest model	ruwikiruscorpora_upos_cbow_300_10_2021	45	7	61248	81%
Tree Decision model	ruwikiruscorpora_upos_cbow_300_10_2021	45	7	61248	70%
Naive Bayesian model	ruscorpora_upos_skipgram_300_5_2018	45	7	61248	75%
K nearest neighbors model	ruscorpora_upos_skipgram_300_5_2018	45	7	61248	85%

(*continued*)

Table 2. (*continued*)

ML model	Vectorizer model	Number of the extracted words, *dim_words*	Number of topics	Size of dataset, number news messages	Result, %
Random Forest model	ruscorpora_upos_skipgram_300_5_2018	45	7	61248	81%
Tree Decision model	ruscorpora_upos_skipgram_300_5_2018	45	7	61248	69%
Naive Bayesian model	**news_upos_skipgram_300_5_2019**	**70**	**7**	**150000**	**80%**
K nearest neighbors model	**news_upos_skipgram_300_5_2019**	**70**	**7**	**150000**	**88%**
Random Forest model	**news_upos_skipgram_300_5_2019**	**70**	**7**	**150000**	**84%**
Tree Decision model	**news_upos_skipgram_300_5_2019**	**70**	**7**	**150000**	**72%**

These results are obtained by using data set of news from lenta.ru. This data set is free for download from https://github.com/natasha/corus#reference. It contains more than 700 thousands of news messages. We also use this data set for modeling news Russian language and then apply the built model for prediction the topic on the message news. To build the news model we use algorithm Continuous Bag of Sentences (CBOS) which is described in Sect. 1. We use the following parameters for construction the model:

- Dimension of embedding space, $dim = 100$
- Word window size, dimension of input layer, $n = 4$
- Number of sentence for sliding window, $m = 1$
- Number of learning epochs = *16000*

1.3 Information Content of Media Text

In media communication, informative texts are distinguished, i.e. conveying information about any fact of reality, and influencing texts, among which manipulative texts stand out, which are characterized by a distortion of the real state of affairs, a shift in the information focus, and a number of other techniques. These texts interpret the facts of reality in order to influence. *Clickbaiting* uses exclusively manipulative techniques to attract the attention of the addressee [9, 10]. Of course, such a text can also report a fact that corresponds to a real situation, but this is not the main thing for him. Speaking of informative, influencing texts, we proceed from their main function. Several functions can be combined in one text. Here we attempt to build the machine learning model for identification of the informative texts. This problem is urgent due to a wide usage in the modern mass media of various phenomena which spring up in response to the challenges of our time.

We build ML model using approach which is present above. For learning we use dataset which is collected by our colleagues from Volgograd State University. Each text of this data set written to the separate file which is located at one of the two directories: for informative and not informative texts. We take as a basis different vector representations of Russian language mentioned above. Further, we add one of those listed in the Table 3 classifier. Below we give the obtained results.

Table 3. The results of the experiments to identify of the information content.

ML model	Vectorizer model	Result, %
Naive Bayesian model	taiga_upos_skipgram_300_2_2018	59%
K nearest neighbors model	taiga_upos_skipgram_300_2_2018	62%
Tree Decision model	taiga_upos_skipgram_300_2_2018	52%
Random Forest model	taiga_upos_skipgram_300_2_2018	61%
Naive Bayesian model	ruwikiruscorpora_upos_cbow_300_10_2021	57%
K nearest neighbors model	ruwikiruscorpora_upos_cbow_300_10_2021	60%
Random Forest model	ruwikiruscorpora_upos_cbow_300_10_2021	61%
Tree Decision model	ruwikiruscorpora_upos_cbow_300_10_2021	55%
Naive Bayesian model	ruscorpora_upos_skipgram_300_5_2018	60%
K nearest neighbors model	ruscorpora_upos_skipgram_300_5_2018	57%
Random Forest model	ruscorpora_upos_skipgram_300_5_2018	62%
Tree Decision model	ruscorpora_upos_skipgram_300_5_2018	53%
Naive Bayesian model	news_upos_skipgram_300_5_2019	61%
K nearest neighbors model	news_upos_skipgram_300_5_2019	61%
Random Forest model	news_upos_skipgram_300_5_2019	68%
Tree Decision model	news_upos_skipgram_300_5_2019	57%
Naive Bayesian model	**Model on the basis CBOS (trained on the lenta.ru news dataset)**	**76%**
K nearest neighbors model	**Model on the basis CBOS (trained on the lenta.ru news dataset)**	**78%**
Random Forest model	**Model on the basis CBOS (trained on the lenta.ru news dataset)**	**81%**
Tree Decision model	**Model on the basis CBOS (trained on the lenta.ru news dataset)**	**72%**

2 Conclusion and Future Work

The results obtained allow us to make the following conclusions. For solving the classification problem it needs to construct corresponding model of the nature language. This model must be trained on the corpus of texts which is semantically closed to the classification problem. In particular, this paper shows that a corpus of news messages is very well suited for revealing the informational content of a text. Also, we note that our proposed algorithm Continuous Bag of Sentences works just as well as the algorithms CBOW and CSG. We plan to construct and study nature language models of various thematic areas. For example: science and technology, art and culture, industry and agriculture and the like.

Acknowledgments. The research is funded by the Russian Science Foundation, № 23–28-00509, https://rscf.ru/project/23-28-00509/.

References

1. Mikolov, T., et al.: Efficient estimation of word representations in vector space arXiv preprint arXiv:1301.3781. (2013)
2. Kryukov, K.V., Pankova, L.A., Pronina, V.A., Suhoverov, V.S., Shipilina, L.B.: Semantic similarity measures in ontology. Rev. Classificat. Problemy Upravleniya **5**, 2–14 (2010)
3. Irkhin, I.Y.A., Bulatov, V.G., Vorontsov, K.V.: Additive regularizarion of topic models with fast text vectorizartion. Comput. Res. Model. **12**(6), 1515–1528 (2020)
4. Grigorieva, E.G.E., Klyachin, V.A.: The study of the statistical characteristics of the text based on the graph model of the linguistic corpus. Izvestiya of Saratov Univ. Mathem. Mech. Informat. **20**(1), 116–126 (2020)
5. Grigoryeva E.G., Kochetova L.A., Pomelnikov Y.V., Popov V.V., Shtelmakh T.V.: Towards a graph model application for automatic text processing in data management. In: IOP Conference Series: Materials Science and Engineering. p. 012077 (2019)
6. Grigoryeva E. G., Klyachin V. A., Pomelnikov Yu. V., Popov V.V.: Algorithm of the key words search based on graph model of linguistic corpus, Vestnik Volgogradskogo gosudarstvennogo universiteta. Seriia 2: Iazykoznanie, vol. 16(2), pp. 58–67 (2017)
7. Kipyatkova, I.S., Karpov, A.A.: Automatic processing and statistic analysis of the news text corpus for a language model of a Russian language speech recognition system. Inform. Control Syst. **4**(47), 2–8 (2010)
8. Voronina I.E., Kretov A.A., Popova I.V.: Algorithms of semantic proximity assessment based on the lexical environment of the key words in a text, Vestnik Voronezhskogo gosudarstvennogo universiteta. Seriia: Sistemnyi analiz i informatsionnye tekhnologii, vol. 1, pp. 148–153 (2010)
9. Gavrikova, O.A.: Clickbait as a factor of false narrative creation in political mass media discourse. Politicheskaya lingvistika **3**(75), 22–30 (2019)
10. Sladkevich, Z.: Headlines in internet media services: between informing and clickbaiting. Media Linguist. **6**(3), 353–368 (2019). (In Russian)

Anxiety Mining from Socioeconomic Data

Fahad Bin Gias(✉), Fahmida Alam, and Sifat Momen

Department of Electrical and Computer Engineering, North South University, Plot 15, Block B, Bashundhara, Dhaka 1229, Bangladesh
{fahad.gias,fahmida.alam18,sifat.momen}@northsouth.edu

Abstract. Covid-19 pandemic has caused an increase in anxiety and stress-related problems. This research explores the detection of anxiety (specifically state anxiety) using socioeconomic data. The dataset, which originally appeared in Turkish, contains 2853 records. Following the translation of the dataset to English using an existing library, extensive exploration and pre-processing were performed. On the pre-processed data, machine learning algorithms were applied. XGBoost has been found to provide the best results after appropriate hyper-parameter tuning. Finally, SHAP is used to interpret the XGBoost model.

Keywords: State Anxiety · Machine Learning · XAI · Model prediction

1 Introduction

COVID-19 pandemic has brought about unprecedented challenges to individuals and societies at large. It has unleashed a wave of anxiety, with studies reporting staggering increases in state anxiety levels at individual and population levels [2,18]. Pieh and colleagues [9] reported that the prevalence of anxiety symptoms has increased by up to three times in some countries. One study [1] indicates that uncertainty and fear due to Covid-19 resulted in mental health exacerbation in US adults. Most (53%) US adults displayed an array of behavioral changes, including greater substance abuse, gambling, and overeating as a means of coping with stress and anxiety. Studies in India, Italy, and China have also found high rates of anxiety among the general population and healthcare workers [7,12, 20]. In one study with Bangladeshi students [17], it has been reported that the pandemic resulted in an increased level of depression and anxiety which in turn had a deleterious impact on their behavioral changes, including poor sleeping and dietary habits. Another study shows that stress and anxiety affect various routine habits (such as sleeping habits), which in turn can be used to detect the stress level of an individual [3].

There are various forms of anxiety disorder. One common anxiety disorder is State Anxiety (SA) which is a transitory emotional state, typically associated with feelings of apprehension, nervousness, and physiological sequelae such as an increased heart rate or respiration [19]. It reflects the temporary reactions

R. Silhavy and P. Silhavy (Eds.): CSOC 2023, LNNS 724, pp. 472–488, 2023.
https://doi.org/10.1007/978-3-031-35314-7_42

directly related to adverse situations in a specific moment [5]. It is different from generalized anxiety disorder (GAD), which is an excessive form of anxiety that is not specific to any particular event or situation and continues for the long term beyond control. Anxiety disorders, e.g., state anxiety, are common mental health conditions and, if left untreated, can lead to negative consequences such as chronic stress, sleep disturbances, fatigue, decreased cognitive function, social isolation, and even severe anxiety disorders and depression. Early detection of state anxiety and providing appropriate treatment can help individuals cope with symptoms and prevent it from aggravating further.

Socio-economic data are comparatively easier to access. This paper utilizes machine learning approaches to predict state anxiety during COVID-19 from socio-economic data. It also applies explainable AI (XAI) techniques to interpret the "black box" model. We have used State Anxiety Inventory (SAI) as the scale for State Anxiety, which is a 20-question subset from the State-Trait Anxiety Inventory (STAI) developed by Spielberger et al. [16].

The rest of the paper is organized as follows: Sect. 2 discusses related works. Following this, we discuss the methodology adopted in this work. Results have been discussed in Sect. 4, and following that, Sect. 5 discusses the key findings of our work. Finally, the paper is concluded in Sect. 6.

2 Literature Review

Perpetuini et al. [8] used machine learning approach to investigate the relationship between state anxiety and various physiological parameters measured by photoplethysmography (PPG) in healthy individuals. As measured by the Receiver Operating Characteristic (ROC) analysis with an area under the curve of 0.88, their findings indicate that it is possible to accurately predict SA.

Fenfen and colleagues [4] used machine learning approach to study the psychological states of 2009 Chinese undergraduate students during the Covid-19 pandemic. They found a prevalence rate of probably anxiety and insomnia of 12.49% and 16.87% respectively. Using XGBoost algorithm, they predicted anxiety and insomina amongst the students with an accuracy of 97.3% and 96.2% respectively.

Priya et al. [11] applied machine learning algorithms to predict stress, anxiety, and depression among employed and unemployed people from various cultural backgrounds using the Depression, Anxiety, and Stress Scale questionnaire (DASS 21). They used five different algorithms to predict the occurrence of anxiety, sadness, and stress on five different severity levels. Random Forest classifier was found to have the highest accuracy (73.3%) among the five applied algorithms.

Study by Roy-Byrne et al. [13] aimed to find out if poor outcomes among individuals with depression and anxiety from low socioeconomic status could be caused by lack or inadequate mental health treatment. The study used data from 1,772 individuals in the National Comorbidity Survey Replication (NCS-R) who met criteria for a mood or anxiety disorder. The study found that socioeconomic status does not significantly impact treatment for depression and anxiety. However, poor outcomes among these individuals may be due to factors other than quality of care, such as ongoing stress.

Sau et al. [15] used machine learning techniques to predict anxiety and depression in elderly patients. They aimed to create an accurate predictive model based on sociodemographic and health-related data to help doctors diagnose these conditions early. They tested a dataset of 510 elderly patients and found that the Random Forest classifier had the highest accuracy of 89%. They also tested the model on an external dataset of 110 elderly patients and found it to be 91% accurate with a false positive rate of 10%.

Nemesure et al. [6] used a novel machine learning pipeline to predict major depressive disorder (MDD) and GAD by re-analyzing data from observational research. 4,184 undergraduate students participated in the study and had a general health checkup and a mental evaluation for MDD and GAD. The pipeline used 59 biological and demographic features from the general health survey and a set of engineering features for model training, eliminating all psychiatric data. The model's performance was evaluated on a held-out test set, and it had an AUC of 0.73 for GAD and 0.67 for MDD. The study found having a satisfying living situation and public health insurance as key predictors of MDD, while up-to-date vaccination records and marijuana usage as the main predictors of GAD.

Pintelas and colleagues [10] reviewed machine learning prediction methods for anxiety disorders and conducted a comparative literature study on research to predict various anxiety disorders. The analysis of 16 studies showed that machine learning methods could effectively identify anxiety disorders. They discussed various supervised machine learning algorithms such as Bayesian networks, Artificial Neural Networks (ANNs), Support Vector Machines (SVM), Decision trees, Linear regression, Neuro-Fuzzy Systems (NFS), ensemble classifiers and many more. They found that ANNs were the best performer for GAD, and the super learner technique received the highest score for predicting PTSD. Hybrid methods and support vector machines were the most popular techniques for PTSD and SAD. Lastly, fMRI tool was only used to generate input data for SAD, PD, and agoraphobia.

3 Methodology

Figure 1 shows the details of our study protocol, which consists of a number of key subparts. The following subsections describe each of the subparts in detail.

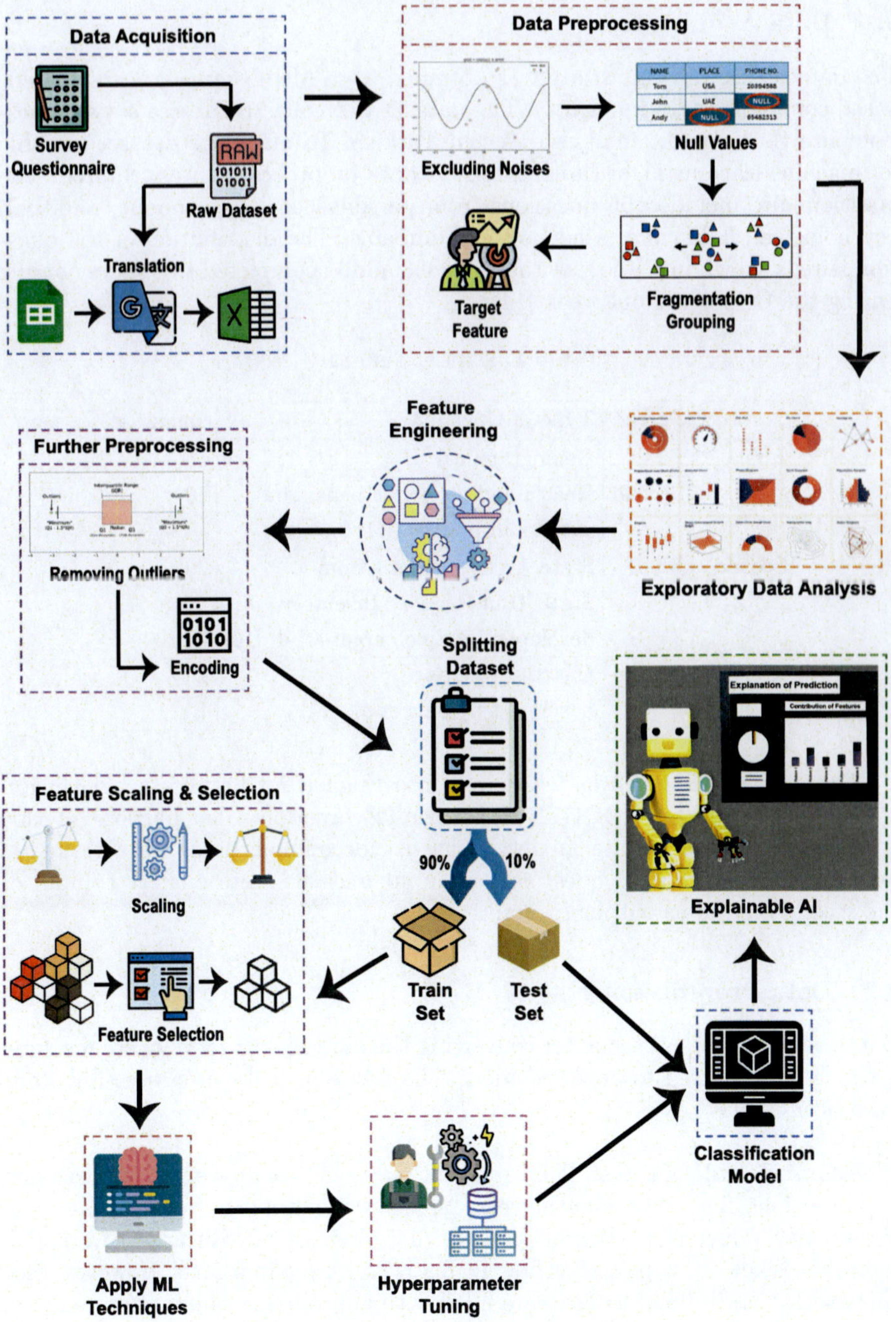

Fig. 1. Methodology adopted in this research

3.1 Data Acquisition

The dataset used in this study was obtained from Mendeley [14], a free cloud-based communal data repository. The dataset was collected from a survey questionnaire that was originally written in Turkish. To make the dataset suitable for machine learning algorithms, extensive data preprocessing was required. The questionnaire included 55 questions from personal, social, economic, habitual, psychological, behavioral, and political domains. The distribution of the questionnaire is shown in Table 1. A total of 2853 adults completed this survey online during the COVID-19 outbreak [14].

Table 1. Dataset Summary

Sections	Parts	Type of Questions	Quantity
First	First	Demographical	15
	Second	Health and economic behavioral, social and political perceptions	10
Second	First	State Anxiety subset from State-Trait Anxiety Inventory developed by Spielberger et al. [16]	20
	Second	Distress Tolerance developed by Sari et al. [14]	10

The Turkish dataset was translated into English for better comprehension using Google Sheets' GOOGLETRANSLATE function. The purpose of this study was to predict state anxiety using socioeconomic data. As a result, it used the first section (25 questions) as input and the state anxiety column of the second section as output column.

3.2 Data Preprocessing

Data preprocessing is crucial for converting the dataset to a form where machine learning can be used to create models. The following data preprocessing steps have been applied.

Noises and Null Values: The irrelevant features- features excluding the survey questions (i.e., day, month, year, time)- were dropped from the dataset. Some noisy values in particular features (e,g., living city, education) were also removed. Likewise, rare null values in features were eliminated. However, features with plentiful null values were filled with appropriate values after analysis.

Fragmentation and Grouping: A categorical feature obtained from a checklist question (that allows multiple selections of choices) can contain numerous unique values. This will result into feature explosion if one-hot encoding is used to provide numeric representation to the categorical features. To prevent this explosion of features, a custom fragmentation method was developed. It broke down a single checklist question into n number of binary features where n is the number of options (Fig. 2).

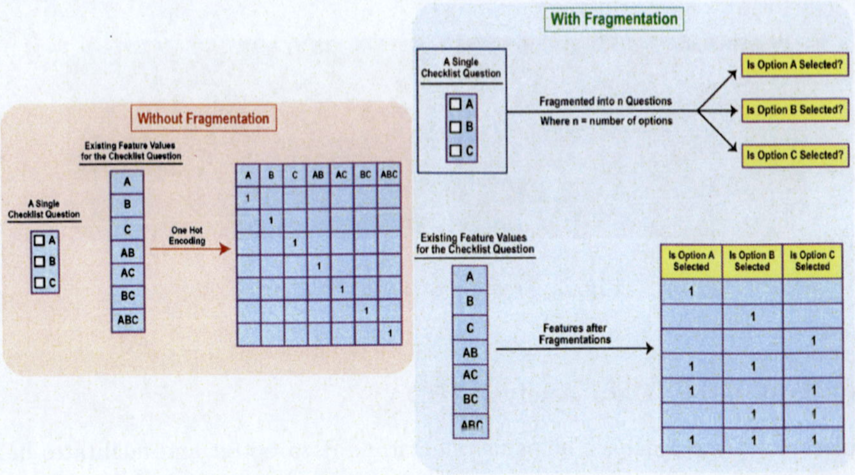

Fig. 2. Number of Features Before and After Fragmentation

On the other hand, grouping was a separate approach taken to make imbalanced features more balanced. Infrequent feature values (e.g., married, divorced and other marital status) were grouped together to provide a better feature representation. Grouping was performed on 4 features, which made them more balanced (Fig. 3).

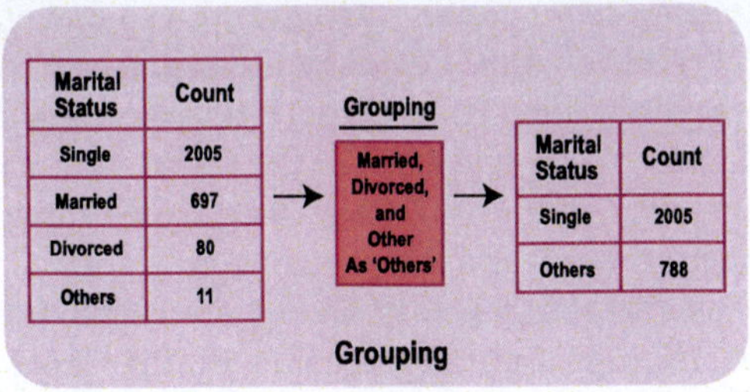

Fig. 3. Grouping

Target Feature: The state anxiety questions, consisting of positive and negative emotions, were on a 4-point scale. These features were rescaled using two different approaches.

The negative feelings were rescaled with 'none' being 1 and 'high' being 4. Conversely, the positive feelings were rescaled in reverse order, with 'none' being

4 and 'high' being 1. The scores for all questions were then combined to create a single feature, "Anxiety_Score", with higher scores indicating higher levels of anxiety. Finally, the binary target feature, the presence of state anxiety or not, was created following a threshold in anxiety score proposed by Sari et al. [14] (Fig. 4).

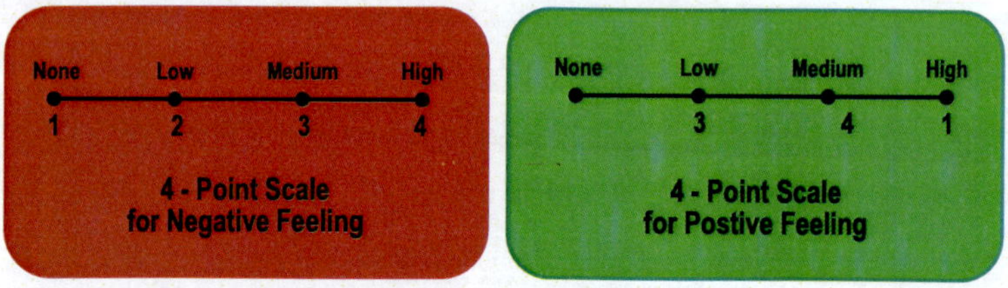

Fig. 4. Creation of Target Feature

3.3 Exploratory Data Analysis (EDA)

Exploratory Data Analysis helps us understand data better and facilitates better feature extractions, engineering, and selection techniques, which in turn positively impacts the quality of prediction.

Since all the input features are categorical and Anxiety_Score (target feature is directly dependent on it) is continuous, EDA was performed using box plot visualization. To perform a comprehensive EDA, all twenty-five input features were plotted with respect to the anxiety score. Analysis of those box plots pointed out some strong and weak co-relations. The vital and decisive co-relations were health status, political views, personal safety, current income, credit card usage, etc.

Current health status, personal safety, and political opinion have a strong linear correlation to anxiety score. Again, age group and current month's income also have a linear correlation but not that strong. Also, feeling uneasy in a crowded place, stockpiling, having a chronic disease, and being affected by COVID-19 have a slightly linear correlation. On the other hand, hygienic behavior, credit card usage, and online shopping have a V-shaped correlation with anxiety scores. Furthermore, marital status, gender, living city type, number of children, and the remote working facility have a noticeable influence on anxiety.

Nevertheless, the average monthly income in the previous year and the mother's education have no influence on anxiety. Occupation and education level seem to have some influence on anxiety score, but no remarks could be drawn. Besides the working sector, the previous occupation does not have any influence on anxiety. However, employment status has a significant contribution to anxiety, which implies regardless of the occupation and the working sector, every individual is anxious about their job security (Fig. 5).

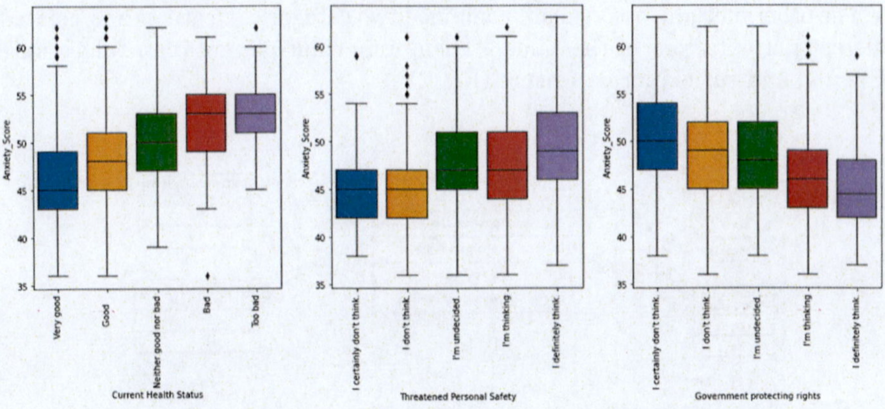

Fig. 5. Box Plot Visualization of Dominant Features

3.4 Feature Engineering

Feature engineering creates new features by manipulating, combining, or extracting existing features. These new features help the model to learn better and give a different insight into data that was not available earlier.

Utilizing the knowledge of EDA, some new features were developed through feature engineering. A basic engineered feature was the change in monthly income obtained by subtracting the monthly income from the preceding month. Besides, fancier features were created with complex feature engineering techniques using two methods- deviance and assortment.

The deviance method determined the deviation of a feature from others based on a category, e.g., how different a person's monthly income was relative to others in the same profession (Fig. 6).

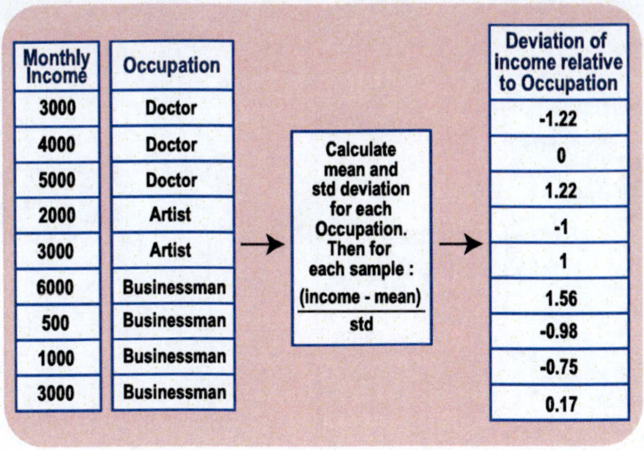

Fig. 6. Complex Feature Engineering with Deviance Method

The other method, assortment, combined two categorical features and created different permutations of the values. Then, important permutation values were extracted and turned into a feature (Fig. 7).

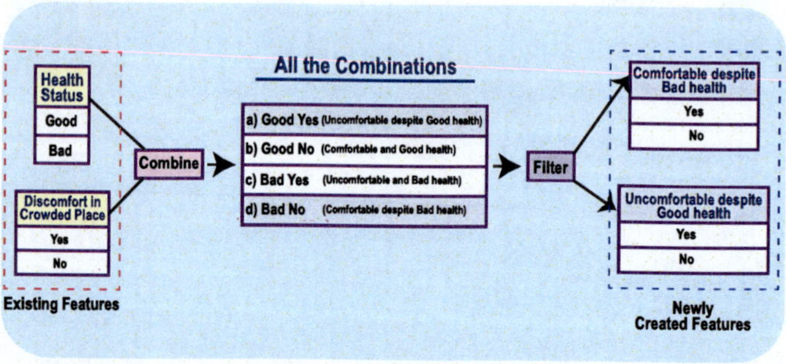

Fig. 7. Complex Feature Engineering with Assortment

3.5 Removing Outliers

Outliers are extreme values that reside outside the normal trend and often make it difficult to interpret or generalize. Tukey's method has been employed to detect outliers. According to Tukey's rule,

$$f(x) = \begin{cases} 1, \text{ if } x < Q1 - 1.5 \times IQR \\ 0, \text{ if } Q1 - 1.5 \times IQR \leq x \leq Q3 + 1.5 \times IQR \\ 1, \text{ if } x > Q3 + 1.5 \times IQR \end{cases} \quad (1)$$

where $f(x) = 1$ indicates that x is an outlier and $f(x) = 0$ indicates that x is not an outlier.

After detecting outliers from the prominent features, they were removed from the dataset (Figs. 8 and 9).

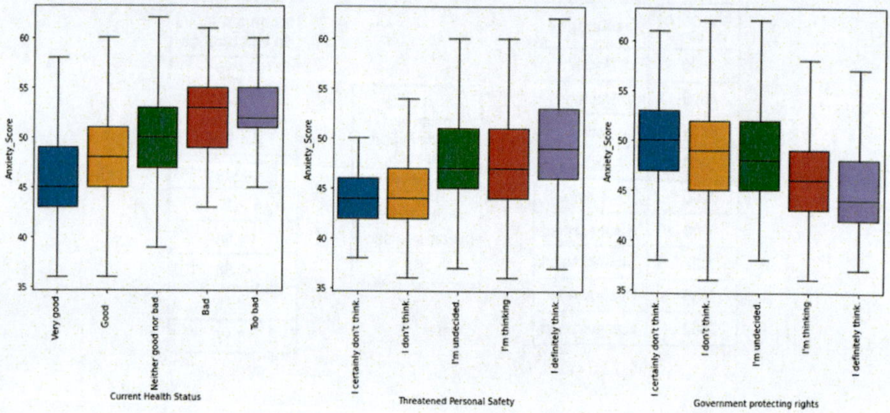

Fig. 8. Outliers Removed from Most Important Features

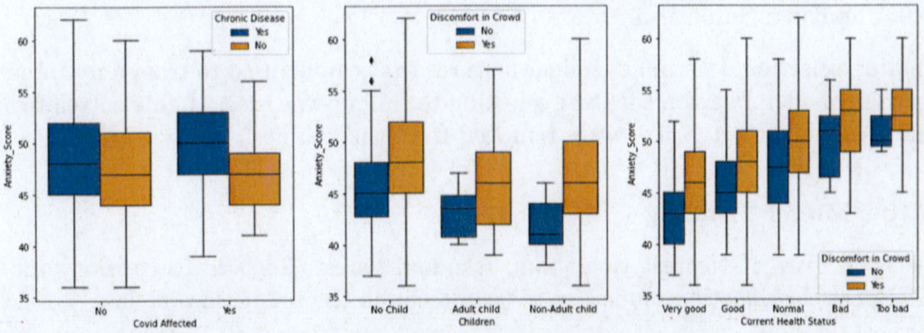

Fig. 9. Outlier Removal from Newly Engineered Features

3.6 Encoding

The dataset had predominantly categorical features. Two types of categorical features- nominal and ordinal- were encoded into numerical features through encoding. One hot encoding was used to convert the nominal categorical features, and Label encoding was applied to ordinal categorical features. This resulted in 8 features being one-hot encoded and 14 features being label encoded (Fig. 10).

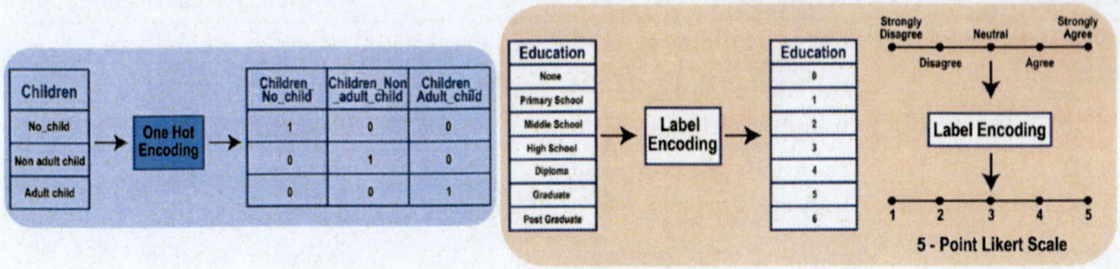

Fig. 10. One Hot and Label Encoding

3.7 Splitting

To evaluate machine learning models, the dataset was split into train and test sets in the ratio 90:10. Stratified train test split was used to preserve the class ratio of the target feature.

3.8 Scaling

Features with different ranges of values create dominance of some features over others. Distance-based machine learning algorithms become biased to features with larger ranges. To overcome this problem, scaling is applied to convert different ranges into a single range. MinMaxScaler is applied to the dataset, which transforms different ranges into a single [0, 1] range.

3.9 Feature Selection

Feature selection determines salient features that can be used to train a machine learning model. Pearson's feature selection technique was used. Highly correlated features (coefficient > 0.9) were removed from the train set.

3.10 Model Training

Decision Tree, K-Nearest Neighbors, Random Forest, Logistic Regression, and XGBoost classification models were used to build the prediction model.

3.11 Hyperparameter Tuning

Hyperparameter tuning was used to identify the best sets of hyperparameters for each model. It was complemented with cross-validation to get a better estimation of the models' future performance. GridSearchCV, along with Stratified K-fold Cross-Validation, was executed to elicit the optimum performing model (Table 2).

Table 2. Hyperparameter Tuning of the Models

Model	Hyperparameter Space	Best Hyperparameter Set
K-Nearest Neighbors	metric: minkowsi, euclidean, manhattan	metric: manhattan
	n_neighbors: 20, 22, 24, ..., 30	n_neighbors: 24
	weights: uniform, distance	weights: distance
Decision Tree	criterion: gini, entropy	criterion: gini
	max_depth: 1–25	max_depth: 4
	splitter: best, random	splitter: random
Random Forest	criterion: gini, entropy, log_loss	criterion: entropy
	n_estimators: 40, 45, 50, ..., 120	n_estimators:110
XGBoost	cols_sample_bytree: 0–1	cols_sample_bytree:0.1
	eta: 0.1–0.01	eta: 0.1
	max_depth: 1–6	max_depth:1
	n_estimators: 30, 35, 40, ..., 100	n_estimators:40

4 Result

4.1 Performance Metrics

The appropriate performance metric for a model depends on the type of problem and the characteristics of the dataset. In this study, since the target feature was almost balanced, accuracy would be an appropriate metric. Additionally, recall value would be important as the problem fits into the medical disease detection spectrum. Besides precision, f1 score and AUC score would also provide valuable insight into how good the model is.

Decision Tree classifier showed better performance compared to other models, followed by Random Forest and Logistic Regression. In contrast, the other models, K-Nearest Neighbors and XGBoost, performed poorly. The decision tree classification has not only produced the best accuracy but also delivered the best recall and f1-score, which are crucial in our problem spectrum (Table 3).

Table 3. Performance of ML Algorithms without Hyperparameter Tuning

MODEL	Accuracy	Precision	Recall	F1 Score	AUC Score
Decision Tree	**0.7127**	**0.7032**	**0.7676**	**0.7340**	**0.7108**
K-Nearest Neighbors	0.6836	0.6923	0.6971	0.6947	0.6831
Random Forest	0.7054	0.7194	0.7042	0.7117	0.7055
Logistic Regression	0.7090	0.7279	0.6971	0.7122	0.7094
XGBoost	0.6509	0.6691	0.6408	0.6546	0.6512

After applying stratified k-fold cross-validation with GridSearchCV hyperparameter tuning, the best model changed. The best hyperparameters set of the models for this problem are as follows (Table 4):

Table 4. ML Models' Score after Hyperparameter Tuning

MODEL	Best Score
K-Nearest Neighbors	0.6642
Decision Tree	0.6701
Random Forest	0.6904
XGBoost	**0.6956**

Retraining the dataset with the best model (XGBoost) and hyperparameters set the following result. Accuracy, precision, and AUC score were better than previous models (Table 5).

Table 5. Best ML Model Performance after Tuning

Best Model	Accuracy	Precision	Recall	F1 Score	AUC Score
XGBoost	0.7200	0.7372	0.7112	0.7240	0.7202

The performance of the XGBoost model for each class was evaluated as follows (Table 6):

Table 6. Performance of the Model on Two Classes

Model	Classification	Precision	Recall	F1 Score
XGBoost	No Anxiety	0.703	0.729	0.716
	Anxiety	0.737	0.711	0.724

4.2 Explainable AI

Machine learning models are considered "black boxes" because their inner workings and decision-making processes are at many times unclear. This often makes AI models difficult to trust, even when they produce good results. Explainable AI addresses this problem by making models more interpretable through various XAI methods. One such method is SHapley Additive exPlanations (SHAP), which is a mathematical method used to explain the predictions of a machine learning model. In this paper, we use SHAP to explain the predictions of the XGBoost model.

SHAP values calculate the contribution of each feature to the final prediction, where the sum of all feature contributions equals the final outcome value. This allows for the creation of bar plots of SHAP values to demonstrate feature importance and summary plots to show the positive or negative influence of a feature on the prediction.

From the bar plot, the most important feature for the prediction is found to be the health status, followed by personal safety and political opinion. They are followed by marital status, gender, *srh_c_anc*, online shopping habit, *newchi_c_anc*, stockpiling habit, and hygienic behavior. *SRH_C_anc41* and *new_Chi_C_anc01* are our engineered features; the first indicates whether a person with good health feels uneasy in a crowded place, while the second indicates how people with no children feel in crowded places (Fig. 11).

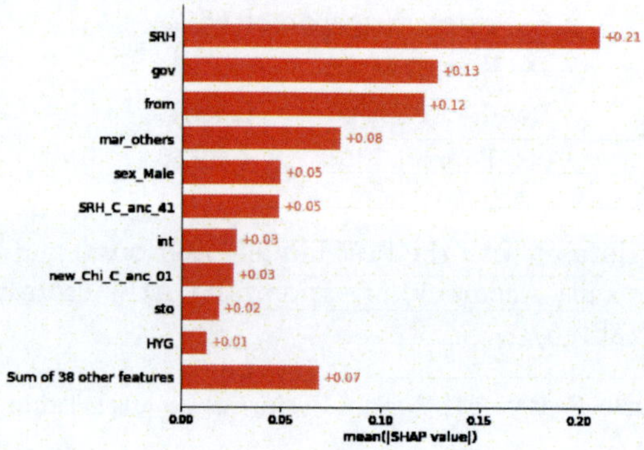

Fig. 11. SHAP Bar Plot of the Final XGBoost Model

The summary plot shows that the worse the health status, the higher the anxiety. Again, for personal safety features, the more unsafe people feel in society, the higher their anxiety. On the contrary, for the political opinion feature, the lower the satisfaction with the government, the lower the anxiety (Fig. 12).

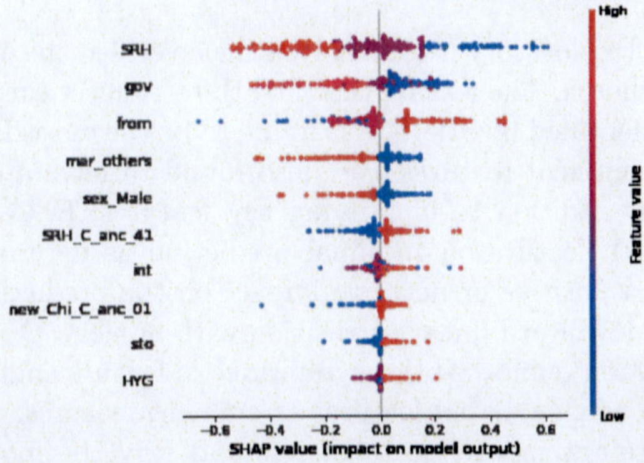

Fig. 12. SHAP Summary Plot of the Final XGBoost Model

Furthermore, SHAP can implement local interpretability by calculating shap values for each individual prediction and illustrating how the features contribute to that single prediction. Some false prediction samples were randomly taken, and their local interpretations were visualized through SHAP waterfall plots (Fig. 13).

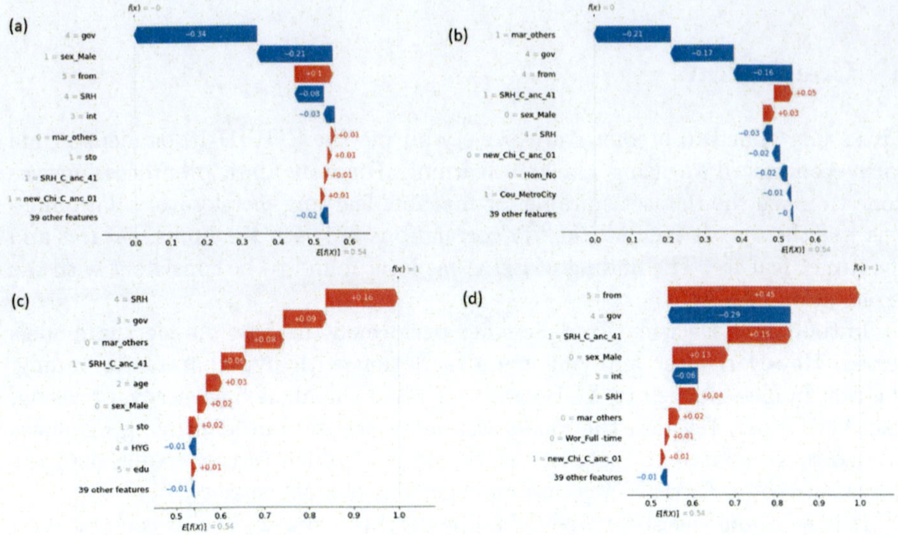

Fig. 13. SHAP Local interpretability of Wrong Predictions

5 Discussion

It is worth noting that the methodology used in this study, although it is based on a specific data source, can be applied to other datasets as well. The preprocessing steps and the EDA can be adapted to suit different datasets with similar characteristics.

The findings of Exploratory Data Analysis have verified the explanations provided by Explainable AI. The feature that had the strongest correlation with the target feature, as identified by SHAP, was found to be the most dominant feature. Similarly, other prominent features were also found to have a clear correlation with the target. In addition to identifying key features, EDA also establishes the influence of each feature on the final prediction as determined by SHAP. Features that had a positive or negative impact on the prediction, respectively, had an upward or downward linear correlation with anxiety. On the other hand, the XAI interpretation reinforced the significance of feature engineering by highlighting two of the newly created features as eminent features.

Further, an attempt was made to understand why the model made incorrect predictions: false positives and negatives. Random false prediction samples were investigated with the help of SHAP local interpretability. However, no generalized pattern was identified as different features contributed differently to particular predictions, and the most important features varied among samples. As a result, this investigation approach could scarcely find any explanation for the wrong predictions. This highlights the need for further research to improve the interpretability of these wrong predictions, which could significantly increase the overall performance of the model.

6 Conclusion

This study aimed to predict state anxiety during the COVID-19 pandemic from socio-economic data using machine learning. Rigorous data preprocessing was done to make the dataset suitable for machine learning. Exploratory data analysis was then performed to identify correlations between the input features and the target feature. The finding of EDA was later found to be consistent with the explanation of XAI.

Initially, the Decision Tree classifier performed the best among the models tested. However, after applying cross-validation with hyperparameter tuning, the best model changed to XGBoost, which had the highest accuracy, precision, and AUC score. Besides, the model was more accurate in identifying instances of anxiety compared to instances of no anxiety, which further emphasizes the importance of early detection and management of state anxiety.

In conclusion, this study highlights the potential of using socio-economic data to predict state anxiety during the COVID-19 pandemic and the importance of hyperparameter tuning and feature engineering in achieving optimal model performance. Nevertheless, it also emphasizes the need for further research to improve the performance and interpretability of the models.

References

1. Avena, N.M., Simkus, J., Lewandowski, A., Gold, M.S., Potenza, M.N.: Substance use disorders and behavioral addictions during the Covid-19 pandemic and Covid-19-related restrictions. Front. Psych. **12**, 653674 (2021)
2. Burkova, V.N., et al.: Predictors of anxiety in the Covid-19 pandemic from a global perspective: data from 23 countries. Sustainability **13**(7), 4017 (2021)
3. Chowdhury, R.N., Hassan, M.F., Arshaduzzaman Fahim, M., Momen, S.: Stress mining from sleep-related parameters. In: Silhavy, R., Silhavy, P., Prokopova, Z. (eds.) Data Science and Algorithms in Systems, CoMeSySo 2022. Lecture Notes in Networks and Systems, vol. 597, pp. 740–750. Springer, Cham (2023). https://doi.org/10.1007/978-3-031-21438-7_62
4. Ge, F., Zhang, D., Wu, L., Mu, H.: Predicting psychological state among Chinese undergraduate students in the Covid-19 epidemic: a longitudinal study using a machine learning. Neuropsychiatr. Dis. Treat. **16**, 2111 (2020)
5. Leal, P.C., Goes, T.C., da Silva, L.C.F., Teixeira-Silva, F.: Trait vs. state anxiety in different threatening situations. Trends Psychiatry Psychother. **39**, 147–157 (2017)
6. Nemesure, M.D., Heinz, M.V., Huang, R., Jacobson, N.C.: Predictive modeling of depression and anxiety using electronic health records and a novel machine learning approach with artificial intelligence. Sci. Rep. **11**(1), 1–9 (2021)
7. Pal, D., Sahu, D.P., Maji, S., Taywade, M.: Prevalence of anxiety disorder in adolescents in India: a systematic review and meta-analysis. Cureus **14**(8), e28084 (2022)
8. Perpetuini, D., et al.: Prediction of state anxiety by machine learning applied to photoplethysmography data. PeerJ **9**, e10448 (2021)
9. Pieh, C., Budimir, S., Delgadillo, J., Barkham, M., Fontaine, J.R., Probst, T.: Mental health during covid-19 lockdown in the United Kingdom. Psychosom. Med. **83**(4), 328–337 (2021)
10. Pintelas, E.G., Kotsilieris, T., Livieris, I.E., Pintelas, P.: A review of machine learning prediction methods for anxiety disorders. In: Proceedings of the 8th International Conference on Software Development and Technologies for Enhancing Accessibility and Fighting Info-exclusion, pp. 8–15 (2018)
11. Priya, A., Garg, S., Tigga, N.P.: Predicting anxiety, depression and stress in modern life using machine learning algorithms. Procedia Comput. Sci. **167**, 1258–1267 (2020)
12. Riello, M., Purgato, M., Bove, C., MacTaggart, D., Rusconi, E.: Prevalence of post-traumatic symptomatology and anxiety among residential nursing and care home workers following the first Covid-19 outbreak in Northern Italy. R. Soci. Open Sci. **7**(9), 200880 (2020)
13. Roy-Byrne, P.P., Joesch, J.M., Wang, P.S., Kessler, R.C.: Low socioeconomic status and mental health care use among respondents with anxiety and depression in the NCS-R. Psychiatr. Serv. **60**(9), 1190–1197 (2009)
14. Sari, E., Kağan, G., Karakuş, B.Ş, Özdemir, Ö.: Dataset on social and psychological effects of Covid-19 pandemic in Turkey. Sci. Data **9**(1), 1–7 (2022)
15. Sau, A., Bhakta, I.: Predicting anxiety and depression in elderly patients using machine learning technology. Healthc. Technol. Lett. **4**(6), 238–243 (2017)
16. Speilberger, C.D., Gorsuch, R., Lushene, R., Vagg, P., Jacobs, G.: Manual for the State-Trait Anxiety Inventory. Consulting Psychologists, Palo Alto, CA (1983)
17. Sultana, J., Quadery, S.E.U., Amik, F.R., Basak, T., Momen, S.: A data-driven approach to understanding the impact of Covid-19 on dietary habits amongst Bangladeshi students. J. Positive School Psychol. **6**, 11691–11697 (2022)

18. Tedaldi, E., Orabona, N., Hovnanyan, A., Rubaltelli, E., Scrimin, S.: Trends in state anxiety during the full lockdown in Italy: the role played by Covid-19 risk perception and trait emotional intelligence. Trauma Care **2**(3), 418–426 (2022). https://www.mdpi.com/2673-866X/2/3/34
19. Wiedemann, K.: Anxiety and anxiety disorders. In: Smelser, N.J., Baltes, P.B. (eds.) International Encyclopedia of the Social & Behavioral Sciences, pp. 560–567. Pergamon, Oxford (2001). https://doi.org/10.1016/B0-08-043076-7/03760-8. https://www.sciencedirect.com/science/article/pii/B0080430767037608
20. Wu, S., Zhang, K., Parks-Stamm, E.J., Hu, Z., Ji, Y., Cui, X.: Increases in anxiety and depression during Covid-19: a large longitudinal study from china. Front. Psychol. **12**, 2716 (2021)

Detection of IoT Communication Attacks on LoRaWAN Gateway and Server

Tibor Horák(✉), Peter Střelec, Szabolcs Kováč, Pavol Tanuška, and Eduard Nemlaha

Institute of Applied Informatics, Automation and Mechatronics, Faculty of Materials Science and Technology in Trnava, Slovak University of Technology in Bratislava, Trnava, Slovakia
{tibor.horak,peter.strelec,szabolcs.kovac,pavol.tanuska, eduard.nemlaha}@stuba.sk

Abstract. Internet of things (IoT) devices are widespread and frequently used for data acquisition. Increasingly, they are exposed to different types of attacks. IoT communications often use the Long-Range Wide Area Network (LoRaWAN) protocol for longer distances. This proposal enables the identification of a Denial-of-Service (DoS) attack on a LoRaWAN gateway and its integration into the monitoring and detection of security incidents. The presented model can be applied to LoRaWAN gateway infrastructures that provide communication coverage, where differences in the distances of IoT device transmitters are exploited. The demonstration is a mechanism for integrating data from the network layer and communication logs of the LoRaWAN gateway. The communication logs were recorded at the LoRaWAN gateway. Each IoT device to which the gateway provides connectivity generates data. These data is collected and used to identify and detect a DoS attack. This model will allow the attack to be detected and add information about the quality of the communication, or it can help identify and shut down the gateways that are targeted by the attack.

Keywords: IoT · LoRaWAN security · LoRa transmit · DoS attacks

1 Introduction

LoRaWAN protocol security is designed to provide the basic conditions of use for IoT devices. The protocol takes into account the IoT device's low power consumption, ease of implementation, wide scalability, and affordable prices [1].

The start part of the connection, when the connection of the IoT device to the LoRaWAN gateway occurs, authentication is implemented. With authentication, it is ensured that only authenticated devices can connect to the network. Application and MAC (Media Access Control) messages are authenticated in origin, and transmission integrity protection is established [2].

The transmitted data are encrypted. This basic type of protection, built on mutual authentication, ensures that the content of network traffic is not altered and originates from a legitimate device. It also serves as a basic protection against eavesdropping on traffic that is intercepted by attackers [3].

R. Silhavy and P. Silhavy (Eds.): CSOC 2023, LNNS 724, pp. 489–497, 2023.
https://doi.org/10.1007/978-3-031-35314-7_43

LoRaWAN security further defines in the protocol part end-to-end encryption for the application payload exchanged between endpoint devices and the application server. LoRaWAN is one of the IoT protocols that use end-to-end encryption [4].

The LoRa Alliance publishes the LoRaWAN protocol standard, which is primarily used to connect end devices where long battery life is required. Multiple specifications v1.0.47 (released in 2020) and v1.12 (released in 2017) have been released. However, available end-devices are still compatible with the previous LoRaWAN standards v1.0.2 released in 2016 and v1.0.38 from 2018. Such devices cannot be upgraded to newer versions of the LoRa specification. The latest v1.1 and v1.0.4 versions differ in the definition of hardware requirements [5].

LoRaWAN's security mechanisms use Advanced Encryption Standard (AES) cryptographic algorithms. LoRaWAN uses AES encryption in concert with several modes that provide integrity and encryption during transmissions. The first is Cipher-based Message Authentication Code3 (CMAC) for integrity protection and Counter Mode Encryption4 (CTR) for encryption. LoRaWAN devices come with a unique 128-bit AES key (AppKey) and a globally unique identifier EUI-64-based (DevEUI). Both of these device values are used in the device authentication process on the network. LoRaWAN networks are identified by a 24-bit, globally unique identifier assigned by the LoRa Alliance [6].

When ABP is enabled, the device address (DevAddr) and network identity (NwkID) for the preselected network are fixed. The end device uses them unchanged and they remain the same throughout the life of the device. The connection procedure is simpler than that with OTAA. In this setting, the device can only work properly in its predefined network [7].

Even if the network allows registration of end devices with DevAddr values other than those defined for the network, the Packet Broker cannot route data from these end devices to the correct network because the DevAddr is different from what is allowed. If the Packet Broker is set to accept such a transfer, a problem with the transfer to the network server may occur again. A server-side error may occur here, where it will not be able to optimize the DevAddr allocation and the data will be ignored. IoT devices that use OTAA are assigned a new DevAddr each time a new session is created. This will ensure and enable easy transfer of devices and registration to different networks [8].

To connect an OTAA, it is necessary to ensure that the terminal equipment is within the coverage range of the network in which it is registered. The OTAA connection procedure requires that the end-device can receive a Join Accept downlink message from the network server. If the end-device does not have network coverage at power-up, the join procedure will fail [9].

Depending on the LoRaWAN firmware and memory used in the device, a failed OTAA connection may have a high probability of attempting to retry the transmission. Thus, it can easily happen that the device does not go to sleep. It will re-execute the connection sequence, thus consuming a lot of power and the lifetime of the device will be low [10].

It is an incorrect practice to use ABP instead of OTAA thinking that this connection method is more reliable. In this respect, ABP is less flexible and allows the device to start transmitting messages immediately after power-up [11].

It does not take into account at all whether it has available network coverage. It is also convenient to use OTAA if the device is rebooting. ABP uses storage to hold the frame counters between restarts. The OTAA stores the OTAA session instead of the frame counters, and thus a seamless communication is guaranteed [12].

2 Background

LoRaWAN uses a radio signal, according to the respective specification that is valid for different regions. As shown in Fig. 1, LoRaWAN is one a Low-power wide-area networks (LPWAN). It is designed to transmit small volumes of data but over long distances. As mentioned, it is precisely the low power requirements and battery life that these IoT devices are often used in areas that must cover distances in the order of kilometers [13].

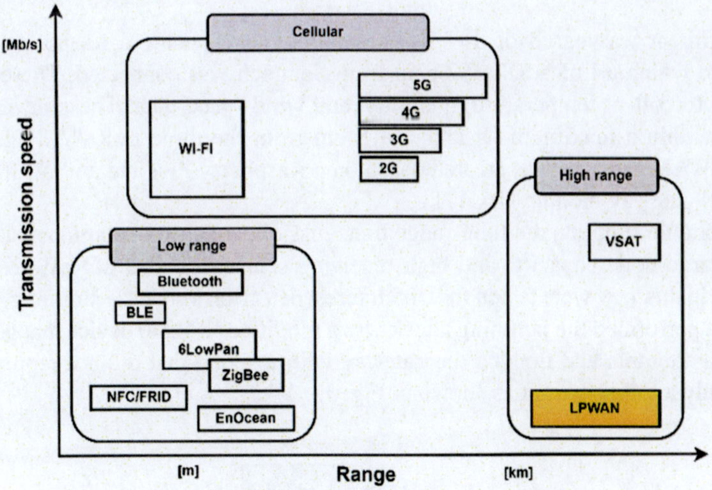

Fig. 1. Protocol transmission speed by range

The radio signals produced by IoT devices can be jammed or intercepted. It is the interference at the gateway or node that can lead to interrupted transmission and result in Denial-of-Service (DoS), preventing legitimate devices from using the service. However, such an attack can be detected [14].

There are, of course, some radio-frequency interferences that are quite difficult to detect, but due to the type of modulation used for LoRaWAN, these interferences can be well observed. An overview of the attacks that are known on LoRaWAN networks can be seen in Fig. 2. To show network identification of attack we choose Jamming DoS attack [15].

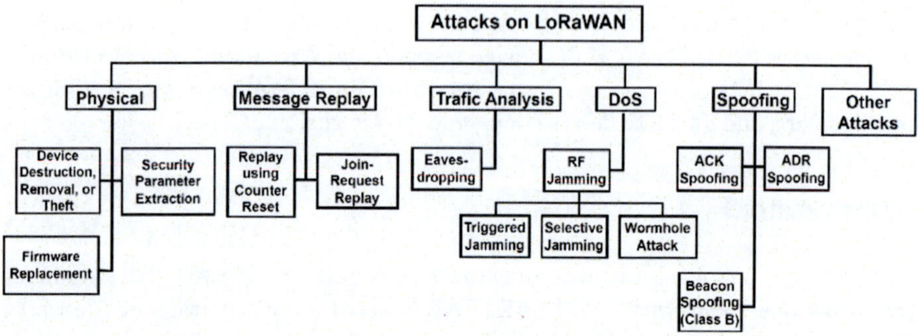

Fig. 2. Attacks on LoRaWAN

3 Materials and Methods

The experiment was based on IoT devices implemented using a Raspberry Pi Pico RP2040, to which a LoRa SX1262 chip from Semtech was connected. These devices were used to collect temperature, humidity, and wind speed data. These devices were deployed within a maximum Distance of 8 km from the three LoRaWAN gateways. The LoRaWAN gateway was created based on a Raspberry Pi 4 and an SX1302 868M LoRaWAN gateway module [16, 17].

This module supports the long-range transmission, concurrent communication, can handle a large node capacity, and high receiving sensitivity. The LoRaWAN devices connected in this way were tested for DoS interest detection known as jamming. The IoT device that performed the jamming attack was a modified RP2040 device that generated connection attempts and flooded the gateway with a signal that did not represent real data but only a login attempt as shown in Fig. 3.

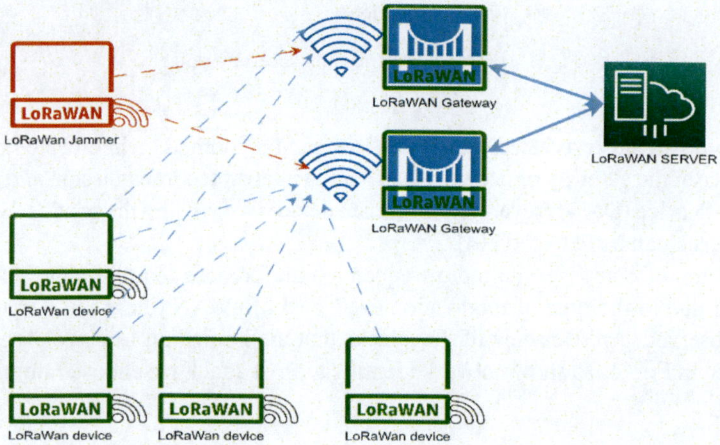

Fig. 3. The block scheme IoT communication with the Jammer device

Jammer is a type of DoS attack, and the goal of the attacker is to prevent or completely shut down the usability of the LoRaWAN network. During this attack, a jammer was planted by the attacker and placed as close as possible to the LoRaWAN gateway. Using a modified end IoT device, on which the maximum transmission power is usually also set and its typical behavior is constant retransmission, which it overloads the transmission line so that other devices cannot communicate with the gateway [18].

The physical location of the gate must be known to achieve the best effect of the attack. Simultaneously, it is necessary to know the frequency band in which the network operates. IoT devices communicated using OTAA. The jamming attack was conducted using a single IoT device, and then the experiment was repeated using multiple devices [19].

The connection sequence was selected for detection. The activation sequence of the transmission using the OTAA method must have the end devices set up. The DevEUI and AppKey values are mandatory, which are known by both the transmitting IoT device and the server. The IoT device broadcasts a join-request where the join-request packet identifier and the DevNonce value are additionally present. DevNonce is a 16-bit, randomly generated number by the IoT device. This number is randomly generated each time a join sequence occurs. It is possible to use the DevNonce value to detect a jamming attack [20].

The DevNonce value is generated using an n-count Least Significant Bit (LSB) read operation of the RegRssiWideband. The value of this register is obtained from the broadband signal strength at the receiver each 1 ms. It is assumed that the LSB value varies constantly and randomly depending on the quality of the signal [21].

4 Experiment Implementation and Result

To illustrate the use of network protocol analysis as a source of information to evaluate the security of an attack, a DoS attack was used. Data. These were logged on the LoRaWAN harness were evaluated with analytical tools and classified according to the type of communication. Subsequently, it was evaluated which gateway was affected by the attack. Figure 4 shows the LoRaWAN network topology and it was mapped to the OSI Layer. In the Gateway and Network Server parts, there can be obtained data that can be used to detect of DoS attack.

Detection of jamming attack can be based on the computation of the Hamming distance between DevNonce during the join sequence. This is the difference between the newly received join-request packet and the DevNonce of the last received join-request packet. This method can be used to determine the baseline values of the hamming distance of consecutive DevNonce values. These values are further used as thresholds to detect the jammer in the network. In addition, join-request packets are signed with an AppKey to ensure integrity. An attacker would need to know the value of the AppKey to pass off his activity as legitimate traffic [22].

Both the network and application server along with the gateway were configured on Raspberry Pi. The open-source solution Chirpstack was used to configure the servers. For proper server operation, it was necessary to install the mqtt broker to start with. Gateway-Bridge communication was then set up and for both the network and application servers.

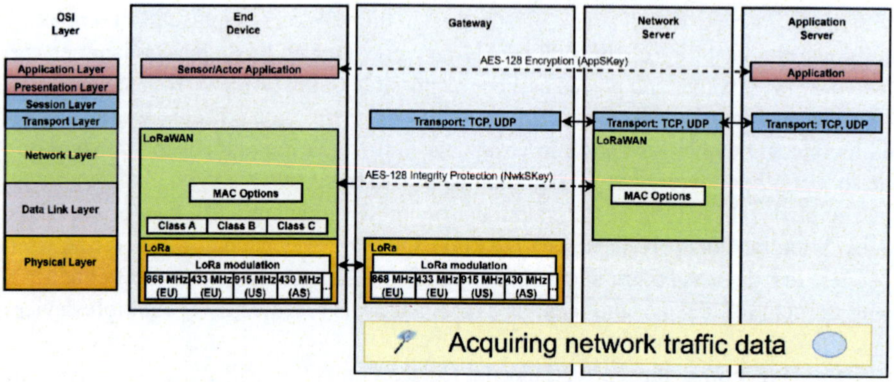

Fig. 4. LoRaWAN Network layers with data acquisition

The communication between the gateway and the server was secured using a packet-forwarder. The packet forwarder is also responsible for logging radio and network traffic information.

Packet forwarder logs were in JSON format. This data is fed to the input of the parser, which detects the packet sent by the gateway to the server with the "up:" flag. Another flag identifying the data packet is the "rxpk" attribute. This packet is then for further processing, only the payload ("data") is selected. The join-request packet is a special civic in that it is exactly 46 hexadecimal characters long. A sufficient condition for identification is the identification of packets with 64 hexadecimal characters. If a join request is identified from the log, the DevNonce value is read, and the Hamming distance is calculated. The Fig. 5 shows the standard connection traffic. The Fig. 6 shows both standard and jammer traffic. The DoS attack is visible between the standard connection process and the process when the jammer attacker is active.

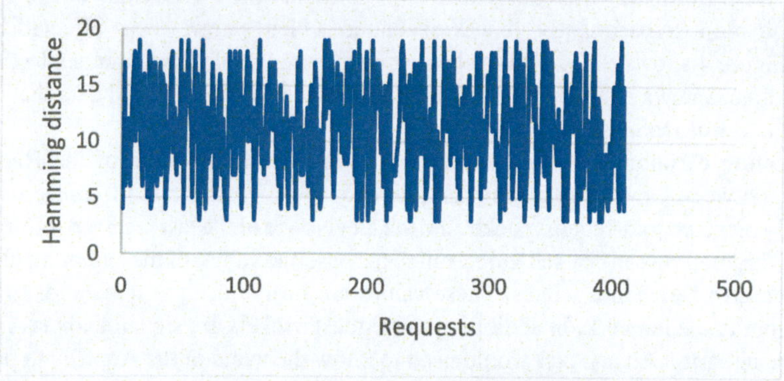

Fig. 5. Hamming distance standard traffic

The jammer-induced DoS attack where the hamming visibility values for DevNonce can be seen in Fig. 6 between requests numbers 250 to 450 and between requests 750 and

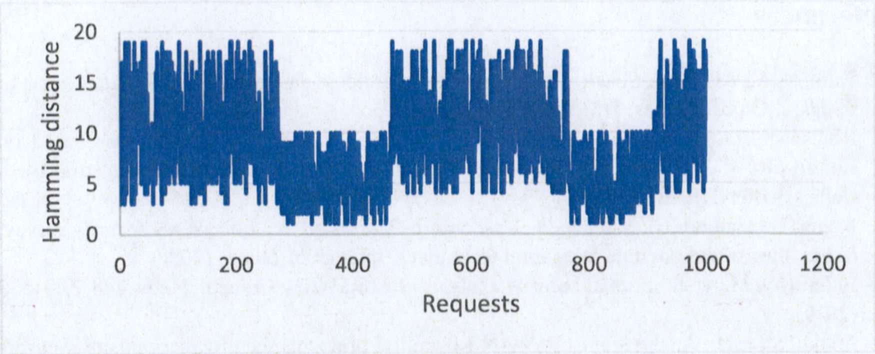

Fig. 6. Hamming distance with jammer activity

920. Calculating and storing the hamming distance is not a computationally intensive operation. By storing this data and mapping it according to packet delivery time, it is easy to obtain a daily overview of the behavior of gateways and servers.

5 Conclusions

IoT devices that use the LoRaWAN protocol are exposed to several types of attacks. The impact of a Jammer attack has been analyzed. Multiple gateways were implemented to identify the network traffic to allow logging of the communication. The type of attack thus identified can be further used to monitor the network traffic. Simultaneously, the information that is generated during the traffic to the log file can be used for further potential attack detection. These data can be further analyzed to provide a better way to monitor traffic or identify problem devices. In the case of deploying multiple gateways, it is possible to prevent attackers from shutting down the communication.

Verification of the detection of potential attacks led to the design of LoRaWAN network protection using multiple LoRaWAN gateways, which can be switched off or on as required. However, more devices need to be secured and operated, which can make the setup more challenging, especially for wide area networks.

Acknowledgments. This research was funded by the Scientific Grant Agency of the Ministry of Education, Science, Research, and Sport of the Slovak Republic and the Slovak Academy of Sciences, grant number VEGA 1/0176/22 Proactive control of hybrid production systems using simulation-based digital twin and VEGA 1/0193/22 Proposal of identification and monitoring of production equipment parameters for the needs of predictive maintenance in accordance with the concept of Industry 4.0 using Industrial IoT technologies. This article was also written thanks to the generous support under the Operational Program Integrated Infrastructure for the project: "Strategic research in the field of SMART monitoring, treatment and preventive protection against coronavirus (SARS-CoV-2)", Project no. 313011ASS8, co-financed by the European Regional Development Fund.

References

1. Falih, M.A., Ali, N.G., Shakir, W.M.R.: Lorawan protocol data communication in internet of things. J. Glob. Sci. Res. **7**(5), 2279–2282 (2022)
2. Khutsoane, O., Isong, B., Abu-Mahfouz, A.M.: IoT devices and applications based on LoRa/LoRaWAN. In: IECON 2017–43rd Annual Conference of the IEEE Industrial Electronics Society. IEEE (2017)
3. Noura, H., Hatoum, T., Salman, O., Yaacoub, J.-P., Chehab, A.: LoRaWAN security survey: issues, threats and possible mitigation techniques. Internet of Things (2020)
4. Eldefrawy, M., et al.: Formal security analysis of LoRaWAN. Comput. Netw. **148**, 328–339 (2019)
5. Loukil, S., et al.: Analysis of LoRaWAN 1.0 and 1.1 protocols security mechanisms. Sensors **22**(10), 3717 (2022)
6. Mamvong, J.N., Goteng, G.L., Zhou, B., Gao, Y.: Efficient security algorithm for power-constrained IoT devices. IEEE Internet Things J. **8**(7), 5498–5509 (2021). https://doi.org/10.1109/JIOT.2020.3033435
7. Hayati, N., et al.: Potential development of AES 128-bit key generation for LoRaWAN security. In: 2019 2nd International Conference on Communication Engineering and Technology (ICCET). IEEE (2019)
8. Thaenkaew, P., Quoitin, B., Meddahi, A.: Evaluating the cost of beyond AES-128 LoRaWAN security. In: 2022 International Symposium on Networks, Computers and Communications (ISNCC). IEEE (2022)
9. Chacko, S., Job, D.: Security mechanisms and Vulnerabilities in LPWAN. In: IOP conference Series: Materials Science and Engineering, vol. 396, no. 1. IOP Publishing (2018)
10. Lestari, R.I., Suryani, V., Wardhana, A.A.: Digital signature method to overcome sniffing attacks on LoRaWAN network. Int. J. Electr. Comput. Eng. Syst. **13**(7), 533–539 (2022). https://doi.org/10.32985/ijeces.13.7.5
11. Seller, O.: LoRaWAN security. J. ICT Stand., 47–60 (2021)
12. Tsai, K.-L., et al.: AES-128 based secure low power communication for LoRaWAN IoT environments. IEEE Access **6**, 45325–45334 (2018)
13. Mikhaylov, K., Petaejaejaervi, J., Haenninen, T.: Analysis of capacity and scalability of the LoRa low power wide area network technology. In: European Wireless 2016; 22th European Wireless Conference. VDE (2016)
14. Tomasin, S., Zulian, S., Vangelista, L.: Security analysis of lorawan join procedure for internet of things networks. In: 2017 IEEE Wireless Communications and Networking Conference Workshops (WCNCW). IEEE (2017)
15. Ingham, M., Marchang, J., Bhowmik, D.: IoT security vulnerabilities and predictive signal jamming attack analysis in LoRaWAN. IET Inf. Secur. **14**(4), 368–379 (2020)
16. Raspberry Pi Documentation https://www.raspberrypi.com/documentation/microcontrollers/rp2040.html. Accessed 20 Jan 2023
17. Semtech SX1262 https://www.semtech.com/products/wireless-rf/lora-core/sx1262. Accessed 22 Jan 2023
18. Kuntke, F., et al.: LoRaWAN security issues and mitigation options by the example of agricultural IoT scenarios. Trans. Emerg. Telecommun. Technol. **33**(5), e4452 (2022)
19. Ruotsalainen, H., et al.: LoRaWAN physical layer-based attacks and countermeasures, a review. Sensors **22**(9), 3127 (2022)
20. Na, S.J., et al.: Scenario and countermeasure for replay attack using join request messages in LoRaWAN. In: 2017 International Conference on Information Networking (ICOIN). IEEE (2017)

21. Kim, J., Song, J.S.: A simple and efficient replay attack prevention scheme for LoRaWAN. In: Proceedings of the 2017 the 7th International Conference on Communication and Network Security (2017)
22. Danish, S.M., Nasir, A., Qureshi, H.K., Ashfaq, A.B., Mumtaz, S., Rodriguez, J.: Network intrusion detection system for jamming attack in LoRaWAN join procedure. In: 2018 IEEE International Conference on Communications (ICC), Kansas City, MO, USA, pp. 1–6 (2018). https://doi.org/10.1109/ICC.2018.8422721

The Application of Multicriteria Decision-Making to the Small and Medium-Sized Business Digital Marketing Industry

Anthony Saker Neto, Marcelo Bezerra de Moura Fontenele, Michele Arlinda Aguiar, Raquel Soares Fernandes Teotonio, Robervania da Silva Barbosa, and Plácido Rogério Pinheiro(✉)

Graduate Program in Administration, University of Fortaleza (UNIFOR), Fortaleza, Ceará, Brazil
placido@unifor.br

Abstract. Marketing is the art of exploring, creating, and delivering value to satisfy a consumer market's needs and/or wishes. It is an industry that focuses on generating value for a business's product, service, or brand, aiming to conquer and retain customers. Since digital marketing is understood as a communication process with a target audience and to externalize knowledge in an adequate and impactful way, many small and medium businesses are faced with the dilemma of what marketing structure to adopt, whether in-house or outsourced. Investing in digital marketing is a need of nearly all of today's organizations; however, choosing the best structure is still challenging for many entrepreneurs. Aiming to tend to (not eliminate) the inherent subjectivity of said decision process, and thus to meet the objectives of this research, the multicriteria decision-aid method was applied through the AHP (Analytic Hierarchy Process) method. Regarding the methodology, the present article is classified as applied, quali-quantitative, exploratory, and bibliographic research. In addition, through the literature review, the scientific knowledge that supported this study was obtained; however, it is worth highlighting that the satisfactory results of this work entail an initial perspective of the theme. As such, further conclusive research may be carried out.

Keywords: Digital Marketing · AHP Method · Decision-making · Small and Medium Businesses

1 Introduction

Companies of all sizes invest in digital marketing to promote and sell their products online, boosting their customer reach regardless of their geographical location, which is an essential strategy for promoting brands to the world, considering that the current trend is for customers to search what they are looking for online.

R. Silhavy and P. Silhavy (Eds.): CSOC 2023, LNNS 724, pp. 498–505, 2023.
https://doi.org/10.1007/978-3-031-35314-7_44

To [12], the main objectives of marketing are related to the focus on sales growth, greater market participation, and increasing sales gross income. Through digital marketing, multiple objectives may be reached, enabling business promotion, and strengthening the quality of customer relationships. Regarding small and medium businesses, the dilemma of adopting the right marketing industry structure is even more relevant since, in many sectors, the resources are few and the costs are thoroughly monitored in order to safeguard some profit while strengthening themselves in the market as well as, in the best-case scenario, aiming to grow further. Therefore, this research's core problem is *what is the best structure for the digital marketing industry in small and medium businesses?* Marketing strategies may be established either through in-house or outsourced teams.

Based on [21], the multicriteria method allows for a greater understanding of the analyzed decision-making context. In this regard, the general objective of this study is to support the decision-making of digital marketing applications in small and medium businesses, based on the decision-making multicriteria methodology approach, the AHP (Analytic Hierarchy Process) method. Its specific objectives are to identify relevant criteria that support decision-making to choose the best structuration model, and to order the criteria according to their own importance through relative weights.

To support the feasibility of the presented research, this article is structured as follows: Sect. 2 introduces the theoretical framework by conceptualizing small and medium businesses and their entrepreneurial context, the basics of marketing and its evolution to digital marketing, the variables concerning the delimitation and planning of marketing in the companies, as well as the comprehension of the decision-taking method AHP. Moreover, Sect. 3 introduces the adopted research methodology and its limitations. Also, Sect. 4 exposes the results and analysis of the application of the AHP method, while Sect. 5 presents this article's final considerations.

2 Theoretical Framework

An insightful entrepreneur must be attentive and flexible to opinions. It is a challenging task since it is common to believe that having a fixed idea equals power. However, according to [26], having a solid goal does not mean being unyielding in their strategies to achieve it. Marketing can be understood as a process beyond the social sphere since it even overlooks needs in a lucrative manner [11]. Meanwhile, [26] conceptualizes it as a social and managerial process through which individuals gain what they need and wish for by creating and exchanging products and values. From a market standpoint, according to [15], it is necessary to understand the consumer and their expectations, needs, and satisfactions or in satisfactions when consuming.

This consumer observation and analysis process has gained a new chapter ever since the spread of the virtual relationship environment. In this sense, according to [3], digital marketing has gained strength by establishing permanent contact with clients, favoring the buying, and selling relationship while building trust. According to [25], marketing is still a managerial tool that encompasses all areas of a company, one that is committed to internal and external investments and its consequential return estimate. The same author portrays the flow of marketing's systemic process, which involves inputs consisting of financial, human, and material resources, as well as throughputs, or resource transformation processes for the product/services, price/remuneration, distribution/sales, and communication output scopes. In this context, management needs, among other decisions, to define whether the marketing structure it will adopt in its company will be in-house or outsourced. Multicriteria analysis methodologies are important tools for decision-making since preference conflicts involving people and factors exist [4].

To aid the resolution of this research's problem, the multicriteria AHP method was employed, as it is effective and presents an advantage by allowing its users to distribute, intuitively, relative weights for multiple criteria while making a pairwise comparison between them. Thus, even though there might be two or more incompatible variables, it allows, through peoples' knowledge and expertise, its users to recognize which criteria are more relevant for decision-making.

2.1 The AHP (Analytic Hierarchy Process) Method

Developed by Tomas L. Saaty in the early 1970s, the AHP is a multicriteria method widely employed and known for supporting decision-making for conflict resolution in multiple criteria problems [13, 27]. The AHP method consists of the following analytic thought steps: i) hierarchical level structuration; ii) weights and priorities definition; iii) logical consistency determination; iv) Decision Matrix [13, 27].

i) *Hierarchical level structuration:* To [6], a great number of complex problems can be simplified by dividing them into hierarchical levels. Generally, the problem's hierarchy is represented by the main objective, its relevant criteria, and decision alternatives. As an auxiliary technique for hierarchization, a conceptual map was used, which, according to [18], represents ideas or concepts in a written or graphic hierarchical diagram scheme by reporting the relations between concepts to reflect the cognitive structure of the chosen subject.

ii) *weight and priorities definition:* in this stage, it is necessary to fulfill the following steps:

a) Pairwise judgements:

Through the established hierarchical model, an evaluation through pairwise comparison is initiated, which means comparing each pair of criteria and their listed subcriteria, if there are any [2]. To perform this comparison, it is essential to use a number scale that shows the relevance of one element over another. Thus, this work has used Saaty's fundamental scale (1987), which contains the intensity of importance of the fundamental scale (1 to 9), its definition, and the explanation behind each item [16].

According to Saaty's [27] methodology, in which i represents a line and column, the comparison matrix adheres to the following cataloged rules:

- $a_{ii} = 1$ for every i. Every criterion, when compared to itself, shall have the same importance of value 1;
- $c_{ij} = 1/c_{ji}$. Suppose that in comparing criterion i to criterion j, the assigned importance was 7; in this case, in comparing criterion j to criterion i, this importance will be, for example, of 1/7 value.

b) Local and Global Priorities

The method's next phase aims to estimate the local and global priorities. It is about obtaining, through mathematical calculations, the relative contribution of each element of the hierarchical structure when compared to the immediate objective, as well as the primary objective. According to [27], the medium local priorities of the elements compared in the judgment matrix can be obtained through matrix operations by calculating its primary eigenvector and normalizing it after.

iii) *Logical Consistency Determination:* According to [6], although AHP incorporates often subjective preferences, the quality of the final decision depends on the consistency of comparison between pairs. The value indicated by [27] as an inconsistency limit is of 0,1. If this value exceeds 0,1, the problem needs to be restudied and the judgments reviewed [7].

3 Methodology

This article presents a literature review based on secondary data, encompassing books, dissertations, national and international published journals, and articles relevant to the theme, this being the application of the digital marketing industry in small and medium businesses and the application of the multicriteria AHP (Analytic Hierarchy Process) method for decision-making, calculated through excel spreadsheets.

Through literature research, which aimed to introduce the researchers to all that had been produced about the subject, scientific knowledge that served as a basis for this study's investigation was obtained. To [23], literature research:

> It is a necessary research strategy for the conduction of any scientific research. Literature research aims to explain and discuss a subject, theme, or problem based on references published in books, journals, magazines, encyclopedias, dictionaries, newspapers, websites, CDs, congress annals etc. It aims to know, analyze, and explain contributions on a specific subject, theme, or problem. Literature research is an excellent way of scientific training when done by itself - theoretical analysis - or as an essential part of any scientific paper, aiming to construct the theoretical study platform.

In addition, this article is based on quali-quantitative research, which turns qualitative judgments into a numeric scale, thus legitimizing the collected information and generating practical knowledge. Considering its purpose, this research is of applied nature to generate practical application knowledge to specific problems.

Regarding its objectives, it is exploratory research since it allows a broad viewpoint regarding a fact. According to [17], exploratory research aims to, among other things, provide greater information about the investigated subject, aid the delimitation of the research's theme, and direct the fixation of goals and formulation of hypotheses. It is worth mentioning that due to the exploratory character of this work, its results are not definitive, providing only an initial perspective about the theme. That means further conclusive research may be carried out.

4 Analysis

Initially, the construction of a cognitive map was carried out by identifying the necessary elements to be previously considered when a small and medium business defines its digital marketing industry structure. It is verifiable that said map is a perfectly adequate tool of organization and structuration of knowledge for the analysis of intentions and reasonings, which allows the presentation of information in an objective and ordered manner to aid the users in the comprehension of a situation/problem.

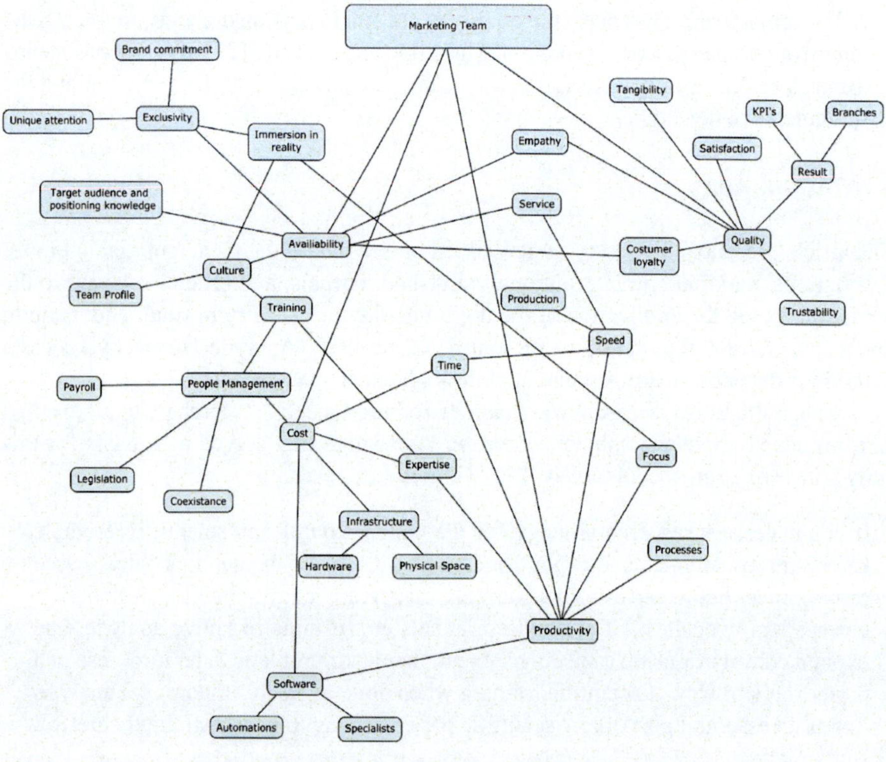

Fig. 1. Cognitive Map. **Source**: created by the authors

Through the cognitive map, built based on the scientific knowledge obtained from literature research of the theme, it was possible to identify the elements with more connections and/or derivatives, which were then adopted as the main criteria to guide decision-making, being: Cost, Availability, Productivity, and Quality (as seen in Fig. 1). To the Cost criterion, elements related to the expenditure necessary to constitute and create an in-house digital marketing team were considered, such as people, software, and infrastructure management. To the Quality criterion, aspects of investment results that could translate into satisfaction were linked. Meanwhile, items that expressed resource efficiency, such as focus and processes, were linked to the Productivity criterion. As for the Availability criterion, it included components that depicted the connection generated by using an ongoing, exclusively-dedicated team.

After that, following the multicriteria AHP method, a pairwise matrix was developed to compare the identified criteria, establishing the importance degree of each criterion to the others, using Saaty's fundamental scale (1 to 9) [27]. Later, calculations were made through matrix operations to obtain this relative contribution, calculating the matrix's primary eigenvector, and then normalizing it. The result obtained in each line represents the total relative percentage of preference between the criteria, considering this research's problem. It is worth noting that the sum of the priority vector's elements equals 1 (100%). According to the judgements carried out, the criteria for the digital marketing structure choice in small and medium businesses are presented in the following importance order: Cost (46,940%) was more important than Productivity, Availability, and Quality. Quality (30,762%) was more important than Productivity and Availability. Productivity (16,992%) was considered superior to Availability. Lastly, Availability was (5,306%) listed as the least relevant criterion.

It is worth mentioning that, to determine the judgments' consistency, the Consistency Ratio (CR) was measured. Considering that the RI value for $n = 4$ is 0,9, the CR value found was 0,057, that is, smaller than 0,10 which is the limit accepted by the AHP method. Thus, it is evident that this work's model calculated weights are satisfactory.

5 Conclusions

Given this article's objectives, it was possible to observe that the multicriteria AHP method can be a relevant ally for entrepreneurs of small and medium businesses when defining the marketing industry structure to be adopted. This support is displayed through the method's steps, from brainstorming for the cognitive map construction, defining criteria and structuring them in hierarchical levels to attributing weights and priorities to the pairwise analysis of the identified criteria while allowing to verify its logical consistency. Though the method can make quantitative analysis more objective, it is based on the knowledge, experience, and living of the individuals who conducted this work. Another relevant factor is the company's scenario when adopting the method, along with the acceptance and sensibility of their products and services to digital marketing [28]. As such, by portraying Cost as the primary criterion, followed up by Quality, Productivity, and Availability, it can be understood that for small and medium businesses, the expenditure amount for digital marketing is relevant and tends to choose alternatives that present the best cost-effectiveness. It is worth noting that the model structured in this article has its limitations and must be considered as a starting point for future research [29, 30]. Among further studies, we can list the incorporation of a larger group of entrepreneurs in the cognitive map review, the development of subcriteria, analysis for its hierarchization by criteria, and finally the proposition of alternatives.

Acknowledgments. The sixth author thanks the National Council for Technological and Scientific Development (CNPq) through grant # 04272/2020–5. The authors thank the Edson Queiroz Foundation/University of Fortaleza for all the support provided.

References

1. de Abreu, F.A.: Outsource Marketing: Efeitos Na Relação Entre Investimentos Em Ações De Marketing Digital E Desempenho Financeiro (2016). https://repositorio.unb.br/handle/10482/20791. Accessed 02 Feb 2023
2. Abreu, L.M.: Escolha de um programa de controle da qualidade da água para consumo humano: aplicação do método AHP. Revista Brasileira de Engenharia Agrícola e Ambiental **4**, 257–262 (2000)
3. Alves, L.S., Do Lago, M.M.: Marketing Digital: A Influência das Mídias Sociais no Comportamento do Consumidor (2019). https://www.aedb.br/seget/arquivos/artigos19/27028244.pdf. Accessed 02 Feb 2023
4. de Oliveira Barreto, T.B., Pinheiro, P.R., Silva, C.F.G.: The multicriteria model support to decision in the evaluation of service quality in customer service. In: Silhavy, R. (ed.) CSOC2018 2018. AISC, vol. 763, pp. 158–167. Springer, Cham (2019). https://doi.org/10.1007/978-3-319-91186-1_17
5. Cittadin, J.: Gestão de Marketing na pequena empresa de confecção de vestuários. Revista Eletrônica de Administração e Turismo **11**(6), 1326–1348 (2017). https://periodicos.ufpel.edu.br/ojs2/index.php/AT/article/view/11905. Accessed 01 Feb 2023
6. Colin, E.C.: Pesquisa Operacional - 170 Aplicações em Estratégia, Finanças, Logística, Produção, Marketing e Vendas, 2.ed. Atlas, Rio de Janeiro (2017)
7. De Araujo, W.C., Gonçalves, I. F., Esquerre, K.P.O.: Aplicação do método AHP para auxílio à tomada de decisão do melhor tratamento para a borra oleosa gerada na indústria petroquímica. Revista Gestão Industrial **16**(4) (2020)
8. Forte, S.A.B., Forte, S.H.A.C., Pinheiro, P.R.: Strategic decision method structured in SWOT analysis and postures based in the MAGIQ multicriteria analysis. In: Silhavy, R., Silhavy, P., Prokopova, Z. (eds.) CoMeSySo 2017. AISC, vol. 662, pp. 227–237. Springer, Cham (2018). https://doi.org/10.1007/978-3-319-67621-0_21
9. Garcia, R.L.P.: Fatores Determinantes do Outsourcing das Atividades de Marketing (2013). https://repositorio-aberto.up.pt/bitstream/10216/69701/2/25322.pdf. Accessed 02 Feb 2023
10. Junior Cardoso, E.C.: A Importância do Marketing Digital para Pequenas Empresas: Uma Revisão Integrativa. Revista Interdisciplinar Pensamento Científico, 5(4) (2019). http://reinpeconline.com.br/index.php/reinpec/article/view/371. Accessed 02 Feb 2023
11. Kotler, P.: Administração de marketing, 10th edn. Prentice Hall, São Paulo (2000)
12. Las Casas, A.L.: Plano de Marketing para Micro e Pequenas Empresas. 6.ed. Atlas, São Paulo (2011)
13. Marins, C.S., de Souza, D.O., da Barros, M.S.: O Uso do Método de Análise Hierárquica na Tomada de Decisões Gerenciais - Um Estudo de Caso. XLI SBPO **1**(49) (2009)
14. Mcgovern, G., Quelch, J.: Outsourcing marketing. Harvard Bus. Rev. (2005). https://hbr.org/2005/03/outsourcing-marketing?language=pt. Accessed 06 Feb 2023
15. Passos, J.C.: Contribuições do Marketing para Micro e Pequenas Empresas do Setor de Serviços: Um Estudo no Brasil. Revista de la Agrupación Joven Iberoamericana de Contabilidad Y Administración De Empresas (AJOICA) (11), 105–116 (2013). http://www.elcriterio.com/revista/contenidos_11/janduhy_camilo.pdf. Accessed 01 Feb 2023
16. Pimenta, L.B.: Processo Analítico Hierárquico em Ambiente SIG: Temáticas e Aplicações Voltadas à Tomada de Decisão Utilizando Critérios Espaciais. Interações (Campo Grande) **20**, 407–420 (2019)
17. Raupp, F.M., Beuren, I.M.: Metodologia da pesquisa aplicável às ciências. Como elaborar trabalhos monográficos em contabilidade: teoria e prática, pp. 76–97. Atlas, São Paulo (2006)
18. Rediske, G., Storch, C.R.R., Nara, E.O.B.: Construção de mapas conceituais do método AHP e Promethee. V Congresso Brasileiro de Engenharia de Produção (2015)

19. de Ribeiro, M.C.C.R., Da Silva Alves, A.: Aplicação do Método Analytic Hierarchy Process com a Mensuração Absoluta num Problema de Seleção Qualitativa. Sistemas Gestão **11**(3), 270–281 (2016)
20. Robaina, M.O.: Análise do Marketing Digital e Mídias Sociais: Estudo Multicasos baseado na percepção de gestores. Revista de Administração da UNIMEP **19** (2022). https://web.s.ebscohost.com/ehost/detail/detail?vid=0&sid=6db15141-433f-4f58-9b72-a5255e51d7fa%40redis&bdata=Jmxhbmc9cHQtYnImc2l0ZT1laG9zdC1saXZl#AN=160970719&db=foh. Accessed 02 Feb 2023
21. Ferreira, C., Nery, A., Pinheiro, P.R.: A multi-criteria model in information technology infrastructure problems. Procedia Comput. Sci. **91**, 642–651 (2016). https://doi.org/10.1016/j.procs.2016.07.161
22. da Silva, P.Ó.G.: A Importância das Organizações Recorre aos Serviços de Outsourcing em Marketing. Tese de Doutorado (2017). http://hdl.handle.net/10400.26/18338. Accessed 01 Feb 2023
23. Soares, S.V., Picolli, R.A., Casagrande, J.L.: Pesquisa bibliográfica, Pesquisa Bibliométrica, Artigo de Revisão e Ensaio Teórico em Administração e Contabilidade. Administração: Ensino e Pesquisa **19**(2), 308–339 (2018)
24. Sousa, T.C.S.: Estratégia de Marketing como Instrumento de Competitividade na Pequena Empresa. Lisboa: ISCTE, Tese de mestrado (2008). http://hdl.handle.net/10071/1100. Accessed 02 Feb 2023
25. Yanaze, M.H., Almeida, E., Yanaze, L.K.H.: Marketing Digital: Conceitos e Práticas. Editora Saraiva (2022). Accessed 04 Feb 2023. https://integrada.minhabiblioteca.com.br/#/books/9788571441408/
26. Zenaro, M., Pereira, M.F.: Marketing Estratégico para Organizações e Empreendedores: Guia Prático e Ações Passo a Passo (2013). ISBN 9788522486380. https://integrada.minhabiblioteca.com.br/#/books/9788522486380/. Accessed 06 Feb 2023
27. Saaty, T.L.: Fundamentals of Decision Making and Priority Theory: With the Analytic Hierarchy Process. RWS Publications, Pittsburgh (1994)
28. Tamanini, I., Carvalho, A.L., Castro, A.K., Pinheiro, P.R.: A novel multicriteria model applied to cashew chestnut industrialization process. In: Mehnen, J., Köppen, M., Saad, A., Tiwari, A. (eds.) Applications of Soft Computing. Advances in Intelligent and Soft Computing, vol. 58. Springer, Heidelberg (2009). https://doi.org/10.1007/978-3-540-89619-7_24
29. Almeida, L.H., Albuquerque, A.B., Pinheiro, P.R.: A multi-criteria model for planning and fine-tuning distributed scrum projects. In: 2011 IEEE Sixth International Conference on Global Software Engineering, Helsinki, Finland, pp. 75–83 (2011). https://doi.org/10.1109/ICGSE.2011.36
30. Pinheiro, P.R., Machado, T.C.S., Tamanini, I.: Dealing the selection of project management through hybrid model of verbal decision analysis. Procedia Comput. Sci. **17**, 332–339 (2013). https://doi.org/10.1016/j.procs.2013.05.043

Opportunities to Improve the Effectiveness of Online Learning Based on the Study of Student Preferences as a "Human Operator"

Yu. I. Lobanova(✉)

St. Petersburg State University of Architecture and Civil Engineering (SPbGASU), 4, 2-ya Krasnoarmeyskaya St., 190005 Saint-Petersburg, Russia
gretta25@list.ru

Abstract. The article explores the possibilities of improving the effectiveness of online learning based on the analysis of a student as a "human operator" engaged in working with information in the "man-machine" system. The results of the empirical research conducted by the author are presented. With the help of the author's questionnaire, the preferences of students of different levels of education and different areas of training in relation to individual electronic didactic tools, their characteristics and methods of organizing interaction with students in theoretical classes (lectures) used by teachers in the educational process were studied. The preferences common to the whole group of respondents were revealed: trainees prefer a variety of templates and slides used, the use of different information visualization tools (figures, tables), the practical orientation of theoretical classes and their interactive nature. Undergraduate students prefer lectures to a greater extent than undergraduates, which include practical tasks, tests, that is, interactive elements. Master's students perceive lectures – videos and abstract-schematic images of people on slides more positively. For students of technical fields of training, the most important factor was the readability of the slides. Slides with color highlighting of formulas and important elements of the text turned out to be attractive for students of the humanities. Men prefer the Canva program for creating presentations and using a light background of slides, women prefer highlighting important information elements (definitions, formulas) in color and font. Conclusions are drawn that when teaching the same subjects, the style of presentation of information itself should differ if the teacher works with different bachelor's and master's degrees, as well as with different areas of training. Both the format of the slides and the organization of interactions with students is important for getting pleasure in the learning process and, accordingly, for the effectiveness of learning. It is indicated that the search for opportunities to convert presentations and interaction formats within the framework of training sessions from those given to those subjectively convenient for the learner is the way of individualization of digital education. #CSOC1120.

Keywords: Information and Communication Technologies · Teams · Moodle · Power Point · Canva · Slide Types · Schemes · Color · Font · Photo · Interaction · Education Level · Humanitarians · Learning Effectiveness · Preferences · Covid-19

R. Silhavy and P. Silhavy (Eds.): CSOC 2023, LNNS 724, pp. 506–519, 2023.
https://doi.org/10.1007/978-3-031-35314-7_45

1 Introduction

The Covid-19 pandemic brought to a new level the use of information and communication and digital technologies in all areas of people's lives - almost everywhere where direct human-to-human contact was assumed, both professionals and clients were forced to try out interaction through websites, computer programs, and messengers.

Around the world, information and communication and digital technologies began to be used (in the learning, for instance) much earlier than in Russia. The interest of scientists in the factors which ensure the effectiveness of online learning has predetermined the main directions of research: the attitude of students and teachers to different learning formats (including dynamics) were studied as well as the possibility of using online learning for training specialists in different fields, the questions of optimal combination of offline and online learning (within the mixed format) were raised [2–4].

Studies that were conducted in the pre-pandemic times showed that the attitude of learners to the use of online learning among learners was mostly cautious, if not negative [2, 3]. Thus, back in 2012 attempts to use certain technologies, such as videoconferencing, to teach a foreign language did not cause much enthusiasm among students [2]. In Germany, at the same time, they thought about the introduction of online learning in the creation of unique undergraduate programs in electrical engineering [3]. Adults who had to combine their studies with work in the relevant industry had to learn from these programs. In order to make better use of distance learning, the opinion of graduate students on the use of distance technology was studied. But at that time they were conservative in their preferences, leaning towards working with printed versions of textbooks and learning face-to-face with a teacher (offline format) [3].

In general, since online learning was not used extremely widely until the pandemic era, the problem of satisfaction-dissatisfaction with its process and results, as well as the factors determining them, were clearly understudied [4–6].

The times of the pandemic made both distance learning itself and research in this field particularly relevant [7–14]. However, the widespread use of distance learning has shown that it may perform with different efficacy in different subject areas. For example, in the field of medical education, the use of technology has proved to be limited, and satisfaction with the learning process only occurred if there were results that correlated with the degree of training of teachers and trainees [16].

Problems of distance learning have touched everyone, so the studies which were conducted by researchers from different countries, despite some commonality of the analyzed problems, had relevance due to cross-cultural differences which determine both the specifics of information and communication and digital technologies [6, 7, 14] and their subjective attractiveness for participants of the educational process.

Studies conducted by scientists from non-European countries (China, India, Brazil) have shown that, in addition to the subject aspects, the effectiveness of distance learning is also related to the cultural aspects. Thus, Chinese researchers [6] found that the main three factors determining satisfaction with online learning are: "Insight and Reliability"; "Stimulation and Attractiveness"; and "Usability and Innovation". By contrast, a regression analysis showed that the strongest predictor of student satisfaction with online learning during COVID-19 disruptions was "Stimulation and Attractiveness." This new finding, according to Chinese researchers, points to the need for hospitality and tourism educational institutions to develop an attractive and motivating visual environment for online courses, as a stimulating online learning environment is crucial in the context of pedagogical failures caused by COVID-19. However, the researchers indicate that the findings may be specific to Chinese students and reflect their satisfaction with learning, which may differ in other contexts.

In Russia, interest in scientific research in the field of distance learning has also increased significantly during the times of covid, when education had to switch urgently to COVID-19, and teachers had to master work with different (including previously almost unused by them) information and communication technologies. Practice has shown that there are sufficient numver of problems in this area. Researchers in Russia are studying the experience of using different information and communication and digital technologies in education, comparing their capabilities and the results achieved, analyzing the use of different social learning networks [15–18].

The author of this paper goes consistently in his research from studying the attitude of students to different learning formats [19–22], the analysis of objectively achieved results and evaluation of student satisfaction when working in different formats to the analysis of digital didactic tools and forms of organizing online classes in order to select the most effective for teaching (taking into account sociodemographic, socio-professional characteristics of groups of students, as well as subjective preferences of students and teachers).

The author of this paper has conducted a series of empirical studies, aimed at studying the subjective perception of distance learning and individual factors that predetermine its effectiveness.

Information overload, fatigue and monotony, as well as hypodynamia among learners [21, 22], as well as among tutors, turned out to be important problems of distance learning. However, the learners' load can be partially relieved by pre-recording the lectures, while the prevention of negative psychophysiological states of the learners, who act as "human operators" in distance learning, is an open question.

In order to approach the problem the model of T. Gallwey [23] - the triangle of efficiency - result, learning, pleasure - was applied.

Based on this model, the effectiveness of learning is achieved when:

1) The learning process itself brings the learner not only results, but also pleasure.
2) Obtained results apparently satisfy the subject and (probably) exceed the invested costs: for example, a skill or ability is formed, and there are no manifestations of fatigue yet.

It is logical to assume that if ideally the process of achieving a goal may bring pleasure, in reality it may not provide any. Negative emotions may arise due to both underachievement of results and discomfort, inconvenience (subjective or objective - it is necessary to understand) - of the process itself.

Two questions arise:

1. How do we reduce costs? Since the learning process is largely a communicative process, how can we make it as efficient as possible?

2. How to make the learning process enjoyable, comfortable, and conducive to achieving its goals?

To find answers to these questions, we can use G. Lasswell's model of the communicative process [24]. Based on Lasswell's diagram, the communication process (as information transfer) includes the source of communication, the addressee of communication, the channel through which the message is transmitted, and the information actually transmitted.

This article will analyze only the process of information transfer and the possibilities of increasing its effectiveness. The content of information itself will be left out of the brackets. As part of the educational process in teaching a particular discipline the content remains unchanged. But the channels of information transmission, symbols and codes through which information is transmitted can change.

If we imagine a usual full lecture in Teams or Zoom, information can be transmitted in these programs using, for example, a screen presentation (showing various content to the students, including slide presentations), interactive whiteboard, voice, image.

When preparing a presentation, the teacher chooses the program with which to visualize the desired content, chooses the types of slides, can use tables, diagrams, pictures, photos, change the size and types of fonts, work with color. The choice of the same photos or pictures can also be specified in terms of specific details.

As part of the lecture itself, the teacher can make the lecture interactive and practical, or concentrate on a purely theoretical, abstract and logical presentation of material.

When making a lecture interactive, different methods and techniques can be used, including questions, assignments, examples and tests.

The specifics of the organization of classes and the choice of the channels and means to be used are chosen by teachers. In the conditions of the educational process in the university, which is sharpened to professional standards, some disciplines are taught according to the same programs in different areas of training, while the classes are taught by the same teachers.

Students of different ages, different qualifications, different levels of education, and different genders may prefer different channels and means of transmitting information and, accordingly, feel pleasure or displeasure when teachers use specific channels, methods, and means. We can even assume that in some cases subjective preferences of a teacher and his desire to achieve his own subjective convenience in conducting classes can lead to negative changes in psychophysiological states of students and development of their negative emotions and experiences. Teaching against the background of unfavorable psychophysiological (in particular case - psychoemotional) state obviously has a negative impact on the learning process and leads to lower results achieved (consequently, it can reduce satisfaction with the results themselves).

In connection with what develops mental fatigue, which in higher education is the main one? Due to the increased expenditure on the part of the central nervous system in the formation of new neural connections when mastering new material. To make the learning process more effective, the costs need to be reduced. How can this be done? It should be understood that within the framework of a learning activity (as well as any other) a subject can use 3 types of resources [25]:

- internal (intra-individual);

- external (extra-individual);
- inter-individual.

Those that can be influenced by the teacher are obviously external. The teacher can recommend students to use a certain technique when working in the personal offices, certain programs (if the university provides such an opportunity), websites, and most importantly - to present information to students so that it is easier for them to understand, that is, to improve the communicative process.

But how exactly should this be done? This task was originally traditional for engineering psychology, in which the object is a human operator: at the dawn of engineering psychology the problems of signal coding were investigated in order to facilitate human-machine interaction. It should be noted that in the 50–80 s "human operators" was a special professional group, rather monogenic in its characteristics, aimed at solving a strictly defined range of professional tasks. Now the same task is becoming traditional for educators as well, since both they and their students are becoming human operators under distance learning conditions, since the interaction process is mediated by information-communication and digital technologies. At the same time, the participants in the educational process (at different levels of education) differ in age, in gender, and in the direction of training.

Research hypothesis: students from groups with different socio-demographic and socio-professional characteristics subjectively perceive the use of digital didactic tools in learning by teachers differently, giving more preference to some and less to others.

The purpose of the study was to axamine students' subjective preferences of digital didactic tools and interactive teaching methods used by teachers when conducting classes in a distance format.

Research objective is to study students' preferences (from groups with different socio-demographic and socio-professional characteristics) regarding:

- quality and design features of presentation slides used by the teacher as illustrative or demonstration material;
- the use of specific software to create presentations for lectures, and certain presentation templates and styles (presentation uniformity);
- application of specific visualization tools (diagrams, charts, tables) to create digital illustrative materials;
- the use of photos with a particular content (such as people) as illustrative material.
- certain methods of organization and formats of lectures, as well as supplementing the theoretical material of lectures with practical tasks and tests.

2 Methods

To conduct the study we used the author's questionnaire designed to collect information about the subjective preferences of students used in online classes (lectures): software for presentation design, templates, illustrative tools, lecture formats and the organization of interaction in their process.

Description of the research sample:
Sixty-eight respondents were interviewed during data collection, including: 37 bachelor students and 31 master students.

Engineering field of study - 38 people.
Humanitarian field of study - 30 people.
Men - 27 people.
Female - 41 people.

3 Results

Table 1 shows the average values of scores, which were given to certain digital didactic means, their individual characteristics, as well as certain forms of organization of lecture classes in online format during questioning. Values are ranked in descending order (from the maximum values to the minimum ones). Thus, the first five items in the table correspond to the most attractive elements for learners, while the last ones correspond to the least preferred ones.

If we analyze the preferences for channels, tools, and features of class organization that were recorded in the surveyed group of respondents as a whole, the following features should be noted:

1. The use of exclusively corporate templates for teachers' preparation of presentations for lectures is rather not welcomed by the learners.
2. The use of different types of slides (with text, figures, photos, diagrams, etc.) is more preferable for them.
3. In terms of clarity of delivery of information, the use of diagrams and figures in the presentation is preferred.
4. Inclusion of questions and practical exercises in the lecture is perceived rather positively by the respondents.
5. Lecture given as a recording of the presentation with voice-over narration - perceived positively by the majority of the respondents.

At the same time, the presentation of lectures exclusively in text format, poor readability of the slides, the use of Powerpoint, and the excessive multi-color slides in the presentation are perceived rather negatively.

Table 2 shows significant differences in the preferences of students at different levels of education in the choice of characteristics of digital didactic tools, as well as forms of organization of classes.

It is clear from the table that bachelor students prefer highlighting meaningful text fragments and formulas on slides in color, using certain templates when creating slides. They like the availability of tasks and tests for the lectures, and they positively perceive the lectures created with the help of software tools.

Master's students prefer slides with light backgrounds, schematic and abstract images of people.

Practical tasks, which are included in the lecture itself or are offered as a supplement to it, are more attractive to bachelor students. Bachelor students are also more interested (as compared to master students) in lecture questions, assignments and tests.

When analyzing the preference for lecture formats, it turns out that bachelor students prefer text lectures in Moodle, and master students prefer video lectures recorded by lecturers.

Table 1. Learners' preferences.

	N	Range	Minimum	Maximum	Total	Average value	Standard deviation
Presentation plus voice	68	8.00	1.00	9.00	493.00	7.2500	2.38387
Practical assignments	68	8.00	1.00	9.00	413.00	6.0735	2.90813
Figures	68	8.00	1.00	9.00	404.50	5.9485	1.86111
Highlighting the main thing in color	68	8.00	1.00	9.00	402.00	5.9118	2.19215
Corporate style	68	8.00	1.00	9.00	391.50	5.7574	2.30265
Variety	68	8.00	1.00	9.00	389.00	5.7206	2.51025
Highlighting questions in color	68	8.00	1.00	9.00	388.50	5.7132	2.74313
Assignments	68	8.00	1.00	9.00	387.00	5.6912	2.81126
Patterns	68	8.00	1.00	9.00	374.00	5.5000	2.36738
Font highlighting	68	8.00	1.00	9.00	369.50	5.4338	2.02024
Highlighting formulas	68	8.00	1.00	9.00	366.50	5.3897	2.42322
Video lectures	68	9.00	0.00	9.00	353.00	5.1912	3.09674
Schemes	68	8.00	1.00	9.00	346.50	5.0956	1.45648
Font	68	8.00	1.00	9.00	346.50	5.0956	2.91581
Canva	68	8.00	1.00	9.00	340.50	5.0074	2.61952
Questions	68	8.00	1.00	9.00	338.50	4.9779	2.67023
Chat without a moderator	68	8.00	1.00	9.00	337.00	4.9559	2.73961
Photo	68	8.00	1.00	9.00	333.50	4.9044	1.56991
Chat with a moderator	68	8.00	1.00	9.00	329.00	4.8382	2.71321
People schematically	68	8.00	1.00	9.00	327.50	4.8162	2.74257
Tables	68	8.00	1.00	9.00	321.50	4.7279	1.35336
Diagrams	68	8.00	1.00	9.00	315.50	4.6397	1.41116
Text	68	8.00	1.00	9.00	311.00	4.5735	1.53629

(continued)

Table 1. (*continued*)

	N	Range	Minimum	Maximum	Total	Average value	Standard deviation
Tests	68	8.00	1.00	9.00	311.00	4.5735	2.94890
Emoji	68	8.00	1.00	9.00	303.50	4.4632	2.74856
MOODLE lectures	68	8.00	1.00	9.00	301.00	4.4265	2.33287
Slide lightness	68	8.00	1.00	9.00	280.00	4.1176	2.69067
Color (variety)	68	8.00	1.00	9.00	275.00	4.0441	2.22894
Photo, people - schematic image	68	8.00	1.00	9.00	252.50	3.7132	1.89523
Powerpoiint	68	8.00	1.00	9.00	244.50	3.5956	2.45445
Readability	68	8.00	1.00	9.00	199.00	2.9265	2.63906
Lectures, text	68	8.00	1.00	9.00	196.00	2.8824	2.65718
N of valid (by the list)	68						

Table 3 shows the significant differences in the preferences of bachelor students in the choice of characteristics of digital didactic tools and forms of class organization.

Comparing the preferences of students of different fields of study (regardless of the level of education), for engineering students the most important factor was the readability of the slides. Color or font highlighting of important points in the text, formulas, questions to the lecture was more important for humanities students. Besides, possibility to discuss some questions in chats during the lecture was also more useful and important for humanities students.

Table 4 shows the significant differences in the preferences of students of different genders in the choice of characteristics of digital didactic tools and forms of lesson organization.

Men prefer presentations made with Canva software, as well as slides made on a light background.

For women, color highlighting important points in the text of the slide, color-coding formulas, using templates, and lecture tests are more attractive.

Table 2. Comparison of average (by levels of education): bachelor students and master students.

Level of education *		N	Average value	Standard deviation	Mean square error of the average	Significance level (2-way significance)
Light background of the slides	1	37	**3.4324**	2.39839	.39429	0. 021
	2	31	**4.9355**	2.82767	.50786	
People schematically	1	37	**4.1351**	2.79048	.45875	0. 024
	2	31	**5.6290**	2.48987	.44719	
Font	1	37	**5.8649**	2.79545	.45957	0. 016
	2	31	**4.1774**	2.83004	.50829	
Formula highlighting	1	37	**5.8514**	2.51900	.41412	0. 083
	2	31	**4.8387**	2.21881	.39851	
Patterns	1	37	**6.0405**	2.25271	.37034	0. 039
	2	31	**4.8548**	2.37414	.42641	
Questions	1	37	**5.5135**	2.64703	.43517	0. 07
	2	31	**4.3387**	2.59611	.46628	
Assignments	1	37	**6.6757**	2.43589	.40046	0. 001
	2	31	**4.5161**	2.81213	.50507	
Video lectures	1	37	**4.5946**	3.00425	.49390	0. 083
	2	31	**5.9032**	3.10220	.55717	
MOODLE lectures	1	37	**4.8919**	2.51422	.41334	0. 072
	2	31	**3.8710**	1.99569	.35844	
Practice assignments	1	37	**6.8108**	2.56946	.42242	0. 021
	2	31	**5.1935**	3.08133	.55342	
Tests	1	37	**5.2973**	3.07196	.50503	0. 026
	2	31	**3.7097**	2.58449	.46419	

*1 - bachelor degree
2 - master degree

Table 3. Comparison of the average (by field of study).

Field of study*		N	Average value	Standard deviation	Mean square error of the average	Significance level (2-way significance)
Readability	1	38	**3.5789**	2.95581	.47950	0. 021
	2	30	**2.1000**	1.91815	.35021	
Schematic-abstract portrayal of people	1	38	**4.3421**	2.58139	.41876	0. 109
	2	30	**5.4167**	2.86502	.52308	

(*continued*)

Table 3. (*continued*)

Field of study*		N	Average value	Standard deviation	Mean square error of the average	Significance level (2-way significance)
Font highlighting of the main thing	1	38	**4.9868**	2.73981	.44446	0. 091
	2	30	**5.2333**	3.16700	.57821	
Highlighting formulas in color	1	38	**4.8816**	2.53205	.41075	0. 051
	2	30	**6.0333**	2.14931	.39241	
Use of patterns	1	38	**5.0789**	2.25570	.36592	0. 099
	2	30	**6.0333**	2.43514	.44459	
Highlighting questions in color	1	38	**5.1447**	2.79698	.45373	0. 054
	2	30	**6.4333**	2.53844	.46345	

* 1 - representatives of engineering fields of study
2 - representatives of non-technical areas of training

Table 4. Comparison of average (by gender) (men and women).

Gender*		N	Average value	Standard deviation	Mean square error of the average	Significance level (2-way significance)
Light background of the slides	1	27	**4.8519**	2.36487	.45512	0.067
	2	41	**3.6341**	2.80852	.43862	
Text	1	27	**4.1667**	1.23257	.23721	0.076
	2	41	**4.8415**	1.66748	.26042	
Font highlighting of the main thing	1	27	**4.8333**	1.79208	.34489	0.046
	2	41	**5.8293**	2.08450	.32554	
Highlighting formulas in color	1	27	**4.4444**	2.18972	.42141	0.008
	2	41	**6.0122**	2.39136	.37347	
Canva	1	27	**5.6667**	2.74563	.52840	0.101
	2	41	**4.5732**	2.47124	.38594	
Patterns	1	27	**4.7407**	2.56594	.49382	0.031
	2	41	**6.0000**	2.11246	.32991	
Tests	1	27	**3.6296**	2.66239	.51238	0.031
	2	41	**5.1951**	2.99349	.46750	

*1 – men
2 – women

4 Discussion

What might be the reasons for the recorded general student preferences and the differences between students of different educational levels, fields of study, and genders? Explanations may include the following:

1. From the point of view of the marketing generational theory [26, 27] the differences can be explained by the fact that bachelor and master students, despite a relatively small difference in age, belong to different generations. Master students still belong to generation "Y", whereas first-year bachelor students are mostly "Z". Hence the preference for lectures created with the help of software tools in Moodle by younger students (as a continuation of their habitual and comfortable stay in a virtual environment) and video lectures by older students (they are more resembling traditional classes in the classroom).
2. A certain role in the formation of preferences, very likely, played personal characteristics of students of different fields of study, which, among other things, determined the engineering and humanitarian orientation, the choice of the nature of future professional activity - rather creative or rather performing.
3. Probably, the peculiarities of the disciplines taught to the students of specific fields of study are also important - the curricula of different fields of study have their own peculiarities. And presentation of information within the framework of these disciplines in its turn requires obligatory use of certain types of coding and channels for transmitting information. For example, with lawyers it is impossible to do without slides with texts, with designers without figures.
4. The recorded differences in preferences for certain didactic means and methods of organizing classes between groups of male and female respondents require additional research, as groups of students in different fields of study are heterogeneous in terms of the prevailing gender.
5. Separate importance, apparently, has such a factor as individual-psychological features and professional-pedagogical competence of the teachers themselves who teach certain disciplines (sometimes the same for students of different fields of study, for example – "Social Communications. Psychology"). Teachers may differ in:
   - the level of instrumental preparation in terms of using certain tools;
   - the availability of different software tools;
   - pedagogical skills and knowledge of different pedagogical technologies;
   - personal characteristics;
   - style of pedagogical communication.

All of the above creates the basis for a certain style of lecturing in a distant format by the teacher, which includes the style of presentation and its combination with the transfer of information through speech. Students' and teachers' preferences may or may not coincide, and it is in the latter case that monotony, information overload and overstrain are more likely to develop in students. Both preferences of students and teachers - in fact - is a human factor, which subjectivizes the learning process.

In the future, on the basis of a number of in-depth studies of the preferences of students and teachers it will be possible to gradually approach the algorithmic process of creating presentations and voice messages, and their transformation (conversion) from a given style in the one which is subjectively better perceived by a particular group of students, and later - and a particular student. In other words, research of this kind can become the basis for individualization of learning with the help of software tools.

However, the problem of transferring personal meanings, personal knowledge [19, 20, 28] in the course of learning (at least in the near future) will remain unsolvable outside of direct human-to-human interaction.

5 Conclusion

The purpose of the study has been achieved. The hypothesis is confirmed. The research objective has been achieved. The differences have been revealed. Different groups of students (with different socio-demographic and socio-professional characteristics) really prefer different digital didactic tools and forms of organizing classes.

Previously, in the pre-computer era, there was little choice: a blackboard and a book, but now teachers have at their disposal a huge variety of information and communication and digital technologies, digital didactic tools, while the competition for students in the educational space is growing. To win in the competition, the educational process must be made effective, which is achievable through a combination of enjoyment in the learning process and achievement of the required results. Analysis of the preferences of the learners in this regard is the cornerstone.

References

1. Singh, V., Thurman, A.: How many ways can we define online learning? a systematic literature review of definitions of online learning (1988–2018). Am. J. Dist. Educ. **33**(4), 289–306 (2019)
2. Candarlia, D., GulruYukselb, H.G.: Students' perceptions of video-conferencing in the classrooms in higher education. Procedia – Soc. Behav. Sci. **47**, 357–361 (2012). https://doi.org/10.1016/j.sbspro.2012.06.663
3. Muller, A.L., Kurz, R., Hoppe, B.: Development of distance learning concept for a Bachelor of Science in electrical engineering programme. Procedia Soc. Behav. Sci. **93**, 1484–1488 (2013)
4. Harsası, M., Sutawıjaya, A.: Determinants of student satisfaction in online tutorial: a study of a distance education institution. Turk. Online J. Dist. Educ. **19**(1), 89–99 (2018)
5. Lee, J.: An exploratory study of effective online learning: assessing satisfaction levels of graduate students of mathematics education associated with human and design factors of an online course. Int. Rev. Res. Open Distrib. Learn. **15**(1), 38–42 (2014)
6. Sun, A., Chen, X.: Online education and its effective practice: a research review. J. Inf. Technol. Educ. Res. **15**, 157–190 (2016). https://doi.org/10.28945/3502
7. Guptaa, S., Dabasb, A., Swarnimc, S., Mishrad, D.: Medical education during COVID-19 associated lockdown: faculty and students' perspective. Med. J. Armed Forces India **77**(1), S79–S84 (2021). https://doi.org/10.1016/j.mjafi.2020.12
8. Limaa, F.B., Lautertb, S.L., Gomesc, A.S.: Contrasting levels of student engagement in blended and non-blended learning scenarios. Comput. Educ. **172**, 104241 (2021). https://doi.org/10.1016/j.compedu.2021.104241

9. Agyeiwaah, E., Baiden, F.B., Gamor, E., Hsu, F.C.: Determining the attributes that influence students' online learning satisfaction during COVID-19 pandemic. J. Hosp. Leis. Sport Tour. Educ. **30**, 100364 (2022). https://doi.org/10.1016/j.jhlste.2021.100364
10. Dhawan, S.: Online learning: a panacea in the time of COVID-19 crisis. J. Educ. Technol. Syst. **49**(1), 5–22 (2020)
11. Dzhangarov, A.I., Hanmurzaev, H.E., Potapova, N.V.: Digital education in the coronavirus era. J. Phys: Conf. Ser. **1691**, 012133 (2020)
12. Faize, F., Nawaz, M.: Evaluation and improvement of students' satisfaction in online learning during COVID-19. Open Praxis **12**(4), 495–507 (2020)
13. Kirscha, C., de Abreua, P.M.J.E., Neumannb, S., Wealer, C.: Practices and experiences of distant education during the COVID-19 pandemic: the perspectives of six- to sixteen-year-olds from three high-income countries. Int. J. Educ. Res. Open **2**, 100049 (2021)
14. Bani Hani, A., et al.: Cross-sectional Study E-Learning during COVID-19 pandemic; turning a crisis into opportunity: a cross-sectional study at the University of Jordan. Ann. Med. Surg. **70**, 102882 (2021)
15. Kvashko, L.P., Aleksandrova, L.G., Shesternina, V.V., Erdakova, L.D., Kvashko, V.V.: Distance learning during self-isolation: comparative analysis. J. Phys: Conf. Ser. **1691**, 012013 (2020). https://doi.org/10.1088/1742-6596/1691/1/012013
16. Makarenya, T.A., Stash, S.V., Nikashina, P.O.: Modern educational technologies in the context of distance learning. J. Phys: Conf. Ser. **1691**, 012117 (2020). https://doi.org/10.1088/1742-6596/1691/1/012117
17. Orishev, A.B., Mamedov, A.A., Kotusov, D.V., Grigoriev, S.L., Makarova, E.V.: Digital education: vkontakte social network as a means of organizing the educational process. J. Phys: Conf. Ser. **1691**, 12092 (2020). https://doi.org/10.1088/1742-6596/1691/1/012092
18. Yarychev, N., Mentsiev, A.: New methods and ways of organizing the educational process in the context of digitalization. J. Phys: Conf. Ser. **1691**, 012128 (2020). https://doi.org/10.1088/1742-6596/1691/1/012128
19. Lobanova, Y.: Distant learning experience reflection during the pandemic of Covid-19 (On the example of teaching in the technical university). J. Phys: Conf. Ser. **1691**, 012152 (2020). https://doi.org/10.1088/1742-6596/1691/1/012152
20. Lobanova, Y.I.: Distance learning advantages and disadvantages: teaching experience analysis at the university with the basis on different informational-communicative technologies. In: Silhavy, R. (ed.) CSOC 2021. LNNS, vol. 229, pp. 499–506. Springer, Cham (2021). https://doi.org/10.1007/978-3-030-77445-5_46
21. Lobanova, Yu.I.: Student or human operator? objective results and subjective perceptions of the process and results of offline and online learning. In: Silhavy, R. (ed.) Cybernetics Perspectives in Systems: Proceedings of 11th Computer Science On-line Conference 2022, vol. 3, pp. 121–127. Springer, Cham (2022). https://doi.org/10.1007/978-3-031-09073-8_11
22. Lobanova, Y.I.: From offline learning to the future: subjective assessment and learning outcomes when using different formats. In: Anikina, Zhanna (ed.) Integration of Engineering Education and the Humanities: Global Intercultural Perspectives: Proceedings of the Conference Integration of Engineering Education and the Humanities: Global Intercultural Perspectives, 20–22 April 2022, St. Petersburg, Russia, pp. 376–385. Springer, Cham (2022). https://doi.org/10.1007/978-3-031-11435-9_41
23. Gallwey, W.T.: The Inner Game of Work: Focus, Learning, Pleasure, and Mobility in the Workplace. Alpina Business Books, Moscow (2005)
24. Lasswell, H.D.: Describing the effects of communications. In: Smith B.L., Lasswell H.D., Casey R.D. (eds.) Propaganda, Communication and Public Opinion, pp. 95–117. Princeton University Press, Princeton (1946)

25. Lasswell, H.D.: The structure and function of communication in society. In: Berelson B., Janowitz M. (eds.) Reader in Public Opinion and Communication, pp. 178–189. The Free Press, New York (1966). (in Eng)
26. Tolochek, V.A.: Styles of Activity: A Resource-Based Approach, p. 366. Institute of Psychology, Russian Academy of Sciences, Moscow (2015)
27. Dynkina, E.D.: Selection of Tools for Effective Training of Generation Y. In: Business Education in Knowledge Economy, no. 1, pp. 50–53 (2015)
28. Polanyi, M.: Personal Knowledge. Progress, Moscow (1985)

Capabilities of the Matrix Method of Teaching in the Course of Foreign Language Professional Training

Victoria Sogrina(✉), Diana Stanovova, Elina Lavrinenko, Olga Frolova, Inessa Kappusheva, and Jamilia Erkenova

MIREA-Russian Technology University, Moscow, Russia
sogrinavika6@gmail.com

Abstract. The article deals with the application of the matrix method of teaching in the course of foreign language professional training in the Russian Technological University (RTU MIREA). Within the framework of the experimental research the authors have evaluated the possibility to improve the communicative skills of intercultural interaction contributing to the solution of professional tasks. #CSOC1120.

Keywords: Matrix Method of Teaching · Educational Space · Communicative Skills · Professional Foreign Language Training

1 Introduction

In today's world, digitalization has influenced all spheres of human professional activity: economic, political and, above all, educational environment [1]. Now, due to the difficult epidemiological situation (COVID-19) and controversies on the world stage, the need to train a highly qualified competent specialist is increasing. That is why training is carried out with the help of different methods, e.g. the reproductive method of training. This method is based on the repetition of the acquired material, thus, the productivity increases [2]. Other innovative methods are also available: lecture and distance format of learning material [3]. With the use of information technology in the learning process, the integration of teaching methods used in technical and humanities disciplines has become relevant. For example, matrix algebra in the applied sense can reflect the cost ratio of production structures; can be used in technical calculations of the design of structures, in the study of systems of m linear equations with n variables, in the calculation of field values near inhomogeneity and statistics and much more [4]. Accordingly, the matrix method of teaching is universal with respect to different fields of application.

2 Methods

For the first time the concept of this method was theoretically grounded and practically proved by N.F. Zamyatkin. According to his theory, the matrix method of teaching a foreign language consists in arranging the step-by-step assimilation of the heard and

R. Silhavy and P. Silhavy (Eds.): CSOC 2023, LNNS 724, pp. 520–525, 2023.
https://doi.org/10.1007/978-3-031-35314-7_46

foreign speech by perceiving, speaking and analyzing the foreign language material on the presented topic [5, 6]. The concept of the matrix method has certain conditions for implementation, such as: the unity of the topics of audio materials in the matrix; time restrictions (the duration of each audio recording from 30 s to a minute and a half); filling the audio materials with lexical and grammatical constructions to consolidate the learned material; absence of external stimuli (noise effects, pauses, inarticulate speech).

English matrix by Zamyatkin method consists of the following mandatory items: listening to audio lessons, with a frequency of at least two hours a day; listening to audio recordings with a parallel viewing of the printed text, this is called verification; read aloud, copying intonation and pronunciation. The matrix is nothing else but a set of dialogues composed in a special way: originally a 3*3 grid of dialogues (9 samples), mastered to perfection by repetition; then comes a 5*5 matrix (25 dialogues) and 6*6 (36 dialogues). The author claims that by "installing" the 3*3 matrix by the "meditative-matrix" method, the learner will master, on average, 25% of the speech, which, when the process of "implanting" the matrix is completed, is perfected to 60% of the language acquisition.

There should be 25–30 dialogues in the matrix. The duration of the first 5 dialogues should be 20–40 s, the next ones should be longer, but no longer than 70 s. Each dialogue must be logically complete, that is, you can't cut it off. Choose either "British English" or "American English," and do not mix them. All dialogues must be spoken by the speakers at a normal speed and with a familiar everyday intonation. Each matrix dialogue should be repeated the more times the better. In the dialogues no extraneous sounds and any special effects are allowed, as well as long pauses. You should preferably choose dialogues with a high-frequency vocabulary, they should contain frequently used words, phrases, grammatical patterns, and it is better not to use monologues.

The same audio passage is listened to until you develop the ability to understand and translate each word without much effort. At the initial stage, each audio file should be listened to for at least 3 weeks in a row. It is better to start with tiny excerpts, lasting no more than a minute. It is necessary to monitor the written text in parallel, and everything becomes clear: you can hear what is "swallowed" and what is pronounced with what intonation. Working with the text and audio at the same time, after a certain time manages to understand and hear every word, understand all the phrases and memorize them in full. If you read English text aloud, copying the pronunciation and intonation of the author of a method, it allows you to quickly start speaking in a foreign language.

With the first 3 dialogues need to work for 14 days, at least 3 h a day. Then, with each dialogue you need to work for 3 h over 3 days. After that you will need to listen to the text with your eyes for 10–13 days per dialogue. The next step is articulation, speaking, 3–5 days for each dialogue, 3 h a day. Sometimes it is necessary to return to old dialogues to refresh their memory (once every 2–3 months). In total, at least three hours a day should be spent on work with the matrix. When the matrix is finished (all dialogues have been worked out), it is necessary to repeat it "in a circle".

The following advantages of this method can be noted: a special style of writing, which is well perceived and has an effective impact on people; the author insists that the language learning is impossible without a sincere and strong desire; the creator of the method motivates to give more time to self-study without sparing for it any effort;

learning the language by method of Zamyatkin does not oblige the user to notch the grammar, numerous rules of spelling, lists of irregular verbs, and so on; the author believes that the human brain in the state of the brain of a human. As you know, if you do not choose the most convenient way to learn languages, then to achieve success will be a long time. In this regard, the methodology is really perfect, and after five to six months of hard work, you can learn to speak fluent English.

3 Results

Within the framework of our research the application of the matrix method of teaching is considered on the basis of the Russian Technological University (RTU MIREA) in professional foreign language training, which will contribute to the development of communicative skills to build intercultural relationships.

To carry out the experimental work we interviewed first-year students of the Institute of Radioelectronics and Informatics, as well as the Institute of Artificial Intelligence, the sample of which amounted to 50 people. During the ascertaining stage of the experiment we found that most respondents were not aware of the existence of the matrix method of teaching in the course of linguistic training (Fig. 1).

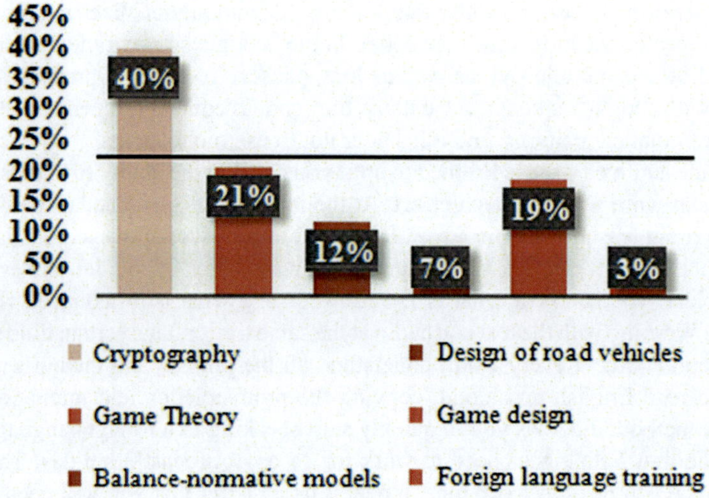

Fig. 1. Applications of matrices.

Within the framework of professional foreign language training of RTU MIREA students, the discipline of foreign language (English), one of the topics of the second semester is "The greatest architectural structure", which is filled with authentic texts, lexical and grammatical tasks and an audio track. On the example of the audio material Architectural styles the matrix method of teaching was implemented, which met all the conditions: duration, content, clarity and structure. Experimental research (formative stage) was conducted in 3 stages, on each of which the respondents were empirically evaluated on the subject: hearing (simple perception of sounds), listening (trying to isolate the main idea from the flow of foreign language information), analysis (to make any notes about the heard audio material), imitation (trying to reproduce what is heard by memory), and understanding (performing the presented task with the ability to retell what is heard). The survey showed that the fragmentation into stages of such a part of language learning as listening has an effective effect on the formation of communicative skills of the student [4]. It is worth noting that the ratio of the "hear" and "understand" stages is very different at the first listening, thus showing that respondents can perceive foreign speech but cannot formulate the main idea on the spot. Repeated listening to the audio track allowed students to make notes to analyze the material presented, thus increasing the number of students who could partially imitate the text they heard. At the final listening stage, students were asked to use an audio script. This allowed them to follow the text visually and to relate the presented material to the topic of this practical lesson "The greatest architectural structure", bringing the understanding of the dialogue to almost 100% of the result (Fig. 2).

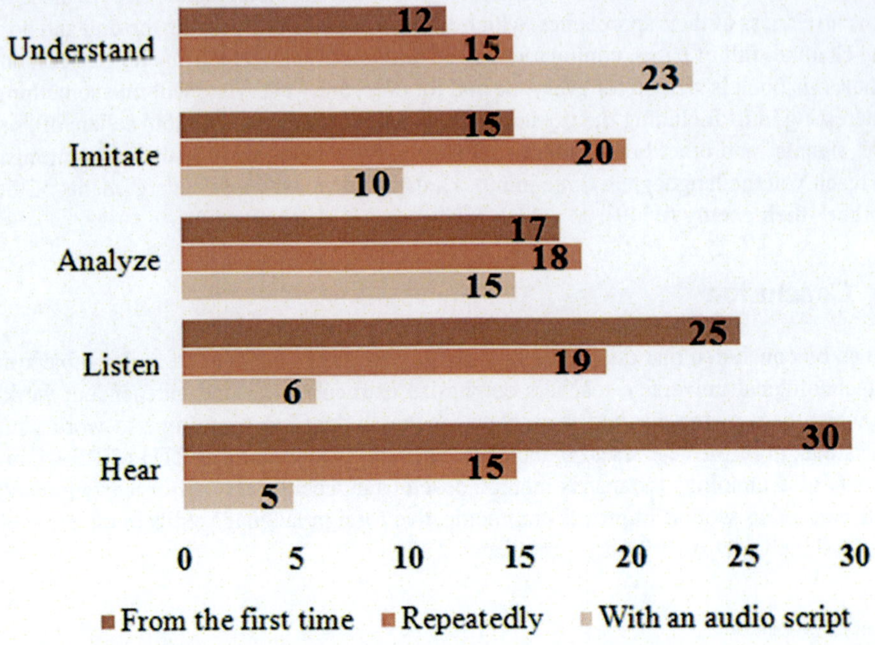

Fig. 2. Applying the matrix method of teaching, using the "Architectural styles" audio track as an example.

Experimental work shows us that the introduction of new methods of teaching a foreign language with the use of information and communication technologies has a positive impact on the results of the students' progress [7, 8]. As it turned out, the range of matrices is so wide that it is possible to use them in almost any discipline presented in the university.

4 Discussion

After the release of N. Zamyatkin's book "You cannot be taught a foreign language", many linguists and foreign language teachers had different opinions. In our paper we wanted to present one of the opinions about Zamyatkin's matrix method.

Honest to the last comma, the book that immediately became a classic of the genre and a must-read for everyone who is at least somewhat interested in languages. A paradoxical book that inexorably destroys myth after myth, fiction after fiction, error after error. A book that frees the reader from the pervasive old misconceptions that keep him from mastering a foreign language. Anyone who studies or intends to study a foreign language simply must read this book, unparalleled either in the accessibility of the author's language or in the quantity and quality of useful tips. Brilliant style and relaxed humor of presentation make this book interesting for those who have already "studied" a foreign language at school or university and as a result finally believed in their "inability" to languages - they will understand why, after all those painfully long years and did not master - and could not master! - the language remaining within the generally accepted format of "learning". Those who speak foreign languages will be glad to be convinced of the correctness of their approaches, which allowed them to escape from the dull and boring chamber full of cases, conjugations and participles that frighten any normal person. Thus, this book is written for everyone and for everyone - everyone will find something interesting in it! Including the organizers of language "scams", the rabid sellers of "secret signals" and other brash authors of "successful" books, who shamelessly promise to teach you the language in three minutes a day: they must know the arguments of the author - their enemy #1! [9].

5 Conclusion

It can be concluded that due to the fact that foreign language is not a major subject in a technological university, teachers emphasize extracurricular and independent work, in which as a matrix method is relevant, as the student has more time to work with tasks than in full-time classes. In the course of linguistic training at RTU MIREA University of Technology the matrix method develops independence, self-organization and self-control, as well as improves communicative foreign language skills in the logic of professional activity of future specialists.

References

1. Kitaygorodskaya, G.A.: Intensive training in foreign languages. Theory and practice. Higher School, Kitaygorodskaya School, Moscow, Russia (2014)
2. Shokanova, R.D., Tarasova, E.N.: Reproductive method in teaching Russian language to foreign students and its innovative aspects. Russ. Technol. J. **9**(3), 98–107 (2021)
3. Sivarajah, R.T., Lee, J.T., Richardson, M.L.: A review of innovative teaching methods. Acad. Radiol. **26**, 101–113 (2019)
4. Monakhov, V.M.: Informatization of educational and methodical support of the integral process of competence formation and technological monitoring of their quality management. Vestnik MGU **4**, 46–59 (2012)

5. Arkusova, I.V.: Modern pedagogical technologies in teaching a foreign language (structural-logical tables and practice of application). Non-governmental educational institution VPOOMPSI, Moscow, Russia (2014)
6. Passov, E.I.: Fundamentals of teaching methodology of foreign languages. Russian language, Moscow, Russia (2015)
7. Shchukin, A.N.: Modern intensive methods and technologies of teaching foreign languages. Filomatis, Moscow, Russia (2014)
8. Pikhart, M., Al-Obaydi, L.: Potential pitfalls of online foreign language teaching from the perspective of the university teachers. Heliyon **9**, 45–51 (2023)
9. Zamyatkin, N.: It is impossible to teach you English. https://ik-ptz.ru/en/dictations-on-the-russian-language--class-3/vas-nevozmozhno-nauchit-angliiskomu-yazyku-zamyatkin-zamyatkin-n-vas.html. Accessed 02 Feb 2023

An Approach for Making a Conversation with an Intelligent Assistant

A. V. Kozlovsky, Ya. E. Melnik(✉), and V. I. Voloshchuk

Institute of Computer Technologies and Information Security, Southern Federal University, St. Chehova, 2, Taganrog 347922, Russia
{kozlovskiy,iamelnik,vvoloshchuk}@sfedu.ru

Abstract. The problems associated with modern text generation methods are considered in the article. The task of making an interactive dialogue with an intelligent assistant is set. Advantages and disadvantages of analogues are revealed. An approach for analyzing of incoming requests and constructing responses to them is proposed. A graph model is chosen as the basis for semantic text conversion. In addition, the possibility of adding vectors to the main model to take into account the time aspect and other additional characteristics necessary for a full-fledged analysis of both a separate sentence and a whole text has been shown. The general system operation algorithm system operation is constructed. Directions for further research have been identified.

Keywords: Text Generation · Neural Networks · Graph · Vector

1 Introduction

The development of anything, in particular humanity, implies simplifying and facilitating interactions and actions performed by certain processes automation. Initially, the processes associated with hard work were automated, then just monotonous ones. The common trait of these two categories is that the actions performed can be unambiguously described algorithmically, which excludes the element of intelligence. With the science and technology development, it has become possible to automate tasks that are not directly amenable to algorithmization. Artificial intelligence made it possible to do this, in particular, the most popular section of machine learning and its subsection deep learning [1].

The artificial intelligence development has made it possible to highlight objects in photos and videos, generate text, process sound and more. In solving problems in these areas has already achieved amazing results, and work on improvements and problems is still underway, which potentially means even more amazing and practically more convenient and thoughtful ideas and tools in the not so distant future. However, at the moment, there are still areas where the presence of a person is necessary so that he can assess the intelligent system operation correctness when it gives a signal or shows a hint. Such control is necessary, since machine learning models simply cannot always show unambiguously accurate results with 100% probability. This should be taken into account

R. Silhavy and P. Silhavy (Eds.): CSOC 2023, LNNS 724, pp. 526–536, 2023.
https://doi.org/10.1007/978-3-031-35314-7_47

in certain areas of such systems and products application [2]. For example, analyzing a video stream to identify atypical behavior [3]. The system may mistakenly perceive some actions as aggressive and react, which is undesirable. That is why additional human control is necessary in such situations. However, this does not in the least detract from the importance and usefulness of such solutions, since they work with high accuracy and in most cases no corrections are required. This allows you to reduce the employees number and optimize work processes.

Also, intelligent assistants can be very useful in report generation, because it often takes a huge amount of time due to the execution subtleties. And the intelligent system, having received the necessary data for input, will be able to make it in a short time, taking into account all features.

In addition, it is worth considering the convenience of working with such a system, since it is aimed at simplifying the work, and for example, an overly complex interface can eliminate this main advantage [4]. The most convenient way of communication on work issues for a person is a dialogue, so the form of interaction in the form of a chatbot seems to be the most preferable [5, 6].

However, it is quite a difficult task to make fully coherent dialogue between a machine and a person, since the system must fully understand the subject area it operates, and also preserve the conversation context and work with it in order to correctly perceive new information. And it is necessary to identify specific problems and study analogues in detail in order to speed up and simplify work on such a system.

2 Highlighting Problems

The main problems when generating text for a user request are an incomplete perception of the words semantics and natural language constructions and loss of context. For example, most popular solutions are based on recurrent neural networks that evaluate the relevance of sentences related components and texts as a whole [7]. This makes it possible to develop texts of a certain level well. Such texts are usually highly specialized and have a fairly strict structure, from which it is not worth departing. However, there is no question about disparity in comparing with the flexibility of human responses to replicas. The problem of preserving semantics almost directly follows from the previous one. The program simply analyzes the words sequence and selects those that in its opinion are most suitable. However, with this approach, the continued use of pronouns (and other universal constructions) may entail unknown consequences, since they are used in completely different areas and situations.

Also, among the making interactive conversations with a computer problems, the following two can be distinguished: the lack of training in new areas and the inability to process large requests. The first of these features would be useful for expanding the certain system applications range for interaction with various users. That is, when finding the best approach to text analysis, there is no need to create a new solution for each new case. This will optimize the work in this research area. And the ability to form large requests will allow people to fully explain their question or task. In addition, people prefer complete and understandable answers, which also imply more than one sentence. Remembering the context will help significantly in solving the latter problem.

Thus, four problems can be identified:

1. incomplete perception of the words semantics and natural language constructions;
2. lack of the context memorization;
3. lack of training in new knowledge areas;
4. lack of ability to work with large requests.

3 Analysis of Analogues

In the search for analogues, one was found that was most suitable for the tasks. ChatGPT is an artificial intelligence trained using a combination of machine learning algorithms and methods. One of the main algorithms used in the learning process is a neural network called a transformer, which is designed to process sequential data such as language [8].

To train the ChatGPT transformer model, the technique of unsupervised learning was used: the model tries to predicting the next word in a sequence based on the words that were in front of it, based on analyzing a large amount of text data and studying the patterns and structure of the language.

In addition to the transformer model, ChatGPT uses other machine learning algorithms and methods. For example, it uses information retrieval algorithms to search training data and find the most relevant information in response to a query, and also uses natural language processing techniques to understand and generate human-like responses. In general, the combination of these algorithms and methods allows ChatGPT to process efficiently and understand large amounts of text data and communicate with users in natural language.

The ability to process and understand large amounts of text data is one of the main technical advantages of ChatGPT, which allows it to answer questions and provide information on a wide range of topics. It's also able to communicate in natural language, which makes it easier for people to interact with ChatGPT.

One of the new solution disadvantages is that it doesn't have the ability to browse the Internet or access new information that goes beyond the one it was taught. This means that ChatGPT may not have the most up-to-date information or may not be able to answer questions about events that occurred after it gained knowledge.

In addition, the ChatGPT reward model, designed with human supervision in mind, may be overly optimized and thus reduce performance according to Goodhart's law. As it's known, when training, reviewers preferred longer answers, regardless of the actual understanding or the actual content, which indicates the algorithmic bias of the ChatGPT data.

Less similar solutions were also found. Among the most popular of the Russian - speaking sector, the following were analyzed:

- YaLM (Another Language Model), in particular "Balaboba";
- GPT-3 (Generative Pre-trained Transformer 3), in particular the Russian-language model GPT-3 Large (ruGPT-3);
- mGPT (multilingual Generative Pre-trained Transformer).

Lesser-known systems and projects were also considered:

- PC Writer 1.0;

- Zinaida Falls;
- the novel "The day when the computer will write a novel";
- Philip Parker Algorithm;
- Narrative Science.

YaLM is a tool developed and used by Yandex in its own products. It is used both in real tasks ("Search" and "Alice") and in entertainment ("Balaboba"). Copes well with its tasks, i.e. searching the Internet, interacting with devices and conducting simple conversations, as well as generating short texts in a certain style [9]. However, it will not be suitable for building something structured and larger. At the same time, there are problems with preserving the context.

GPT-3 is the OpenAI project. It is the third generation of the natural language processing algorithm. The following applications are distinguished: generating articles and texts, conducting a conversation in particular, using chatbots [10, 11]. Open air gave access to its model to large companies so that they could create something of their own based on it. One such example is ruGPT-3 Large. This model was created so that Russian-speaking users could fully appreciate the capabilities of the model. Other application methods, such as, for example, writing poetry and prose, performing translations, are also possible. Such a wide range of supported tasks became possible due to deep learning. This means that the neural network was not trained specifically for each task. However, the problems of preserving the context and the small size of requests and responses remain [12].

mGPT – development of the Russian-language GPT-3 Large model. Sber continued its work in this direction and the result was this multilingual model, which supports 61 languages. The problems remain the same as the predecessor. However, in this case, the task was complicated by the fact that there is not so much data for less commonly used languages. But it is worth noting the fact that the model manages to take into account cultural peculiarities when constructing the text [13].

PC Writer 1.0 is a project that can be called an "artificial writer". This program was able to write a book that was even published. When studying the passage, it became obvious that it is quite difficult to perceive what is happening described in the book, since the details pile up on top of each other. It seems that the program does not fully grasp the context of what is happening and the semantics of words and text in general. This is also confirmed by the fact that the lines of reasoning are not fully understood and can jump over. However, it is worth considering the fact that this result was achieved in 2008, which is impressive.

Zinaida Falls is an older project of Yandex, developed for writing poetry. In the materials on this program, it is noted that it was able to convey feelings and sensations, like a person. It can also imitate the styles of various famous poets. These are the undeniable advantages of this solution, but there is no work with large requests. Also, words are not always used appropriately, which indicates an incomplete semantics perception.

It is also worth mentioning the novel "The Day when the computer will write a novel". This work made it to the final of the Japanese novel contest, which is the best result for works of this kind. The jury, in general, praised this work, but noted that cliches were constantly used in it.

Philip Parker's algorithm is not aimed at writing works of fiction, but is aimed at more routine tasks. However, the author of the algorithm himself also highlights such an application as writing specific texts that real authors can't or don't want to undertake.

The Narrative Science project was created for writing reports, in particular for writing articles about baseball games [14]. In one work, this program was even able to do a better job than a real writer, because it used a more appropriate method of constructing a text. This solution is very good, but it is applicable to a rather limited range of spheres.

Thus, the following can be noted in relation to the considered analogues. A part of solutions don't have the possibility of interactive communication, i.e. they receive data and begin generation. Other solutions, on the contrary, have this advantage and also offer a wider range of tasks to be solved, however, they have problems with the semantics perception and working with large queries. Also, most analogues have a problem with retaining the context for its further use. In addition, no information was found about possibilities of these solutions to adaptation to new information.

Thus, the problems highlighted in Sect. 2 are confirmed and become more evident.

4 Proposed Approach

First of all, it is necessary to study the text analysis methods, based on which it is possible to build a full-fledged conversion algorithm, including a transitional format for storing text data semantics.

The very first way to transform text structures into something other than a sequence of words was a method based on recursive neural networks [15]. Such neural networks work with variable length data and use hierarchical sample structures in training [16]. The most striking example of using such architecture is images that are composed of scenes that include many objects [17]. Identifying the scene structure and deconstructing it is a non–trivial task. At the same time, it is necessary to identify both individual objects and the entire structure of the scene. Most often, this architecture is used with some improvement for sequential decryption of an image natural scenes [18] or for structuring sentences of a natural language [19].

The key difference from recurrent neural networks is that the latter work exclusively within the framework of a linear progression in time, which is clearly visible on the branch expressed in the form of an LSTM model [20].

Such models have a good ability to see the context within a small number of "iterations". An example is ChatGPT [21], which was trained to predict the next word using text sent to it as a dataset. However, such an example is limited in the semantics perception: the model may separate the request from the unrelated text, but post-processing will be limited in terms of response formation and will more often respond with cliched phrases. That is, the semantics itself has been revealed, but is limited to the rough concepts of the basic topics embedded in the model.

It is worth saying that, in contrast to the above, the recursive approach is not too widespread. However, it showed himself very well in natural language processing tasks, which is just necessary to solve the task. Recursiveness itself is expressed in the fact that the one module output signal is often fed to the input of the same type another module. Thus, it is possible to make full-fledged constructions from the identified semantics, including meaningful and structured sentences.

The key problem is that this approach does not work in the opposite direction. This means that the algorithm is designed more for post-processing of textual information and works exclusively with hierarchical trees [22]. As part of the task, there is even a dataset for training, which stores a lot of sentences converted into the described constructions. However, an algorithm is needed that turns the text model into an intermediate one, possible for analysis by recursive neural networks. In such a situation, knowing the format that is used for postprocessing will help: since semantically, the data is a tree format, a graph model that has proven itself well on many similar scope tasks is best suited for the intermediate state [23].

Thanks to the structure built on the concept of vertices and edges, it becomes possible to implement scalable text structures, such as the source text, the request, the information necessary for the request within a single system.

It is worth noting that it is impossible to preserve absolutely all semantics, however, due to a multi-level approach, most of it can be preserved. At the same time, the lowest level will include an exceptionally primitive structure of material objects explicitly expressed in the text. In this case, the subject (material objects) are represented as vertices. In turn, the sentences elements are used as edges, often expressed as predicates, but they can also represent circumstances (adverbial turns).

Based on this, a directed graph is obtained, which can be processed algorithmically and mathematically.

Example:

A serious bear attacked a poor hare running past this forest two days ago.

The main objects will be a bear, a hare and a forest in this case.

The interactions of the bear with the hare and the hare with the forest will be the edges. If we classify the forest as a place of action, then the bear also refers to the forest, but this is not considered at this stage.

This example can be represented in the form of a graph shown in Fig. 1.

The classification of "this" in this case will depend on the context of the proposals standing before it, otherwise it is simply not taken into account.

Then the resulting structure is filled with values for the formalized processing possibility. Each subject will be automatically assigned a score in the form of a coefficient reflecting the degree of the vertices on each other influence. Each vertex is somehow connected to any other, but the influence degree may vary.

In the example above, there is an obvious connection between a bear and a hare and a hare with a forest, which indicates their strong mutual influence, on the basis of which a relatively high coefficient can be put to these connections. On the other hand, despite the fact that the bear is in the forest, their relationship is indirect at the moment, which is estimated by a weaker coefficient.

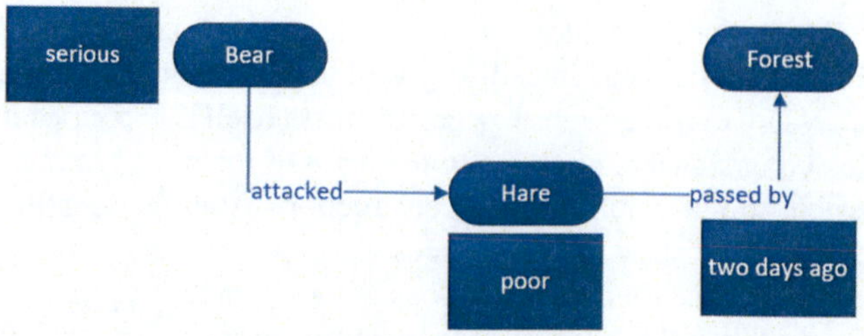

Fig. 1. Graph semantically reflecting the sentence presented as example

If additional actions are added in the current history, a significant drawback of the selected model can be noticed.

Example:

A serious bear attacked a poor hare running past this forest two days ago. However, the hare's speed was great, and it ran away into the thicket.

Two vertices are added in this case: speed and thicket. Speed is related to the hare, as is the thicket.

The graph corresponding to this situation is shown in Fig. 2.

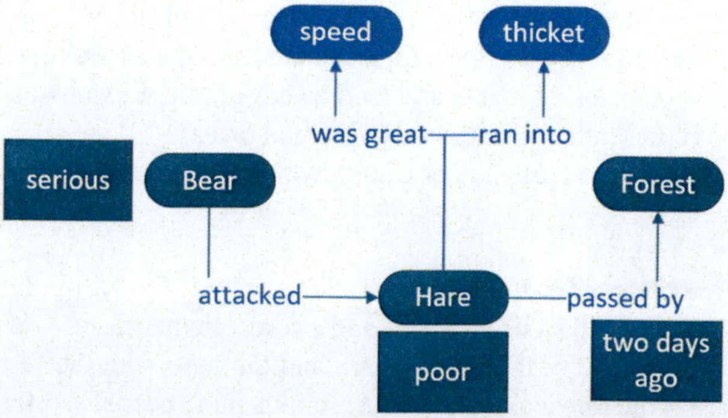

Fig. 2. The graph corresponding to the complicated example

However, it is now impossible to determine which action occurred earlier. In order for the sequence to remain relevant, it is necessary to take into account such an important factor as time, since without it the entire text structure will turn into an indivisible block devoid of sequence information. In order to preserve the graph model taking into account the new parameter, it was decided to combine it with the coordinate plane and add a third dimension to reflect time. This results in a full-fledged graph structure in vector space [24].

The presence of a graph is due to the need to structurally link objects with each other and to designate their relationships. Vector space, in turn, allows to place objects globally from the general history point of view, based on several characteristics. One of them is time.

It should also not be forgotten the fact that some words can be interpreted differently depending on the context. This problem suggests the need to complicate the original model by adding several levels of semantic perception.

Thus, the method of textual information converting into an intermediate vector-graph model was chosen, which allows adapting to any language in which it was trained storing and allocating, including information about the request, which subsequent processing stages carefully delete.

The proposed solution should eliminate the shortcomings of existing analogues. It also provides more flexible work with the graph, which will provide opportunities for fairly flexible query generation before conversion (Fig. 3).

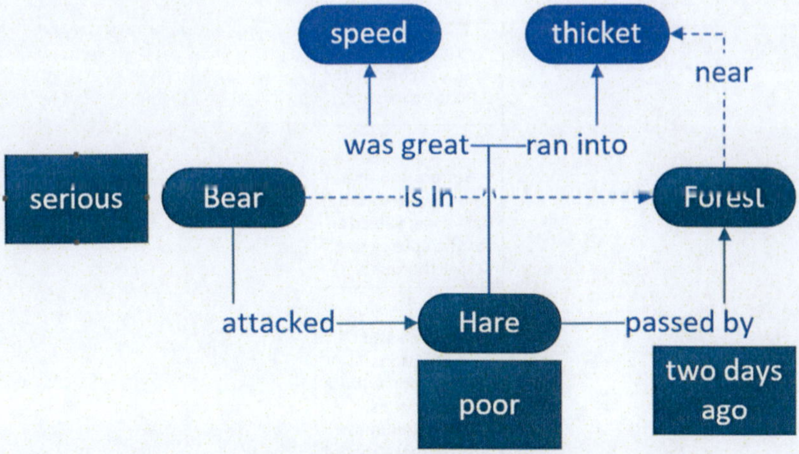

Fig. 3. A complicated graph describing indirect connections

5 The Structure of the Algorithm Implementing the Proposed Approach

The following algorithm has been developed to implement the approach.

First of all, it is necessary to analyze the original sentence and identify semantic points. To do this, a morphological analysis is performed and an intermediate model is built based on a graph structure and a vector levels representation to take into account the time concept.

Next, the system determines a number of post-processing actions that need to be performed based on a request from the user. After that, the stack LSTM model transforms the intermediate model into query-optimized trees. As a result, recursive neural networks "mix" data into single text blocks and form a full-fledged, semantically conceptual image.

For clarity, a generalized algorithm implementing the proposed approach is shown in Fig. 4. Each stage includes a specific neural network that solves the task assigned to it.

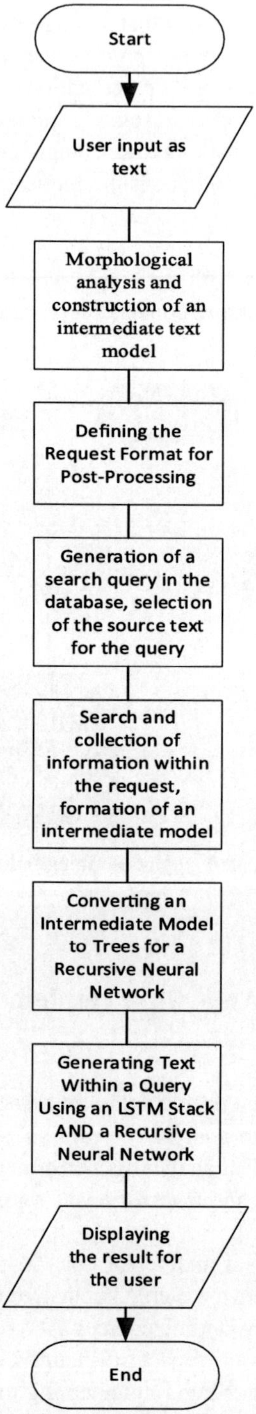

Fig. 4. A generalized algorithm implementing the proposed approach

In the first layer, the model recognizes a sequence of words, forming an intermediate vector-graph format.

The second layer plays a key role in the system. First of all, the model identifies instructions from the intermediate model aimed at the system itself. For this purpose, both pre-agreed constructions and figurative representations can be used, which are read by the model automatically.

In the third layer, all priority will be given to the LSTM model [25], which selects a query from the text and leaves information that is extremely important for semantics. This layer is combinational, so after the LSTM model is working, the system searches the database for the information necessary for the query, including using the classification problem solution [26], and converts it into a tree that can recognize the next layer.

The fourth, final layer, is fully expressed through recursive neural networks. The model, taking into account the generated parameters, forms a sequence of trees, represented as a text block, the size and structure of which are formed depending on the user's request.

6 Conclusion

In this paper, the problems associated with the issues of text generation were highlighted, the available analogues were analyzed and a solution based on a new approach to text processing that takes into account the context and logical relationships of objects was also proposed.

This solution should help improve formalized text processing, in particular, maintaining an interactive and fully coherent dialogue with an intelligent assistant.

The proposed solution allows to get away from such problems as semantics incomplete perception, the inability to work with large queries, as well as loss of context and the inability to replenish the system's knowledge base.

Acknowledgments. This approach will be implemented on the base of the presented algorithm as a part of the project of the prototype development of a text processing help web service.

Project No. 925ГУЦЭС8-D3/82999 of 27.12.2022.

References

1. Kolesnikova, G.I.: Artificial intelligence: Problems and prospects. Videonauka **2**(10), 1–6 (2018)
2. Kalyaev, I.A., Melnik, E.V.: Trusted control systems. Mekhatron. Avtomatiz. Upravlen. **22**(5), 227–236 (2021)
3. Ryabchikov, I.A.: A method for automatic recognition of deviant behavior of people based on the integration of computer vision and knowledge management technologies to support decision-making by operators of video monitoring systems. Inf. Technol. Telecommun. **87**(3), 21–36 (2022)
4. Gorelova, G., Melnik, E., Safronenkova, I.: The problem statement of cognitive modeling in social robotic systems. Lect. Notes Comput. Sci. **12998**, 62–75 (2021)

5. Matveev, A., et al.: Virtual dialogue assistant for remote exams. Mathematics **9**(18), 2229 (2021)
6. Budzianowski, P., Vulić, I.: Hello, it's GPT-2 - How can i help you? Towards the use of pretrained language models for task-oriented dialogue systems. In: 3rd Workshop on Neural Generation and Translation, pp. 15–22. Association for Computational Linguistics, Stroudsburg (2019)
7. Benderskaya, E.N., Benderskaya, E.N.: Recurrent neural network as dynamical system and approaches to its training. Comput. Telecommun. Control **4**(176), 29–40 (2013)
8. Rudolph, J., Tan, S., Tan Sh.: ChatGPT: Bullshit spewer or the end of traditional assessments in higher education? J. Appl. Learn. Teach. **6**(1), 1 (2023)
9. Bubyakin, M.Y., Il'ina, Y.A.: The possibilities of neural network language models in modern realities. Sci. Educ. **44** (2022)
10. Menal, D.: A tool of conversation: Chatbot. Int. J. Comput. Sci. Eng. **5**, 158–161 (2017)
11. Borodin, A.I., Veynberg, R.R., Litvishko, O.V.: Methods of text processing when creating chatbots. Humanit. Balkan Res. **3**(5) (2019)
12. Li, L., Bamman, D.: Gender and representation bias in GPT-3 generated stories. In: 3rd Workshop on Narrative Understanding, pp. 48–55. Association for Computational Linguistics, Stroudsburg (2021)
13. Shliazhko, O., Mikhailov, V., Fenogenova, A., Kozlova, A., Tikhonova, M., Shavrina, T.: mGPT: Few-shot learners go multilingual. arXiv preprint arXiv:2204.07580 (2022)
14. Ghuman, R., Kumari, R.: Narrative science: A review. Int. J. Sci. Res. **2**(9), 205–207 (2013)
15. Androsova, E.E.: Application of recursive recurrent neural networks. New Inf. Technol. Automat. Syst. **19**, 107–114 (2016)
16. Proshina, M.V.: Modern methods of natural language processing: Neural networks. Constr. Econ. **5**, 27–42 (2022)
17. Myznikov, P.: Application of recursive neural networks in the problem of hierarchical image segmentation. In: 59th International Scientific Student Conference, p. 152, Novosibirsk National Research State University, Novosibirsk (2021)
18. Falomkin, I.I.: Generalized algorithm of adaptive morphological image filtering. In: 9th International Conference "Intelligent Systems and Computer Science", pp. 291–294. Faculty of Mechanics and Mechanics of MGU, Moscow (2006)
19. Socher, R., et al.: Recursive deep models for semantic compositionality over a sentiment Treebank. In: Conference on Empirical Methods in Natural Language Processing, pp. 1631–1642 (2013)
20. Hochreiter, S., Schmidhuber, J.: Long short-term memory. Neural Comput. **9**(8), 1735–1780 (1997)
21. Lund, B.D., Wang, T.: Chatting about ChatGPT: How may AI and GPT impact academia and libraries? Library Hi Tech News (2023)
22. Shen, Y., Tan, Sh., Sordoni, A., Courville A.: Ordered neurons: Integrating tree structures into recurrent neural networks. arXiv preprint (2018)
23. Ore, O.: Graphs and their Uses. The Mathematical Association of America, Washington, D.C. (1996)
24. Pennington, J., Socher, R., Christopher, D.: Manning glove: Global vectors for word representation. In: Proceedings of the 2014 Conference on Empirical Methods in Natural Language Processing, pp. 1532–1543. Association for Computational Linguistics. Stroudsburg (2014)
25. Heryadi, Y., Warnars, H.L.H.S.: Learning temporal representation of transaction amount for fraudulent transaction recognition using CNN, Stacked LSTM, and CNN-LSTM. In: 2017 IEEE International Conference on Cybernetics and Computational Intelligence (CyberneticsCom), pp. 84–89. IEEE. Red Hook (2017)
26. Vorobev, E.V., Puchkov, N.V.: Classification of texts using convolutional neural networks. Young Res. Don **6**(9), 2–7 (2017)

Fraud Detection in Mobile Banking Based on Artificial Intelligence

Derrick Bwalya(✉) and Jackson Phiri

Department of Computer Science, University of Zambia, Lusaka, Zambia
{derrick.bwalya,Jackson.phiri}@cs.unza.zm

Abstract. The advent of modern technological advances has led to traditional Banks and other financial institutions to reposition their financial services and introduce new transaction channels and modes of payments. Mobile Banking (M-Banking) usage has been on the rise in the recent years and has provided opportunities for growth and financial inclusion. However, this substantial increase in the volume of mobile banking transactions, has had a corresponding increase in fraud cases recorded in millions of dollars every year by banks and financial institutions. Managing vast amounts of mobile banking transactional data and making choices about risk, client retention, fraud detection, and marketing management are crucial tasks in the banking industry. This study explores an effective application of Artificial Intelligence data mining algorithms fraud detection in mobile banking transactions. This research aims to provide a robust, cost effective, efficient yet accurate data mining-based solution to detect mobile banking fraud. We propose Artificial Intelligence (AI) model using K-Means clustering and Anomaly detection algorithms to detect "fraudulent" and "genuine" transactions in near- real time. The results of applying K-means clustering and Isolation Forest anomaly detection produced a detection rate of 5% of the volume of transactions analyzed.

Keywords: Mobile Banking · K-means · Data Mining · Clustering · Anomaly detection

1 Introduction

Traditional banking has been around for a while and is still the most popular way to conduct bank transactions in both developed and developing nations [1]. However, the conventional banking approach is increasingly giving way to the modern digital banking approach in the twenty-first century. Information technology (IT) has been steadily improving over the past several years to aid in streamlining corporate processes around the globe, particularly in the banking sector where the usage of Automated Teller Machines (ATMs) was developed to make cash withdrawals for clients easier. Following the rise of online banking, mobile phone banking finally emerged. A bank's long-term success—or even survival—depends on an efficient digital banking experience. The secret to success lies in being able to adopt and utilise new technologies and skills. A modular microservices-led design has become the standard since modern digital banking platforms need to be flexible, adaptive, and scalable without affecting performance.

R. Silhavy and P. Silhavy (Eds.): CSOC 2023, LNNS 724, pp. 537–554, 2023.
https://doi.org/10.1007/978-3-031-35314-7_48

The development of the mobile banking ecosystem in some African nations has been accelerated by adoption of advanced technological infrastructure, the interoperability of financial systems, and developments in regulatory framework. The transaction and payment options accorded by the introduction of mobile banking have been instrumental in facilitating trade and commerce amidst the COVID-19 pandemic. The mentioned advent of modern technological advances has led to traditional Banks and other financial institutions to reposition their financial services and introduce new transaction channels and modes of payments through digital banking. Digital Banking is the automation of traditional banking services, enabling a bank's customers to access banking products and services via an electronic/online platform. Mobile Banking is the use of mobile gadget (mobile phones, tablets, Personal digital Assistants) to conduct a financial transaction [2]. Put it another way, Mobile banking is an electronic banking innovation that connects mobile phones and other mobile devices to banking systems and enables access to a variety of financial services via the mobile interface. Mobile banking is a subset of Digital banking. The primary goal of mobile banking is financial inclusion, ensuring that the unbanked and rural population are provided with appropriate financial services.

Mobile banking allows access to bank services which includes access to mini-statements; alerts on account activity or passing of set thresholds; access to loan statements; status on cheque, stop payment on cheque; ordering cheque books; account balance inquiry; change of PIN; funds transfers; mobile airtime and bundle purchase; utility bill payment; and mobile agency deposits [3].

In a bid to stay competitive and offer conveniency to its customers, most financial institutions in Africa, and Zambia in particular, have introduced mobile banking services. This has been supported by the rapid growth and penetration of mobile phone usage by the general populace. According to the Bank of Zambia (BOZ) 2020 Annual payments systems report, the national financial Switch (NFS) recorded an increase of 61% in volume and 67% in value of transactions [4]. With all the opportunities for development and growth presented by this rapid growth, it has also attracted cybercriminals leading toa sharp rise in the number of mobile banking fraud cases affecting both the end users and financial institutions. According to [5], fraud has diverse effects which include reduced industry confidence, destabilizing customer savings, and affects the cost of living for the affected customers. The Association of Certified Fraud Examiners (ACFE) defines fraud as an any intentional or deliberate act of depriving another of property or money by cunning, deception, or other unfair acts [6]. In the white paper titled "Understanding Credit Card fraud", the authors define fraud as, "When an individual uses another individuals' credit card for personal reasons while the owner of the card and the card issuer are not aware of the fact that the card is being used". They further state that, "the individual using the card has no connection with the cardholder or issuer, and has no intention of either contacting the owner of the card or making repayments for the purchases made" [7]. In another context, fraud may be defined as the purposeful and deliberate activities made by participants in the mobile financial services ecosystem with the objective to make money, deny other participants in the ecosystem revenue, or harm the reputation of other stakeholders. The Federal Bureau of Investigations (FBI) defines fraud as fraudulent conversion and acquisition of money or property by false pretense [8]. Chimobi et al. [9] stated that fraud is defined by law as the dishonest act of denying someone something they may otherwise be entitled to or would have received but for the fraud that was committed.

With the ever-increasing threats in the cyberspace, fraud detection requires training a machine learning algorithm to identify concealed observations from any normal observations. In this study, we apply AI machine learning algorithms to cluster data and identify anomalies on mobile banking transactions. We focus on debits transactions on customer accounts, that ultimately reduces the balance.

2 Literature Review

This section seeks to analyse pertinent research papers related AI and machine learning in fraud detection and conclusions from the theoretical and empirical literature in order to strengthen the research and its ability to understand the issues it intends to address.

2.1 Artificial Intelligence

Artificial intelligence systems are driven by algorithms, applying techniques such as machine learning, deep learning and rules developed by domain experts. AI is a general term that addresses the need to build and use computer systems to model intelligent human behavior with minimal human intervention [10] Several scholars have endeavored to give definition to artificial intelligence. Poole et al. [11] define AI as "the field that studies the synthesis and analysis of computational agents that act intelligently." They further go on to state that an agent is said to be intelligent when: its actions are consistent with the prevailing circumstances; it has the ability to learn from experience, its able to adapt to changing environments; and given its perceptual and computational limitations, it is capable of making the right decisions. AI is further described as the ability of machines and computer programs to mimic human thinking processes and behavior patterns, such as the capacity to learn from experience and make judgments about and respond to unplanned events [12, 13].

As a branch of computer science AI, focuses on developing machines that can do activities that typically require human intelligence [14]. Loosely speaking, this involves incorporating human intelligence into machines. The main goal of artificial intelligence development is to replicate the human mind through the use of computer programs that can comprehend human behavior to investigate the behavior of human intelligence [15]. AI is divided in various sub-domains of intelligence, which includes machine learning (ML), deep learning (DL), Neural Networks (NN), Natural language processing (NLP), computer vision (used in visual perception and pattern-recognition), and cognitive computing in which computing algorithms mimic a human brain.

2.2 Machine Learning

Machine Learning is a sub-domain of AI [16] Arthur Samuel [17] defines machine learning as "the field of study that gives computers the ability to learn without being explicitly programmed." He pioneered research in machine leaning by developing a checkers game playing program. Tom Mitchell [18] gave a more refined and formal definition of ML stating that, "*A computer program is said to learn from experience E with respect to some class of tasks T and performance measure P, if its performance at tasks in T, as measured by P, improves with experience E.*" He further goes on to state the central question that ML seeks to answer [19]: "*How can we build computer systems that automatically improve with experience, and what are the fundamental laws that govern all learning processes?*" The main focus of ML learning, he states, is how to make computers have the capability to program themselves.

ML teaches a computer how to draw conclusions and decide based on prior knowledge gained from past experience. Without relying on human experience, ML recognizes patterns and examines historical data to deduce the significance of these data points and arrive at a potential conclusion. Machine learning has been applied across industries and various disciplines, such as Speech Recognition, Computer vision in the US Post Office to automatically sort letters with handwritten addresses, and robot control. Research shows several applications of machine learning in financial statement fraud detection [20].

2.3 Data Mining

Fraud detection has long been accomplished using conventional/traditional data analysis techniques. However, these techniques require complex and time-consuming investigations to discover patterns in data. Managing vast amounts of transactional data and making choices about risk, client retention, fraud detection, and marketing management are crucial tasks in the banking industry. In their study on Data mining and its application in Banking sector, Chitra et al. [21] defines data mining as a process of analysing vast data repositories to find useful information and gain insightful knowledge to address important business concerns. This process also known as Knowledge Discovery from Data (KDD) involves extraction of interesting, previously unknown and potentially useful patterns or knowledge from huge amount of data. It reveals hidden to human analysis implicit correlations, trends, patterns, exceptions, and oddities. In this process, Algorithms are applied on the data set to find relationships and patterns in the data, and then this information about the patterns is used to make forecasts and predictions. Pulakkazhy et al. [22] states that the discovered knowledge from the data must be new, obscure (not obvious), and must be relevant and can be applied in the domain where this knowledge has been obtained.

Data Mining involves various logical process flows and knowledge discovery as shown in Fig. 1. The data mining process can be broken down into two main phases namely: data preparation/data preprocessing, and data mining. The data preparation involves the first four processes, namely: data selection, data cleaning, data integration, and data transformation. Data mining phase involves the last three processes including data mining, pattern evaluation and knowledge representation. There are several techniques to data mining based on the kind and amount of the data that is being used, as well as on the nature of the business challenge being addressed.

Machine Learning (ML) algorithms are generally categorized into three, namely: supervised learning, unsupervised learning, and reinforcement learning.

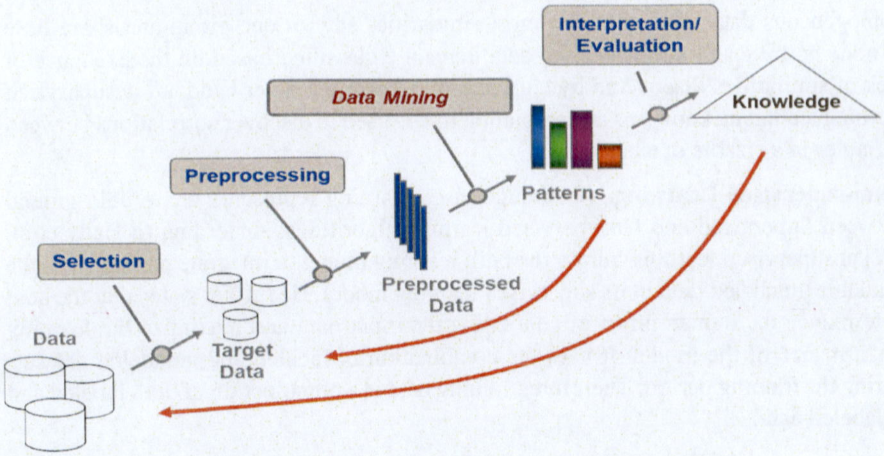

Fig. 1. Knowledge Discovery process in Databases [23]

Supervised Learning. Supervised learning is a machine learning technique in which models are trained using labeled data features that define the meaning of the data. Typically, supervised learning starts with a pre-existing set of data and a predetermined understanding of how that data is classified. The goal of supervised learning is to identify data patterns that may be used in an analytics process. A supervised algorithm can either be a regression or classification. When the label is continuous, it is a regression; when the data comes from a finite set of values, it known as classification. In the case of supervised learning for fraud detection, the data is labelled as fraudulent (1) or non-fraudulent (0).

Supervised machine learning models are trained on labeled data and examine historical transactions to mathematically determine how a typical fraudulent transaction looks and assigns a fraud probability score to each transaction [24]. The built model needs to find the mapping function to map the input variable (X) with the output variable (Y) as illustrated in Eq. 1 below.

$$Y = f(X) \tag{1}$$

The most popularly applied supervised learning algorithms are, the Artificial neural networks (ANN), support vector machines (SVMs), Naïve Bayes (NB), Random Forests (RF) as well as decision trees (DT) [25].

Unsupervised Learning. This is a machine learning technique in which patterns are deduced from unlabeled input data. Unlike supervised learning, unsupervised learning models find the hidden patterns and insights from the given dataset. Unsupervised leaning removes the aspect of classification feedback that is inherent in supervised learning; instead, the target classification is unknown during the training process [26]. The goal of unsupervised learning is to find the underlying structure of a dataset, classifying the data into clusters based on similarities, and representing the dataset in a compressed manner. Unsupervised learning can be used for two types of problems: Clustering and Association. Using the clustering technique, items are grouped into clusters so that homogeneous data items with the most similarities stay in one group and share little to none with those in another. The data items are classified based on the existence or lack of similarities discovered by cluster analysis. On the other hand, an unsupervised learning technique known as an association rule is used to discover correlations between variables in a sizable database.

Semi-supervised Learning. Semi-supervised learning represents the middle ground between Supervised and Unsupervised learning algorithms. According to Hady et al. [27] unsupervised learning allows the built learning model to integrate part or all of the available unlabeled data in its supervised learning model. He further states that the goal is to reduce the human effort of annotating the input data and maximize the learning performance of the model. It uses the combination of labeled and unlabeled datasets during the training period. Therefore, training data is a combination of both labeled and unlabeled data.

Reinforcement Machine Learning. Unlike the other machine learning techniques, reinforcement learning is a feedback-based approach in which agents learn by interacting with the environment. This machine learning technique involves various steps, such taking action, changing state, and getting feedback from the environment. It therefore includes repeatedly interacting with the environment while employing the most recent learnt policy, gathering knowledge that is then used to refine the policy [28].

2.4 Related Works

Gian Antonio et al. [29] proposed a Machine Learning-based decision Support System to automatically associate a risk factor to each transaction performed through mobile banking system. The proposed approach has a hierarchical architecture: The first step involved the application of unsupervised Machine Learning module to detect abnormal patterns or wrongly labeled transactions. The second step applies a supervised module risk factor for the transactions that were not marked as anomalies in the previous step.

In a conference paper on Credit Card Fraud detection by [30] using machine learning methods, the authors compared Logistic Regression, Random Forest, Naive Bayes and Multilayer Perceptron. The results showed that each of these algorithms can be used for credit card fraud detection with high accuracy, but that Random Forest had the best performance. The authors went on to highlight the importance of using sampling methods to address class imbalance issues prevalent in fraud datasets. In a journal entitled "Use of Hidden Markov Model as Internet Banking Fraud Detection" Sunil. S. Mhamane et al. [31] proposed a methodology for fraud detection in internet banking using Hidden Markov Model (HMM). They modelled the sequence of operations in online internet banking using HMM. The solution is split between the training and detection phases. In the training phase, the train model is created based on the historical set of transactions to determine the behavioural profile of the account holder. The detection and prevention phase looks for the deviation from the historical behavioural profile to detect fraud. Comparative studies indicated the proposed methodology had an accuracy of over 72% and scalable to handle large transaction volumes.

In a similar research work, [32] proposed bank anti-fraud model based on K-means and Hidden Markov Model. They applied the K-means algorithm to symbolize the transaction amount and frequency sequency of the of a bank account. An HMM is initially trained with the normal behaviour of an account. With enough historical transactions, the results show that the model performs well for low, medium frequency and amount of user groups.

3 Methodology

In their publication on research design, [33] describes a research methodology as the overall method selected to include the many study components in a logical sequence, ensuring that the research topic is adequately handled. Several literature reviews were conducted to have an in-depth understanding of the subject area and unearth gaps in the existing literature on mobile banking fraud in the Zambian context.

This paper proposes an approach that uses machine learning algorithms to detect anomalous transactions. Figure 2 depicts the architecture of the proposed mobile banking fraud and anomaly detection transaction flow system design.

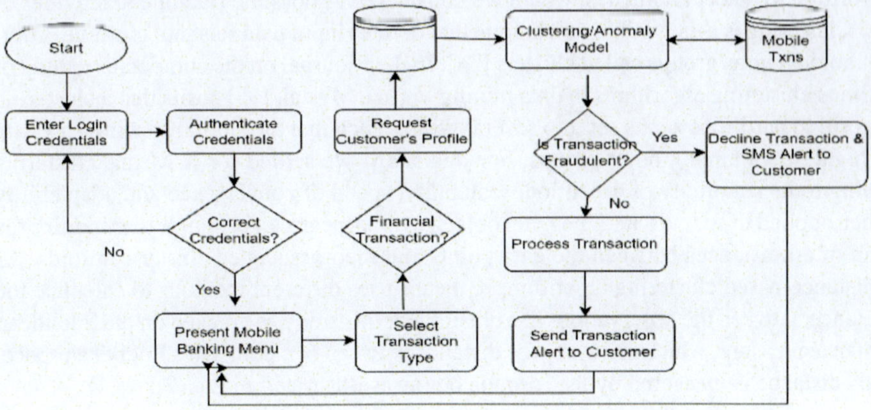

Fig. 2. Proposed System Architecture/Flow Design

As stated in Sect. 2.3, data mining can is broken down into 2 phases.

3.1 Data Collection and Preparation

The first step in this phase involved collection of data for the study from a named Commercial Bank's (name withheld for confidentiality purposes) Oracle relational database using Oracle PL/SQL. This dataset was derived from a combination the historical mobile transactions tables. The collected raw data covered the period 1st January 2022 to 31st 2022.

The second step involved exploratory Data analysis (EDA), using machine learning visual techniques to gain an in-depth characteristic of data and discover trends and patterns in the data. This was achieved using python's inbuilt machine learning libraries (pandas, numpy, matplotlib and seaborn). Using these libraries and associated functions, we were able to obtain the descriptive information about the dataset as well as getting information about the data types, null values and identify duplicate values. Like in many data mining problems, the collected data required cleaning and pre-processing to remove anomalies such as null/missing values, and duplicates.

In the data transformation stage, the categorical features from the raw mobile banking transaction data were removed, and thus we worked with the 8 numerical features. The transformation process also involved renaming some columns so as to hide sensitive personal identifiable information (PII) such as account number and transaction branch.

3.2 Data Mining

This phase involved the application of unsupervised Machine learning algorithms, namely Clustering and Anomaly detection algorithms, to the preprocessed data to gain hidden insights into the data. The data mining methodology applied involved applying K-means clustering algorithm to the preprocessed data, grouping the data into clusters. The clustered data was then fed into unsupervised Isolation Forest (IF) anomaly detection algorithm to detect anomalies, which we considered as possible fraudulent activities.

Clustering is a data mining technique that divides input data sets into multiple different homogenous groups called clusters [34]. In their journal on the comparative study of various clustering algorithms in data mining Verma. M et al. [35] posits that a clustering algorithm partitions a data set into several groups such that the similarity within a group is higher than among other groups. For this study, we settled for K-Means clustering owing to its simplicity of execution, scalability, speed of convergence and adaptability when applied to sparse data. The main objective of k-means algorithm is to minimize the sum of all distances between the data points and their associated cluster centroids. As a distance-based clustering algorithm, K-means uses different methods to calculate the distance between the observations in a cluster. For this study, we picked on the Euclidean distance measure, which measures the distance between two points in a Euclidean space. This distance is measured by the formula below as illustrated in Eq. (2):

$$d(\mathrm{x}, \mathrm{y}) = \sqrt{\sum\nolimits_{k=0}^{n}}(x_i - y_i)^2 \tag{2}$$

where, x and y are two vectors of length n.

The steps involved in K-means clustering are:

1. Define the number of clusters (k) that will be generated in the final solution.
2. Randomly selecting k objects from the data set to serve as the initial centers for the clusters (centroid)
3. Assign each observation to its closest centroid, based on the Euclidean distance between the object and the centroid (Eq. 1).
4. For each of the k clusters update the cluster centroid by calculating the new mean values of all the data points in the cluster.
5. Iteratively repeat the cluster assignment and centroid update until the cluster assignments stop moving (convenience is achieved).

The steps outlined above are depicted in Fig. 3.

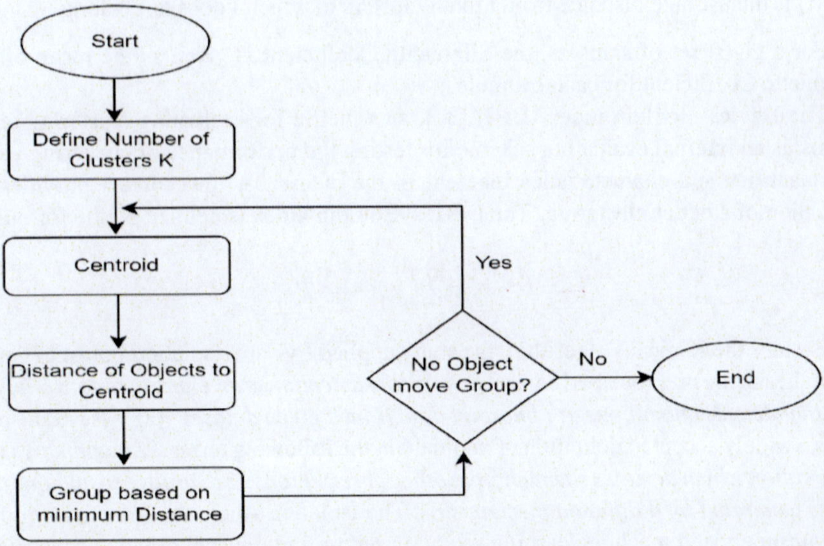

Fig. 3. K-Means Clustering Process

One of the drawbacks of K-means algorithm is the need to assign the number of clusters beforehand. To overcome this challenge in selecting the best K value for the optimal number of clusters, different approaches are applied. For this study, we picked on the elbow method, which is used to determine the optimal number of clusters by looking at the percentage of the comparison between the number of clusters that will form an elbow at a point [36]. The idea behind this approach is to start with K = 2 and on increasing it in each step by 1 to calculate the clusters and the costs associated with it. When the cost drops dramatically at some value of K, and it reaches a plateau, to give you the optimal K value. Finding the optimal number of clusters can be achieved using several techniques. We employed the elbow method to find the "elbow" point, where additional data does not alter cluster membership significantly.

To interpret and check consistency across data clusters, we applied the Silhouette technique, which provides a graphical representation of how well each point in the dataset has been classified. The resultant silhouette provides a measure of the cohesion between a datapoint and a cluster when compared to other clusters, ranging from −1 to +1. The silhouette coefficient is calculated using the average intra-cluster distance (a) and the average nearest-cluster distance (b) for each sample. The silhouette ratio is therefore represented mathematically as:

$$S(i) = \frac{b(i) - a(i)}{max\{a(i), b(i)\}} \tag{3}$$

where:

- S(i) is the silhouette coefficient of the data point i.
- a(i) is the average distance between i and all the other data points in the cluster to which i belongs.

- b(i) is the average distance from i to all clusters to which i does not belong.

For a given set of samples, the Silhouette Coefficient is given as the mean of the Silhouette Coefficient for each example.

The Davies–Bouldin index (DBI) [37], as a metric for evaluating clustering algorithms, is an internal evaluating scheme for testing the performance of clustering using the quantities and characteristics inherent in the dataset. A lower index provides an indication of a better clustering. The Davies–Bouldin index is defined by the formulae:

$$DB = \frac{1}{k}\sum\nolimits_{i=1}^{k} maxRij \tag{4}$$

Besides the K-means algorithm, the study applied PyCaret Isolation Forest to detect anomalies in the preprocessed dataset. *"Anomaly detection is an important data analysis task that detects anomalous or abnormal data from a given dataset"* [38]. Hawkins provides a widely accepted definition of anomaly in the following terms: "*An anomaly is an observation which deviates so much from other observations as to arouse suspicions that it was generated by a different mechanism*". The Isolation Anomaly Detection Module is an unsupervised machine learning module used to find unusual events, occurrences, or observations that raise questions by deviating sharply from the bulk of the data. Of the several unsupervised anomaly detection algorithms, this study applied Isolation Forest (IF) to detect the anomalous transactions in the mobile banking dataset. The idea behind Isolation Forest is that, on average, anomalies will be closer to the root node than normal instances. Conceptually, a lower average depth (closer to the root node) implies a higher likelihood of a data point being an outlier. Anomaly detection algorithms detect fraud by providing a ranking to the datasets that reflect the degree of anomaly. The data points are sorted according to their path length or anomaly score. Path Length h(x) of a point x is measured by the number of edges x traverses an iTree from the root node until the traversal is terminated at an external node Given a dataset with n instances, the average path of unsuccessful search is illustrated by Eq. 5 below:

$$c(n) = 2H(n-1) - (2(n-1)/n) \tag{5}$$

where H(i) is the harmonic number estimated by ln(i) + 0.577215649

The anomaly score of an instance x is defined by Eq. (6) given by:

$$\mathrm{S(x, n)} = 2^{-\frac{E(h(x))}{c(n)}} \tag{6}$$

where n is the number of instances, h(x) is the depth at which the data point is found in a particular tree. E(h(x)) is its average over different trees, and H(i) is the harmonic number. s(x, n) is a number between 0 and 1, where the higher the score the more likely it is an outlier.

We used principal component analysis (PCA), an unsupervised technique used in machine learning to reduce the dimensionality of data by compressing the feature space by identifying a subspace that captures most of the information in the complete feature matrix. We applied Data mining algorithms to determine the optimal number of clusters and detect the anomalies in the data.

The preprocessed data was divided into what we termed as seen (95%) and unseen (5%) data. The seen data was used to build and train the model, while the unseen data was used for prediction.

4 Analysis and Results

The results obtained from applying clustering and anomaly detection algorithms are presented in the sub sections that follow.

4.1 Data Extraction

As indicated in Sect. 3.1, the dataset for this study was extracted from an Oracle Database using PL/SQL. The extracted raw data consisted of 310,362 observations and 13 features.

4.2 Data Analysis

We applied various EDA techniques to explore the raw data and gain meaningful insights from the dataset. Applying the various python functions, we removed the categorical features and reduced the feature list to 8, taking into account the important features needed for clustering. We explored the distribution of each variable and we discovered that most of the variables were either bimodal or multimodal, with the exception of PR_CODE. TR_AMT, AC_CLASS and AC_CATEGORY (Fig. 4).

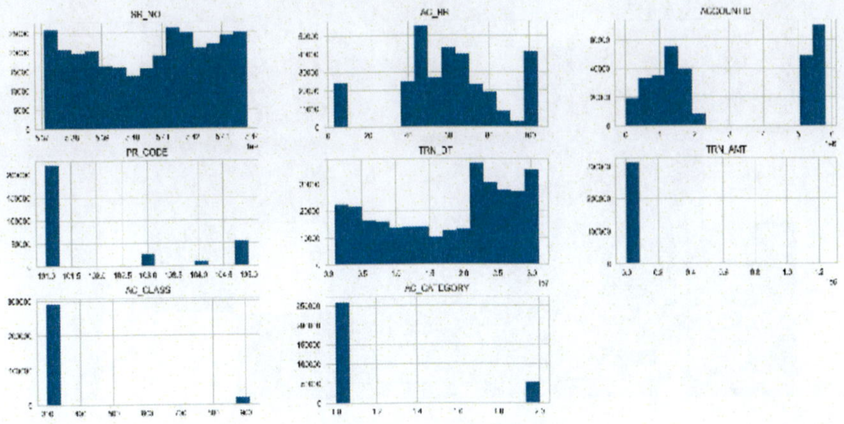

Fig. 4. Distribution of Variables

The collected and preprocessed data was distributed into the following categories of mobile banking transaction types. We noted that the bulky of the transactions involved local funds transfer between accounts within the same financial institution. This represents 70.29% of the total data set collected for analysis. Water utility bill payment represents a small share of the total transaction processed, contributing only 0.52% of total transactions (Table 1).

Table 1. Data Distribution Per Product

Product code	Description	Count
101	Local Transfers	218,164
102	Water Utility Bill payment	1,616
103	Pay TV Bill payment	26,466
104	RTGS Transfers	10,444
105	Electricity Purchase	53, 672

We used the corr() pandas function in combination with heatmap() Seaborn function to create a heatmap that visualized the correlation of all variable values. Figure 5 shows the correlation matrix obtained during the EDA. The figure depicts a high correlation (0.99) between the serial number (SR_NO) and the transaction date (TRN_DT).

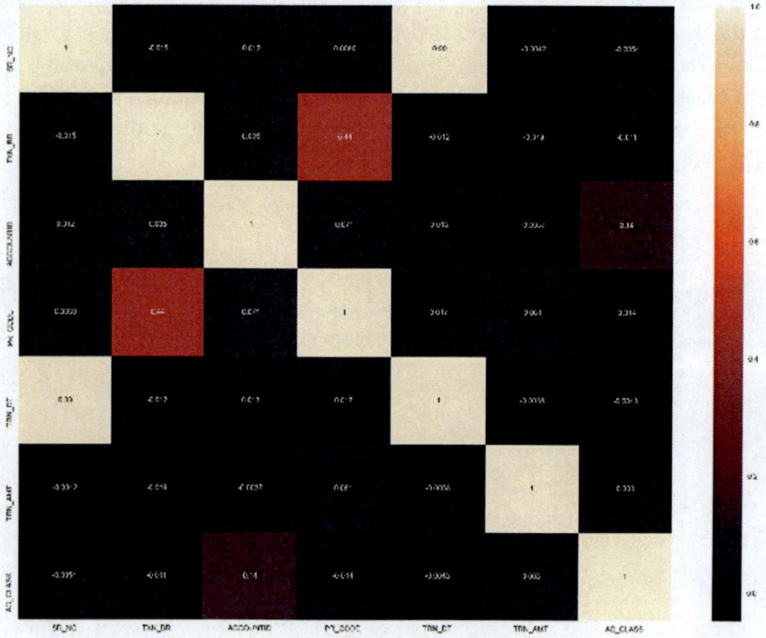

Fig. 5. Correlation Matrix

The elbow method was used to determine the optimal number of clusters. For this study, the optimal number of clusters was obtained with a K-value of 5 as depicted in Fig. 6, and a corresponding Silhouette coefficient of 0.2953.

The silhouette plot as shown in Fig. 7 displays a measure of how close each point in one cluster is to points in the nearby clusters. The figure shows that all the data points in all the 5 clusters are above the average silhouette score of 0.2953. This result makes the

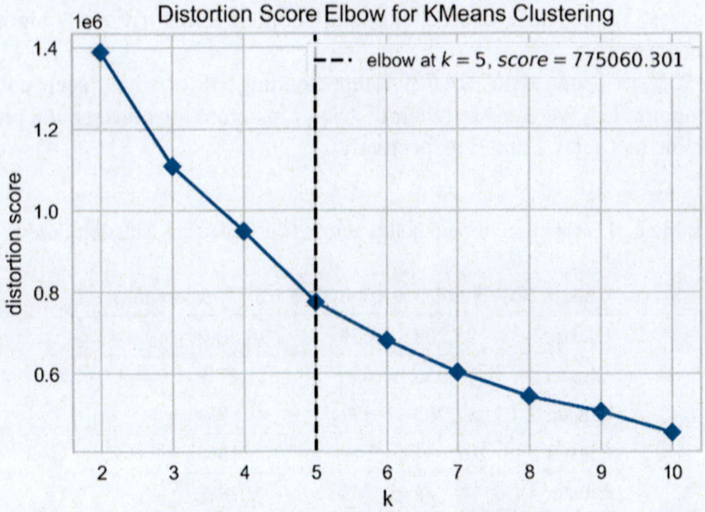

Fig. 6. K-value using Elbow method.

choice of k = 5 an acceptable number of clusters. However, we noted a wide fluctuation in the thickness of the cluster plots, with cluster 2 been smaller in thickness compared to the clusters.

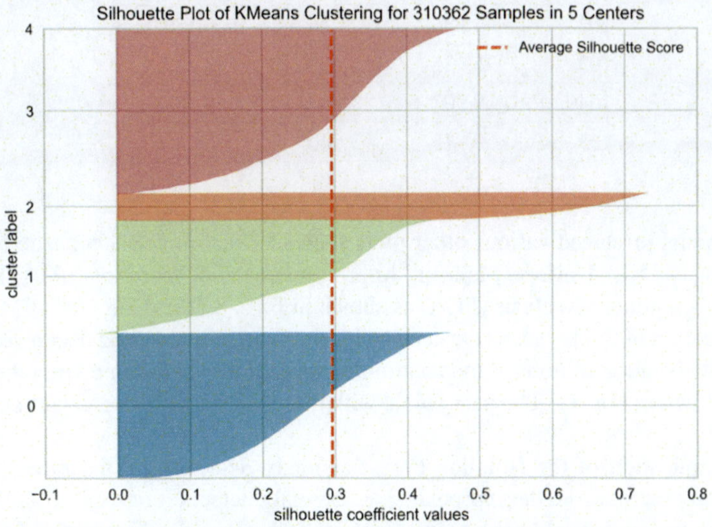

Fig. 7. Silhoutte Analysis of Mobile Transactions.

After assigning the data into 5 respective clusters, we separated the clustered data into 95% "seen Data" (294,844) and 5% "unseen data" (15,518). The 95% of the modelling data, termed as "seen data" was fed into the Isolation Forest (iForest) algorithm for anomaly detection.

Of the 95% of "seen data" 14,739, (representing 5% of total "seen data") were labelled as anomalies. We analyzed their distribution across the clusters and products as depicted below in Tables 2 and 3 respectively.

Table 2. Distribution of Anomalies across Clusters using Isolation Forest

Cluster No	Number	Anomalies	% Of Anomalies
Cluster 0	95,208	5,675	38.50%
Cluster 1	73,956	1,690	11.47%
Cluster 2	17,703	6,572	44.59%
Cluster 3	107,934	759	5.15%
Cluster 4	43	43	0.29%

Table 3. Distribution of Anomalies across Products.

Product Code	Data Across Product	Anomalies	% Of Anomalies
101	207,296	6,368	43.21%
102	1,536	26	0.18%
103	25,135	3,420	23.20%
104	9,902	985	6.68%
105	50,975	3,940	6.73%

The model produced various other plots such as T-distributed Stochastic Neighbor Embedding (t-SNE), Uniform Manifold Approximation and Projection (uMAP), and the Principal Component Analysis (PCA) as shown in Figs. 8, 9 and 10.

Application of the K-means clustering algorithm on preprocessed data revealed that Cluster 4 had highest minimum and maximum transactions values for all mobile transactions as shown in Fig. 11. The data points with these values were deemed as candidates for fraud.

The application of the Isolation Forest anomaly detection on the modelling data (95% of the preprocessed data) produced an anomaly detection rate of 5.0%. When we applied the trained model to the holdout data (5% "unseen data") the model produced an anomaly detection rate of 4.9%, which is consistent with the 5% detection rate in the modelling stage. These results validate the accurate performance of the model to detect mobile banking transactions.

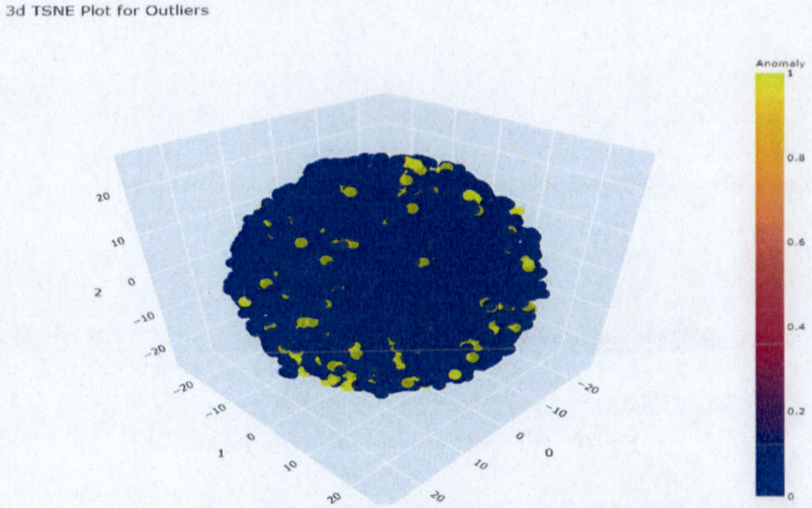

Fig. 8. T-distributed Stochastic Neighbor Embedding (t-SNE)

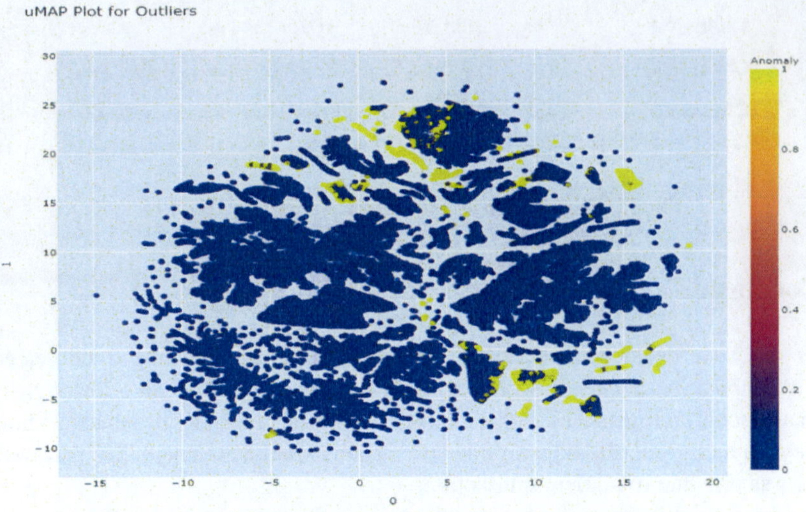

Fig. 9. Uniform Manifold Approximation and Projection (uMAP)

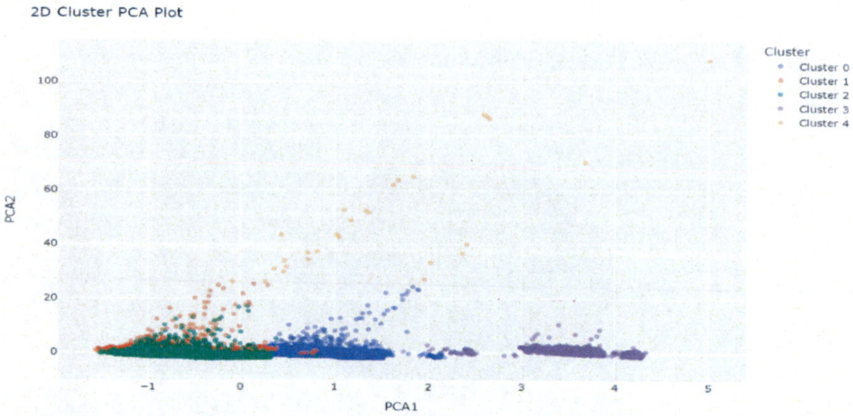

Fig. 10. Principal Component Analysis (PCA)

Cluster	min	max	mean	median
Cluster 0	7.0	200000.0	3840.401059	400.0
Cluster 1	7.0	222051.0	4886.334266	400.0
Cluster 2	7.0	101500.0	2793.941326	1000.0
Cluster 3	8.5	200000.0	10008.109812	520.0
Cluster 4	230000.0	1250000.0	461409.360000	378589.0

Fig. 11. Distribution of Transaction Amounts in Clusters

5 Conclusion

In this study, we developed a data mining model for mobile banking fraud detection using a combination of clustering and anomaly detection algorithm algorithms, namely K-means, Isolation forest. The study showed that we can apply K-means algorithms to mobile data and group the transactions into clusters and detect anomalies that we can consider as possible fraudulent transactions.

For future work, we recommend that the various clustering and anomaly detection be compared to establish the algorithm with the best results when run on the same dataset. We further recommend the implementation of mobile transaction fraud detection system that uses NoSQL databases to store historical customer mobile transactions and profiles. Use of NoSQL model increases the performance and speed of data storage and extraction from historical data.

With the increasing integration between Financial Service Providers (FSPs) and Mobile Network Operators (MNOs) through the National Financial Switch (NFS), driven by the Central Bank (Bank of Zambia), we recommend that this study be extended to fraud perpetuated through NFS.

References

1. Luarn, P., Lin, H.-H.: Toward an understanding of the behavioural intention to use mobile banking. Comput. Hum. Behav. **21**, 873–891 (2005). https://doi.org/10.1016/j.chb.2004.03.003
2. Sabharwal, M., Swarup, A.: Banking by the use of handheld devices & gadgets like Smartphones, Tablets (Using Banking Applications & Widgets that are Based on Mobile Operating Systems like Android etc.)
3. Mohammad, A.B.: E-banking of economical prospects in Bangladesh. J. Internet Bank. Commer. **15**, 1–10 (2010)
4. https://www.boz.zm/2020NPSAnnualReport.pdf
5. Sadgali, I, Sael, N., Benabbou, F.: Performance of machine learning techniques in the detection of financial frauds. In: Second International Conference on Intelligent Computing in Data Sciences (ICDS 2018)
6. http://www.acfe.com/uploadedfiles/acfewebsite/content/documents/rttn-2010.pdf
7. Bhatla, T.P., Prabhu, V., Dua, A.: Understanding credit card frauds. Cards Bus. Rev. **1**(6), 1–15 (2003)
8. https://ucr.fbi.gov/crime-in-the-u.s/2010/crime-in-the-u.s.-2010/offense-definitions
9. Chimobi, E.C., Jude, E.I., Livinus, E.I.: Business frauds in Nigeria: Underlying causes, effects and possible remedies: case study of banking sector
10. Hamet, P., Tremblay, J.: Artificial intelligence in medicine. Metabolism **69**, S36–S40 (2017)
11. Poole, D.L., Mackworth, A.K.: Artificial Intelligence: Foundations of Computational Agents. Cambridge University Press (2010)
12. Huang, M.-H., Rust, R.T.: Artificial intelligence in service. J. Serv. Res. **21**(2), 155–172 (2018)
13. Cioffi, R., Travaglioni, M., Piscitelli, G., Petrillo, A., De Felice, F.: Artificial intelligence and machine learning applications in smart production: Progress, trends, and directions. Sustainability **12**(2), 492 (2020)
14. Jakhar, D., Kaur, I.: Artificial intelligence, machine learning and deep learning: Definitions and differences. Clin. Exp. Dermatol. **45**(1), 131–132 (2020)
15. Aggarwal, K., et al.: Has the future started? The current growth of artificial intelligence, machine learning, and deep learning. Iraqi J. Comput. Sci. Math. **3**(1), 115–123 (2022)
16. Kühl, N., Goutier, M., Hirt, R., Satzger, G.: Machine learning in artificial intelligence: Towards a common understanding. arXiv preprint arXiv:2004.04686 (2020)
17. Samuel, A.L.: Machine learning. Technol. Rev. **62**(1), 42–45 (1959)
18. Mitchell, T.M.: Machine Learning. McGraw-Hill, New York (2007)
19. Mitchell, T.M.: The discipline of machine learning. Carnegie Mellon University, School of Computer Science, Machine Learning (2006)
20. Perols, J.: Financial statement fraud detection: An analysis of statistical and machine learning algorithms. Audit. J. Pract. Theory **30**(2), 19–50 (2011)
21. Chitra, K., Subashini, B.: Data mining techniques and its applications in banking sector. Int. J. Emerg. Technol. Adv. Eng. **3**(8), 219–226 (2013)
22. Pulakkazhy, S., Balan, R.S.: Data mining in banking and its applications-a review (2013)
23. Fayyad, U., et al.: From knowledge discovery to data mining: An overview. In: Fayyad, U., et al. (eds.) Advances in Knowledge Discovery and Data Mining. AAAI/MIT Press (1995)
24. Green, B.P., Choi, J.H.: Assessing the risk of management fraud through neural network technology. Auditing **16**, 14–28 (1997)
25. Gao, J., Zhou, Z., Ai, J., Xia, B., Coggeshall, S.: Predicting credit card transaction fraud using machine learning algorithms. J. Intell. Learn. Syst. Appl. **11**, 33–63 (2019). https://doi.org/10.4236/jilsa.2019.113003

26. Siddique, N., Adeli, H.: Computational Intelligence: Synergies of Fuzzy Logic, Neural Networks and Evolutionary Computing. John Wiley & Sons (2013)
27. Hady, M.F.A., Schwenker, F.: Semi-supervised learning. In: Handbook on Neural Information Processing, pp. 215–239 (2013)
28. Levine, S., Kumar, A., Tucker, G., Fu, J.: Offline reinforcement learning: Tutorial, review, and perspectives on open problems. arXiv preprint arXiv:2005.01643 (2020)
29. Susto, G.A., Terzi, M., Masiero, C., Pampuri, S., Schirru, A.: A fraud detection decision support system via human on-line behavior characterization and machine learning. In: 2018 First International Conference on Artificial Intelligence for Industries (AI4I), pp. 9–14 (2018). https://doi.org/10.1109/AI4I.2018.8665694
30. Varmedja, D., Karanovic, M., Sladojevic, S., Arsenovic, M., Anderla, A. (2019). Credit card fraud detection - machine learning methods. In: 2019 18th International Symposium INFOTEH-JAHORINA (INFOTEH), pp. 1–5 (2019). https://doi.org/10.1109/INFOTEH.2019.8717766
31. Mhamane, S.S., Lobo, L.M.R.J.: Use of Hidden Markov model as internet banking fraud detection. Int. J. Comput. Appl. **45**(21) (2012)
32. Wang, X., Wu, H., Yi, Z.: Research on bank anti-fraud model based on K-means and hidden Markov model. In: 2018 IEEE 3rd International Conference on Image, Vision and Computing (ICIVC), pp. 780–784 (2018). https://doi.org/10.1109/ICIVC.2018.8492795
33. Creswell, J.W.: Research Design: Qualitative, Quantitative, and Mixed Methods Approaches. Sage Publications, Thousand Oaks (2013)
34. Dubey, A., Choubey, A.: A systematic review on k-means clustering techniques. Int. J. Sci. Res. Eng. Technol. **6**(6), 624–627 (2017)
35. Verma, M., Srivastava, M., Chack, N., Diswar, A.K., Gupta, N.: A comparative study of various clustering algorithms in data mining. Int. J. Eng. Res. Appl. (IJERA) **2**(3), 1379–1384 (2012)
36. Syakur, M.A., Khotimah, B.K., Rochman, E.M.S., Satoto, B.D.: Integration k-means clustering method and elbow method for identification of the best customer profile cluster. In: IOP Conference Series: Materials Science and Engineering, vol. 336, no. 1, p. 012017. IOP Publishing (2018)
37. Milligan, G.W., Cooper, M.C.: An examination of procedures for determining the number of clusters in a data set. Psychometrika **50**(2), 159–179 (1985)
38. Ahmed, M., Mahmood, A.N., Hu, J.: A survey of network anomaly detection techniques. J. Netw. Comput. Appl. **60**, 19–31 (2016)

A Comparative Study on the Correlation Between Similarity and Length of News from Telecommunications and Media Companies

Yougyung Park(✉) and Inwhee Joe

Department of Computer Science, Hanyang University, Seoul 04763, South Korea
eunicepark83@gmail.com
http://wm.hanyang.ac.kr/

Abstract. This study is based on the increasing damage caused by fake news, abusing articles, plagiarism, and similar news included in online news, and activities and systems for monitoring news copyright violations are emerging to prevent this. Copyright can be detected based on the similarity of news content. However, it is very difficult to check the similarity of news that hundreds of thousands of cases are distributed every day and contains similar topics and contents. As we pay attention to the tendency of news length to be shortened due to mobile, we would like to investigate the correlation between news similarities. To this end, news from five sections registered in the first half of 2022 was collected on the Naver portal, measured the length of the news, and analyzed the correlation by measuring the news similarity. News similarity was analyzed for news showing similarity of 0.7 or higher after obtaining similarity using cosine similarity and TF-IDF algorithm. In this study, two approaches were attempted to analyze the correlation between news length and news similarity. The first is to analyze the correlation between news length and similarity by classifying it by news section. The second is a method of analyzing the correlation by first classifying news into news producers, telecommunication companies, and media companies. The first approach was less correlated, but applying the second method showed a negative correlation result that the shorter the news length, the higher the news similarity. It was found that telecommunication companies had a higher correlation between news length and similarity than media companies.

Keywords: News length · News similarity · cosine similarity · TF-IDF · correlation analysis

1 Introduction

More than 300,000 news stories are produced every day. The produced news is mainly distributed through the portal news platform. Naver News is the mainstream in Korea. Recently, new content using news is also being distributed on SNS such as blogs, YouTube, Instagram, and Facebook. Above all, activities for economic benefit are increasing.

R. Silhavy and P. Silhavy (Eds.): CSOC 2023, LNNS 724, pp. 555–569, 2023.
https://doi.org/10.1007/978-3-031-35314-7_49

1.1 Motivation

There's also a lot of fake news, plagiarism news like an audit article, and similar news. Similar news refers to news with the same title and content. According to the results of the newspaper industry survey released by the Korea Press Foundation on December 29, 2022, there are 5,397 domestic newspapers. ? News producers are divided into news agencies and media companies, and there are about 5,397 news producers. Yonhap News Agency, a representative news agency that writes news and supplies it to newspapers and broadcasting stations, has a structure that can be used by other news media companies that have signed contracts. There are 28 news agencies, including GNN News Communications, NSP Communications, Newsis, News1, and Newsfim. The rest are classified as media companies.

1.2 Objective

However, news that can be easily read online produces new content using news on SNS as well as media companies. On blogs, YouTube, Instagram, and Facebook, activities that make economic profits from news-based paid content are increasing. Therefore, recently, online platforms such as Google and Facebook have been preparing for each country, such as paying for news usage fees. Korea also regulates and monitors news. (Re)The Korea Press Foundation is monitoring news copyright violations that quote and copy news on SNS.

Against this background, research on the correlation between news similarity and news length is expected to be beneficial in discussing news quality such as plagiarism and copyright of news in the future.

2 Related Work

2.1 Study on the Length of News

Research on news length is being conducted in several studies. According to the "Study on the Sentence Length of Sports Articles," [3] the sentence length of newspaper articles is in the order of politics, economy, editorial, and sports, and there is a study that "short sentences are read well" in terms of readability. On the other hand, there is a study that calculated the news evaluation index while suggesting that the length of the news is highly related to the items of stakeholders and transparent reporters. A study found that 633 characters were appropriate for ease evaluation, 1,368 characters for information volume evaluation, and 346 characters for readability evaluation, based on two manuscript sheets (400 characters).

2.2 Study on News Similarity

Research on news similarity is a methodology 5 using supervised learning in 'similarity detection and contextual data study of time-sensitive online news articles based on RSS feeds', and there is a study that shows a correlation of 0.89.

2.3 Cosine Similarity and TF-IDF Weight Model

Cosine similarity, which is most commonly used in document similarity studies, is a measure of how similar the two vectors are by calculating the angle between the two vectors, and interprets that the closer they are to 1, the more similar they are [1,2] The equation for cosine similarity is as follows:

$$\cos\theta = \frac{A \cdot B}{\| A \| \| B \|} = \frac{\sum_{i=1}^{n} A_i \times B_i}{\sqrt{\sum_{i=1}^{n} (A_i)^2} \times \sqrt{\sum_{i=1}^{n} (B_i)^2}} \tag{1}$$

TF-IDF (Term Frequency-Inverse Document Frequency) is a statistical word expression method that weights each word in a Document Term Matrix (DTM) in consideration of its importance. The method of calculating the weight is as follows. After DTM is generated, TF-IDF weights are given for each word by multiplying the frequency of word appearance (TF) and inverse document frequency (IDF), just like the TF-IDF name (Table 1). The formula is as follows:

$$TF(d,t) \times IDF(d,t) = TF - IDF(d,t) \tag{2}$$

Table 1. TF-IDF Symbol Explain

Symbol	Description
$TF(d,t)$	Number of appearances of the word t in d in the document
$DF(t)$	The number of documents in which the word t appeared
$IDF(d,t)$	The IDF value decreases as the number $DF(t)$ increases inversely proportional to $DF(t)$

3 Proposed Method

3.1 Data Crawling

Data crawling was done using the Python-based BeautifulSoup library at Google colab. The news data was targeted at Naver News. The news has 11 sections. Among them, only five sections were collected: politics, economy, society, life/culture, and IT/science. The period is the first half of 2022 (January 1 to June 30) (Fig. 1). Each news section has a subsection. One was selected from each subsection. In the political section, North Korea, the economy is a medium-term venture, society is an environment, life/culture is a health information, and IT/science is a mobile section (Fig. 2).

There are 84,672 news collected.

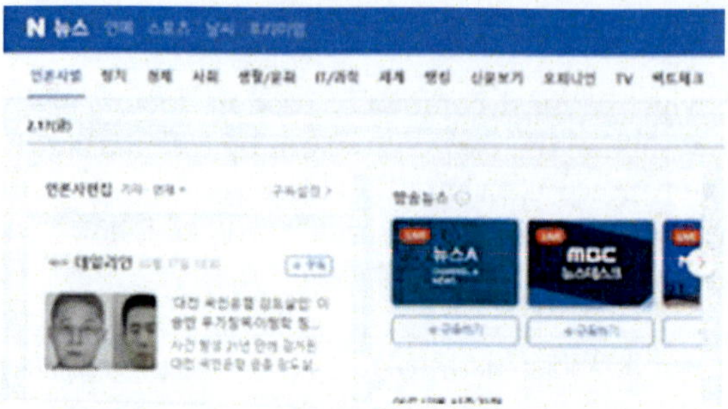

Fig. 1. Naver News Section

Fig. 2. Naver News Political Subsection

3.2 Data Classification

There are two categories of news. There are categories by section and media classification. Sectional classification is indicated as the basis for news classification. They are the political, economic, social, life/culture, IT/science sections. Looking at the amount of data collected, there were 19,122 political sections, 26,760 economic cases, 15,636 social cases, 12,976 living cultures, and 10,178 IT/science cases (Table 2).

Table 2. Total News Collected

Volume of news	
Politics (North Korea)	23,207
Economy (medium-term venture)	26,760
Society (environment))	15,636
Life/Culture (Health Information)	21,118
IT/Science (Mobile)	10,178
Sum	96,719

The second classification is classified into news producers, telecommunications companies and media companies. The telecommunications companies were designated as three (Yonhap News, Newsis, and News1), and the rest were classified as media companies. The amount of news from three telecommunications companies and the total amount of news from several media companies are as follows (Table 3).

Table 3. Total News by Media

Volume of news		News Agency	Media
Politics (North Korea)	23,207	15,822	7,205
Economy (medium-term venture)	26,760	12,511	14,249
Society (environment)	15,636	12,799	2,837
Life/Culture (Health Information)	21,118	1,983	19,135
IT/Science (Mobile)	10,178	4,226	5,952
Sum	96,719	47,341	49,378

3.3 Measure the Length of the News

News is divided into titles and content. The length of the news was measured by dividing it into title and content. The number of characters in the news ranges from at least two to a maximum of 75 titles, with an average of 19.9. The content of the news averaged 327.4 from a minimum of 4 to a maximum of 5,782.

As a result of examining the length of the news by news section, the section with the lowest average of the titles is society (environment), and the section with the highest average is life/culture. In terms of content, the social (environment) section was the lowest, and the life/culture section was the highest (Tables 4 and 5).

Table 4. Value of News Title and Content

Title			Content		
Min	Max	Average	Min	Max	Average
2	75	19.9	4	5782	327.4

Table 5. News Length by Section

volume of news		Title Average/Per article	Content average/per article
Politics(North Korea)	23,027	21.1	310.6
Economy (medium-term venture)	26,760	22.3	541.9
Society (environment)	15,636	16.5	243.3
Life/Culture (Health Information)	21,118	24.6	555.5
IT/Science (Mobile)	10,178	21.1	331.7
Sum	96,719	19.9	327.4

The results of classification as media showed that the average length of the title of one news was 25.2 for media companies and 19.1 for telecommunication companies, 731.9 for media companies and 267.8 for telecommunication companies (Table 6).

3.4 Measurement of News Similarity

As a result of measuring cosine similarity and TF-IDF similarity, news with a news similarity of 0.7 or higher was extracted and the subject of analysis. Of the 96,719 news collected by crawling, 18,269 (18.78%) were news with a similarity of 0.7 or higher. The most similar news section was 33.49% in society and 32.71% in politics, higher than the average (17.18%) (Table 7).

Table 6. News Length by Media

volume of news		Title Average/Per article	Content average/per article
Media Companies	23,027	25.2	731.9
Telecommunications	26,760	19.1	267.8
Sum	96,719	19.9	327.4

Table 7. News Similarity 0.7 or higher in Section

volume of news		similarity 0.7 or more	similarity the rate of news
Politics	23,027	7,531	0.327051
Economy	26,760	2,098	0.078401
Society	15,636	5,236	0.334868
life/culture	21,118	487	0.23061
IT/Science	10,178	1,265	0.124288
Sum	96,719	16,617	0.171807

Similarities are very different in the media. Media companies accounted for 4.32 percent with 2,133 and telecommunications companies with 14.848, representing 30.60% (Table 8).

Table 8. News similarity 0.7 or higher by Media

volume of news		similarity 0.7 or more	similarity the rate of news
Media Companies	23,027	2,133	0.043197
Telecommunications	26,730	14,484	0.305950
Sum	96,719	16,617	0.171807

3.5 Comparing Data Standardization to Three Methods

News consists of titles and content. The target of data preprocessing is the content of the news.

The length of the news content ranges from 4 to 5,782 characters. Since the similarity is in the range of 0.7 to 1, it is difficult to compare (apply). Therefore, it is necessary to standardize news data. In this study, we tried three methods for data standardization.

Ratio Scale. The ratio scale is a measure of equal spacing between lengths and the presence of absolute values. The length of the news content was classified into 200 characters, and the minimum and maximum sections were designated as ratios. The ratio scale is classified as 1–6, and the number of news included in this scale is 15,893 (95.65%) (Table 9).

Table 9. Ratio scale per length of news

length of the news		Ratio scale	Volume of the news
Minimum	Maximum		
4,500	6,000		8
3,000	4,500		15
1,500	3,000		299
1,200	1,500		405
1,000	1,200	6	580
800	1,000	5	657
600	800	4	708
400	600	3	656
200	400	2	4,473
0	200	1	8,819

Data were searched using descriptive statistics with the news length and news similarity of the data in the 1–6 section of the ratio scale. In the derived statistics, if the median and average values are similar, the probability of a variable being close to a normal distribution is high, and it is considered to form a normal distribution when $|dwarfism| < 3$, $|kurtosis| < 8$ (Table 10).

Table 10. Technical statistics

		N=15,893
	length of the news	News Similarity
Average	1.886742591	0.892638607
a standard error	0.010764048	0.001303889
median value	1	1
the most frequent value	1	1
standard deviation	1.356996039	0.164377982
Dispersion	1.84143825	0.027020121
kurtosis	1.833513609	−0.05623851
skewness	1.685329428	−1.212803079
Range	5	0.51
Minimum value	1	0.49
Maximum value	6	1
The sum	29986	14186.70539
an observation number	15893	15893
Confidence level (95.0%)	0.021098754	0.002555771

Correlation analysis was conducted to examine the correlation between the two variables based on the above results. Correlation analysis is an analysis method for determining whether a linear relationship exists between the two variables, and can determine how close it is linearly. The correlation coefficient derived through correlation analysis is a numerical value of the degree of correlation, and does not explain the causal relationship between the two variables. The closer the correlation coefficient is to 1(−1), the stronger positive (negative) correlation is shown, and 0 is considered to have no correlation between data. The correlation coefficient between the news length and news similarity of the ratio scale 1–6 section data is -. It can be said that it is 029438, and it shows a significant negative correlation as it appears as a significant probability (0.000<.05). However, it is difficult to secure the reliability of the data because it does not arbitrarily include data (727) with a length of 6,000 or more in the analysis between the length of news content and news similarity of the data in sections 1–6. In addition, smooth comparison was difficult because the length and similarity of the news content were very different. As a way to overcome this point, the data outliers were removed after classifying the entire data (n=16,617) into media companies and telecommunication companies, and data standardization was performed through natural log conversion for the length variable of the news content (Fig. 3, Tables 11 and 12).

Table 11. Correlation between news length and news similarity of media companies

	length of the news	News Similarity
length of the news	1	
News Similarity	−0.029483177***	1

The reason for taking natural logs for data analysis is that nonlinear data can be linearly converted, and the accuracy of the analysis results can be improved if the normal distribution assumption is established through natural log conversion. First of all, we visualized it with BoxPlot to check the outliers of the news content length of the entire data (n = 16,617). In order to remove outliers identified in BoxPlot, the results were compared by applying the following two methods (Fig. 4).

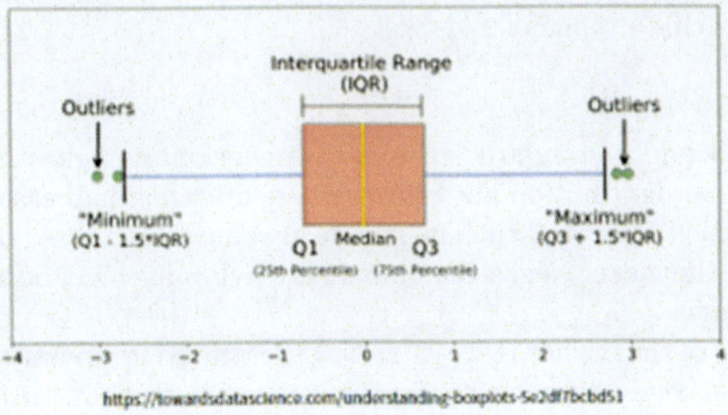

Fig. 3. IQR Structure

Table 12. Number of news by Media

	total amount of news
Media Companies	2,133
Telecommunications	14,484
Sum	16,617

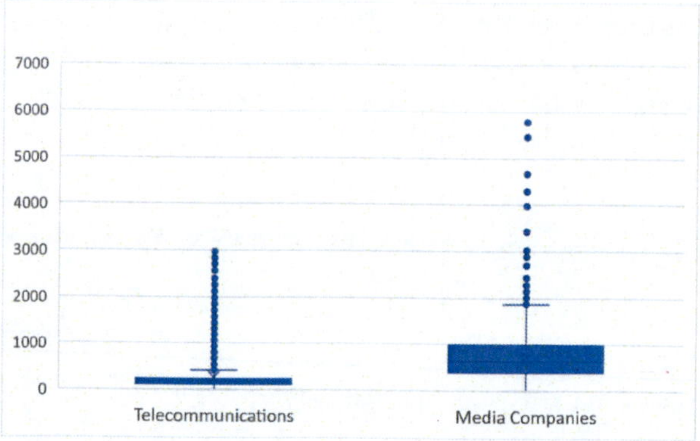

Fig. 4. Boxplot by Media

Data Standardization. In this study, two methods were tried to standardize data, that is, eliminate outliers. It was difficult to compare because the length of the news content ranged from 4 to 5,782 and the similarity ranged from 0.7 to 1 (Table 13).

First: Inter Quactile Range (IQR) IQR is a technique that uses the deviation of quartile values to detect outliers (Outliers) that deviate from the entire data pattern and specify a range of 1/4–3/4.

$$z = \frac{x - \mu}{\sigma} \tag{3}$$

Second: Z-Score A standard score is a dimensionless figure that creates a statistical normal distribution and shows where an individual case occupies the standard deviation, Z-Scorefkrh. The standard value Z indicates how far the raw value x is from the mean. Negative numbers are below average, positive numbers are above average.

In the Z score, the outlier is < -3 or > 3 is removed as an outlier (Table 14).

As a result of conducting two methods to remove outliers on media and carrier data, it was found that the outlier removal method through IQR conservatively determined outliers than the Z-SCORE outlier removal method (Table 15).

The process from data collection to processing analysis is illustrated (Fig. 5).

Table 13. Explain about formula (3)

X is the number of normalized raw materials
σ is the standard deviation from the population
μ is average in the population

Table 14. Comparison of IQR and Z-Score results by Media

	length of the news	News Similarity
length of the news	1	
News Similarity	−0.045132022**	1

Table 15. Comparison of IQR and Z-Score results by Media

	IQR		Z-SCORE	
	Data after outlier removal	an outlier	Data after outlier removal	an outlier
Media Companies	2,063	70	2,102	31
Telecommunications	12,671	1,813	14,088	396
Sum	14,734	1,883	16,190	427

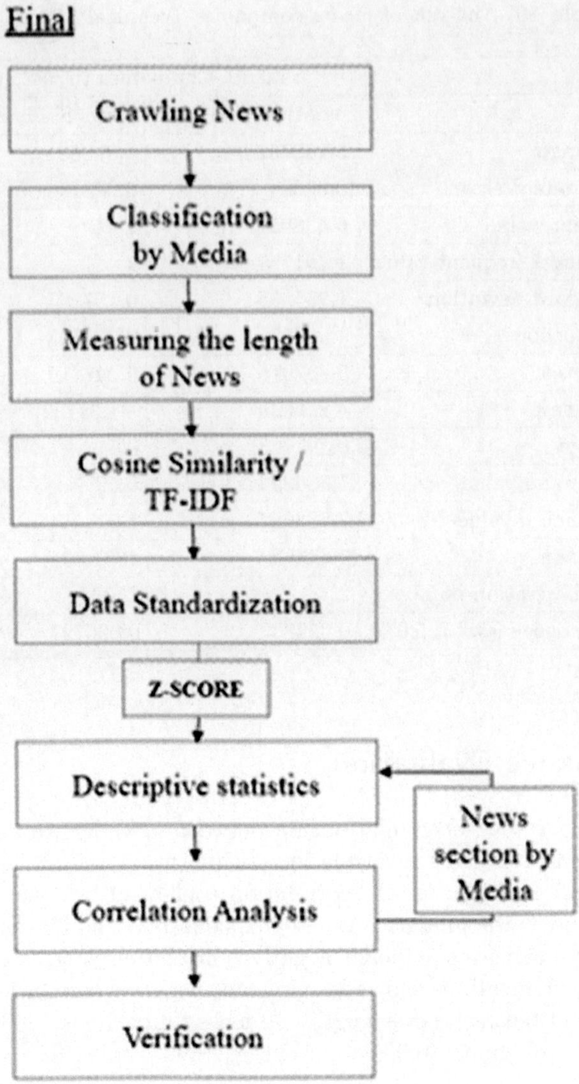

Fig. 5. Final Process

3.6 Technical Statistics

Therefore, descriptive statistics were examined to confirm the normality of the data by taking each result data. The results of comparing the descriptive statistics according to the outlier removal method confirmed that both methods satisfied normality. In this study, Z-SCORE was adopted to minimize data loss after outliers were removed (Tables 16 and 17).

Table 16. Analysis of media companies Technical Statistics

	Media Companies (n = 2,102)	
	length of the news	News Similarity
Average	6.150762	0.899631
a standard error	0.02694	0.00343
median value	6.436951	1
the most frequent value	1.791759	1
standard deviation	1.235135	0.157263
Dispersion	1.525558	0.024732
kurtosis	5.887075	0.247319
skewness	−2.34196	−1.31732
Range	6.397763	0.51
Minimum value	1.386294	0.49
Maximum value	7.784057	1
The sum	12928.9	1891.024
an observation number	2102	2102
Confidence level (95.0%)	0.052832	0.006727

4 Performance Evaluation

In order to analyze the linear relationship between news length and news similarity, it was measured using a correlation coefficient.

In Media Companies news, the correlation coefficient between the length of news content and news similarity is −.045132022 and significance probability ($0.0385 < .05$), showing a significant negative correlation (Fig. 6 and Table 18).

The Telecommunications shows a significant negative correlation as the correlation coefficient between news length and news similarity is −.086397825, and the significance probability ($0.000 < .05$) (Fig. 7 and Table 19).

Table 17. Statistical analysis of telecommunications

	Telecommunications (n'=14,088)	
	length of the news	News Similarity
Average	5.150627	0.891371
a standard error	0.006615	0.001394
median value	5.187386	1
the most frequent value	1.386294	1
standard deviation	0.78511	0.165436
Dispersion	0.616398	0.027369
kurtosis	7.246856	−0.10532
skewness	−1.05081	−1.19448
Range	5.715382	0.51
Minimum value	1.386294	0.49
Maximum value	7.101676	1
The sum	72562.04	12557.63
an observation number	14088	14088
Confidence level (95.0%)	0.012966	0.002732

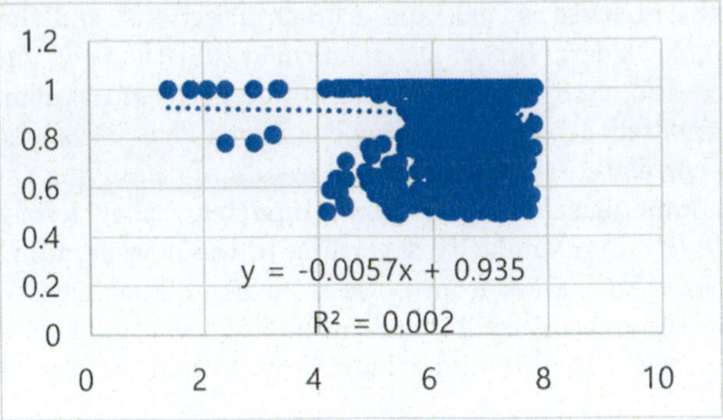

Fig. 6. Scatterplot of media companies

Table 18. media companies correlation results

Media Companies	the length of the news	News Similarity
length of the news	1	
News Similarity	−0.045132022**	1

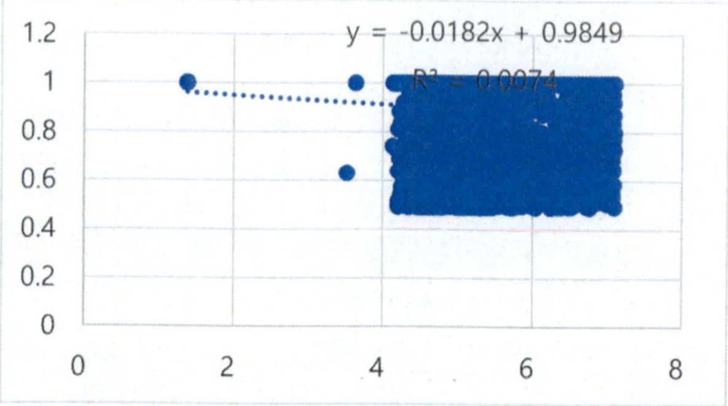

Fig. 7. telecommunications scatterplot

Table 19. telecommunications correlation analysis results

Media Companies	the length of the news	News Similarity
length of the news	1	
News Similarity	−0.086397825***	1

5 Limitations of the Study

As a result of this study, it can be seen that online news has the highest news production in the political, social, and economic sections, and the news length is mainly 4–400 characters. It was also confirmed that the news similarity of 0.7 or more reached about 18% in all news. News length and similarity were also found to be correlated. Above all, it was confirmed that news length and news similarity were correlated according to the results of classifying them into news producers, telecommunications and media companies, rather than the correlation between news length and similarity according to the news section. The method of separating news from news from news from news from media companies is thought to be a study necessary for correlation analysis. I hope it will be a basic study necessary for the quality and reliability of news in the future.

6 Conclusion

This study is a study that can confirm news similarity through correlation analysis between news length and news similarity. In particular, it is considered a meaningful study as a study that measures the difference according to the section and media of online news. It is expected to be used as basic research data for future research on news. Implementing a system that can separate news with high similarity amidst the pouring news will be beneficial to news-related industries and people in charge. The amount of news produced every day is huge. The limitations of this study are limited from the start. This is because the amount of news collected for research is extremely small. It was very difficult to process, standardize, and analyze from the stage of collecting about 10 news stories. Above all, it was difficult to have less research on news. There are many activities to increase the reliability of information by using news through SNS. There are many inappropriate activities among them. Therefore, I think there is a need for a system that can screen news. We hope that the study of news length and news similarity will be the starting point for these systems.

References

1. Arabi, H., Akbari, M.: Improving plagiarism detection in text document using hybrid weighted similarity. Exp. Syst. Appl. **207**, 118034 (2022)
2. Nasim, Z., Haider, S.: Evaluation of clustering techniques on urdu news head-lines: A case of short length text. J. Exp. Theor. Artif. Intell. 1–22 (2022)
3. Yoo, B.C., Lee, J.Y.: A study on the sentence length of sports news in the era of the convergency. J. Digit. Converg. **15**(6), 505–511 (2017)

A Survey of Bias in Healthcare: Pitfalls of Using Biased Datasets and Applications

Bojana Velichkovska[1(✉)], Daniel Denkovski[1], Hristijan Gjoreski[1], Marija Kalendar[1], and Venet Osmani[2]

[1] Faculty of Electrical Engineering and Information Technologies, Ss. Cyril and Methodius University, Skopje, North Macedonia
{bojanav,danield,hristijang,marijaka}@feit.ukim.edu.mk
[2] Fondazione Bruno Kessler, Trento, Italy
vosmani@fbk.eu

Abstract. Artificial intelligence (AI) is widely used in medical applications to support outcome prediction and treatment optimisation based on collected patient data. With the increasing use of AI in medical applications, there is a need to identify and address potential sources of bias that may lead to unfair decisions.

There have been many reported cases of bias in healthcare professionals, medical equipment, medical datasets, and actively used medical applications. These cases have severely impacted the quality of patients' healthcare, and despite awareness campaigns, bias has persisted or in certain cases even exacerbated. In this paper, we survey reported cases of different forms of bias in medical practice, medical technology, medical datasets, and medical applications, and analyse the impact these reports have in the access and quality of care provided for certain patient groups. In the end, we discuss possible pitfalls of using biased datasets and applications, and thus, provide the reasoning behind the need for robust and equitable medical technologies.

Keywords: healthcare bias · attitudes of healthcare professionals · biased datasets · medical applications

1 Introduction

With the rapid development of artificial intelligence (AI) and the efficiency which AI offers, there are entire industries that rely on its application in solving everyday challenges. Advantageously, AI is an important part of a vast range of actively-used applications, spanning from social media and personalized recommender systems, all the way to smart homes, smart cars, surveillance, and so on. With that, the benefits offered by AI are manifold. However, there are risks that social media and recommender systems can use information to influence user's opinions, especially for high-stakes events, such as elections.

These issues then raise the question, "With AI having a high error margin, what happens when that same AI is employed to sensitive areas such as medicine?" There

R. Silhavy and P. Silhavy (Eds.): CSOC 2023, LNNS 724, pp. 570–584, 2023.
https://doi.org/10.1007/978-3-031-35314-7_50

is a plethora of actively-used medical applications based on artificial intelligence [1]. Namely, AI is widely used in medical setting, with applications in everything from patient care and maintaining medical records to billing. Therefore, the use of AI in medicine directly, and sometimes indirectly, determines the treatments given to patients, and consequently, patients' outcomes. One of the many applications of AI in medicine nowadays is the identification and diagnosis of different medical conditions [2]. A sub-branch of AI in diagnostic capacity is medical imaging diagnosis, where AI algorithms are taught to recognize complex patterns in imaging data and provide health assessments of the patients' conditions, which as a diagnostic tool has had large success in past years [3]. Other applications include, but are not limited to, personalised medicine [4], smart health records [5], clinical trial and research [6], drug discovery and manufacturing [7], etc.

However, technology is prone to different forms of malfunction, so it is expected that issues can arise with actively-used applications. Of all potential issues, there are some which happen as an unintentional and unexpected by-product of AI, and these issues are therefore more serious than others, because they are "silent" or "hidden". With this, the use of AI in medicine and healthcare is full of potential for clinical, social, and ethical conflicts. Namely, there is a risk of patient harm due to prevailing errors in AI models, influenced by biased inequities in the health system, exacerbated by lack of transparency in patient selection when creating medical datasets, as well as the evident lack of transparency in development of AI-based medical applications [8].

With these risks in mind, it becomes important to understand healthcare biases in medical practice, and address the potential pitfalls of using biased datasets and biased AI applications. Furthermore, increasing awareness would lead to improving application of AI in clinical practices and consequently better outcomes.

This paper is organised into four sections. Section 2 gives the definition of bias and the different types of bias possible in medical AI. The next section gives a survey of reported cases of bias in medical datasets and medical AI-based applications. Section 4 gives a summary of the different types of bias and the lessons learned. The paper concludes in Sect. 5.

2 Bias and Different Types of Bias

According to the Merriam Webster dictionary, bias is defined as:

- a personal and often unreasoned judgment for or against one side;
- an unreasoned and unfair distortion of judgment in favour of or against a person or thing;
- a settled and predictable leaning in one direction and connotes unfair prejudice;
- a systematic error introduced into sampling or testing by selecting or encouraging one outcome or answer over others.

Therefore, the meaning of the word bias can be surmised into a propensity to show partiality towards certain persons or groups in favour of others, which often negatively impacts the marginalised group. From this definition, we can infer that the presence of bias in medical setting would mean overlooking a group of patients discriminatively, which can result in significantly poorer health services to certain patient groups that might lead to long-term complications, impairments, and in worst case scenarios even deaths which could have been prevented.

According to a recent presentation on AI in healthcare delivered by Marzyeh Ghassemi, bias is already present in the clinical landscape [9]. The bias present in the healthcare system can: one, exist consciously and be exhibited through prejudiced ideas such as racism and sexism; two, be unconscious, but cemented thoughts based on learned stereotypes; or, three, can happen mistakenly by being based on conclusions drawn only from working with a uniform portion of the population. With either of the listed, there are different forms of bias which can occur: racial bias, bias based on sex, gender or sexual identity, socioeconomical bias, educational bias, as well as bias arising from geographical location, overweight and obesity, or age (see Fig. 1). Some forms of bias are more dominant than other, but each have manifold consequences that can negatively impact certain patient groups.

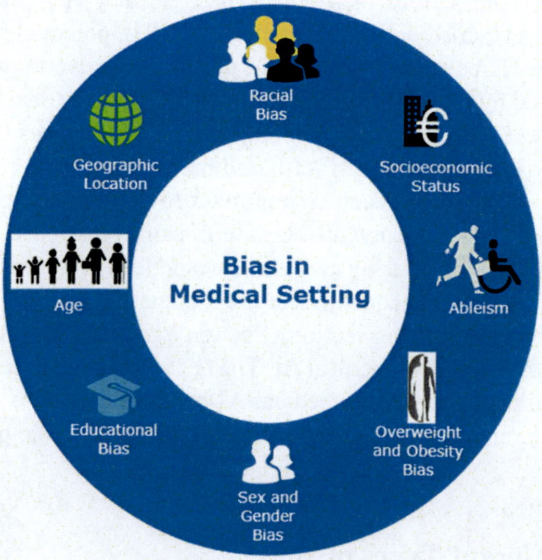

Fig. 1. Bias in Medical Setting: Different types of bias present.

3 Bias in Medicine

AI-related bias in medical setting can be dissected from four distinct aspects: data-driven, algorithmic, technological, and human [10]. Essentially, all these aspects are, directly or indirectly, linked to the human aspect in medicine, wherein biased medical decisions, attitudes, and behaviour are a daily occurrence, and unfortunately also get propagated and taught to medical students [11].

3.1 Human Aspect of Bias

The first direction we decided to investigate was the human aspect of bias. According to the key findings presented in the U.S. National Healthcare and Disparities Report [12] conducted in 2019, it was observed that in spite of efforts disparities in medicine have persisted and some have even worsened, mostly for poorer and uninsured populations. The report also showed disparities based on residence location. Additionally, there were racial and ethnic disparities, i.e., White patients were found to have received better care compared to other racial and ethnic groups: Blacks, American Indians, Alaska Natives, Hispanics, Asians, and Native Hawaiians (Pacific Islanders). In [13] the authors discussed how people of colour in the U.S. face disparities in access to healthcare, the quality of the care received, which consequently impacts the health outcome of these patients. According to [14], people of colour face more barriers in accessing care, which means less access to preventive services and treatments, and even when access to care is secured, patients of colour tend to have unsatisfactory interactions with healthcare providers. Racial bias has extended even to patient recommendations for bypass surgery [15]. Namely, some physicians were more likely to recommend White patients over Black patients, because they believed that Black patients would not adhere to the necessary physical activity needed after surgery.

There have also been reported cases of sex and gender bias. A study [16] showed that medical professionals were more likely to dismiss chronic pain in women than men, expressing it as difference of "brave men" compared with "emotional women". These biases can lead to the silencing of patients in addressing important health problems. Such is also the case with transgender people, who feel reluctant to receive proper healthcare due to expected unfair treatment [17].

Other types of bias are also present in behaviour, attitudes, and opinions of medical professionals. Namely, medical professionals find working with older patients and their families as challenging, and have described these patients as demanding and offensive, and wanting to manage their own treatment [18]. Furthermore, disabled patients have limited access to certain areas in healthcare centres. The authors of [19] discussed how over 80% of medical professionals would rather work with people without disabilities. People with obesity are also likely to receive poor treatment by their health provider, as well as have their symptoms attributed to their weight [20]. People with lower socioeconomic status are more likely to experience delays in testing and treatment, which creates issues with regular and quality preventive treatments [21].

Altogether, the presented types of bias result in reduction of healthcare access and quality for certain patient groups based on prejudiced opinions and beliefs deeply rooted in the behaviour of medical personnel. Therefore, these groups are exposed to different serious risks which directly affect their health, due to delayed or non-existent treatment, incorrect diagnosis, which might overlook serious conditions and complications.

3.2 Bias in Medical Technology

Reporting of bias has also been extended to medical technology actively used in practice to evaluate patients. Inherent in the medical technology itself, due to the fact that the device's testing did not include diverse population, this type of bias has been long overlooked, whilst medical technologies incorporating it were being constantly used and were directly influencing patient statistics.

The first reported bias in medical technology dates back to 1968 [22]. Namely, guidelines given by manufacturers indicated the need for higher radiation exposure for Black patients, which was absorbed into the medical practice and resulted with X-ray technicians routinely exposing Black patients to higher doses of radiation compared to those received by White patients. Naturally, the guidelines were based on the false belief that Black patients have denser bones and thicker skin.

A study [23] found that forehead thermometers, which measure temperature through the skin using infrared technology, were 26% less likely to detect fever for Black patients compared to oral thermometers. Another study [24], which gained attention during the COVID-19 pandemic, found that pulse oximeters were three times less likely to detect low oxygen levels in Black patients, which delays necessary treatment and puts patients at risk.

In actuality, bias in medical equipment, and one that is being used on daily bases, can have unforgivable consequences for underserved patients. The problems with the equipment lead to penalisation of certain population groups, which can contribute to delay in diagnosis or treatment administration, that might lead to fatalities.

3.3 Bias in Medical Datasets

If left unchecked, systemic human biases, stigmatic opinions and bias in medical technology can be incorporated in medical datasets and AI-based medical algorithms, and heighten the presence of bias in a wide spectre of applications developed with the aim of assisting and bettering the healthcare process and experience for sick patients.

Biased datasets are either full of biased markers or have underrepresentation of certain patient groups, which can stem from one or more in a list of reasons:

- systematic discrimination arising from unequitable treatment of patients from medical personnel due to racial, socioeconomic, or additional aspects;
- bias imbedded during the data collection process;
- lack of diversity and interdisciplinarity in accessing medical equipment and technology quality;
- or simply, lack of detailed quality investigations in obtained technological, clinical, and scientific research.

Collecting data from medical institutions without mitigating biased opinions, practices, and treatments can effectively lead to bias in medical datasets. Considering the already presented issues in behaviour of medical professionals, it is understandable why there have been many reported cases of bias in medical datasets. Biased datasets can come from transference from implicit bias of medical professionals, or implicit bias during the data selection process and underrepresentation of diverse patient groups.

Data has to be representative of population variety, otherwise it can reinforce lack of generalisation and different forms of bias [10, 25]. In [26] the authors investigated the history and physical notes from 18,259 patients that were collected in an urban academic medical centre. The study analysed the presence of negative descriptors, like noncompliant or resistant, regarding patients and their behaviour. The results showed presence of racial bias in the analysed electronic health records; namely, Black patients were 2.5 times more likely to be given negative descriptors compared with White patients.

Medical imaging datasets have also been scrutinised for being ladened with different forms of bias. The authors in [27] address the importance of gender balance in medical imaging datasets by showing consistent decrease of performance for underrepresented genders. The authors also investigated the influence which imbalanced datasets had on model performance. They conclude that when working with 25%/75% imbalance ratio between classes model performance across the minority class is significantly lower compared to the majority class. On the other hand, that difference was not observed in balanced datasets.

Medical imaging datasets have been investigated for presence of racial and ethnic imbalance. The National Lung Screening Trial collected data from 53,000 smokers to investigate lung cancer diagnosis [28], however from the selected patients only 4% were Black. Another dataset targeted for its biased data is the International Skin Imaging Collaboration. The dataset is one of the most used open-access dataset on skin lesions in the diagnostic process of melanoma which the most serious form of skin cancer; however, the data was collected from mostly fair-skinned patients [29]. Another form of bias is geographic bias, as pointed by the researchers in [30]. The study was conducted in 2020 and it showed that 71% of studies in U.S. where geographic location was present were using data only from three states: California, Massachusetts, and New York. Additionally, they found that conducted studies used data from only 16 countries, whereas there were no datasets available from the remaining 34 countries.

Altogether, the presented cases in which bias has been noted in medical datasets show that open-access data available for researchers can be significantly limited on patients from a certain geographical area, and belonging to one dominant race or gender. This impacts uniform patient representation; thus, datasets carry limited knowledge which does not allow for thorough understanding of medical conditions and subtle changes which might occur across different patient populations.

3.4 Bias in AI-Based Medical Applications

When AI algorithms use biased datasets during the training process, the algorithms have a limited view into the problem and have better understanding of the problem from the perspective of the dominantly present group. Therefore, the models can learn the bias which the data incorporates, and have lower performance accuracy over certain patient groups. Consequently, the trained algorithms reinforce inequities in healthcare in everything from cancer-detection algorithms which are less effective for Black patients [31] to cardiac risk-scores that underestimate the amount of care needed by Black patients. There have been reports of bias in algorithms used for maternal health. Namely, a widely-used Vaginal Birth after Caesarean (VBAC) algorithm contributed to higher rates of c-section among women of colour because it was predicting lower successful VBAC rates for pregnant women of colour [32].

A case study [33] showed evidence of racial bias in an actively used algorithm, which carried decisions for more than 200 million people in the U.S. The origin of the bias came from using health costs as a proxy for health needs. Since less money was being spent on Black patients who have the same level of need as White patients, the algorithm learned this discrepancy, and therefore, assigned the same level of risk scores to White and Black patients, even though the Black patients were in worse medical condition compared to the White patients. According to an estimation made by the authors of the study, the number of Black patients identified for extra care was reduced 2.5 times compared with what it should have been.

Another study [34] evaluated the performance of state-of-the-art algorithms in detecting abnormalities (e.g., pneumonia, lung nodules, lesions, fractures, etc.) in chest X-rays. The results showed that young females had the highest rate of underdiagnosis, followed by Black patients, then by patients with public health insurance due to low income. This was even further pronounced with patients with fusion of more than one of the listed criteria, i.e., a Black woman with public insurance and low-income background had the highest rate of underdiagnosis.

An investigation into an AI-based tool for early detection of sepsis, actively used by more than 170 hospitals, showed the model's inability in predicting this life-threatening illness in 67% of patients who developed it [35]. Furthermore, the model also generated false sepsis alerts on thousands of patients who did not develop the illness.

Another algorithm was criticised for [36] suggesting extreme cuts to in-home care of disabled patients, which caused extreme disruption of patients' lives and resulted with increased hospitalisation. Another study which reported a form of socioeconomic bias aimed to assess the degree to which data quality of electronic health records related to socioeconomic status [37]. The machine learning models investigated in the study aimed to predict asthma exacerbation in children. The results of the study showed worse predictive model performance in patients with lower socioeconomic status. An AI-based model for Alzheimer's diagnosis from audio data, built in Canada, underperformed for patients with certain accents because the training process included speech samples from one accent, therefore making the application unusable for everyone else in the country [38].

Altogether, AI-based medical applications with biased performance across different patient groups have been widely reported only after actively being used in medical setting and severely impacting the quality of care offered to patients. Many reported cases have endangered patients' lives by missing disease diagnosis in life-threatening situations. Other cases show undue stress inflicted to patients by inaccurate diagnosis of illnesses which are later proven as non-existent. Furthermore, the algorithms' flaws are more expressed in patients which have a diminished access to healthcare therefore creating severe difficulties for them, i.e., patients with low income and limited access to medical care cannot afford a second opinion; this makes erroneous diagnosis in these cases a heavy-handed and punitive action towards the patient.

4 Summary of Bias and Lessons Learned

The allotted sections and the papers discussed in them are summarised in Table 1. For each of the papers referenced, the table lists the type of bias, a brief description of the paper, the source of bias (medical practice, datasets, AI applications, with a more detailed and concrete scope), and the implications which can be drawn from the observed.

From the ample examples referenced, it is evident bias is present in medical practice and technology. In addition, that bias can easily transfer from medical practice to medical datasets, and eventually, to AI-based applications unless adequate actions are taken. Briefly, data presumed to reflect different population groups equitably has on many occasions failed to do just that, AI-based algorithms presumed to have equal performance across different population groups have proven biased in vast scope, and understandably so, when they are impacted by human influence that embeds societal prejudices against patients of different race, gender, appearance, socioeconomic status, etc. Therefore, from these lessons learned, the question of mitigating bias emerges.

The primary problem from which all others derive is the human aspect in medical practice. For that reason, the most important step to take is countering biased practices by imposing mandatory training of all medical personnel with the purpose of imprinting fundamental understanding of bias in health and consequences. Non-functional medical equipment in the 21st century is another huge problem. Diversity in testing new medical equipment before releasing it for mass production and ensuring that equipment cannot do any inadvertent harm to some populations is another must.

Creating datasets which have blind spots across race, gender, socioeconomic status, and so on, should not be available to everyone around the world. Enabling diverse and multi-disciplinary teams, when creating medical datasets, can be beneficial in reducing, and even maybe eradicating, those cultural or academic blind spots, and thus allowing for fair and equitable datasets.

Every stage after the data collection can also succumb to bias. In order to provide responsible algorithm development, steps must be taken in the pre-processing, in-processing, and post-processing stages. Namely, when working on a model the development teams should require maximising the accuracy of the model while at the same time minimising the influence of biased markers. Post-processing mitigation also helps, in that, through thorough analysis of the model performance across different population groups issues with AI-based applications can be detected in the early stages of production. That would allow models to be retrained or retuned to work equitably for everyone.

In summary, ensuring accurate medical equipment and adequate data gathering with wide representation and accurate labelling is extremely important, since with faulty data little can be done to prevent bias transfer to the AI application. Furthermore, regulations must be followed when creating the application. Teams must be equipped to handle different aspects of a problem, which is why a vast array of diversity, knowledge, and understanding is a must. In the end, even after all precautions are taken, the model must be rigorously analysed for bias before being put into practice. Once all requirements for fairness are met, sharing details on how the model was developed is essential, for several reasons: one, it allows the research community to better understand the steps which should be taken in order to develop unbiased applications; two, it would account for how the model should be used; three, additional bias assessments can be conducted by impartial teams; four, transparency would help patients with trusting the process; and more.

The brevity of it all is, there are fundamental issues to be considered and corrected, and with urgency, as they impact lives all over the world. And, that change should come from us all.

Table 1. Summary of bias in Healthcare.

Ref	Bias Type	Description	Source	Implications
Human Aspect of Bias				
[12]	socio-economic, location, racial and ethnic	worsened care for poorer populations; subpar care for some racial and ethnic groups	medical practice	bias mitigation strategies not entirely effective
[13]	racial and ethnic	people of colour facing disparities in healthcare access and quality	medical practice	reduced preventive care, impact on outcomes for patients of colour
[14]	racial and ethnic	people of colour facing barriers in accessing care	medical practice	reduced preventive care, impact on outcomes for patients of colour
[15]	racial and ethnic	patient recommendations for bypass surgery based on skin colour	medical practice	Black patients were denied bypass surgery
[16]	sex and gender	dismissal of chronic pain in women	medical practice	can lead to silencing patients on important health problems
[17]	sex and gender	transgender people feel reluctant to receive proper healthcare	medical practice	transgender people might avoid visiting a healthcare professional unless urgent
[18]	age	medical professionals find working with older patients demanding	medical practice	unfair treatment of older patients
[19]	ableism	over 80% of medical professionals would rather work with people without disabilities	medical practice	favouring working with able-bodied people can create an issue for disabled patients and access to quality, objective healthcare

(*continued*)

Table 1. (*continued*)

Ref	Bias Type	Description	Source	Implications
[20]	obesity	people with obesity have their symptoms attributed to weight	medical practice	medical conditions can be overlooked
[21]	socio-economic	poorer people are more likely to experience delays in testing and treatment	medical practice	chances of worsening medical conditions and complications due to wait time
Bias in Medical Technology				
[22]	racial and ethnic	higher radiation exposure for Black patients	equipment	absorbed in medical practice and routinely applied
[23]	racial and ethnic	forehead thermometers were 26% less likely to detect fever in Black patients compared with oral thermometers	equipment	missed fevers could lead to delays in diagnosis and treatment, and possibly cause an increased death rate in Black patients
[24]	racial and ethnic	pulse oximeters were three times less likely to detect low oxygen levels in Black patients	equipment	could lead to delays in diagnosis treatment, and possibly cause an increased death rate in Black patients
Bias in Medical Datasets				
[26]	racial and ethnic	physical notes from 18,259 patients showed negative descriptors for certain racial and ethnic groups	medical practice transferred to medical datasets	2.5 times more negative descriptors for Black patients compared with White patients can lead to AI applications learning that discrepancy and operating on that bias

(*continued*)

Table 1. (*continued*)

Ref	Bias Type	Description	Source	Implications
[27]	gender	decrease of model performance in case of underrepresented classes in datasets	medical datasets transferred to AI applications	working with imbalanced ratio significantly overlooks the minority of the population and can result with AI applications performing worse for certain population groups
[28]	racial and ethnic	only 4% of selected patients for cancer diagnosis dataset were Black	medical datasets	underrepresentation can lead to AI applications performing well only for certain ethnic and racial groups
[29]	racial and ethnic	dataset for skin cancer collected from mostly fair-skinned people	medical datasets	underrepresented populations that will very likely lead to AI applications unfamiliar with different skin colours and understanding cancer only for White patients
[30]	location	71% of studies with geographic location included came from only three states	medical datasets	underrepresentation of patients coming from certain areas can lead to AI models operating accurately only for the areas in the data
Bias in AI-based Medical Applications				
[31]	racial and ethnic	cancer detection algorithm less effective for Black patients	medical applications	loss of lives (in Black patients) which could have been prevented

(*continued*)

Table 1. *(continued)*

Ref	Bias Type	Description	Source	Implications
[32]	racial and ethnic	VBAC algorithm predicts higher rates of c-section among women of colour	medical applications	higher rates of potentially unnecessary procedures for women of colour
[33]	racial and ethnic	same risk scores were assigned to White and Black patients, even though Black patients were in worse medical condition	medical application	number of Black patients identified for extra care was reduced 2.5 times compared with what it should have been
[34]	socio-economic, sex and gender, racial and ethnic	chest X-ray showed young females had the highest rate of underdiagnosis, followed by Black patients, then by patients with public health insurance	medical application	disregard for serious illnesses
[35]	ableism	extreme cuts to in-home care of disabled patients	medical application	extreme disruption of patients' lives which resulted with increased hospitalisation
[36]	socio-economic	researched the degree to which data quality of electronic health records related to socioeconomic status	medical dataset transferred to medical application	worse predictive model performance in patients with lower socioeconomic status
[37]	linguistic	Alzheimer's diagnostic tool underperformed for patients with certain accents	medical dataset transferred to medical application	the application was unusable for a large population of people

5 Conclusion

Certain patient groups are marginalised due to different aspects pertaining to their gender and sexuality, the colour of their skin, their socioeconomic status, etc., which affects the quantity and quality of care which they are offered. Biased practices impact patients' healthcare, and patients are subjected to opinions and behaviour which negatively influence their quality of life. Therefore, detecting the presence of different types of bias in the healthcare system, namely biases in medical technology, behaviour of medical professionals, datasets collected from patients, and AI-based medical applications, as well as understanding the sources of existing bias are important and needed steps for improving healthcare access and the quality of care offered to different patient groups.

Therefore, in this work we illustrated different types of bias present in the healthcare systems, focusing on surveying papers which illustrate four different types of bias: in healthcare professionals, in the technology used for medical procedures, in the datasets collected in medical setting, and in AI-based medical applications for wide use. Our survey showed cases of different forms of bias which have had significant impact on patient lives around the world.

With this reflective analysis on AI technologies, we wish to raise awareness for the need of creating clinically robust and safe medical applications, built on widely-representative datasets, which successfully address ethical complaints and are transparent all throughout the development process, and therefore can be successfully and safely integrated in healthcare practices.

Acknowledgements. Part of the study was supported by WideHealth project – Horizon 2020, under grant agreement No 95227.

References

1. Park, C.W., Seo, S.W.: Artificial intelligence in health care: Current applications and issues. J. Korean Med. Sci. **35**(42), 379 (2020)
2. Kumar, Y., Koul, A., Singla, R., Ijaz, M.F.: Artificial intelligence in disease diagnosis: a systematic literature review, synthesizing framework and future research agenda. J. Ambient. Intell. Humaniz. Comput. (2021). https://doi.org/10.1007/s12652-021-03612-z
3. Oren, O., et al.: Artificial intelligence in medical imaging: switching from radiographic pathological data to clinically meaningful endpoints. Lancet Digit. Health **2**(9), 486–488 (2020)
4. Johnson, K.B., et al.: Precision medicine, AI, and the future of personalized health care. Clin. Transl. Sci. **14**(1), 86–93 (2021)
5. Rayan, Z., Alfonse, M., Salem, A.B.M.: Machine learning approaches in smart health. Procedia Comput. Sci. **154**, 361–368 (2019)
6. Weissler, E.H., et al.: The role of machine learning in clinical research: Transforming the future of evidence generation. Trials **22**, 537 (2021)
7. Paul, D., Sanap, G., Shenoy, S., Kalyane, D., Kalia, K., Tekade, R.K.: Artificial intelligence in drug discovery and development. Drug Discov. Today **26**(1), 80–93 (2021)
8. Bernal, J., Mazo, C.: Transparency of artificial intelligence in healthcare: Insights from professionals in computing and healthcare worldwide. Appl. Sci. **12**(20), 10228 (2022)

9. Ghassemi, M.: Exploring healthy models in ML for health. In: AI for Healthcare Equity Conference, AI & Health. MIT (2021)
10. Norori, N., Hu, Q., Aellen, F.M., Faraci, F.D., Tzovara, A.: Addressing bias in big data and AI for health care: A call for open science. Patterns **2**(10), 100347 (2021)
11. Brooks, K.C.: A piece of my mind. A silent curriculum. JAMA **313**(19), 1909–1910 (2015)
12. 2019 National Healthcare Quality and Disparities Report. Rockville, MD. Agency for Healthcare Research and Quality. AHRQ Pub. No. 20(21)-0045-EF (2020)
13. Fiscella, K., Franks, P., Gold, M.R., Clancy, C.M.: Inequality in quality: Addressing socioeconomic, racial, and ethnic disparities in health care. JAMA **283**(19), 2579–2584 (2000)
14. 2013 National Healthcare Disparities Report. Rockville, MD. Agency for Healthcare Research and Quality, US Dept of Health and Human Services. AGRQ Pub. No. 14-0006 (2014)
15. Dovidio, J.F., Eggly, S., Albrecht, T.L., Hagiwara, N., Penner, L.: Racial biases in medicine and healthcare disparities. TPM-Test. Psychom. Methodol. Appl. Psychol. **23**(4), 489–510 (2016)
16. Samulowitz, A., Gremyr, I., Eriksson, E., Hensing, G.: "Brave Men" and "Emotional Women": A theory-guided literature review on gender bias in health care and gendered norms towards patients with chronic pain. Pain Res. Manag. **2018**, 6358624 (2018)
17. Casey, L.S., et al.: Discrimination in the United States: Experiences of lesbian, gay, bisexual, transgender, and queer Americans. Health Serv. Res. **54**(Suppl 2), 1454–1466 (2019)
18. Ben-Harush, A., et al.: Ageism among physicians, nurses, and social workers: Findings from a qualitative study. Eur. J. Ageing **14**(1), 39–48 (2016). https://doi.org/10.1007/s10433-016-0389-9
19. VanPuymbrouck, L., Friedman, C., Feldner, H.: Explicit and implicit disability attitudes of healthcare providers. Rehabil. Psychol. **65**(2), 101–112 (2020)
20. Phelan, S.M., Burgess, D.J., Yeazel, M.W., Hellerstedt, W.L., Griffin, J.M., van Ryn, M.: Impact of weight bias and stigma on quality of care and outcomes for patients with obesity. Obes. Rev. **16**(4), 319–326 (2015)
21. Arpey, N.C., Gaglioti, A.H., Rosenbaum, M.E.: How socioeconomic status affects patient perceptions of health care: A qualitative study. J. Prim. Care Commun. Health **8**(3), 169–175 (2017)
22. Bavli, I., Jones, D.S.: Race correction and the X-ray machine—The controversy over increased radiation doses for black Americans in 1968. N. Engl. J. Med. **387**(10), 947–952 (2022)
23. Bhavani, S.V., Wiley, Z., Verhoef, P.A., Coopersmith, C.M., Ofotokun, I.: Racial differences in detection of fever using temporal vs oral temperature measurements in hospitalized patients. JAMA **328**(9), 885–886 (2022)
24. Sjoding, M.W., Dickson, R.P., Iwashyna, T.J., Gay, S.E., Valley, T.S.: Racial bias in pulse oximetry measurement. N. Engl. J. Med. **383**(25), 2477–2478 (2020)
25. Tasci, E., Zhuge, Y., Camphausen, K., Krauze, A.V.: Bias and class imbalance in oncologic data-towards inclusive and transferrable AI in large scale oncology data sets. Cancers (Basel) **14**(12), 2897 (2022)
26. Sun, M., Oliwa, T., Peek, M.E., Tung, E.L.: Negative patient descriptors: Documenting racial bias in the electronic health record. Health Aff. **41**(2), 203–211 (2022)
27. Larrazabal, A.J., Nieto, N., Peterson, V., Milone, D.H., Ferrante, E.: Gender imbalance in medical imaging datasets produces biased classifiers for computer-aided diagnosis. Proc. Natl. Acad. Sci. **117**(23), 12592–12594 (2020)
28. Ferryman, K., Pitcan, M.: Fairness in Precission Medicine (2018)
29. Adamson, A.S., Smith, A.: Machine learning and health care disparities in dermatology. JAMA Dermatol. **154**(11), 1247–1248 (2018)
30. Kaushal, A., Altman, R., Langlotz, C.: Geographic distribution of US cohorts used to train deep learning algorithms. JAMA **324**(12), 1212–1213 (2020)

31. Sourlos, N., Wang, J., Nagaraj, Y., van Ooijen, P., Vliegenthart, R.: Possible bias in supervised deep learning algorithms for CT lung nodule detection and classification. Cancers (Basel) **14**(16), 3867 (2022)
32. Taylor, L.M.: Race-Based Prediction in Pregnancy Algorithm Is Damaging to Maternal Health (2021)
33. Obermeyer, Z., Powers, B., Vogeli, C., Mullainathan, S.: Dissecting racial bias in an algorithm used to manage the health of populations. Science **366**(6464), 447–453 (2019)
34. Seyyed-Kalantari, L., Liu, G., McDermott, M., Chen, I.Y., Ghassemi, M.: CheXclusion: Fairness gaps in deep chest X-ray classifiers. Pac. Symp. Biocomput. **26**, 232–243 (2021)
35. Wong, A., et. al.: External validation of a widely implemented proprietary sepsis prediction model in hospitalized patients. JAMA Intern. Med. **181**(8), 1065–1070 (2021)
36. Lutz, R.: Incident Number 110. In: McGregor, S. (ed.) Artificial Intelligence Incident Database. Responsible AI Collaborative. incidentdatabase.ai/cite/110. Accessed 16 Feb 2023
37. Juhn, Y.J., et al.: Assessing socioeconomic bias in machine learning algorithms in health care: A case study of the HOUSES index. J. Am. Med. Inform. Assoc. **29**(7), 1142–1151 (2022)
38. Fraser, K.C., Meltzer, J.A., Rudzicz, F.: Linguistic features identify Alzheimer's disease in narrative speech. J. Alzheimers Dis. **49**(2), 407–422 (2016)

Professional Certification for Logistics Manager Selection: A Multicriteria Approach

Fabiano Porto de Aguiar, José Pereira da Silva Filho, Paulo de Melo Macedo, Plácido Rogério Pinheiro(✉), Rafael Albuquerque Cavalcante, and Rodrigo Bastos Chaves

University of Fortaleza, Fortaleza, CE, Brazil
placido@unifor.br, rodbascha@bol.com.br

Abstract. The research aims to identify and classify the certification of an administrative professional in road freight transport, using the verified multicriteria for decision making psychological, theoretical, practical, and evaluation of alternatives for professionals with experience in the area, researched on the social media platform LinkedIn, randomly drawn four professionals in the transport logistics department. In decision-making, the multicriteria methods add values of paramount significance. Therefore, they are not treated by empirical intuition processes used but have a conference with transparency and clarity. As a decision-making objective, it searches for an alternative with a performance, evaluation, and quality agreement between processes expected by the user of the decision relating to the components. The classification of people is often a subjective process that can be reduced by the method proposed in this research and can be applied to classify other professional categories. This research applied the Analytic Hierarchy Process (AHP) as an evaluative multicriteria decision-making method. In several areas of knowledge, we can use the AHP as a valuable mechanism of the multicriteria methodology to support decision making. Peer analysis establishes the priorities that will give possibilities for making highly complex decisions.

Keywords: Administration · Decision Making · Multicriteria · AHP Decision Analysis · Personnel Certification · Personnel Selection

1 Introduction

In Brazil, the road freight transport activity is one activity that grows according to the economy due to the need to transport the products of transactions between companies from different sectors within the country [1]. The transport of cargo through the road system in Brazil has a good structure and is responsible for the flow, which ranges from entire crops to simple orders. The road freight transport activity crosses several Brazilian states, that is, an interstate and intercity activity, being a demand that heats the road cargo transport companies and makes investors in this segment have a vision of growth and renewal of the action. In this sense, there is also an initial investment, which is the purchase of a truck, trailer, or other transport aggregates which cannot remain idle, and

R. Silhavy and P. Silhavy (Eds.): CSOC 2023, LNNS 724, pp. 585–596, 2023.
https://doi.org/10.1007/978-3-031-35314-7_51

need to generate profit for the company, also aiming at the complexity of carrying out the transport activity. Above all, maintain operating cost efficiencies with fuel variables, parts, labor, maintenance, etc.

Because of this, we face the need for professionals to work in the complex and challenging segment. It involves several varieties of costs and risky operations and demands the transport of third-party goods with added value. It is observed in the cognitive map the Criterion of Professional Certification of Administrator in an area of Road Cargo Transport, being this a professional in the logistics department since it is from this professional that, among other places, the efficiency of transporting and demanding a cost will be developed. Reference found [1] states that such professionals have the knowledge to carry out the operation plans and are fundamental to the company.

In another reference [2], it is textually stated:

"The operational logistics strategy is developed through three basic and distinct activities: storing, transporting, and distributing. The sum of them needs comprehensive and integrated management, which forms logistics. The phases must be integrated at synchronized moments, as a rupture or information mismatch causes problems for the entire operation. In this case, it is clear the need to choose, within criteria, a professional with work performance, theoretical experience and knowledge, to carry out, develop and innovate the operation."

As shown in one of the studies carried out [1]:

"Employing a technical analysis, it can be stated that the problems to be solved in the planning and programming of the operation are of high complexity and involve related issues, for example, routing, construction of lines, allocation of fleets, and scheduling of crew. This reconciles rationalization, economy, and safety with delivery deadlines, labor legislation, timetables, available vehicles, etc."

Therefore, the cognitive map of the Transport Area Administrators is observed. We will focus on the Multicriteria decision in the logistics department, certifying the professional who will lead the complexity of the operations, but with macro knowledge of the activity within the company, aligned with planning strategic, leading with tactical planning and focused on operational planning, knowing that the decisions taken have an impact on several areas. In the view of [2], the theory is based on classification according to the American school. It assumes that all states are comparable (there is no incomparability); there is transitivity in the preference relation and transitivity in the indifference relation.

On the other hand, the fact that a professional completes a specific course at an institution does not guarantee companies that he can work in the sector with the desired productivity. At this point, a gap emerges that can be filled with an assessment of specific skills aimed at the individual combination of experiences and professional training in the sector of activity. Several possible combinations can be considered appropriate, and as an example, we have a professional trained in Computer Science obtaining a certification to work in Project Management. Another positive example is a certified English language teacher from a recognized US university. This study will seek evaluation criteria for an administrative professional to act as a road freight transport company manager in this context.

2 Methodology

One qualitative research was used as a methodology for this study, which aimed to choose the appropriate person for the position that was the object of this work, that is, the identification and certification classification of an administrative professional in the specific area of road transport of loads.

The steps of the method used follow the order: discovery of the problem, precise formulation of the problem, search for relevant knowledge or instruments of the problem, attempt to solve the problem with the aid of the multicriteria analysis tool, obtaining the solution, investigation of the consequences of the solution obtained and the proof of the solution. As for the purposes and objectives, this research is characterized as exploratory research, which aims to develop, clarify, and modify concepts and ideas and may constitute the first stage of broader and descriptive research with characteristics of specific populations or phenomena [3].

The proposed study is characterized by its applied nature, seeking to find a solution to a practical problem and procedure of documentary character using primary sources, that is, materials that have not yet received an analytical treatment [3]. The form of data collection at work is also worth mentioning, consisting of observation and document access. After analyzing the cognitive map, from which the assumptions of the work were treated and diagnosed, we proceeded to the decision analysis through the multicriteria method, where the psychological, theoretical, and practical aspects were chosen for study and placed in order of importance, applying the AHP method as a multicriteria evaluative method in decision making.

Candidates for vacancies for the position we are looking for were extracted from LinkedIn at random but within the criteria already established by the tools already described. The selection process was concluded by carefully analyzing the chosen aspects and their importance within our proposed context and situation.

3 Multicriteria Decision Analysis Method

Daily we make decisions in our lives; often, they are not noticeable, without intention. Decision-making is related to a problem, and the solution to this problem consists in finding paths and options that arrive at solutions. Complex issues in the decision-making process are usually linked to multicriteria analysis [4]. Still, by the same author [4], the criteria serve as a norm for a judgment and differentiate what is wrong and right, that is, discernment. These criteria specify the conditions that must be taken as a choice to be considered. Obtain a listing of criteria for judging the alternatives provides advantages, namely:

- Assist in the thinking of the entire decision-making process;
- Provide practical alternatives for solving the problem;
- Present clarity in the structure that is interconnected with the evaluation process;
- Helping the decision-making group agree with what they are expecting;
- Making the line of thought you choose presents clarity, precision, and visibility, which will be extremely important in decision-making subject to open scrutiny.

With decision-making, choices arise among the various alternatives for us to consider. The different options resolve the problems in question more consistently or even manage to solve the facts. The other options are a series of choices with exclusivity that serve as a tool to solve the problems. In decision-making, the greater the number of alternatives, the greater the probability and veracity of success in solving problems. According to [1], multicriteria decision analysis is formalized as the exercise of decision subject or decision analysis, which helps to obtain alternative answers to questions during a process. These alternatives have the mission of clarifying each decision. It is based on behavior that increases the coherence between the evaluation of the process, the tasks, the values, and the exercise whose judgment the analyst seeks to position.

Over a long time, at least four phases are identified, namely:

- Summarize the content of the process;
- Modeling in the form of an explanation of behavioral and functional characteristics;
- Choose the alternatives;
- Review the content of the facts.

Continuing with [1] in multicriteria decisions, there is the so-called Hierarchical Analysis Method, also called the AHP method (Analytic Hierarchy Process). In this methodology, the problem is broken down into several steps in a hierarchical way in a facilitating and comprehensive manner. This method summarizes the decision maker's values in a generalized form for each alternative, selecting and ordering at the end of information processing. In the AHP method, we will present elements that are of fundamental importance, namely:

- Attributes and properties: the alternatives have a comparison concerning properties;
- Binary correlation: it is the comparison of two elements having a preference for one component to another;
- Fundamental scale: association of a priority value that has a scale reading of positive numbers;
- Hierarchy: separation of sets of elements in the hierarchy.

There are two ways to compare the process of calculating the AHP method:

- Absolute comparison: the alternatives are compared with a pre-established standardization over a certain period. This whole scale is used to rank alternatives concerning the criteria. Right after this step, it compares the alternatives to place in the order of each one. Finally, the sum of values is performed, and there will be a new scale for the alternatives soon after they are normalized.
- Comparative in a relative way: comparing alternatives in a way that each one has in joint with another.

In [1], according to the AHP method, the attributions in the criterion weights will be determined by questions addressing the most relevant criteria to be examined and categorized according to their significance level. The decision analyst will respond to this classification with a numerical value that positions the action. This methodology uses a scale parameter from 1 (one) to 9 (nine) as an alternative scale so that the AHP method is based on the ratio scale as the beginning. This methodology will only be used in criteria that have their importance assigned in a ratio scale. In the comparison process of the AHP method, we identified that the most important alternative is always used as an integer value of the scale and of the most minor importance as the inverse of this unit.

4 Multicriteria Method Applied to Hiring a Logistics Manager

Undoubtedly, the labor market has undergone profound changes recently, and it is necessary to adapt the Human Resources systems of the various institutions. That said, some questions challenge these systems, such as:

- What is the qualification level of the institution's employees?
- Are there any employees who have the potential for promotion?
- How to reduce subjectivity in personnel selection processes?

In this sense, professional certification can be a good strategy to answer these questions and support decision-making within the universe of Human Resources.

One of the objectives of this publication is to present a model and application of the AHP Method to support the certification process that can be used in practice. Therefore, it is essential to recognize some concepts that have already been considered [5]:

- Professional certification is "the formal recognition of the worker's knowledge, skills, attitudes and competences, required by the productive system and defined in terms of previously agreed standards or norms, regardless of how they were acquired";
- It is an "effective resource for organizing the labor market and promoting productivity";
- *"Certification has been identified as an adjustment tool to a flexible form of production (capable of adapting to frequent changes in demand)."*
- Public and private institutions are mobilizing in the search for professional certification. Several institutions have their certification processes:

a) The Federal Council of Administration (CFA - association that regulates the profession of administrator in Brazil) launched the Professional Certification Program in Administration of the CFA/CRA System, as can be seen on the institution's website [6];
b) The Brazilian Association of Financial and Capital Market Entities (AMBIMA) has specific certification programs for investment professionals [7];
c) The Brazilian Association of Financial and Capital Market Entities (AMBIMA) has specific certification programs for investment professionals [7];
d) In Brazil, the Ministry of Education and the Ministry of Labor and Employment established the "National Professional Certification Network (CERTIFIC Network), created in 2009, it is intended to serve workers who seek formal recognition of knowledge, knowledge and professional skills developed in formal and non-formal learning processes and life and work, through professional certification processes" [8];

The proposal of professional certification is reinforced with the recognition that the activity of managing is associated with practical application and not with the simple accumulation of knowledge of a professional, as mentioned below [9]:

"Unlike many other occupations, which bring together in their knowledge the use of a specific technique restricted only to members of their class, Management contemplates, in the exercise of its activity, the application of practical knowledge, which corresponds to a performance rather than a robust body of knowledge."

4.1 Construction of the Decision-Making Model

With all the examples and concepts above, the definition of the certification model for management professionals in a road freight transport company started with a brainstorming session and the design of a cognitive map presented in the Initial Cognitive Map (Fig. 1).

The Initial Cognitive Map summarizes the authors' knowledge on relevant topics and areas of expertise, considering the target audience of the certification. This target audience can be occupants of positions or candidates for leadership positions within the road freight transport segment selected for this study. Some necessary actions for evaluating personnel qualifications were also raised in this map.

Note the importance of knowledge of the fundamentals of quality management, processes, and maintenance, always focusing on cost reduction.

As previously mentioned, cost reduction as an objective is justified due to the high investment in trucks and the consequent need to occupy idle production capacity.

To optimize resources in a company, the human resources team uses many assessment tools in the processes of recruitment: curriculum evaluation, interview (structured and unstructured), social media consultation, knowledge tests, group dynamics, and psychological tests [10].

Considering the costs that could be involved in the creation and implementation of a complete model and creating a simpler model that summarizes the most relevant dimensions, the criteria below were selected to assess the competencies that may be required for the manager position that represents the knowledge, skills, and attitudes [11–13]:

- Psychological Capacity: the best decisions for resource allocation must be taken by leadership positions even in times of pressure to obtain results;
- Theoretical knowledge: knowledge of management tools and topics related to the sector allows growth and improvement of the company's processes;
- Practical Ability: reflects the effective Ability to solve problems, transform concepts and implement management tools to obtain results.

With the Initial Cognitive Map and the concepts presented above, it was possible to arrive at the map of the model to be developed (see Fig. 2).

Following the AHP Method, the next step is comparing the pair-by-pair criteria and their contribution to the professional classification presented in Table 1 according to the authors' judgment [14, 15].

Evaluation of Criteria and Model Balancing. From the Criteria Comparison Matrix, the Weight Vector [0.790; 0.320; 1.946] was calculated, and the Consistency Vector [3.033; 3.011; 3.072], according to the AHP Method, reaching a result of λ_{max} equal to 3.039. Subsequently, the Consistency Index (CI), equal to 0.019, and the Random Index (RI), equivalent to 0.58, were obtained. With these data, the Consistency Ratio (RC) was 0.033, less than 0.10, and we can conclude that the relative priority values are consistent and can be used for decision making.

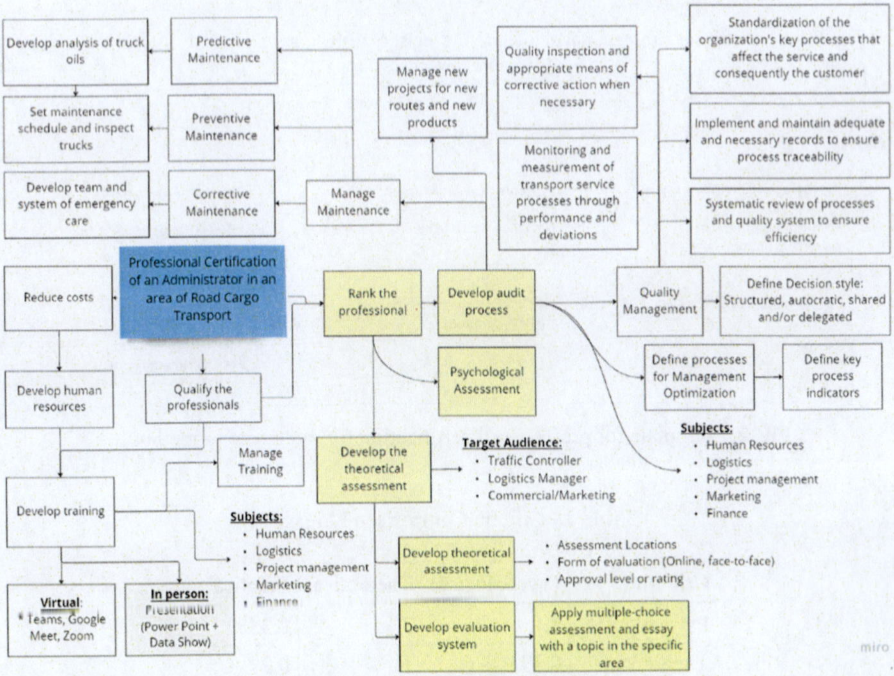

Fig. 1. Initial Cognitive Map

4.2 Model Experimentation

To try out the model, we used the LinkedIn website to obtain profiles of professionals working in the logistics field randomly. The objective here was to simulate a certification process that could be used to "rank" candidates aiming, for example, to fill the position of Logistics Manager in a company in the road freight transport sector.

The above criteria were used with a simplified detailing considering the experimentation conditions summarized in [18]. The performance time in leadership positions was supposed to measure the Psychological Criterion because it was not possible to apply psychological tests that would give greater accuracy. Yet, this alternative was recommended since professionals in leadership roles are subject to pressure, and the to remain in these positions illustrates their medium- and long-term resiliency. For the Theoretical Criterion, it was considered that the completion of academic stages measures the professional's content in the training area and, therefore, would have the knowledge of the tools the market has already developed.

And for the Practical Criterion, working in the logistics department represents the knowledge for operational achievements in the segment (Table 2).

As the situation created is hypothetical, the following conditions were used in the selection of LinkedIn profiles:

- Belong to or work in the logistics department for at least one year;
- Have consistent academic training information containing the names of schools and courses with completion dates and.

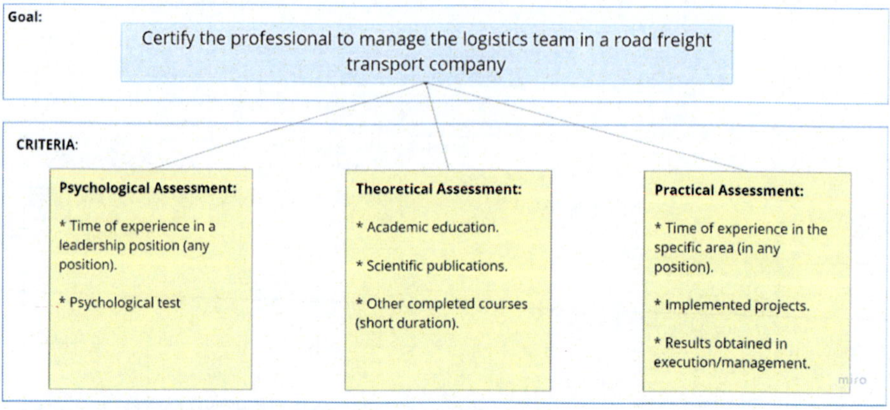

Fig. 2. Template for professional certification for multicriteria method.

Table 1. Criteria Comparison Matrix.

Criterion	Psychological	Theoretical	Practical
Psychological	1,	3,	0.333
Theoretical	0.333	1,	0.2
Practical	3,	5,	1,
Total	4.333	9,	1,533

Source: Authors' calculations

Table 2. LinkedIn profile evaluation criteria.

Criteria	Assessment
Psychological	Time spent in leadership positions in any sector of activity
Theoretical	Academic background and courses are taken in the field of Logistics
Practical	Time of professional experience in the field of Logistics

Source: Authors' definition

- Have information consistent with professional experience (start and end dates and company names with some description of responsibilities).
- With these characteristics, four profiles were randomly selected below:
- Candidate 1: Graduated in Logistics with an MBA, eight years in Logistics, all in leadership positions.
- Candidate 2: Graduated in Logistics and Economic Sciences, 31 years of experience in the area, and 13 in leadership positions.
- Candidate 3: Graduated in IT, Polytechnic course in Mechanical Maintenance, and MBA in Transport Management, 20 years of experience in the area, 14 of which in a leadership position.

- Candidate 4: Graduated in Business Administration with two MBAs in Logistics Production and Business Management, 15 years of experience in the area, with two leadership positions.

By observing the profiles, Candidate two stands out for their experience and Candidate 4 for their academic background. To avoid subjectivities and elaborate a fair score, the times of experience in the sector and leadership positions are in Table 3.

With these times, it is possible to normalize the results and calculate the relative position (rank) for the Practical criteria (associated with the time of experience in the logistics department) and Psychological (linked to the time of experience in leadership positions) of each candidate, rounding the result concerning the candidate who has less time.

Table 3. Survey of candidate experience times

Candidate	Experience in the area	Experience as a Leader
1	8.3 years	8.3 years
2	31.3 years	13.1 years
3	20.3 years	14.7 years
4	15.0 years	2.4 years
Total	74.9 years	38.5 years

Source: Authors' calculations

For the Theoretical Criterion, the authors established the weights of each course level taken in Logistics Management and the like according to Table 4. These weights produced the scores of each candidate were also normalized, generating a comparison relative to the smallest value obtained. From the rankings calculated above, comparisons and analyses between pairs fed the Comparison Matrices for each Criterion.

Similarly to the sequence of steps for the Theoretical Criterion, the Table 5 presents the values for the Psychological Criterion. As a final calculation step, the Preferences Matrix was multiplied by the Preferences Vector, obtaining the result shown in Table 7. In this table, the best-ranked professional is Candidate 2, with the most professional experience (as Practical Criterion) and appropriate time in leadership positions (as Psychological Criterion), these two criteria being those with the most significant weight in the proposed model (Table 6).

Table 4. Weights for academic training

Completed course	Weight
Technician or Technologist	1
Graduation	2
MBA	3
Master's or Doctorate	4

Source: Authors' definition

Table 5. Comparison Matrix - Psychological Criterion

Candidate	1	2	3	4
1	1,	0.6	0.5	3,
2	1,667	1,	0.833	5,
3	two,	1.2	1,	6,
4	0.333	0.2	0.167	1,

Source: Authors' calculations

Table 6. Preference Matrix

Candidate	Psychological	Theoretical	Practical
1	0.2	0.273	0.111
2	0.333	0.091	0.444
3	0.4	0.273	0.222
4	0.067	0.364	0.222

Source: Authors' calculations

Table 7. Final ranking result by the AHP Method

Candidate	Result
1	0.151
2	0.378
3	0.274
4	0.197

Source: Authors' calculations

5 Final Considerations

Considering the above, the proposed model is replicable for situations of certification and/or assessment of the qualification level of personnel in a company in the logistics sector. It is also possible to assume the application of this model in other sectors of the economy with the due balancing of weights according to the company/sector and the eventual inclusion of new criteria that were not addressed here [16].

This model proposal minimizes the subjectivity of the judgments by discussing the personnel evaluation criteria. This model does not end the discussion on evaluating performance and potential in human resources. Still, it indicates a path for future studies by adopting other criteria that would make the model more complete. For example, performance in an interview is a criterion that could be added. It is important to note that the model does not guarantee that the best-scoring candidate is the most suitable for the company [17]. Still, it does indicate future discussion of the selection and weighting of the adopted criteria.

During the elaboration of this work, the authors identified the possibility of using the model to select personnel considering the different demands of a company's strategic, tactical, or operational levels. This perception opens the discussion for new studies in that, for example, a ranked professional with good academic background can be indicated for more strategic work, planning, or product development. On the other hand, another example is hiring a professional with more experience who could be hired for critical short-term situations, which may be tactical and/or operational, even if this professional does not have an excellent academic background. For this application, it is necessary to reassess the weights applied in the Criteria Comparison Matrix (Table 1). This model also makes it possible to obtain the ranking of logistics professionals using only curriculum data and create a basis for companies to assess their positioning concerning the market, bringing insights into the development of human resources. Finally, with the new IT tools (Python computer language, for example), it is possible to propose the automation of this model for the recruitment of professionals from data from professional social networks such as LinkedIn, where the curriculum can be transformed and ranked in a relative way to support the selection processes in human resources such as the development and promotion of employees.

Acknowledgments. The fourth author thanks the National Council for Technological and Scientific Development (CNPq) through grant # 04272/2020–5. The authors thank the Edson Queiroz Foundation/University of Fortaleza for all the support provided.

References

1. Gomes, L.F.A.M., Gomes, C.F.S.: Princípios e métodos para tomada de decisão - Enfoque Multicritério. Editora Atlas, São Paulo (2019)
2. GOMES, L.F.A.M.: Tomada de decisões em cenários complexos: introdução aos métodos discretos do apoio multicritério à decisão. Cengage Learning, São Paulo (2011)
3. Gil, A.C.: Como elaborar projetos de pesquisa. Editora Atlas, São Paulo (2008)
4. Jones, D.: Tomada de decisão para leigos. Alta Books, Rio de Janeiro, RJ (2015)

5. Alexim, J.C., Lopes, C.L.E.: A Certificação Profissional Revisitada. Boletim Técnico do SENAC. **29**, 2–15 (2003)
6. Conselho Federal de Administração: Programa de Certificação CFA | FGV. https://certpessoas.fgv.br/cfa/. Accessed 06 Feb 2023
7. AMBIMA - Associação Brasileira das Entidades dos Mercados Financeiro e de Capitais: Certificação. https://www.anbima.com.br/pt_br/educar/. Accessed 06 Feb 2023
8. Ministério da Educação, B.: Rede Certific. http://portal.mec.gov.br/rede-certific. Accessed 13 Feb 2023
9. Cicmanec, E.R., Nogueira, E.E. da S.: O Corpo de Conhecimentos da Profissão do Administrador no Brasil: contribuições do Sistema CFA/CRAs para sua Legitimação. Revista Eletrônica de Ciência Administrativa. **17**, 9–34 (2018). https://doi.org/10.21529/recadm.2018001
10. Santos, F.P.R., Lima, M.C., Almeida, N.C.: Os instrumentos de avaliação no recrutamento e seleção de pessoas. http://ric.cps.sp.gov.br/handle/123456789/7017. Accessed 08 Feb 2023
11. Chiavenato, I.: Planejamento, Recrutamento e Seleção de Pessoal - Como Agregar Talentos à Empresa. Grupo GEN, São Paulo (2021)
12. de Oliveira Barreto, T.B., Pinheiro, P.R., Silva, C.F.G.: The multicriteria model support to decision in the evaluation of service quality in customer service. In: Silhavy, R. (ed.) CSOC2018 2018. AISC, vol. 763, pp. 158–167. Springer, Cham (2019). https://doi.org/10.1007/978-3-319-91186-1_17
13. de Lima Silva Junior, R.F., Pinheiro, P.R.: A multicriteria structured model to assist the course offering assertiveness. In: Silhavy, R. (ed.) CSOC2018 2018. AISC, vol. 763, pp. 464–473. Springer, Cham (2019). https://doi.org/10.1007/978-3-319-91186-1_48
14. De Andrade, S.J.M., et al.: Prioritising maintenance work orders in a thermal power plant: a multicriteria model application. Sustainability **15**, 54–73 (2023)
15. Capistrano, J.R., Pinheiro, P.R., da Silva Júnior, J.G., de Sousa Nogueira, P.: Structuring A Multicriteria Model to Optimize the Profitability of a Clinical Laboratory. In: Silhavy, R., Silhavy, P., Prokopova, Z. (eds.) CoMeSySo 2019 2019. AISC, vol. 1047, pp. 41–51. Springer, Cham (2019). https://doi.org/10.1007/978-3-030-31362-3_6
16. Leite, G.S., Albuquerque, A.B., Pinheiro, P.R.: A multi-criteria model application in the prioritization of processes for automation in the scenario of intelligence and investigation units. In: Silhavy, R., Silhavy, P., Prokopova, Z. (eds.) Software Engineering Perspectives in Intelligent Systems: Proceedings of 4th Computational Methods in Systems and Software 2020, Vol.1, pp. 947–965. Springer International Publishing, Cham (2020). https://doi.org/10.1007/978-3-030-63322-6_82
17. Filho, E.G., Pinheiro, P.R., Pinheiro, M.C.D., Nunes, L.C., Gomes, L.B.G., Farias, P.P.M.: Support to Early Diagnosis of Gestational Diabetes Aided by Bayesian Networks. In: Silhavy, R. (ed.) CSOC 2019. AISC, vol. 985, pp. 360–369. Springer, Cham (2019). https://doi.org/10.1007/978-3-030-19810-7_36
18. Tamanini, I., Carvalho, A.L., Castro, A.K., Pinheiro, P.R.: A Novel Multicriteria Model Applied to Cashew Chestnut Industrialization Process. In: Mehnen, J., Köppen, M., Saad, A., Tiwari, A. (eds.) Applications of Soft Computing, pp. 243–252. Springer Berlin Heidelberg, Berlin, Heidelberg (2009). https://doi.org/10.1007/978-3-540-89619-7_24

Evaluation of Artificial Intelligence-Based Models for the Diagnosis of Chronic Diseases

Abu Tareq, Abdullah Al Mahfug, Mohammad Imtiaz Faisal, Tanvir Al Mahmud, Riasat Khan(✉), and Sifat Momen

Department of Electrical and Computer Engineering, North South University, Bashundhara, Dhaka 1229, Bangladesh
{abu.tareq,abdullah.mahfug,imtiaz.faisal03,tanvir.mahmud15, riasat.khan,sifat.momen}@northsouth.edu

Abstract. Recent advancements in computer vision and artificial intelligence have facilitated improved healthcare diagnosis, which plays a critical role in saving lives, detecting life-threatening diseases, and prolonging life expectancy. This study focuses on the prediction of six common life-threatening conditions, namely diabetes, coronavirus, viral pneumonia, cardiac disorder, liver, and chronic kidney diseases. Supervised machine learning and deep learning techniques were utilized to accomplish the task. The automatic disease prediction system has subsequently been implemented as a website and an Android smartphone application. In the final model that has been deployed, pathology reports and/or chest X-ray images are taken as inputs, followed by models to predict whether an individual suffers from the diseases listed above. Random forest outperformed other models and achieved accuracies of 85%, 97%, 75%, and 85% for diabetes, cardiac disorder, liver, and chronic kidney diseases, respectively. Covid-19 and pneumonia were identified using chest X-ray images. An attention-based customized CNN model has been employed to accomplish the task with a validation accuracy of 92%. The F1-macro of the attention-based CNN model is found to be 92.67% with individual F1 scores of predicting covid-19, viral pneumonia and normal healthy cases of 91%, 97% and 90%, respectively. Finally, LIME and Grad-CAM-based explainable AI techniques have been used to interpret the prediction made by the black box models.

Keywords: Attention Module · Convolutional Neural Network · COVID-19 · Disease Prediction · Explainable AI · Machine Learning

1 Introduction

According to the World Health Organization (WHO), the healthcare sector in the least developed countries (LDCs) suffers from a shortage of physicians and nurses, as well as administrative issues such as absenteeism, inefficiency, and

R. Silhavy and P. Silhavy (Eds.): CSOC 2023, LNNS 724, pp. 597–626, 2023.
https://doi.org/10.1007/978-3-031-35314-7_52

poor management [3]. Heart disease, stroke, diabetes, and other chronic non-communicable diseases are anticipated to account for half of all deaths worldwide [16]. As per a contemporary study by the Centers for Disease Control and Prevention (CDC) [13], over 35 million people in the United States have been affected by chronic kidney failure and almost half of these are not even diagnosed. Recent public health research finds a high prevalence of non-alcoholic fatty liver diseases amongst the general population in developed countries. This has been attributed mainly to the unconstrained food habits in the masses [8]. The recent coronavirus pandemic has resulted in respiratory problems and pneumonia and has taken millions of lives worldwide [2]. In most countries, coronary heart disease is the second leading cause of mortality, followed by diabetes in fifth place and liver and chronic kidney diseases in eighth and ninth place, respectively. Diagnosing these disorders in advance will facilitate effective therapy and lessen their resulting impacts.

Early detection of complex life-threatening diseases can lead to a lower mortality rate, enhance the quality of life and lessen the financial hardship of the mass people. However, ineffective healthcare sectors in developing countries make it more challenging to do so. This complex scenario has inspired us to create a data-driven solution to predict chronic diseases based on user inputs. Such a system is effective in assisting healthcare professionals in making a quick diagnosis of diseases. This is particularly useful in developing countries where there exist shortages of nurses and health professionals. These diseases have been considered in this work because they have a severe impact worldwide, are responsible for approximately 75% of deaths, and have been investigated in recent articles to make preliminary reference judgments for health professionals.

As the pathogeny of chronic disorders is incredibly complex, it is challenging to diagnose a broad spectrum of diseases for clinicians. With the availability of medical data in the public domain and the parallel growth of computer vision and artificial intelligence techniques, it is now possible to build a predictive model with a low error margin for diagnosing diseases.

Researchers employed a wide variety of deep learning approaches for automatic disease prediction. For instance, Wang et al. [41] proposed a procedure that systematically predicts multiple disease detection by manipulating patients' diagnostic medical information to assess future disease risks. Men et al. [23] proposed an approach that performs multi-disease prediction using deep learning frameworks. They utilized the long-short-term-memory network (LSTM) and broadened it to two independent appliances, time-aware and attention-based frameworks. Ampavathi and his team [4] focused on a system that deploys multi-disease prediction using a deep learning approach. They used various datasets related to lung cancer, hepatitis, Alzheimer's disease, liver tumor, diabetes, Parkinson's, and cardiac illness. The proposed automatic disease detection system is divided into three parts: normalization of data, prediction and weighted normalized feature extraction. The Jaya method-based multiverse optimization algorithm (JAMVO) was used to optimize the weighting function by combining two metaheuristic algorithms. Haque and colleagues [17] utilized a CNN-based approach for predicting COVID-19 disease. The proposed model accomplished 95.34% and 97.56% accuracy and precision, respectively. Dubey [11] proposed a technique to predict multiple diseases using deep learning. The neurons in the hidden layer count of DBN and NN are carefully tuned using the same technique in both prediction systems.

Many researchers have used machine learning techniques for different disease predictions. Pranto et al., for instance, [29] trained machine learning algorithms on the PIMA Indian Diabetes dataset to create models which were then tested on a local dataset (obtained from a local medical center of Dhaka, Bangladesh) that the authors have collected. The highest validation accuracy and F1-score of 81.2% and 88.0% were achieved using the KNN classifier. It was later extended to investigate the impact of cost-sensitive learning and oversampling techniques on the reliability of the models [30]. Empirical results show the best accuracy of 81% achieved by employing cost-sensitive learning and synthetic minority oversampling techniques. Yaganteeswarudu [43] presented a systematic view that could hold multiple disease predictions in a single platform using Flask API and machine learning techniques. This work attained the best prediction results using the logistic regression model with 0.92 and 0.95 accuracies for cardiac disease and cancer detection, respectively. Jackins et al. [20] used three datasets, NIDDk, Framingham heart study, and Wisconsin breast cancer dataset, to detect heart disease, cancer, and diabetes using the random forest and naive bayes algorithms. Accuracy of the naive bayes algorithm accuracy was 82.35% for coronary heart disease, 74.46% for diabetes, and 63.74% for cancer. In contrast, the accuracy of the random forest model was 83.85%, 92.40%, and 74.03% for coronary heart disease, cancer data, and diabetes, respectively. For all three diseases, the random forest model surpassed the naive bayes method in terms of accuracy. In [10], the researchers utilized the R statistical software along with logistic regression and naive bayes techniques for the early identification of heart disease. Empirical results show the highest accuracy 91.61% using logistic regression. Arumugam and his coauthors [6] employed multiple machine learning techniques to predict heart disease in diabetic patients. They used the naive bayes, SVM, and decision tree models for prediction purposes. The decision tree model gave the best accuracy with 90.10% of the three models. Harimoorthy and Thangavelu [18] proposed a technique to predict diabetes, CKD, and heart disease. Among all the machine learning algorithms, the SVM-radial bias kernel method gives the best accuracy in the CKD, diabetes, and heart disease datasets with 98.3%, 98.7%, and 89.9% precision, respectively. Mohit and his colleagues [24] created a disease prediction web application that can predict the diseases such as diabetes, cardiac disease, and breast cancer. The application uses machine learning algorithms to predict the conditions. Logistic regression gave the best prediction for diabetes and breast cancer with 77.60% and 94.55% accuracies, and KNN for heart disease prediction with 83.84% accuracy. Tasin and his team [40] implemented semi-supervised machine learning techniques for diabetes prediction. The authors employed a small-scale, privately curated dataset of which insulin feature was estimated from pseudo-labeling of the open-source Pima Indian dataset. The authors demonstrated that the pseudo-labeling applied by various machine learning techniques outperformed the statistical imputation techniques. The XGBoost approach with synthetic oversampling accomplished 0.81 accuracy for the merged dataset.

Many of the previously carried out works did not address some key issues: (1) We have noticed that a large amount of work in the literature did not consider class imbalance issues which may contribute to model overfitting, (2) Most of the work were limited to the use of the black-box model in making the prediction. However, this is less trustworthy as it is unclear how these models are making the predictions, and (3) Most of the models in the literature have not been deployed and hence cannot be used in real-time by the end users.

This work focuses on using machine learning and deep learning techniques from users' data to predict whether the user is infected with a specific disease. The contributions of this work are depicted below:

- Machine learning and deep learning models are deployed on a website and as an Android application that will assist end users in quickly checking whether a patient has one of the diseases.
- Machine learning models have been created to predict various life-threatening diseases, i.e., COVID-19, viral pneumonia, diabetes, heart, liver and kidney diseases.
- An attention-based CNN model from the chest X-ray image dataset has been used to predict COVID-19 and pneumonia.
- SMOTE technique has been used to overcome class imbalance issues. Hyper-parameter tuning has been used to determine the optimal SMOTE ratio and the best set of parameters for the used classifiers.
- Explainable AI libraries LIME and Grad-CAM were used to interpret the detection results and determine the main contributing factors that influence the prediction.

The rest of the paper is structured as follows: The methodology of the work has been presented in Sect. 2. Results and discussion of the proposed multi-disease prediction system are given in Sect. 3. Finally, we conclude the paper in Sect. 4.

2 Methodology

This article aims to build reliable predictive models for detecting diabetes, COVID-19, cardiac, liver and chronic kidney diseases.

Figure 1 shows the overview of the system. Datasets were obtained from credible open-source repositories, which were then preprocessed and fed to the training step. After a series of experiments, the final chosen model is uploaded to a server. In the case of the web application, after the profile of a patient is given, the server is communicated with the patient profile. The model is then used to generate a result which is provided back to the web application.

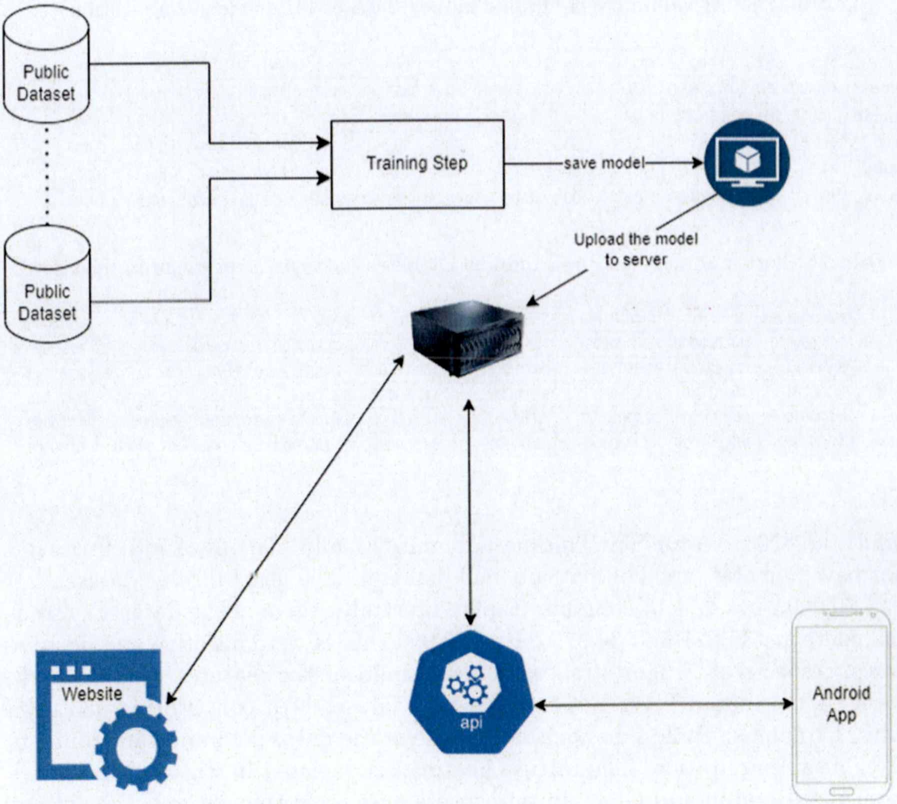

Fig. 1. Overview of various steps of the proposed disease prediction system.

2.1 Dataset and Preprocessing Data

In this work, we have used public datasets collected from various reliable sources on the internet. The chronic kidney disease dataset [39] comprises 400 rows and 26 columns. In Table 1, we can observe the primary features of the used CKD dataset after mean imputation. Figure 2 shows the heatmap of these features. The dataset required preprocessing as it consisted of missing values. For example, 152 entries from the red blood cells (RBC) column, 130 entries from the red blood cell count (RC) column, and 105 from the white blood cell count (WC) were missing. We dropped the id column as it had nothing to do with the prediction. We filled the missing entries with the mean values and converted the categorical values to numerical values. 250 of 400 rows had harmful chronic kidney disease, but 150 had positive kidney disease. Next, for diabetes detection, we used the Pima Indians Diabetes dataset. It contains a total of 2,768 rows and 9 columns. Table 2 demonstrates some of the dataset's essential attributes after mean imputation. Figure 3 shows the heatmap of the features. In this database, 1,330 entries were missing from the Insulin column, 1,816 from the outcome col-

Table 1. Short summary of chronic kidney dataset (after mean imputation).

	id	age	bp	sg	al	su	bgr	bu	sc	sod	pot	hemo	classification
count	400.00	391.00	388.00	353.00	354.00	351.00	356.00	381.00	383.00	313.00	312.00	348.00	400.00
mean	199.50	51.48	76.47	1.02	1.02	0.45	148.04	57.43	3.07	137.53	4.63	12.53	0.38
std	115.61	17.17	13.68	0.01	1.35	1.10	79.28	50.50	5.74	10.41	3.19	2.91	0.48
min	0.00	2.00	50.00	1.01	0.00	0.00	22.00	1.50	0.40	4.50	2.50	3.10	0.00
max	399.00	90.00	180.00	1.03	5.00	5.00	490.00	391.00	76.00	163.00	47.00	17.80	1.00

Table 2. Short summary of Pima Indians Diabetes dataset (after mean imputation).

	Pregnancies	Glucose	BloodPressure	SkinThickness	Insulin	BMI	DPF	Age	Outcome
count	765.000000	765.000000	765.000000	765.000000	765.000000	765.000000	765.000000	765.000000	765.000000
mean	3.840523	121.425490	72.350327	29.076471	159.045098	32.483529	0.471570	33.223529	0.347712
std	3.370364	30.481869	12.225928	9.903924	111.597578	6.892014	0.331823	11.767311	0.476556
min	0.000000	44.000000	24.000000	7.000000	14.000000	18.200000	0.078000	21.000000	0.000000
max	17.000000	199.000000	122.000000	99.000000	846.000000	67.100000	2.420000	81.000000	1.000000

umn, and 800 from the SkinThickness column. Finally, 500 out of 768 entries did not have diabetes, and 268 patients had diabetes. The heart disease dataset [21] had 70,000 rows and 13 columns. Table 3 illustrates some of the dataset's essential contents. 35,021 and 34,979 out of 70,000 entries were negative and positive cases, respectively. Figure 4 shows the heatmap of the features. In this work, we used the Indian Liver Patients Records dataset [27], constituting 583 rows and 11 columns. Table 4 demonstrates some of the dataset's important contents after mean imputation. The feature heatmap is depicted in Fig. 5. Four entries from the Albumin and Globulin ratios were missing in this dataset. We did not have to drop any columns, as features were not correlated. The dataset had 416 entries for the liver patient and 167 for the non-liver patient.

For COVID-19, we used COVID-19 Radiography Database [32]. It contains 6,012 X-ray images of lung opacity, 1,345 of viral pneumonia, 3,616 of COVID-19, and 10,192 of regular patients. Since this dataset is imbalanced, we take 1,345 X-ray images for each class. We divide the entire dataset into training and validation with an 85:15 ratio. The models are trained in batches of 32. Three subcategories are created from each training and validation dataset, i.e., 'COVID,' 'Normal,' and 'Viral Pneumonia,' containing various chest X-ray pictures. All photographs are transformed to 256×256 pixels to simultaneously retain unanimity and image quality. Next, the chest X-ray images are shuffled and converted to RGB color space. Figure 6 shows sample images from the dataset of three categories. All the basic steps of supervised machine learning are summarized in a block diagram in Fig. 7.

2.2 Splitting Data

The data has been divided into two parts using a stratified holdout split approach. 85% of the data were kept for training purposes. The remaining 15% was used for testing.

Table 3. Short summary of heart disease dataset (after mean imputation).

	id	age	gender	height	weight	ap_hi	ap_lo	cholesterol	glucose	smoke	alco	active	cardio
count	70000.00	70000.00	70000.00	700000.00	70000.00	70000.00	70000.00	70000.00	70000.00	70000.00	70000.00	70000.00	70000.00
mean	49972.42	19468.87	1.35	164.36	74.21	128.82	96.63	1.37	1.23	0.09	0.05	0.80	0.50
std	28851.30	2467.25	0.48	8.21	14.40	154.01	188.47	0.68	0.57	0.28	0.23	0.40	0.50
min	0.00	10798.00	1.00	55.00	10.00	-150.00	-70.00	1.00	1.00	0.00	0.00	0.00	0.00
max	99999.00	23713.00	2.00	250.00	200.00	16020.00	11000.00	3.00	3.00	1.00	1.00	1.00	1.00

Table 4. Short summary of Indian Liver Patients Records dataset (after mean imputation).

	Age	Total_Bilirubin	Direct_Bilirubin	Alkaline Phosphotase	Alamine Aminotransferase	Aspartate Aminotransferase	Total Proteins	Albubin	Albumin and_Globulin_Ratio	Dataset
count	583.00	583.00	583.00	583.00	583.00	583.00	583.00	583.00	579.00	583.00
mean	44.75	3.30	1.49	290.58	80.71	109.91	6.48	3.14	0.95	1.29
std	16.19	6.21	2.81	242.94	182.62	288.91	1.09	0.80	0.32	0.45
min	4.00	0.40	0.10	63.00	10.00	10.00	2.70	0.90	0.30	1.00
max	90.00	75.00	19.70	2110.00	2000.00	4929.00	9.60	5.50	2.80	2.00

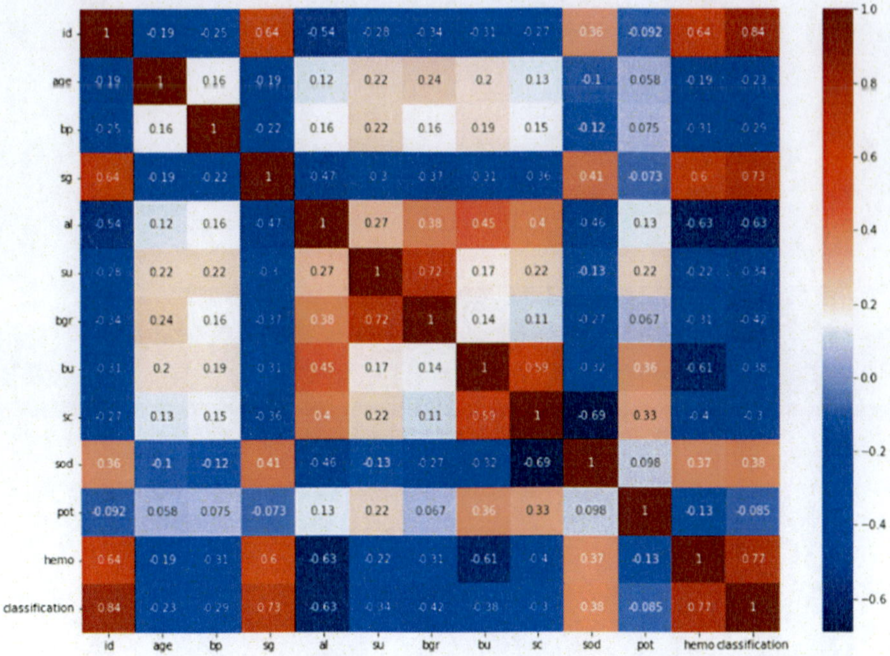

Fig. 2. Heatmap of chronic kidney dataset.

2.3 Machine Learning Algorithms

In this paper, several machine learning techniques have been used for predicting diabetes, cardiac and kidney diseases, which have been briefly described below.

- Logistic regression: It is a classification algorithm that works on probability theory to predict the dependent variable [14]. This machine learning model

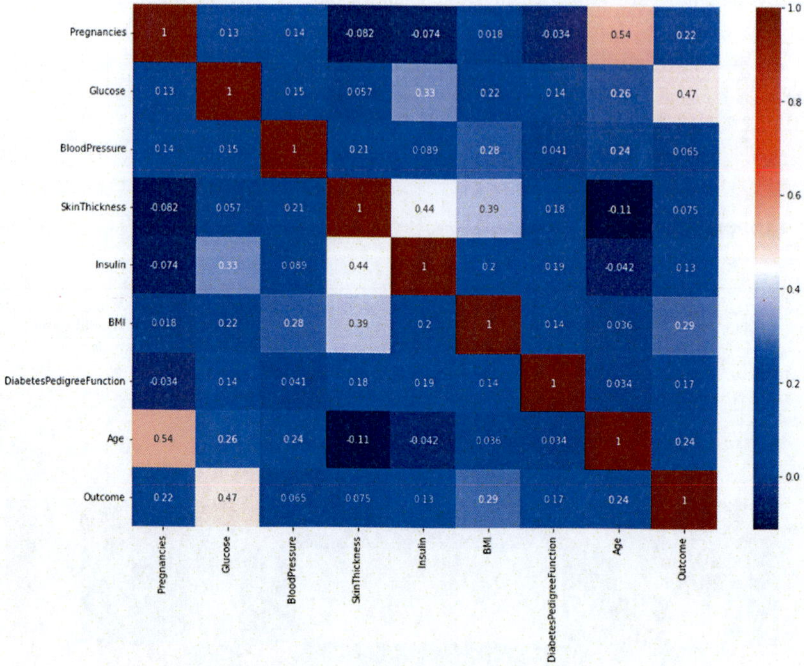

Fig. 3. Heatmap of Pima Indians Diabetes dataset.

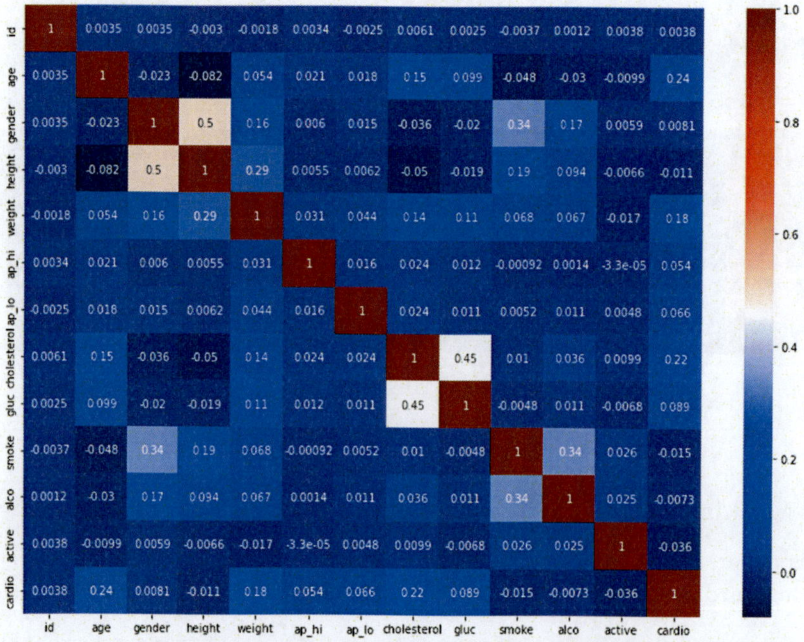

Fig. 4. Heatmap of heart disease dataset.

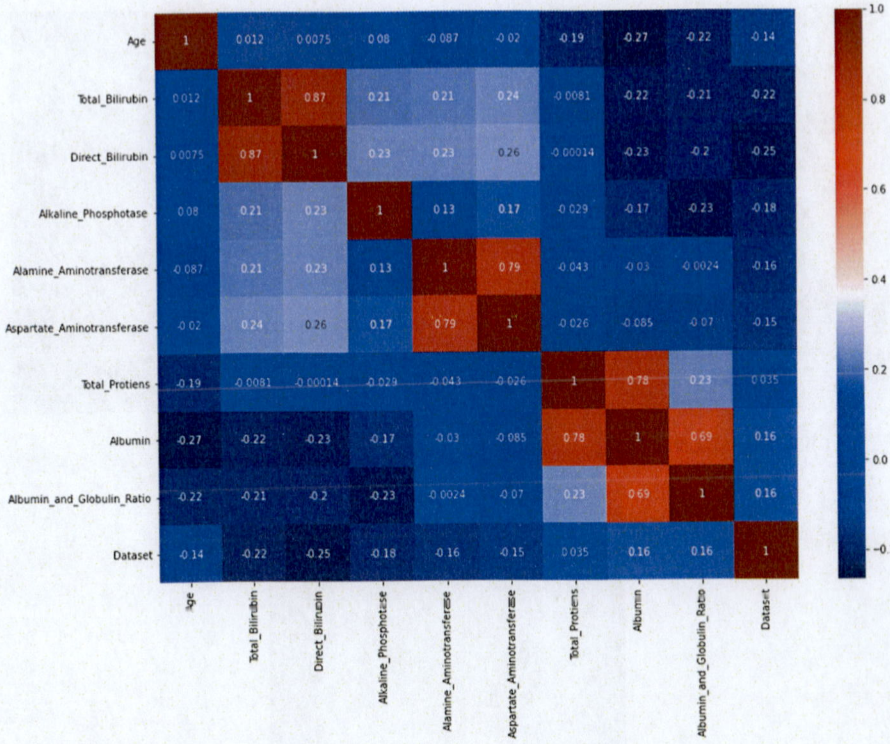

Fig. 5. Heatmap of Indian liver patients records dataset.

uses an S-shaped logistic function that is applied to the linear regression model to produce the prediction.

- SVM: SVM is a unique algorithm that works on classification problems in machine learning. It works well on practical issues and can solve linear or nonlinear problems. This algorithm plots a hyperplane for each attribute, and quality is represented as a coordinate in the dataset [14]. Hyperplane divides the class from one to another, and the main reason is that we will set the latest data to the best category in the future.
- KNN: KNN is a supervised machine learning algorithm that classifies a record based on the votes provided by the record's nearest neighbors that need to be classified.
- Decision tree: A robust classification technique in machine learning is the decision tree. To determine the best attribute and new branch node, this model uses the feature selection coefficients and recursive partitioning technique [14].

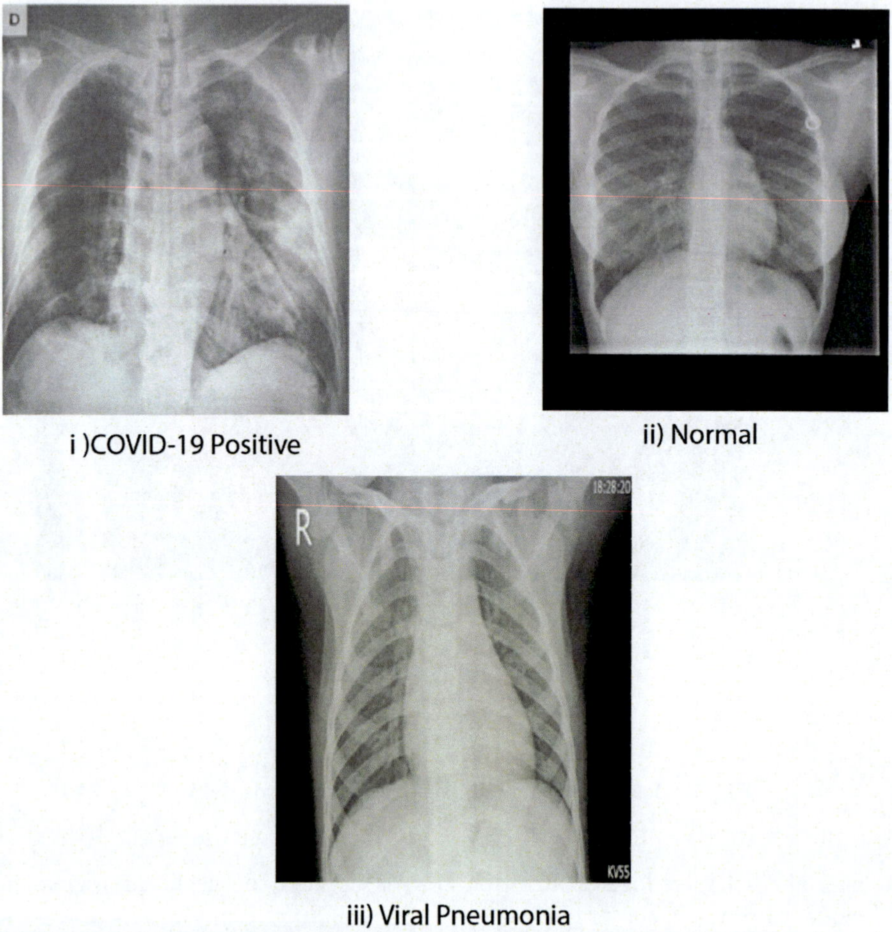

Fig. 6. Sample images from COVID-19 dataset: i) COVID-19 Positive, ii) Normal and iii) Viral Pneumonia.

- Random forest: Random forest is an ensemble machine learning algorithm that uses many decision trees, row, and feature samplings to create the model.
- Extra trees classifier: This ML model is similar to a random forest classifier, but the decision trees are built differently. It can contain randomization while still optimizing, and this technique utilizes averaging to increase predictive accuracy and avoid overfitting problems [14]. This algorithm has various advantages, including a short execution time and a low computational cost.

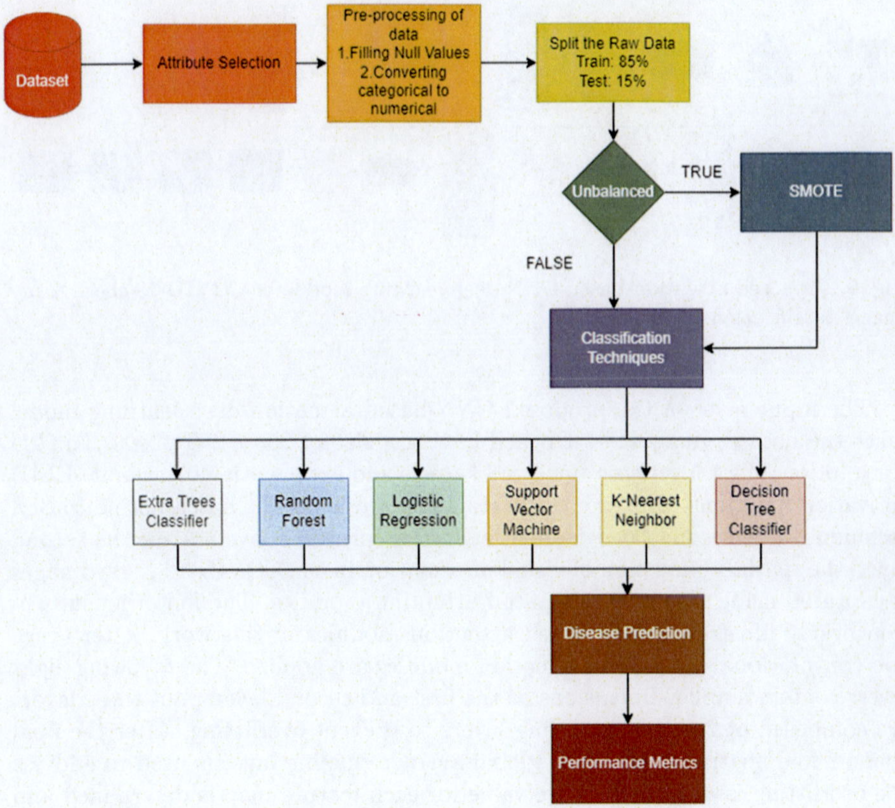

Fig. 7. Methodology of the supervised machine learning techniques.

2.4 Attention Based CNN Model

This work uses a CNN-based attention deep learning model for coronavirus and pneumonia detection from chest X-ray photographs. CNN has been extensively applied in classifying images, particularly in healthcare and disease detection. The proposed COVID-19 detection model is a custom CNN model based on channel and spatial attention, with ten convolutional layers, four max-pooling layers, and one fully connected layer. Figure 8 depicts a general block diagram for the proposed CNN-based attention deep learning model. This model was trained on a total of 1,345 images comprising chest X-rays of healthy, COVID-affected and pneumonia-affected individuals.

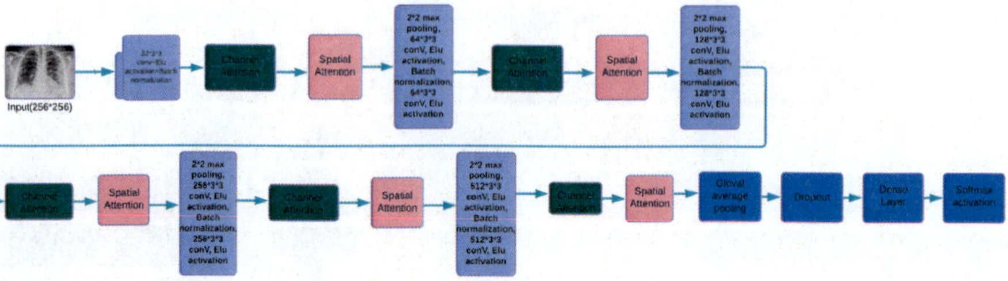

Fig. 8. Proposed attention-based CNN deep learning model for COVID-19 chest X-ray image classification.

The input layer of the proposed CNN-based attention deep learning model takes the dataset images – with each having a size of 256×256 pixels. The 2D convolutional first layer contains 3×3 kernels and exponential linear unit (ELU) activation functions. The inputs are standardized using the batch normalization technique. The second layer is identical to the previous layer. After the second layer, the spatial connection of visual information in the COVID-19 X-ray images is captured using channel and spatial attention modules. The following subsection briefly illustrates the utilized attention modules in this work. After every two convolutional layers, the attention modules are applied. The following eight layers contain similar dimensions as the first and second layers, but these layers are composed of 2×2 max-pooling layers to prevent overfitting. After the final channel and spatial attention, a global average pooling layer is used to address the overfitting issue. The average value for each feature map is determined and a dropout layer of 0.50 is used. Finally, a dense layer receives the input data and identifies the class the image belongs to. Since the dataset contains three categories, softmax activation is applied to determine the corresponding classes.

2.5 Attention Module

The channel attention and spatial attention frameworks proposed by Woo et al. [42] have been applied in this study. Figure 9 shows a schematic diagram of the employed attention modules. The channel attention accepts normalized convolution layers tensors. The outputs of the channel attention module are then passed to the spatial attention module, which forwards them to a convolution layer. This approach is repeated twice on every convolution layer.

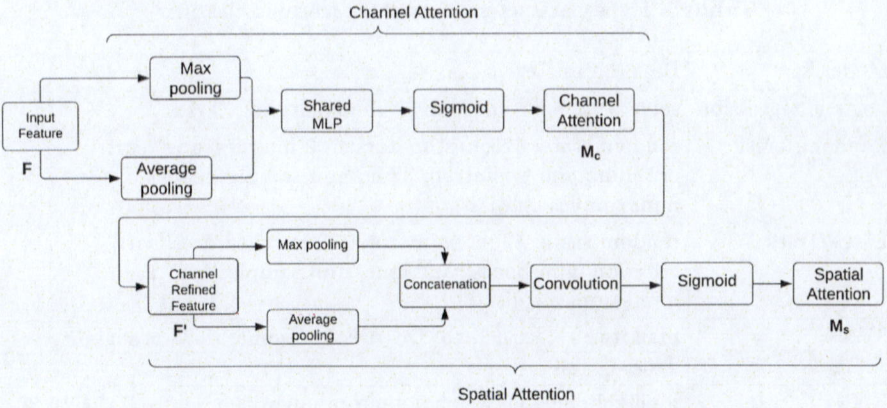

Fig. 9. Block diagram of the employed attention module.

Channel Attention Module. We conduct max and average pooling on the input tensor from a normalized convolution layer in the channel attention module. Following that, max pooling and average pooling are sent to a shared multilayer perceptron (MLP) network. The resultant feature vectors are then combined using element-wise summation. The channel attention is determined using Eq. 1.

$$M_c(F) = \sigma(W_1(W_0(F^c_{avg})) + W_1(W_0(F^c_{max}))) \quad (1)$$

Here, σ denotes the sigmoid function, W_1 and W_0 denote the parameter matrices for MLP and average pooling networks, respectively.

Spatial Attention Module. The input tensor from the channel attention module is subjected to max pooling and average pooling networks in the spatial attention module. These resulting tensors are concatenated to construct a convolution of filter size (f) of 7×7 using the sigmoid function (σ). The output tensor is given in Eq. 2.

$$M_s(F) = \sigma(f^{7\times7}([F^c_{avg}; F^c_{max}])), \quad (2)$$

2.6 Hyperparameter Tuning

A hyperparameter is a parameter that is defined before the start of the learning process. Before the learning algorithm starts training the model, the ML user selects and sets the value of various hyperparameters. These values are adjustable and directly impact the model performance [28]. The hyperparameters used in this study include the number of epochs, the number of branches in a decision tree, learning rates, the number of layers, etc. Hyperparameters with their corresponding best values obtained from the GridSearchCV technique for various machine learning models applied in this work are listed in Table 5.

Table 5. Hyperparameters of machine learning classifier

Model	Best Parameters
Logistic regression	solver: liblinear, max_iter: 100 penalty: l2, C: 1.0
Random forest	random_state: 42, n_estimators: 100 max_features: sqrt, criterion: gini bootstrap: True, min_samples_leaf: 1 min_samples_split: 2
Extra Trees	random_state: 42, n_estimators: 28 max_features: sqrt, criterion: gini bootstrap: True, min_samples_leaf: 1 min_samples_split: 2
SVM	max_iter: -1, kernel: rbf Degree: 3, gamma: scale probability: True, C: 1.0
KNN	n_neighbors: 100, weights: uniform algorithm: auto, leaf_size: 30
Decision tree	random_state: 42, max_depth: 16 max_features: 8, criterion: gini splitter: best, min_samples_leaf: 1 min_samples_split: 2

2.7 Over Sampling Technique

Unbalanced datasets are common in practice, and they influence the performance of machine learning algorithms. In this work, we address the problem of data imbalance by using SMOTE technique [28]. SMOTE does not duplicate data points but rather creates synthetic ones that deviate slightly from the original data points. The SMOTE function has different hyperparameters which can be tweaked to obtain the optimum parameters that result in the best performances, SMOTE ratio and k neighbors value, i.e., the number of nearest neighbors utilized in the creation of synthetic samples.

2.8 Explainable AI

This work uses the explainable AI framework to interpret the pattern identification and final disease prediction results.

- LIME: Local interpretable model agnostic explanation (LIME) is a technique that can explain any ML model's prediction [28]. LIME creates a simpler model around the data points that helps to explain the prediction made by the machine learning model.
- Grad-CAM: Selvaraju et al. [35] used gradient-weighted class activation mapping (Grad-CAM) to highlight the salient regions in the image. For each class designation, Grad-CAM creates a heatmap visualization that can be used to confirm where the CNN is working in the image visually.

3 Results and Discussion

The performances of implemented machine learning models have been measured using various metrics, accuracy, precision, recall, F1 score, and area under the ROC curve.

Table 6. Performance evaluation of various algorithms for diabetes prediction

Classifier	Sampling technique	Accuracy	Precision	Recall	F1 Score	AUC
Logistic regression	Without SMOTE	0.76	0.79	0.70	0.71	0.845
	With SMOTE	0.76	0.76	0.70	0.71	0.847
Random Forest	Without SMOTE	**0.80**	**0.80**	**0.76**	**0.77**	**0.855**
	With SMOTE	**0.85**	**0.84**	**0.84**	**0.84**	**0.868**
Extra Trees	Without SMOTE	0.77	0.77	0.73	0.74	0.853
	With SMOTE	0.84	0.84	0.83	0.84	0.864
SVM	Without SMOTE	0.74	0.76	0.68	0.69	0.817
	With SMOTE	0.80	0.79	0.81	0.79	0.850
KNN	Without SMOTE	0.74	0.77	0.67	0.68	0.838
	With SMOTE	0.74	0.77	0.67	0.68	0.825
decision tree	Without SMOTE	0.68	0.66	0.64	0.64	0.637
	With SMOTE	0.80	0.81	0.77	0.78	0.763

Table 7. Performance evaluation of various algorithms for liver disease prediction

Classifier	Sampling technique	Accuracy	Precision	Recall	F1 Score	AUC
Logistic regression	Without SMOTE	0.71	0.62	0.55	0.53	0.747
	With SMOTE	0.71	0.44	0.48	0.43	0.748
Random Forest	Without SMOTE	0.68	0.59	0.56	0.57	0.690
	With SMOTE	**0.75**	**0.71**	**0.67**	**0.69**	**0.763**
Extra Trees	Without SMOTE	**0.72**	**0.65**	**0.61**	**0.62**	**0.733**
	With SMOTE	0.74	0.70	0.68	0.68	0.748
SVM	Without SMOTE	0.71	0.36	0.50	0.42	0.715
	With SMOTE	0.71	0.35	0.50	0.41	0.725
KNN	Without SMOTE	0.70	0.48	0.50	0.43	0.690
	With SMOTE	0.70	0.60	0.51	0.44	0.690
decision tree	Without SMOTE	0.65	0.59	0.59	0.59	0.592
	With SMOTE	0.70	0.66	0.66	0.66	0.649

Table 8. Performance evaluation of various algorithms for cardiac disease detection

Classifier	Sampling technique	Accuracy	Precision	Recall	F1 Score	AUC
Logistic regression	Without SMOTE	0.64	0.64	0.64	0.64	0.696
Random Forest	Without SMOTE	**0.72**	**0.72**	**0.72**	**0.72**	**0.778**
Extra Trees	Without SMOTE	0.71	0.71	0.71	0.71	0.762
SVM	Without SMOTE	0.64	0.64	0.63	0.63	0.696
KNN	Without SMOTE	0.63	0.63	0.63	0.63	0.692
decision tree	Without SMOTE	0.69	0.69	0.69	0.69	0.692

Tables 6, 7, 8 and 9 show the performance of different ML classifiers for detecting diabetes, liver, heart and kidney disease predictions. For diabetes prediction, random forest surpasses the other classifiers. Without SMOTE, this classifier

Table 9. Performance evaluation of various algorithms for CKD prediction

Classifier	Sampling technique	Accuracy	Precision	Recall	F1 Score	AUC
Logistic regression	Without SMOTE	0.91	0.91	0.93	0.91	0.981
Random Forest	Without SMOTE	**0.97**	**0.97**	**0.98**	**0.97**	**0.993**
Extra Trees	Without SMOTE	0.95	0.95	0.95	0.95	0.993
SVM	Without SMOTE	0.89	0.88	0.90	0.89	0.977
KNN	Without SMOTE	0.93	0.92	0.93	0.92	0.959
decision tree	Without SMOTE	0.90	0.90	0.89	0.90	0.892

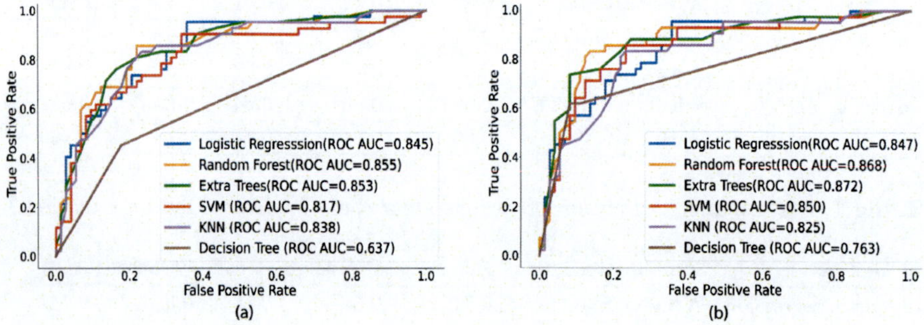

Fig. 10. ROC curves of six algorithms for diabetes disease. (a) Without SMOTE and (b) With SMOTE.

achieved an accuracy of 80%, and after using SMOTE, accuracy increased to 85%. For liver disease, the extra trees classifier achieved an accuracy of 72% before using SMOTE and after applying SMOTE, the random forest classifier's accuracy increased to 75%, the highest among the classifiers.

As the heart and kidney disease datasets are balanced, there is no need to use SMOTE in this scenario. The best performance is accomplished by the random forest classifier, which is 72% accuracy and 0.778 AUC for heart disease prediction. We also see the same pattern for the kidney dataset as the dataset was balanced. With a 97% accuracy and 0.993 AUC, the random forest classifier surpassed the other ML classifiers.

The ROC curve (true positive rate vs. false positive rate) summarizes a classifier's ability to distinguish across different classes. The ROCs of logistic regression, random forest, extra trees, decision tree, SVM, and KNN algorithms for diabetes, liver, heart, and kidney disease are presented in Fig. 10, Fig. 11, Fig. 12, and Fig. 13, respectively. From the ROC curve, we can observe why our proposed model is more competent in making predictions than other classifiers.

Figure 14 shows radar plots that compare the performance of six different machine learning algorithms in diagnosing of four diseases. Results indicate that

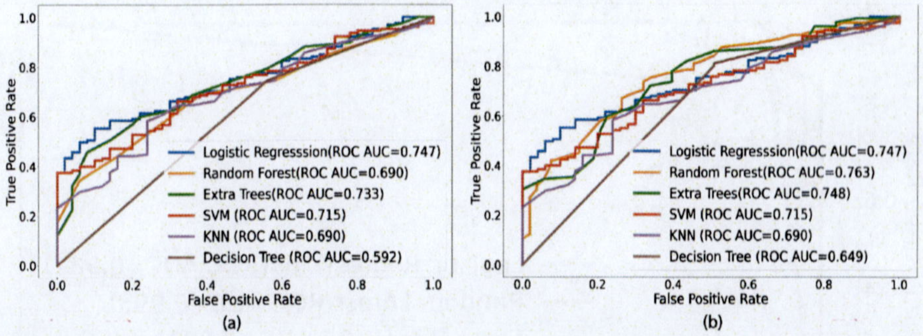

Fig. 11. ROC curves of six algorithms for liver disease. (a) Without SMOTE and (b) With SMOTE.

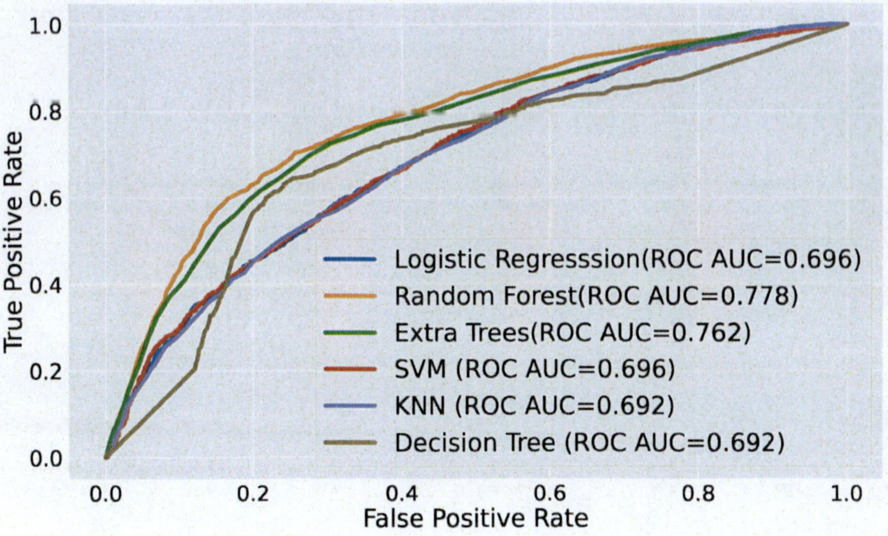

Fig. 12. ROC curves of six algorithms for heart disease without SMOTE.

cardiac and diabetes can be diagnosed with relatively high performance (i.e. precision, recall, F1 score, AUC-ROC and accuracy). However, detecting chronic kidney disease can be predicted with good but relatively lower performance compared to the other three diseases.

In Table 10, we analyze the performance of our proposed ML models with other existing works for diabetes, liver, heart, and kidney disease prediction in terms of the employed dataset and classifier, accuracy and various metrics. The proposed random forest model performance is better than other contemporary proposed models concerning the classification accuracy and F1 score.

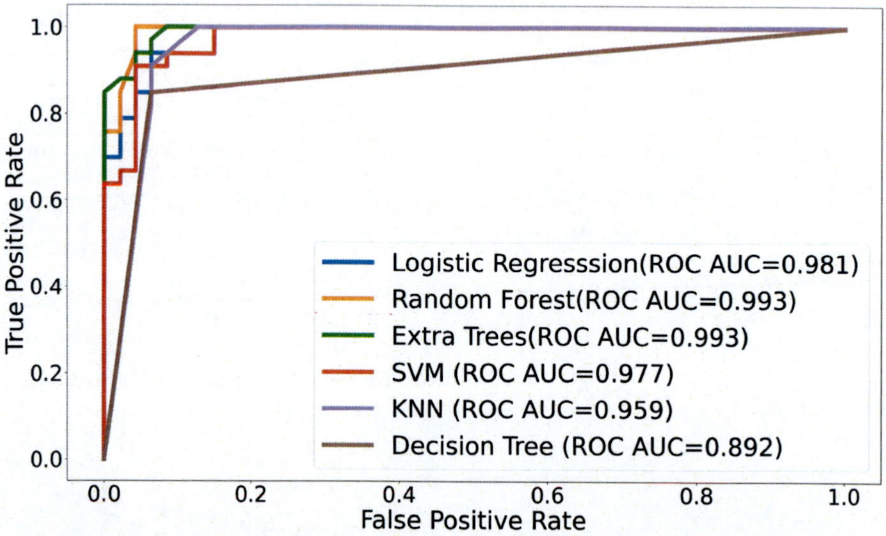

Fig. 13. ROC curves of six algorithms for kidney disease without SMOTE.

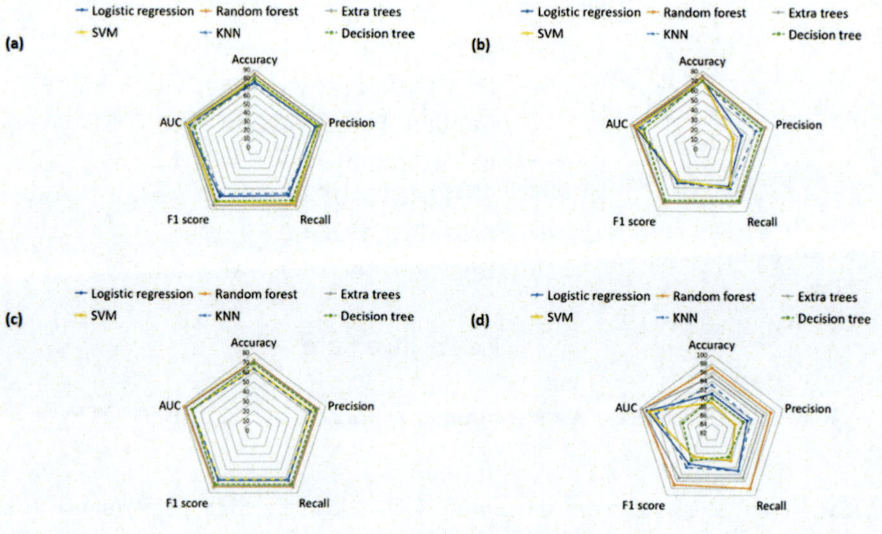

Fig. 14. Comparison of performance metrics of six algorithms for diagnosing (a) diabetes, (b) liver, (c) cardiac and (d) kidney diseases.

For COVID-19 detection, an attention-based CNN model has been used. This model was trained for 18 epochs and in batches of 32. The proposed attention-based CNN model's training and validation accuracies and losses with the change of corresponding epochs are displayed in Fig. 15.

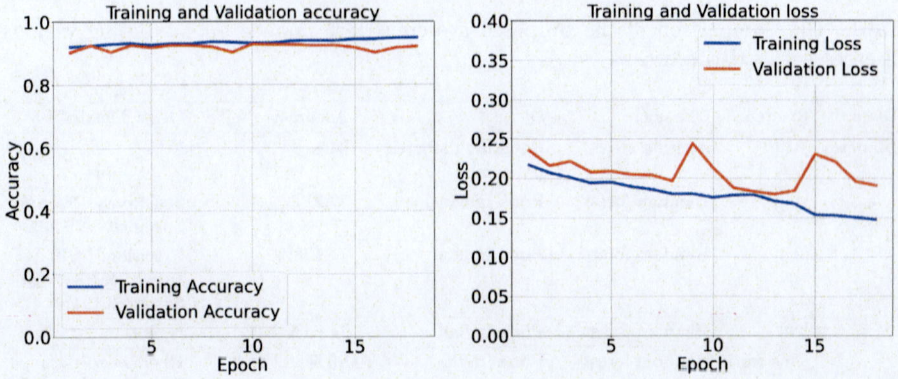

Fig. 15. Training and validation accuracies and losses vs. epochs for the attention-based CNN model.

Figure 15 reveals the validation accuracy of percent, with a loss of 0.1906. The training accuracy rate is 94.86% and a loss of 0.1476.

Table 11 provides the class-wise F1 score, recall, and precision score of the COVID-19 dataset. The F1 score, recall, and precision in each class are greater than or equal to 90%. The best result has been obtained for the viral pneumonia class.

Table 12 compares our work with others in the context of COVID-19 prediction. The comparison indicates that our work outperformed other notable works in the literature in all performance metrics.

3.1 Explainable AI Interpretations

In this section, we aim to interpret how a random forest model makes a prediction as it achieved the best classification accuracy and F1 score.

For diabetes disease, a random record has been selected from the test data for the purpose of interpretation. Figure 16 depicts an illustration of a patient affected by diabetes and summarizes the patient's medical history, symptoms and the final result. The probability of being diagnosed with diabetes in this situation is 0.96. The orange bars represent the medical history and symptoms with much weight in favor of the prognosis, whereas the blue bars represent data that contradicts it. According to the explanation, glucose, diabetes pedigree function, BMI, age, insulin, and skin thickness are the patient's significant symptoms and medical records that primarily contribute to the prediction. Figure 17 describes the explanation of a non-liver disease-affected patient. It was based on their diagnosis values.

Since we chose a random test case for the explanation, we can observe that the test case was of a negative healthy patient with 87 percent confidence. As the blue bar represents the vital feature, and the orange bar for less critical ones, we can see there are two features with orange marked and the rest of them are

Table 10. Comparison of the proposed system's performance for predicting various diseases with existing works

Disease	Reference	Classifier	Dataset	Accuracy	Other Metrics
Diabetes	[34]	decision tree	Egyptian diabetes patients	84%	N/A
	[44]	Random forest	Pima Indian	77.21%	Sensitivity: 74.58% Specificity: 79.58%
	[25]	Random forest	Pima Indian	71.428%	Precision: 64.102% Recall: 45.454% F1 Score: 53.191%
	[20]	Random forest	Pima Indian	74.03%	N/A
	This work	Random forest	Pima Indian	**85%**	**Precision: 84% Recall: 84% F1 score: 84% ROC AUC: 0.868**
Liver	[19]	PSO	Egyptian NCCVH database	66.4%	N/A
	[9]	Naïve Bayes and FT Tree	WEKA dataset	Naïve Bayes: 75.54% FT Tree: 72.66%	N/A
	[38]	SVM and Backpropagation	(UCI) Machine Learning Repository	SVM: 71% Backpropagation: 73.2%	N/A
	[33]	Modified Rotation Forest	UCI liver dataset and Indian dataset	74.78%	N/A
	This work	Random forest	(UCI) Machine Learning Repository	**75%**	**Precision: 71% Recall: 67% F1 score: 69% ROC AUC: 0.763**
Heart	[26]	Random forest	(UCI) Machine Learning Repository	84.1604%	ROC AUC: 0.9018
	[12]	Decision tree	Hungarian heart disease	67.7%	N/A
	[36]	Random forest	Cleveland Heart Disease dataset	85.81%	N/A
	[20]	Random forest	Framingham heart study	83.85%	N/A
	This work	Random forest	(UCI) Machine Learning Repository	**85%**	**Precision: 72% Recall: 72% F1 score: 72% ROC AUC: 0.778**
Kidney	[31]	RBF	(UCI) Machine Learning Repository	95.84%	Precision: 90% Recall: 91.14% F-measure: 0.9057
	[5]	Decision tree	(UCI) Machine Learning Repository	93%	Precision: 92.5% Recall: 93% F1 score: 92.7%
	[15]	Random forest	(UCI) Machine Learning Repository	94.16%	Precision: 95.12% Recall: 96.29%
	This work	Random forest	(UCI) Machine Learning Repository	**97%**	**Precision: 97% Recall: 98% F1 score: 97% ROC AUC: 0.993**

Table 11. Performance metrics of three classes of the COVID-19 dataset

Class	Precision	Recall	F1 score
COVID-19	0.91	0.91	0.91
Normal	0.90	0.90	0.90
Viral Pneumonia	**0.97**	**0.97**	**0.97**

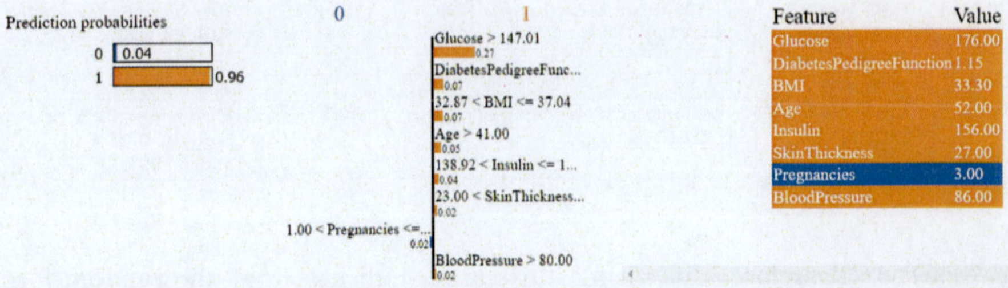

Fig. 16. Explanation of the prediction for diabetes using LIME.

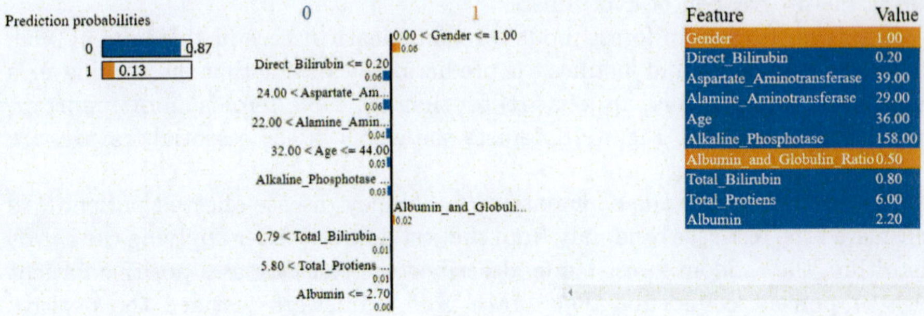

Fig. 17. Explanation of the prediction for liver disease using LIME.

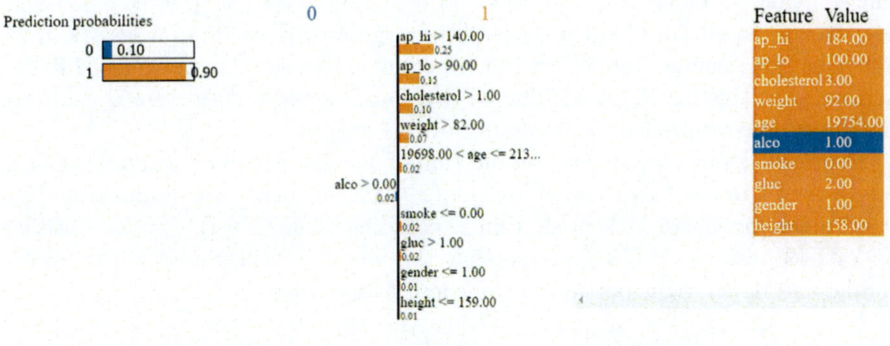

Fig. 18. Explanation of the prediction for heart disease using LIME.

Table 12. Comparison of the proposed system's performance for COVID-19 prediction with existing works

Reference	Techniques	Dataset	Validation Accuracy	Other Metrics
[7]	CNN (ResNet101)	5,982 chest X-ray images (1,765 COVID)	71.9%	Sensitivity: 77.3%, Specificity: 71.8%
[22]	ResNet50	15,478 chest X-ray images (473 COVID)	93%	Sensitivity: 90.1% Specificity: 89.6%
[1]	DeTraC	1,768 chest X-ray images (949 COVID)	93.1%	N/A
[37]	VGG-16 with both attention and convolution module	1,125 chest X-ray images (125 COVID)	79.58%	Precision: 84% Recall: 78% F1 score: 80%
This work	Attention-based CNN model	4,035 chest X-ray images(1345 COVID)	**92%**	**Precision: 92% Recall: 92% F1 score: 92%**

blue marked. Therefore, this visual illustration indicates that the random forest model considers the eight features more important than others. As a result, LIME discovered direct bilirubin, age, total bilirubin, alkaline_phosphatase, alanine_aminotransferase, aspartate_aminotransferase, and total_proteins all had a role in the progression of liver illness.

The applied random forest model predicts heart disease in this patient with 90 percent confidence and justifies the prediction by stating that the ap_hi level is more significant than 140, ap_lo is greater than 90, cholesterol is more important than 1.0, and so forth. Figure 18 depicts the worth of the essential features for the patient.

Figure 19 describes the explanation of a kidney disease-affected patient. We considered the test case randomly from the test dataset. After applying the LIME algorithm, the random forest model described the test case as a positive patient with 97 percent confidence. The LIME XAI framework justified the positive prediction by stating that the hemoglobin is higher than 12.70 and less than or equal to 14.35, hypertension (htn) is less than or equal to 0, diabetes mellitus (dm) is less than or equal to 0, red blood cell count (RC) is greater than four and less than or equal to 5, white blood cell count (WC) is greater 6,525 and less than or equal 79, blood pressure (bp) is greater than 70, and less than or equivalent 80, anemia (ane) is less than or equal 0, and age is in between 52 to 61. According to this figure, blood glucose (bgr) and appetite class were considered less important for predicting the final positive output.

Figure 20 depicts a visual representation of how the proposed attention-based deep learning model takes some portion of a CXR image as more important than other areas for COVID-19 prediction. According to Fig. 20(i), for the positive COVID-19 case, Grad-CAM shows that the model concentrates on the lower-middle and upper-right portion of the chest X-ray image.

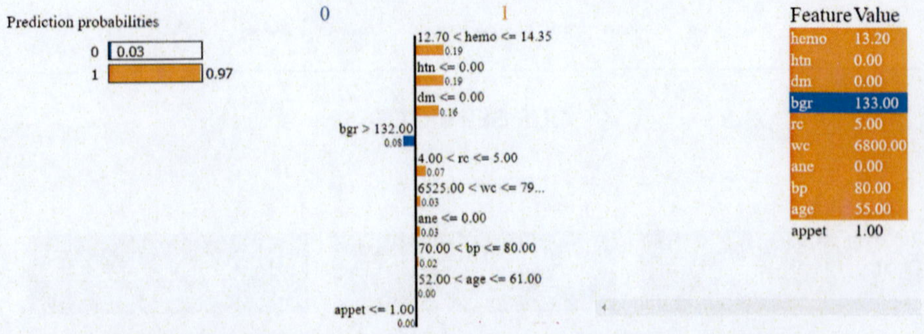

Fig. 19. Explanation of the prediction for kidney disease using LIME.

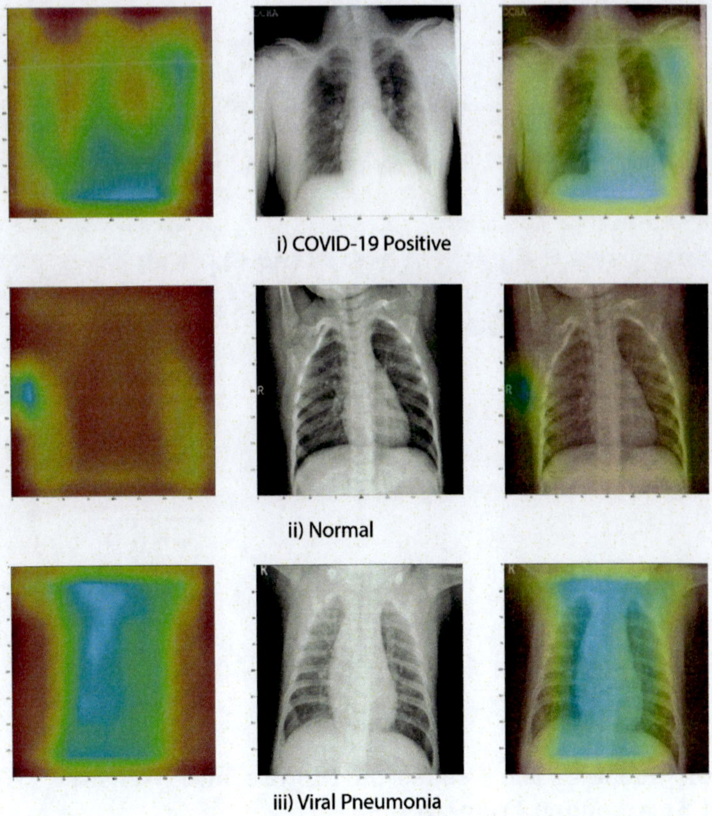

Fig. 20. Grad-CAM visualization for COVID-19 prediction.

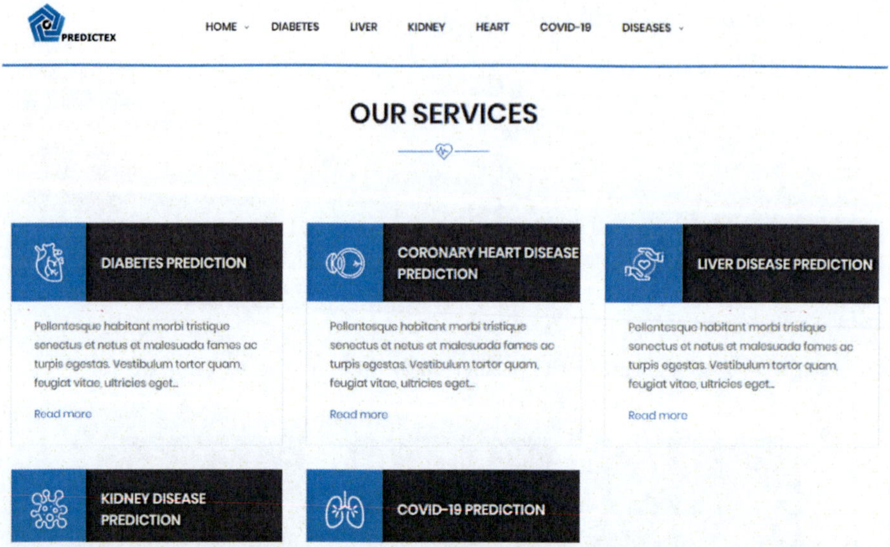

Fig. 21. The system's front-end view.

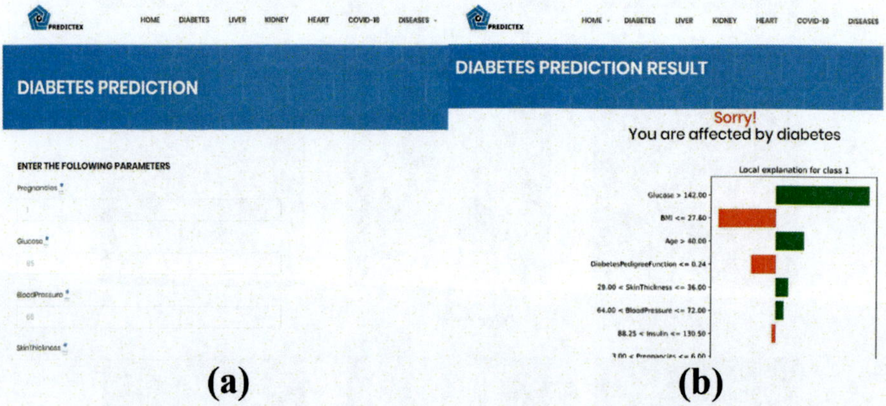

Fig. 22. Disease prediction: (a) Input textual data; (b) Output of the detected diseases

3.2 Deployment of the Prediction Models into Website and Smartphone Frameworks

The user interface of the proposed web application has been created in such a way that the user can utilize it efficiently. The user interface design for the multi-disease prediction system is shown in Fig. 21.

Figure 21 shows the home page of the designed website system, and users have an interface consisting of multiple disease prediction choices. From the

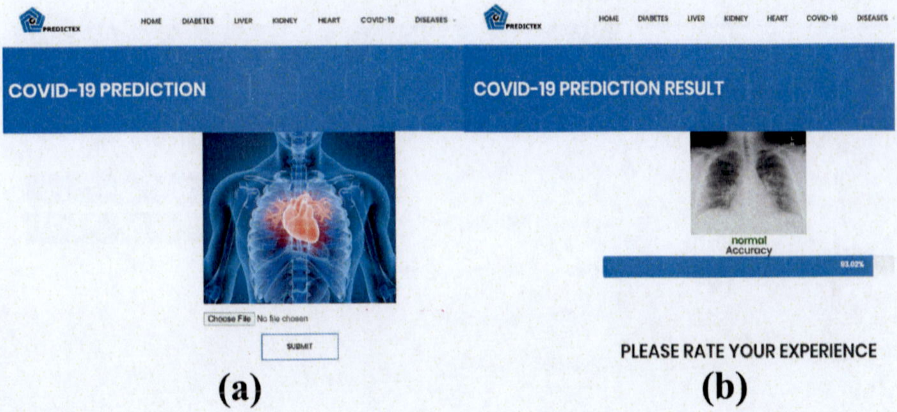

Fig. 23. COVID-19 detection: (a) CXR image upload; (b) Final prediction result.

home screen, users can select predicting various diseases, i.e., diabetes, kidney, liver, coronary heart disease, and COVID-19 or pneumonia.

Figure 22 shows that the website's interface will appear on the home page after selecting a particular disease. For example, after choosing diabetes prediction, this page takes various parameters related to this disease. For diabetes, it takes pregnancies, glucose, blood pressure, skin thickness, etc.

Figure 22 shows that after submitting the expected values, the random forest algorithm processes the values and gives an output whether the user is infected with the disease or not. If the user is infected, the system provides a result with a message: "Sorry! You are affected by Diabetes." It will also show an explanation of prediction using explainable AI.

Users can upload and submit a chest X-ray image for COVID-19, as shown in Fig. 23.

For COVID-19 or pneumonia detection, users first upload and submit a chest X-ray image on the website, as shown in Fig. 23(a). Figure 23(b) illustrates that the imaging system gives the prediction result with the classification accuracy output. Figure 24 shows the designed Android application user interface. The proposed Android application work through an API. When a user uses the designed Android application to request a disease prediction, the request is sent to the central server, which computes the result and returns it as a response. The Android application receives the corresponding response and then displays the final prediction result.

Finally, a comprehensive survey has been performed to rate the designed disease prediction Android smartphone application. A total of 24 participants participated in the survey, of which four were female and others were male. All of the individuals were aged between 18 and 26. The focus group was asked to rate the Android application based on five features on a scale from 0 to 5. Criteria included the productivity of the application, consistency of the results, the engagement level of the application, the design of the application, and the sim-

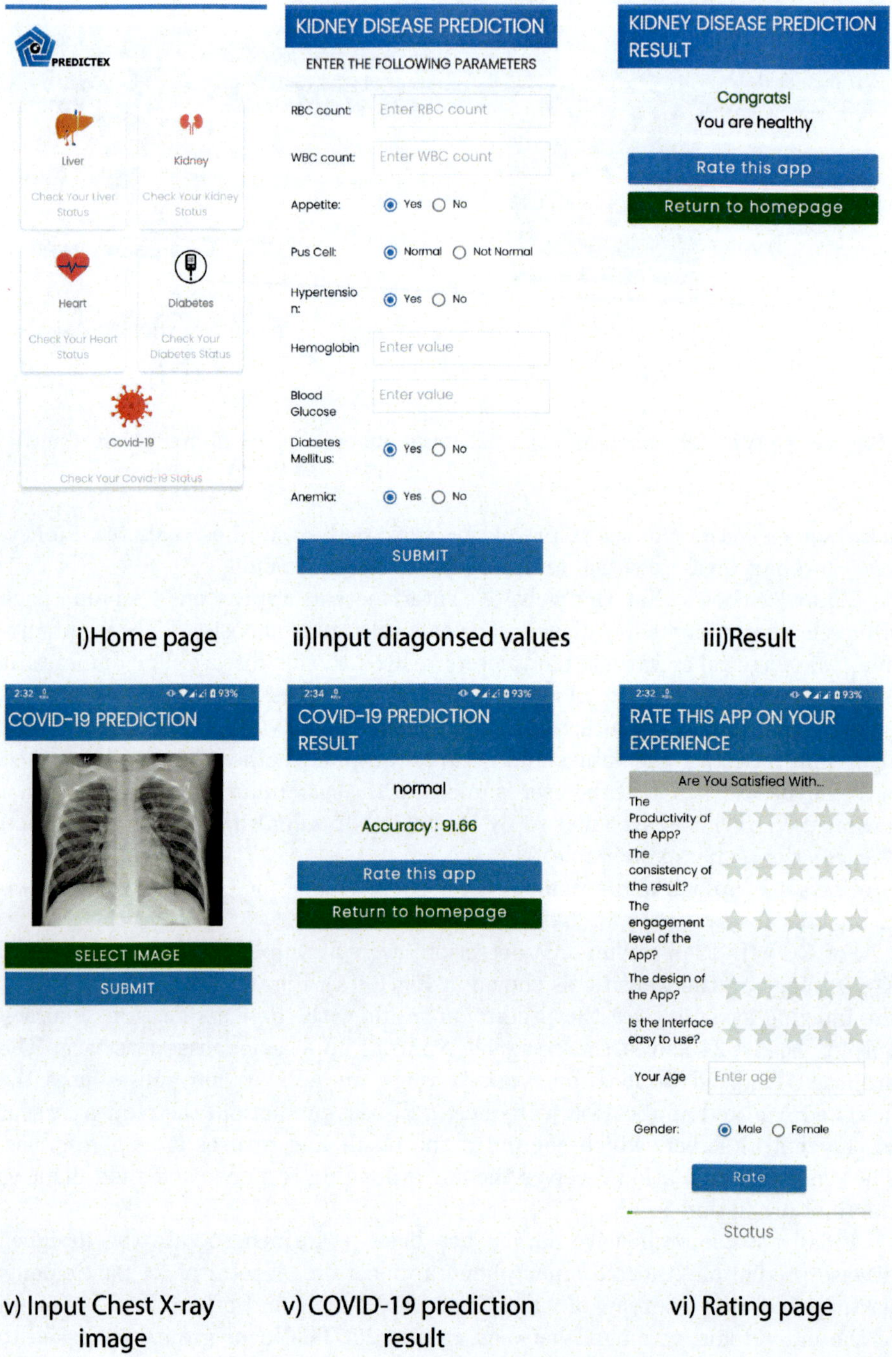

Fig. 24. Android application user interface.

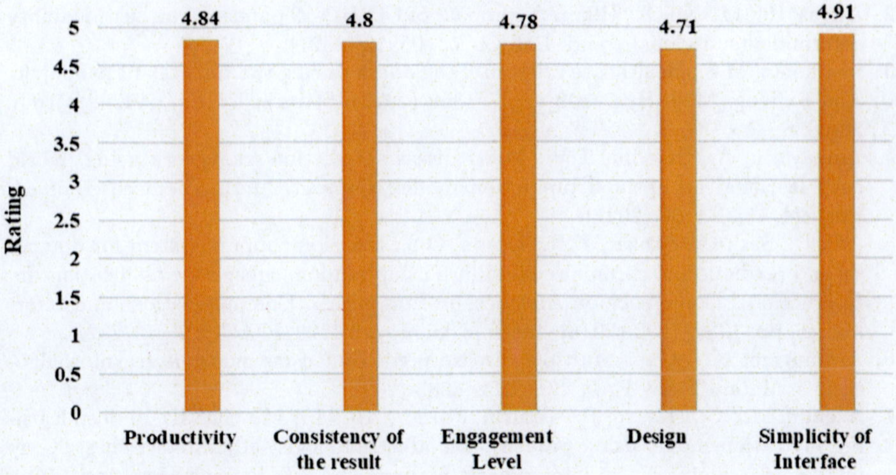

Fig. 25. Android application features review survey results.

plicity of the interface. The findings of the survey are shown in Fig. 25. According to this figure, the average rating of all the features is 4.81, with a maximum and minimum of 4.91 and 4.71, respectively.

4 Conclusion

The prognosis of a wide range of diseases is extremely difficult due to their complex patterns and intricate variations. In this work, an automatic multi-disease prediction system has been developed using different supervised machine learning algorithms for diabetes, cardiac, liver and kidney disorders and a deep learning approach for COVID-19 and pneumonia detection. Among all the ML frameworks, the random forest classifier and the attention-based CNN model offer the most accurate prediction of these diseases. Explainable AI models have been employed to provide interpretability and clarity to disease predictions. Following that, a website and an Android smartphone application have been developed based on the proposed models to forecast whether a person is affected by these diseases or not in real time. In the future, we intend to improve the performance of the deep learning framework the performance using image segmentation and adversarial learning techniques. Advanced meta-heuristic approaches can be utilized for parameter and coefficient tuning of the machine learning models. Furthermore, we plan to improve the existing set of disease datasets and add more medical disorders to the existing pipeline.

References

1. Abbas, A., Abdelsamea, M.M., Gaber, M.M.: Classification of COVID-19 in chest X-ray images using DeTraC deep convolutional neural network. Appl. Intell. **51**, 854–864 (2021)

2. Abbas, H.S.M., Xu, X., Ullah, A.: Impact of COVID-19 pandemic on sustainability determinants: a global trend. Heliyon **7**, e05912 (2021)
3. Al-Zaman, M.S.: Healthcare crisis in Bangladesh during the COVID-19 pandemic. Am. J. Trop. Med. Hyg. **103**, 1357–1359 (2020). https://doi.org/10.4269/ajtmh.20-0826
4. Ampavathi, A., Saradhi, T.V.: Multi disease-prediction framework using hybrid deep learning: an optimal prediction model. Comput. Meth. Biomech. Biomed. Eng. **24**, 1146–1168 (2021)
5. Ani, R., Sasi, G., Sankar, U.R., Deepa, O.S.: Decision support system for diagnosis and prediction of chronic renal failure using random subspace classification. In: International Conference on Advances in Computing, Communications and Informatics, pp. 1287–1292 (2016). https://doi.org/10.1109/ICACCI.2016.7732224
6. Arumugam, K., et al.: Multiple disease prediction using machine learning algorithms. Mater. Today Proc. **80**, 3682–3685 (2021)
7. Azemin, M.Z.C., Hassan, R., Tamrin, M.I.M., Ali, M.A.M.: COVID-19 deep learning prediction model using publicly available radiologist-adjudicated chest X-ray images as training data: preliminary findings. Int. J. Biomed. Imaging **2020**, 8828855 (2020)
8. Cholongitas, E., et al.: Epidemiology of nonalcoholic fatty liver disease in Europe: a systematic review and meta-analysis. Ann. Gastroenterol. **34**, 404–414 (2021)
9. Dhamodharan, S.: Liver disease prediction using Bayesian classification. In: National Conference on Advanced Computing, Application and Technologies (2016)
10. Dinesh, K.G., Arumugaraj, K., Santhosh, K.D., Mareeswari, V.: Prediction of cardiovascular disease using machine learning algorithms. In: International Conference on Current Trends towards Converging Technologies, pp. 1–7 (2018)
11. Dubey, A.K.: Optimized hybrid learning for multi disease prediction enabled by lion with butterfly optimization algorithm. Sādhanā **46**(2), 1–27 (2021). https://doi.org/10.1007/s12046-021-01574-8
12. Ekız, S., Erdoğmu, P.: Comparative study of heart disease classification. In: Electric Electronics, Computer Science, Biomedical Engineerings' Meeting, pp. 1–4 (2017). https://doi.org/10.1109/EBBT.2017.7956761
13. Elshahat, S., Cockwell, P., Maxwell, A.P., Griffin, M., O'Brien, T., O'Neill, C.: The impact of chronic kidney disease on developed countries from a health economics perspective: a systematic scoping review. PLoS ONE **15**, 1–19 (2020)
14. Geron, A.: Hands-On Machine Learning with Scikit-Learn, Keras, and TensorFlow: Concepts, Tools, and Techniques to Build Intelligent Systems, 2nd edn. O'Reilly Media Inc. (2019)
15. Gupta, R., Koli, N., Mahor, N., Tejashri, N.: Performance analysis of machine learning classifier for predicting chronic kidney disease. In: International Conference for Emerging Technology, pp. 1–4 (2020). https://doi.org/10.1109/INCET49848.2020.9154147
16. Hajat, C., Stein, E.: The global burden of multiple chronic conditions: a narrative review. Prev. Med. Rep. **12**, 284–293 (2018)
17. Haque, K.F., Haque, F.F., Gandy, L., Abdelgawad, A.: Automatic detection of COVID-19 from chest X-ray images with convolutional neural networks. In: International Conference on Computing, Electronics & Communications Engineering, pp. 125–130 (2020)
18. Harimoorthy, K., Thangavelu, M.: Multi-disease prediction model using improved SVM-radial bias technique in healthcare monitoring system. J. Ambient. Intell. Humaniz. Comput. **12**, 3715–3723 (2021)

19. Hashem, S., et al.: Comparison of machine learning approaches for prediction of advanced liver fibrosis in chronic hepatitis C patients. IEEE/ACM Trans. Comput. Biol. Bioinf. **15**, 861–868 (2017)
20. Jackins, V., Vimal, S., Kaliappan, M., Lee, M.Y.: AI-based smart prediction of clinical disease using random forest classifier and Naive Bayes. J. Supercomput. **77**, 5198–5219 (2021)
21. Janosi, A., Steinbrunn, W., Pfisterer, M., Detrano, R.: Heart Disease. UCI Machine Learning Repository (1988)
22. Lin, T.C., Lee, H.C.: COVID-19 chest radiography images analysis based on integration of image preprocess, guided grad-CAM, machine learning and risk management. In: International Conference on Medical and Health Informatics, pp. 281–288 (2020)
23. Men, L., Ilk, N., Tang, X., Liu, Y.: Multi-disease prediction using LSTM recurrent neural networks. Exp. Syst. Appl. **177**, 1–11 (2021)
24. Mohit, I., Kumar, K.S., Reddy, U.A.K., Kumar, B.S.: An approach to detect multiple diseases using machine learning algorithm. J. Phys. Conf. Ser. **2089**, 1–7 (2021)
25. Orabi, K.M., Kamal, Y.M., Rabah, T.M.: Early predictive system for diabetes mellitus disease. In: Industrial Conference on Data Mining, pp. 420–427 (2016)
26. Pahwa, K., Kumar, R.: Prediction of heart disease using hybrid technique for selecting features. In: International Conference on Electrical, Computer and Electronics, pp. 500–504 (2017). https://doi.org/10.1109/UPCON.2017.8251100
27. Parvatikar, S.: Indian liver patients records (2021). https://doi.org/10.21227/rtpv-rc68
28. Patil, A., Framewala, A., Kazi, F.: Explainability of SMOTE based oversampling for imbalanced dataset problems. In: International Conference on Information and Computer Technologies, pp. 41–45 (2020)
29. Pranto, B., Mehnaz, S.M., Mahid, E.B., Sadman, I.M., Rahman, A., Momen, S.: Evaluating machine learning methods for predicting diabetes among female patients in Bangladesh. Information **11**, 374 (2020)
30. Pranto, B., Mehnaz, S.M., Momen, S., Huq, S.M.: Prediction of diabetes using cost sensitive learning and oversampling techniques on Bangladeshi and Indian female patients. In: International Conference on Information Technology Research, pp. 1–6 (2020)
31. Rady, E.H.A., Anwar, A.S.: Prediction of kidney disease stages using data mining algorithms. Inf. Med. Unlocked **15**, 1–7 (2019)
32. Rahman, T., et al.: Exploring the effect of image enhancement techniques on COVID-19 detection using chest X-ray images. Comput. Biol. Med. **132**, 104319 (2021)
33. Ramana, B.V., Babu, M.P., Venkateswarlu, N.: Liver classification using modified rotation forest. Int. J. Eng. Res. Dev. **6**, 17–24 (2012)
34. Sahoo, S., Mitra, T., Mohanty, A.K., Sahoo, B.J.R., Rath, S.: Diabetes prediction: a study of various classification based data mining techniques. Int. J. Comput. Sci. Inf. **25**, 100605 (2022)
35. Selvaraju, R.R., et al.: Grad-CAM: visual explanations from deep networks via gradient-based localization. In: International Conference on Computer Vision, pp. 618–626 (2017)
36. Singh, Y.K., Sinha, N., Singh, S.K.: Heart disease prediction system using random forest. In: International Conference on Advances in Computing and Data Sciences, pp. 613–623 (2016)

37. Sitaula, C., Hossain, M.B.: Attention-based VGG-16 model for COVID-19 chest X-ray image classification. Appl. Intell. **51**, 2850–2863 (2021)
38. Sontakke, S., Lohokare, J., Dani, R.: Diagnosis of liver diseases using machine learning. In: International Conference on Emerging Trends & Innovation in ICT, pp. 129–133 (2017)
39. Taal, M.: Chronic kidney disease. In: Landmark Papers in Nephrology, vol. 2, pp. 326–328 (2013)
40. Tasin, I., Nabil, T.U., Islam, S., Khan, R.: Diabetes prediction using machine learning and explainable AI techniques. Healthc. Technol. Lett. **10**, 1–10 (2022)
41. Wang, T., Tian, Y., Qiu, R.G.: Long short-term memory recurrent neural networks for multiple diseases risk prediction by leveraging longitudinal medical records. IEEE J. Biomed. Health Inform. **24**, 2337–2346 (2019)
42. Woo, S., Park, J., Lee, J.-Y., Kweon, I.S.: CBAM: convolutional block attention module. In: Ferrari, V., Hebert, M., Sminchisescu, C., Weiss, Y. (eds.) ECCV 2018. LNCS, vol. 11211, pp. 3–19. Springer, Cham (2018). https://doi.org/10.1007/978-3-030-01234-2_1
43. Yaganteeswarudu, A.: Multi disease prediction model by using machine learning and flask API. In: International Conference on Communication and Electronics Systems, pp. 1242–1246 (2020)
44. Zou, Q., Qu, K., Luo, Y., Yin, D., Ju, Y., Tang, H.: Predicting diabetes mellitus with machine learning techniques. Front. Genet. **9**, 515 (2018)

Enhancing MapReduce for Large Data Sets with Mobile Agent Assistance

Ahmed Amine Fariz(✉), Jaafar Abouchabaka, and Najat Rafalia

Faculty of Sciences, LaRIT, IbnTofail University ofKenitra, Kenitra, Morocco
a.amine.fariz@gmail.com

Abstract. The paper goes on to explain that as the amount of data being generated and collected continues to grow at an unprecedented rate, traditional database and software techniques are struggling to keep up. This is where the MapReduce Framework comes in. It is a programming model designed for processing large datasets in parallel across a cluster of computers. However, as the volume of data increases, the limitations of this framework become more apparent. The proposed approach utilizes agent technology to improve the speed and efficiency of the MapReduce programming model. The use of agents allows for the dynamic and adaptive distribution of tasks, resulting in a significant increase in processing speed.

The case study presented shows how this approach was able to achieve a 70% increase in speed when processing 52 Mb of data. The paper concludes by highlighting the potential of this approach to revolutionize the way we process big data and the benefits it can bring to businesses and organizations. It also encourages further research and development in this area to further improve the performance of the MapReduce Framework.

Keywords: Agents; Big Data; Map-Reduce · Hadoop · Jade

1 Introduction

Big Data has created a significant opportunity for businesses to gain an advantage and improve service delivery. However, with the large amount of unstructured data, a major challenge is how to process and sort it quickly. Map-Reduce has become a vital tool in the processing of Big Data, due to its ability to parallelize computation and distribute data. In this paper, we propose a multi-agent system for parallel execution of Map-Reduce on multiple databases. We have also integrated a Regex System in the Map-Reduce technique, based on matching a pattern from a string. Our proposed method was validated by experimental results and achieved a 70% speedup compared to traditional Map-Reduce techniques for 52 Mb of data. In addition, the paper discussed how authors in the literature approached the challenges of big data processing using different modifications of Hadoop Map-Reduce framework and other techniques such as online aggregation and message passing interface.

In addition to the benefits of parallelization and distribution, our proposed multi-agent system for Map-Reduce also allows for increased flexibility and autonomy. Each agent

R. Silhavy and P. Silhavy (Eds.): CSOC 2023, LNNS 724, pp. 627–639, 2023.
https://doi.org/10.1007/978-3-031-35314-7_53

is able to set its own goals and actions, and interact and collaborate with other agents through a communication protocol. This allows for more efficient and effective data processing, as well as the ability to adapt to changing circumstances and requirements. Overall, our proposed system is a valuable solution for organizations dealing with large amounts of data and seeking to gain a competitive advantage through faster and more efficient data processing. We believe that this system can greatly improve the ability of organizations to extract valuable insights and make data-driven decisions in real-time.

2 MapReduce

2.1 MapReduce Description

Map-Reduce is a powerful programming model that enables parallel and distributed processing of large data sets in a distributed environment. This approach is particularly useful for big data processing and can significantly reduce the time required for data processing by utilizing the power of parallel architecture. The Hadoop framework is one of the most widely used implementations of the Map-Reduce model, which enables the execution of the "Map" and "Reduce" tasks. The data processing starts by distributing the data among different nodes called "mappers". These mappers process the input data and generate a set of intermediate key/value pairs as output. These intermediate key/value pairs then enter the "Reduce" phase, where data with the same key is bundled together and processed by the "reducer" node, In detail, input data can be given by a list of records or words that are represented by (key, value) pairs, as shown in Fig. 1. The reducer combines different values with the same key and sums up the intermediate results from the mappers to produce the final output. The MapReduce methodology utilizes parallel processing control through a network of cluster workers. The platform is divided into four main components: the MapReduce framework, data loading, worker nodes, and analytics or output. Data is first loaded into a datastore, where it is divided into smaller chunks for easier processing through the MapReduce model. These chunks of data are then preprocessed through averaging techniques or other preprocessing methods to optimize the performance of the Map-Reduce model. The worker nodes play a crucial role in data processing by communicating with the server and coordinating the execution of the Map and Reduce phases. The output from the Reduce phase is then consolidated to produce the final output, which can be used for analytics or other downstream applications.

Additionally, Map-Reduce is a powerful programming model that enables parallel and distributed processing of large data sets in a distributed environment. This approach is particularly useful for big data processing and can significantly reduce the time required for data processing by utilizing the power of parallel architecture. Additionally, MapReduce is a popular choice for big data analysis because it allows for fault tolerance and scalability, making it suitable for processing large and complex data sets. It is also widely used in various fields such as natural language processing, computer vision, and machine learning. The use of a distributed environment and parallel processing makes it possible to process and analyze large amounts of data in a relatively short amount of time. Furthermore, the MapReduce model is easy to implement and can be integrated with other big data technologies such as Hadoop and Spark for even more efficient data processing. Overall, MapReduce is a powerful tool for big data processing and can

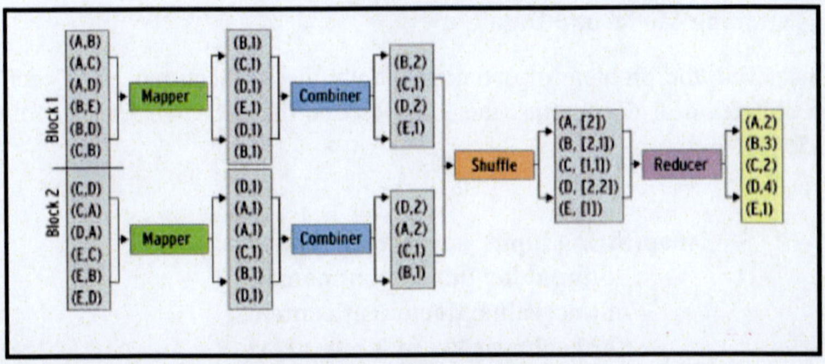

Fig. 1. Map-Reduce example [6].

greatly aid organizations in extracting valuable insights and making data-driven decisions. The Hadoop framework is one of the most widely used implementations of the Map-Reduce model, which enables the execution of the "Map" and "Reduce" tasks. The data processing starts by distributing the data among different nodes called "mappers". These mappers process the input data and generate a set of intermediate key/value pairs as output. These intermediate key/value pairs then enter the "Reduce" phase, where data with the same key is bundled together and processed by the "reducer" node. The reducer combines different values with the same key and sums up the intermediate results from the mappers to produce the final output.

The MapReduce methodology utilizes parallel processing control through a network of cluster workers. The platform is divided into four main components: the MapReduce framework, data loading, worker nodes, and analytics or output. Data is first loaded into a datastore, where it is divided into smaller chunks for easier processing through the MapReduce model. These chunks of data are then preprocessed through averaging techniques or other preprocessing methods to optimize the performance of the MapReduce model. The worker nodes play a crucial role in data processing by communicating with the server and coordinating the execution of the Map and Reduce phases. The output from the Reduce phase is then consolidated to produce the final output, which can be used for analytics or other downstream applications.

2.2 Background: Map-Reduce

Map-Reduce is a widely-used programming model that allows for parallel and distributed processing of large data sets in a distributed environment. It is a core component in an ecosystem of tools for distributed, scalable, and fault-tolerant data storage, management, and processing. Map-Reduce is essentially a distributed grep-sort-aggregate or, in database terms, a distributed execution engine for select-project via sequential scan, followed by hash partitioning and sort-merge group-by. It is particularly well-suited for data already stored on a distributed file system, which offers data replication as well as the ability to execute computations locally on each data node. There are two key aspects of Map-Reduce: the programming model and the distributed execution framework. We will explore these aspects in more detail later, after introducing a simple example.

2.3 Programming Model and Data Flow

When considering the problem of counting the number of occurrences of each word in a large collection of documents, the Map-Reduce the user will use the following pseudocode [5, 9]:

```
map(String input_key, String input_value):
//input key: document name;
//input_value: document contents;
for each word w in input_value:
EmitIntermediate(w,n);
reduce (String intermediate_key, Iterator
intermediate_values):
// intermediate_key : word;
//intermediate_values:????;
int result =0;
for each v in intermediate_values:
result +=v;
EmitFinal(intermediate_key,result);
```

The function referred to as "reduce" takes all of the counts generated for a specific word [5], and combines them by summing them up. This process is based on the number of occurrences associated with that word.

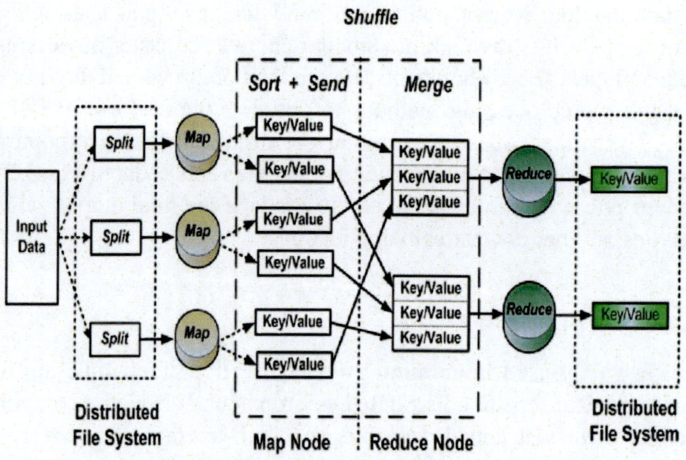

Fig. 2. Overview of the Map-Reduce execution framework.

Map-Reduce is a programming model that is based on a well-established abstraction in functional programming. The model consists of two functions, the mapper and the reducer, which operate on key-value pairs. The mapper function, MAPPER, takes in key-value pairs as input and produces a set of intermediate key-value pairs as output.

The reducer function, REDUCER, takes in the intermediate key-value pairs and groups them together based on the key, and applies a reduction operation to produce the final output.

To execute this computation on a large cluster, the data is partitioned into a number of input splits. Each split is assigned to a separate map task, which processes the input and produces the intermediate key-value pairs. The intermediate key-value pairs are then partitioned among a number of reduce tasks, by hashing on the intermediate key. Each reducer receives a part of the intermediate key space and merges, sorts, and applies the reduction operation on the key-value pairs to obtain the final results. This data flow model is illustrated in Fig. 2.

2.4 Scalable Distributed Processing: The Map-Reduce Framework

The Map-Reduce framework allows for the transparent execution of a wide range of computations on a distributed cluster architecture. One important feature of this framework is that the storage cluster may partially or completely overlap with the compute cluster, even when more advanced distributed data stores are used. This means that computation tasks can be executed on machines that host local copies of the input data. This allows for efficient data processing and reduces the need for data transmission over the network.

Figure 2b illustrates a possible placement of the elements from Fig. 2a, including data chunks and computation tasks, onto cluster machines. In this example, it is possible to place all map tasks on machines that host a local copy of their input split. If this is not possible, then the data will be transmitted over the network from a remote storage node. Additionally, an intermediate combiner can be inserted between each mapper and the final reducer. This combiner combines all local map outputs using either the same or a separate reducer function before they are sent to the reducers. This can further optimize the data processing and reduce the amount of data transmitted over the network.

The distributed execution model in Map-Reduce also handles load balancing and fault tolerance in an effective manner. One of the key benefits of Map-Reduce is its ability to scale and use any number of machines, as the volume of output data is usually much smaller than the volume of input data. This leads to a performance improvement that is almost proportional to the number of nodes used.

3 Distributed Mining Process

The process of distributed data mining, as illustrated in Fig. 3, includes several stages including data gathering, pre-processing, analysis, and post-processing. Many of these stages involve distributed processing through a data storage layer, such as GFS, HDFS, or KFS, or through a higher-level data access and job description language, such as Sawzall, Pig, or Cascading.

The first step in the distributed data mining process is data gathering, which involves identifying the sources of data and obtaining it. This can include tasks such as crawling millions of web pages, querying heterogeneous databases, performing large-scale scientific simulations, or monitoring distributed systems. These tasks are often performed in a distributed manner and can be expressed as Map-Reduce jobs.

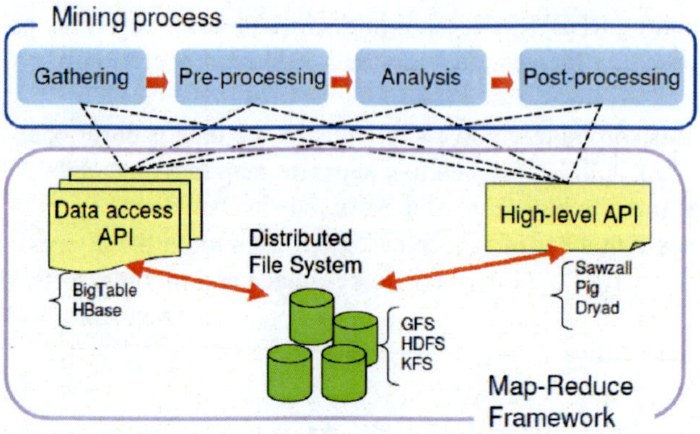

Fig. 3. Distributed mining process.

After obtaining the raw data, it is important to perform data pre-processing to transform the data into a format suitable for analysis. This step, which typically involves data cleaning and can consume the majority of time for exploratory data mining tasks, has been largely ignored in the research literature despite calls for its importance from established researchers. However, it is crucial to not overlook this step as the amount of data continues to grow.

Researchers, including the authors of this text, are finding that they increasingly need to handle large amounts of data, sometimes in the form of gigabytes or even terabytes. For example, it was reported that parsing 4.5 terabytes of compressed text logs for 30 days worth of MSN instant messaging data took five full days on an eight-processor machine with fast local disks (reference [27]). Similarly, the authors recently experienced a five-hour process to extract source/destination IP pairs from a 350 gigabyte raw network event log (similar to an example in Sect. 2.1), even though the data was accessed over a 2 Gbps Fibre Channel link to a SAN. The TREC data, which is 100GB of text, also took several days to pre-process on a single powerful machine (four cores and 32GB RAM). However, they found that they were able to achieve better performance on a few commodity nodes running Hadoop, with minimal effort required for setup (about two to three hours for a moderately experienced person). Other approaches, such as the DPH (desperate Perl hacker) approach and the traditional database management system approach, were less efficient and required more effort. The authors believe that Hadoop offers a better relative effort-to-benefit ratio. For co-clustering specifically, there are two main pre-processing tasks.

1. Constructing the graph from raw data
2. Preparing the transpose in advance

The first step in building the graph from raw data is primarily focused on extracting the graph, such as source-destination or document-term pairs, and may also include other related tasks like stemming and stopword removal. To optimize co-clustering, it is necessary to iterate over both rows and columns, thus the need for pre-computing the adjacency lists for both the original graph and its transpose. Transposition is similar to computing an inverted index, which is one of the original applications for Map-Reduce

[12]. On average, this step takes a few minutes. In Sect. 5, we provide detailed information on the actual times for processing real-world data.

Procedure 3 CCROWREDUCER (k, V)

Row group label is $p \equiv k$
Initialize $\mathbf{g}_p \leftarrow 0$, $I_p \leftarrow \emptyset$
for each map value $(\mathbf{g}, I) \in V$ **do**
  $\mathbf{g}_p \leftarrow$ COMBINESTATISTICS$(\mathbf{g}_p, \mathbf{g})$
  $I_p \leftarrow I_p \cup I$
emit $(p, (\mathbf{g}_p, I_p))$

Procedure 4 COLLECTRESULTS

Initialize $\mathbf{G} \leftarrow 0$, $\mathbf{r} \leftarrow 0$
for reduce output $(p, (\mathbf{g}_p, I_p))$ **do**
  $g_{p:} \leftarrow \mathbf{g}_p$
  $r(i) \leftarrow p$, for all $i \in I_p$
return G and **r**

The process of splitting a cluster into smaller groups during co-clustering requires some modifications. The decision of which cluster to split can be made in a similar way as described in [6], but the method of determining which rows or columns should be assigned to the new group is not easily parallelizable. A strategy that has been found to be effective is using a random split of 50–50, where half of the rows or columns are placed in the new cluster and half remain in the original cluster. This approach has been observed to yield better results than the criterion described in [6], and it can be further improved by conducting multiple random trials to take advantage of cluster resources.

4 Methodology

4.1 System Overview

We propose a system that uses parallel execution of the Map-Reduce algorithm on multiple databases through the use of intelligent agents. These agents, called Map-Reduce Agents, were developed using the JADE open source platform for peer-to-peer agent-based applications. The goal of the system is to create an architecture with autonomous agents that can set their own goals, actions, and interact and collaborate with each other through a communication protocol. The system is composed of several components.

1. The Load Data Agent retrieves the name of a selected input file by taking in its full path.
2. The Parallel Map-Reduce Agents process the dataset from the file name received from the Load Data Agent to obtain the desired results.
3. The Search module within the Map-Reduce Agent allows users to specify a word to search for, the column to search in, and the level of detail desired in the output. If a parallel search is conducted, a third Map-Reduce Agent can be activated to search through a second file by sending its location to the second agent.
4. The Result export module outputs the final results as text files for further processing by other systems.

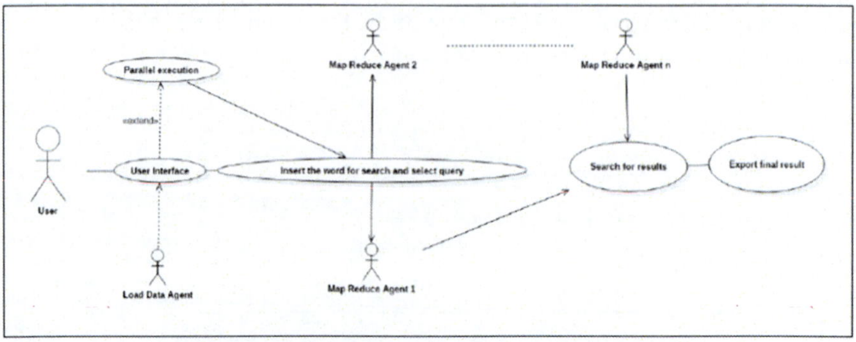

Fig. 4. Figure 2 System architecture.

5. The User interface simplifies the user's interaction with the system.

The procedures carried out by the Regex Map-Reduce method for an individual agent are detailed below.

1. The user inputs one or more words to search for and determine their frequencies in the input dataset.
2. The dataset to be searched is selected by the user, which can be in the form of.xlsx,.xls,.csv, or.txt file with a header describing the attributes. The user can also specify a subset of columns to search within.
3. The input words and subset of data are applied to the search process.
4. The Map-Reduce technique is enhanced by a Regex System and applied to the dataset. An example of the proposed Regex is provided as follows: (words[switch1].matches(".?\b" + WordUtils.capitalize(try.toLowerCase()) + "\b.?")||word s[switch1].matches("." + try + "."))
5. The results of the Map-Reduce Agent are displayed.
6. The final results are output in the User Interface.
7. The user can choose to delete any temporary files used during the process.

The diagram in Fig. 3 illustrates the interactions between the User, User Interface, Choose Options, Run Map-Reduce, Word Count, and Map-Reduce Agent Output objects. It is evident that the user only inputs keywords and selects search options, and the remaining components then exchange intermediary and final results. Lastly, the user can delete the temporary file utilized to store the outcome, to initiate a new search.

4.2 Improved Map-Reduced Methodology

The following is a summary of the algorithm for parallel execution of Map-Reduce:

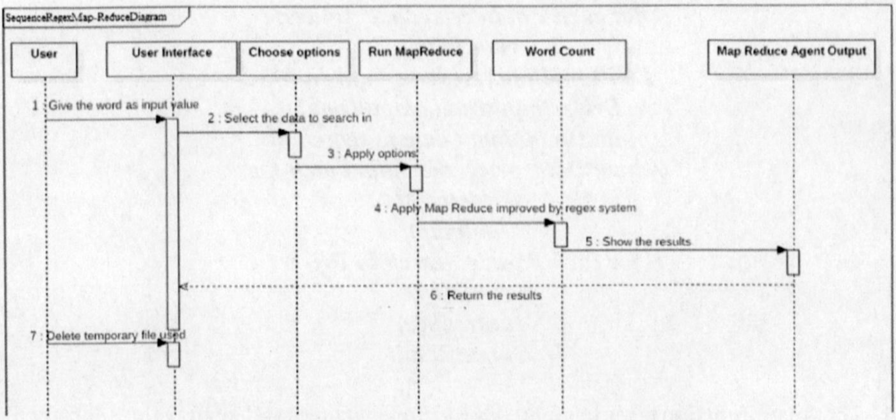

Fig. 5. Figure 3 Regex Map-Reduce sequence diagram.

map(String input_key, String input_value):
//input key: document name;
//input_value: document contents;
for each input searched word (w) in input_value:
EmitIntermediatevalue(w,1);
reduce (String intermediate_key, Iterator intermediate_values):
// intermediate_key : word;
//intermediate_values:????;
int result =0;
for each v in intermediate_values:
result +=v;
EmitFinal(intermediate_key,result);
map1(String input_key, String input_value):
//input key: different document names;
//input_value: document contents;
for each input searched word (w) in input_value:
EmitIntermediatevalue(w,1);
reduce (String intermediate_key, Iterator intermediate_values):
// intermediate_key : word;
//intermediate_values:????;
int result =0;

for each v in intermediate_values:
result +=v;
EmitFinal(intermediate_key,result);
Driver(inputData,outputData)
//inputData: input data to search in
//outputData: select the output path for data
run(map)
for each ParallelSearch in Driver
run(map)
run(reduce)
OutputResult

1. The "map" function takes in input_key (document name) and input_value (document contents) as parameters.
2. For each searched word (w) in input_value, it emits an intermediate value (w,1).
3. The "reduce" function takes in intermediate_key (word) and intermediate_values (iteration of values) as parameters.
4. It initializes a result variable to 0, then iterates through each value (v) in intermediate_values and adds it to the result.
5. It then emits the final result (intermediate_key, result).
6. The "map1 " function is similar to the first "map" function, but with different input_key and input_value.
7. The "reduce" function is also the same as the first one.
8. The "Driver" function takes in inputData (input data to search) and outputData (selected output path for data) as parameters.
9. It runs the "map" function, followed by "reduce" function.
10. The result is outputted.

Note that the exact meaning of intermediate_values is not specified in the provided code snippet, so it is unclear what it is referring to.

The Map class reads each raw from the input file, applies the sort method, and sends the result to the Reduce class. The diagram illustrates the relationship of the proposed Driver class with the other classes.

The proposed Regex System uses a pattern-matching approach to find data, even if it is incomplete or case-sensitive. The system performs a search in the dataset and attempts to find the best match for the input string. The MapReduce algorithm is utilized in conjunction with the Regex System to execute user requests and search for the same pattern in a column or string within the dataset. (Figs. 4 and 7).

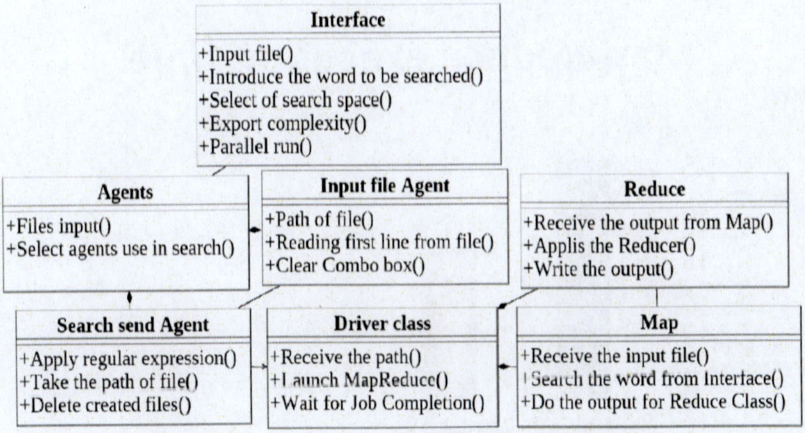

Fig. 6. Figure 4 Class diagram for proposed method.

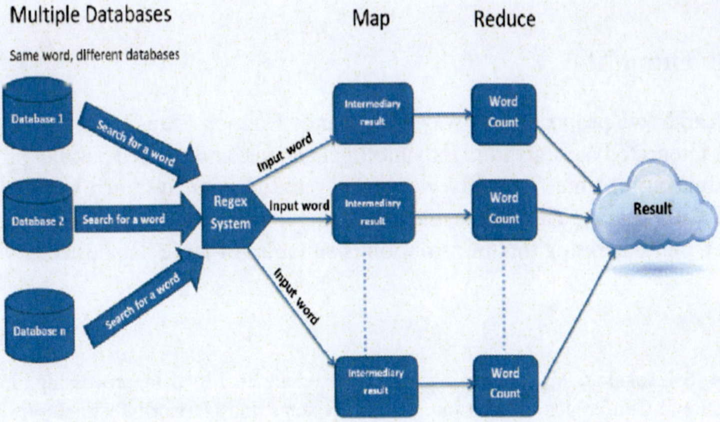

Fig. 7. Figure 5 Parallel search with Map-Reduce Regex System.

5 Assessment of Performance

The experiments were conducted on two databases that consist of 52 Megabytes of data and 163066 records each, which can be found in references [13, 14]. These databases contain information about the hospital-specific charges for more than 3,000 U.S. hospitals that receive Medicare Inpatient Prospective Payment System (IPPS) payments for the top 100 most frequently billed discharges, paid under Medicare based on a rate per discharge using the Medicare Severity Diagnosis Related Group (MS-DRG) for Fiscal Years 2011, 2012, and 2013. The parallel search was performed on these databases, and it yielded results in 16 ms. Additionally, we implemented the Map-Reduce algorithm on both databases, and it generated results in 27 ms (Fig. 6) as shown in [13, 14] (Fig. 8).

The use of parallel search agents allowed for querying of both databases simultaneously, thus reducing the Map-Reduce execution time.

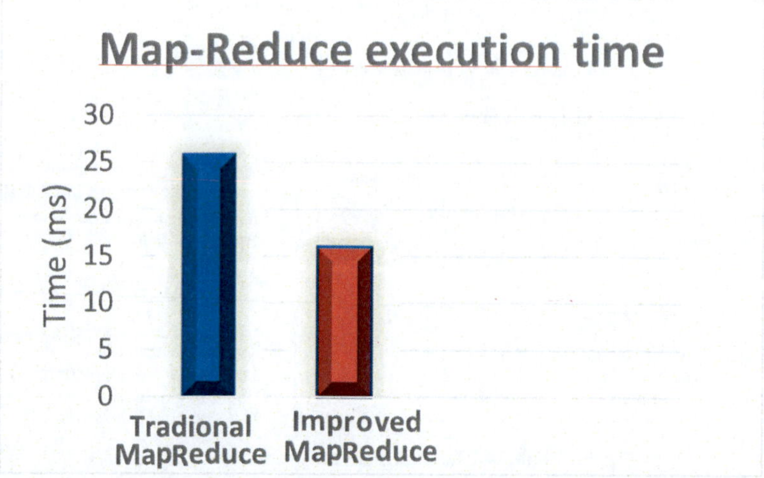

Fig. 8. Figure 6 Map-Reduce run time.

6 Final Thoughts

In this research, we proposed two ways to enhance Hadoop Map-Reduce performance. Firstly, we integrated the algorithm into intelligent agents and applied parallel processing to improve execution time. Secondly, we incorporated a Regex System to boost algorithm speed and result clarity. Through testing, we achieved a 70% increase in speed for a 52 Mb dataset, demonstrating that our solution is suitable for large databases.

References

1. Pandey, S., Tokekar, V.: Prominence of MapReduce in big data processing. In: Fourth International Conference on Communication Systems and Network Technologies (CSNT), pp. 555–560 (2014)
2. Condie, T., Conway, N., Alvaro, P., Hellerstein, J.M.: MapReduce Online. In: Proceedings of the 7th USENIX Symposium on Networked Systems Design and implementation (NSDI 2010), pp. 1–15 April 2010
3. Fariz, A., Abouchabka, J., Rafalia, N.: Improving MapReduce process by mobile agents. Adv. Intell. Syst. Comput. **1295**, 851–863 (2020)
4. Mohamed, H., Marchand-Maillet, S.: Enhancing mapReduce using MPI and an optimized data exchange policy. In: 2012 41st International Conference on Paralle Processing Workshops, pp. 11–18, Pittsburgh, PA, 10–13 September 2012
5. El Fazziki, A., Sadiq, A., Ouarzazi, J., Sadgal, M.: A multi-agent framework for a hadoop based air quality decision support system. In: Krogstie, Juel-Skielse, Kabilan (Eds.) Proceedings of the CAiSE-2015 Industry Track co-located with 27th Conference on Advanced Information Systems Engineering (CAiSE 2015), Stockholm, Sweden, pp. 45–59. 11 June 2015
6. Jeffery, D., Ghemawat, S.: MapReduce: simplified data processing on large clusters. In: Appeared in OSDI 2004: Sixth Symposium on Operating System Design and Implementation, San Francisco, CA, December, 2004, Google Research Publication, pp. 1–13 (2004). http://research.google.com/archive/mapreduce.html

7. Leskovec, J., Rajaraman, A., Ullman, J.: Mining of Massive Datasets: Map-Reduce and the New Software Stack. Stanford Computer Science course. http://www.mmds.org/mmds/v2.1/ch02-mapreduce.pdf
8. http://hadoop.apache.org/
9. Altamirano, A., Lukas Forer, S.S.: Analyzing Big Data using Hadoop MapReduce. In: Marshall Plan Scholarship Paper, Utah State University University of Innsbruck (LFU) in cooperation with Medical University of Innsbruck, p. 7 (2014)
10. Howe, B.: Introduction to Data Science: MapReduce Pseudocode. University of Washington course. https://class.coursera.org/datasci-001/lecture/73
11. JAVA Agent DEvelopment Framework. http://jade.tilab.com/
12. Nasri, M., Hossain, M.R., Ginn, H.L., Moallem, M.: Agent-based real-time coordination of power converters in a DC shipboard power system. In: 2015 IEEE Electric Ship Technologies Symposium (ESTS), Alexandria, VA, pp. 8–13 (2015)
13. Ahmad, I., Kazmi, J.H., Shahzad, M., Palensky, P., Gawlik, W.: Co-simulation framework based on power system, AI and communication tools for evaluating smart grid applications. In: 2015 IEEE Innovative Smart Grid Technologies - Asia (ISGT ASIA), Bangkok, pp. 1-6 (2015)
14. https://data.cms.gov/Medicare/Inpatient-Prospective-Payment-System-IPPS-Provider/97k6-zzx3

The Effect of E-Learning Quality, Self-efficacy and E-Learning Satisfaction on the Students' Intention to Use the E-Learning System

M. E. Rankapola(✉) and T. Zuva

Vaal University of Technology, Vanderbijlpark 1911, Republic of South Africa
merankaps@gmail.com

Abstract. The study investigated the extent to which e-learning quality, self-efficacy and e-learning satisfaction influenced the students' intention to use/use the e-learning system amid the Covid-19 pandemic and level five locdown restrictions. A total of 857 students took part in the study. A response rate of 63.36% was achieved. A positivist research philosophy reinforced by a quantitative research approach was employed in carrying out the study. Structural Equation Modelling (SEM), Confirmatory Factor Analysis (CFA) and Path Analysis (PA) were employed to analyse data. Cronbach alpha and composite reliability scores for the latent constructs used in this study were greater than 0.7, indicating that they were all reliable. Path analysis was used to investigate and verify the hypothesised relationships between the variables. According to the findings, the effect of technical service quality on user satisfaction is statistically significant and positive [$\beta = 0.031$; p-value > 0.05], whereas the effects of content and information quality, educational service quality, and service quality are statistically significant and positive [$\beta = 0.072$, p-value 0.05; $\beta = 0.203$, p-value 0.05; $\beta = 0.137$, p-value 0.05]. Furthermore, user self-efficacy exhibited a statistically significant beneficial influence on user satisfaction [$\beta = 0.711$, p-value 0.01]. The results show that user self-efficacy moderates the relationship between technical service quality [$\beta = 0.04$, p-value 0.10], educational service quality [$\beta = 0.067$, p-value 0.05], and service quality [$\beta = 0.065$, p-value 0.05].

Keywords: e-learning quality · e-learning satisfaction · self-efficacy

1 Introduction

The COVID-19 epidemic has had an impact on all aspects of life, including education. Universities needed to adapt to online teaching and learning modality. Many parties (students and teachers) were unprepared for this rapid and unprecedented paradigm shift. In South Africa, the hasty transition from traditional face-to-face interaction had proven to be more challenging and problematic because most students come from unstable economic circumstances which influence access to the necessary devices and the ability to

R. Silhavy and P. Silhavy (Eds.): CSOC 2023, LNNS 724, pp. 640–653, 2023.
https://doi.org/10.1007/978-3-031-35314-7_54

use those devices effectively. Untrained lecturers, under-prepared learners, unsupported internet networks, and other academic and social factors, added to the likelihood of unsuccessful e-learning implementation.

However, some institutions in the country had already introduced multi-modal teaching and learning strategies e.g. UNISA, UP, TUT etc. Nevertheless, the coronavirus disaster required that institutions implement a full-online teaching and learning strategy as contact lectures were completely suspended. Other institutions, faculties and departments were obliged to introduce e-learning for the first time.

This study was conducted to investigate the influence of e-learning quality on e-learning quality on self-efficacy and e-learning satisfaction. This study was triggered by the rapid transition from the face-to-face modality to the e-learning modality, lack of appropriate resources, underprepared learners and untrained lecturers as well as other academic and social factors that have a bearing on the success of e-learning implementation. We believe that it is critically important to uncover and understand the quality of an e-learning system's effect on social factors such as computer self-efficacy and e-learning satisfaction which predicts and determines the success of an e-learning system.

The rest of the paper is arranged as follows; literature review, conceptual framework and hypotheses development, research methodology, findings and analysis and discussions and conclusions, limitations and future research work and lastly the references.

2 Review of Existing Literature

2.1 E-Learning Content and User Satisfaction

E-learning is a modern learning system inspired by information technology and has become the norm in higher education due to its technology resources and benefits [1] It is a digital delivery system that allows learners to control instructional content through features such as accessibility and navigation [2]. The enjoyment of the user is essential in online learning. Learner satisfaction is influenced by time flexibility, learning material, and ease of use [3]. Student satisfaction is increased by well-designed course content that includes dynamic content and is easy to use [4]. Practical e-learning course content improves learner skills and knowledge. Effective use of learning materials is recognised as a dynamic way to distribute learning content to stakeholders [5].

2.2 E-Learning Content and User Intention

Previous research has found that using technology improves e-learning. The COVID-19 epidemic altered the world and how it worked, compelling the education industry to shift from traditional to online learning [6]. This has not only kept education going during the COVID-19 pandemic, but it has also decreased time and distance problems, allowing users to enhance their skills at a lesser cost and with greater flexibility [7]. This e-learning technique saves money and makes learning easier. Several UTAUT investigations discovered that performance expectation has a significant influence on behavioural intention to use e-learning [8]. E-learning components like collaborative learning, personalisation, cost, and performance have a positive relationship with the behavioural intention [2].

2.3 Self-efficacy and User Satisfaction

How confident a student feels in their ability to study is important, especially when taking classes online. A person's confidence in his or her abilities to complete a task in an information and communications technology setting. Directly, it affects one's freedom of choice [9]. One's confidence in their ability to make effective use of educational technology is directly tied to their level of self-efficacy in that setting. An individual's sense of competence and satisfaction increases when they can successfully utilise technology [10, 11]. Users' confidence and satisfaction with their learning interactions both increase with increased self-efficacy [12]. At long last, contentment. An individual's level of satisfaction depends on more than just how well a product works. Based on their research, [13] propose a paradigm in which customer happiness is the result of high-quality systems, data, and service that ultimately benefits their education. Having confidence in one's abilities is said to have a positive impact on one's performance and enjoyment in the classroom [14].

2.4 Self-efficacy and User Intention

Having trust in one's ability to use a computer successfully to accomplish a task is what we call "computer-related self-efficacy" [14]. Previous studies have shown that students' perceptions of their computing abilities affect their approach to schoolwork and performance[15]. Users' confidence in their ability to make effective use of technology is known as self-efficacy. Improved users' self-efficacy increases their pleasure and motivation [16]. Users' intentions may be influenced by their level of computer self-efficacy while using a collaborative e-learning platform that requires some familiarity with technology. At last, deliberate action fosters an amiable setting for learning [17].

2.5 User Satisfaction and User Intention

When attempting to decipher user intent, customer happiness is paramount. Because of how happy its stakeholders are, online education has become increasingly common. An individual's level of satisfaction with a service can be seen as an indicator of how well their educational experience with that service was overall [18]. During a formal evaluation process, it might be challenging to please all of the relevant parties (users, teachers, and institutions). Customer happiness is a key factor in promoting repeat business [19]. Without future use, satisfaction is meaningless. Users will continue with their intended purpose if their needs are met [20].

Information, communication, and technology support users' online learning intentions [21]. ICT in the learning system includes the behavioural intention to utilise and technological utilisation. The intention to engage in a specific behaviour is a behavioural intention [22]. As the IS success model developed by [13] demonstrates, various factors affect users' decisions to engage with an information system [13]. One of the best predictors of continued use is how happy the customer is with the product or service. According to the aforementioned claims, all parties involved, notably the course's participants, will benefit by engaging in high-quality course learning. Consistently happy users are more likely to return [23].

3 Conceptual Framework and Hypothesis Development

According to theories of e-learning success, such as Social Cognitive Theory (SCT), DeLone and McLean IS success model; many elements contribute to its success in schools. For example, [24] applying the SCT found that compatibility (environmental factor) and personal innovativeness (personal factor) strongly influenced student learning performance and intentions to use. The results accord with those of [25], who proposed that knowledge and attitude influence behaviour by acting as a "mediator" between the two. The current analysis supports Seddon's three concept categories: system and information quality, net benefit perceptions, and IS conduct [26].

The MELSS model emphasised e-learning measurements such as System Quality and Service Quality (technical level), educational quality, content and information quality (semantic level), user satisfaction, intention to use/use, loyalty to the system, benefits of using the system, and goal achievement (influence level) [27]. [28] claimed that result anticipation and self-efficacy influence behaviour. The ability to attain a goal or complete a task depends on self-efficacy and outcome expectancy [29]. This research adds to the understanding of the topic by focusing on the moderating effect of user self-efficacy. The theoretical framework used in this study is depicted in Fig. 1.

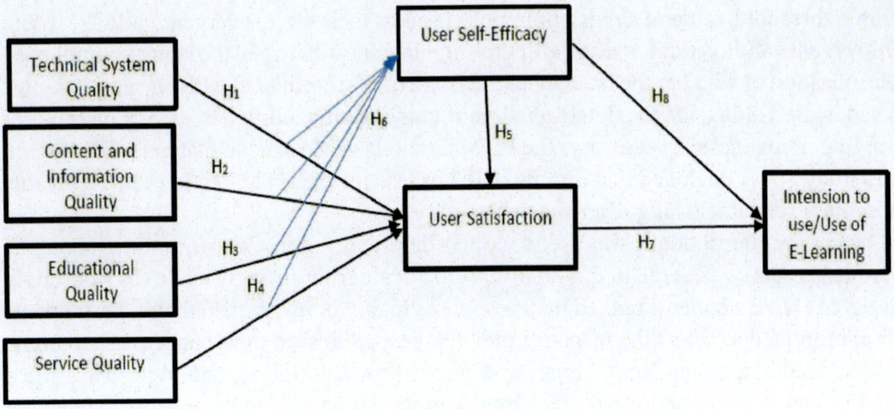

Fig. 1. Conceptual Model

From Fig. 1, the derived hypothesis that the study sought to test are as follows:

H1 Alt: Technical service quality has a significant impact on user satisfaction.

H2 Alt: Content and information quality has a significant impact on user satisfaction.

H3 Alt: Educational service quality has a significant impact on user satisfaction.

H4 Alt: Service quality significantly influences user satisfaction.

H5 Alt: User self-efficacy significantly influences user satisfaction.

H6a Alt: The association between technical service quality and use satisfaction is moderated by user self-efficacy.

H6b Alt: The association between educational service quality and user satisfaction is moderated by user self-efficacy.

H6c Alt: The association between content and information quality and user satisfaction is moderated by user self-efficacy.

H6d Alt: The relationship between service quality and user satisfaction is moderated by user self-efficacy.

H7 Alt: User satisfaction has a significant impact on E-Learning Intention to Use/Use.

H8 Alt: User self-efficacy significantly influences E-Learning Intention to Use/Use.

4 Research Methodology

The Structural Equation Modelling [30] analysis, which comprised Confirmatory Factor Analysis (CFA) and path analysis using Amos 28 was employed as the primary statistical approach. SEM is a multivariate statistical approach that is used to investigate structural correlations [31]. A combination of factor analysis and regression analysis describes the approach taken. CFA and path analysis are two popular statistical methods for analysing data [32]. A summary of the proposed hypothesis and the quality of the model is also included in the data analysis method.

Cronbach's alpha was used in the study to measure the instrument's dependability. According to [33], a variable must have a Cronbach Alpha value greater than 0.7 to be considered statistically reliable. [34]showed that average variance extracted [24] with a threshold value of 0.5 is commonly used to measure convergent validity, where convergent validity is defined as the degree of relatedness between the latent components. The question of whether the latent constructs are overly correlated with one another must be addressed alongside the determination of convergent validity. As such, a method for gauging uniqueness is necessary; the HTMT ratio is used to make that determination in this study. [35] concluded that a value of 0.9 or less for the HTMT ratio is an acceptable minimum for establishing discriminant validity.

A survey questionnaire was used to collect data, and within it were questions about the various factors considered in this analysis. On a scale from strongly agree (5) to strongly disagree (1), respondents ranked their responses to the survey questionnaire. Participants were prompted to select the response they felt best addressed their concerns. Lecturers, tutors, academic managers, alumni, and instructional designers (among others) filled out the survey over two months. The total sample size was 857 respondents to meet the minimum number of participants needed to get reliable results from a quantitative study. Large sample size was used to produce credible data that accurately reflected the whole picture.

The research was conducted with input from a wide range of university personnel who work closely with the online learning platform regularly. Due to the importance of ethical considerations in conducting research, the researcher had also guaranteed that all ethical guidelines were adhered to. The survey was made available online after participants gave their consent, and an ethical form detailing all the rules was provided. They were also briefed on the research goals, as well as the significance of the survey data they will be providing. Information provided by respondents was not shared with any third parties., and everything was stored safely in encrypted devices.

5 Research Findings

This section of the study presents survey results. The study had a 63.36% response rate suggests a reliable sample population, according to [36], who argued that a response rate between 50% and 92% is suitable for quantitative research. More so, 55.2% (n = 300) of respondents were females, 44.2% (n = 240) were males, and 0.6% did not share their gender, indicating that females dominated the study sample population. Most of the respondents were young adults, with 66.1% aged 19 to 25 years, 11.6% aged 26 to 30 years, 8.7% aged 41 and above, 8.1% aged 31 to 35 years, and 5.5% aged 36 to 40 years.

5.1 Confirmatory Factor Analysis

Confirmatory factor analysis primary purpose is to evaluate the constructs in a study and establish their validity and reliability [37]. In this respect, the reliability of the factors was evaluated using factor loadings. In their study,[38] postulated that factor loadings needed to be above 0.6 to be considered significant. That being said, values greater than the 0.6 threshold are acceptable. Table 1 demonstrates that all indicators are above 0.6, so none need to be eliminated. However, [33] found that a composite reliability of 0.7 or higher indicated that a variable could be considered reliable for statistical inferences. Thus, all Cronbach alpha and composite reliability scores for the latent constructs used in this study are greater than 0.7, indicating that they are all reliable. All latent constructs were found to be statistically and convergently valid within one decimal place.

Table 1. Reliability and Convergent Validity of the Constructs

Construct	Indicators	Factor Loadings	Cronbachs Alpha	Composite Reliability	AVE
Technical System Quality	TSQ1	0.808***	0.730	0.801	0.474
	TSQ2	0.813***			
	TSQ3	0.796***			
Content & Information Quality	C&IQ1	0.878***	0.869	0.714	0.691
	C&IQ2	0.898***			
	C&IQ3	0.895***			
Educational System Quality	ESQ1	0.818***	0.795	0.726	0.566

(*continued*)

Table 1. (*continued*)

Construct	Indicators	Factor Loadings	Cronbachs Alpha	Composite Reliability	AVE
	ESQ2	0.858***			
	ESQ3	0.851***			
Service Quality	SQ1	0.827***	0.825	0.817	0.631
	SQ2	0.915***			
	SQ3	0.838***			
User Self-Efficacy	USE1	0.885***	0.875	0.721	0.696
	USE2	0.916***			
	USE3	0.882***			
User Satisfaction	US1	0.920***	0.891	0.761	0.734
	US2	0.896***			
	US3	0.903***			
Intention to use/use	ITU/U1	0.864***	0.823	0.796	0.610
	ITU/U2	0.910***			
	ITU/U3	0.806***			

Source: Author Estimations; *** indicates statistical significance at 1%

In addition to determining convergent validity, it is essential to determine whether the latent constructs are overly interconnected. In this regard, it is necessary to assess the degree of distinctiveness using the HTMT ratio. The results reveal that none of the constructs has a strong correlation with the other, confirming the constructs' discriminant validity as indicated in Tables 2 and 3.

5.2 Model Quality

The data were validated using CFA prior to SEM. The requirements for convergent validity were satisfied. Among the indices used to evaluate the quality of a measurement model's fit are CMIN/DF (χ 2/Df), the Goodness of Fit Index (GFI), the Adjusted Goodness of Fit Index (AGFI), the Normed Fit Index (NFI), the Tucker-Lewis Index (TLI), the Comparative Fit Index (CFI), and the Root Mean Square Error of Approximation (RMSEA). A model's χ 2/Df should be between 0 and 5, with lower values indicating a better match [39]. Furthermore, a satisfactory fit is suggested when the values of GFI, AGFI, NFI, TLI, and CFI are near to 1, as well as when the RMSEA is between 0.05–0.10. [40]. The measurement model provided a reasonable match with (χ 2/Df = 3.197;

Table 2. Discriminant validity determination using HTMT ratio

	TSQ	CIQ	ESQ	SQ	USE	US	ITU/U
TSQ							
CIQ	0.150						
ESQ	0.170	0.822					
SQ	0.159	0.742	0.878				
USE	0.082	0.668	0.770	0.804			
US	0.088	0.676	0.821	0.844	0.828		
ITU/U	0.077	0.593	0.716	0.737	0.820	0.871	

Source: Author Estimations

Table 3. Hypotheses Assessment Results

	Hypotheses Relationship	Estimate	Remark
H1	Technical Service Quality --- > User Satisfaction	0.031	Rejected
H2	Content and Information Quality --- > User Satisfaction	0.072**	Accepted
H3	Educational Service Quality --- > User Satisfaction	0.203 **	Accepted
H4	Service Quality --- > User Satisfaction	0.137 **	Accepted
H5	User Self Efficacy --- > User Satisfaction	0.711 ***	Accepted
H6	User Self Efficacy moderates Technical Service Quality and User Satisfaction	0.04 *	Accepted
	User Self Efficacy moderates Content and Information Quality and User Satisfaction	0.016	Rejected
	User Self Efficacy moderates Educational Service Quality and User Satisfaction	0.067 **	Accepted
	User Self Efficacy moderates Service Quality and User Satisfaction	0.065 **	Accepted
H7	User Satisfaction --- > Intention to Use E-learning	0.803 ***	Accepted
H8	User Self Efficacy --- > Intention to Use E-learning	0.107	Rejected

Source: Author Estimations; * ** *** indicates statistical significance at 10%, 5%, and 1%

GFI = 0.904; AGFI = 0.865; NFI = 0.92; TLI = 0.926; CFI = 0.944 and RMSEA = 0.064).

5.3 Path Analysis

After establishing the constructs and indicators' factor structure, reliability, and validity, we tested the hypotheses about the path they predicted. Path analysis was used to investigate and verify the hypothesised relationships between the included variables, as described by [41]. The results in this regard are shown in Fig. 2 below. According to the findings, the effect of technical service quality on user satisfaction is statistically significant and positive [β = 0.031; p-value > 0.05], whereas the effects of content and information quality, educational service quality, and service quality are statistically significant and positive [β = 0.072, p-value 0.05; β = 0.203, p-value 0.05; β = 0.137, p-value 0.05].

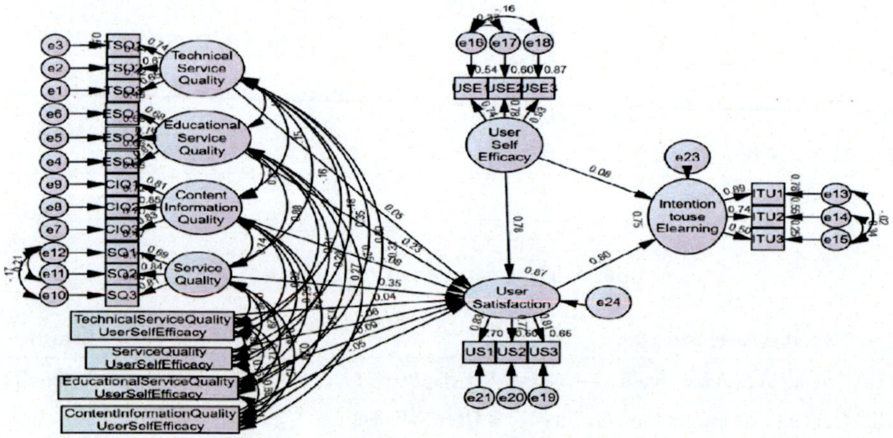

Fig. 2. Path Analysis

Furthermore, user self-efficacy exhibited a statistically significant beneficial influence on user satisfaction [β = 0.711, p-value 0.01]. The positive coefficients for technical service quality, educational service quality, content and information quality, service quality, and user self-efficacy imply that these characteristics contribute to higher levels of user satisfaction. Furthermore, user self-efficacy had a statistically positive but negligible effect on e-learning intention, whereas user satisfaction had a statistically significant positive effect on e-learning intention.

However, the study's main contribution is that it examines whether user self-efficacy is a predictor of user satisfaction, and the findings show that it does. The results show that user self-efficacy moderates the relationship between technical service quality [β = 0.04, p-value 0.10], educational service quality [β = 0.067, p-value 0.05], and service quality [β = 0.065, p-value 0.05]. The presence of a positive coefficient indicates that user self-efficacy improves the impact of technical service quality, educational service quality, and service quality on e-learning user satisfaction.

6 Discussion and Conclusion

The study was carried out in South Africa to explore the association between self-efficacy, interaction aspects (technical service quality, educational service quality, content and information quality, and service quality), user satisfaction, and intention to utilise e-learning. Previous research has identified a number of factors that act as drivers towards e-learning, including perceived usefulness, ease of use, network externality [20, 42], user satisfaction, information quality [43, 44], and real-time access to information [44].

In the context of this research, "interaction" refers to the ways in which a user of an e-learning platform can communicate with the software's many constituent parts, including instructors, students, support staff, and course materials (that is, technical service, educational service, content and information and service quality). This finding is significant because it suggests that other factors, such as the availability of technical support, the simplicity of facilitating teacher-student communication, and the convenience of accessing online learning contents, all play a role in determining whether or not users are satisfied with and willing to adopt e-learning. Previous studies have suggested that user participation is crucial to user satisfaction and e-learning adoption [45, 46]. As a result, hypotheses H2 (content and information quality improves user satisfaction), H3 (educational services quality significantly affects user satisfaction), and H4 (services quality significantly affect user satisfaction) were accepted.

User self-efficacy was found to have a significant positive effect on user satisfaction and e-learning intention to use, and it was validated in the role of moderator between technical service quality, educational service quality, service quality, and user satisfaction. Self-efficacy in this study refers to an individual's confidence in his or her ability to use digital tools proficiently. These results corroborate prior studies that found a strong link between user confidence in their own abilities and their satisfaction with the experience [47, 48]. This means that H5 and H6 (a, c, d) are correct in their predictions that higher levels of self-efficacy led to greater user satisfaction and that self-efficacy acts as a moderator between the quality of technical services, educational services, content and information, and overall service quality and user satisfaction.

The findings of this study show that self-efficacy has a significant positive relationship with user satisfaction, which may explain the rise of self-efficacy in South Africa. This may be due to the greater availability of the internet and the greater openness and preference of the country's younger population for using digital tools. Therefore, governments need to acknowledge that boosting e-learners' sense of competence is essential to online education's long-term success. It may also be necessary to launch awareness campaigns and work to improve accessibility, training, and the availability of competent trainers, as well as adequate infrastructure.

Furthermore, the statistically substantial positive link between user happiness and intention to use indicates that user pleasure is a primary and critical element for user intention. As a result, the study findings are consistent with [23] showing that a continuous degree of satisfaction develops users' intention. Stakeholder satisfaction has attracted a great deal of attention to online learning. Satisfaction is also defined as a factor that drives people to intend to return in the future [19]. Furthermore, this is expanded upon in that users will continue with their desired aim provided their prerequisites are met satisfactorily [20].

This study's findings show that there is no significant positive relationship between technical service quality and user satisfaction; that user self-efficacy does not moderate the relationship between technical service quality and user satisfaction; and that self-efficacy and user intention to adopt e-learning are not related. One of the most frequently mentioned shortcomings of e-learning is the lack of opportunity for peer-to-peer learning and meaningful connection between the learner and mentor [49]. This study's findings back up previous findings. As a result, hypotheses H1 (the quality of technical services has a significant impact on user satisfaction), H6(b) (user self-efficacy moderates the relationship between content and information quality and user satisfaction), and H8 (self-efficacy significantly influences intention to use) are rejected.

Instead, the data indicate that significant revisions, alterations, and upgrades to the e-learning content are required to increase engagement. Because faculty members are primarily responsible for developing e-learning content, teachers may find this duty extremely draining, as not all professors are camera-savvy and comfortable filming lectures. Furthermore, few people may have the experience and expertise to create appropriate e-learning content. As a result, legislators and administrators must ensure that content creators receive appropriate training, equipment, and infrastructure to generate meaningful and dynamic e-learning content.

[50] argues that a person's sense of efficacy has a significant impact on whether or not they are able to persist in the face of adversity, whether they are able to meet the challenges posed by their environment without succumbing to stress and depression, and whether or not they are able to achieve their goals. This research showed that user satisfaction and intention to use with e-learning can be affected by both the learner's perception of their own abilities and the quality of the e-learning itself. Students' satisfaction with e-learning, in turn affecting their intention to use e-learning, was shown to be influenced by both students' self-efficacy and the quality of the e-learning system.

7 Implications, Limitations and Future Studies

7.1 Study Implications

Students' satisfaction with and commitment to e-learning are impacted by factors including their own sense of competence and the quality of the e-learning opportunities available to them, as shown by SEM. Students' satisfaction and future plans to use e-learning were found to improve when they had high levels of self-efficacy and access to high-quality technical service, educational service, content and information, and service. It showed that universities need to offer a range of services to back up the quality of e-learning if they want students to enjoy using it. Since both students and instructors use e-learning, instructors must develop strategies to boost students' confidence in their ability to learn online. Due to the lack of immediate, personal interaction, e-learning can be challenging to put into practice. While some students may be more receptive to online courses, others may benefit more from interacting with a live instructor in a classroom setting. Some students have high levels of self-efficacy regarding their ability to learn to use e-learning [51] because they believe online learning systems are easier to use and more valuable. However, it is argued that students' achievement in online courses is affected by personal traits like students' perceived self-efficacy [52]. Because of the challenges often encountered in e-learning, students' self-efficacy must be developed and improved for them to adopt this system voluntarily.

Users' satisfaction with e-learning is strongly influenced by both the e-learning quality and their own sense of competence in using it (self-efficacy). Universities should be able to foster and improve e-learning given their emphasis on quality e-learning and self-efficacy in delivering modules with tasks. Important because the university is responsible for providing and encouraging instructors to provide instructional materials that are effective for students. Self-efficacy learning's can be gauged by looking at students' grades and how happy they are with the experience.

The quality of e-learning and self-efficacy have a substantial impact on the satisfaction of e-learning users. The importance of e-learning quality and self-efficacy in delivering modules with tasks, then universities should be able to nurture and improve e-learning. It is critical since one of the university's roles is to supply and encourage professors to provide material that effectively teaches students. Determining the efficacy of e-learning is possible by analysing student achievement and satisfaction.

7.2 Limitations and Future Studies

As with other studies, this one has its caveats. At first, researchers will focus on the dimensions that are constrained by each variable; later, researchers will expand their scope to include all dimensions. Furthermore, the finding opens up several opportunities for further research. First, the current study does not investigate the influence of culture on e-learning adoption. Future research into the impact of culture may provide a more in-depth understanding. Second, the analysis suggests a single path to e-learning adoption, mediated through user pleasure. As a result, future research could concentrate on the influence of independent elements (technical service quality, content and information quality, educational service quality, and service quality) on user intention.

References

1. Moore, J.L., Dickson-Deane, C., Galyen, K.: E-Learning, online learning, and distance learning environments: are they the same? Internet High. Educ. **14**(2), 129–135 (2011)
2. Asvial, M., Mayangsari, J., Yudistriansyah, A.: Behavioral intention of e-Learning: a case study of distance learning at a junior high school in indonesia due to the COVID-19 pandemic. Int. J. Technol. **12**(1), 54–64 (2021)
3. Jiang, H., Islam, A.Y.M.A., Gu, X., Spector, J.M.: Online learning satisfaction in higher education during the COVID-19 pandemic: a regional comparison between Eastern and Western Chinese universities. Educ. Inf. Technol. **26**(6), 6747–6769 (2021). https://doi.org/10.1007/s10639-021-10519-x
4. Almaiah, M.A., Alyoussef, I.Y.: Analysis of the effect of course design, course content support, course assessment and instructor characteristics on the actual use of E-learning system. IEEE Access **7**, 171907–171922 (2019)
5. Khan, N.U.S., Yildiz, Y.: Impact of intangible characteristics of universities on student satisfaction. Amazonia Investiga **9**(26), 105–116 (2020)
6. Liu, X., et al.: Research on the effects of entrepreneurial education and entrepreneurial self-efficacy on college students' entrepreneurial intention. Front. Psychol. **10**, 869 (2019)
7. Zhao, Y., et al.: Do cultural differences affect users'e-learning adoption? A meta-analysis. Br. J. Edu. Technol. **52**(1), 20–41 (2021)
8. Nurkhin, A.: Analysis of factors affecting behavioral intention to use e-learning uses the unified theory of acceptance and use of technology approach. KnE Social Sciences, pp. 1005–1025–1005–1025 (2020)
9. Bandura, A.: Reflections on self-efficacy. Adv. Behav. Res. Ther. **1**(4), 237–269 (1978)
10. Alqurashi, E.: Predicting student satisfaction and perceived learning within online learning environments. Distance Educ. **40**(1), 133–148 (2019)

11. Chan, E.S., et al.: Self-efficacy, work engagement, and job satisfaction among teaching assistants in Hong Kong's inclusive education. SAGE Open **10**(3), 2158244020941008 (2020)
12. Dash, G., Chakraborty, D.: Transition to E-learning: By choice or by force â€"A cross-cultural and trans-national assessment. Prabandhan: Indian J. Manag. **14**(3), 8–23 (2021)
13. DeLone, W.H., McLean, E.R.: The DeLone and McLean model of information systems success: a ten-year update. J. Manag. Inf. Syst. **19**(4), 9–30 (2003)
14. Puška, E., Ejubović, A., Đalić, N., Puška, A.: Examination of influence of e-learning on academic success on the example of Bosnia and Herzegovina. Educ. Inf. Technol. **26**(2), 1977–1994 (2020). https://doi.org/10.1007/s10639-020-10343-9
15. Fathema, N., Shannon, D., Ross, M.: Expanding the Technology Acceptance Model (TAM) to examine faculty use of Learning Management Systems (LMSs) in higher education institutions. J. Online Learn. Teach. **11**(2), 210–233 (2015)
16. Rahmi, B., Birgoren, B., Aktepe, A.: Identifying factors affecting intention to use in distance learning systems. Turk. Online J. Distance Educ. **22**(2), 58–80 (2021)
17. Alzahrani, L., Seth, K.P.: Factors influencing students' satisfaction with continuous use of learning management systems during the COVID-19 pandemic: an empirical study. Educ. Inf. Technol. **26**(6), 6787–6805 (2021). https://doi.org/10.1007/s10639-021-10492-5
18. Albelbisi, N.A.: Development and validation of the MOOC success scale (MOOC-SS). Educ. Inf. Technol. **25**(5), 4535–4555 (2020). https://doi.org/10.1007/s10639-020-10186-4
19. Rienties, B., Toetenel, L.: The impact of learning design on student behaviour, satisfaction and performance: a cross-institutional comparison across 151 modules. Comput. Hum. Behav. **60**, 333–341 (2016)
20. Cheng, Y.-M.: How does task-technology fit influence cloud-based e-learning continuance and impact? Educ. Training **61**(4), 480–499 (2019)
21. Maheshwari, G.: Factors affecting students' intentions to undertake online learning: an empirical study in Vietnam. Educ. Inf. Technol. **26**(6), 6629–6649 (2021). https://doi.org/10.1007/s10639-021-10465-8
22. Ngai, E.W., Poon, J., Chan, Y.H.: Empirical examination of the adoption of WebCT using TAM. Comput. Educ. **48**(2), 250–267 (2007)
23. Pozón-López, I., et al.: Perceived user satisfaction and intention to use massive open online courses (MOOCs). J. Comput. High. Educ. **33**, 85–120 (2021)
24. Panigrahi, R., Srivastava, P.R., Panigrahi, P.K.: Effectiveness of e-learning: the mediating role of student engagement on perceived learning effectiveness. Inf. Technol. People **34**(7), 1840–1862 (2021)
25. Adefolalu, A.O.: Cognitive-behavioural theories and adherence: application and relevance in antiretroviral therapy. South. Afr. J. HIV Med. **19**(1), 1–7 (2018)
26. Seddon, P.B.: A respecification and extension of the DeLone and McLean model of IS success. Inf. Syst. Res. **8**(3), 240–253 (1997)
27. Hassanzadeh, A., Kanaani, F., Elahi, S.: A model for measuring e-learning systems success in universities. Expert Syst. Appl. **39**(12), 10959–10966 (2012)
28. Bandura, A.: The self system in reciprocal determinism. Am. Psychol. **33**(4), 344 (1978)
29. Akbari, M., et al.: The effect of E-learning on self-efficacy and sense of coherence of cancer caregivers: application of the bandura and antonovsky social cognitive theory. Curr. Health Sci. J. **47**(4), 539 (2021)
30. Fatahi, S., Kazemifard, M., Ghasem-Aghaee, N.: Design and implementation of an e-Learning model by considering learner's personality and emotions. Adv. Electr. Eng. Comput. Sci. **39**, 423–434 (2009)
31. Afshan, S., et al.: Internet banking in Pakistan: an extended technology acceptance perspective. Int. J. Bus. Inf. Syst. **27**(3), 383–410 (2018)
32. Bowen, N.K., Guo, S.: Structural Equation Modeling. Oxford University Press, Oxford (2011)

33. Latan, H., Noonan, R. (eds.): Partial least squares path modeling. Springer International Publishing, Cham (2017). https://doi.org/10.1007/978-3-319-64069-3
34. Avkiran, N.K., Ringle, C.M. (eds.): Partial least squares structural equation modeling. Springer International Publishing, Cham (2018)
35. Brown, T.A.: Confirmatory factor analysis for applied research. Guilford publications, New York (2015)
36. De Vaus, D., de Vaus, D.: Surveys in social research. Routledge, London (2013)
37. Sharif Nia, H., et al.: A second-order confirmatory factor analysis of the moral distress scale-revised for nurses. Nurs. Ethics **26**(4), 1199–1210 (2019)
38. Hair, J.F., Ringle, C.M., Sarstedt, M.: The use of partial least squares (PLS) to address marketing management topics. J. Mark. Theory Pract. **19**(2), 135–138 (2011)
39. Gunday, G., et al.: Effects of innovation types on firm performance. Int. J. Prod. Econ. **133**(2), 662–676 (2011)
40. Makanyeza, C., Mabenge, B.K., Ngorora-Madzimure, G.P.K.: Factors influencing small and medium enterprises' innovativeness: evidence from manufacturing companies in Harare Zimbabwe. Global Bus. Organ. Excellence **42**(3), 10–23 (2023)
41. Valenzuela, S., Piña, M., Ramírez, J.: Behavioral effects of framing on social media users: how conflict, economic, human interest, and morality frames drive news sharing. J. Commun. **67**(5), 803–826 (2017)
42. Cheng, Y.M.: Antecedents and consequences of e-learning acceptance. Inf. Syst. J. **21**(3), 269–299 (2011)
43. Cidral, W.A., et al.: E-learning success determinants: Brazilian empirical study. Comput. Educ. **122**, 273–290 (2018)
44. Phutela, N., Dwivedi, S.: A qualitative study of students' perspective on e-learning adoption in India. J. Appl. Res. High. Educ. **12**, 545–559 (2020)
45. Harrati, N., et al.: Exploring user satisfaction for e-learning systems via usage-based metrics and system usability scale analysis. Comput. Hum. Behav. **61**, 463–471 (2016)
46. Haryaka, U., Agus, F., Kridalaksana, A.H.: User satisfaction model for e-learning using smartphone. Procedia Comput. Sci. **116**, 373–380 (2017)
47. Dash, G., et al.: COVID-19 and E-Learning adoption in higher education: a multi-group analysis and recommendation. Sustainability **14**(14), 8799 (2022)
48. Daultani, Y., et al.: Perceived outcomes of e-learning: identifying key attributes affecting user satisfaction in higher education institutes. Measuring Bus. Excellence **25**(2), 216–229 (2021)
49. Oyediran, W.O., et al.: Prospects and limitations of e-learning application in private tertiary institutions amidst COVID-19 lockdown in Nigeria. Heliyon **6**(11), e05457 (2020)
50. Bandura, A.: Social cognitive theory: an agentic perspective. Annu. Rev. Psychol. **52**(1), 1–26 (2001)
51. Lee, J.-W., Mendlinger, S.: Perceived self-efficacy and its effect on online learning acceptance and student satisfaction. J. Serv. Sci. Manag. **4**(03), 243 (2011)
52. Yavuzalp, N., Bahcivan, E.: The online learning self-efficacy scale: its adaptation into Turkish and interpretation according to various variables. Turk. Online J. Distance Educ. **21**(1), 31–44 (2020)

A Constructivist Approach to Enhance Student's Learning in Hybrid Learning Environments

M. E. Rankapola(✉) and T. Zuva

Vaal University of Technology, Vanderbijlpark 1911, Republic of South Africa
merankaps@gmail.com

Abstract. The study used two case studies over a period of 12 months to develop and implement a constructivist approach for using lecture podcasts to enhance student learning in a hybrid learning environment. Two groups of undergraduate students enrolled at the Tshwane University of Technology participated in the study. The university's Learning Management System (LMS), MyTUTor which is based on D2L Brightspace was used as a server to host the podcasts. In the first semester (case study 1), the podcasts were merely utilized to complement the traditional teaching and learning mechanisms. In the second semester (case study 2), the podcasts were made compulsory (i.e., linked to progression requirements). Semi-structured interview and MyTUTor system log files were used to collect research data. The MyTUTor log files provided insight into frequency access and use of the podcasts. Semi-structured interview was conducted to gain insight into the students' experiences of podcasts use thereby establishing a relationship between the use of podcasts and learning facilitation. The findings posit that lecture podcasts are conducive to learning if they are student-centred and are integrated in teaching and learning design. The findings also postulate that student-generated podcasts accorded students an opportunity to become actively involved in the learning discourse thereby constructing their own knowledge and improving their problem-solving skills. The study concluded that student-generated podcasts led to increased podcast use and therefore enhanced learning if integrated in the teaching and learning design (i.e., if it is linked to progression requirements).

Keywords: Hybrid · Podcasts · Constructivist · Student-centered · LMS

1 Introduction

Mobile technologies have become ubiquitous among university students globally. In South African universities, 98 percent of students own mobile devices such as smart-phones, androids and iPods [1]. [1] examined the use of mobile technologies in the quest to enhance learning in the South African university environment. His study revealed that the ubiquitous presence of mobile phones with advanced features such as MP3/MP4 players; e-mail system and Internet have not been exploited for educational purposes. Similarly, another study conducted by [2] revealed that at the University of Cape Town in South Africa (UCT) over 98 percent of students own mobile phones and other mobile

R. Silhavy and P. Silhavy (Eds.): CSOC 2023, LNNS 724, pp. 654–663, 2023.
https://doi.org/10.1007/978-3-031-35314-7_55

technologies, although the use of these devices in teaching and learning is limited. Based on anecdotal evidence, almost all the students at Tshwane University of Technology (TUT) own modernised mobile devices with advanced features listed above. It is therefore pivotal to incorporate these devices which are commonly available to students in hybrid learning environments.

Hybrid learning environment involves the use of face-to-face lecture sessions and online-lecture sessions to facilitate learning. [2] suggested that the emphasis, when using ubiquitous technologies for educational purposes, should be on how computing and mobile devices share, distribute and enhance engagement (learning) with both content and knowledgeable human agents (lecturers and students). The current study described how student learning can be facilitated and enhanced in a hybrid learning environment by employing a constructivist approach.

In a constructivist approach students play an active role in the construction of knowledge and are not perceived as passive recipients of information and knowledge. Students used mobile devices to create, upload, listen to podcasts and engage with the learning content thereby constructing their own knowledge in a hybrid learning environment. Hein [3] informed us that constructivism refers to the idea that students construct knowledge for themselves - each student individually (and socially) constructs meaning as he or she learns. [4] noted that a constructivist type of learning requires a multiplicity of perspectives so that students will have a full range of options from which to construct their own knowledge. This is in accordance with the current study which aims to develop a constructivist approach to enhance learning in a hybrid learning environment.

2 Justification of the Study

The study exploited the opportunity of using the pervasive availability of mobile devices among university students for educational purposes. The study further explored how these mobile devices can be used to facilitate and enhance learning in a hybrid learning environment using a constructivist tactic. Thus, mobile devices such as cell phones were used to extend the learning space beyond the authentic formal learning contexts to informal learning settings such as home, on transport and leisure places. [5] contended that because hybrid learning environments contain both online and face-to-face components, they (hybrid formats) have the potential to offset deficiencies of traditional, large lectures while retaining positive aspects of classroom setting. The main source of motivation for this study is derived from the embedded disadvantages in both online and traditional learning environments and the opportunity to explore advantages of both in a constructivist methodology.

Literature [1, 2, 4–11] in the area of HLEs has provided empirical evidence about various learning environments and student learning preferences. Thus it has become apparent that online learning environments would become ineffective if they are designed independent to traditional face-to-face learning environments. The opposite of the preceding statement is also true. With the rapid developments in technology, learning environments ought to adapt and conform to the current standards and meet the needs of the digital age student. Adoption of hybrid teaching and learning approach does not automatically imply that students' learning will be enhanced and that students will be actively involved in the learning activity. A constructivist approach is necessary to be incorporated in hybrid learning environment.

3 Theoretical Framework

Constructivism is a theory of learning and knowledge closely associated with the research work of prominent authors [12, 13]. According to [13], the most fundamental and radical epistemological principle of constructivism embraces the idea that knowledge does not and cannot have the purpose of producing representations of an independent reality, but instead has an adaptive function. When people actively participate in real-world activities and problem solving activities, learning occurs [9]. According to the constructivist perspective, learning is determined by the complex interplay among students' prior knowledge, the social context (learning space), and the problem to be solved (assignments/projects). [14] echoed [15] who understood Constructivist Learning Environments (CLEs) as technology based spaces in which students 'explore, experiment, construct, converse and reflect on what they are doing so that they learn from their experiences'. This implies that students have to play an active role in constructing knowledge for themselves by exploring (using internet resources and other tools), experimenting/reflective learning (putting into action what has been learnt) and be able to communicate their own understanding to peers and lecturers. Active learning is a central tenet of constructivist learning theory and widely accepted as a hallmark of effective instruction [9]. Students in CLEs ought to be given a platform to discuss ideas, analyse, and reflect their thoughts towards enhancing cognitive and metacognitive outcomes. The constructivist paradigm views the 'context' in which learning occurs as central to the activity of learning itself, and this has proved to be a useful theory for designing and developing e-learning programs [16].

Hybrid Learning Environments are designed to integrate the best features of regular face-to-face learning with technology-based online-learning by dichotomizing the total class time into a web-based learning portion and an in-class or face-to-face meeting portion [2]. Hybrid learning environments blend traditional learning platforms and online or e-learning platforms. [6] described hybrid learning as learning systems that combine face-to-face instruction with technology mediated instruction. The hybrid learning approach warrants the use of computers or mobile devices and connectivity to support out-of-class independent learning activity and also motivate students to actively participate in class discussions. According to [17], learning quantity and quality suffers when students are solely and completely immersed in technology-based instructional delivery methods. Therefore, to eliminate the deleterious effects associated with the sole use of technology-based learning methods, many colleges and universities have adopted various forms of blended/hybrid instruction to more effectively deliver instructional content to students and promote their learning.

Podcasting is one of the platforms used in mobile learning or e-learning environments whereby a mobile device like a cell phone, laptop or even a stationery desktop computer is used to listen to an audio podcast or watch a video podcast [11]. [18] described podcasting as a blend of two words, namely iPod, the popular digital music player from Apple, and broadcasting. According to [19], podcasting is an audio content delivery approach based on web syndication protocols such as an Really Simple Syndicate (RSS) feed and secondly, podcasting intends to distribute data to mobile devices such as iPods, Moving Picture Experts Group (MPEG)-2 audio layer III (MP3) players, Personal Digital Assistants (PDAs) and mobile phones. Mobile devices which may be used for m-learning include, digital media players, notably, iPods and MP3 players; smartphones such as Blackberry and iPhone as well as PDAs like palmtops and pocket Personal Computers (PCs). Figure 1 below depicts the relationship between Constructivist Learning Theory (CLT) and Hybrid Learning Environment (HLE) empowered by the use of podcasting technology.

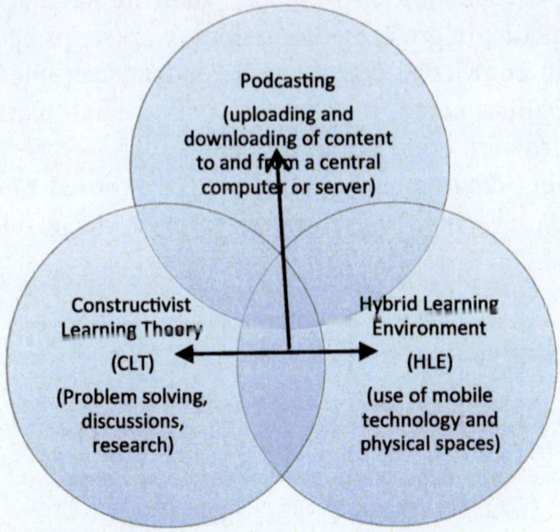

Fig. 1. Relationship between CLT, HLE & Podcasting.

4 Related Studies

[14] described the design of constructivist learning environment on pedagogical, social and technological perspectives. He suggested that the effectiveness of the constructivist learning environments in knowledge construction can be examined by further research in the field. [2] conducted a study to develop an approach for using podcasting to enhance student learning. Their findings suggested that students were confident in using podcasts and that lecture podcasts work effectively if they are tightly coupled to curriculum.

[16] investigated the influences of constructivist learning environments through the use of laptops. The findings of their study revealed different aspects of students' learning outcomes and enforcement to use creative thinking in building students' knowledge within constructivism learning context. The [7] case study investigated how blending of different instructional approaches with technology affects student engagement. The study reported that active learning was achieved not due to student individual differences but rather the learning environment provided in the problem-based blended learning.

5 Methodology and Description of Study

Different research paradigms and models are based on varying philosophical foundations and conceptions of reality [20, 21]. [22] posits that the positivist paradigm holds that knowledge is absolute and objective and that a single objective reality exists external to human beings. Interpretivist, by contrast, aims to find new interpretations or underlying meanings and adheres to the ontological assumption of multiple realities, which are time-and context dependent [22]. The current study followed an interpretivist paradigm because the study's fundamental aim is to design and develop a constructivist approach to enhance learning by putting the student at the centre of the system (student centeredness). In a constructivist learning environment, students have to construct their own knowledge by participating in projects, discussions, workshops etc. Through participation or involvement in knowledge construction, students can interpret and make their own meanings. As [20] has noted, in the process of research participants often create new meanings and make new connections of ideas.

In a hybrid learning environment, students were exposed to e-learning sesstions delivered through the universityd's LMS in the form of videos and traditional face to face sessions.

5.1 Description of Study:

Case Study 1: Semester 1 (Table 1).

Table 1. Case study 1

Number of participants	Number of contact weeks	Teaching & learning strategy	Pedagogy
20	15	Wikis Blog Face to face lectures Group work	Lecturer generated podcasts: complemental Student generated podcasts: complemental

Case Study 2: Semester 2 (Table 2).

Table 2. Case study 2

Number of participants	Number of conctact weeks	Teaching & learning strategy	Pedagogy
20	15	Wikis Blog Face to face lectures Group work	Lecturer generated podcasts: compulsory Student generated podcasts: compulsory

6 Data Analysis

Research data was collected by using semi-structured interviews and access logs from the MyTUTor Learning Management System (LMS). MyTUTor system logs provided insight into the frequency of podcast access or use. These data assisted in establishing whether or not students were engaged in learning. Interview data was analysed using themes that were developed from the data collected during the interviews (Table 3).

6.1 Access Logs

In case study 1, the lecturer-generated podcast indicates highest access or use at the interval of 31–35 (11). Moreover, the lecturer-generated podcast achieved an average of 59% as compared to 41% of the student-generated podcast. Although there is only 8% difference in case study 1 podcast accesses, it appears as if students were more receptive to the lecturer-generated podcast than to the student-generated podcast. Some of the reasons that could have contributed to this scenario are; a) some students were first time users of the podcasting software (Camtasia, audacity) and also the MyTUTor system, therefore were not able to produce good quality podcasts, b) students still believed that the lecturer is the knowledge agent with superior knowledge than students, and c) some of the student-generated podcasts did not comply with the principles of good podcast production in terms of length (some were 20 min long). In case study 2, the student-generated podcast achieved 58% access rate whereas the lecturer-generated podcast achieved 42% access rate. The results of case study 2 are slightly the opposite of the case study 1 results. In this case, students seem to be more receptive to the student-generated podcast than the lecturer-generated podcast. The main reason that could have attributed to this outcome may be that students were to be graded on podcast-production as part of their progression requirement. Subsequently, all students generated excellent podcasts that were shared amongst their colleagues which were then used as valuable learning resources (Figs. 2 and 3).

Table 3. Access logs

	Case study 1		Case study 2	
Interval	Lecturer podcast	Student podcast	Lecturer podcast	Student podcast
0 - 5	0	1	0	2
6 - 10	1	3	2	4
11 - 15	3	0	2	3
16 – 20	4	1	3	4
21 - 25	2	3	1	2
26 – 30	7	2	4	6
31 – 35	11	4	6	3
36 - 40	2	5	1	5
41 - 45	1	0	0	2
46 - 50	2	2	0	2
51 – 55	3	0	6	3
56 - 60	1	2	2	4
61 - 65	0	1	1	0
66 - 70	0	0	0	1
71 - 75	0	2	3	2
Total	37	26	31	43
Percentage	59%	41%	42%	58%

6.2 Interviews

Semi-structured interviews were conducted at the end of case study 2. All participants (n = 20) were invited to participate in the interview. However only 14 students were available to be interviewed on their experience in using podcasts for learning. Thematic analysis method was used to analyse the interview data. The following themes emerged from the interview (Table 4):

Technical System Quality
The majority of students (n = 9) reported that they experienced difficulties in accessing lecture and/or student podcast away from the university premises because of internet connectivity constraints. Only a small number of students (n = 5) reported that they were able to access the podcasts at home. Therefore, many students were obliged to access the podcast at the university and downloaded to their mobile devices for offline use.

Collaborative Learning
Students were receptive to collaborative strategies such as group discussions in class and online interactions that were facilitated through the use of a wiki and a forum. The

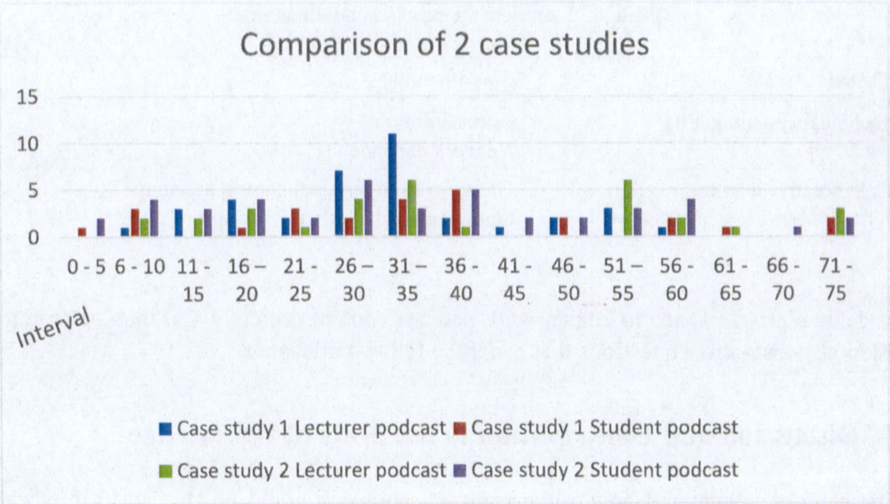

Fig. 2. Comparison of case study 1 and case study 2

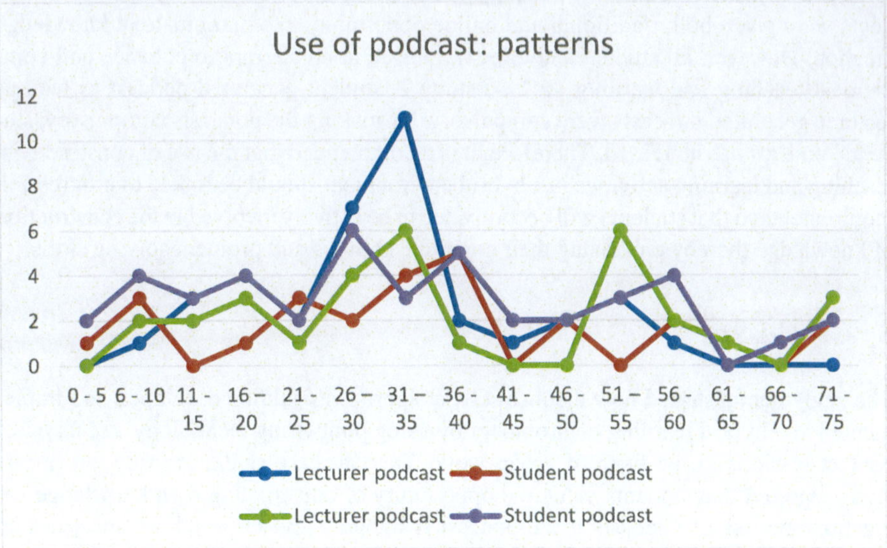

Fig. 3. Podcast access patterns

interview results indicated that students were gratified by been given an opportunity to construct knowledge and share it with colleagues. As the generation of the podcast by the students was compulsory, students were highly motivated to perform better knowing that successful completion of this task will have a bearing on their final average. It also emerged that the use of a wiki to post questions to the lecturer and colleagues, accorded

Table 4. Interview themes and classifications

Theme	Classification
Technical system quality	Accessibility Network dependency
Collaborative learning	Group work in physical contact sessions Online interactions (wiki, forum)

students a private space to engage with podcast content critically and thus be able to ask/post constructive questions that benefited fellow colleagues.

7 Discussion and Contribution to the Body of Knowledge

The study demonstrated how a constructivist approach could be applied in a hybrid learning environment by conducting two case studies. Teaching and learning strategies and pedagogy designs should be student-centred thereby giving students an opportunity to be actively involved in constructing and acquiring knowledge. In case study 1, students were given both traditional and online opportunities to participate in knowledge creation. However, all student activities were used in an attempt to enhance and complement teaching and learning. In case study 2, student-generated podcast as well as lecturer-generated podcasts were compulsory. By making the podcasts compulsory, students' podcast use increased. Therefore, it is recommended that the use of constructivist teaching and learning activities in a hybrid environment should be linked to progression requirements so that students will be motivated to be actively involved in the construction of knowledge thereby enhancing their cognitive abilities and problem solving skills.

8 Conclusion

The study demonstrated how a constructivist approach could be developed and implemented in a hybrid learning environment by using podcasting technology and physical contact sessions in the form of group work. The two case studies carried out in this study reported that students value the opportunity of constructing own knowledge i.e. creation and use of podcasts if the podcast is integrated in the teaching and learning design. As seen in case 1, a complemental podcast was used and students appeared to be less receptive in creating and using the podcasts. Case study 2 reported increased podcast creation and use because the podcasts formed part of progression requirements. The study recommends and concludes that constructivist approach in a hybrid learning environment activates and stimulates students' cognitive and social competences. The extent to which this approach is applied successfully, depends on various environmental and pedagogical factors. As seen in this study, some challenges which include technical system quality may hamper the successful development and implementation of this approach. It is therefore important to be aware of environmental challenges before embarking on this approach.

References

1. Foko, T.: The use of mobile technologies in enhancing learning in South Africa and the challenges of increasing digital divide. In: ED-MEDIA 2009-World Conference on Educational Multimedia. Honolulu, Hawaii: Hypermedia & Telecommunications (2009)
2. Ng'ambi, D., Lombe, A.: Using podcasting to facilitate student learning: a constructivist perspective. Educ. Technol. Soc. **15**(4), 181–192 (2012)
3. Hein, G.E.: The museum and the needs of people. In: International Committee of Museum Educators. Jerusalem Israel: Lesley College (1991)
4. Noodrink, M.: Different ways of teaching, different pedagogical approaches: pedagogies for flexible learning supported by technology, in e-learning. Ootmarsum, Netherlands (2010)
5. Riffel, S.K., Sibley, D.H.: Learning online: student perceptions of a hybrid learning format. J. Coll. Sci. Teach. **32**(6), 394–399 (2003)
6. Bonk, C.J., Graham, C.R.: The handbook of blended learning: global perspectives. local designs. Acad. Manag. Learn. Educ. **7**(1), pp. 132–137 (2008)
7. Delialioğlu, Ö.: Student engagement in blended learning environments with lecture-based and problem-based instructional approaches. Educ. Technol. Soc. **15**(3), 310–322 (2012)
8. Dewhurst, D.G., Macleod, H.A., Norris, T.A.M.: Independent student learning aided by computers: an acceptable alternative to lectures? Comput. Educ. **35**(3), 223–241 (2000)
9. Oakleaf, M., VanScoy, A.: Instructional strategies for digital reference: methods to facilitate student learning. Comput. Educ. **49**(4), 380–390 (2010)
10. Ostiguy, N., Haffer, A.: Assessing differences in instructional methods uncovering how students learn best. J. Sci. Teach. **30**(6), 370–374 (2001)
11. Rankapola, M.E.: The effect of podcasting revision lectures in improving the learners' academic performance. Online J. Distance Educ. e-Learning **2**(4), 81–94 (2014)
12. Piaget, J.: Sociological studies. J. Coll. Sci. Teach. 1–10 (1989)
13. Vygotsky, L.S.: Mind in society: the development of higher psychological processes.. J. Coll. Sci. Teach. 34–45 (1978)
14. Wang, Q.: Designing a web-based constructivist learning environment. Interact. Learn. Environ. **17**(1), 1–13 (2007)
15. Taber, K.: Constructivism as education theory: contingency in learning, and optimally guided instruction. Educ. Theor. 39–61 (2011)
16. Sultan, W.H., Woods, P.C., Koo, A.-C.: A constructivist approach for digital learning: Malaysian schools case study. Educ. Technol. Soc. **14**(4), 149–163 (2011)
17. Lim, D.H., Morris, M.L.: Learner and instructional factors influencing learning outcomes within a blended learning environment. Educ. Technol. Soc. **12**(4), 282–293 (2009)
18. Evans, C.: The effectiveness of m-learning in the form of podcast revision lectures in higher education. Comput. Educ. **50**(2), 491–498 (2008)
19. Dale, C.: Strategies for using podcasting to support student learning. J. Hospitality Leisure Sports Tourism Educ. **6**(1), 646–657 (2007)
20. Cohen, L., Manion, L., Morrison, K.: Research Methods in Education (5th Edn.) Br. J. Educ. Stud. **48**(4), 446–468 (2000)
21. Du Poy, E., Itlin, L.N.: Introduction to Research: Understanding and Applying Multiple Strategies (2nd ed.). Mosby Inc: St Louis (1998)
22. De Villiers, M.R.: Three approaches as pillars for interpretive Information Systems research: development research, action research and grounded theory. In: Proceedings of SAICSIT 2005. South Africa: UNISA, pp. 112 (2005)

Edge Detection Algorithms to Improve the Control of Robotic Hands

Ricardo Manuel Arias Velásquez(✉)

Universidad Tecnológica del Perú, Lima, Peru
ricardoariasvelasquez@hotmail.com, rariasv@utp.edu.pe
https://utp.edu.pe/

Abstract. This paper improves the remote control of a robotic hand through a comparison of two artificial vision algorithms, by using the "Edge Detection" and "Find Hand" respectively, for both cases the programming uses the Python language. In addition, the system features a wireless connection with the robotic hand's servo motor controller, making it easy to manipulate the hand from a far point with Kalman modified, Butterworth filters, and control movement prediction with two feedbacks loops. In response to the tests carried out, this article determined an optimal operation for the automatic and remote-control of the hand and fingers, with high accuracy, with the precision of 78.7% to 94.44%, and F1 measurement from 77.38% to 92.30%; with an improvement to reduce the noise and mechanical influence in the electrical circuit.

Keywords: Artificial intelligence · edge detection · robotic hand

1 Introduction

Currently, robotics has presented improvements in its control and the communication channels that are used by users for its manipulation, highlighting among these the use of remote control for its remote and wireless management. On the other hand, the algorithms and programming used have been improved, making these machines do not require so much manipulation from the operators [1].

Additionally, in the last 3 years the authors in [1,2]; they indicate that robotics is commonly used in areas such as studies, industries, exhibitions, research fields and the like. On the one hand, the International Federation of Robotics (IFR) points out that since 2010 the request for robots has increased annually by 15% to date [2]. In recent times, the use of remote control of robots for research and teaching has been marked, and in the same way that in industries they present problems for users who want a reduced response time or a wider control range.

Due to this, other studies propose teleoperated systems to maneuver elements and dangerous materials through wireless devices such as virtual reality controls or artificial vision. Some establish a prototype of the robotic hand considering

R. Silhavy and P. Silhavy (Eds.): CSOC 2023, LNNS 724, pp. 664–678, 2023.
https://doi.org/10.1007/978-3-031-35314-7_56

the use of neural networks, achieving a degree of accuracy of 85% [3]. All this developed to minimize accidents in dangerous operations or to support people.

After reviewing the research related to the remote control of a robotic hand, it was shown that they use different devices to control the system. However, these systems hardly include artificial vision or if they do include it, they only focus on objects or position. For this reason, it is necessary to carry out a comparative analysis between image processing algorithms that allow improving the remote-control system through artificial vision.

This research article has the following content: Sect. 2, the systematic review for the evaluation of the artificial vision and control of robot hand with the limitations and contributions. The Sect. 3 has the development of the Butterworth and Kalman modified filter and the algorithms for the visual recognition and feedback to the closed loop control, based on the representation of the Denavit Hartenberg parameters for the hand and fingers integrated. The Sect. 4 has the discussion and constraints in the real implementation with a robotic hand. Finally, the last section has the conclusions and recommendations for future works.

2 Systematic Review for the Evaluation of Artificial Vision in Robotic Applications

The last decade has increased the research articles associated to robotics hands and artificial vision with new technological advances that allow robots to be manipulated remotely to replace human participation in dangerous situations or on occasions in which attendance is an impossible alternative. Among them, the papers in [3–5] introduce the manipulation and movements of the robotics hands through the use of kinematic methods and equations. On the other hand the papers in [4,5] introduce similarity the use of a haptic control, with a master - slave architecture with two and one arms, with kinematic equations. In the same way, the Ref. [3] carries out its research in remote control using virtual reality gloves (VRFree) together with a virtual hand to manipulate objects with the robot's fingertips, allowing the comparison of the precision of each one for the results. On the other hand, research linked to the control of robots in a wired or wireless way also presents the use of artificial vision as the main theme. In addition, they use artificial vision to improve grip in robotic hands or arms in order to optimize the manipulation of objects. The evolution and limitation of several authors associated to the methodologies in the Table 1. In [3], the methodology is It uses two data gloves, the VRFree and the Manus VR, and a vision-based system, the Leap Motion Controller. The methodology is based on synergy. Besides in [4], the kinematics equations consider the amount of DoF of the robot. Therefore, they use a bilateral teleoperation system composed of a master interface and a speed-controlled [5], with a synchronization in real time with the slave robot. As a complementary, the PD controller applied to robot dynamics [6], as an example, the haptic interfaces enable end-effector matching and force feedback with a Panda arms, it uses an SL robotics simulator for real-time control [6], the evaluation by using an architecture of multilevel convolutional neural networks associated to vision algorithms with two cameras to recognize grip points on the object and uses of Corner Grasping to obtain data. On the other hand, the use of clamp-mounted RGB-D camera to avoid the use of additional sensors, in this case, the segmentation is used to improve accuracy and grip. It uses a neural network capable of detecting different grip poses. 4'400 images were used in training. 50 layers of residual neural networks [7]. In other words, through a camera the element is captured using artificial vision, for this purpose, a neural network is used to identify the object and make an association with the required movement, using robot kinematics algorithms for joint control [8]. Consequently, through a camera the information of the position in the 3D Cartesian plane is captured, for which they use kinematics equations to control the robot. In this, a non-linear neural network is used to compare patterns and calibrate the images. With the homogeneous transformation, the authors have a computational time of 0.142 and a precision of 1.715 mm. While with the neural network it has a computational time of 2.355 and a precision of 0.923 mm.

Table 1. Methodology, KPI and constraints in the last research articles.

Author (Year)	KPI	Constraints
C. Mizera et al. [3] (2019)	Robotic hand teleoperation to manipulate objects with the tips of the fingers in a virtual way. The system resolution is 12 MP. The cameras are located 10 to 2 m from the target, and it has a margin of error of less than 0.4 mm	The slides between the user's wrists produce errors, the sensors detect non-existent movements that should be filtered. Depending on the controller used, there are limitations on its average flex, the lowest being 7.5° and the highest being 16.5°
R. Rahal et al. [4] (2020)	Remote control of a robotic arm through haptic control. The human workload Hw presented an average of 60.7%, with the variation in NASA TLX it is adjusted to 60.3%	Time and placement error degrade by 14% and 10%, respectively. So the robot deviates from the shortest path in exchange for greater comfort for the user
R. Muthusamy et al. [10] (2021)	Use of artificial vision by events for the control of a robotic arm. Five tests were carried out for each of the three shapes of objects and an average holding coefficient was obtained: triangle 1.22 cm, for rectangle 0.48 cm and for pentagon 0.44 cm	Depending on the processing rate of the event, a corner detection algorithm should be selected to improve the speed. Caution should be taken with lighting and variable speeds

During 2021, an event-based control method (EBVS) is used, for which a UR10 robot is used as a base, and to this it applies artificial vision control based on event-based corner detector algorithms. Finally, in Table 1, the stereo method is used to capture events, for which it uses kinematic models and makes D-H tables, using two cameras to capture the position of the object, the success is based on the hidden layer of the neural network has 5 neurons, which achieves a learning rate of 0.05 and several iterations of 10,000. For its part, research [10] also uses artificial vision, but instead of using a common registration method like other studies, this research uses an event-based control method (EBVS). In addition, it is important to mention that the study works based on the UR10 robot to which it adds the artificial vision tool. It is also necessary to detail that it focuses on events for the detection of corner features, through heat maps it locates and generates data for the tracking and alignment of the clamping. Likewise, depending on the intensity of the light used, they present as a result errors ranging from 8.39% to 26.02%, comparing intensities of 365 lumens and 10 lumens. On the other hand, the study [11], like the previous study, focuses on capturing events but unlike the previous one, it uses a stereo method for which it requires two cameras that capture the position of the object to be manipulated. In addition, as an additional value, it uses kinematic models, and it makes tables of homogeneous matrices with Denavit Hartemberg parameters. In addition, because of 192 data to which the stereo method was applied, the average error was approximately 2.5 cm.

3 Methodology to Improve the Hand Control for Robotic Applications

In recent years, the methods used to control robotic hands have various applications such as: Control of robotic prostheses [9,11], rehabilitation [10], teleoperation [12,13], detection of diseases such as rheumatoid arthritis [14], etc. And depending on your application some methods are more suitable, being the most used the one that makes use of inertial measurement units (IMUs), such as accelerometers, gyroscopes and magnetometers, and this due to features such as: A high accuracy, high usability, high portability and low cost. Said results seven come reflected in its average error that ranges from 3 to 3.5°, and standard deviations of 1°.

3.1 Kalman Modified Filter

The Kalman modified filter consists of two steps (prediction according to the actual position in Eq. 1 with the gain k in the Eq. 2, and the associated covariance of the error for the next moment and update in the Eq. 3 and Eq. 4 and aims to minimize future value error by performing a linear combination of an estimate compared to the difference between the actual measurement and the measurement predicted by the Kalman gain [15].

$$x_n = x_{n-1} + k(y - C_{x_{new}}) \tag{1}$$

$$k = \frac{p_{n-1}C^T}{CP_{n-1}C^T + R} \tag{2}$$

$$P_n = P_{n-1} + k(C - P_{n-1}) \tag{3}$$

$$y_{n+1} = C_{X_{n-1}} \tag{4}$$

where: X: It is the position of the process during the process (real position of the hand). y: It is the image read in the cam (with a delay in the lecture). k: It is the Kalman gain. n: It is the period 0, in real time. n − 1: It is the previous movement. n + 1: It is the future estimated. P: It is covariance matrix, associated to the error in the state. R: It is the covariance of the noise.

3.2 Butterworth Filter

The Butterworth filter is one of the most widely used frequency domain filters (in ECG, EEG and EMG signals) as it provides a frequency response that is maximally flat thanks to the use of Butterworth polynomials and, compared to other filters do not present ripples in the pass band, but the attenuation is the slowest [16]. The Butterworth filter is a recursive filter that depending on its order the curve can achieve a smooth decay or may resemble a step function [17]. The complementary filter is a filter used for data fusion such as that of the accelerometer and gyroscope, which can be considered as a simplification of the filter of Kalman that completely dispenses with statistical analysis.

$$\theta = k \times (\theta_{n-1} + \theta_g) + (1 + k) \times \theta_a \tag{5}$$

where: θ: Angle of moment in the curve of each hand and fingers. θ_g: Gyroscope value in real time. θ_a: Accelerometer value in real time.

The implementation of the sensors has been used with an evaluation of a real hand in the Fig. 1 A) and the robotic hand with the control cables.

3.3 Artificial Vision Algorithms

It shows the text editor with the OpenCV, CVZone, MediaPipe and Serial Tools machine vision libraries as input. Besides, the lines identified and recognize the hand through a webcam. Besides, a pre-defined variables are shown that allow capturing the video, in this case the variable in charge of that is 'cap', followed by the variable 'detector' stores the information of the detection of the hand by means of vectors, likewise allows the configuration of how many hands can be detected and, also works with a value of 0.8 in the second variable since being equal to or greater than 1.

Fig. 1. A) Real hand for the analysis of the movements and, B) The robotic hand control, motors for the finger movements and hand.

3.4 Control Integrated with PID, Artificial Vision and Filters

Finally, the block diagram was elaborated as shown in Fig. 2, where the flow and feedback of this system is explained, starting with the desired angle setting, to then calculate the error through the comparison of the desired angle and the angle obtained by the IMU sensors through your DMP module. Once the error is calculated, it will go through the PID controller and then by a saturation block, which will command the power circuit and the servomotors, updating the reading of the IMU sensors, thus returning to the loop initial. In this case the change in the position identified by the artificial vision algorithm, with the noise in the signal of the accelerometers and gyroscope and the electronics devices of the hand and fingers caused by the motor actions.

3.5 Key Performance Index (KPI) for the Evaluation.

Table 2. Confusion matrix evaluation for the KPI.

Reality	Prediction 0	Prediction 1
0	TN	FP
1	FN	TP

According to the Table 2, the paper has the following KPI:

- Precision: With the precision metric we can measure the quality of the machine learning model in classification tasks [19,20], in the Eq. 6

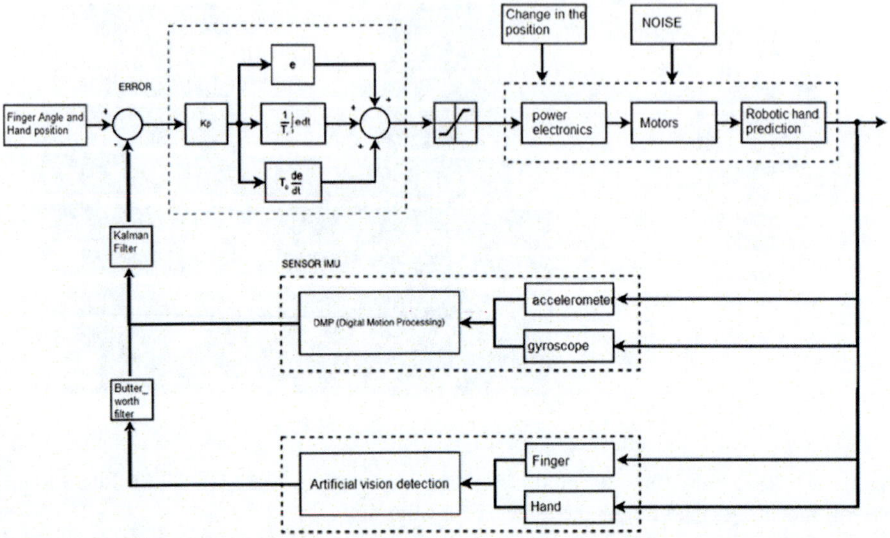

Fig. 2. Methodology for the control and prediction with Kalman and Butterworth filter, sensor IMU and artificial vision.

- Recall: The completeness metric will inform us about the amount that the machine learning model is able to identify [19,20], in the Eq. 7.
- F1 measurement: The F1 value is used to combine the precision and recall measures into a single value. Therefore, it compares the combined performance of accuracy and completeness between various solutions; it is evaluated by taking the harmonic mean between precision and completeness [19,20], in the Eq. 8.
- Accuracy: it evaluates the percentage of cases that the model has succeeded [19,20], in the Eq. 9.

$$Precision = \frac{TP}{TP + FP} \tag{6}$$

$$recall = \frac{TP}{TP + FN} \tag{7}$$

$$F1 = \frac{2 \times precision \times recall}{precision + recall} \tag{8}$$

$$accuracy = \frac{TP + TN}{TP + TN + FP + FN} \tag{9}$$

4 Discussion

The first step in calibrating the power driver is to configure, programmatically, its internal clock to match the refresh signal of its PWM outputs with the control frequency of the servomotors (which 50 Hz according to manual), for which the digital oscilloscope Tektronik model TBS 1102BEDU, according to the Fig. 3.

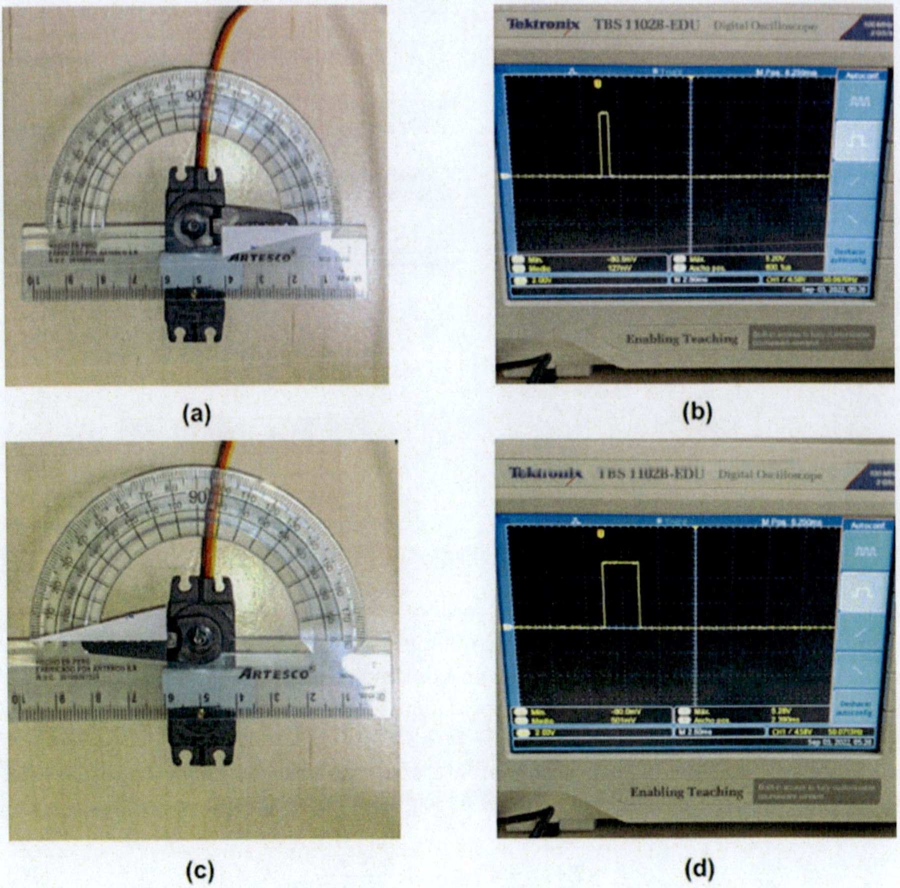

Fig. 3. Calibration of the Power driver PCA9685 with oscilloscope with A): 0° and B) detection of 0°, and C) 180° and D) detection of 180°.

To interpret the measured data, it is necessary to make a referencing. In this section explains the calibration process and provides the necessary knowledge for the conversion of IMU sensors, starting from its internal structure. The MPU 6050 sensor has an internal digital processing engine of movement, also known as DMP, which is used to merge the data of the gyroscope and accelerometer, then said data will be stored in the FIFO buffer, reducing the data acquisition

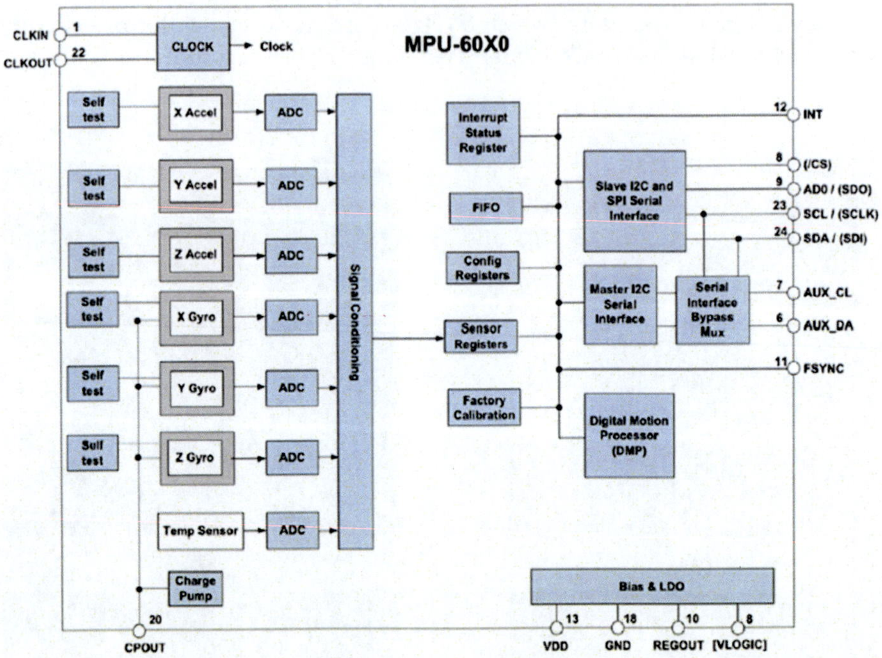

Fig. 4. Block diagram MPU6050.

time, by reading bursts of data and not one by one, in Fig. 4. The evaluation with a step input of the GP system in continuous time using a PID controller in its parallel form. Where the proportional gain P = 0.692, the integral gain I = 0.692/0.863, the derivative gain D = 0.21575 * 0.692 and the derivative filter coefficient N = 100; in order to evaluate the response curve of the embedded PID controller before a step entrance in Eq. 10.

$$G_c(s) = P + \frac{1}{s} + \frac{D_s}{N_s + 1} \tag{10}$$

Therefore, the behavior of the embedded PID controller over time discrete and we notice a random alteration in the measured angular position, this is due to the noise that is generated by merging the gyroscope and accelerometer signals for the calculation of the angle of rotation in the Y axis of the sensor, in the Fig. 5. It also evaluates the controller reacts to said spasm, correcting its position in 1.2 s, in Fig. 6 A), in this case the response of the embedded PID controller to noise or external movement. Besides, in the Fig. 6. B) with the evaluation of the

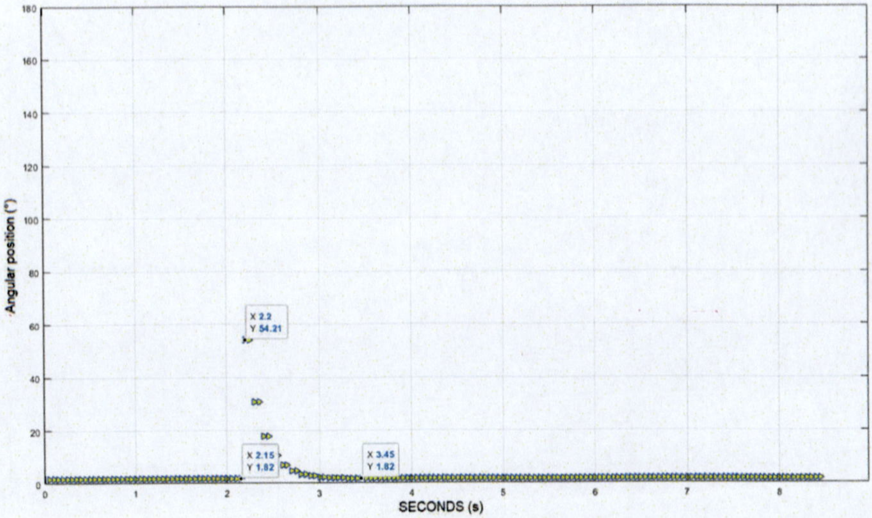

Fig. 5. Regulation of a step pulse in the motors.

Fig. 7, practical experiments show that the controller has the ability to generate configurations, but spasms may occur that momentarily alter the position of a finger, the evaluation is improved with the artificial vision. In this paper in the Fig. 8, a comparative analysis of the edge detection and Find Hand algorithms was developed through the use of python and artificial vision through a webcam. For which, a prototype of an artificial hand that can react to the data processed by the software was implemented and for this, an Arduino Uno controller was used that drives the servomotors connected to each finger of the hand prototype. On the other hand, to obtain quantitative results, 120 tests were carried out with Python during the day and at night. Likewise, a confusion matrix was used for which 120 tests were carried out to find specific accuracy metrics of the algorithms. Then, a comparison of the results of the two algorithms was made to determine the percentages of accuracy and error in the Table 3. The main indicators have the following results with the PDI control and the PDI with the methodology proposed, the improvement with the traditional method indicated in the Table 4, with 120 hand movements:

- Precision from 78.70% to 94.44%.
- Recall from 76.10% to 90.26%.
- F1 measurement from 77.38% to 92.30%.
- Accuracy from 70.83% to 85.83%.

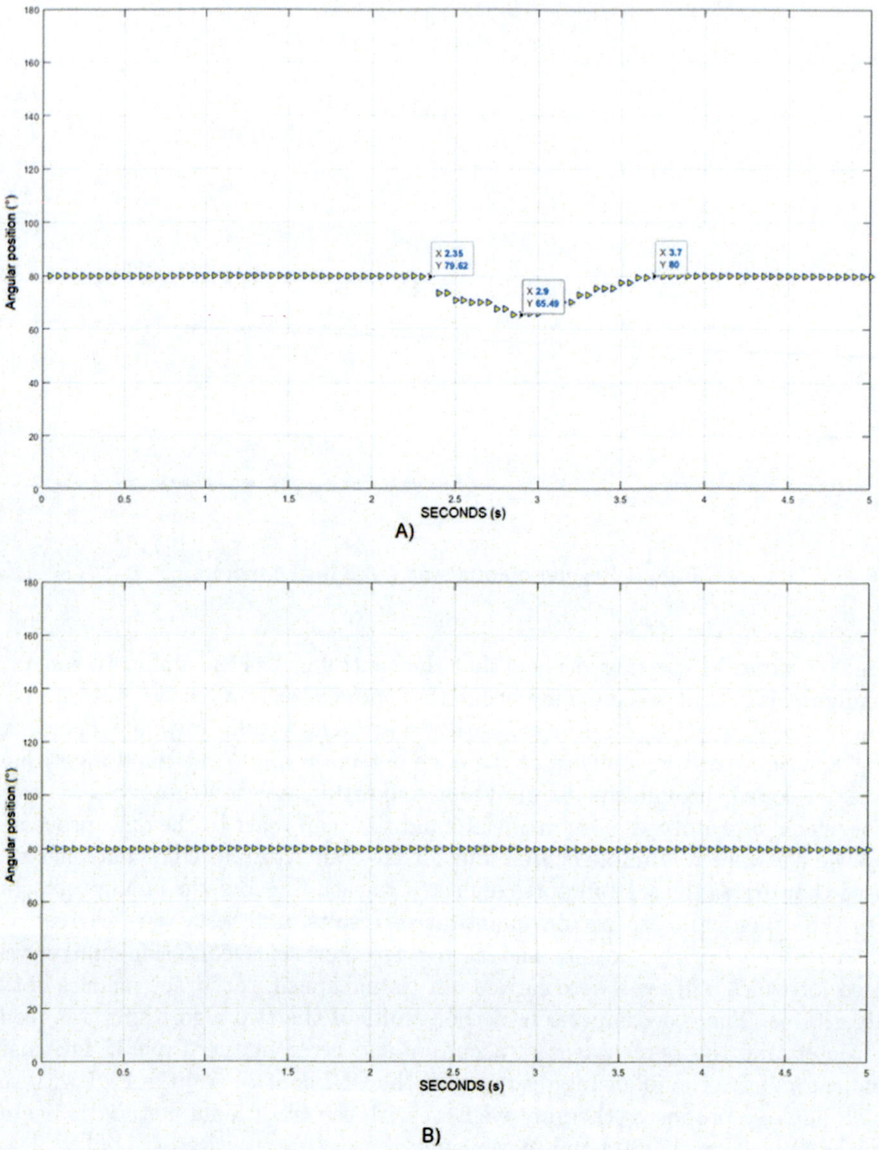

Fig. 6. A) Noise or external movement influence. B) Same condition of the noise but with the feedback of filters and artificial vision.

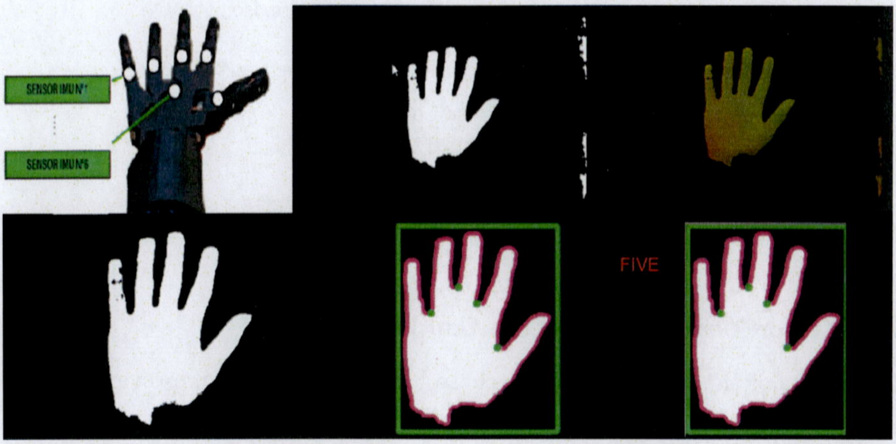

Fig. 7. Reduction of the noise with the artificial vision of the robot.

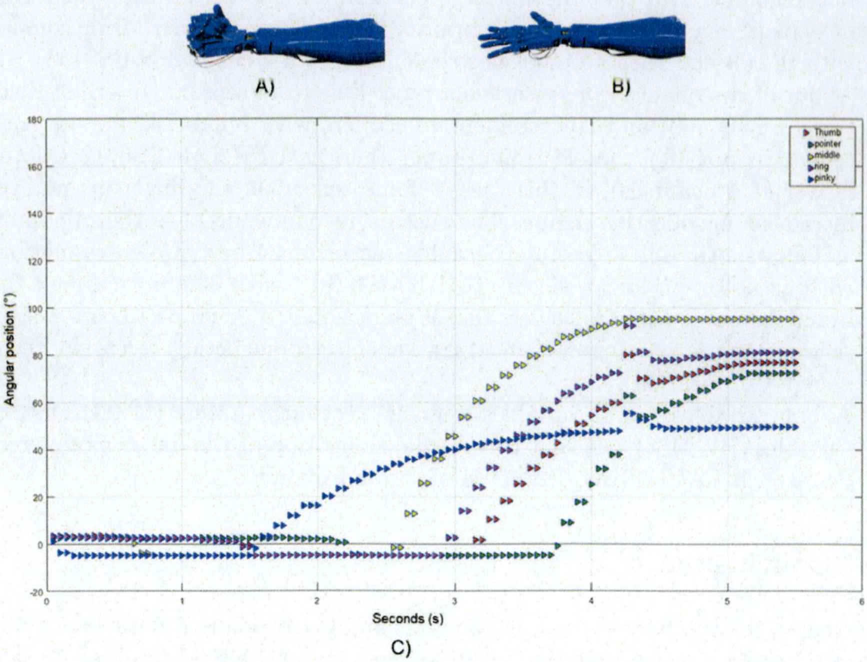

Fig. 8. A) Hand closed, B) hand open, C) angle position evaluated.

Table 3. Confusion matrix results

Reality	False	Positive
Incorrect movement	1	6
Correct movement	11	102

Table 4. Key performance index comparison

Description	With PDI control	With PDI + filters + artificial vision
Precision	78.70%	94.44%
Recall	76.10%	90.26%
F1	77.38%	92.30%
Accuracy	70.83%	85.83%

4.1 Lesson Learned About the Control

The results of the hands and finger control has the following improvements:

- The limitations of the Control technique: It set the initials parameters in the position of the joints, in this case, the error of the force and position cannot converge to zero at the same time, when holding objects deformable. It is recommended that the converge position error to zero during the finger movement in a free space and give priority to regulation of force during contact with the object; with a power electronic MOSFTET channel N BSS138.
- About of the quantity of sensors and type: Five force sensors (Interlink Electronics FSR 400) and 10 incremental encoders with Micro DC motors with reduction box, for the electronic power there is DFRobot TB6612 (Motor Driver of 4 channels), in this case, a force sensor on the fingertips and an increment encoder by engine. The sensors recommended are the minimum as follows: An ASIC electromyographic sensor of 24 bits of own resolution, with actuators: 5 linear sensors (L12-30-100-6-I), with 30mm races and the brand Actuonix Motion Devices Inc; as an optional, it could be 1 Leap Motion sensor, 3 motion sensors flex and 4 trackers electromagnetic (Ascension Technology Corporation).
- About variables: Input variables: Signals electromyography (EMG) Output variables: PWM signals that control the elongation of the linear motor, recommended LAUNCHXL-F28069M of Texas Instruments.

5 Conclusion

Currently, the feedback with accelerometers and gyroscope in robotic hands and fingers has limitations precision 78.70%, accuracy of 70.83%, recall of 77.38% and F1 of 77.38% with the implementation of Butterworth, Kalman modified filter and the artificial vision the results increase: The key indicators are the precision 94.44%, accuracy of 85.83%, recall of 90.26% and F1 of 92.30%. The improvement in the control is possible with the representation of the Denavit Hartenberg in the finger control and binarized image emit with an RGB image and the edge recognition for the prediction of the next movement; in this case, the complementary filter has a correction for the data fusion such as that of the accelerometer and gyroscope; besides, minimize future value error by performing a linear combination of an estimate compared to the difference between the actual measurement and the measurement predicted. For future works, it is possible to reduce the velocity of the feedback with extension of the system that allows the control of the entire arm robotic as the bicep, wrist and shoulder.

Acknowledgements. Universidad Tecnológica del Perú.

References

1. Sivaraman, P., et al.: Humanoid gesture control ARM with manifold actuation by embedded system. Mater. Today Proc. **37**(Part 2), 2749–2758 (2020). https://doi.org/10.1016/j.matpr.2020.08.545
2. Seitz, M.: Qué países tienen más robots en sus fábricas y cuán cierto es que nos están robando los puestos de trabajo. BBC NEWS, March 2017. https://www.bbc.com/mundo/noticias-39267567
3. Mizera, C., Delrieu, T., Weistroffer, V., Andriot, C., Decatoire, A., Gazeau, J.-P.: Evaluation of hand-tracking systems in teleoperation and virtual dexterous manipulation. IEEE Sens. J. **20**(3), 1642–1655 (2020). https://doi.org/10.1109/JSEN.2019.2947612
4. Rahal, R., et al.: Caring about the human operator: haptic shared control for enhanced user comfort in robotic telemanipulation. IEEE Trans. Haptics **13**(1), 197–203 (2020). https://doi.org/10.1109/TOH.2020.2969662
5. Singh, J., Srinivasan, A.R., Neumann, G., Kucukyilmaz, A.: Haptic-guided teleoperation of a 7-DoF collaborative robot arm with an identical twin master. IEEE Trans. Haptics **13**(1), 246–252 (2020). https://doi.org/10.1109/TOH.2020.2971485
6. Yu, Q., Shang, W., Zhao, Z., Cong, S., Li, Z.: Robotic grasping of unknown objects using novel multilevel convolutional neural networks: from parallel Gripper to Dexterous hand. IEEE Trans. Autom. Sci. Eng. **18**(4), 1730–1741 (2021). https://doi.org/10.1109/TASE.2020.3017022
7. Rosenberger, P., et al.: Object-Independent human-to-robot handovers using real time robotic vision. IEEE Robot. Autom. Lett. **6**(1), 17–23 (2021). https://doi.org/10.1109/LRA.2020.3026970
8. Santina, C.D., et al.: Learning from humans how to grasp: a data-driven architecture for autonomous grasping with anthropomorphic soft hands. IEEE Robot. Autom. Lett. **4**(2), 1533–1540 (2019). https://doi.org/10.1109/LRA.2019.2896485
9. Enebuse, I., Foo, M., Ibrahim, B.S.K.K., Ahmed, H., Supmak, F., Eyobu, O.S.: A comparative review of hand-eye calibration techniques for vision guided robots. IEEE Access **9**, 113143–113155 (2021). https://doi.org/10.1109/ACCESS.2021.3104514
10. Muthusamy, R., et al.: Neuromorphic eye-in-hand visual servoing. IEEE Access **9**, 55853–55870 (2021). https://doi.org/10.1109/ACCESS.2021.3071261
11. Hsieh, Y.-Z., Lin, S.-S.: Robotic arm assistance system based on simple stereo matching and Q-learning optimization. IEEE Sens. J. **20**(18), 10945–10954 (2020). https://doi.org/10.1109/JSEN.2020.2993314
12. Li, S., Rameshwar, R., Votta, A.M., Onal, C.D.: Intuitive control of a robotic arm and hand system with pneumatic haptic feedback. IEEE Robot. Autom. Lett. **4**(4), 4424–4430 (2019). https://doi.org/10.1109/LRA.2019.2937483

13. Sekhar, R., Musalay, R.K., Krishnamurthy, Y., Shreenivas, B.: Inertial sensor based wireless control of a robotic arm. In: 2012 IEEE International Conference on Emerging Signal Processing Applications, ESPA 2012 - Proceedings (2012). https://doi.org/10.1109/ESPA.2012.6152452
14. Connolly, J., Condell, J., O'Flynn, B., Sanchez, J.T., Gardiner, P.: IMU sensor-based electronic goniometric glove for clinical finger movement analysis. IEEE Sens. J. **18**(3), 1273–1281 (2018). https://doi.org/10.1109/JSEN.2017.2776262
15. Ghadrdan, M., Grimholt, C., Skogestad, S., Halvorsen, I.J.: Estimation of primary variables from combination of secondary measurements: comparison of alternative methods for monitoring and control. Comput. Aided Chem. Eng. **31**(2012), 925–929 (2012). https://doi.org/10.1016/B978-0-444-59506-5.50016-X
16. S. Edition, The Authoritative Dictionary of IEEE Standards Terms (2000)
17. AlHinai, N.: Introduction to biomedical signal processing and artificial intelligence. In: Zgallai, W. (ed.) Biomedical Signal Processing and Artificial Intelligence in Healthcare, 1st edn., pp. 1–28. Elsevier (2020). https://doi.org/10.1016/b978-0-12-818946-7.00001-9
18. Li, Z., Schicho, J.: A technique for deriving equational conditions on the Denavit-Hartenberg parameters of 6R linkages that are necessary for movability. Mech. Mach. Theory **94**, 1–8 (2015). https://doi.org/10.1016/j.mechmachtheory.2015.07.010
19. Velásquez, R.M.A., Lara, J.V.M.: Converting data into knowledge with RCA methodology improved for inverters fault analysis. Heliyon **8**(8), e10094 (2022). https://doi.org/10.1016/j.heliyon.2022.e10094
20. Lara, J.V.M., Velásquez, R.M.A.: Low-cost image analysis with convolutional neural network for Herpes zoster. Biomed. Signal Process. Control **71**(Part B), 103250 (2022). https://doi.org/10.1016/j.bspc.2021.103250

Detection of Variable Astrophysical Signal Using Selected Machine Learning Methods

Denis Benka(✉), Sabína Vašová, Michal Kebísek, and Maximilián Strémy

Institute of Applied Informatics, Automation and Mechatronics,
Faculty of Materials Science and Technology in Trnava,
Slovak University of Technology in Bratislava, Bratislava, Slovakia
{denis.benka,sabina.vasova,michal.kebisek,maximilian.stremy}@stuba.sk

Abstract. Machine learning methods are widely used to identify specific patterns, especially in image based data. In our research we focus on quasiperiodic oscillations (QPO) in astronomical objects known as cataclysmic variables (CV). We work with very subtle QPO signals in the form of a power density spectrum (PDS). The confidence of detection of the latter using some common statistical methods could yield less significance than the reality is. We work with real and simulated QPO data and we use sigma intervals as our main statistical method to get the confidence levels. As expected, most of our observed QPO fell under 1-σ and based off this method, such QPO is not significant. In our work, we would like to propose and subsequently evaluate two machine learning algorithms with different lengths of training data. Our main goal is to testify the accuracy and feasibility of the selected machine learning methods in contrast to the sigma intervals. The aim of this paper is to summarise both the theory needed to understand the problem and the results of our conducted research.

Keywords: quasi-periodic oscillation · simulation · support vector machine · long-short term memory network

1 Introduction

Signal is a certain phenomenon, a process or a function that carries coded information. It can be in various formats, e.g. electromagnetic radiation, energy, impulse, optical, mechanical, acoustic, etc. It is transmitted between the transmitter and the receiver, and these may also be diametrically different (in structure, essentially, in the way it is formed, received and processed) [1].

Astrophysical signals are observed in the form of electromagnetic radiation either by means of devices on the ground surface that can record low to medium frequencies (GHz to THz - optical area) or satellites placed in Earth's orbit due to the space radiation filtering of Earth's magnetic field and atmosphere (THz and higher - UV, gamma, beta radiation, etc.). We study low-frequency

R. Silhavy and P. Silhavy (Eds.): CSOC 2023, LNNS 724, pp. 679–691, 2023.
https://doi.org/10.1007/978-3-031-35314-7_57

and low-energy signals observed using satellites. These detect the radiation of objects and phenomena in the universe and their detectors convert it to a signal that we can analyze. The common output signal from satellites is the number of photons (or the amount of detected energy) emitted from the observed source per unit of time [2,3].

Nowadays, astronomical satellites are experiencing an exponential development as well as the growth of everyday data collected. Every day the satellites spew an enormous amount of data. Extensive data archives are being built (astrophysical scientific archives for high-frequency data Heasarc currently archives data from 20 satellites in the range of 30 years) [4]. We will work with Kepler satellite data of a binary star system called MV Lyr. A binary star system (cataclysmic variable, CV) consists of a main star which accretes matter from its host, a secondary star. There are many phenomena to study, we chose one specific which does not have a strict period of occurrence, but rather manifests stochastically. Such phenomena are quasi-periodic oscillations of the accretion disc [3].

For the purposes of the analysis of our QPO signals, various software tools are currently used [5,6], as the manual processing of this signal would be very difficult for scientists. This is where the space for artificial intelligence elements is and this could relieve scientists from a lengthy signal analysis and estimation. Machine learning (ML) provides a suitable tool for analyzing a large volume of data. It is more and more used in everyday life. There are various methods of ML for processing and analyzing the signal. ML could be utilized e.g. in medicine to identify tumors, for weather forecasting, stock prices forecasting, voice recognition, for generating 3D objects, etc. The input into such a system can be in the form of an image, numerical matrix or text strings. Depending on the specific problem, a ML algorithm can learn to recognize certain characteristics, or patterns and generate or predict new outputs [7–9].

We will work in the Python programming language as its libraries provide a wide range of functions and data analysis tools available without a license. This language is used by many companies and offers complex data analysis options.

2 Methodology

2.1 Astrophysical Signal Processing Background

Ultra violet and X-ray radiation come from high energy objects or phenomena in space. These include CV, and exotic objects such as clusters of galaxies, black holes, neutron stars, active galactic nuclei. Regarding the latter, scientists thrive to find out their origin, structure, chemical composition, their behavior, life cycle, etc. In this work we focus on the study of a CV, the binary star system MV Lyr. Such star systems are widely occuring in the universe although it may not seem so. In our work, we will focus on the photometric signal captured by the Kepler space telescope. The spacecraft has cameras operating on the principle of a charge-bound structure (CCD from the Charged Couple Device). It is basically a connected network of 42 photometers capturing photons in the field of view.

These are detected on the chip in the form of an electric charge, which is recorded with time data [2,3,10].

The observational data are stored in the Mikulski Archive for Space Telescopes (MAST). The raw image data need to be processed with the help of a suitable software (SaS, Heasoft, Fits View, etc.) to extract the necessary data, i.e. the number of photons captured by the CCD detector and the time in which they were captured. From these two informations we can create a very important product for our future signal processing research called a light curve (Fig. 1).

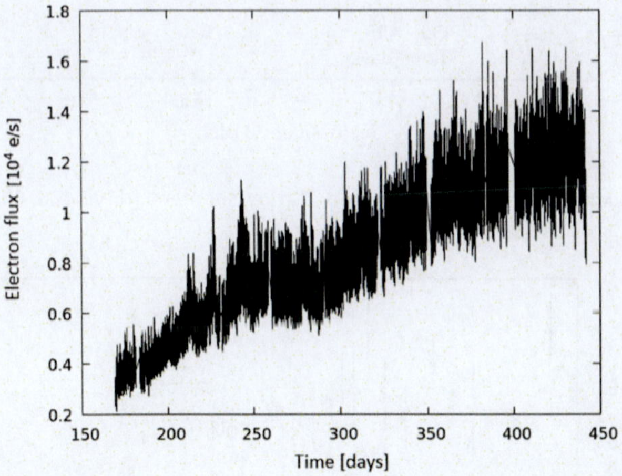

Fig. 1. Complete 272-day long light curve of our observed binary star system MV Lyr.

2.2 Data Processing

Experienced astronomers could see some variability in our observed light curve but the period of its occurence is not strict. First of all we need to process our signal. We divided the 272-day long light curve into equal 10-day segments. This needs to be done in order to confirm the presence of a QPO. The flux from each day of the segment was then fed to a Lomb-Scargle [11] algorithm to obtain the power of the signal at a respective frequency (Fig. 2). In order to see the variability more clearly, we used log-log scale of our periodograms. We used Bartlett's [12] method to average the 10-day periodograms and to obtain a PDS (Fig. 3). Each bin in the PDS represents the power of the signal at that specific frequency and also a standard deviation of the 10 averaged values used [13].

In the created PDS (Fig. 3) we can see hump profiles, these are the so-called QPO. Some are rather obvious like the one at $\log(f/\mathrm{Hz}) \simeq -2.9$. Our area of interest is the QPO at frequency $\log(f/\mathrm{Hz}) \simeq -3.4$. This QPO is very subtle, manifesting only on three power bins. Based on our previous research [14–16],

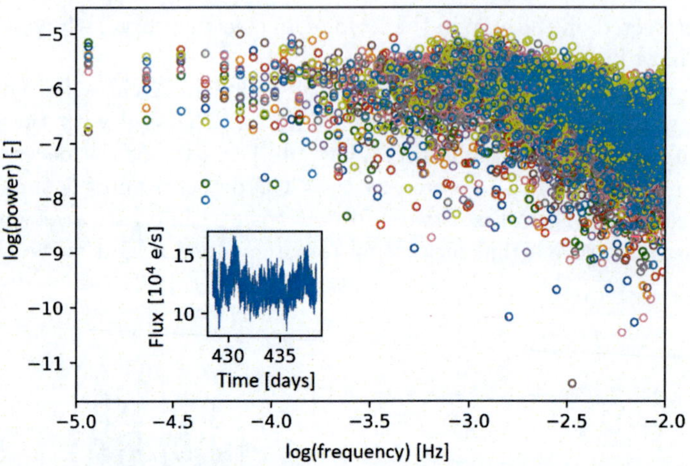

Fig. 2. Lomb-Scargle periodogram from a 10-day segment, days 428–438, used part of the light curve is in the zoom.

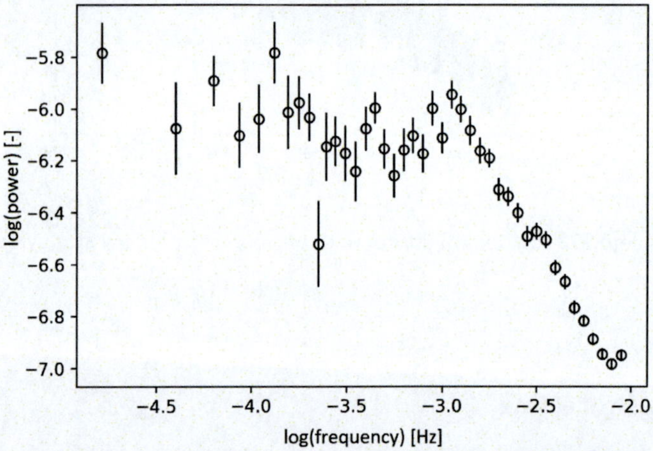

Fig. 3. PDS of the 10-day periodogram as shown in Fig. 2.

we believe, that most of such binary star systems manifest such a subtle QPO at the latter frequency. First of all we need to find a model which will correctly describe our data. For this purpose Lorentz models are used to describe the QPO peaks [17]. We used a four Lorentz profile model to create a representative fit to our data. To estimate the confidence of such QPO we created Timmer & Koenig [18] simulations and estimated the sigma confidence intervals (Fig. 4). It is clear that using confidence intervals to estimate the confidence of our observed QPO, the QPO fell under the 1-σ interval, meaning the confidence is under 68%. This is not high enough to be taken as significant.

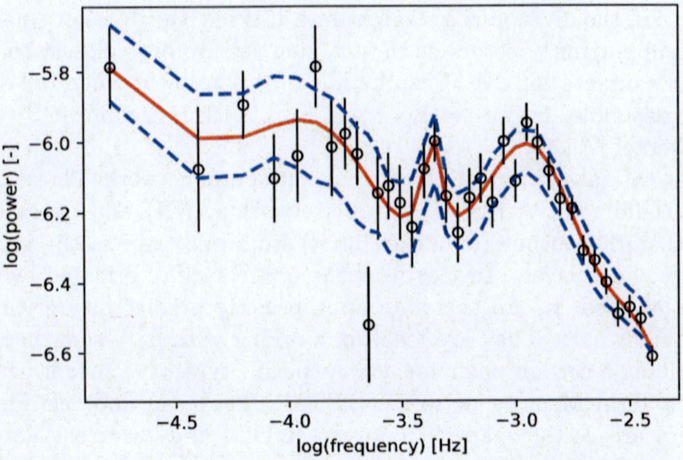

Fig. 4. PDS with Timmer & Koenig confidence intervals (blue dashed line) and Lorentz fit (red solid line).

In our research we want to try to apply several ML methods to identify our QPO to see whether the algorithm could yield a higher level of confidence. For training purposes of our ML models, we will work with simulated PDS data. We will have two categories, one with the QPO at our frequency of interest, one without it. We will train the algorithms using datasets with different sizes (i.e. 50, 100, 250, 500, 1000).

2.3 Machine Learning Methods

The aim of a machine learning algorithm is to optimize the created model, i.e. minimize the loss function (1) expressing inaccuracies between the predictions of data from the training sets and test data. To be more precise, it expresses the degree of uncertainty of input-output relation [19].

$$J = \frac{1}{n} \sum_{i=0}^{n} (y_p - y)^2 . \tag{1}$$

In (1) n expresses the total number of training samples in the training dataset, y_p is a predicted output, y is the real output. If we want to optimize this function, we need to find minimal loss and we can do so with gradient descent.

In our work, one of the ML algorithms we used is a Supporting Vector Machine (SVM). They are one of the most robust supervised learning models. The main optimization problem of this model is to find a plane that distributes the N-dimensional space characterizing the training set (N is the number of features or categories). The search focuses on maximizing distance between categories of the training data. This ensures the robustness of the algorithm in such a way that data that is still unknown will be easier to identify, e.g. if we have the two categories, the division is a straight line, if three, the division "plane" is 2D, etc. The "supporting" vectors of the training set are ones closest to the plane and affect its orientation. SVM can be linear or non-linear when the data is not well distinguishable. In our work we will work with a non-linear SVM using a Gaussian kernel [20,21].

Our next ML algorithm will be a recurrent neural network. Those are derived from the feedforward Artificial Neural Network (ANN) and operate on such principle that the output (or a part of it) from one layer is the input to the next and previous layers. In this way, the occurrence of oriented cycles in the network is allowed. It can therefore appropriately predict future values based on the previous ones. They are ideal for working with data sequences, number rows describing a certain phenomenon. A specific type of recurrent networks are Long-Short Term Memory networks (LSTM). They can find relations in data where there are certain gaps in between. LSTM and recurrent networks use a back-propagation algorithm to calculate the weights and a gradient descent algorithm to optimise its loss function [22,23].

2.4 Related Works

The research regarding machine learning and astronomy is flourishing. [24] and [25] used three CNN's for the detection of radio galaxies in noisy telescope image data. Their approach yielded more than 90% accuracy. [26] applied ML methods to the problem of gravitational lensing. They used simulated data to train a pre-trained CNN. His research yielded a promising 77% prediction accuracy. They also stated that the accuracy could be improved using more accurate calculations of the simulations used, as the CNN identified many false positive cases. A CNN can be used to reconstruct damaged image data. [27] created a two-layer CNN to reconstruct incomplete images of galaxies from satellite data. The resulting images were good in both quality and scientifical correctness. In comparison to commonly used methods in astronomy, they proved that ML methods could be computationally efficient and also scientifically precise. [28] used five CNNs each working with specific band-based images. The input from these was then passed to a final decision based neural network yielding binary output. They also proved that multi-epoch training is more fruitful rather than training with a small number of epochs. Their approach had hit 99.95% testing accuracy. There were several attempts to detect binary black holes using CNNs. [29] used simulated images of binary black holes to train a CNN. Using the ADAM (adaptive momentum estimation) optimization algorithm with a fixed learning rate, they proved that a well-trained CNN can be efficient in binary black hole identification and a close match to the methods used in common astronomical approaches. [30] and [31] studied the possibility of using a CNN for the forecasting of Earth's wavefronts based on images obtained from satellites. They also used a pretrained CNN called Inception and trained it using their simulated datasets. The results of their research proved that this approach is even more precise in comparison with some commonly used weather forecast models. Noise removal in astronomical images can be a pesky task using traditional methods.

[32] created two CNN classifiers using a supervised approach for the detection of the source of contaminants of astronomical images such as radiation rays, hot pixels, nebula clouds and diffraction spikes. The CNNs were trained for 30 epochs with 50000 samples. The average accuracy was 97%. Mixing the types of training data (the contaminants) did not bring any significant impact to the performance of the proposed method proving that CNNs are able to find the necessary features if trained correctly. Supernova outbursts detection is a highly used application of CNNs in astronomy nowadays. [33] proposed a CNN trained on observational image data of such events. The prediction accuracy of 99.32% shows the success of their research.

All of the above used image-based data in their research regarding the use of their ML method in astronomical tasks. Only a few use any other types of training data. Some examples of the latter include [5], who created an autoencoding RNN to detect variability in light curves of selected supernovae. They used raw data meaning the data was in numerical form. The testing accuracy was about 80%. The method as proposed by [34] yielded a 75% accuracy in most of their prediction classes using also a RNN. Their dataset consisted of variable star data in the form of light curves, just like our area of research. They did use only necessary preprocessing steps to be able to feed the RNN with data. [6] developed a trustworthy method for the detection of supernova eruptions. The algorithm consists of a RNN trained to classify light curves using photometric data. Their approach yielded an extraordinary accuracy of 99%, They also tried to classify incomplete light curves with a 86% accuracy.

All of the works above prove that ML algorithms can be valid assets not only in image-based detection. The aim of our research is to use several different ML approaches to tackle the problem of subtle QPO detection and to compare our results to the commonly used method of estimating confidence - confidence intervals.

2.5 RNN Training

Using the Python 3.9 *Keras* package, all RNNs were created. One input layer (42 $\times$ 1), three hidden layers with a *sigmoid* activation function (we want smaller learning steps), a dense layer with a sigmoid activation function, and one unit for the binary classification output (for each category) make up each trained network. The number of 10 units per hidden layer was experimentally determined after testing the performance of the LSTM network with a range of 2–15 neurons. For each hidden layer, L2 normalization was utilized. A batch normalization layer was utilized to normalize the output of the current batch in between the hidden layers. We employed several batch sizes for different sizes of training datasets, ranging from 5 to 15. 300 epochs were used for each LSTM training. The starting learning rate used was 0.003, with a declination on plateau. After 150 epochs of no progress in validation accuracy, the learning process was likewise early terminated. A binary cross-entropy loss function was utilized to validate each training epoch, and the Adam optimizer was employed to build each model [22]. The architecture of the used RNN is shown in Fig. 5.

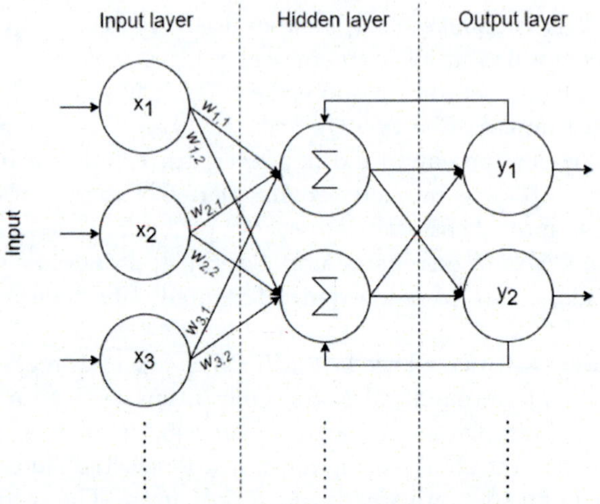

Fig. 5. Proposed RNN architecture.

Each input neuron is connected to a neuron in hidden layer. Here are the weights and input features summed and processed to be sent to the last, output layer. Some of the output of the last layer is used as an input to the hidden neurons, this should be beneficial for obtaining higher accuracy of the model [21–23].

2.6 SVM Training

SVM models were created using the *sklearn* library of Python 3.9. One training sample is also a 2D numerical matrix (42 $\times$ 2) but this time we include the frequency in addition to the power of the signal in log scale. There were also several dataset sizes tested, including 50, 100, 250, 500, and 1000 samples. We employed a validation training dataset as well as k-fold cross validation set. The SVM kernel was a radial basis function $e^{\gamma\|x-x'\|^2}$, which was optimized using a grid search. This type of kernel is non-linear and it is a good choice to tackle our main objective based on the profile of the fit function for our data. The optimised parameters for each size of the training dataset are summarized in Table 1. The first value of each hyperparameter was optimised using normal training validation, i.e. 20% of the training dataset, we also used a 10-fold cross-validation set for comparison, this was also 20% of each training dataset [8]. We were using supervised approach and the data were divided into two categories (with/without the QPO) as shown on a sample in Fig. 6, where the orange line depicts the class where a QPO is present at $\log(f/\mathrm{Hz}) \simeq -3.4$ and blue line is the sample without the QPO.

Table 1. Optimized hyperparameters of our SVM for each dataset size. Values obtained using normal/10-fold cross-validation set in the training process

Validation	Parameter	50	100	250	500	1000
Normal/k-fold	C	10/1	2/10	10/3	5/5	5/5
	γ	10/10	2/5	10/10	10/10	1/1
	Accuracy	0.60/0.61	0.59/0.61	0.59/0.63	0.60/0.63	0.59/0.63

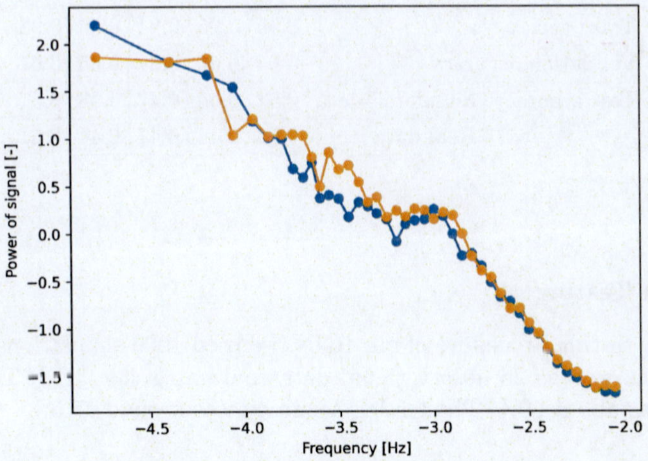

Fig. 6. Sample from each training class. Simulated PDS with a QPO at our researched frequency (orange) and without the QPO (blue).

3 Results

The dataset for both the RNN's and SVM's were separated into three sections: training (80%), validation (10% of the training data), and testing (10% of the training data). We also used k-fold cross-validation with 10 folds. We also tested one case, where 5000 training samples were used. The results were quite similar to other dataset sizes. For testing, we fed both our ML algorithms with simulated data (1000 samples) and real QPO data in numerical form (26 samples).

3.1 RNN Testing

First of all, we applied z-normalization to each dataset size, which resulted in zero mean and a standard deviation of one. Each dataset was reshaped into a N $\times$ 42 $\times$ 1 matrix, with N being the number of training data in each set. Because of the validation split, the dataset was shuffled before each training epoch. Positive integers were used to normalize the class labels to 0, 1 [19,22]. A sample of each training class was depicted in Fig. 6. We tested the RNN's with the data as mentioned above. In all cases, training accuracy was about

95%. However, the average validation accuracy was 65.2%. The average testing accuracy with simulated data was 47,48%, while the testing accuracy with real data was 47,59%. The results are summarized in Table 2.

Table 2. Dataset sizes, training accuracies, validation and testing accuracies (both simulated and real data used) for each trained RNN

Dataset size		50	100	250	500	1000
Train accuracy		0.95	0.95	0.94	0.96	0.98
Validation accuracy		0.64	0.66	0.64	0.65	0.67
Test accuracy	Simulated data	0.52	0.44	0.47	0.43	0.51
	Real data	0.53	0.44	0.47	0.44	0.51

3.2 SVM Testing

Same as the testing procedure of our RNN, we used 1000 simulations from both categories, as well as 26 observational data containing the QPO. The training accuracy was about 60%. The testing accuracy was around 45.5%.

Table 3. Dataset sizes, training accuracies, validation and testing accuracies (both simulated and real data used) for each trained SVM

Dataset size		50	100	250	500	1000
Train accuracy		0.65	0.61	0.59	0.57	0.59
Validation accuracy		0.63	0.50	0.49	0.55	0.57
Test accuracy	Simulated data	0.41	0.40	0.42	0.39	0.42
	Real data	0.52	0.48	0.53	0.51	0.52

Increasing the amount of the training dataset had no effect on testing or training accuracy. Using 10-fold cross validation only slightly increased accuracy by about 1%. Testing with simulated data yielded a 41% accuracy, whereas testing with real data yielded 51%. Results from testing our SVM's are summarized in Table 3.

4 Discussion

Our SVM did not perform as we anticipated. The average accuracy of tests was 51%. Compared with the confidence level of Timmer & Koenig [18], our method does not improve the reliability of QPO detection. Our rate of positive detection or SVM recall was higher when using a larger data set. Increasing the number of training samples also slightly improves accuracy or true positive rate. Using 10-fold cross-validation increased the accuracy of the test by 1%. We tried to use a training dataset of 5000 samples (2500 samples from both training categories). The training took two weeks to complete and the results were even worse than the former (42% accuracy of tests with real data). In our case, we should have focused more on the smaller number of training samples, as large data sets turned out to perform worse. The truth is that the data in each target class overlap in the sense that the data have similar properties. This produces unwanted local optima due to the nature of the optimization kernel used. This can cause our low bias kernel we used to have trouble distinguishing between ground-truth and noise. We also tried using quadratic polynomial kernel. The results were not reasonable because the accuracy of the tests was just over 30%.

The testing results of our RNN's were also not plausible. We decided to pre-process the data to create a more identifiable data set. Instead of this methodology, if we wish to use RNN architectures in the future, we should use fairly raw photometric data of the MV Lyr system to achieve higher classification accuracy. We also trained the RNN models on a dataset of size 5000. The training time was about 5 h and the resulting test accuracy was not reasonable (62%). We used a batch size of 15 and trained for 500 epochs. The learning rate and its decay remained the same as the other training process of the RNN. The accuracies while using simulated data for testing were as follows: training accuracy (96%), validation accuracy (67%), and testing accuracy (52.5%) on simulated data and (52.7%) on real data.

5 Conclusion

We processed the raw light curve data of the binary system MV Lyr. Data are taken from the MAST. Our time-series analysis confirms the existence of multiple QPOs. The target QPO is $\log(f/\mathrm{Hz}) \simeq -3.4$. Using Timmer & Koenig [18] simulated data we estimated sigma confidence intervals, the latter is within 1-σ, giving our QPO confidence of 68%. Our goal was to improve this score using a RNN and a SVM architecture that detects such QPO's with higher accuracy. The data for training our models were simulated from the Lorentz model used to fit our PDS. The data were simulated in two categories, with and without the target QPO. The training dataset was in numerical form, where values of flux were used (also with respective frequencies for SVM models). We tested with both simulated and real data. The final classification accuracy was not as good as expected. The overall validation and testing accuracy was over 50%. Compared to statistical methods used by astronomers, the latter value is below the 1-σ confidence interval for the observed QPO signal.

Our future work includes using the raw light curve data instead of the processed data (i.e. we will not create PDS but rather use a light curve) to see if this is the problem. We believe that this approach will yield better results than this study. In our future work, we also want to tune our SVM using a training dataset with less samples as the results were quite similar with different dataset sizes. We are also thinking of changing the training data type from numerical to image and see if the results will change.

References

1. Sharma, A., Jain, A., Kumar Arya, A., Ram, M. (eds.): Artificial Intelligence for Signal Processing and Wireless Communication. De Gruyter (2022)
2. Shore, S.N.: The Tapestry of Modern Astrophysics. Wiley-Interscience, Hoboken (2011)
3. Bode, M.F., Evans, A. (eds.): Classical Novae. Cambridge University Press (2008)
4. HEASARC: NASA's Archive of Data on Energetic Phenomena. https://heasarc.gsfc.nasa.gov/
5. Tsang, B.T.-H., Schultz, W.C.: Deep neural network classifier for variable stars with novelty detection capability. Astrophys. J. **877**, L14 (2019). https://doi.org/10.3847/2041-8213/ab212c
6. Möller, A., de Boissière, T.: SuperNNova: an open-source framework for Bayesian, neural network-based supernova classification. Mon. Not. R. Astron. Soc. **491**, 4277–4293 (2020). https://doi.org/10.1093/mnras/stz3312
7. Jain, S., Pandey, K., Jain, P., Seng, K.P.: Artificial Intelligence, Machine Learning, and Mental Health in Pandemics A Computational Approach. Academic Press, London (2022)
8. Joshi, A.V.: Machine Learning and Artificial Intelligence. Springer, Cham (2020). https://doi.org/10.1007/978-3-030-26622-6
9. Ramírez, D.S., Santamaría, I., Scharf, L.: Coherence in Signal Processing and Machine Learning. Springer, Cham (2023). https://doi.org/10.1007/978-3-031-13331-2
10. Collins, G.W.: The Fundamentals of Stellar Astrophysics. W.H. Freeman, New York (1989)
11. Lomb, N.: Least-squares frequency analysis of unequally spaced data. Astrophys. Space Sci. **39**, 447–462 (1976)
12. Bartlett, M.S.: On the theoretical specification and sampling properties of autocorrelated time-series. Suppl. J. R. Stat. Soc. **8**, 27–41 (1946). https://doi.org/10.2307/2983611
13. Vapnik, V.N.: The Nature of Statistical Learning Theory. Springer, New York (2010). https://doi.org/10.1007/978-1-4757-3264-1
14. Dobrotka, A., Orio, M., Benka, D., Vanderburg, A.: Searching for the 1 mHz variability in the flickering of V4743 SGR: a cataclysmic variable accreting at a high rate. Astron. Astrophys. **649**, A67 (2021). https://doi.org/10.1051/0004-6361/202039742
15. Orio, M., et al.: Nova LMC 2009a as observed with XMM-Newton, compared with other novae. Mon. Not. R. Astron. Soc. **505**, 3113–3134 (2021). https://doi.org/10.1093/mnras/stab1391
16. Dobrotka, A., Ness, J.-U., Bajčičáková, I.: Fast stochastic variability study of two SU UMa systems V1504 Cyg and V344 Lyr observed by Kepler satellite. Mon. Not. R. Astron. Soc. **460**, 458–466 (2016). https://doi.org/10.1093/mnras/stw1001
17. Bellomo, N., Preziosi, L.: Modelling Mathematical Methods and Scientific Computation. CRC Press, Boca Raton (1995)

18. Timmer, J., Koenig, M.: On generating power law noise. Astron. Astrophys. **300**, 707–710 (1995)
19. Bishop, C.M.: Pattern Recognition and Machine Learning. Springer, New York (2006)
20. Rasmussen, C.E., Williams, C.K.I.: Gaussian Processes for Machine Learning. MIT Press, Cambridge (2006)
21. Khan, A.I., Al-Habsi, S.: Machine learning in computer vision. Procedia Comput. Sci. **167**, 1444–1451 (2020). https://doi.org/10.1016/j.procs.2020.03.355
22. Clarke, B., Fokoué, E., Zhang, H.H.: Principles and Theory for Data Mining and Machine Learning. Springer, New York (2011). https://doi.org/10.1007/978-0-387-98135-2
23. Kononenko, I., Kukar, M.: Machine Learning and Data Mining: Introduction to Principles and Algorithms. Horwood Publishing, Chichester (2007)
24. Lukic, V., de Gasperin, F., Brüggen, M.: ConvoSource: radio-astronomical source-finding with convolutional neural networks. Galaxies **8**, 3 (2019). https://doi.org/10.3390/galaxies8010003
25. Aniyan, A.K., Thorat, K.: Classifying radio galaxies with the convolutional neural network. Astrophys. J. Suppl. Ser. **230**, 20 (2017). https://doi.org/10.3847/1538-4365/aa7333
26. Davies, A., Serjeant, S., Bromley, J.M.: Using convolutional neural networks to identify gravitational lenses in astronomical images. Mon. Not. R. Astron. Soc. **487**, 5263–5271 (2019). https://doi.org/10.1093/mnras/stz1288
27. Flamary, R.: Astronomical image reconstruction with convolutional neural networks. In: 2017 25th European Signal Processing Conference (EUSIPCO), pp. 2468–2472. IEEE, Kos, Greece (2017)
28. Kimura, A., Takahashi, I., Tanaka, M., Yasuda, N., Ueda, N., Yoshida, N.: Single-epoch supernova classification with deep convolutional neural networks. In: 2017 IEEE 37th International Conference on Distributed Computing Systems Workshops (ICDCSW), pp. 354–359. IEEE, Atlanta, GA, USA (2017)
29. Gabbard, H., Williams, M., Hayes, F., Messenger, C.: Matching matched filtering with deep networks for gravitational-wave astronomy. Phys. Rev. Lett. **120**, 141103 (2018). https://doi.org/10.1103/PhysRevLett.120.141103
30. Andersen, T., Owner-Petersen, M., Enmark, A.: Neural networks for image-based wavefront sensing for astronomy. Opt. Lett. **44**, 4618 (2019). https://doi.org/10.1364/OL.44.004618
31. Andersen, T., Owner-Petersen, M., Enmark, A.: Image-based wavefront sensing for astronomy using neural networks. J. Astron. Telesc. Instrum. Syst. **6**, 1 (2020). https://doi.org/10.1117/1.JATIS.6.3.034002
32. Paillassa, M., Bertin, E., Bouy, H.: MAXIMASK and MAXITRACK: two new tools for identifying contaminants in astronomical images using convolutional neural networks. Astron. Astrophys. **634**, A48 (2020). https://doi.org/10.1051/0004-6361/201936345
33. Cabrera-Vives, G., Reyes, I., Forster, F., Estevez, P.A., Maureira, J.-C.: Supernovae detection by using convolutional neural networks. In: 2016 International Joint Conference on Neural Networks (IJCNN), pp. 251–258. IEEE, Vancouver, BC, Canada (2016)
34. Becker, I., Pichara, K., Catelan, M., Protopapas, P., Aguirre, C., Nikzat, F.: Scalable end-to-end recurrent neural network for variable star classification. Mon. Not. R. Astron. Soc. **493**, 2981–2995 (2020). https://doi.org/10.1093/mnras/staa350

Land Cover Detection in Slovak Republic Using Machine Learning

Sabina Vasova(✉), Denis Benka, Michal Kebisek, and Maximilian Stremy

Institute of Applied Informatics, Automation and Mechatronics, Faculty of Material Science and Technology in Trnava, Slovak University of Technology in Bratislava, Bratislava, Slovakia
{sabina.vasova,denis.benka,michal.kebisek,maximilian.stremy}@stuba.sk

Abstract. There are 44 classes of landscape covers in Slovakia, which are grouped in the Corine Landcover list. Accurate detection of classes of agricultural areas and wetlands (2.3.1. Permanent grasslands, meadows and pastures, 3.2.1 Natural meadows, 4.1. Inland wetlands) is important, but the division of these classes with the help of common classification methods is not possible, because they have the same reflectivity as grasslands. The use of deep learning methods can be one of the solutions to the problem of detection. To this day, the differentiation of 2.3.1 subcategories in Slovakia is detectable only empirically. The goal of this article is to compare supervised learning methods such as decision tree, neural networks, support vector machine, which are used to detect land cover in different countries. We will analyze the used methods and propose a suitable method for solving detection in the above-mentioned areas in Slovak republic.

Keywords: detection · land cover · deep learning · sentinel · copernicus · slovakia · corine land cover

1 Introduction

Detection of land cover change is important from the point of view of collecting information for multiple systems. Detection is used to support decision-making, for example for the protection of countries, in the management of water resources, in land usage and various other activities.

The Copernicus program, whose original name was Global Monitoring for Environment and Security (GMES), is an European Commission project that aims to create a multi-level operational infrastructure for Earth observation. The European Commission in cooperation with the European Space Agency (ESA) launched six missions called Sentinel 1–6 for land cover detection, Meteosat Third Generation (MTG) satellite and the Metop Second Generation (MetOp-SG) satellite of European operational satellite agency (EUMETSAT) for monitoring weather [1,2].

Each of the missions is intended for a different observation. The basic contribution is observations of surface areas, land cover mapping, landscape management, forestry, global disaster control, risk mapping, security, humanitarian

R. Silhavy and P. Silhavy (Eds.): CSOC 2023, LNNS 724, pp. 692–702, 2023.
https://doi.org/10.1007/978-3-031-35314-7_58

operations etc. Sentinel missions use an array of sensors that collect image data for further processing [3–6].

In our case, we want to examine the Earth's surfaces,i.e. land covers, so we want to use the Sentinel 2 mission combined Sentinel 1 and Sentinel 3 images. Sentinel 1–3 will provide us with detailed information about surface temperature, ozone parameters, water parameters, sea level pressure and various other parameters. Combining the data from these three missions will give us more detailed data of the observed area because the first two missions work only based on indices while Sentinel 3 works with infrared images [4–6].

The parameters are used to create a detailed analysis. We will use the images prepared in this way to create an input image training dataset. In the last phase, we will determine the machine learning method with which we want to detect the subclasses of Corine class 2.3.1, 3.2.1 and 4.1 in the territory of the Slovak Republic [7].

The Sentinel 1 mission operates on the principle of radar imaging. It is described using Synthetic Aperture Radar (SAR). The scanning of individual satellites is based on different methods [5].

The Sentinel 2 mission is an optical instrument that samples 13 spectral bands, out of which 4 bands have 10 m spatial resolution, 6 bands have 20 m spatial resolution, and 3 bands have 60 m spatial resolution. Bands are measured in different wavelengths (443–2190 mm) from the visible spectrum (Visible Infrared Spectrum, VIS), short wavelengths of the infrared spectrum (Near Infrared, NIR) to the medium infrared spectrum (Short Wave Infrared, SWIR). The Sentinel 2 mission consists of two satellites phased at 180° (twin satellites), whose measurement frequency is 5 days. The satellites move along the Earth's equator, while the movement follows the same orbit (290 km). One cycle lasts 10 days, i.e. 143 orbits with a cadence of 10 s. It means, one observation lasts 10 days and sampling frequency is 10 s. Using the Universal Transverse Mercator system (UTM), the Earth is divided into 60 zones. The lifetime of the telescope is estimated from 7.5 years to 12 years, while the satellite was placed in the Earth's orbit in 2015 [2,8].

The principle of the Sentinel 2 satellite lies in capturing the light reflected from the Earth using a three-mirror telescope marked M1, M2, M3. Using a beam splitter, the mirrors are focused into two Focal Plane Arrays (FPA). The first is intended for 10 m NIR resolution and the second for SWIR wavelengths [9].

Radiometric calibration is achieved using components located on the inside of the Calibration and Shutter Mechanism (CSM). Image accuracy for SWIR and NIR systems is influenced by 12 detectors located in two horizontal rows, including bandpass filters located on the detectors [4].

Sentinel 2 data products are divided into several levels. Those include levels 0, 1A, 1B, 1C, and 2A. Level 0 is compressed format, which contains telemetry analysis, cloud mask. After decompression we get Level 1A, which contains radiometric corrections and geometric viewing model refinement. Level 1B contains additional resampling images, conversion to reflectances and preview image with masks generation. Level 1C captures data from the top of the atmosphere. Levels 0, 1A, 1B are not publicly available, only ministries and research projects have access. An ordinary user can download the image database using DIAS systems, databases e.g. Data Integration and Analysis System (DIAS) etc., or using the *scihub* tool from the ESA only at level 1C or 2A. Both levels contain images of $100 \times 100\,\text{km}^2$. The images contain an orthorectified image, i.e. do not contain spatial distortion, use atmospheric correction and various other adjustments. The images are stored in the so-called tiles (in the file Granules). From a level 1C image, we can create a level 2A image enriched with all the parameters of both levels using the *Sen2Cor* processor [4,10,11].

The basic difference between these levels is the data is captured off different parts of the atmosphere. Level 2A captures data (reflectivity of surfaces) from the bottom of the atmosphere (Bottom Atmospheric Product, BOA), while product 1C, on the other hand, captures the top of the atmosphere (Top Atmospheric Product, TOA). We can process products of different levels using the ESA's Sentinel Application Platform (SNAP) and its *Qgis* (Semi-Automatic classification) plugins. SNAP is an open-access Earth observation analysis tool [12].

2 Literature Review

In [13] Italy's landscape and its changes were observed. They created EAGLE, a classification system, utilizing a decision tree. The training dataset contained spectral and backscatter features of images obtained from Sentinel 1 and Sentinel 2. They proposed two new indices, the Normalized Difference Chlorophyll Index (NDCI) and Burned Index (BI). BI is used to identify burned areas (either in nature or urban), while NDCI can be used to distinguish between broad-leaved and needle-leaved trees. Their method of classification used yielded very good results and is used to this day.

RF, Maximum Likelihood (MLE) and Support Vector Machine (SVM) were used to map the region of Sahel, a region of Nigeria. [14] created a mapping algorithm based off the combination of the above-mentioned machine learning methods. However, the best-in-case accuracy yielded the RF classifier. In comparison to the method used by [13] on the area of Italy, the resulting accuracy was lower.

SVM as well as a neural network were used by [15] to detect the landscape of Egypt. As a training dataset, they combined several spectral bands obtained from Sentinel 1 and extracted main features using Principal Component Analysis (PCA). This approach yielded the best results in comparison to [13,14].

Convolutional Neural Networks (CNN) were used by [16] to detect the landscape of Australia. They especially aimed to detect cloudiness. They used raw image data from Sentinel 1 and Sentinel 2 and with similar results of classification as [13].

3 Methodology

The images are from the Sentinel 1 and Sentinel 2 databases [3]. We will have to modify the obtained data files into our required form, with which we can continue to work. Processing will be done using common methods in SNAP and Qgis [12]. The basic operations include atmospheric correction, resampling, noise removal from the tilling grid (GRID), temperature noise removal, radiometric calibration, terrain correction, cloud removal, and last but not least, background scattering. The variance is converted to dB units [4,11].

The images are of type 2A (BOA) or 1C (TOA, they are already orthorectified, i.e. treated from spatial distortion), which already contain ECMWF fundamentals values (ozone, water vapor, average sea pressure). Images, processed in this way, are used for mapping and monitoring of water bodies - water mapping. Type 1C is divided into so-called Tiles. Using the *Sen2Cor* (SNAP) processor, we will create a combination of 1C and 2A frames. The result of combination is the L2A type, which is enriched with all the properties of both types, such as aerosols and so on [8,9].

We will resample the created images to the highest spatial resolution of 10 m and limit them only to spectral bands B2, B3, B4 and B8, which are used for vegetation observation. Spectral bands are in B1 - B12 channels. B2 is blue, B3 is green, B4 is red, and near-infrared is B8 channels have 10 m resolution. In ground sampling distance is 20 m resolution, i.e. red edge (B5), near-infrared are B6, B7, B8A, and short-wave infrared (SWIR) are B11, B12. In the end, B1 is coastal aerosol and B10 is a cirrus band have in 60 m spatial resolution. Using the *Subset* function (SNAP), we will divide the image into specific parts [17].

The images will be processed using the methods included in SNAP. Indices (see in Table 1), vegetation index (NDVI), BI, water index (NDWI), snow detecting index and others are applied to the images. Images prepared in this way will serve as an input dataset for machine learning methods [13].

Table 1. List of the most frequently used indices in land cover detection [13]

Satellite	Index Name	Brief Index Formula
Sentinel-2	Normalized Difference Vegetation Index	$NDVI = \frac{(B8-B4)}{(B8+B4)}$
	Normalized Burn Ratio	$NBR = \frac{(B8-B12)}{(B8+B12)}$
	Normalized Difference Water Index	$NDWI = \frac{(B2-B8)}{(B2+B8)}$
	Normalized Difference Snow Index	$NDSI = \frac{(B3-B11)}{(B3+B11)}$
	Normalized Difference Coniferous Index	$NDCI = \frac{(B6-B12)}{(B6+B12)} * \frac{(B8-B11)}{(B8+B11)}$
	Burned Index	$BI = \frac{(1-(B3-B11))}{(1+(B3+B11))}$

We will classify input data using machine learning methods. We know that supervised learning methods (SVM, Decision trees, Neural networks etc.) are used e.g. for feature recognition. Another technique is unsupervised learning

(PCA, K-means clustering, Anomaly detection etc.), where the method adapts to new inputs. This is essential for deriving characteristic features in categories or finding hidden patterns [18].

Unsupervised learning uses clustering, i.e. the input is a single set of data based on characteristic properties, sorted into the categories that will be assigned to it. We also want to test K-means clustering, Anomaly detection, PCA and other methods [19]. Supervised learning can use classification and regression.

During classification, the image dataset is the input, and the output is a class, which we use for the classification of specific items. We would like to compare obtained results from SVM, Discriminant analysis, Naive Bayes and Neural Networks In regression, the output is continuous. It shows similarities in the data. It is used in forecasting and data development. This includes Linear regression, Hierarchical, Ensembles, Decision trees, Neural Networks and other mentioned methods [19,20].

3.1 Results of Analysis

The classification method using the decision tree method was applied to the detection of the territory of Italy (see in Fig. 1) [13]. The method is suitable as long as we can define the exact differences between the given areas.

The method used to detect landscape covers is a decision tree in combination with common methods (indices). Data from Sentinel 1 and Sentinel 2 were used as input. The overall accuracy reaches 83% (see in Table 2) [13]. The highest accuracy from the point of view of users is for the water bodies class. The method was used for automatic classification, which was adapted to additional changes such as burnt areas, landslides and others.

Another area under study was the Sahel region in Nigeria [14]. The input set again consisted of combinations of Sentinel 1 and Sentinel 2 images that have been indexed. Of the surveyed land covers, the areas with rice were detected with the highest accuracy using the SVM. In the case of the MLE, the highest accuracy for water bodies dominates, compared to the Classifier ensemble (CE) and RF, where scattered vegetation dominates (see in Fig. 2) [14]. Among the used methods were the SVM, which achieved an accuracy of approx. 61% (see in Table 3) [14].

The second use case input database was created by combining inputs from Sentinel 1, Sentinel 2 and Google Earth database. The method was used to detect populated areas, streets, deserts and areas with living vegetation. Neural Networks and SVM methods were used for detection.

The input data were combined using the PCA method used to combine the Sentinel 1 radar and Sentinel 2 spectral inputs including cloud removal. Inputs were orthorectified and atmospheric correction was performed using Envi 5.1. It is a tool used for creating high-quality seamless mosaics by combining georeferenced scenes [21,22].

They used the ERDAS IMAGINE system to resample the spatial resolution from 10 m to 5 m. ERDAS is used for imaging and geospatial analysis [23]. They used different data to train the network, e.g. differences in differently polarized

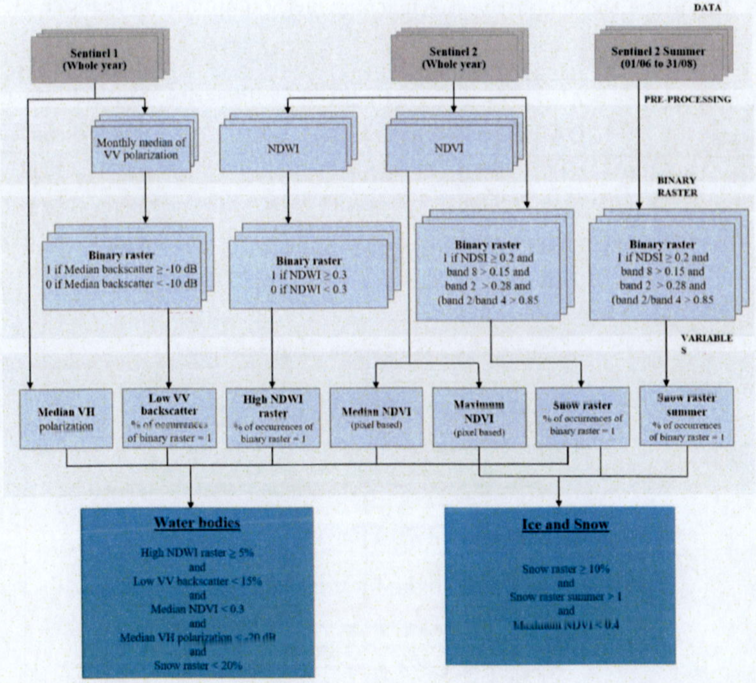

Fig. 1. Conditions for the classification tree method of land cover detection in Italy [13]

Table 2. Overview of accuracies from the classification tree method applied to land cover in the territory of Italy [13]

Overall Accuracy		
0.83		
Land Cover		
Class name	**User's accuracy**	**Producer's accuracy**
Artificial abiotic surfaces	0.92	0.88
Natural abiotic surfaces	0.83	0.78
Water bodies	0.98	0.90
Permanent snow and ice	0.86	1.00
Broad-leaved	0.87	0.81
Needle-leaved	0.90	0.88
Permanent herbaceous	0.92	0.62
Periodic herbaceous	0.86	0.81
Land Cover Change	**User's accuracy**	**Producer's accuracy**
Class name		
Other disturbance	0.35	0.71
Burned areas	0.67	1.00
Land consumption	0.81	1.00
Restoration	0.50	1.00

Table 3. Results of the support vector machine method applied to the area of Nigeria and Sahel, Africa [14]

Input features		All		Reduced Collinearity		Mutual Information	
Accuracies		OA [%]	95% CI	OA [%]	95% CI	OA [%]	95% CI
Classifier	RF	73.3	3.8	67.4	4.1	64.7	4.1
	ML	31.7	3.3	54.6	4.0	58.6	4.2
	SVM	60.8	4.1	62.1	4.1	43.4	3.9
	CE	72.0	3.9	-	-	-	-

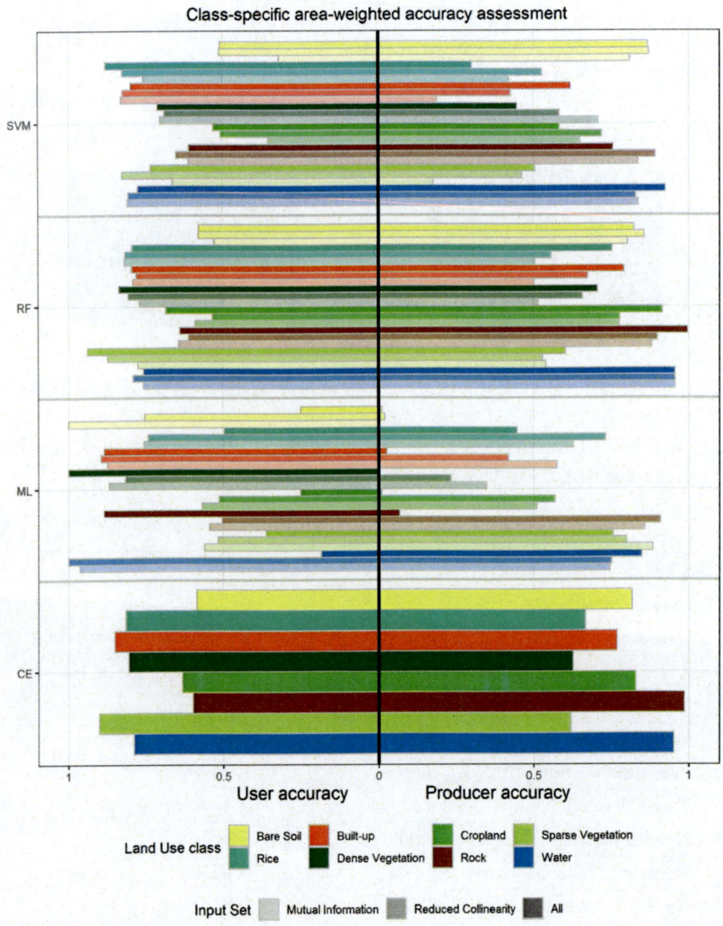

Fig. 2. Classification accuracies of specific types of land cover methods applied to the area of Nigeria and Sahel, Africa [14]

inputs, i.e. co-polarization (VV) and cross-polarization (VH) dB. Neural Networks and SVM were applied to the same input data. The overall accuracy was the highest, approximately 95%–97%, (see in Table 4) in both methods when combining inputs from Sentinel 1 and 2 (processed by the PCA method). The data was confident enough and tested using the kappa index [15,24].

Table 4. Accuracy results using neural networks and support vector machine applied to the area of Egypt [15]

Description of variables	Overall accuracy [%] SVM	Kappa index SVM	Overall accuracy[%] Neural Networks	Kappa index Neural Network
VV	73.38	0.75	71.59	0.70
VH	70.73	0.72	68.31	0.69
Sentinel-2	86.20	0.82	83.1	0.74
merged image using PCA	96.81	0.95	95.22	0.93
VV, VH, (VV-VH)	92.66	0.90	90.26	0.89
VV, VH, (VV-VH)/2	90.22	0.88	89.91	0.85
VV, VH, Texture	93.02	0.92	92.27	0.91

Australia is another of the researched areas [16]. The input data were collected from Sentinel 2 with a resolution of 10 m and images from PlanetScope with a resolution of 3 m. Convolutional Neural Networks (CNN) was used as one of the test methods for detecting clouds, land cover, and the like. The highest accuracy of the network for land cover application reached approx. 88% in the agricultural area (see in Table 5). The accuracies are comparable to the most used *Sen2Cor* methods [25] in the SNAP program [16].

4 Discussion

The use of CNN in our case could be only a partial solution, despite the fact that it achieves a high detection accuracy in the area of Egypt (approx. 95.22%) and in Australia (approx. 88%) [14,16]. Similar to the SVM methods, which achieved an accuracy of approximately 97% in Egypt and 61% in Africa. The random forest method was used in Africa with the accuracy of approximately 74% [14,15]. Detection of classes 2.3.1, 3.2.1 and 4.1 is only partially satisfactory. If we want to distinguish the subclasses of wetlands and grasslands, it is advisable to use one of the supervised classification methods, i.e. CNN, SVM. These classes show characteristic differences that we can also detect using NDVI and NDWI, similarly when using deep learning methods. We conclude that fundamental differences in reflectivity will contribute to higher accuracy when detecting using SVM, CNN. A clear proof is the detection results from Egypt, where the accuracy reached approx. 97% [14]. The problem arises if we want to distinguish between classes 2.3.1 and 3.2.1. Based on our tests, we could not distinguish between those

Table 5. Accuracy results using the CNN method for the area of Australia [16]

			DenseNet201	ResNet50	VGG10	Ensemble
Per-label F_2 (OA)	Cloud labels	Clear	0.88 (0.74)	0.87 (0.72)	0.92 (0.85)	0.91 (0.82)
		Haze	0.33 (0.73)	0.42 (0.87)	0.30 (0.45)	0.00 (0.95)
		Partly cloudy	0.73 (0.64)	0.73 (0.64)	0.69 (0.62)	0.73 (0.67)
		Cloudy	0.91 (0.90)	0.89 (0.90)	0.90 (0.92)	0.90 (0.91)
	Shade labels	Unshaded	0.93 (0.76)	0.93 (0.75)	0.93 (0.75)	0.93 (0.76)
		Partly shaded	0.67 (0.68)	0.62 (0.60)	0.68 (0.68)	0.69 (0.69)
		Shaded	0.50 (0.70)	0.47 (0.69)	0.57 (0.75)	0.45 (0.90)
	Land cover labels	Agriculture	0.88 (0.70)	0.87 (0.69)	0.88 (0.75)	0.88 (0.71)
		Bare ground	0.69 (0.68)	0.71 (0.66)	0.70 (0.57)	0.70 (0.72)
		Habitation	0.69 (0.68)	0.72 (0.78)	0.77 (0.82)	0.76 (0.83)
		Forest	0.85 (0.64)	0.83 (0.66)	0.85 (0.67)	0.84 (0.65)
		Water	0.72 (0.92)	0.71 (0.93)	0.66 (0.89)	0.72 (0.93)
Per-sample F_2 (OA)			0.81 (0.74)	0.80 (0.74)	0.81 (0.73)	0.82 (0.80)

categories as they show the same reflectivity. It is necessary to keep in mind other factors, for example (pastures can be detected as uncultivated parcels that have turned into grassland after a period of at least 3 years of non-use) [26]. Human intervention in nature is also a significant factor, which we can see using 3 annual records of the investigated area.

In this case, we would additionally choose the sorting tree method, where we know exactly what factors make up the differences between the investigated areas. A typical example is soil detection in Italy, where they achieved accuracy of 92% [13]. We see the potential in the clustering method, followed by the use of regression, which could reveal several features of these areas, which we could use with a combination of CNN and classification tree methods. By combining CNN methods, sorting tree and common classification methods in SNAP, Qgis, we assume that we can achieve high detection accuracy.

Based on the results of the previous analysis, the initial experiments were carried out using data from Sentinel 1, Sentinel 2, Sentinel 3, which are intended for the observation of land covers, because we found out that the detection of our mentioned Corine land cover subcategories in Slovakia are still detected only empirically [3]. We see potential in the use of machine learning, which we want to apply to the above-mentioned areas. The detection has implications for global environmental and security monitoring.

5 Conclusion

We introduced the Copernicus program, including the Sentinel 1–6 missions, from which we can obtain images of Earth's surface. These cover urban area management, environmentally friendly development and wildlife preservation, regional and local planning, agriculture, forestry, and fisheries, as well as health, civil defense, infrastructure, transit, and mobility. Tourism is also included.

Nowadays, the subcategories of permanent grasslands and wetlands are distinguished only empirically, we will analyze selected machine learning methods and propose our solution to this problem in order to see if the classification is more reliable.

We have described how we have to process the input images, and the software needed to use, i.e. SNAP and the Qgis plugin. Those are used for basic image processing and basic classification using indices. We clarified the overall process of processing the input dataset for machine learning methods. In brief, we explained which machine learning methods exist, how they are used, and we divided them according to their purpose. Next, we went through the procedures and methods of machine learning that were used to detect the land cover.

We evaluated the results of land cover detection from five regions of the world. In the territory of Italy, the decision tree method was used in combination with standard surface identification methods (indices), where they achieved an accuracy of about 83% [13]. In the territory of Nigeria in Africa, SVM methods were used, which achieved an accuracy of approximately 61%, similarly, the MLE and others were used, which achieved an accuracy of 72% [14]. The SVM and Neural Networks method were used to detect the urban area of Egypt. The overall accuracy was the highest, approximately 95%–97%, of both methods when combining inputs from Sentinel 1 and 2 (processed by the PCA method) [15]. Australia is another of the researched areas. The CNN method was used. The highest accuracy of the network for agricultural area detection reached approx 88% [16].

We evaluated all the results and used methods from several areas of the world and came to the conclusion that the solution to our problem can be a combination of these methods. We see the potential in the clustering method, followed by the use of regression, which will reveal the appropriateness and additional properties that we could include in the combination of CNN and classification tree methods. We can achieve high detection accuracy by combining the methods of CNN, classification tree, and basic classification methods in SNAP and Qgis. The detection can also be increased by finding new features in the data using clustering.

Acknowledgments. This publication is the result of implementation of the project VEGA 1/0176/22: "Proactive control of hybrid production systems using simulation-based digital twin" supported by the VEGA.

References

1. Copernicus and Sentinel missions. https://www.eumetsat.int/copernicus
2. Sentinel 2 mission overview. https://sentinels.copernicus.eu/web/sentinel/missions/sentinel-2/overview
3. Sentinel databases. https://sentinels.copernicus.eu/web/sentinel/sentinel-data-access
4. Definitions of Granules, data dictionary. https://sentinels.copernicus.eu/web/sentinel/user-guides/sentinel-2-msi/definitions

5. Application of Sentinel 1 mission. https://sentinels.copernicus.eu/web/sentinel/user-guides/sentinel-1-sar/applications
6. Definitions of Granules, data dictionary. https://sentinels.copernicus.eu/web/sentinel/user-guides/sentinel-3-olci/applications
7. Copernicus Programme. Copernicus Land Cover. Retrieved from CORINE Land Cover. https://land.copernicus.eu/pan-european/corine-land-cover
8. Applications of Sentinel 2 mission. https://sentinels.copernicus.eu/web/sentinel/user-guides/sentinel-2-msi/applications
9. Sentinel 2 - products and types. https://sentinels.copernicus.eu/web/sentinel/user-guides/sentinel-2-msi/product-types
10. Processing data - Sentinel 2. https://sentinels.copernicus.eu/web/sentinel/user-guides/sentinel-2-msi/processing-levels
11. Sentinel products and algorithms. https://sentinels.copernicus.eu/web/sentinel/technical-guides/sentinel-2-msi/products-algorithms
12. Congedo, L.Q.: Retrieved from Semi-Automatic Classification Tutorial. https://semiautomaticclassificationmanual.readthedocs.io/en/latest/tutorial_1.html#download-the-data
13. De Fioravante, P., et al.: Multispectral Sentinel-2 and SAR Sentinel-1 integration for automatic land cover classification. Land. **10**(6), 611 (2021). https://doi.org/10.3390/land10060611
14. Schulz, D.Y.: Land use mapping using Sentinel-1 and Sentinel-2 time series in a heterogeneous landscape in Niger, Sahel. ISPRS J. Photogrammetry Remote Sens. **178**, 97–111 (2021). https://doi.org/10.1016/j.isprsjprs.2021.06.005
15. Taha, L.G., Ibrahim, R.E.: Land use land cover mapping from Sentinel-1, Sentinel-2 and fused Sentinel images based on machine learning algorithms. Int. J. Comput. Appl. Math. Comput. Sci. **1**, 12–23 (2021)
16. Shendryk, Y.R.: Deep learning for multi-modal classification of cloud, shadow and land cover scenes in PlanetScope and Sentinel-2 imagery. ISPRS J. Photogrammetry Remote Sens. **157**, 124–136 (2019). https://doi.org/10.1016/j.isprsjprs.2019.08.018
17. Band combinations. https://gisgeography.com/sentinel-2-bands-combinations/
18. Troiano, L., et al. (eds.): Advances in Deep Learning, Artificial Intelligence and Robotics: Proceedings of the 2nd International Conference on Deep Learning, Artificial. Springer, Heidelberg (2022)
19. Pinheiro Cinelli, L.A.: Variational Methods for Machine Learning with Applications to Deep Networks. Springer, Cham (2021). https://doi.org/10.1007/978-3-030-70679-1
20. Buduma, N., Buduma, N., Papa, J.: Fundamentals of Deep Learning: Designing Next-Generation Machine Intelligence Algorithms. O'Reilly Media (2022)
21. Definition of ENVI 5.1 tool. https://sdu.sk/iKLn
22. Jolliffe, I.T.: Principal Component Analysis. Springer, New York (2010). https://doi.org/10.1007/b98835
23. Overview about ERDAS. https://hexagon.com/products/erdas-imagine
24. Jordan, R.: Complementing optical remote sensing with synthetic aperture radar observations of hail damage swaths to agricultural crops in the central United States. J. Appl. Meteorol. Climatol. **59**, 665–685 (2020). https://doi.org/10.1175/JAMC-D-19-0124.1
25. European Space Agency. ESA Documentation SNAP. Retrieved from ESA Snap tutorials. http://step.esa.int/main/doc/tutorials/
26. Corine land cover. https://land.copernicus.eu/user-corner/technical-library/corine-land-cover-nomenclature-guidelines/html/index-clc-231.html

Leveraging Synonyms and Antonyms for Data Augmentation in Sarcasm Identification

Aytuğ Onan(✉)

Department of Computer Engineering, Faculty of Engineering and Architecture, İzmir Katip Çelebi University, 35620 İzmir, Turkey
aytug.onan@ikcu.edu.tr

Abstract. Sarcasm identification is a challenging task in natural language processing due to its complex and nuanced nature. This study addresses the challenge of sarcasm identification in social media text, which has significant implications for sentiment analysis, opinion mining, and social media monitoring. Due to the lack of annotated data and the intricate nature of sarcastic expressions, building accurate sarcasm identification models is a challenging task. To overcome this problem, we propose a novel text data augmentation method that employs WordNet lexical database to generate high-quality training examples by swapping words with their synonyms or antonyms. We compare our proposed approach with conventional text data augmentation techniques and evaluate it on five state-of-the-art classifiers, including CNN, RNN, LSTM, GRU, and bidirectional LSTM. The results demonstrate that our proposed scheme outperforms other methods, achieving the highest F1-score and accuracy values across all classifiers. Our findings suggest that this text data augmentation approach can significantly improve the performance of sarcasm identification models, particularly in scenarios with limited annotated data.

Keywords: Sarcasm Identification · Text Data Augmentation · Deep Learning

1 Introduction

Sarcasm is a form of communication in which the speaker intends to convey the opposite or different meaning than the literal interpretation of their words. Sarcasm can be expressed in various ways, such as negation, hyperbole, rhetorical questions, and irony. Sarcasm is often used to convey negative feelings, criticism, or humor, and is prevalent in daily communication, particularly in social media [1]. However, sarcasm identification is a challenging task in natural language processing (NLP) due to its complex and nuanced nature. Sarcasm detection requires understanding the context, semantics, and social cues of the text. The sarcastic utterances are often expressed with a high degree of ambiguity, which makes it challenging to identify the sarcastic intent [2].

Despite its complexity, sarcasm identification has gained significant attention in recent years due to its potential applications in sentiment analysis, opinion mining, and social media analysis. In sentiment analysis, sarcasm detection can help to identify the

R. Silhavy and P. Silhavy (Eds.): CSOC 2023, LNNS 724, pp. 703–713, 2023.
https://doi.org/10.1007/978-3-031-35314-7_59

sentiment of the text accurately. In opinion mining, sarcasm detection can help to identify the true opinion of the author. In social media analysis, sarcasm detection can help to identify the user's intent and the potential impact of the post.

Existing approaches for sarcasm identification can be broadly categorized into three groups: rule-based, feature-based, and deep learning-based. Rule-based approaches rely on handcrafted rules and linguistic features to identify sarcasm [3]. Feature-based approaches use machine learning algorithms to learn features from the data [3]. Deep learning-based approaches use neural networks to automatically learn the features from the data [4]. However, these approaches require a large amount of labeled data to achieve high accuracy, which is often not feasible due to the limited availability of labeled data.

Data augmentation is a common technique used to overcome the problem of limited data in machine learning [5]. Data augmentation involves creating new examples by making slight modifications to the original data. In NLP, data augmentation methods such as synonym replacement, paraphrasing, and back-translation have been proposed to expand the training data for various NLP tasks, such as machine translation, text classification, and sentiment analysis [6]. In NLP, data augmentation methods have also been proposed to expand the training data for various NLP tasks, such as machine translation, text classification, and sentiment analysis [7]. These methods often involve making small modifications to the text to create new examples, such as replacing words with synonyms or antonyms, paraphrasing the text, or generating new sentences with back-translation [6]. One of the advantages of data augmentation is that it can help to reduce overfitting and improve the generalization ability of the model. Overfitting occurs when the model learns to fit the training data too well and fails to generalize to new, unseen data. Data augmentation can introduce more variability into the training data, which helps to prevent the model from memorizing the training data and encourages it to learn more robust and generalizable patterns [5]. Data augmentation can also help to address the problem of class imbalance, which is a common issue in many NLP tasks. Class imbalance occurs when the distribution of the classes in the training data is highly skewed, with one class having much fewer examples than the other(s). This can lead to biased models that perform poorly on the minority class. Data augmentation can generate new examples for the minority class, thereby balancing the distribution of the classes and improving the model's performance on the minority class [6, 7].

In this paper, we propose leveraging synonyms and antonyms based data augmentation for sarcasm identification. The proposed approach generates new examples by replacing words in the original sentences with their synonyms and antonyms, which can help to expand the training data and improve the performance of the sarcasm identification model. The use of synonyms and antonyms provides a natural way to generate new examples that are similar or opposite in meaning to the original sentences, respectively. We believe that our proposed approach can be easily integrated into existing sarcasm identification models and can significantly improve their performance by augmenting the training data with synonyms and antonyms.

The paper is structured as follows. Section 2 provides a comprehensive review of related work on sarcasm identification and data augmentation methods. Section 3 describes the proposed approach in detail, including the data augmentation techniques and the model architecture. Section 4 presents the experimental procedure, including the datasets, evaluation metrics, and implementation details, and reports the experimental results. Finally, Sect. 5 concludes the paper and suggests directions for future research in sarcasm identification and data augmentation.

2 Related Work

Sarcasm identification is a well-studied problem in natural language processing, and various techniques have been proposed for its identification. González-Ibánez et al. [8] conducted one of the most significant studies in the field of natural language processing (NLP) focused on using machine learning techniques for sarcasm identification. They evaluated the predictive performance of various linguistic feature sets, including unigrams, dictionary-based lexical and pragmatic factors, and their frequency, on a Twitter sarcasm corpus with two conventional supervised learners: support vector machines and logistic regression. Reyes et al. [9] evaluated feature sets based on structural, morph syntactic and semantic ambiguity, polarity, unexpectedness, and emotional scenarios for figurative language processing. Barbieri et al. [10] considered frequency-based features, written and spoken style uses, intensity of adjectives and adverbs, punctuation marks, and emoticons for sarcasm identification. Kunneman et al. [11] evaluated the reliability of user-generated hashtags on Twitter for sarcasm identification. Rajadesingan et al. [12] proposed a behavioral modeling scheme for sarcasm detection. Bouazizi and Ohtsuki [13] introduced a pattern-based approach for sarcasm identification using sentiment-related, punctuation-related, syntactic and semantic features, and pattern-based features. Mishra et al. [14] evaluated lexical, implicit incongruity based features, explicit incongruity based features, textual features, simple gaze based features, and complex gaze based features.

Deep learning has been applied in the field of NLP, specifically in sarcasm identification, with successful results. Ghosh and Veale [15] presented a deep learning architecture for sarcasm identification using convolutional neural networks (CNN), long short-term memory (LSTM), and deep neural networks (DNN). Their empirical analysis on Twitter messages showed that their architecture outperformed conventional classifiers, with an F-measure of 0.92. Similarly, Naz et al. [16] utilized a CNN-based architecture for sarcasm identification, and Kumar et al. [17] proposed a sarcasm identification framework based on bidirectional LSTM with CNN. Onan [2] evaluated the predictive performance of four neural language models and traditional feature sets in conjunction with CNN for sarcasm identification, while Ren et al. [18] introduced a weighted word embedding scheme and multi-level memory network for text sentiment classification and sarcasm expression features. Jain et al. [17] presented a deep learning framework for sarcasm identification in English and Hindi languages, utilizing pre-trained GloVe model and bidirectional LSTM with soft-attention mechanism to extract semantic feature vectors. Recently, Onan and Toçoğlu [19] presented an effective sarcasm identification framework for social media data using neural language models and deep neural networks. A three-layer stacked bidirectional long short-term memory architecture is introduced to identify sarcastic text documents, and an inverse gravity moment based term weighted word-embedding model with trigrams is used to represent text documents.

3 Methodology

This section describes the proposed approach in detail, including the data augmentations techniques and the deep neural network architectures.

3.1 Conventional Text Transformation Functions

Typos are a type of data augmentation that can be used to generate new training examples from existing ones [20]. This approach involves introducing spelling errors or typos into the original text, thereby creating new variations of the same sentence. The typos can be introduced randomly, or by using specific rules to mimic common typing mistakes, such as substituting one letter for another, omitting letters, or adding extra letters. By introducing these errors, the resulting augmented dataset can help improve the robustness and accuracy of natural language processing models, particularly those that rely on large amounts of training data.

Back Translation is another data augmentation technique. This technique involves translating a text document from one language to another and then back to the original language. The idea is to introduce variations in the text while preserving the original meaning. This technique can be especially useful in cases where large amounts of data are not available in the target language [21].

3.2 The Proposed Data Augmentation Scheme

Swap Antonomy-WordNet is a text data augmentation technique that swaps words in a sentence with their antonyms, using a lexical database called WordNet. WordNet is a large electronic lexical database of English words and their meanings that is organized into synsets, or groups of synonyms and antonyms. This technique can be used to generate new training examples from existing ones, by swapping out words with opposite meanings. The resulting augmented dataset can be used to train and improve natural language processing models, especially those that rely on large amounts of training data. An example of Swap Antonomy-WordNet text data augmentation could involve replacing the word "happy" with its antonym "sad" in a sentence expressing a positive sentiment. For instance, the sentence "I am so happy to see you!" could be transformed into "I am so sad to see you!" to create an augmented example with the opposite sentiment. This approach can be particularly useful for improving the robustness of sentiment analysis models by exposing them to more diverse and challenging examples. However, it should be noted that the class label of the original example needs to be changed accordingly to reflect the opposite sentiment. Moreover, care should be taken to ensure that the resulting augmented examples are still grammatically correct and semantically meaningful.

Swap Synonym-WordNet is another text data augmentation technique that swaps words in a sentence with their synonyms, using the same WordNet database. This technique can also be used to generate new training examples from existing ones, by swapping out words with similar meanings but different spellings. The resulting augmented dataset can be used to train and improve natural language processing models, especially those that rely on large amounts of training data. For example, the sentence "The food was terrible" can be augmented to "The food was dreadful" by replacing the word "terrible" with its synonym "dreadful."

Algorithm 1. The general structure for the proposed text data augmentation scheme.

Input: Original dataset *D*
Number of augmented examples to generate *n*
Threshold value *t* for the similarity measure
WordNet lexical database
Output: Augmented dataset *D_aug*
Algorithm:

1. Initialize *D_aug* as an empty dataset.
2. For each example *x* in *D*, do the following:
   a. Find all synonyms of each word in *x* using WordNet and store them in a list, *S*.
   b. Find all antonyms of each word in *x* using WordNet and store them in a list, *A*.
   c. For each synonym *s* in *S*, do the following:
      i. Compute the similarity score between *x* and *s* using the similarity measure (i.e., cosine similarity) and check if it is greater than or equal to the threshold *t*.
      ii. If the similarity score is greater than or equal to *t*, replace the corresponding word in *x* with s and add the resulting example to *D_aug* with the same class label.
   d. For each antonym *a* in *A*, do the following:
      i. Compute the similarity score between *x* and *a* using a similarity measure (i.e., cosine similarity) and check if it is greater than or equal to the threshold *t*.
      ii. If the similarity score is greater than or equal to *t*, replace the corresponding word in *x* with *a* and change the class label of the resulting example to the opposite of the original class label.
      iii. Add the resulting example to *D_aug*.
3. Repeat steps 2 and 3 until *n* augmented examples have been generated.
4. Return the augmented dataset by adding *D* and *D_aug*.

Both techniques leverage the vast resources of WordNet, which includes a large collection of English words and their synonyms and antonyms, to increase the diversity and richness of the training data. This can help to improve the robustness and generalization of natural language processing models. The general structure of the proposed text data augmentation scheme has been summarized in Algorithm 1. The value of *n* should be chosen such that it results in a reasonable amount of augmented data while avoiding overfitting. The threshold value should be chosen such that it results in a sufficient amount of variation in the augmented data while avoiding creating unrealistic examples. These values can be tuned through experimentation and analysis of the resulting augmented dataset.

3.3 Text Classification Architectures

The proposed text data augmentation approach has been evaluated empirically using five state-of-the-art text classification models to assess its predictive performance. The rest of this section briefly explains the models.

Convolutional Neural Networks (CNNs) have been shown to be effective in text classification tasks due to their ability to capture local and global relationships between words in a sentence. In CNN-based text classification, the input sentence is first converted into a sequence of word embeddings, which are then fed into a convolutional layer with multiple filters of varying sizes. The filters slide over the input sequence and compute feature maps, which are then passed through a max-pooling layer to extract the most salient features. The resulting feature vectors are concatenated and fed into a fully connected layer for classification. Several studies have demonstrated the effectiveness of CNNs in text classification tasks, including sentiment analysis, topic classification, and sarcasm detection [22].

Recurrent neural networks (RNNs) are a type of neural network architecture that is particularly well-suited for sequential data such as text [23]. RNNs are designed to take into account the temporal dependencies between inputs in a sequence, which makes them a powerful tool for natural language processing (NLP) tasks such as language modeling, machine translation, and sentiment analysis. Unlike traditional feedforward neural networks, RNNs have feedback connections that allow them to use information from previous time steps to inform their current predictions. This makes them particularly effective for tasks where the meaning of a particular word or phrase depends on the context in which it appears. RNNs have been successfully applied to a wide range of NLP tasks.

Long Short-Term Memory (LSTM) is a type of recurrent neural network (RNN) architecture that is designed to overcome the vanishing gradient problem in standard RNNs [24]. The LSTM cell is comprised of multiple gating mechanisms that selectively control the flow of information through the cell. The input gate controls the flow of new input into the cell, the forget gate controls the retention of previously stored information, and the output gate controls the output of the cell. The LSTM architecture is particularly effective for tasks that involve long-term dependencies and sequential input, such as language modeling, machine translation, and speech recognition. LSTMs have been used extensively in natural language processing (NLP) tasks, including sentiment analysis, text classification, and named entity recognition.

Gated Recurrent Unit (GRU) is a type of recurrent neural network architecture that is similar to the LSTM network in that it is designed to address the vanishing gradient problem that occurs in traditional recurrent neural networks, which hinders their ability to capture long-term dependencies in sequential data. The GRU network uses gating mechanisms to selectively update and reset the hidden state at each time step, which allows it to maintain relevant information over longer sequences. The gating mechanisms in GRU are simpler than those in LSTM, as it only has two gates: a reset gate and an update gate. The reset gate determines how much of the previous hidden state should be forgotten, while the update gate determines how much of the new input should be added to the current hidden state.

Bidirectional LSTM (BiLSTM) is an extension of the standard LSTM network for text classification tasks [26]. It processes the input sequence both forwards and backward to capture the contextual information from both directions. This is achieved by using two separate LSTMs, one that reads the sequence from beginning to end, and another that reads it from end to beginning. The output of each LSTM is then concatenated to form the final output sequence. BiLSTMs have shown promising results in various NLP tasks, including sentiment analysis, text classification, and machine translation. They are particularly useful in tasks where context plays a critical role, such as understanding the sentiment of a sentence or determining the meaning of a word based on its context.

4 Experimental Procedure and Results

This section presents the experimental procedure, including the dataset, evaluation metrics, and implementation details, and reports the experimental results.

4.1 Dataset

We collected a dataset of approximately 40,000 tweets in English using self-annotated tweets from Twitter users with hashtags of "sarcasm" or "sarcastic" for sarcastic tweets, and hashtags about positive and negative sentiments for non-sarcastic tweets [2]. After manual annotation, we obtained a balanced corpus of 15,000 sarcastic tweets and 15,000 non-sarcastic tweets. We preprocessed the corpus by tokenizing the tweets using the Twokenize tool, removing unnecessary items, mentions, replies, URLs, and special characters. The dataset was constructed using the framework in [1, 8].

4.2 Experimental Procedure

To evaluate the proposed data augmentation approach, we conducted experiments on the sarcasm identification task using five state-of-the-art text classification architectures: CNN, RNN, LSTM, GRU, and Bidirectional LSTM. We utilized the word2vec word-embedding scheme. We implemented the models using Python and the Keras library, with TensorFlow as the backend.

For data augmentation, we set the number of augmented examples to generate (n) to 10,000 and the similarity threshold (t) to 0.8, based on empirical analysis. We used WordNet as the lexical database for both swap synonym and swap antonym methods.

To evaluate the performance of our proposed data augmentation approach, we compared the accuracy, and F1 score of each model with and without data augmentation. We used 10-fold cross-validation to validate the results and reported the average performance across all folds. The parameters of the models utilized in the empirical analysis have been summarized in Table 1.

4.3 Experimental Results and Discussion

In Table 2, the classification accuracy values obtained by the text data augmentation methods on deep neural networks have been presented. The empirical results indicate

Table 1. The parameter values for the deep neural network architectures.

Model	Embedding Dimension	Hidden Layer Size	Dropout Rate	Batch Size
CNN	300	100	0.5	64
LSTM	300	100	0.5	64
GRU	300	100	0.5	64
BiLSTM	300	100	0.5	64
RNN	300	100	0.5	64

that the proposed scheme outperforms all other data augmentation techniques across all classifiers, achieving an accuracy of 82.46% with CNN. The "Swap Synonym- WordNet" configuration is the second-best performing technique, with an accuracy of 81.06%. The "Swap Antonomy-WordNet" configuration is the third-best performing technique, with an accuracy of 80.32%. The "back translation" technique performs slightly better than the baseline configuration, while the "typos" configuration performs worse than the baseline.

Table 2. The classification accuracies obtained by text data augmentation schemes.

Configuration	CNN	RNN	LSTM	GRU	BiLSTM
Without Data Augmentation	78.235	80.178	81.214	79.976	83.089
Typos	77.625	79.563	80.687	79.259	82.301
Back Translation	79.147	80.854	82.032	80.741	83.467
Swap Antonomy-WordNet	80.318	81.802	82.785	81.636	84.125
Swap Synonym-WordNet	81.058	82.137	83.069	82.098	84.571
Proposed Scheme	82.459	83.258	83.758	83.021	85.362

In terms of classifiers, the BiLSTM architecture performs the best across all configurations, with an accuracy of 85.36% in the proposed scheme configuration. The LSTM and GRU architectures perform similarly, with accuracies in the low 80s across all configurations. The RNN architecture performs slightly worse than LSTM and GRU, while the CNN architecture performs the worst, with accuracies in the high 70s to low 80s across all configurations. The results suggest that the proposed data augmentation scheme, which combines multiple techniques including synonyms, and antonyms, can significantly improve the performance of sarcasm detection models, particularly when used in conjunction with the BiLSTM architecture.

The Table 3 shows the F1-score values obtained by applying different text data augmentation schemes on the sarcasm detection dataset. Table 3 shows that data augmentation techniques generally outperform the baseline (without data augmentation) for all classifiers, indicating the effectiveness of data augmentation in improving the performance of sarcasm detection models. The proposed scheme achieves the highest F1-scores

Table 3. F1-score values obtained by text data augmentation schemes.

Configuration	CNN	RNN	LSTM	GRU	BiLSTM
Without Data Augmentation	0.780	0.802	0.812	0.800	0.831
Typos	0.776	0.795	0.807	0.794	0.823
Back Translation	0.791	0.808	0.821	0.810	0.835
Swap Antonomy-WordNet	0.802	0.817	0.827	0.818	0.841
Swap Synonym-WordNet	0.811	0.825	0.835	0.826	0.848
Proposed Scheme	0.825	0.833	0.840	0.835	0.861

for all classifiers, followed by Swap Synonym-WordNet and Swap Antonomy-WordNet. The results indicate that the proposed scheme can effectively leverage the strengths of the techniques to generate diverse and high-quality augmented data. In terms of classifiers, the BiLSTM consistently outperforms the other classifiers for all data augmentation techniques, followed by LSTM and GRU. CNN shows the lowest performance, indicating that it may not be suitable for the sarcasm detection task, at least without extensive data augmentation. To summarize the main findings of the empirical analysis, Fig. 1 presents the main effects plot for accuracy values.

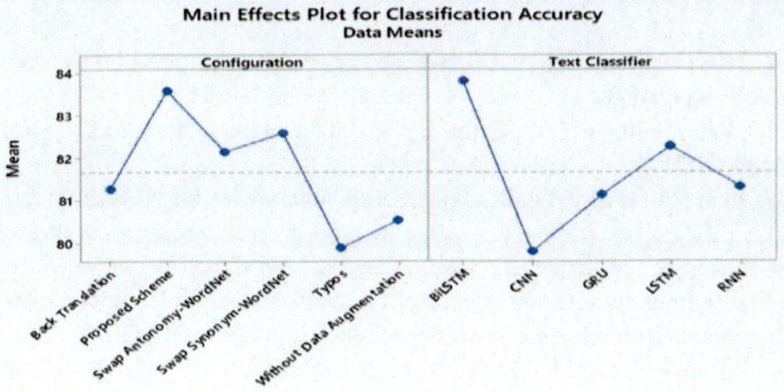

Fig. 1. The main effects plot for classification accuracy.

5 Conclusion

This work focuses on the problem of sarcasm identification in social media text, which has important applications in sentiment analysis, opinion mining, and social media monitoring. Due to the scarcity of annotated data and the high variability and complexity of sarcastic expressions, building accurate sarcasm identification models remains a challenging task. To address this problem, we propose a novel text data augmentation scheme based on WordNet lexical database, which can generate diverse and high-quality training examples by swapping words with their synonyms or antonyms. The proposed approach is compared with several conventional text data augmentation methods and evaluated on five state-of-the-art classifiers, including CNN, RNN, LSTM, GRU, and bidirectional LSTM. The experimental results demonstrate that the proposed scheme outperforms the other methods, achieving the highest F1-score and accuracy values across all classifiers. Our findings suggest that the proposed text data augmentation approach can effectively enhance the performance of sarcasm identification models, especially in low-resource scenarios where annotated data is limited.

References

1. Paredes-Valverde, M.A., Colomo-Palacios, R., Salas-Zarate, M., Valencia-Garcia, R.: Sentiment analysis in Spanish for improvement of product and services: a deep learning approach. Sci. Program. **2017**, 1–12 (2017)
2. Onan, A.: Topic-enriched word embeddings for sarcasm identification. In: Silhavy, R. (ed.) Software Engineering Methods in Intelligent Algorithms, pp. 293–304. Springer, Cham (2019). https://doi.org/10.1007/978-3-030-19807-7_29
3. Eke, C.I., Norman, A.A., Shuib, L., Nweke, H.F.: Sarcasm identification in textual data: systematic review, research challenges and open directions. Artif. Intell. Rev. **53**, 4215–4258 (2020)
4. Eke, C.I., Norman, A.A., Shuib, L.: Context-based feature technique for sarcasm identification in benchmark datasets using deep learning and BERT model. IEEE Access **9**, 48501–48518 (2021)
5. Shorten, C., Khoshgoftaar, T.M.: A survey on image data augmentation for deep learning. J. Big Data **6**, 1–48 (2019)
6. Shorten, C., Khoshgoftaar, T.M., Furht, B.: Text data augmentation for deep learning. J. Big Data **8**, 1–34 (2021)
7. Feng, S.Y., et al.: A survey of data augmentation approaches for NLP. arXiv preprint arXiv: 2105.03075 (2021)
8. González-Ibánez, R., Muresan, S., Wacholder, N.: Identifying sarcasm in Twitter: a closer look. In: Proceedings 49th Annual Meeting of the Association for Computational Linguistics: Human Language Technologies, Short Papers, vol. 2, pp. 581–586 (2011)
9. Reyes, A., Rosso, P., Buscaldi, D.: From humor recognition to irony detection: the figurative language of social media. Data Knowl. Eng. **74**, 1–12 (2012)
10. Barbieri, F., Saggion, H., Ronzano, F.: Modelling sarcasm in Twitter a novel approach. In: Proceedings 5th Workshop Computational Approaches to Subjectivity, Sentiment and Social Media Analysis, pp. 50–58 (2014)
11. Kunneman, F., Liebrecht, C., van Mulken, M., van den Bosch, A.: Signaling sarcasm: from hyperbole to hashtag. Inf. Process. Manage. **51**, 500–509 (2015)
12. Rajadesingan, A., Zafarani, R., Liu, H.: Sarcasm detection on Twitter: a behavioral modeling approach. In: Proceedings 8th ACM International Conference Web Search Data Mining, pp. 97–106 (2015)
13. Bouazizi, M., Otsuki, T.: A pattern-based approach for sarcasm detection on Twitter. IEEE Access **4**, 5477–5488 (2016)
14. Mishra, A., Kanojia, D., Nagar, S., Dey, K., Bhattacharyya, P.: Harnessing cognitive features for sarcasm detection. arXiv:1701.05574 (2017)

15. Ghosh, A., Veale, T.: Fracking sarcasm using neural network. In: Proceedings of the 7th Workshop on Computational Approaches to Subjectivity, Sentiment and Social Media Analysis, pp. 161–169 (2016)
16. Naz, F., et al.: Automatic identification of sarcasm in tweets and customer reviews. J. Intell. Fuzzy Syst. **37**(5), 6815–6828 (2019)
17. Jain, D., Kumar, A., Garg, G.: Sarcasm detection in mash-up language using soft-attention based bi-directional LSTM and feature-rich CNN. Appl. Soft Comput. **91** (2020)
18. Ren, H., Zeng, Z., Cai, Y., Du, Q., Li, Q., Xie, H.: A weighted word embedding model for text classification. In: Li, G., Yang, J., Gama, J., Natwichai, J., Tong, Y. (eds.) Database Systems for Advanced Applications, pp. 419–434. Springer, Cham (2019). https://doi.org/10.1007/978-3-030-18576-3_25
19. Onan, A., Toçoğlu, M.A.: A term weighted neural language model and stacked bidirectional LSTM based framework for sarcasm identification. IEEE Access **9**, 7701–7722 (2021)
20. Gui, T., et al.: TextFlint: unified multilingual robustness evaluation toolkit for natural language processing. arXiv preprint arXiv:2103.11441 (2021)
21. Hayashi, T., et al.: Backtranslation-style data augmentation for end-to-end ASR. In: Proceedings of the IEEE Spoken Language Technology Workshop (SLT), pp. 426–433 (2018)
22. Elman, J.L.: Finding structure in time. Cogn. Sci. **14**(2), 179–211 (1990)
23. Zhang, L., Wang, S., Liu, B.: Deep learning for sentiment analysis: a survey. Wiley Interdisc. Rev. Data Min. Knowl. Discov. **8**(4), e1253 (2018)
24. Rojas-Barahona, L.M.: Deep learning for sentiment analysis. Lang. Linguist. Compass **10**(12), 701–719 (2016)
25. Cho, K., et al.: Learning phrase representations using RNN encoder-decoder for statistical machine translation. arXiv preprint arXiv:1406.1078 (2014)
26. Onan, A.: Bidirectional convolutional recurrent neural network architecture with groupwise enhancement mechanism for text sentiment classification. J. King Saud Univ.-Comput. Inf. Sci. **34**(5), 2098–2117 (2022)

Knowledge Management Methodology to Predict Student Doctoral Production

Ricardo Manuel Arias Velásquez(✉)

Universidad Tecnológica del Perú, Lima, Peru
ricardoariasvelasquez@hotmail.com, rariasv@utp.edu.pe
https://utp.edu.pe/

Abstract. The scientific production of engineering doctoral students is a challenge in several universities, in the last decade a new approach of knowledge management was proposed with a qualitative evaluation. In this research article, an evaluation of the production of 188 doctoral students are investigated with a qualitative and quantitative design, with an approach of knowledge management approach based on capitals (Human, structural, relational), and skills (technology management and innovation). Our findings are only 2.6% of the set has increased their production with an average of thirty-four research articles compared with the remaining students; the most important aspect are the following: i) A strong influence of the knowledge management with internal organization in the university, ii) External communication with authors in others universities, iii) innovation, training and activities; iv) Specific statistical access; v) specific engineering system; vi) operative management software; vii) Availability of resource for new design and pilots; viii) decision-making process and ix) proactivity in the university for research projects. This analysis has evaluated the impact and the number of articles published; it has a strong correlation in doctoral engineering students; the results are a mean absolute error, associated to research articles published (Y1) with 0.8074 correlation coefficient and 1.5203 of mean absolute error (MAE); and h-index (Y2) with 0.8213 correlation coefficient and MAE of 0.545.

Keywords: Knowledge · management skills · research community

1 Introduction

Universities have an important challenge in the knowledge generation, the value in the knowledge chain allows to create patents and innovation in all the industrial sectors, with emphasis in the engineering doctoral program, in this case the major responsibility is on the supervisors and directors. Therefore, they could contribute to administration and improve the results of a specific research field. A particular aspect is the publication of research articles according to the thesis and research of its students. Engineering doctoral students requires technology

R. Silhavy and P. Silhavy (Eds.): CSOC 2023, LNNS 724, pp. 714–732, 2023.
https://doi.org/10.1007/978-3-031-35314-7_60

skills [1] and the appropriate environment to develop the knowledge management (KM) aspects: Human, structural, and relational [2]. For example, In Peru has increased the international certification in all the universities, based on government administration and its National Superintendence of Higher University Education called SUNEDU. It has a total of 51 public universities authorized to operate and 92 private universities in Peru. However, according to the report of the National Institute of Statistics and Informatics-INEI in 2018 [3], only a total of 42 thousand of professors were working, between Peruvian public and private universities, and in 2020, about 74 thousand of professors were incorporated to the research and develop field in Peruvian public and private universities. On the other hand, only two universities have the engineering doctoral program without an international certification. Both have the same challenge about the quality of the new knowledge generated in its program, based on the private regulation. Besides, it is a challenge due to the current situation regarding the pandemic and COVID-19 [2] and the situations to adapt its process, efficiently manage economic, material resources and especially human capital (Professors, supervisors, and administrative workers), after the last peak of COVID-19. In this country, the publishing rate is low, in the others doctoral programs, for example the authors of [5], they had an important conclusion with just 54 research articles in medicine doctoral program evaluated in 2017. According to the national science center (CONCYTEC) [6], Peru has 31% of the universities with a research information management active; therefore, the universities need to increase the knowledge management and skills evaluation to improve the production of doctoral students, due to constrains of an emerging country after the COVID-19, the global high inflation, restrictions of electronics and materials around the world

1.1 Motivation

Doctoral degree should consider managing the change focused on the talent of the experts who work at the university, the availability of laboratories, materials, and knowledge. In such a way that it allows them to find innovative and unique strategies to maintain themselves during the current COVID context [7], and to be a differentiating brand in the market that they develop with the highest knowledge generation demonstrated with research articles published and h index. For example, in Peru, the low rate of research articles is a challenge in the knowledge management of the universities, therefore, this research article allows to understand the characteristics of its problems and an important difference in 2.6% of the total population of one university, to improve the implementation of the knowledge management in the doctoral education. Although the innovation and infrastructure are important, one of the biggest problems is associated to the 3Ws ("What to publish, when to publish and where to publish" [4]); without a particular analysis of the environment and its methodology, the next step could be important for the next generation of the engineering doctoral program. The question associated to the motivation is the following: Could the technology skills and knowledge management map increase the knowledge generation by increasing the publishing average and h index impact?

This paper has the following content: Sect. 2, about the methodology, instrument, data description with the systematic review as knowledge management application. Likewise, the Sect. 3 develops a case study associated to an action-research study with qualitative research of 188 doctoral students, during the years 2012 to 2021. Later in Sect. 4, the discussion of the results, and finally the conclusion and future works.

2 Systematic Review for the Knowledge Management Applied to Skill Evaluation

In the evaluation of the systematic review, we have used the Population, intervention, comparison, outcome, and context (PICOC) methodology. In the Table 1 the search strings considered the aspects and filters for the selection of the references. In the last 24 years, the results were 5,779 research articles associated to doctoral programs based on the population 4,860 papers, in the intervention, comparison and outcome stage found 607 papers and with the context filters, it had 312 papers in the Table 1; however, only 87 research articles were associated to engineering doctoral programs as a category with relevant contribution after the inclusion and restrictions questions:

- Does it have a real contribution in the field?
- Does it have a statistic and metrics to establish the model or methods to other universities of industries?
- Did it have an evaluation in a Q1 or Q2 international journal indexed in Scopus and/or Web of Science?

As a results, only 42 research articles included the human capital contribution, 21 research articles in structural capital influence and 15 research articles associated to relational capitals. Finally, only nine research articles associated to skills and knowledge generation in the engineering doctoral programs, in Fig. 1 Besides, the evolution of this research is in the Fig. 2, it increases since 1997 to 2021. During the evolution of the papers in the Fig. 2, the contribution associated to doctoral students were as follows:

- In [9]: It is an exploratory and descriptive research with a quantitative approach. The objective is to validate, the change rate from offline to online system, to replace the face-to-face system in education. Because of the forced social distancing and restrictions imposed by the Coronavirus has affected the "relationships and performance" of research team (supervisors and engineering doctoral students), as well how to evaluate the technologies and instructions and process "adopted by them to innovate and achieve sustainable education" [9]. The Sample is about teachers in education from Brazil, as well as teachers from the 197 best Brazilian universities. The database analysis and using computer tools (SPSS).

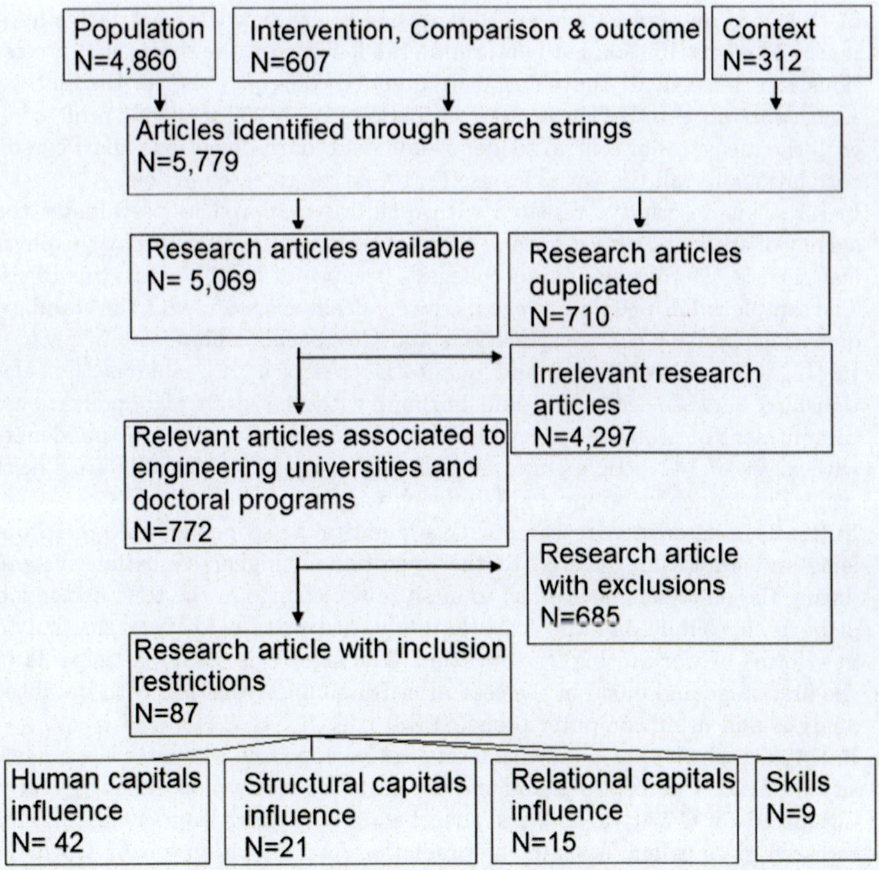

Fig. 1. Evaluation with PICOC evaluation.

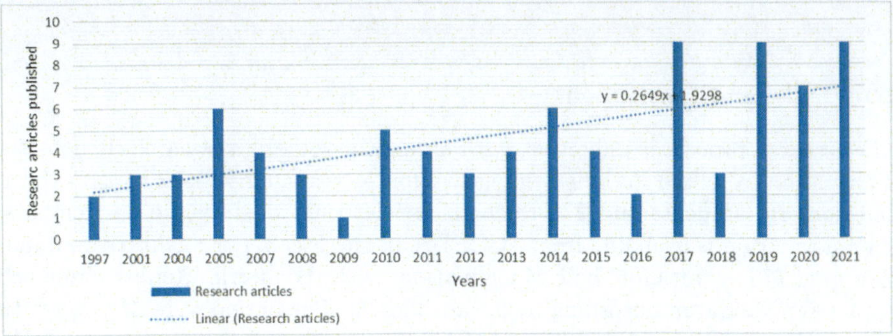

Fig. 2. Evolution of the research articles with inclusion and exclusions criterion.

- In [10], it is an exploratory and descriptive research with quantitative approach. The contribution is to determine the features of the professors' professions and consequently have "some unintended consequences" for the motivational work potential of the professors. The Sample is 202 academic professors; with the measures of central tendency and "standard deviations and Pearson correlations for all the variables used" [10] in the research article.
- In [11], it is descriptive research with qualitative approach; it evaluates the agency of academic professors and alternative responses, "caused by the physical closure of universities and colleges" because of the Corona virus crisis. The sample is 171 people surveyed, entirely academic staff, with the standard deviations and Pearson's correlations and Cronbach's alpha.
- In [12] it is a quantitative and qualitative research, it evaluates the relationships between leadership and burnout with the levels of organizational commitment of professor and supervisors during the Corona virus pandemic; with a set of 147 people surveyed (all professors); the Methods are Likert scale, Pearson correlations and Cronbach's alpha.
- In Russia, a descriptive research with a quantitative scope identifies the problems and limitations related to the operation in higher education system, during the pandemic, as well as to analyze the change in the tone and mood of both the public and the teaching body, in response to the transition to new forms of work in higher education. The sample is 3,431 participants in the first stage and 6,066 in the second in Russian universities, with database analysis and using computer tools (SPSS) [13].
- In [14], descriptive research with a qualitative approach evaluates the university professor's perceptions and the potential for motivation; during e learning forced by Covid -19 is lower than before e learning from covid-19. This research gives us emphasis in the prediction for motivation during Covid-19 with burnout aspects, job satisfaction, job commitment and the attitude. The author has evaluated three institutions with 202 participants: with standard deviations and Pearson's correlations and Cronbach's alpha.

2.1 Data and Instrument

The research process is inductive, with a qualitative design research. It uses the information from one Peruvian university for 24 years, since 2012 to 2021, from 188 doctoral students and 270 research articles published by the participants; the information is available from SCOPUS, in the Fig. 3. For qualitative evaluation used 773 surveys for the KM evaluation; with the metric: Robust Cronbach Alpha as a validator, according to the authors [8]. The period were 24 years with 773 results. It has identified 37 factors in the knowledge management and 110 skills identified during the surveys and the application of these skills for knowledge generation. The doctoral students have responded on a Licker one to five scale ranging from one as "Not applied for me", two as "very rarely true for me", three "rarely for me", four "sometimes true for me" to five "almost always true for me", in the Table 2. In the Fig. 4, it links the factors with the problem and

Table 1. PICOC Methodology questions and evaluation.

Criterion	Questions	Evaluation	Search strings
Population	Who is our main population?	Engineering doctoral students in private and public universities	(knowledge)AND(doctoral) AND(student) AND(engineering) NOT (nursing)NOT(art) NOT(medicine)
Intervention	What approach could it be used?	Qualitative design, inductive process, action-research, university, higher education. Dimensions and factors associated to knowledge generation	model)AND(knowledge) AND(doctoral)AND (community)NOT(medicine) NOT(nurse)AND(action)
Comparison	Could it compare with other specialties?	Engineering programs for doctoral degree. Publication influence	(Research)AND(inductive) NOT(foundation)
Outcome	What does it evaluate?	Knowledge generation, publishing, and impact of the engineering doctoral students	
Context	What is the context?	Normal context and constraints of the COVID in doctoral degree associated to the supervisors	(knowledge)AND ((doctoral)OR(PHD)) AND(supervisor)NOT (medicine)NOT(nurse) AND(COVID)

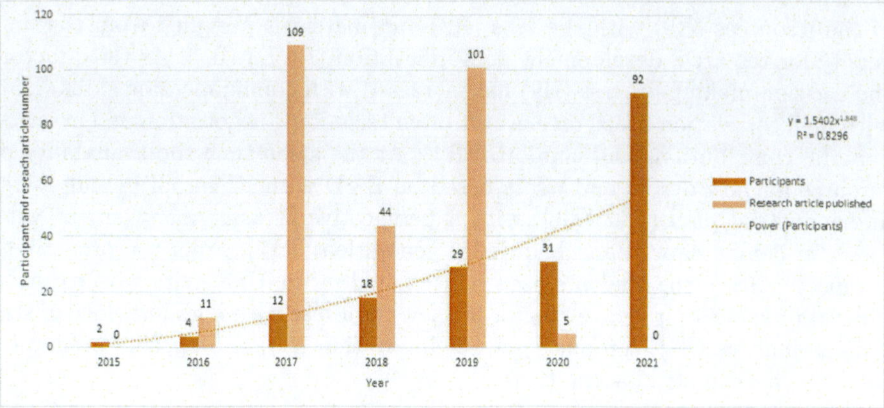

Fig. 3. Database analysis for the case study.

Table 2 indicates the code for each factor, in combination, the Table 2 and Fig. 4 could describe the relation with the factors and the main problem, to improve the knowledge generation in engineering doctoral studies: It has nine core factors with direct influence, the knowledge in the international organization depends

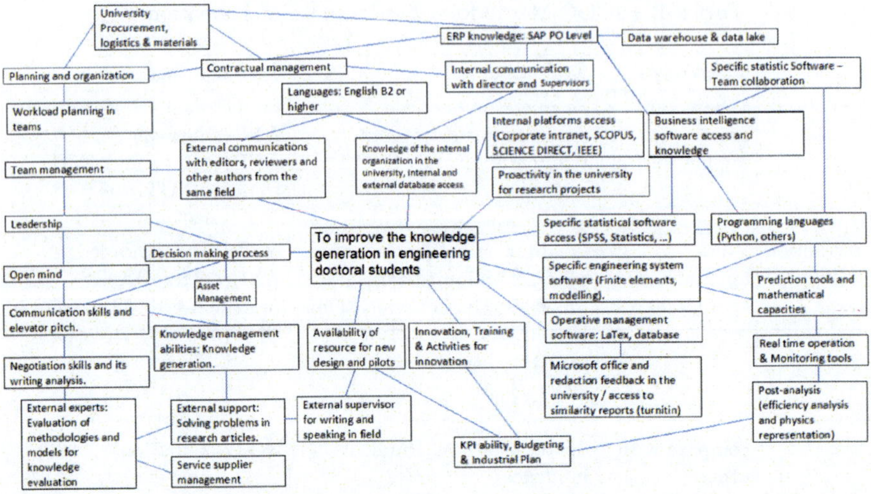

Fig. 4. Core factors and relationship.

on the database access (H1), based on the English level (H4) and the communication with the director and supervisors (H14). Furthermore, the external communications with editors, reviewers, and other authors from the same field (H3) and the external support (R2) is important in the feedback and validation of the perspective (R2). It is therefore the supervisor and doctoral student should develop the team management (H7), and the planning (H12)/organization (H8) of the resources (R9) with the labs (R8) and materials provided from the university for research, development, and innovation (R&D) (R7). As third factor, the decision-making process (H9) has increased with communication skills (H5), with high correlation based on elevator pitch technique, negotiation and external experts' contributions and supports (R4). At the same time the availability of resource for new design and pilots based on R&D with KPI ability, budgeting, and industrial plan (R12) [36], with a perspective of asset management with focus on human capital and knowledge generation (R11). From the proactivity perspective (S1), the students have participated in the University with external innovation contest; it is a key factor for new resources with high influence of statistical analysis (S2), computer science innovation (S3) and specific prediction tools for mathematical scope (R3).

The main tool resources were specific statistical software access and specific engineering software knowledge are important. The programming languages have enhanced the skills and facility in the generation of new knowledge though Python, Julia, and Tensor flow for the data analysis (S4). The Cronbach's Alpha indicator in Human capital is 0.918, the structural capital is 0.870 and relational capital with 0.883. It validated the instrument for the research article analysis [15].

Table 2. Instrument used for the knowledge management evaluation.

Item	Human Capitals	Structural Capitals	Relational capitals
1	H1: Knowledge of the internal organization in the university, internal and external database access	S1: Proactivity in the university for research projects	R1: External experts: Evaluation of methodologies and models for knowledge evaluation
2	H2: Negotiation skills and its writing analysis	S2: Business intelligence software access and knowledge	R2: External support: Solving problems in research articles.
3	H3: External communications with editors, reviewers, and other authors from the same field	S3: Specific statistical software access (SPSS, Statistics, ...)	R3: Prediction tools and mathematical capacities
4	H4: Languages: English B2 or higher	S4: Programming languages (Python, others)	R4: External supervisor for writing and speaking in field
5	H5: Communication skills and elevator pitch	S5: ERP knowledge: SAP PO Level	R5: Real time operation & Monitoring tools
6	H6: Knowledge management abilities: Knowledge generation	S6: Specific engineering system software (Finite elements, modelling)	R6: Service supplier management
7	H7: Team management	S7: Data warehouse & data lake	R7: Availability of resource for new design and pilots
8	H8: Workload planning in teams	S8: Internal platforms access (Corporate intranet, SCOPUS, SCIENCE DIRECT, IEEE)	R8: Innovation, Training & Activities for innovation, labs
9	H9: Decision making process	S9: Specific statistic Software - Team collaboration	R9: University Procurement, logistics & materials
10	H10: Leadership	S10: Operative management software: LaTeX, database	R10: Post-analysis (efficiency analysis and physics representation)
11	H11: Contractual management	S11: Microsoft office and redaction feedback in the university / access to similarity reports (Turnitin)	R11: Asset management
12	H12: Planning and organization		R12: KPI ability, Budgeting & Industrial Plan
13	H13: Open mind		
15	H14: Internal communication with director and supervisors		

2.2 Methodology for Knowledge Management and Skill Evaluation

It develops a qualitative approach for the model, five students selected in random process during the validation process has been evaluated with an action-research design from 188 doctoral students, the complete candidates from 2012 to 2021. The action-research design has been implemented according to the flux diagram in the Fig. 5. The commitment of the doctoral students to follow the process has been developed over 3 years and the publications made through this process. The case study with the five students started with a deep understanding of the

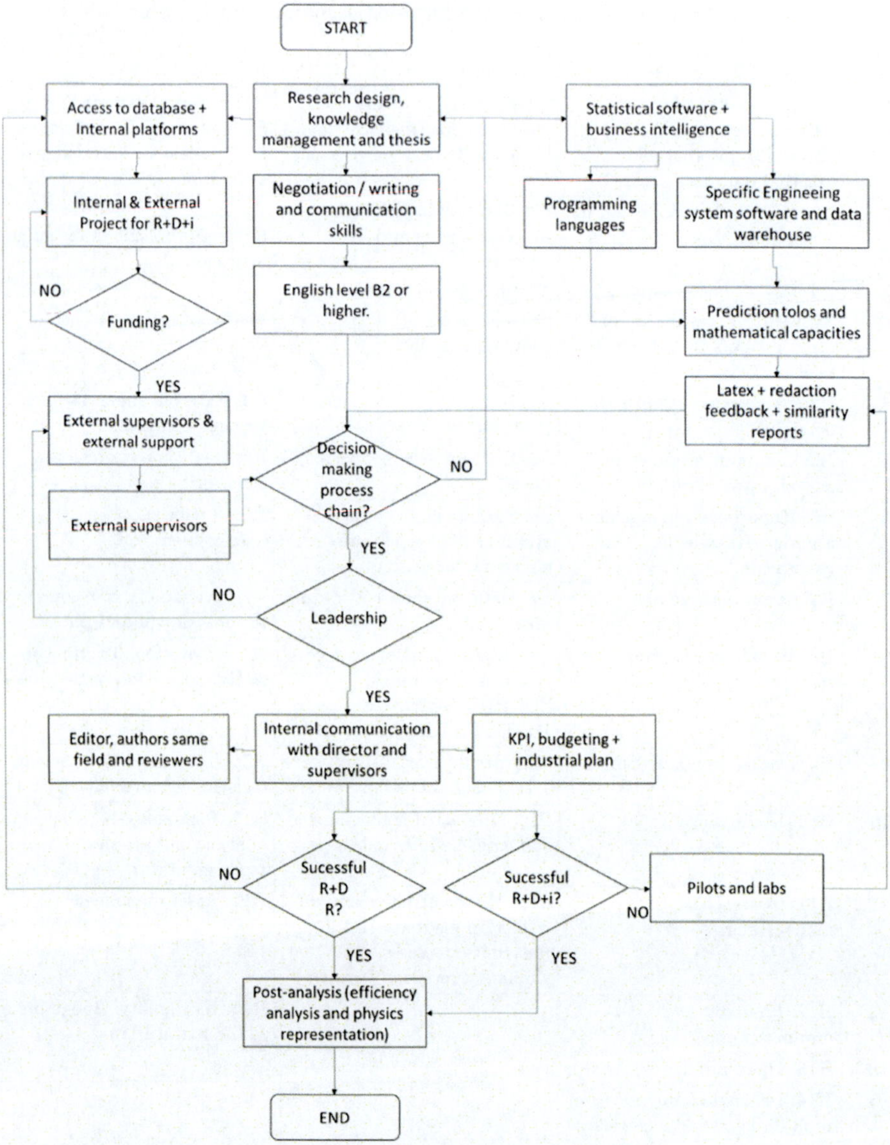

Fig. 5. Model proposed for Knowledge management improvement in engineering doctoral students.

research design for transformative learning process [16], and knowledge management contribution for the research article structure and thesis [17]. The next step was the access to the internal platform and external data base, for the systematic review, the recommended systematic review is the PICOC and Preferred "Reporting Items for Systematic Reviews and Meta-Analyses PRISMA methodology" [18]. It allows to incorporate R&D in internal and external contest for funding [19]; a particular case is the government participation, in New Zealand, it is an important factor for 91% of the doctoral research [20], however, in Peru was a restrictive participation, without adequate funds for engineering [3]. In the first column of the model, the open feedback practices from internal and external supervisors are the first contribution in the decision-making process, suggested by authors in [21]; and the development of skills through the follow-up meetings of the research articles or presentation of the results of the case study, prior to their submission to the journal caused by delay in the supervision of their manuscripts [22]. At the same time, the negotiation, writing, and communications skills are the second pillar for the construction of the model, according the thesis steering model [23]; the best result was achieved with English language for the international journals for a specific purpose of increase the h index [28]. The third pillar is the mathematical and statistical skills associated to business analysis and programming languages for data science; it has been considered according the results in Africa, USA and Europe for doctoral students associated to electrical, mechanical, chemical engineering programs [24,25]. Likewise, the UK and US students have identified the mathematical contribution as an important aspect in the doctoral studies for a successful publication process [26,27]. Finally, with the decision-making process and the leadership for the internal for the selection of the international journal and the field, the leadership is the fundamental key in the external contest [29]; due to KPI target and the university process for the achievement of the committed goals for the use of the funds and the resources of the university, under the internal and external administrative procedure. Later, the perception of the process in the last stage should be evaluated, to incorporate the continuous improvement for the self-perception [30], and the innovation with Agile methodology approach, for the lesson learning diffusion [31], especially from the student to the supervisor, to incorporate the lesson learned for future doctoral students and post doctorate researchers [32]. Finally, we have evaluated with a quantitative research with four independent variables: Human capital (X1), structural capital (X2), relational capital (X3), and technology management and skill (X4) with 2 dependent variables: number of research articles (Y1), h-index impact (Y2); the correlation design is based on four algorithms: Multilayer perceptron (MP), random forest (RF), metaheuristic random subspace (MRS) and metaheuristic additive regression (MAR); with the analysis of coefficient correlation, mean absolute error and the root mean squared error; for the classifier selection.

2.3 Case Study

The case study uses the data in the Sect. 2.2, with doctoral students from 2012 to 2021. It has 773 surveys from 188 students. They have published according to the Fig. 3, with 270 research articles. The case study is based on the application of the methodology, in the Fig. 5, on five doctoral students. With the application of this model, the doctoral students have published 162 research articles (60% of the total of research articles from March 2012 to June 2021), the h-index

Table 3. Doctoral students in the case study.

Last cycle in university	Position	Research article	h index
2019-2	21	42	5
2019-1	86	18	3
2017-2	4	47	15
2017-2	25	34	15
2017-1	30	21	4

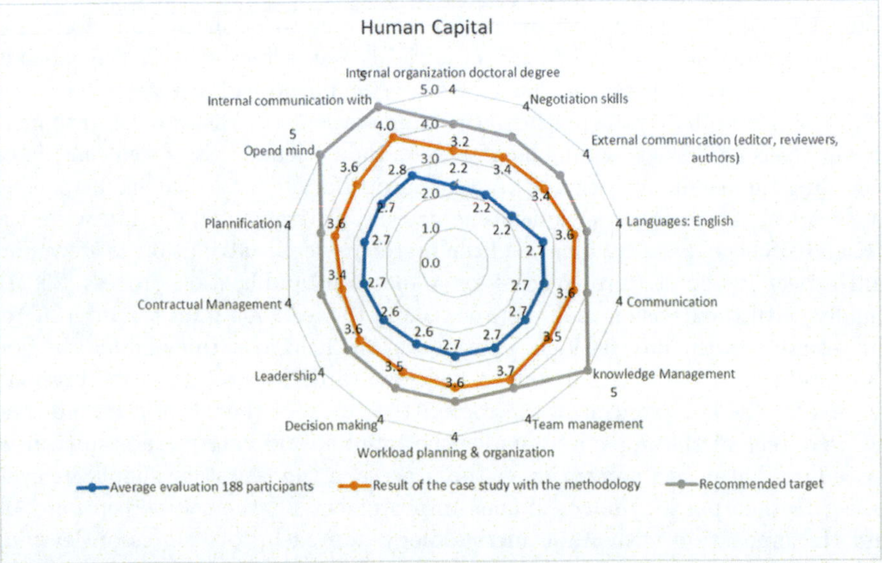

Fig. 6. Human capitals evaluated with 188 participants, the 5 case studies in average and the recommended target.

are from 3 to 15 and research articles per each student from 18 to 47, detailed in the Table 4. The results in the knowledge management are organized in the Fig. 6, for Human Capital (Table 3, from H1 to H14), in the blue line is the average of the 188 doctoral students, and the orange line is for the five doctoral students. The lowest value for all the population is 2.2 for: Internal organization doctoral degree, negotiation skills, external communication. In the case study, the five doctoral students have the lowest values 3.2 in Internal organization doctoral degree. It is related to the COVID restrictions for the laboratory and the funding process; besides, the certification in the last part of the process.

The highest values in the population are 2.8, it is for internal communication with the supervisors and the directors. The same factor is improved to 4.0. Likewise, the recommended target is proposed with a recommendation of the supervisors dedicated to these five students in the case study. The action-

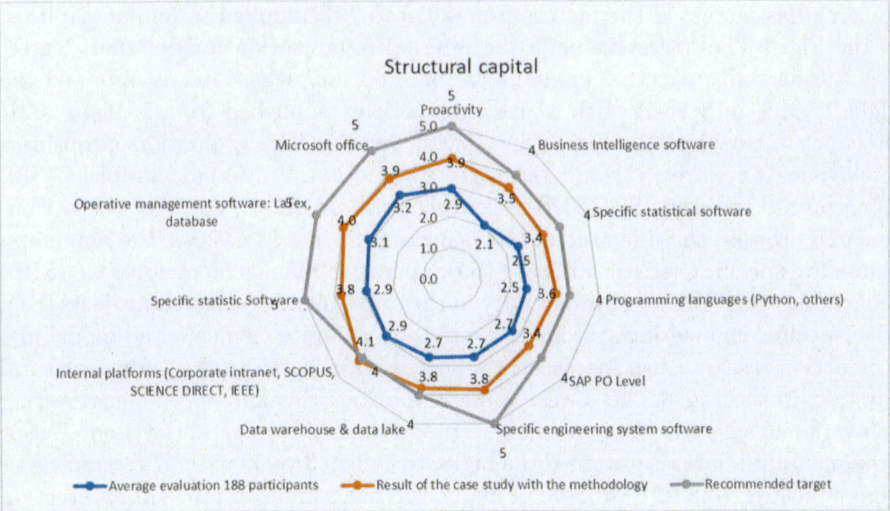

Fig. 7. Structural capitals evaluated with 188 participants, the 5 case studies in average and the recommended target.

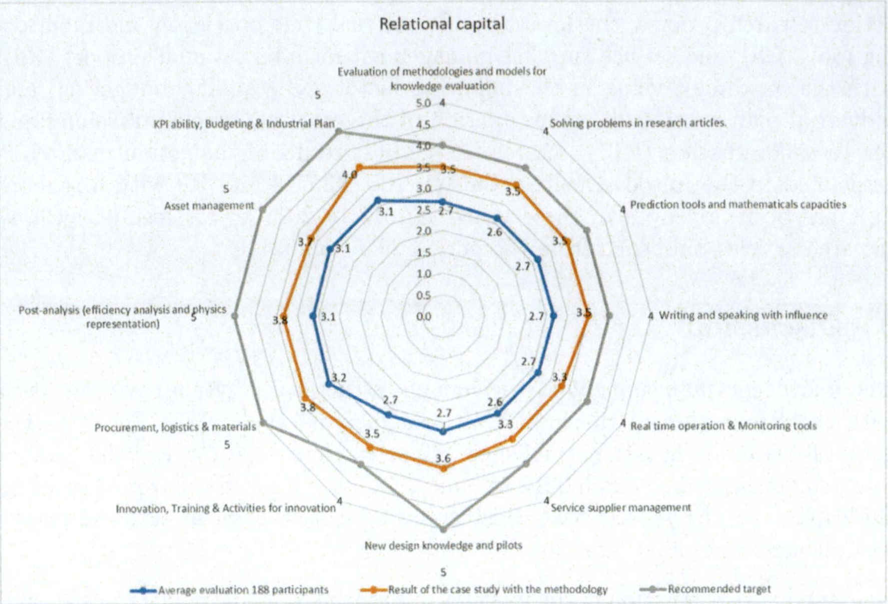

Fig. 8. Relational capitals evaluated with 188 participants, the 5 case studies in average and the recommended target.

research has increased the factors from 2015 to 2017 applied to human capitals, in the Fig. 6 The lowest value is the internal organization for doctoral degree, negotiation skills, external communication, and language, later is obtained the highest value in 3 years with 47 research articles published from 2016 to 2021. In structural capitals in Fig. 7, the lowest value in the evaluation is business intelligence software (S2) and the highest value is the internal platforms (S8), as corporate intranet, SCOPUS, ScienceDirect and IEEE. After of the action-research process, these factors are S2 is 3.2 and S8 is 4.1. Besides, the maximum values are specific statistic software (S9), operative management software (S10), redactions tools (S11) and proactivity in the university for research projects (S1), with specific engineering system software (S6), as finite elements and modelling. The action-research has increased the factors from 2015 to 2017 for structural capitals, in the Fig. 8 The lowest value is the statistics and specific engineering software and systems (finite elements), programming languages (Python, Tensor flow and Julia), later is obtained the highest value in 3 years with 47 research articles published from 2016 to 2021. Finally, the relational capital as an interaction with the stakeholders has the lowest values as solving problems in research articles (R2), with the supervisors with internal and external engineers associated, and the maximum values for the university procurement, logistics and materials for the projects associated in the engineering doctoral programs (R9). After the action-research process, the lowest values are real time operation and monitoring tools (R5), and service supplier management for international support (R6); with the maximum value in the following factor: KPI ability, budgeting, and industrial plan control for the evaluation of the research article publishing and the thesis evaluation (R12). Therefore, the best result in the action research is associated to the improvement in the R1, R2, R3, R8 and R9 with a scale of 2; it has been improved in three years to 2015 to 2017 to 5.0 results, with 47 research articles published from the period 2015 to 2021.

3 Discussion

The knowledge management in the human, structural capitals have improved with the application of the model with engineering skill in the Fig. 5. In the action-research design, they published 162 research articles (60% of the total of research articles from March 2012 to June 2021), the h-index are from 3 to 15, in the Table 4. All the aspects have considered eleven skills associated to technology and management skills, as follows:

- - Operative instruction in the university: Associated to the policy research and publication procedures in the university, with the statistical software (SPSS, Statistics as the most used software), besides, specific engineering software for green design as PV system, PIM and Matlab.
- Communication, innovation ecosystem and legislation: It is associated to the skill for innovation environments with Agile methodology (recommended by the doctoral students with SCRUM and CANVAS), communication protocol for internal and external supervisors.

- Ethics, laboratory, equipment, and statistical analysis: In this perspective, the ethics recommendations with guidelines, to report on achievements on labs. Besides the operative instructions, knowledge of quality, execution process and the commissioning procedure for new equipment used in the doctoral thesis.
- Leadership, decision making process: It is evaluated with the teams with internal and external students and supervisors, with a decision-making process associated to approval with nine gateways.
- Asset management, electrical and electronic engineering: It is associated to PASS 55 and ISO 55001, and a perspective of control and protection in electrical and electronics systems.
- Health, security, environmental and quality in the research and labs: With specific procedures and policies; and planification.
- Reliability and mechanical engineering: With FMEA and reliability techniques with RCM plus and RCM two.
- Agile methods: With the certification according the Project Management Institute with the PMP or Agile Method certification. The main skill is the open mind, and the application of the knowledge operation in innovative projects.
- Programming and computer science: The main aspect is Python, Tensor Flow and Julia software for the analysis. Besides, special consideration for finite elements analysis software, in this case, the COMSOL software is used. On the other hand, the analysis of curves and trends with business intelligence software is used with inspection capacities.
- Redaction: The use of Microsoft word and Latex are common tools, according the requirement of the international journals and the university.
- Circular economy, renewable energy, inventory, green design: The consolidated evaluation of the Supply chain management with circular economy, in order to incorporate the sustainability perspective and industrial application in the research articles.

In the Fig. 9, it presents two box plots with the results of the comparison between the doctoral students and the results of the action research participants. The research productivity of the doctoral students is described, on the left of the Fig. 9, it represents the number of research articles published by students who pursue their PhD in the same university as the one without the action-research influence the median is one research articles published. On the right, it represents the research article productivity of students with a median of thirty-four research articles and an average of thirty-three, the most productive students are the ones who were trained with the new model proposed, which the supervisor draws her coauthors with an average of forty-five research articles published.

In the Table 4 described four methodologies, for the independent variable Human capital, structural capital and structural capitals with skills [33], and the dependent variable of number of research article published (Y1). The best correlation coefficient obtained with Metaheuristic additive regression methodology [34] is 80.74%; with a mean absolute error of number of research article

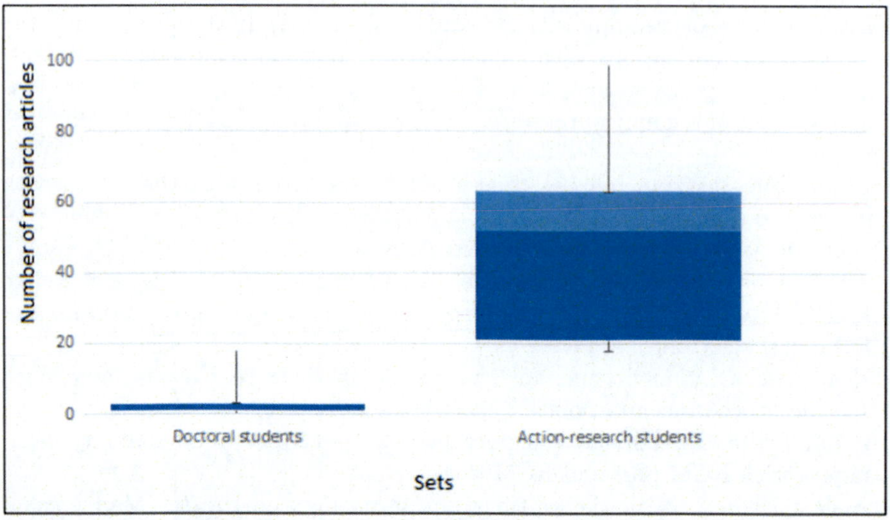

Fig. 9. Knowledge management improvements for the action-research set.

is 1.5; and the root mean squared error is 3.8. In the impact perspective, the h-index is analyzed with the same independent variables and a new dependent variable of h-index (Y2). The Metaheuristic additive regression of 82.13% as correlation coefficient and the mean absolute error is 0.545 with root mean squared error (Table 5).

Table 4. Algorithms and correlation for the prediction: Research articles published from doctoral students

Description	Multilayer perceptron	Random Forest	Metaheuristic Random Subspace	Metaheuristic Additive regression
Correlation coefficient	57.33%	60.37%	70.23%	80.74%
Mean absolute error	2.8105	1.9966	1.9799	1.5203
Root mean squared error	5.6723	4.7366	4.6494	3.849

Table 5. Algorithms and Correlation for the prediction: h-Index from doctoral students

Description	Multilayer perceptron	Random Forest	Metaheuristic Random Subspace	Metaheuristic Additive regression
Correlation coefficient	64.20%	65.72%	15.83%	82.13%
Mean absolute error	0.8927	0.5957	0.7178	0.545
Root mean squared error	1.5608	1.3546	1.7339	1.129

The accuracy - learning process is evaluated in ten models for the Y1 and Y2 variables, the 3 sigma is considered since the model four to ten, in the Fig. 10.

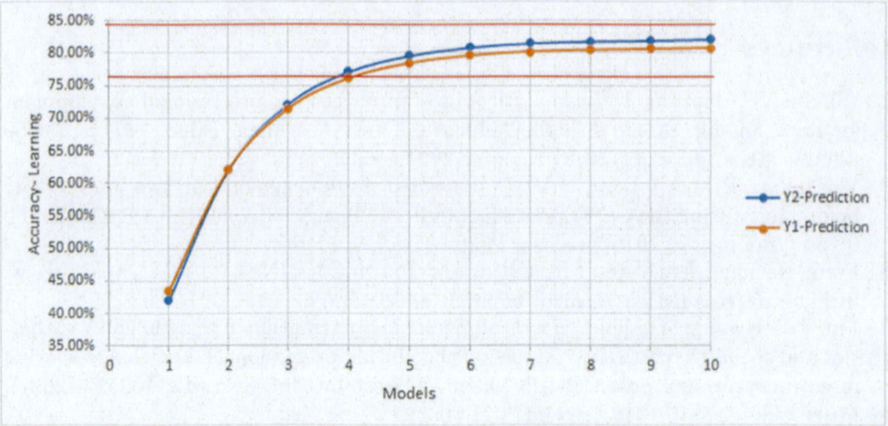

Fig. 10. Accuracy for learning according the models structured.

4 Conclusion

In this paper, it compared the scientific productivity of 188 doctoral students with an action research design in 5 doctoral students during 2012 to 2021 in a research community; with the knowledge management application based on human capitals (14 factors), structural capitals (11 factors), relational capitals (12 factors) and the incorporation of technology and management skills (110 skills with 11 dimensions) in the students. With the incorporation of the new knowledge management approach, the 5 doctoral students published 162 research articles (60% from March 2012 to June 2021) with the h-index 3 to 15. Without this new approach, 183 doctoral students have published 108 research articles. These outcomes have a straight influence for the references on the outstanding productivity of engineering doctoral students and provide evidence of the confident effects of capitals: Human, structural, and relational based on skills. In a comparable way the engineering doctoral students, and post-doctoral students have autonomy in choosing their trainees. Our findings could be extended to several other locations and institutions, as research centers in public and private universities, in which knowledge generation is a main key and in which the director researcher is confronted with the problem of searching for "highly skilled employees across national and international borders" [35]. These results are in the context with and without COVID-19 from 2012 to June 2021 in a Latin American culture context, the core factors and the problem are represented in the Fig. 4. In future research, the need to increase the KM and skills should increase the productivity with specific technology skill and more culturally diverse context.

Acknowledgements. Universidad Tecnológica del Perú.

References

1. Wilkins, S., Hazzam, J., Lean, J.: Doctoral publishing as professional development for an academic career in higher education. Int. J. Manag. Educ. **19**(1), 100459 (2021). https://doi.org/10.1016/j.ijme.2021.100459
2. Velásquez, R.M.A., Lara, J.V.M.: Knowledge management in two universities before and during the COVID-19 effect in Peru. Technol. Soc. **64**(1), 101479 (2021). https://doi.org/10.1016/j.techsoc.2020.101479
3. Peru, National Institute of Statistics and Informatics INEI (2018). https://www.inei.gob.pe/estadisticas/indice-tematico/education/
4. Lin, H., Hwang, Y.: The effects of personal information management capabilities and social-psychological factors on accounting professionals' knowledge-sharing intentions: pre and post COVID-19. Int. J. Account. Inf. Syst. **42**, 100522 (2021). https://doi.org/10.1016/j.accinf.2021.100522
5. Gonzales-Saldaña, J., et al.: Producción científica de la facultad de medicina de una universidad peruana en SCOPUS y Pubmed. Educación Médica **19**(Supplement 2), 128–134 (2018). https://doi.org/10.1016/j.edumed.2017.01.010
6. Melgar, A., Brossard, I., Olivares, C.: Current status of research information management in Peru. Procedia Comput. Sci. **146**, 220–229 (2019). https://doi.org/10.1016/j.procs.2019.01.096
7. Gunasekera, G., Liyanagamage, N., Fernando, M.: The role of emotional intelligence in student-supervisor relationships: implications on the psychological safety of doctoral students. Int. J. Manag. Educ. **19**(2), 100491 (2021). https://doi.org/10.1016/j.ijme.2021.100491
8. Christmann, A., Van Aelst, S.: Robust estimation of Cronbach's alpha. J. Multivariate Anal. **97**(7), 1660–1674 (2006). https://doi.org/10.1016/j.jmva.2005.05.012
9. Sokal, L., Trudel, L.E., Babb, J.: I've had it! Factors associated with burnout and low organizational commitment in Canadian teachers during the second wave of the COVID-19 pandemic. Int. J. Educ. Res. Open **2**(2), 100023 (2021). https://doi.org/10.1016/j.ijedro.2020.100023
10. Kulikowski, K., Przytuła, S., Sułkowski, Ł: The motivation of academics in remote teaching during the covid-19 pandemic in polish universities-Opening the debate on a new equilibrium in e-learning. Sustainability (Switzerland) **13**(5), 1–16 (2021). https://doi.org/10.3390/su13052752
11. Damşa, C., Langford, M., Uehara, D., Scherer, R.: Teachers' agency and online education in times of crisis. Comput. Hum. Behav. **121**, 106793 (2021). https://doi.org/10.1016/j.chb.2021.106793
12. Velásquez, R.M.A., Lara, J.V.M.: Forecast and evaluation of COVID-19 spreading in USA with reduced-space Gaussian process regression. Chaos Solitons Fract. **136**(109924), 1–9 (2020). https://doi.org/10.1016/j.chaos.2020.109924
13. Aleshkovski, I.A., Gasparishvili, A.T., Krukhmaleva, O.V., Narbut, N.P., Savina, N.E.: Russian higher school: forced distance learning and planned switch to distance learning during pandemic (experience of sociological analysis). Vysshee Obrazovanie v Rossii **30**(5), 120–137 (2021). https://doi.org/10.31992/0869-3617-2021-30-5-120-137

14. Narbut, N.P., Aleshkovski, I.A., Gasparishvili, A.T., Krukhmaleva, O.V.: Forced shift to distance learning as an impetus to technological changes in the Russian higher education. RUDN J. Sociol. **20**(3), 611–621 (2020). https://doi.org/10.22363/2313-2272-2020-20-3-611-621
15. Leontitsis, A., Pagge, J.: A simulation approach on Cronbach's alpha statistical significance. Math. Comput. Simul. **73**(5), 336–340 (2007). https://doi.org/10.1016/j.matcom.2006.08.001
16. Bergeå, O., Karlsson, R., Hedlund-Åström, A., Jacobsson, P., Luttropp, C.: Education for sustainability as a transformative learning process: a pedagogical experiment in EcoDesign doctoral education. J. Clean. Prod. **14**, 1431–1442 (2006). https://doi.org/10.1016/j.jclepro.2005.11.020
17. Davis, M.: Why do we need doctoral study in design? Int. J. Des. **2**(3), 71–79 (2008)
18. Puljak, L., Sapunar, D.: Acceptance of a systematic review as a thesis: survey of biomedical doctoral programs in Europe. Syst. Rev. **6**(1), 1–7 (2017). https://doi.org/10.1186/s13643-017-0653-x
19. Fomunyam, K.G.: Post-doctoral and non-faculty doctorate researchers in engineering education: demographics and funding. Int. J. Educ. Pract. **8**(4), 676–685 (2020)
20. Sampson, K.A., Comer, K.: When the governmental tail wags the disciplinary dog: some consequences of national funding policy on doctoral research in New Zealand. Higher Educ. Res. Dev. **29**(3), 275–289 (2010). https://doi.org/10.1080/07294360903277372
21. Bitchener, J.: The content feedback practices of applied linguistics doctoral supervisors in New Zealand and Australian universities. Aust. Rev. Appl. Linguist. **39**(2), 105–121 (2016). https://doi.org/10.1075/aral.39.2.01bit
22. Fulgence, K.: A theoretical perspective on how doctoral supervisors develop supervision skills. Int. J. Doct. Stud. **14**, 721–739 (2019). https://doi.org/10.28945/4446
23. Heldal, I., et al.: Supporting communication within industrial doctoral projects: the thesis steering model. In: ITICSE 2014 - Proceedings of the 2014 Innovation and Technology in Computer Science Education Conference, vol. 325 (2014). https://doi.org/10.1145/2591708.2602680
24. Kitagawa, F.: Collaborative doctoral programmes: employer engagement, knowledge mediation and skills for innovation. Higher Educ. Q. **68**(3), 328–347 (2014). https://doi.org/10.1111/hequ.12049
25. Cross, M., Backhouse, J.: Evaluating doctoral programmes in Africa: context and practices. Higher Educ. Policy **27**(2), 155–174 (2014). https://doi.org/10.1057/hep.2014.1
26. Bancroft, S.F.: Toward a critical theory of science, technology, engineering, and mathematics doctoral persistence: critical capital theory. Sci. Educ. **102**(6), 1319–1335 (2018). https://doi.org/10.1002/sce.21474
27. Alabdulaziz, M.S.: Saudi mathematics students' experiences and challenges with their doctoral supervisors in UK Universities. Int. J. Doct. Stud. **15**, 237–263 (2020). https://doi.org/10.28945/4538
28. Negretti, R., McGrath, L.: English for specific playfulness? How doctoral students in science, technology, engineering and mathematics manipulate genre. Engl. Spec. Purp. **60**, 26–39 (2020). https://doi.org/10.1016/j.esp.2020.04.004
29. Hallinger, P.: A review of three decades of doctoral studies using the principal instructional management rating scale: a lens on methodological progress in educational leadership. Educ. Adm. Q. **47**(2), 271–306 (2011). https://doi.org/10.1177/0013161X10383412

30. Khuram, W., Wang, Y., Khan, S., Khalid, A.: Academic attitude and subjective norms effects on international doctoral students' academic performance self-perceptions: a moderated-mediation analysis of the influences of knowledge- seeking intentions and supervisor support. J. Psychol. Africa **31**(2), 145–152 (2021). https://doi.org/10.1080/14330237.2021.1903188
31. Krahe, J.A.E., Lalley, C., Solomons, N.M.: Beyond survival: fostering growth and innovation in doctoral study-a concept analysis of the ba space. Int. J. Nurs. Educ. Sch. **11**, 1–8 (2014). https://doi.org/10.1515/ijnes-2013-0020
32. Velásquez, R.M.A., Lara, J.V.M.: Converting data into knowledge with RCA methodology improved for inverters fault analysis. Heliyon **8**(8), e10094 (2022). https://doi.org/10.1016/j.heliyon.2022.e10094
33. Szelényi, K., Bresonis, K.: The public good and academic capitalism: science and engineering doctoral students and faculty on the boundary of knowledge regimes. J. Higher Educ. **85**(1), 126–153 (2014). https://doi.org/10.1080/00221546.2014.11777321
34. Trindade, Á.R., Campelo, F.: Tuning metaheuristics by sequential optimisation of regression models. Appl. Soft Comput. **85**(105829), 1–16 (2019). https://doi.org/10.1016/j.asoc.2019.105829
35. Baruffaldi, S., Visentin, F., Conti, A.: The productivity of science & engineering PhD students hired from supervisors' networks. Res. Policy **45**(4), 785–796 (2016). https://doi.org/10.1016/j.respol.2015.12.006
36. Gómez-Marín, N., et al.: Sustainable knowledge management in academia and research organizations in the innovation context. Int. J. Manag. Educ. **20**(1), 100601 (2022). https://doi.org/10.1016/j.ijme.2022.100601

Neural Network Control of a Belt Conveyor Model with a Dynamic Angle of Elevation

Alexey A. Petrov[1(✉)], Olga V. Druzhinina[1,2], and Olga N. Masina[1]

[1] Bunin Yelets State University, Yelets, Russia
xeal91@yandex.ru

[2] Federal Research Center "Computer Science and Control" of Russian Academy of Sciences, Moscow, Russia

Abstract. This paper proposes a generalized description of a switched belt conveyor model with a dynamic change in the angle between the horizontal plane and the plane of the conveyor belt. The model is given by a four-dimensional nonlinear differential equation with control functions. The search for control functions according to a given quality criterion is carried out using a neural network controller. Switching in the proposed model is taken into account in the form of factors of smooth loading and instant unloading of cargo. A neural network structure is developed to control the linear speed and elevation angle of the conveyor. To implement the neural network controller, a control algorithm for the conveyor model was developed taking into account feedback. Graphs of the linear and angular speed of the conveyor, as well as graphs of control signals, are obtained. A comparative analysis with the results of the trajectory dynamics of the model with instantaneous loading is carried out. To study the constructed model, methods for the numerical solution of ordinary differential equations, numerical optimization methods and intellectual analysis methods are used. The software is developed in the Julia language with the involvement of the DifferentialEquations, Plots, BlackBoxOptim libraries, as well as the original neural network computing library. The obtained results can be used in the problems of designing intelligent control systems for production lines and in the stabilization problems of multidimensional dynamic models #CSOC1120.

Keywords: switched dynamic models · differential equations · belt conveyor · optimization · artificial neural networks · intelligent control

1 Introduction

Design, automation and monitoring of conveyor transport systems are current areas of research [1–8]. The range of important problems includes, for example, stabilization problems of the conveyor traction factor, monitoring the dynamic load of conveyor belts, conveyor control parameters optimization, designing a

R. Silhavy and P. Silhavy (Eds.): CSOC 2023, LNNS 724, pp. 733–746, 2023.
https://doi.org/10.1007/978-3-031-35314-7_61

smart conveyer belt, and creating multifunctional continuous transport systems. The solution of these problems is associated with the need to use methods of control theory with elements of artificial intelligence [9–12].

The artificial intelligence tools such as fuzzy control, artificial neural networks and machine learning are used in mathematical modeling of conveyor transport systems [13–21]. In [13] the issues of constructing a fuzzy system for tracking the state of the conveyor are considered. In [14] the aspects of pipeline control optimization using artificial neural networks are studied. In [15] the application of machine learning models for high-precision classification of types of loads on a rubber conveyor belt is considered.

This paper is a continuation of [22–24]. In [22] a model of a belt conveyor with a dynamic change in the angle between the horizontal plane and the plane of the conveyor belt is presented. The issues of optimal control of this model are studied using a combined controller based on the sliding mode and a fuzzy inference system in the Mamdani form. In [23], methods for studying a belt conveyor model based on the construction of PID controllers and neural network controllers are proposed. In [24], such a modified mathematical model of a conveyor belt is developed, which takes into account additional dissipative effects. In addition, in this paper, a comparative analysis of various types of intelligent control of a conveyor belt is carried out.

Refinement of the models considered in [22–24] can take into account various factors, among which the effect of transient processes on the system dynamics should be noted when the conveyor load modes change. In this paper, we solve the problem of synthesizing a neural network controller for a new model of a belt conveyor with a dynamic change in the angle between the horizontal plane and the plane of the conveyor belt. The proposed model takes into account the factors of smooth loading and instant unloading. We conduct a comparative analysis of the results of a computer study of this model with the results for the model considered in [24].

Here is a summary of the paper by sections. Section 2 considers the basic model of a single-drive conveyor with a dynamic change in the angle between the horizontal plane and the plane of the conveyor belt. In Sect. 3, a switched model of a belt conveyor is synthesized, taking into account the smooth loading. In Sect. 4, for a new model of a belt conveyor, a solution to the optimal control problem based on a neural network controller is proposed. Graphs of the linear and angular speed of the conveyor, as well as graphs of control signals, are obtained. A comparative analysis of the research results for various models of a belt conveyor with a dynamic change in the angle of elevation is carried out. Section 5 discusses the results of the paper and considers promising areas of research.

2 Basic Model of Conveyor with Dynamic Angle of Elevation

Consider the basic model of a single-drive belt conveyor with a dynamically variable angle of elevation [22] under the following conditions: the conveyor belt is inextensible, the loads on the belt have a significant effect on the momentum and angular momentum of the conveyor, the friction force of the loads on the conveyor belt is not taken into account. Taking into account these conditions, the differential equations of the model can be written in the form

$$\begin{aligned} \dot{x}_0 &= x_1, \\ \dot{x}_1 &= \frac{u_p(t) - kx_1 - m_1 g \sin(\alpha_0)}{m_1 + m_0}, \\ \dot{\alpha_0} &= \alpha_1, \\ \dot{\alpha_1} &= \frac{u_\alpha(t)}{(m_0 + m_1)c\varepsilon^2} - \frac{g\cos(\alpha_0)}{\varepsilon}, \\ & u_1, u_2 \in U, \varepsilon \in E, m_1 \in M, \end{aligned} \tag{1}$$

where x_0 is the movement of the conveyor belt, m_0 is the mass of the conveyor belt, α_0 is the lifting angle of the conveyor, α_1 is the angular conveyor velocity, m_0 is the mass of the conveyor belt, m_1 is the total mass of loads on the conveyor, s is the coefficient of the moment of conveyor inertia, ε is the position of the conveyor mass center, k is the coefficient of rolling friction, u_1 is the magnitude of tho linoar force of the conveyor translational movement, u_2 is the magnitude of the torque to control the angle of conveyor elevation, U is the set of control vectors, M is the set of masses on the conveyor belt, E is the set of positions of the conveyor mass centers.

Given the description of the model, two types of actions are possible: loading (action 1) and unloading (action 2). It should be noted that the presence of these actions refers model (1) to the class of switched models [22].

Let us first describe the features of the action 1. The action 1 is as follows: let $m = m_0 + m_1$, $x_1(t_{i-1})$ is the speed before the moment of loading, $x_1(t_i)$ is the speed after loading. Since at the time of loading the equality $mx_1 = \text{const}$ is true, we get

$$x_1(t_i)m = x_1(t_{i-1})(m + \Delta m), \quad x_1(t_i) = \frac{x_1(t_{i-1})m}{m + \Delta m}.$$

It should be noted that the position of the mass center also changes according to the equality

$$\varepsilon = f(x_1, m + \Delta m),$$

where $f : x,\ m + \Delta m \to \varepsilon$. Thus, the introduction of the function f allows to calculate the position of the mass center. Features of action 2 are that in the process of unloading the total mass of loads decreases at a constant linear speed.

For model (1), the optimal control problem is formulated and solved in a number of special cases. Specialized software was developed, a number of computational experiments are carried out to stabilize the conveyor model. It should be noted that model (1) contains assumptions that reduce the efficiency of practical use. In particular, the loading and unloading occurs instantly. In this paper, we propose such a modified model that takes into account the presence of transient processes during cargo loading.

3 Switched Belt Conveyor Model Taking into Account Smooth Loading

Next, we propose such a model of a single-drive belt conveyor with a dynamic angle of elevation which is described by differential equations of the form

$$\begin{aligned} \dot{x} &= \frac{p}{m}, \\ \dot{p} &= u_p(t) - k\frac{p}{m} - (m - m_0)g\sin(\alpha_0), \\ \dot{\alpha_0} &= \alpha_1, \\ \dot{\alpha_1} &= \frac{u_\alpha(t)}{mc\varepsilon^2} - \frac{g\cos(\alpha_0)}{\varepsilon}, \\ &u_p, u_\alpha \in U, \varepsilon \in E, m \in M, \end{aligned} \qquad (2)$$

where x is the linear movement of the conveyor belt, p is the momentum of the system, α_0 is the lifting angle of the conveyor relative to the zero position, α_1 is the speed of the conveyor angular rotation, m is the total mass of the system, m_0 is the total mass of loads on the conveyor, $u_p(t)$ is the conveyor traction control function, $u_\alpha(t)$ is the belt lift angle control function, ε is the position of the conveyor mass center relative to the lower roller, c is the coefficient that determines the moment of conveyor inertia, k is the rolling friction coefficient. The sets M, E, U include all possible values of the total mass of loads, center of mass and controls, respectively.

Changes in the modes of operation in model (2) correspond to the choice of m_1 and ε from the sets M, E according to a given law. System (2) belongs to systems with switching.

Note that model (2) is a modification of model (1) taking into account changes in the mechanism for loading and unloading. We assume that the process of engagement of the loaded cargo on the moving belt has a fixed duration. This process is set by changing the mass of loads in the system based on the following switching mode. This switching mode is determined by the algorithm

$$\begin{aligned} &\text{if } t_i \notin T: \quad m(t_i) = m(t_{i-1}) \\ &\text{else:} \quad m(t_i) = m(t_i) + \frac{(t_i - t_{i-1})(m_e - m_b)}{t_e - t_b}. \end{aligned}$$

The following designations are used in the description of the algorithm: T is the set of loading intervals, i is the step number of the algorithm for solving ordinary differential equations, m_b is the mass of loads on the belt at the beginning of loading, m_e is the mass of loads on the conveyor at the end of loading, t_e and t_b are the start time of loading.

We also assume that the unloading process, as for model (1), occurs instantly. Taking into account this assumption, we introduce a switching mode that determines the change in the mass and momentum of the system. This switching mode is determined by the algorithm

$$\begin{aligned}&\text{if } t_i \notin T_u: \quad m(t_i) = m(t_{i-1}),\ p(t_i) = p(t_{i-1})\\ &\text{else:} \quad m(t_i) = m(t_{i-1}) + m_u,\ p(t_i) = m(t_{i-1}) + \Delta x m_u.\end{aligned}$$

In the description of the algorithm, the following designations are adopted: T_u is the intervals for unloading cargo, Δx is the velocity of the conveyor belt at the time of unloading, m_u is the mass of the unloaded cargo. It should be noted that, in contrast to model (1), model (2) describes the momentum of the system, since m is given by a continuous piecewise linear function.

We consider the following formulation of the optimal control problem for model (2). It is necessary with the using of controls u_p, u_α to implement the transition mode and stabilize the phase state of the system (2) near the target point $E(\dot{x}_1, \alpha_{01}, \alpha_{11})$. The coordinates of the target point correspond to the "optimal" values of the linear velocity, elevation angle and angular velocity. An important circumstance is that when solving the problem of optimal control one should take into account the invariance with respect to the initial conditions. We propose the following optimality criterion

$$\lim_{t_n \to \infty} \frac{1}{t_n} \int_0^{t_n} \|E - X(t)\|\, dt \to \min, \tag{3}$$

where $t \in (0, t_n)$, $X = (\dot{x}(t), \alpha_0(t), \alpha_1(t))$. The meaning of this criterion is to implement the fastest possible transition process and maintain the required phase state of the system (2).

4 Construction of a Neural Network Controller and Computer Study of the Model

To implement optimal control in the system (2), a combined controller is used, which has two components: i) sliding mode control for generating u_p; ii) neural network control for generating u_α. The sliding mode control is described by the switching law of the form

$$\begin{aligned}&\text{if } \frac{p}{m} < s: \ u_p = c,\\ &\text{else}: \ u_p = -c,\end{aligned}$$

where c is a given control constant.

For model (2), as well as for model (1), the implementation of the sliding mode to control the elevation angle does not give the expected results, due to the fact that instability occurs in the models. To implement the conveyor elevation angle control for model (2), we propose a neural network controller, the topology of which is shown in Fig. 1.

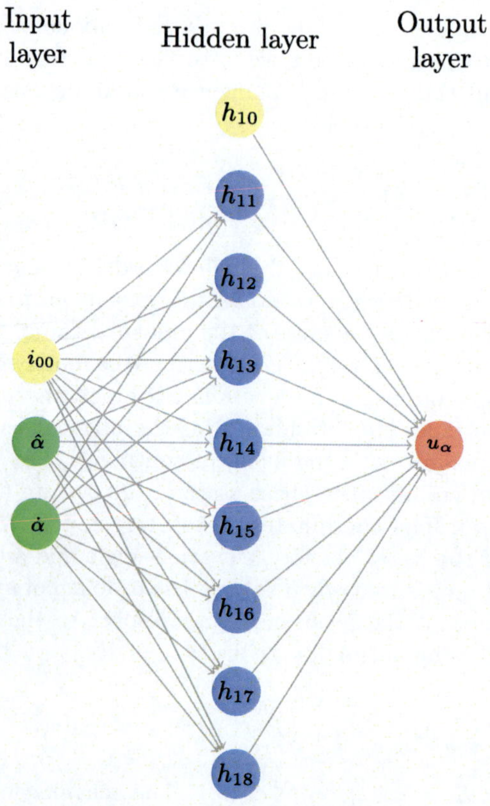

Fig. 1. Topology of an artificial neural network to control the elevation angle in the model (2).

Figure 1 shows the topology of the neural network designed to construct control actions for u_{α}. The input given to the input layer is the angular velocity $\dot{\alpha}$ and the error $\hat{\alpha}$. The hidden and output layers use a tangential activation function. Bias neurons are used in the input and hidden layers.

The result of the neural network working at each step is the value of the control function. The proposed controller implements the principle of feedback control, taking into account the error and its time derivative. For the selection of weight coefficients we use reinforcement learning. The basic principle of the learning algorithm is as follows. It is necessary to minimize the weighted value of n numerical estimates of criterion (3). To do this, at each stage of the optimization algorithm, n numerical solutions of equations (2) are performed, followed by calculation of the value

$$H = \text{mean}(\|E - X(t)\|). \tag{4}$$

The value of H is used as a loss function in neural network training. The choice of reinforcement learning is related to the specifics of the problem of constructing a neural network controller to stabilize the conveyor elevation angle modeled by system (2). We used a neural network controller with a similar topology when construsting a controller for a simpler conveyor model with instantaneous loading and unloading [23].

It should be noted that model (1) takes into account the existence of the additive part of the control function described in elementary functions, using the right parts of the system turn to zero. The specified additive part (balancing variable) has the form

$$u_{0\alpha} = c\varepsilon gm \cos(\alpha_0).$$

When obtaining trajectories, we take into account the fact that the system is in equilibrium with respect to control. As a result, the trajectories enter the oscillating mode, or the mode of a rather smooth change of trajectories with time [23]. However, in practice, the calculation of the value of the balancing variable is not always possible, since this requires accurate values of m and ε, which can only be obtained by accurately measuring the masses and coordinates of the loads on the conveyor belt. Next, in relation to model (2), we consider a more complex problem in which the values of m and ε are assumed to be unknown.

To carry out computational experiments with model (2), a program in the Julia language [25] is developed. We perform a comparative analysis of the trajectories of model (2) with the trajectories of model (1), taking into account the use of similar conditions and model parameters. Note that the computer program for calculating model trajectories (1) is developed in Python3 using the numpy, scipy, matplotlib libraries [26]. Figure 2 and 3 present the results of a computational experiment on calculating the linear velocity for cases where stabilization is carried out by means of a sliding mode controller. For computational experiments, the following set of parameters $m = 2.5$, $m_0 = 0.1$, $k = 0.5$, $g = 9.8$, $c = 0.5$, $\varepsilon = 1$ is used.

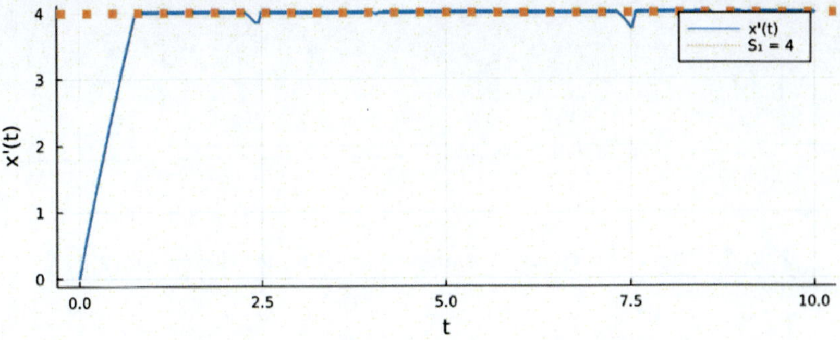

Fig. 2. Linear velocity graph for model (2).

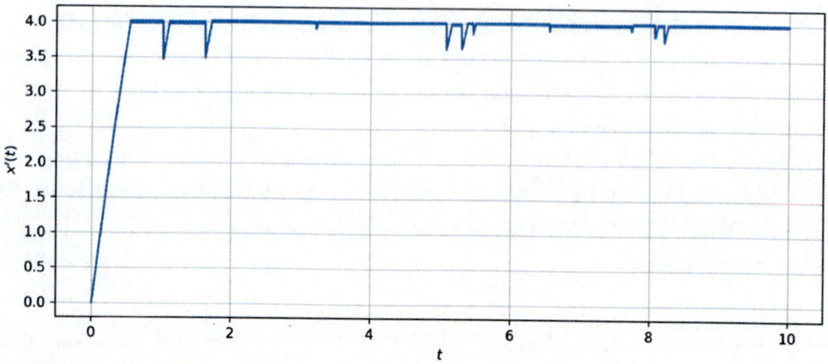

Fig. 3. Linear velocity graph for model (1).

In Fig. 2, 3, the linear velocity (the phase variable $\dot{x}(t)$) is marked with a solid line, the required belt velocity is marked with a dotted line. It can be noted that for the linear velocity in Fig. 2, the achieved quality of control is consistent with the problem statement. A short-term decrease in velocity at certain points in time is due to the loading. Compared to Fig. 3, there are no sharp changes in velocity in the graph, which are associated with instantaneous loading of cargo for model (1).

Figure 4 and Fig. 5 shows graphs of the conveyor angular position in model (2) and model (1), respectively.

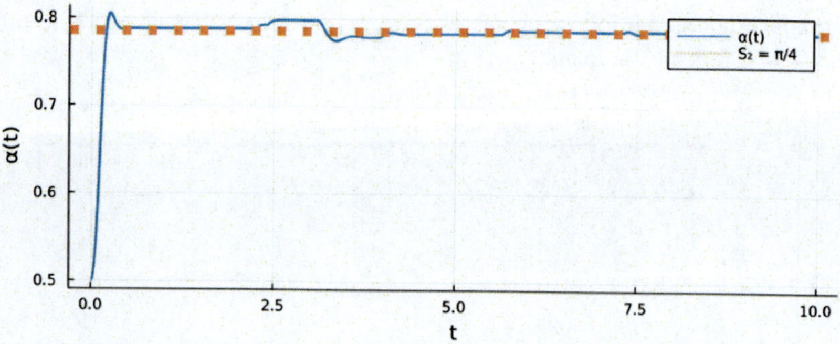

Fig. 4. Graph of the conveyor angular position for the model (2).

Figure 4 shows the results of the stabilization of the angular position by means of a neural network controller with a 2–8–1 topology and with bias neurons (see Fig. 1).

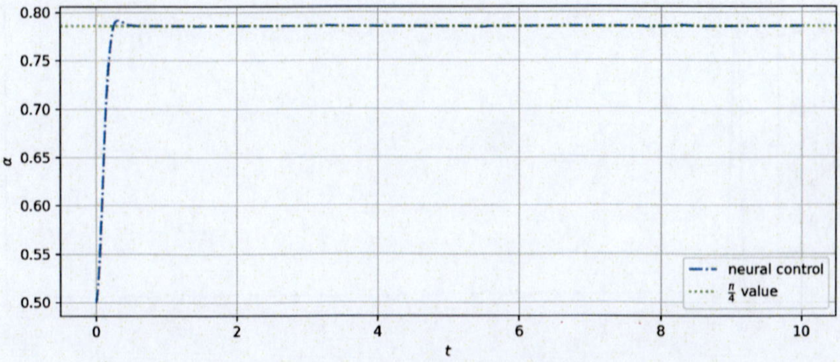

Fig. 5. Graph of the conveyor angular position for the model (1).

With the considered set of parameters, the trajectories corresponding to the angular position for the model (2) are close to the target value $S_1 = \frac{\pi}{4}$. Compared with the angular position graph for model (1), it can be noted that the quality of stabilization of the angular position of model (2) is lower. This result is explained by the fact that the system with smooth loading operator under conditions of uncertainty of the parameters m, ε.

In Fig. 6, 7 are graphs of the conveyor angular velocity. According to Fig. 6,

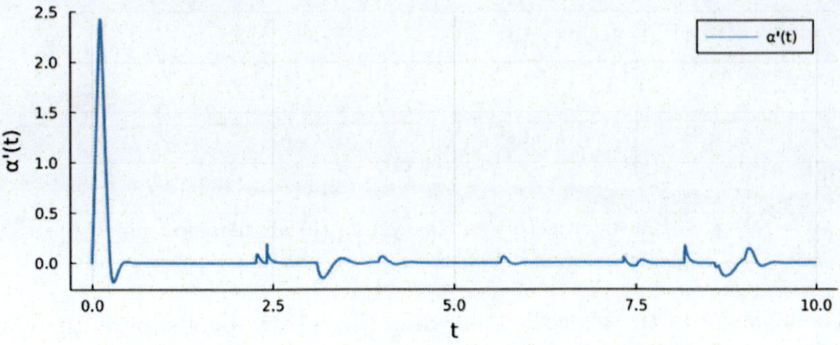

Fig. 6. Graph of conveyor angular velocity for model (2).

the angular velocity in the steady state is close to zero, which is consistent with the problem under consideration. Note that for model (2), in comparison with model (1), the neural network training time increases. Thus, the efficiency of the neural network algorithm (with respect to time) for model (2) is lower, but the information content is higher.

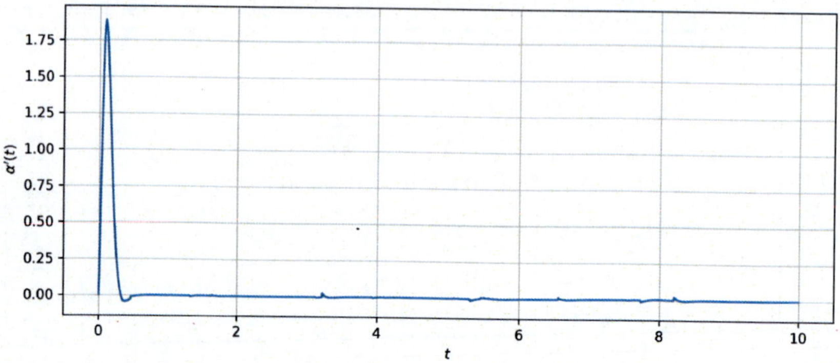

Fig. 7. Graph of the angular velocity for the model (1).

Figure 8, 9 show the phase trajectories of the conveyor angular position for model (2) and model (1), respectively.

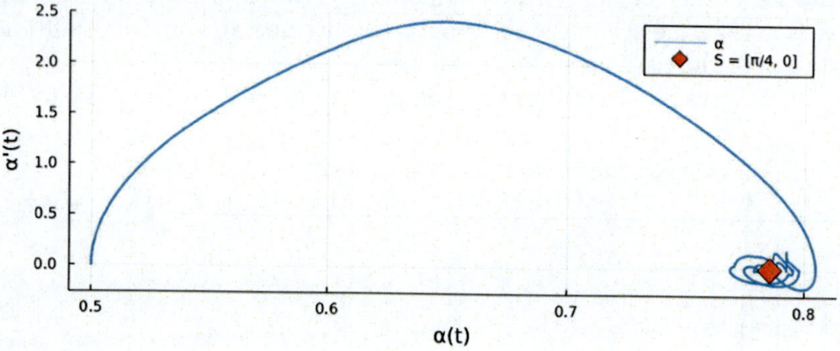

Fig. 8. Phase trajectory of the system (2) angular position.

The behavior of the trajectory shown in Fig. 8 corresponds to multiple focus lying in the neighborhood of the true focus, determined by the conditions for setting the optimal control problem (see markers in Fig. 8 and Fig. 9). It should be noted that the trajectory in Fig. 8 is consistent with condition (3), which specifies the optimality criterion. Comparison with a similar trajectory of model (1) (Fig. 9) shows a similar character of stabilization with a difference at the final stages of movement.

Figure 10 shows a graph of the change in the system (2) momentum.

According to Fig. 10, model (2) with the selected parameters provides an exit to the stabilizing mode with respect to the phase variable of the momentum. Presented in Fig. 10 jump transitions are associated with a change in the mass of loads on the conveyor, which leads to response regulation.

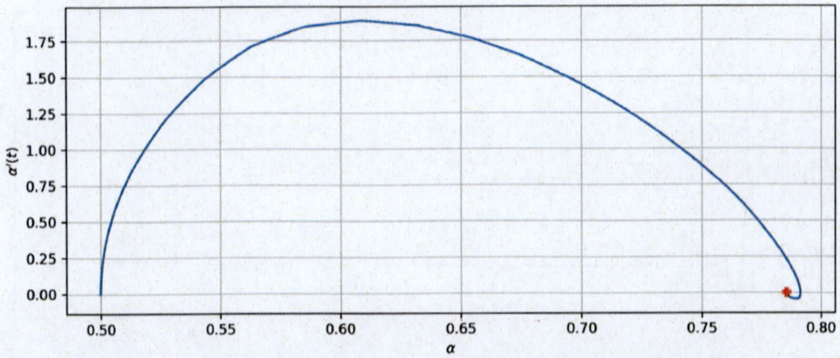

Fig. 9. Phase trajectory of the system (1) angular position.

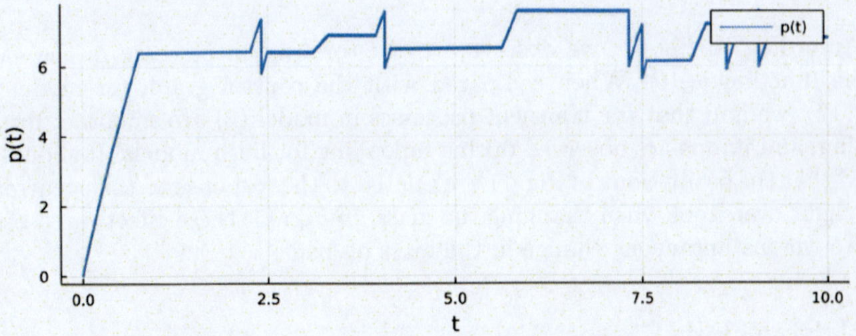

Fig. 10. Graph of the change in the system (2) momentum.

Figure 11 and Fig. 12 show the conveyor angle control graph for models (2) and (1), respectively.

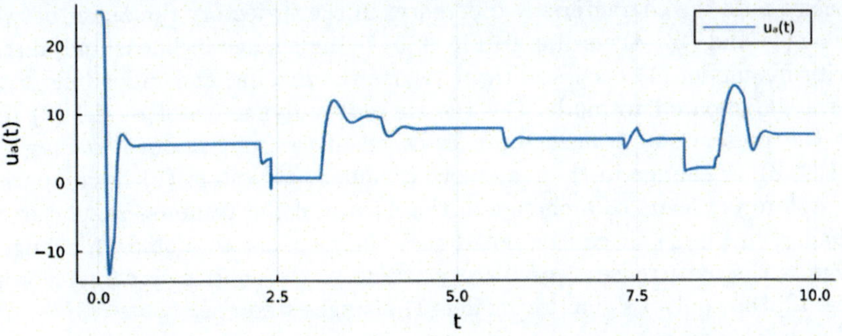

Fig. 11. Conveyor elevation angle control graph for model (2).

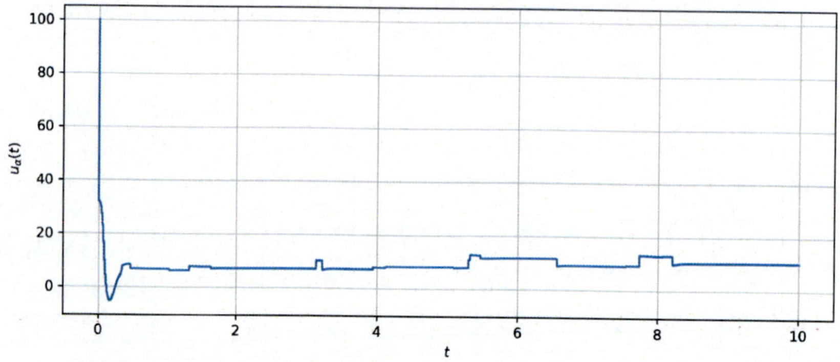

Fig. 12. Conveyor elevation angle control graph for model (1).

According to Fig. 11, we observe a trend towards an increase in the value of the function $u_\alpha(t)$. When compared with the control graph for model (1) (Fig. 12), we find that the transient processes in model (2) are smoother during loading, and jumps are observed during unloading for both models. It should be noted that the oscillations in Fig. 11 are related to the response to the occurrence of a slight overshoot when changing the mass. In Fig. 11, these effects are absent due to the instantaneous change in the mass of loads.

5 Discussion

Verifying the model, we obtained that the sliding mode and the mode of using the neural network controller have similar characteristics of achieving system destabilization. Note that system (2) loses stability when the control switching period is more than 0.01 s. In [23], a similar conclusion is reached when comparing the sliding mode and the fuzzy controller mode for a "simpler" system (1), while the switching period is up to 0.1 s.

Let us note the characteristic differences in the trajectory dynamics between systems (1) and (2). According to Fig. 3, 5, 7, the change in linear and angular velocity for model (1) occurs abruptly, since the loading and unloading occurs instantly. Compared to model (1), the trajectory dynamics of model (2) indicates the presence of transient processes resulting from smooth loading (see Fig. 2, 4, 6). A change in the trajectory dynamics of system (2) in comparison with system (1) leads to a change in the nature of the response regulation. In particular, in Fig. 11, it can be noted that the response to a smooth change in weights is the occurrence of oscillatory effects in the control signal. According to Fig. 12, the control signal for model (1) changes stepwise.

The implementation of the sliding mode for linear speed control leads to effective results in the stabilization of the belt conveyor, both in the case of the model with instantaneous loading and unloading, and in the model with smooth loading of the load and instantaneous unloading. Note that model (2) takes into account a fairly wide range of physical effects, but at the same time it is rather complicated for optimal trajectories search.

6 Conclusion

The analysis of the basic and modified models of a belt conveyor with intelligent control demonstrated the effectiveness of the applied neural network modeling method. It should be noted that the modified conveyor model takes into account the smooth loading and the ability to control the elevation angle of the conveyor belt. Based on the quality criterion, a loss function is proposed that takes into account the software implementation of the artificial neural network training algorithm. Graphs of trajectories and a graph of the control function are obtained, taking into account the selected parameters of the models. The method of searching for optimal control based on the implementation of an artificial neural network is tested to solve the stabilization problem of elevation angle for a belt conveyor model. Qualitative effects associated with the features of the proposed modified belt conveyor control model are revealed. The performed comparative analysis of the neural network control effectiveness for various models is aimed at obtaining a universal assessment of the algorithms applicability for various operating conditions of systems. The use of the Julia language in combination with Jupyter system and Plots, DifferentialEquations, BlackBoxOptim libraries made it possible to achieve high performance indicators of the developed software. The use of the developed algorithmic and instrumental support is associated with the possibilities of implementing technologies for intelligent control of conveyor transport systems. As a prospect for further research, it should be noted the construction and analysis of hybrid dynamic models of a belt conveyor with a dynamic elevation angle.

References

1. Dmitriev, V.G., Verzhanskiy, A.P.: Grounds of the Belt Conveyor Theory. Gornaya kniga, Moscow (2017)
2. Subba Rao, D.V.: The Belt Conveyor: A Concise Basic Course. CRC Press, New York (2020) https://doi.org/10.1201/9781003089315
3. Zhao, L., Lyn, Y.: Typical failure analysis and processing of belt conveyor. Procedia Eng. **26**, 942–946 (2011)
4. Andrejiova, M., Grincova, A., Marasova, D.: Monitoring dynamic loading of conveyer belts by measuring local peak impact forces. Measurement **158**, 107690 (2020)
5. Andrejiova, M., Grincova, A., Marasova, D.: Measurement and simulation of impact wear damage to industrial conveyor belts. Wear **12**, 368–369 (2016)
6. Dmitrieva, V.V., Sizin, E.P.: Continuous belt conveyor speed control in case of reduced spectral density of load flow. Mining Inf. Anal. Bull. **2**, 130–138 (2020)
7. Listova, M.A., Dmitrieva, V.V., Sizin, E.P.: Reliability of the belt conveyor bed when restoring failed roller supports. In: IOP Conference Series: Earth and Environmental Science, p. 012002 (2021)

8. Zyuzicheva, Y.E.: Model of a belt conveyor located at an angle to the horizon determination of the optimal inclination angle for the transition process. Mining Inf. Anal. Bull. **7**, 212–216 (2006)
9. Kumar, R., Singh, V.P., Mathur, A.: Intelligent Algorithms for Analysis and Control of Dynamical Systems. Springer, Singapore (2021). https://doi.org/10.1007/978-981-15-8045-1
10. Wen, Y.: Recent Advances in Intelligent Control Systems. Springer, Heidelberg (2009). https://doi.org/10.1007/978-1-84882-548-2
11. Kim, H.: Intelligent control of vehicle dynamic systems by artificial neural network. PhD thesis (1997)
12. Zaitceva, I., Andrievsky, B.: Methods of intelligent control in mechatronics and robotic engineering: a survey. Electronics **11**(15), 2443 (2022)
13. Aliworom, C., Uzoechi, L., Olubiwe, M.: Design of fuzzy logic tracking controller for industrial conveyor system. Int. J. Eng. Trends Technol. **61**, 64–71 (2018)
14. Khalid, H.: Implementation of artificial neural network to achieve speed control and power saving of a belt conveyor system. East.-Eur. J. Enterp. Technol. **2**, 44–53 (2021)
15. Žvirblis, T., et al.: Investigation of deep learning models on identification of minimum signal length for precise classification of conveyor rubber belt loads. Adv. Mech. Eng. **14** (2022)
16. Lee, D., Seo, H., Jung, M.W.: Neural basis of reinforcement learning and decision making. Ann. Rev. Neurosci. **35**(1), 287–308 (2012)
17. Kozhubaev, Y.N., Semenov, I.M.: Belt conveyor control systems. Sci. Tech. Bull. St. Petersburg State Polytech. Univ. **2**(195), 181–186 (2014)
18. Ma, M.X., Gao, X.X.: Coal belt conveyor pid controller parameter regulation with neural network. Appl. Mech. Mater. **319**, 583–589 (2013)
19. Farouq, O., Selamat, H., Noor, S.: Intelligent modeling and control of a conveyor belt grain dryer using a simplified type 2 neuro-fuzzy controller drying. Technology **33**(10), 1210–1222 (2015)
20. Lv, Y., Liu, B., Liu, N., Zhao, M.: Design of automatic speed control system of belt conveyor based on image recognition. In: IEEE 2020 3rd International Conference on Artificial Intelligence and Big Data (ICAIBD), Chengdu, China, pp. 227–230 (2020)
21. Lutfy, O.F., Selamat, H., Mohd Noor, S.B.: Intelligent modeling and control of a conveyor belt grain dryer using a simplified type 2 neuro-fuzzy controller. Drying Technol. **33**, 1210–1222 (2015)
22. Masina, O.N., Druzhinina, O.V., Igonina, E.V., Petrov, A.A.: Synthesis and stabilization of belt conveyor models with intelligent control. Lect. Notes Netw. Syst. **228**, 645–658 (2021)
23. Druzhinina, O.V., Masina, O.N., Petrov, A.A.: Modeling of the belt conveyor control system using artificial intelligence methods. J. Phys. Conf. Ser. **2001**, 012011 (2021)
24. Masina, O.N., Druzhinina, O.V., Petrov, A.A.: Controllers synthesis for computer research of dynamic conveyor belt model using intelligent algorithms. Lect. Notes Netw. Syst. **502**, 462–473 (2022)
25. Bezanson, J., Edelman, A., Karpinski, S., Shah, V.B.: Julia: a fresh approach to numerical computing. SIAM Rev. **59**(1), 65–98 (2017)
26. Mckinney, W.: Python for Data Analysis, 2e: Data Wrangling with Pandas, Numpy, and Ipython. OReilly, Boston (2017)

Detection of Vocal Cords in Endoscopic Images Based on YOLO Network

Jakub Steinbach[1(✉)], Zuzana Urbániová[2(✉)], and Jan Vrba[1(✉)]

[1] Department of Mathematics, Informatics and Cybernetics Technická 1905/5, University of Chemistry and Technology in Prague, 166 28 Praha 6, Prague, Czech Republic
{jakub.steinbach,jan.vrba}@vscht.cz

[2] 3rd Faculty of Medicine Department of Otorhinolaryngology University Hospital Královské Vinohrady, Charles University, Prague, Czechia
zuzana.urbaniova@lf3.cuni.cz

Abstract. This article presents an application of the YOLOv5 object detection algorithm to detect vocal folds in laryngoscopic videos without any additional image enhancement. Accuracy in the form of mean average precision, precision, and recall of architectures are evaluated and compared. Results suggest that the YOLO detection algorithm could be used to locate the region of vocal folds for a further objective evaluation that could potentially be implemented in clinical practice.

Keywords: vocal cords · vocal folds · YOLO · neural networks · machine learning

1 Introduction

Vocal cords play an important role in phonation, breathing, and swallowing. The impairment of their mobility causes patients to have a fundamental clinical problem with basic physiological functions. Vocal cord movement disorder can be caused by trauma, arytenoid articulation disorder (arytenoid arthritis, ankylosis, or luxation of the arytenoid joint), myogenic disease (systemic myopathies such as myasthenia gravis), or neurogenic lesions. Neurogenic dysfunctions include the most common innervation disorder of the recurrent laryngeal nerve or superior laryngeal nerve, as well as rare lesions of the vagus nerve. Pathology of the central nervous system (stroke, Parkinson's disease, multiple sclerosis) with the resulting central paresis of the vocal cords cannot be omitted, and voice disorders related to spasmodic dysphonia must also be included in the neurogenic group [6,12].

1.1 Clinical Visualization of Vocal Cords

Patients with vocal cord impairment often suffer from dysphonia. The essential step in physical evaluation and follow-up therapy is to visualize the larynx and vocal cords [20].

R. Silhavy and P. Silhavy (Eds.): CSOC 2023, LNNS 724, pp. 747–755, 2023.
https://doi.org/10.1007/978-3-031-35314-7_62

In daily otorhinolaryngological practice, clinicians have variations of visualization methods from indirect mirror laryngoscopy to high-speed kymography. The most common and relevant clinical methods for functional evaluation of the vocal tract are stroboscopic visualization of the vocal mucosal wave and dynamic evaluation of the voice with flexible laryngoscopy. Combining these techniques is desirable for detailed information on the vibration activity of the vocal cord mucosa and also the function of the entire vocal tract. With rigid laryngostroboscopy, we obtain a greater magnification of the vocal cords and excellent optical image quality. On the other hand, the main disadvantage is the slightly restricted position of the patient with protrusion of the tongue. Rigid laryngoscopy also has limitations in the absence of a comprehensive functional evaluation of the entire vocal tract [18].

The mentioned visualization methods come with a certain degree of subjectivity. The final finding is influenced by the experience of the otorhinolaryngologist. Until now, there has been no practical objective method that would reliably assist in evaluating the outcome of patient treatment. In clinical practice, the assessment of impaired mobility of the vocal cords is qualitative and standardized terminology is lacking for a clear definition of movement disorders. Therefore, it can be difficult for clinicians to accurately convey important clinical information.

1.2 Vocal Cords Detection with Machine Learning

To utilize machine learning methods for objective evaluation of vocal cords, first, it is necessary to detect the location of vocal cords. The YOLO [16] algorithm, introduced in 2016, is one of the state-of-the-art detectors. As a one-stage detector, it solves object detection as a regression problem. Both the bounding box parameters and the corresponding label are predicted with a single computation, hence the name *You Only Look Once*. YOLO, namely its version YOLOv5 [9], is utilized in many studies, where, usually, static images or non-shaky videos are processed.

There are applications related to COVID-19, where YOLO is used for covid mask recognition [11] or COVID-19 detection using CT images [15]. The study [13] proposes a modified YOLOv5 network for the detection and classification of breast tumors, while the study [5] aims at the detection and classification of brain tumors. The suitability of YOLOv5 for real-time detection is demonstrated in the detection and localization of lung nodules from low-dose CT scans [8]. There are also studies based on YOLOv5 dedicated to the detection of ulcers [3], stroke lesion [4], colorectal polyps [23] or type of white blood cells [17]. YOLO is also a prospective tool for improving existing work on vocal cord detection, such as detection of vocal fold paralysis from clean images in [1] or detection of laryngeal cancer from cropped images in [2] by localizing the region of vocal folds in the image before using classification algorithms.

In this study, we introduce the YOLOv5 application for vocal cord detection, in the shaky video obtained during flexible and rigid laryngoscopy.

2 Materials and Methods

2.1 Recorded Data Preprocessing

We performed the video analysis of the three vocal cord recordings taken at the Department of Otorhinolaryngology, 3rd Faculty of Medicine, Charles University, and the University Hospital Královské Vinohrady in Prague. The first two recordings were taken with a flexible Olympus rhino-laryngoscope, and the third recording was taken with a rigid laryngoendoscope. Stroboscopic examination using the Highlight plus Invisia diagnostic video chain was used for all recordings. During the videolaryngoscopy examination, patients were asked to vocalize "ee" followed by a deep breath.

The first recording represents a patient with left-sided vocal cord paralysis. The second recording shows the patient on the second day after total thyroidectomy with slight partial paresis of the right vocal cord, and the third recording represents a healthy patient with normal vocal cord function.

The videos are all less than a minute long. The sampling rate of the videos 25 Hz and the frame size is 960 × 720 pixels. You can see the information on utilized videos in the Table 1.

Table 1. Data Source Information

Diagnosis	left-sided vocal cord paralysis	partial right vocal cord paresis	healthy patient
Framerate	25 fps		
Frame Size	960 × 720		
Total Length	49 s	38 s	19 s
Total Frames	1237	965	499
Frames Used	487	418	201

First, all three videos were divided into individual frames. After obtaining a total of 2 701 images, the position of the vocal cords in each image was manually marked using the Label Studio Data Labeling Platform [21]. During the labeling process, 1,595 images were discarded either due to missing vocal cords or due to increased blurriness, resulting in a total of 1,106 labeled images that were used for training and validation.

The labeled images were then downloaded with the information on the labels in a JSON file and manually transformed into a YOLO format.

2.2 Object Detection

The YOLO algorithm consists of three steps: (i) image resizing, (ii) using the convolutional network to detect bounding boxes for all classes, and (iii) tresholding the bounding boxes based on the box confidence.

For feature extraction, YOLOv5 uses the CSP-Darknet53 convolutional network, also used in previous versions of the YOLO algorithm.

The general formula of the loss function is described as follows:

$$L = L_{box} + L_{conf} + L_{cls} \tag{1}$$

L_{box} (defined by Eq. 2) represents the regression loss due to misposition of the bounding box. The misposition is given by the Complete Intersection over Union ($CIoU$) which, compared to the other variations of the intersection over union metric, includes the information on the position of the real and predicted bounding box centers, and the information on the ratio of the width and height of the two boxes. This reduces the number of iterations needed to train the model.

$$L_{box} = 1 - CIoU = 1 - IoU - \frac{\rho^2(A^p, A^g)}{c^2} + \alpha v \tag{2}$$

IoU denotes the Intersection over Union, ρ^2 denotes the Euclidean distance between the two boxes, c^2 represents the diagonal length of the smallest concave box enclosing both bounding boxes, and αv define the aspect ratio consistency of the two boxes.

If an object is detected, the classification loss at each cell is the squared error of the class conditional probabilities for each class:

The classification loss sums the cross entropy of the conditional probability of each detected object c for each grid cell i of the image. It is defined by:

$$L_{cls} = \sum_{i=0}^{K^2} I_i \sum_{c=1}^{C} E(\hat{p}_i(c), p_i(c)) \tag{3}$$

The I takes 1 if the object exists in the cell i, otherwise, it takes 0. The predicted and true probability of an object belonging to class c are denoted $\hat{p}_i(c)$ and $p_i(c)$, respectively. E describes the cross-entropy function.

The confidence loss evaluates the rate of correct and incorrect bounding box predictions by calculating the binary cross-entropy value of predicted and real confidence values of each detected and existing bounding box. It is defined by:

$$L_{conf} = \sum_{i=0}^{K^2} \sum_{j=0}^{M} I_{i,j} E(\widehat{C}_i, C_i) + \lambda_{noobj} \sum_{i=0}^{K^2} \sum_{j=0}^{M} (1 - I_{i,j}) E(\widehat{C}_i, C_i) \tag{4}$$

Similarly to classification loss, $I_{i,j}$ takes 1 for each presence of j-th bounding box in i-th cell; otherwise, it takes 0. The predicted and real confidence values are denoted $\widehat{C}_i$ and C_i, respectively. The λ_{noobj} parameter allows to decrease the dependence of the loss function on correct classification of unlabeled images and is usually set to 0.5.

3 Results

3.1 Model Validation

To estimate the performance of various YOLOv5 architectures, we evaluated the mean average precision (threshold 0.5) [7], mean average precision over multiple thresholds (from 0.5 to 0.95 with step 0.05) [10], recall and precision. Recall and precision metrics are calculated as follows:

$$Precision = \frac{TP}{TP + FP} \tag{5}$$

$$Recall = \frac{TP}{TP + FN} \tag{6}$$

We also evaluated the training time (TT, see Table 2).

Due to the limited size of our data set, we performed a 5-fold validation. First, all images were randomized, regardless of diagnosis. Then, 5 folds were established and training, respectively, validation was performed. The cross-validation process is shown in Fig. 1

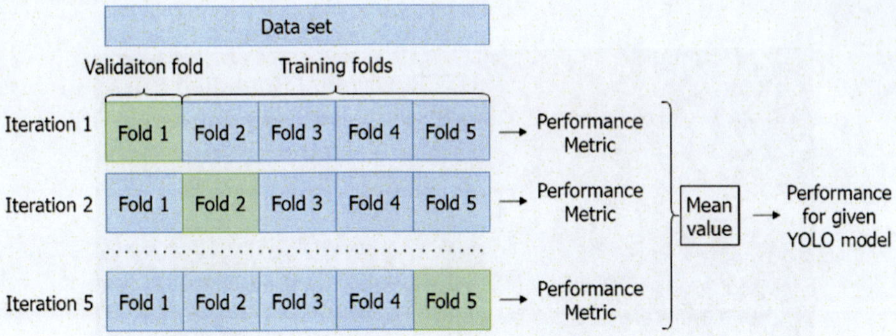

Fig. 1. Schematics of the 5-fold cross-validation used in this study

3.2 Implementation Specification

In this experimental study, we used Pytorch[14] YOLOv5 release 7.0 implementation that is publicly available at https://github.com/ultralytics/yolov5. We conducted experiments with different YOLO models, namely YOLOv5x, YOLOv5l, YOLOv5m, YOLOv5s, and YOLOv5n. The number of weights (#Params.) for each model is in Table 2. All models utilize input images with a size of 640×640 (pixels). To train the models, the stochastic gradient descent optimizer [19] with a learning rate of 0.01 was used. All experiments were carried out with a batch size of 16. The computational time was measured for every single YOLOv5 model. The experiments were performed on a PC using an AMD Ryzen 5900X

12 core CPU running at 3701 MHz MHz with 128 GB RAM. The YOLOv5 was trained using MSI GeForce RTX 3090 Ti GAMING TRIO 24G with CUDA 11.6 driver. The operating system was Windows 10 Pro 64-bit version 10.0.19044 and the code was written in Python 3.9.13 [22] with Pytorch-1.13.1. The results obtained are summarized in Table 2. Note that TT, mAPs, recall, and precision are average values due to a 5-fold validation. Examples of successful detection are shown in Fig. 2 and Fig. 3.

Table 2. Results of Model Validation for Selected Models

Model	Params (M)	Train time	mAP[0.5]	mAP[0.5:95]	Recall	Precision
YOLOv5x	86.7	122.68	0.995	0.736	0.999	0.999
YOLOv5l	46.5	79.52	0.995	0.738	0.999	0.999
YOLOv5m	21.2	56.96	0.995	0.745	0.999	0.999
YOLOv5s	7.2	40.34	0.995	0.748	0.999	0.999
YOLOv5n	1.9	37.43	0.995	0.738	0.999	0.999

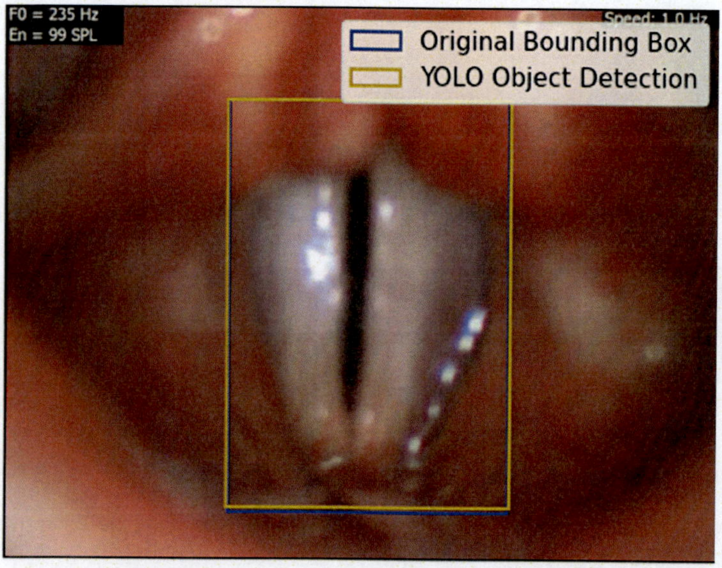

Fig. 2. Original Label and the trained YOLO inference in a validation sample. Rigid laryngoscopy - healthy patient. Ground truth bounding box coordinates: $x_{min} = 303.46, y_{min} = 121.15, x_{max} = 679.62, y_{max} = 658.85$. Predicted bounding box coordinates: $x_{min} = 300.16, y_{min} = 121.68, x_{max} = 681.51, y_{max} = 652.12$, $IoU = 97.33\%$

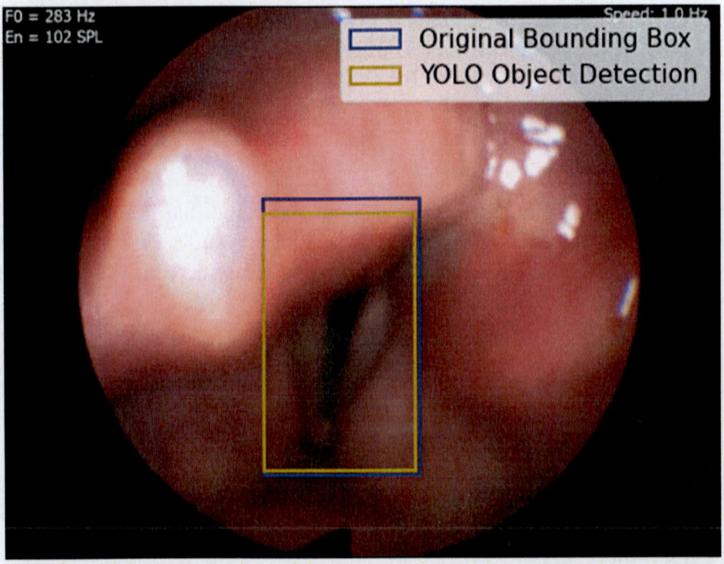

Fig. 3. Original Label with the trained YOLO inference in a validation sample. Flexible laryngoscopy - thyroidectomy patient. Ground truth bounding box coordinates: $x_{min} = 350.77, y_{min} = 251.54, x_{max} = 556.15, y_{max} = 612.69$. Predicted bounding box coordinates: $x_{min} = 350.75, y_{min} = 271.56, x_{max} = 549.91, y_{max} = 606.72$, $IoU = 97.33\%$

4 Conclusion

This article introduces the use of the YOLOv5 object detection algorithm to detect and find the location of vocal folds in videos taken with a rhino-laryngoscope and rigid laryngoendoscope without additional enhancement to image quality.

For training, the evaluated videos were split into individual frames and manually labeled with the Label Studio Data Labeling Platform. Five different YOLOv5 architectures with different numbers of parameters were used for detection and their performance was evaluated with the following metrics: five-fold cross-validation using the train time, mean average precision with threshold 0.5 and over multiple thresholds from 0.5 to 0.95, recall, and precision.

Based on the results, there are no significant differences between the selected YOLOv5 architectures in precision, recall, or mean average precision for the 0.5 threshold. All architectures were able to detect the correct region with an accuracy of >99%. The results show that it is possible to successfully detect vocal folds in unprocessed images using a neural network with a relatively low number of parameters. We see the potential benefit of objective assessment of vocal cord function mainly in facilitating the description of objective findings and also the evaluation of patient treatment results.

Acknowledgements. Jakub Steinbach acknowledges his specific university grant (IGA) A1_FCHI_ _2023_003. Zuzana Urbániová acknowledges her specific research project of Charles University COOPERATIO 43 - Surgical disciplines.

References

1. Adamian, N., Naunheim, M.R., Jowett, N.: An <span style="font-variant:small-caps;">Open-Source</span> Computer Vision Tool for Automated Vocal Fold Tracking From Videoendoscopy. Laryngoscope **131**(1) (2021), https://onlinelibrary.wiley.com/doi/10.1002/lary.28669
2. Azam, M.A., et al.: deep learning applied to white light and narrow band imaging videolaryngoscopy: toward real-time laryngeal cancer detection. Laryngoscope **132**(9), 1798–1806 (2022), https://onlinelibrary.wiley.com/doi/abs/10.1002/lary.29960
3. Brüngel, R., Friedrich, C.M.: Detr and yolov5: exploring performance and self-training for diabetic foot ulcer detection. In: 2021 IEEE 34th International Symposium on Computer-Based Medical Systems (CBMS), pp. 148–153. IEEE (2021)
4. Chen, S., et al.: Automatic detection of stroke lesion from diffusion-weighted imaging via the improved yolov5. Comput. Biol. Med. **150**, 106120 (2022)
5. Dipu, N.M., Shohan, S.A., Salam, K.: Deep learning based brain tumor detection and classification. In: 2021 International Conference on Intelligent Technologies (CONIT), pp. 1–6. IEEE (2021)
6. Dršata, J.: Foniatrie - hlas. Medicína hlavy a krku, Tobiáš, Havlíčkův Brod, 1 edn. (2011). http://arl.uhk.cz/arl-hk/cs/detail-hk_us_cat-0014865-Foniatrie-hlas/
7. Everingham, M., Van Gool, L., Williams, C.K., Winn, J., Zisserman, A.: The pascal visual object classes (voc) challenge. Int. J. Comput. Vis. **88**(2), 303–338 (2010)
8. George, J., Skaria, S., Varun, V., et al.: Using yolo based deep learning network for real time detection and localization of lung nodules from low dose ct scans. In: Medical Imaging 2018: Computer-Aided Diagnosis, vol. 10575, pp. 347–355. SPIE (2018)
9. Jocher, G., et al.: ultralytics/yolov5: v7.0 - yolov5 sota realtime instance segmentation (2022). https://zenodo.org/record/7347926
10. Lin, T., et al.: Microsoft COCO: common objects in context. CoRR abs/arXiv: 1405.0312 (2014). http://arxiv.org/abs/1405.0312
11. Loey, M., Manogaran, G., Taha, M.H.N., Khalifa, N.E.M.: Fighting against covid-19: A novel deep learning model based on yolo-v2 with resnet-50 for medical face mask detection. Sustain. Urban Areas **65**, 102600 (2021)
12. Merati, A.L., Heman-Ackah, Y.D., Abaza, M., Altman, K.W., Sulica, L., Belamowicz, S.: Common movement disorders affecting the larynx: a report from the neurolaryngology committee of the AAO-HNS. Otolaryngology-Head Neck Surgery **133**(5), 654–665 (2005). https://onlinelibrary.wiley.com/doi/10.1016/j.otohns.2005.05.003
13. Mohiyuddin, A., Basharat, A., Ghani, U., Abbas, S., Naeem, O.B., Rizwan, M.: Breast tumor detection and classification in mammogram images using modified yolov5 network. In: Computational and Mathematical Methods in Medicine 2022 (2022)
14. Paszke, A., et al.: Pytorch: An imperative style, high-performance deep learning library. In: Advances in Neural Information Processing Systems 32, pp. 8024–8035. Curran Associates, Inc. (2019). http://papers.neurips.cc/paper/9015-pytorch-an-imperative-style-high-performance-deep-learning-library.pdf

15. Qu, R., Yang, Y., Wang, Y.: Covid-19 detection using ct image based on yolov5 network. In: 2021 3rd International Academic Exchange Conference on Science and Technology Innovation (IAECST), pp. 622–625. IEEE (2021)
16. Redmon, J., Divvala, S., Girshick, R., Farhadi, A.: You only look once: unified, real-time object detection. In: 2016 IEEE Conference on Computer Vision and Pattern Recognition (CVPR), pp. 779–788. IEEE, Las Vegas, NV, USA (Jun 2016). http://ieeexplore.ieee.org/document/7780460/
17. Rohaziat, N., Tomari, M.R.M., Zakaria, W.N.W.: White blood cells type detection using yolov5. In: 2022 IEEE 5th International Symposium in Robotics and Manufacturing Automation (ROMA), pp. 1–6. IEEE (2022)
18. Rosen, C.A., et al.: Nomenclature proposal to describe vocal fold motion impairment. European Arch. Oto-Rhino-Laryngology **273**(8), 1995–1999 (2016). http://link.springer.com/10.1007/s00405-015-3663-0
19. Ruder, S.: An overview of gradient descent optimization algorithms. arXiv preprint arXiv:1609.04747 (2016)
20. Stachler, R.J., et al.: Clinical practice guideline: hoarseness (Dysphonia) (update). Otolaryngology-Head Neck Surgery **158**(S1) (2018). https://onlinelibrary.wiley.com/doi/10.1177/0194599817751030
21. Tkachenko, M., Malyuk, M., Holmanyuk, A., Liubimov, N.: Label Studio: Data labeling software (2020–2022), open source software available from https://github.com/heartexlabs/label-studio
22. Van Rossum, G., Drake, F.L.: Python 3 Reference Manual. CreateSpace, Scotts Valley, CA (2009)
23. Wan, J., Chen, B., Yu, Y.: Polyp detection from colorectum images by using attentive yolov5. Diagnostics **11**(12), 2264 (2021)

A Proposal of Data Mining Model for the Classification of an Act of Violence as a Case of Attempted Femicide in the Peruvian Scope

Sharit More[1] and Wilfredo Ticona[1,2](✉)

[1] Universidad ESAN, Lima, Peru
16100961@ue.edu.pe
[2] Universidad Tecnológica del Perú, Lima, Peru
wmamani@esan.edu.pe

Abstract. Nowadays, femicide is one of the biggest problems worldwide in which the human rights of the victims are violated. In addition, it also constitutes a public health concern, with serious physical and psychological consequences. The objective of this research is to implement a data mining model to classify an act of violence as a case of attempted femicide in Peru. This study used public data of 2021 of the statistics portal National Aurora Program of the Ministry of Women and Vulnerable Populations (MIMP). The applied methodology was based on 5 phases: Data collection, data understanding, data preprocessing, data mining and model evaluation. Results obtained with Balanced Random Forest and Logistic Regression models demonstrated the best performances with a Recall of 0.88 and 0.86, respectively. Furthermore, the application of SMOTE improved the performance of both models. This investigation will contribute to find patterns related to the characteristics of aggressors and victims, that can help to put into action new instruments based on Data Mining to prevent more murders of women.

Keywords: Attempted Femicide · Balanced Random Forest · Classification Model · Data Mining · Feature selection · Light GBM · Logistic regression · SMOTE

1 Introduction

Attempted femicide [24] is a terrible crime that occurs when an agent initiates the execution for murdering a woman, but the victim survives. Unfortunately, in every society in the world, violence against women has marked a tragic milestone because it affects millions [6]. On the other hand, according to the United Nations Office on Drugs and Crime (UNODC), there is no global, standardized or consistently recorded data on femicide [2] where it was reported that 87,000 women worldwide were intentionally killed and more than half of them (50,000) by intimate partners or family members. Also, the European Institute for Gender Equality (EIGE) claims that data gaps mask the true scale of violence [2]. In

R. Silhavy and P. Silhavy (Eds.): CSOC 2023, LNNS 724, pp. 756–772, 2023.
https://doi.org/10.1007/978-3-031-35314-7_63

2020, gender-based violence in Peru was no stranger to the covid 19 pandemic. In the first two months, 32 cases of femicide and 120 cases of attempted femicide were detected [13]. In addition, the 100 line of the Ministry of Women and Vulnerable Populations registered nearly 39,000 calls, of which 77% (more than 30,000) involved a woman as a victim. On the other hand, the virtual portal "Mujeres Desaparecidas Peru" states that between March 20th and April 20th 2020, 228 reports of women's disappearances were reported [6]. In the light of the above, the increase in homicide cases of women in Peru in recent years has led to more research to find patterns that lead to these behaviors in order to strategies for later. That is why, it is important to develop strategies based on the characteristics of the victims and aggressors in order to determine the recurrence of gender-based violence to something more serious, which is femicide. The aim of this research is to present a new method to detect if a woman who suffers current violence is at serious risk of losing her life based on the data from the statistical portal of the National Aurora Program from the MIMP.

2 Related Work

In recent years, with the continuous development of many techniques of data mining [4], much efforts of researchers have been made to apply it to classify and predict crimes against women [17]. In 2019, the author [23] carried out a study to develop a predictive model for recognizing physically abused Peruvian women. Then, the results demonstrated that the Random Forest Classifier was the best classifier whit the average score was 0.7182. On the other hand, in 2021 Guerrero et al. [17] proposed the use of Chi-squared as feature selection techniques and implemented models like Multinomial Logistic Regression, Naive Bayes, Random Forest and Support Vector Machines in order to find the most significant variables for the prediction of IPV in Peru.

Otherwise, in 2020 [37] did a research to assess and predict the crime against women in India analyzing data of crime reporting for the last ten years. The results indicated that the highest accuracy score of 0.80769 with the algorithm Logistic Regression. Furthermore, [12] proposed to evaluate the efficacy and application of Machine learning algorithms such as KNN, Decision Trees, Naïve Bayes, Linear Regression, CART and SVM algorithms to diagnose the crime rate against Women. At the same time, in [13] the purpose was to build and Intimate partner violence (IPV) perpetration triage tool which could be built and implemented in the field to identify young people in Los Angeles who are at high-risk for engaging in violence perpetration.

Then, in 2019 [5], by objectively analyzing various techniques of data mining which were applied to many studies, they discovered that few authors have used the Indian Crime dataset. Also, algorithms like Naive Bayes, decision tree Bayesnet, J48, JRip & OneR were considered as the most frequent algorithms for the analysis and predicting crime or violence against women. The authors in [3] found that the risk of IPV in South Africa was associated with the husband or partner's characteristics more than the woman's, based on data from the 2016

South African Demographic and Health Survey. Finally, [35] studied the application of fuzzy-rough oscillation concept in the field of data mining especially in the subject of crime against women, they found that the maximum crime is occurring due to domestic violence/family discord. The main reason for such a crime is either economic pressure or increase of women participation in the workplace.

3 Methodology

3.1 Proposed Method

The proposed methodology was based on [17] and [37]. It consists of five phases: Data collection, Data Understanding, Data preprocessing, Data Mining and Model evaluation. Then, each stage has general activities that are of paramount importance for the development of the research, see Fig. 1.

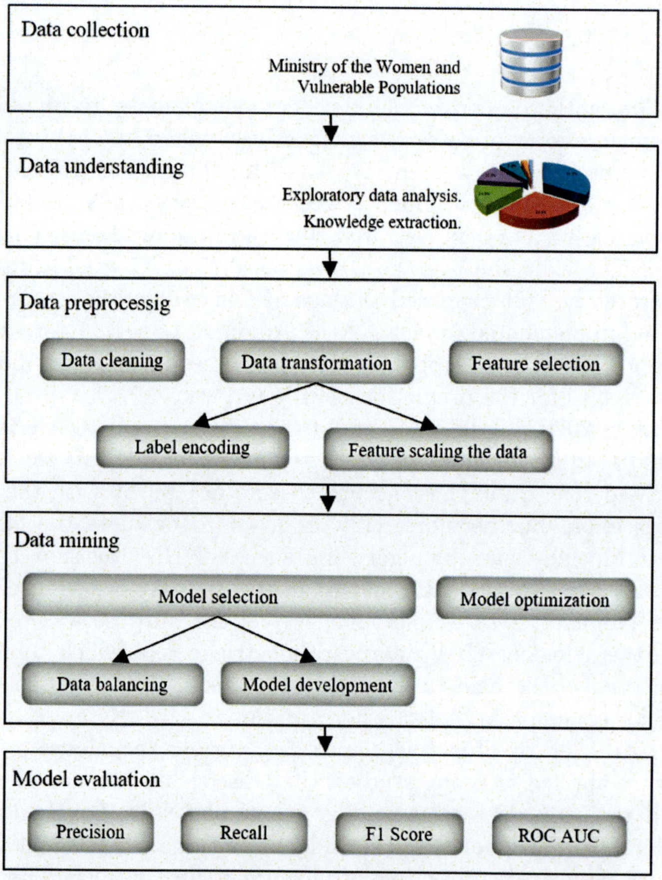

Fig. 1. Proposed Methodology. Source: Own elaboration

3.2 Data Collection

Dataset. The Dataset collection stage consisted of collecting data related to acts of violence against women [17], committed by their partners or ex-partners and linked as attempted femicide. Therefore, the dataset used was obtained from the "Ministerio de la Mujer y Poblaciones Vulnerables" (Ministry of the Women and Vulnerable Populations, MIMP) of Peru (https://portalestadistico.aurora.gob.pe/bases-de-datos-2019/), which contains 181,885 records and 156 variables of complaints made by people who have been victims of domestic and sexual violence, considered in some cases as attempted femicide, throughout 2019 across the country. Table 1 shows some variables and their respective meanings, according to the glossary of terms of the CEM. Also, the dataset [17] was composed of more than 97% of nominal categorical features of the whole dataset.

Table 1. Description of Dataset variables [2]

Variable	Description
Age_Victim	Age at which a woman was a victim of the last assault
Attempted_Feminicide	When the aggressor in his violent attack put the life of the person attacked at risk (1 = YES) (0=NO)
Couple_Bond	It is the bond of familiarity of the aggressor with the person assaulted (1 = YES) (0=NO)
Educational_victim_level	The integrity of the victim is at risk of death (1 = PRIMARY EDUCATION), (2 = SECONDARY EDUCATION) (3 = HIGHER EDUCATION)
Personal_life_Integrity_Risk	The integrity of the victim is at risk of death (1 = YES) (0 = NO)
Place_Ocurrence Aggression	Place where victim assault occurred: Victim's house aggressor's house, both of them, street, desolate place, or other
Type_Violence	Type of violence affecting the person attacked (1 = PSYCHOLOGICAL VIOLENCE), (2= PHYSICAL VIOLENCE), (3 = SEXUAL VIOLENCE),

3.3 Data Understanding

The Data Understanding stage seeks to explore the nature of data to understand what can be found out of them, in order to extract knowledge. There are different visualization techniques, which allows a better understanding of the behavior of the variables, in this study applied Exploratory Data Analysis (EDA), we performed Univariate and Bivariate analysis, most of the data were categorical, so we used bar charts and pie charts [5]. The above Fig. 2 shows that physical violence, categorised as 2, is the most frequent for this type of case, characterised by the use of knives or guns to attempt to kill their couples. Besides, it was found that the average age at which a woman suffered from some act of violence exercised by a relative, partner or acquaintance was 29 years. It should be noted that girls and elderly women are also prone to suffer aggression being in the range of 0 to 80 years. Among aggressors, alcohol consumption as a factor accounted for 24.3%, while death threats accounted for 11.4%, the latter is regarded as a means by which aggressors inflict fear and submission [24].

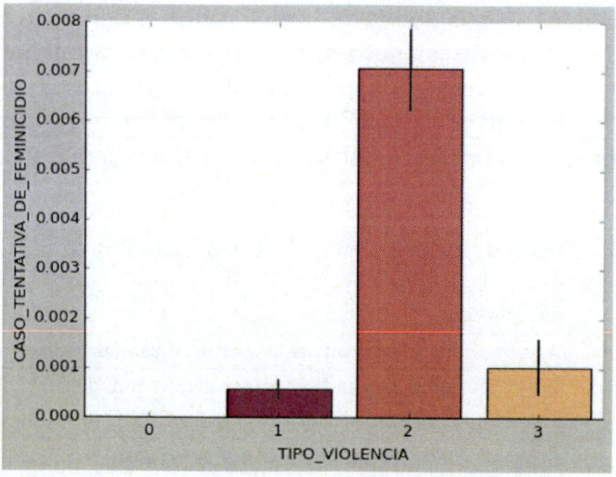

Fig. 2. Relationship of a case of attempted femicide and the type of violence. Source: Own elaboration

3.4 Data Preprocessing

Data preprocessing [30] is a data mining technique that is used to transform the data and make it suitable for another processing use (e.g. classification, clustering). It is a preliminary step that can be done with several techniques among which data cleaning and feature selection. The following stage consisted of developing the processes of data cleansing and feature selection. For that, the first step was to export [17] the original dataset that was in SPSS into a CSV format.

Data Cleaning. The following treatment [23] was applied with the main purpose to remove the "dirty data" and improved the development of the algorithms [5]. Firstly, we identified that the missing values were in empty string format and on first analysis the reports reflected that all the fields were filled in, continuing with a second analysis we transformed the values into empty objects (NaN) and on reviewing the reports we found the high dimension of these in the dataset. Also, a test was carried out to verify whether a conversion of these values was really needed, using models such as Decision Tree, Random Forest and Logistic

Regression, where difficulties were found for the completion of the execution. Then, for categorical data, most of them presented more than 80% of null values [17]. The treatment was to be filled up with zeros, the features containing binary categorical variables and for the features with ordinal category, we applied the mode of them [16]. On the other hand, for numerical features like "The age of the aggressor", the method applied was the imputation by KNN which has been successfully applied to a broad range of applications [35], so we decided to implement it, the sklearn library was used, and by default the number of neighbours was 5. Also, the weighting of the weights was equal, and the distance metric was Euclidean.

Data Transformation. Data transformation [37] is the method of converting data from one format to another, in order to transform it in appropriate forms suitable for the mining process [9]. Firstly, we applied a categorical data encoding technique which was Label Encoding technique [7], the importance of this treatment is that almost all algorithms tend to work much better when the values are numerical. So, we applied it for categorical data, such as the gender of victims and the rest of them, by variables like the educational level of aggressors. We tried as far as possible to maintain the order of importance. Therefore, this type of data was manually converted into numerical. After that, we applied the method called Feature Scaling the Data, to standardise the experiments. StandardScaler [37] was used to improve the quality of our data after rescaling the attribute to all have a similar scale, between 0 and 1. Consequently, the formula is defined in Eq. 1:

$$Z = \frac{X - \mu}{\sigma} \tag{1}$$

Steps for finding Standard deviation [37]
Step1: Subtract the mean value from the value X. X is an observations.
Step 2: Divide the result values from Step1 by σ.

Feature Selection. One of the contributions of this paper is to find the optimum number of features that can provide the best classification performance in a real dataset [30]. To identify the most relevant variables, the LightGBM model was used. LightGBM [26] is a powerful implementation of boosting method that is similar to XGBoost but varies in a few specific ways, specifically in how it creates the tree or base learners and similar to XGBoost, the importance of each feature can be obtained from the feature_importances_ attribute embedded in the algorithm. Firstly, we decided to apply three tree-based models, which were Light GBM, XGBoost and Random Forest. However, after their implementation, the last two were affected by over adjustment in the first six variables so we decided to just use the first one. Then, the variables were divided into subsets of 15, 20, 30 and 35. Figure 3 shows the results of the most relevant variables of the LightGBM model with subsets of 15 and Fig. 4 the subset of 20 variables.

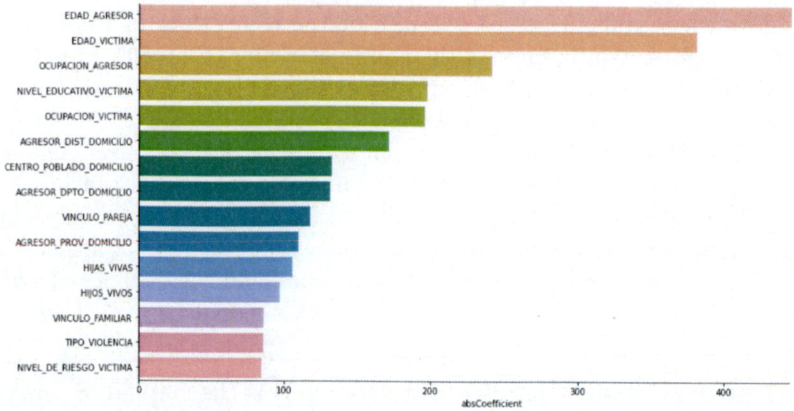

Fig. 3. Ranking of most important variables with subset of 15 variables after Light GBM. Source: Own elaboration.

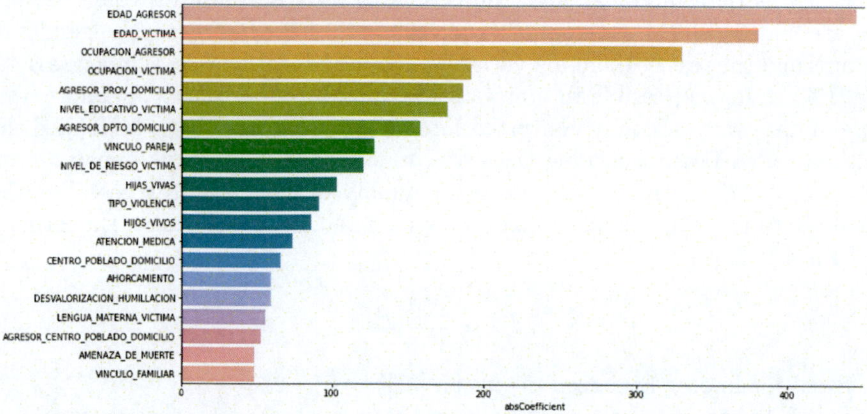

Fig. 4. Ranking of most important variables with subset of 20 variables after Light GBM. Source: Own elaboration.

Moreover, all variables are ranked in order of importance. According to previous reports, they were the age of the offender, the age of the victim, the occupation of the offender, the occupation of the victim, the educational level of the victim and the relationship.

3.5 Data Mining

Data Mining [13] is the application of specific algorithms for extracting patterns from data. This stage consisted of selecting and developing models to classify an act of violence against women as a case of attempted femicide.

Model Selection. Firstly, we split our data in Y contains "Attemped femicide Class" column and X contains all other columns and their corresponding row entries. Then, we applied a randomized distribution, we used to train the model, 80% for the validation and 20% for testing. Then, during the analysis of the target variable we discovered that the positive class was just represented at around 10% of total. In this scenery, to avoid that the results of the DM algorithms of DM could present overfitting. As [16] where they proposed the use of the Synthetic Minority Oversampling Technique (SMOTE) along with a random under-sampling to compensate for the minority and majority classes, respectively [17]. This technique oversamples the instances of the minor class through the introduction of new samples by linking all or any of the closest. The next step was Model Development. For this stage, we performed our experiments with six supervised learning models in order to obtain the predictive model related to the attempted femicide cases of Peruvian women. The following models were implemented effectively in [3,4,13,17,23] and [35]. Several types of DM algorithms were used in this work.

1. Balanced Random Forest. This model consists of a modification of RF [19], by iteratively extracting a bootstrap sample with equal proportions of minority and majority class data points. It implements a subsampling technique for each decision tree formation process in the Random Forest algorithm and combines a sampling technique with an ensemble idea [1]. Based in [27], its algorithm is shown below.

Algorithm 1. Balanced Random Forest (BRF)

1: Start RFC Iteration
2: From the under sample class selection of bootstrapped data sample.
3: From the oversampled class random selection of cases.
4: Apply replacement of data samples from oversampled to under sampled data set.
5: Tree Induction to maximizing tree size
6: Random Selection of Variable
7: Repeat step 2 to 6 as per the requirement.
8: Use of ensemble method for prediction of class

2. Decision Tree. Decision trees learn from data to approximate a sine curve with a set of if-then-else decision rules [23]. It decides which value of the target variable will be the node that will become the root of a subtree [33].
3. Logistic Regression. A data analysis technique [39] that uses mathematics to find relationships between two data factors. It then uses this relationship to predict the value of one of those factors based on the other. It predicts [34] the "probability value" by means of a linear combination of the given characteristics within an logistic function [5] which is given by Eq. 2:

$$S(X) = \frac{1}{1 + e^{-x}} \tag{2}$$

4. Naive Bayes. Is one of the simplest but effective classifiers [17], it works on the probability of the events which have happened in the past. This model [5] is a linear classifier based on Bayes' theorem, which is represented in Eq. 3.

$$P(A/B) = \frac{P(A)\ X\ P(B/A)}{P(B)} \tag{3}$$

   where:
   P(A/B): The posterior probability of A of the data
   P(B): The previous probability
   P(B/A): The probability of data given a hypothesis
5. Random Forest. An ensemble model [38] which takes a decision tree as a basic classifier and contains several decision trees trained by the method of Bagging in order to identify the class label for unlabeled instances [4]. It is based on K decision trees. The description of K decision trees is as follows in Eq. 4:

$$h(X, \theta_k), k = 1, 2, ..., K \tag{4}$$

   where K is the number of decision trees contained in random forest θ_k [38]
6. Support Vector Machines. The folklore view of this learning machine model is that they find an "optimal" hyperplane as the solution to the learning problem [18]. Feature of SVM is that it minimizes an upper bound of generalization error through maximizing the margin between separating hyperplane and dataset [32]. The loss function that helps maximize the margin is hinge loss [14] its represented in Eq. 5:

$$c(x, y, f(x)) = (1 - y * f(x))_+ \tag{5}$$

Model Optimization. To improve the effectiveness of predictions, we seek to find the best parameters for six models by tuning their hyper-parameters [30]. A 5-fold cross-validation was used [25]. On the other hand, in order to find the right configuration for each algorithm, we applied the GridSearchCV function of the Scikit-learn library with different modifications of the hyper parameters to improve its performance [4].

3.6 Model Evaluation

To assess the capability of the models created and identify relevant patterns, we implemented metrics such as AUC, specificity, precision, f1-score, ROC curve because of they were used in [3,4,13,17], and [23]. Besides, through the Confusion Matrix it was possible to visually assess the effectiveness of the models in classifying cases as attempted femicide.

Precision. It is the number of correct positive results divided by the number of positive results predicted by the classifier [25] in Eq. 6:

$$Precision = \frac{TruePositives}{TruePositives + FalsePositives} \tag{6}$$

Recall. It is the number of correct positive results divided by the number of all relevant samples (all samples that should have been identified as positive) [25] in Eq. 7:

$$Recall = \frac{TruePositives}{TruePositives + FalseNegatives} \tag{7}$$

F1 Score. Allows the balance between precision and recall [25] in Eq. 8:

$$F1Score = \frac{Precision * Recall}{Precision + Recall} \tag{8}$$

ROC Curve. It stands for area under receiver operator characteristics (ROC) curve. It is a graph plotting true positive versus false positive [7] view Fig. 5.

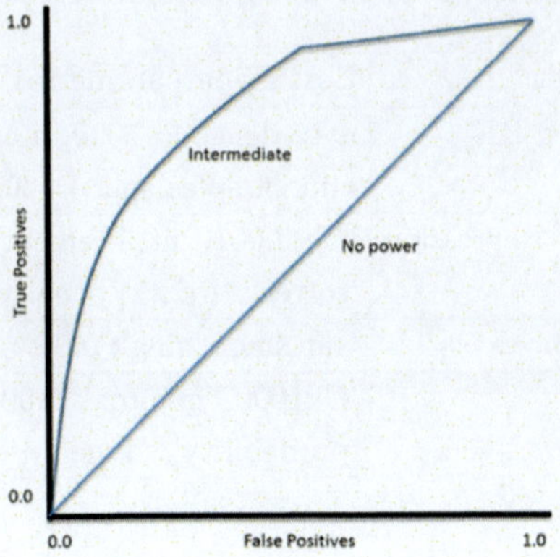

Fig. 5. ROC curve (Source: AISC 811, p. 497)

4 Results

The comparative performance for the classification models is presented in Table 2, which illustrates the results obtained from the different sets of 15, 20, 30 and 35 variables selected by using the method Light GBM technique. In this table, the columns contain the metrics analyzed, and the rows represent the models employed. On the other hand, we observed that the set of 20 variables had the best scores for all development models. The first best model is Logistic Regression, in which Precision value is 0.88, Recall value is 0.87 and F1 score is 0.87. The second best model is Balanced Random Forest, whose Precision value is 0.85, Recall value is 0.85 and F1 score of 0.85. In the same way, for all other models, their scores have a stall point with no meaningful increase or decrease in the evaluated metrics, except for SVM which shows a considerable increase in their Recall score from 0.57 to 0.83 and Precision score from 0.76 to 0.85 with 35 variables.

As we mentioned before the most important class is where the fact of violence was qualified as a case of attempted femicide so we try to minimize the false negatives using Grid Search. Table 3 shows the optimal hyper parameters obtained from the best models with the different set of variables where they had better results.

Table 2. Choice predictive models and best parameters

# Feature	Classiffier	Best Hyperparametros
35	Balanced RF	max_depth = 1000, n_estimators = 300 min_samples_split = 20, criterion = 'gini'
20	Logistic Regression	'C': 0.001, 'fit_intercept': True 'solver': 'lbfgs', 'penalty': 'l2'
15	Naive Bayes	'var_smoothing': 0.0
35	SVM	'C':1000, 'gamma': 0.001 'probability': 'True'

Based on the results of the performance of Logistic Regression model, the confusion matrix showing that and 29,313 (90.60%) attempted femicide and 27,222 (84.13%) Case of non-attempted femicide cases have been classified correctly. Otherwise, around 3041 attempted femicide cases have been classified as non-attempted (False negative). At this point, we can theorize that with 20 features, the LR reaches a stall point because after this set the evaluated metrics will begin to decrease in the evaluated metrics. Finally, the precision score of this positive class was 85%. See Fig. 6.

Table 3. Comparing the Results of Data Mining Models

Feature Selection	15 variables			
Technique	*Classifier*	*Precision*	*Recall*	*F1*
Light GBM	Balanced Random Forest	0.82	0.82	0.82
	Decision Tree	0.66	0.66	0.66
	Logistic Regression	0.87	0.87	0.87
	Naive Bayes	0.86	0.86	0.86
	Random Forest	0.77	0.77	0.77
	SVM	0.75	0.54	0.47
Feature Selection	20 variables			
Technique	*Classifier*	*Precision*	*Recall*	*F1*
Light GBM	Balanced Random Forest	0.85	0.85	0.85
	Decision Tree	0.72	0.74	0.74
	Logistic Regression	0.88	0.87	0.87
	Naive Bayes	0.85	0.84	0.84
	Random Forest	0.82	0.82	0.82
	SVM	0.76	0.57	0.47
Feature Selection	30 variables			
Technique	*Classifier*	*Precision*	*Recall*	*F1*
Light GBM	Balanced Random Forest	0.86	0.86	0.88
	Decision Tree	0.75	0.75	0.75
	Logistic Regression	0.87	0.87	0.87
	Naive Bayes	0.85	0.85	0.85
	Random Forest	0.83	0.83	0.83
	SVM	0.77	0.61	0.54
Feature Selection	35 variables			
Technique	*Classifier*	*Precision*	*Recall*	*F1*
Light GBM	Balanced Random Forest	0.88	0.88	0.88
	Decision Tree	0.71	0.70	0.70
	Logistic Regression	0.88	0.88	0.88
	Naive Bayes	0.85	0.85	0.85
	Random Forest	0.85	0.85	0.85
	SVM	0.85	0.83	0.82

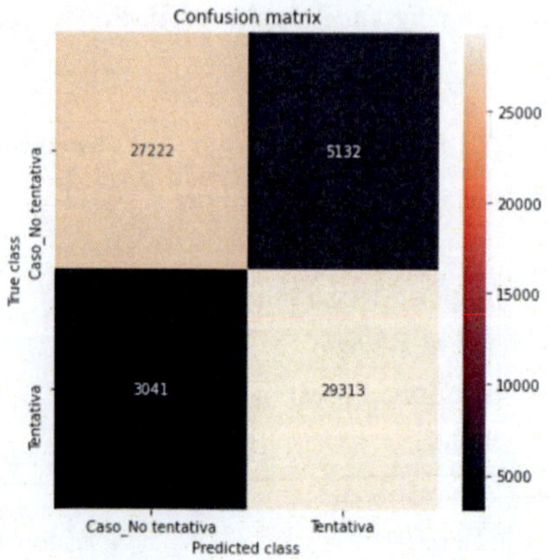

Fig. 6. Confusion Matrix of Logistic Regression Model 20 variables

The second one was Balanced Random Forest with its Confusion matrix showing that 29,970 cases have been classified correctly as an attempted femicide. See Fig. 7. Otherwise, the sensitivity for this model to identify positives was much more effective in performance than the Logistic Regression model. We can perceive that the set of 35 features reached the best parameters on the side of the recall scores, which was 93.0% on the side of Peruvian women who will suffer an aggression that could cause death. We prioritize the best scores of recall since as we know is more serious when a severity case has been ignored, which can lead to the death of the victim.

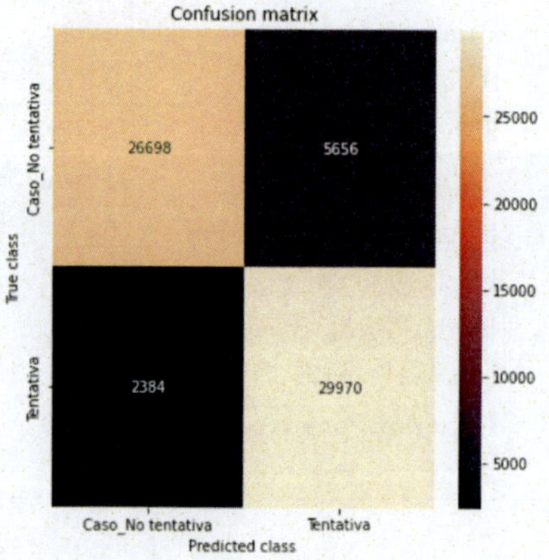

Fig. 7. Confusion Matrix of Balanced Random Forest Model 35 variables

The above Fig. 8 and Fig. 9 show that the AUROC reaches to 0.9172 and 0.94 using each model. These results demonstrated that both models were able to distinguish between a positive case and a negative one.

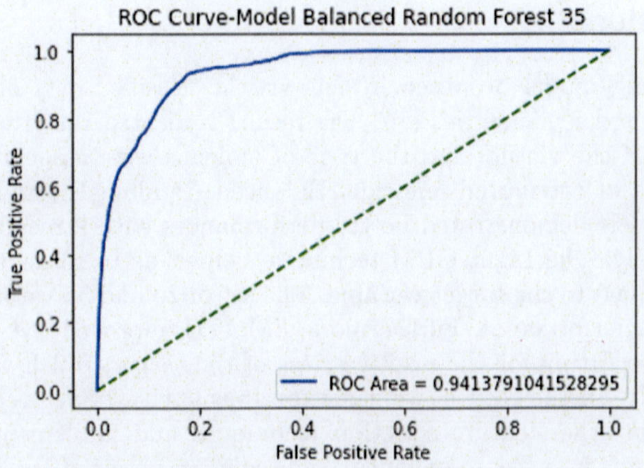

Fig. 8. Curve Roc of Balanced Random Forest Model

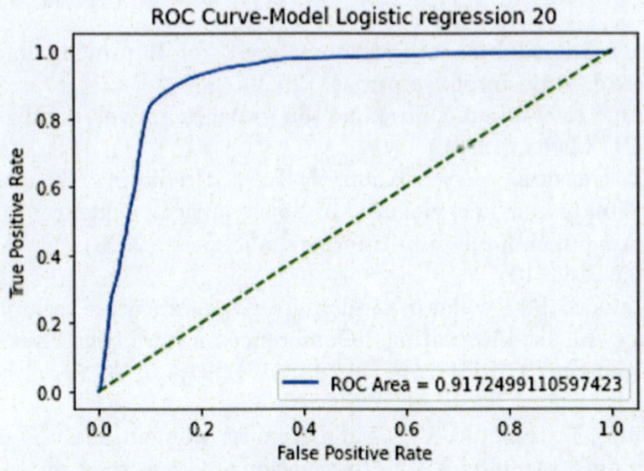

Fig. 9. Curve Roc of Logistic Regression Model

5 Discussion

In this study, it was identified that the most appropriate models to classify an act of violence as a case of attempted femicide are: Balanced Random Forest and Logistic Regression [17], [23] and [37]. On the other hand, through feature selection, variables were identified that influenced the characterization of a case as an attempted femicide or not, when reviewing what was published in the MIMP, they resemble: place of the events, frequency of aggressions, use of weapons, age of the victim and recurrent consumption of alcoholic beverages by the aggressors.

It is recommended to unify the databases of official sources such as the complaints made to the Peruvian National Police, given that the Women's Emergency Center only has a small proportion of the actual magnitude of cases.

6 Conclusions

The data mining model proposed in this article classified acts of violence as cases of attempted femicide in Peru. The results indicated that the ages of the aggressor and of the victim, and the type of violence were associated with the committed case of attempted femicide. Balanced Random Forest and Logistic Regression models demonstrated better performances with a recall of 0.88 and 0.86, respectively. The Light GBM technique helped us to select the variables most closely linked to the target variable. The set of 20 and 35 variables demonstrated a good performance. Furthermore, SMOTE improved the performance and avoided overfitting for the positive class of the target, which were of interest to the study. In the future, we can experiment with newer versions of the dataset, include other feature selection techniques and implement more data mining algorithms in order to enhance the predictive power of the model.

References

1. Agusta, Z.P.: Modified balanced random forest for improving imbalanced data prediction. Int. J. Adv. Intell. Informat. (2019)
2. Amaya, S.: ¿Qué es el feminicidio y qué tan grave es a nivel mundial? Cable News Network (CNN) Mexico (2022)
3. Amusa, L.B., Bengesai, A.V., Khan, H.T.A.: Predicting the vulnerability of women to intimate partner violence in south africa: evidence from tree-based machine learning techniques. J. Interpers Violence (2020). https://doi.org/10.1177/0886260520960110
4. More, A.S., Rana, D.P.: Review of random forest classification techniques to resolve data imbalance. In: 1st International Conference on Intelligent Systems and Information Management (ICISIM), pp. 72–78 (2017). https://doi.org/10.1109/ICISIM.2017.8122151
5. Kaur, B., Ahuja, L., Kumar, V.: Crime Against women: analysis and prediction using data mining techniques. In: International Conference on Machine Learning, Big Data, Cloud and Parallel Computing (COMITCon), pp. 194–196 (2019). https://doi.org/10.1109/COMITCon.2019.8862195
6. Centro de la Mujer Peruana (CMP) Flora Tristan.: La violencia contra la mujer: Feminicidio en el Perú (2005)
7. Chatterjee, S., Das, S., Banerjee, S., Biswas, U.: An approach towards development of a predictive model for female kidnapping in india using R programming. In: International Ethical Hacking Conference (2018). https://doi.org/10.1007/978-981-13-1544-2_40
8. Corzo, S.: The other pandemic. On gender-based violence in the midst of quarantine, Legal Defense Institute (2020)
9. Data preprocessing in data mining. GeeksforGeeks (2019)
10. Díaz, K.A.A., de Almada, G.M.B., González, L.B.D., Rolón, M.M.V., Toledo, G.D.: Perfil de víctimas de violencia de género en pacientes del hospital regional de Alto Paraná, aplicando minería de datos. FPUNE Scientific (2020)
11. Díaz, M.: Violence against women in times of quarantine. La Ley (2020)
12. Dr, V., Anbarasu, P.D.S.: Analysis and prediction of crime against woman using machine learning techniques. Annals Romanian Soc. Cell Biol. **25**(6), 5183–5188 (2021)

13. Fayyad, U., Piatetsky-Shapiro, G., Smyth, P.: From data mining to knowledge discovery in databases. AI Mag. **17**(3), 37 (1996). https://doi.org/10.1609/aimag.v17i3.1230
14. Gandhi, R.: Support Vector Machine - Introduction to Machine Learning Algorithms (2022)
15. Ghosh, D.: Predicting vulnerability of Indian women to domestic violence incidents. Res. Pract. Soc. Sci. **3**(1), 48–72 (2007)
16. Gupta, A., Mohammad, A., Syed, A., Halgamuge, M.N.: A comparative study of classification algorithms using data mining: crime and accidents in Denver City the USA. Education **7**(7) (2016)
17. Guerrero, A., Cárdenas, J. G., Romero, V., Ayma, V.H.: Comparison of classifiers models for prediction of intimate partner violence. In: Arai K., Kapoor S., Bhatia R. (eds.) Proceedings of the Future Technologies Conference (FTC), pp. 469–488. Advances in Intelligent Systems and Computing (2021) https://doi.org/10.1007/978-3-030-63089-8_30
18. Kecman, V.: Support vector machines-an introduction. In: Support Vector Machines: Theory and Applications, pp. 1–47. Springer, Heidelberg (2005). https://doi.org/10.1007/10984697_1
19. Kobyliński, Ł, Przepiórkowski, A.: Definition extraction with balanced random forests. In: Nordström, B., Ranta, A. (eds.) GoTAL 2008. LNCS (LNAI), vol. 5221, pp. 237–247. Springer, Heidelberg (2008). https://doi.org/10.1007/978-3-540-85287-2_23
20. Longadge, R., Dongre, S.: Class imbalance problem in data mining review. arXiv preprint arXiv:1305.1707 (2013)
21. Loinaz, I., Marzabal, I., Andrés-Pueyo, A.: Risk factors of female intimate partner and non-intimate partner homicides. Europ. J. Psychol. Appli. Legal Context **10**, 49–55 (2018). https://doi.org/10.5093/ejpalc2018a
22. Abdulkareem, L.R., Karan, O.: Using ANN to predict gender-based violence in iraq: how AI and data mining technologies revolutionized social networks to make a safer world. In: International Symposium on Multidisciplinary Studies and Innovative Technologies (ISMSIT), pp. 298–302 (2022) https://doi.org/10.1109/ISMSIT56059.2022.9932831
23. Nemias Saboya, A., Sullon, A., Loaiza, O.L.: Predictive model based on machine learning for the detection of physically mistreated women in the peruvian scope. In: Proceedings of the 2019 3rd International Conference on Advances in Artificial Intelligence (ICAAI 2019). Association for Computing Machinery, New York, pp. 18–23 (2022) https://doi.org/10.1145/3369114.3369143
24. Ministerio de La Mujer y Poblaciones Vulnerables (MIMP).: Actualizacion del protocolo Interinstitucional de Accion Frente al Feminicidio, Tentativa de Feminicidio de Frente Accion al Feminicidio, Tentativa de Feminicidio y Violencia de Pareja de Alto Riesgo, Lima (2017)
25. Dahouda, M.K., Joe, I.: A deep-learned embedding technique for categorical features encoding. IEEE Access **9**, 114381–114391 (2021). https://doi.org/10.1109/ACCESS.2021.3104357
26. Mohammad, F., Golsefid, S.: Evaluation of feature selection methods WHITE PAPER (2020)
27. More, A.S., Rana, D.P.: Review of random forest classification techniques to resolve data imbalance. In: 2017 1st International Conference on Intelligent Systems and Information Management (ICISIM) (2017). https://doi.org/https://doi.org/10.1109/icisim.2017.8122151

28. Patel, B., Zala, M.C.: Crime against women analysis & prediction in india using supervised regression. In: First International Conference on Electrical, Electronics, Information and Communication Technologies (ICEEICT), pp. 1–5 (2021) (2022 February)
29. Petering, R., Um, M.Y., Fard, N.A., Tavabi, N., Kumari, R., Gilani, SN.: Artificial intelligence to predict intimate partner violence perpetration. In: Artificial Intelligence and Social Work, vol. 195 (2018)
30. Aouedi, O., Piamrat, K., Parrein, B.: Performance evaluation of feature selection and tree-based algorithms for traffic classification. IEEE International Conference on Communications Workshops (ICC Workshops) (2021) https://doi.org/10.1109/ICCWorkshops50388.2021.9473580
31. Cumbicus-Pineda, O.M., Abad-Eras, T.E., Neyra-Romero, L.A.: Data mining to determine the causes of gender-based violence against women in ecuador. In: IEEE Fifth Ecuador Technical Chapters Meeting (ETCM), pp. 1–6 (2021). https://doi.org/10.1109/ETCM53643.2021.9590664
32. Rejani, Y., Thamarai Selvi, S.: Early detection of breast cancer using SVM classifier technique. ArXiv preprint (2009)
33. Sehra, C.: Decision Trees Explained Easily - Chirag Sehra. Medium (2020)
34. Sharma, A.: Logistic Regression Explained from Scratch (Visually, Mathematically and Programmatically). Medium (2022)
35. Khandelwal, S.K.: CRIME AGAINST WOMEN : CAUSES AND COMPULSIONS.: A Socio-legal Study of Crimes Against Women: A Critical Review of Protective Law, chap. 2 (2015)
36. Suthar, B., Patel, H., Goswami, A.: A survey: classification of imputation methods in data mining. Int. J. Emerg. Technol. Adv. Eng. **2**(1), 309–312 (2012)
37. Tamilarasi, P., Rani, R.U.: Diagnosis of crime rate against women using k-fold cross validation through machine learning. In: International Conference on Computing Methodologies and Communication (ICCMC), pp. 1034–1038. IEEE (2020)
38. Tan, X., et al.: Wireless sensor networks intrusion detection based on smote and the random forest algorithm. Sensors (Basel) (2019)
39. What is logistic regression? - Explanation of the logistic regression model. AWS (2021)

On the Characterization of Digital Industrial Revolution

Ndzimeni Ramugondo(✉), Ernest Ketcha Ngassam, and Shawren Singh

University of South Africa, Florida Campus, 28 Pioneer Ave Florida Park Roodepoort, Johannesburg 1709, South Africa
Ndzimeni.ramugondo@gmail.com, eketcha@gmail.com, singhs@unisa.ac.za

Abstract. The world is, amid rapid technological disruptions, characterized by the adoption and usage of advanced digital technologies now forming part of our daily activities. These technology disruptions forms part of the Digital Industrial Revolution which marks the fourth phase of the Industrial Revolution. The Industrial Revolution refers to the fast-paced changes in the manufacturing, business, technology, or engineering enterprises that were or are enabled by technology and capable of driving societal and economic changes. To date, the world has witnessed three phases of Industrial Revolution and now in the middle of the fourth phase. In this paper, we explore the literature on the evolution of Industrial Revolution as well as the characterization of each phase. The characterization form part of the foundation towards the development of a comprehensive body of knowledge on the Digital Industrial Revolution as well as its evolution over the years in tandem with Industrial Revolution. The outcome of the body of knowledge is to establish a foundation for a broader study on the socio-economic dynamics of Digital Industrial Revolution and how such dynamics can be exploited for socio-economic development in Africa.

Keywords: Industrial Revolution · Digital Industrial Revolution · Digital Technologies · Socio-economy

1 Introduction

The digital disruptions currently underway as part of the Digital Industrial Revolution (DIR) captures an all-encompassing reality of today's progressive usage of digital technologies in our society that are driving changes in the expectations and behaviors of both consumers and businesses [1]. Technology was a central element of advancement with prior phases of the Industrial Revolution (IR), and remain central for the DIR with its distinct feature being the rapid rate at which technologies are changing and impacting virtually all sectors and industries. IR is defined as a "period of technological change with a high impact on society" [2]. To date, the world has witnessed three (3) phases of IR and now in the middle of the fourth phase with phases widely known as 1IR, 2IR, 3IR and 4IR. Each phase of IR has been identified by its dominant technologies with the digital technologies becoming the dominant technology as from the 3IR to

R. Silhavy and P. Silhavy (Eds.): CSOC 2023, LNNS 724, pp. 773–782, 2023.
https://doi.org/10.1007/978-3-031-35314-7_64

date. These technologies are driving the DIR and are advancing from the technology achievements of the 3IR through addition of intelligence capability to enable them to perform functions that are known to be performed by humans with hyper-connectivity comprised of machines, sensors, intelligence with its scope much wider, ranging from gene sequencing to nanotechnology and renewables to quantum computing [3].

This paper seeks to present a comprehensive body of knowledge on DIR by unpacking its evolution over the years and characterization in tandem with IR phases. DIR is mostly presented as a distinct phase or a continuation of 3IR with no direct linkages to technologies from the 1IR and 2IR. Although these technologies emerged as analog and mechanical, the development of the digital versions during the 3IR marked the beginning of a digital revolution. The adoption and usage of new technologies represents a distinctive characteristic of IR which remains the same characteristic for DIR with efficient integration and interoperability across value chain and socio-economic elements [4]. The high adoption rate of digital technologies was witnessed with the response to Covid-19 pandemic which propelled high number of organizations to leapfrog their technology adoption to deliver digital enabled products and services while complying with lockdown regulations.

The development of the body of knowledge is critical for the establishment of a foundation towards a broader study on the socio-economic dynamics of DIR and how such dynamics can be exploited for socio-economic development in Africa. This is critical noting that the developed countries managed to attain the developmental (socio-economic) benefits from the technologies adopted prior and during the prior IR phases with the developing countries having missed the opportunity. With rapid speed and impact, a comprehensive understanding of DIR, its evolution over the period of IR and associated characteristics is critical to ensure successful adoption of digital technologies. Such an understanding is critical to enable countries who missed the developmental benefits of technology in the prior IR phases to leapfrog and adopt usage of digital technologies towards inclusive socio-economic and sustainable development. This paper is comprised of the following sections: Sect. 2 defines the Digital Industrial Revolution, followed by phases and characteristics of the Industrial Revolution in Sect. 3; Sect. 4 outline the characterisation of DIR and Sect. 5, covering the conclusion of the paper together with envisage future research.

2 Defining Digital Industrial Revolution

IR resulted in development of major technological innovations that were used to simplify production processes such as the flying shuttle, spinning jenny, loom and others which changed the complexion of industries while signalling the beginning of an era of industrialization. The digital technologies have long emerged as part of the IR, and it is unfortunate that such technologies did not feature amongst the main technologies identified as part of the 1IR and 2IR and only became dominant during the 3IR. The developments in the communication and computation sector included the invention of telegraph, phonograph, facsimile, telephone, and computation machines which all occurs during the 1IR and 2IR setting a strong foundation for digital technologies. Although these technologies were analog and mechanical, the emergence of their digital versions marked the beginning of the digital revolution. As described by [5], the distinct features of DIR to lie in its impact on systems with speed that has no historical precedent compared to the previous revolutions while evolving at an exponential manner rather than linear. Given the broad nature of DIR, the following concepts and their definition are essential to simplify DIR understanding:

- **Computation** – The early versions of computers were designed to simplify complex calculation problems with the genesis of digital computers being the computing machines that were programable and capable of performing different computing tasks, mainly used for military technical and scientific computations [6].
- **Digital** – digital refers to an outlook that aims to leverage technology, data, and ways of working to establish new business and service models for efficient operations that delivers value [7]. It is about electronic technology that generates, stores, and processes data in terms of two state binary code. The two states being on and off symbolized by 1 and 0.
- **Digital Technology** – digital technology on the other hand, presents a combination of information, computing, communication, and connectivity technologies that are fundamental to the transformation of business strategies, business processes, capabilities, products and services [8].
- **Digitization** – digitization refers to the process of digitally enabling analog or physical artifacts for the purpose of implementing the artifacts into business processes with the aim of acquiring newly formed knowledge while creating new value for the stakeholders [9].
- **Digitalization** – digitalization is defined by [9] as "a fundamental changes made to business operations and business models based on newly acquired knowledge gained through value-added digitization initiatives". Therefore, digitalisation is about the implementation and usage of digital technologies to simplify business processes while creating value.
- **Digital Transformation** – digital transformation is defined as a business or organisational change that is enabled by digital technologies with a view of implementing essential innovative changes or improvements within an organisation to create value for its stakeholders through leveraging available resources and capabilities [10]

Based on the foregoing and for the purpose of this paper, we defined Digital Industrial Revolution (DIR) as "a continuous implementation of digital transformation initiatives enabled by digital technologies amplifying for fast-paced organizational, societal and economic transformation across value chain". Core to digital transformation is the usage digital technologies to enable the transformation of business processes, capabilities, products, and services while creating value.

3 Industrial Revolution (IR) Phases and Characteristics

To date, we have witnessed three phases of IR and currently in the fourth phase characterised by fast-paced changes enabled by digital technologies. The digital technologies that are now driving the current phase resulted from the maturity and advancement of some technologies that emerged along the prior IR phases.

3.1 First Industrial and Digital Revolution

The invention of steam engine, mechanization of simple tasks and construction of railroads are some of the developments that triggered the beginning of First Industrial Revolution (1IR) reported to have taken place between 1760 and 1840 [11]. The 1IR is regarded as one of the important advancements in humanity through technology developments of the time [12]. The beneficiaries of the 1IR are countries that industrialize earlier and took full advantage of the available technologies and most of them are amongst developed countries. Until the beginning of the 1IR, most of the operations were manual and executed by trained craftsmen with goods transported on horseback before invention of steam trains [13] and there were no efficient communication systems until the invention of telegraph and related communication systems.

Certainly, the emergence of 1IR advancements simplified production processes, transportation including communication, and computation. It is unfortunate that the 1IR has been broadly defined as the era of mechanisation and steam power with less reference to early versions of communication and computation systems. There are several communication and computation related technologies that can be traced to the period of the 1IR which amongst them includes the Jacquard's loom, Telegraph, Fax, Typewriter(s) and computation machines. The computation machines invented during the 1IR is the Difference and Analytical Engines. Collectively, these technologies are critical as they laid a strong foundation for the current digital technologies and constitutes the first phase of digital industrial revolution (1DIR).

3.2 Second Industrial and Digital Revolution

The usage of technologies powered by electricity with the advent of mass production and clear division of labour signaled the beginning of 2IR [12]. The 2IR is reported to be characterized by mass production, advent of electricity and assembly line production system. The phenomenal issue about technology developments of the 2IR lies on its profound impact on the improvement of living standards with developments such as electricity, internal combustion engine, telephones, radio's, indoor plumbing including petroleum and chemicals [14]. The same period also witnessed improvements and new inventions in communication and computing technologies constituting the second phase of DIR (2DIR).

The technologies that constitute 2DIR include telephones, phonograph, radio frequency, television, computation, and data processing machines. The different computation and data processing machines developed during the period of 2DIR includes the Tabulating Machine, Differential Analyzer, Turing Machines, Atanasoff-Berry Computer (ABC) and Mark I. The ABC is regarded as the first electronic digital computer with Mark I regarded as the first programmable digital computer. The invention of these two computers during the 2DIR was a turning point towards the advancement of digital revolution.

3.3 Third Industrial and Digital Revolution

Some of the major advancements that occurred during the period of the 3IR includes the development of nuclear power and wide usage of electronics [15]. During this phase, the wide usage of electronics and computers became mainstream with digital technologies becoming dominant a technology. During this period, major technology developments included the birth of internet through the Advanced Research Projects Agency Network (ARPANET) in 1969 [16], semiconductors which were central for the development of a transistor in 1947 [13], mainframe computing around the 1960s, personal computing around the 1970's and 80's including the emergence of wide usage of internet in the 1990's.

The above cumulatively describes the 3DIR innovations which enabled the adoption of electronics including the Information and Communication Technology (ICT) in the production environments [12]. There were several technology developments that emerged during this period in which [17] categorized them into three (3) technology waves as described in Table 1 below:

Table 1. Technology Waves

Technology Wave	Technology Description
First Wave Technologies	Computers, Broadband, Mobile and Telecommunications
Second Wave Technologies	Internet, Social Networks and Cloud Computing
Third Wave Technologies	Internet of Things (IoT), Artificial Intelligence, Big Data, Robotics, Machine Learning, etc.

3.4 Fourth Industrial and Digital Revolution

We are in the middle of the fourth phase popularly known as the Fourth Industrial Revolution (4IR) or DIR in the context of this paper. This phase is based on the expansion of ICT based digital technologies of 3IR which set a strong foundation for the current phase. The DIR is reported to be building on top of the technology achievements of 3IR with addition of intelligence capability to enable them to perform functions that are known to be performed by humans with hyper-connectivity comprised of machines, sensors, intelligence with its scope much wider, ranging from gene sequencing to nanotechnology and renewables to quantum computing [3]. While these technologies are based on the expansion and or maturity of 3IR technologies, there is an opposite view that indicates that there is no fourth phase but a continuation of 3IR as there is no clear evidence of major new innovations other than evolution of the 3IR technologies [18]. The speed at which these technologies are transforming virtually all sectors which is at much higher level than 3IR, such is one factor that qualifies for this phase to be categorized as another phase and collectively referred as the DIR.

The distinct features of DIR to lie in its impact on systems with speed that has no historical precedent compared to the previous revolutions while evolving at an exponential manner rather than linear [5]. This phase is distinguished by the availability of multiple digital technologies which are fusing the physical, digital, and biological (CPS) impacting virtually all sectors and industries while contributing to socio-economic development. The digital technologies that are driving this phase includes amongst others, the Artificial Intelligence (AI), Internet of Things (IoT), Big Data Analytics, Blockchain and Cloud Computing. These technologies are fundamental towards the development of intelligent digital technology capability comprised of devices, machines, production modules and products with the capability to independently exchange information, trigger actions, and control each other, enabling an intelligent manufacturing or production environment [19].

4 Characterization of DIR

The technological advancements experienced with each of the prior IR phases as described above were significant and contributed greatly to socio-economic development while setting a foundation for the current and future technology developments. The DIR is characterized by the application of new and diverse digital technologies that has a broad impact on social, economic, and cultural change. The new digital technologies are digitally transforming virtually all sectors and dubbed to be "disruptive". The digital disruption captures an all-encompassing impact of today's fast-advancing digital technologies while driving changes in the expectations and behaviours of both consumers and businesses [1]. Such disruptions are setting a new technology paradigm while rendering old technologies obsolete [20].

The description of DIR across various literature does not provide the development, evolution and maturity of the technologies which are now at the centre of DIR from the prior IR phases and mostly, the 1IR and 2IR. The DIR technologies gets presented as technologies that possibly first emerged during the 3IR with no direct references or linkages to the technologies from first two IR phases. Although the DIR technologies were not dominant during the 1IR and 2IR, it's important to note that the communication and computation technologies emerged parallel with other dominant technologies across IR phases. As described in Sect. 3 above, parallel to each IR phase, there is a corresponding DIR phase with technologies now evolved to the current DIR technologies. Such technologies first emerged as analog and mechanical and now advanced to digital technologies that became mainstream during the 3IR. Although they are diverse in nature, the digital technologies have now advanced and converged with one technology capable of delivering multiple capabilities. Table 2 below presents the DIR phases and their associated technologies in tandem with IR phases.

There are several technologies referred as digital technologies and categorized into two main categories (Core and Facilitating technologies) by [4] as indicated in Table 3 below. The core technologies refer to the modern and iconic digital technologies that have become commercially available within the last decade with complexity and integrability while the Facilitating technologies, refer to a wide variety of mature and sometimes called legacy information and operations technologies that allow the core technologies to integrate, operate, and function properly [4].

While the 3DIR phase provided the digital technologies and associated environment, the DIR enables for the transformation of the digital environment through digital transformation or digitalization using digital technologies. The above-mentioned technologies have demonstrated their diffusive nature as they are easily adopted and adaptable for wide usage across different sectors of the economy. The diffusive nature represents a

Table 2. IR and DIR Phases

First Phase		**Second Phase**		**Third Phase**	
1IR	**1DIR**	**2IR**	**2DIR**	**3IR**	**3DIR**
• Steam engine • Mechanization. • Flying shuttle • Spinning jenny • Loom	• Jacquard's loom • Telegraph. • Fax. • Typewriter(s). • Computation machines – Difference and – Analytical Engines.	• Electricity. • Assembly line • Mass production. • Indoor Plumbing • Combustion Engines • Petroleum. • Chemicals	• Telephones. • Phonograph. • Radio Frequency. • Television. • Computation and Data Processing Machine. – Tabulating Machine. – Differential Analyzer. – Turing Machines. – Atanasoff-Berry Computer (ABC) and – Mark I.	• Nuclear Power. • Electronics.	• **First Wave Technologies.** – Computers. – Broadband. – Mobile. – Telecommunications. • **Second Wave Technologies.** – Internet. – Social Networks. – Cloud Computing. • **Third Wave Technologies.** – Internet of Things (IoT). – Artificial Intelligence. – Big Data. – Robotics. – Machine Learning.

Table 3. Core and Facilitating Technologies

Core Technologies	**Facilitating Technologies**
• Artificial Intelligence (AI).	• Industrial Actuator and Sensor.
• Cloud Computing.	• Machine and Process Controller.
• Robotics.	• Automated Guided Vehicles.
• Data Analytics.	• Intelligent Enterprise Resource Planning.
• Virtual Reality (VR) and Augmented Reality (AR).	• Communication Interface.
• Digital Twin Technology.	• High-Performance Computing.
• Big Data.	• Smart Wearable Gadget.
• Simulation.	• Predictive Analytics.
• Additive Manufacturing.	• Smart Manufacturing Execution System.
• Cybersecurity.	• Industrial Embedded System.
• Internet of Things (IoT).	• Computer Numerical Control System.
• Semantic Technologies.	
• Blockchain.	
• Three-Dimensional (3D) and Four-Dimensional (4D) Printing.	
• Nanotechnology.	

Source:[4]

distinctive characteristic of IR which remain a critical characteristic of DIR to enable digital transformation across value chain. Based on reviewed literature, we defined Digital Transformation (DT) as a general process of profound improvements in society, business and industries enabled by digital technology for efficient execution of business processes.

Digital transformation processes provide improvements by making use of digital technologies (Core and Facilitating) through an improvement process called digitization or digitalization. Digitization is defined as "a business model driven by the changes associated with the application of digital technology in all aspects of human society" [21]. Without no doubt, the impact of digital transformation is now central to business and people's daily lives as propelled by adoption of technology in response to Covid-19 and its associated lockdown regulations. The response to Covid-19 pandemic enabled majority of organizations to leapfrog their technology adoption and digitally transform their organizations to enable them to deliver products and services while complying to lockdown regulations.

To enable an efficient digital transformation across organisations value chains, it is essential that the systems are highly integrated and interoperable. A value chain represents a way of conceptualising the actions needed in the production of a product or service [22]. Simply, it's about the account of all business activities involved towards the production of a product or service from design to completion. The value chain model is comprised of two main activities which include primary and support with technology being part of the support activities [23]. Technology form part of the core deliverable within the support activities enabling all activities within the value chain resulting in an integrated value chain. An integrated value chain provides visibility across all activities from product/service design, production, marketing, delivery channels, support and throughout the product lifespan. As contained in Table 4 below, [24] described three (3) types of integration that are essential to ensure an integrated value chain.

Table 4. Integration Types

Integration Types	Description
Horizontal Integration	This refers to the integration of several IT systems, processes, resources, and information flows within an organization and between other organizations [19]
Vertical Integration	It refers to the integration of elements through the departments and hierarchical levels of an organization, from product development to manufacturing, logistics and sales [19]
End-to-end Integration	End-to-end Integration enables for product customization and reducing operational costs using CPS to digitally integrate the whole value chain [19]

5 Conclusion and Future Work

The characterization of DIR is essential towards formation of the basis for the development of a comprehensive body of knowledge on DIR, its evolution over the years and characteristics in tandem with IR phases. This body of knowledge is required to establish a foundation for a broader study on the socio-economic dynamics of Digital Industrial Revolution and how such dynamics can be exploited for socio-economic development in Africa. The main characteristic noted with prior IR phases relates to the developmental role of new technologies and their contributions towards socio-economic development and that remain a critical characteristic for DIR.

Given the strategic nature of DIR with high potential to transform society at large, DIR should not be just another phase characterized by the usage of technologies. DIR should be a phase characterized by adoption of digital technologies that empower human society for their development and their usage for socio-economic development of the majority. The developed countries managed to achieve the developmental benefits associated with technologies adopted during the prior IR phases with the developing countries missing the opportunity. For the developing countries to leapfrog from the missed opportunities, a comprehensive understanding of DIR and its developmental benefits is critical to ensure development and upliftment of societies towards eradication of poverty, unemployment, and inequality.

References

1. Kim, B.: Digital Disruption and the Fourth Industrial Revolution (2020)
2. Klingenberg, C.O., Borges, M.A.V., do Vale Antunes, J.A.: Industry 4.0: What makes it a revolution? A historical framework to understand the phenomenon (2022). https://doi.org/10.1016/j.techsoc.2022.102009
3. Schwab, K.: The Fourth Industrial Revolution. World Economic Forum (2016)
4. Ghobakhloo, M., Fathi, M., Iranmanesh, M., Maroufkhani, P., Morales, M.E.: Industry 4.0 ten years on: a bibliometric and systematic review of concepts, sustaina-bility value drivers, and success determinants (2021). https://doi.org/10.1016/j.jclepro.2021.127052
5. Caruso, L.: Digital innovation and the fourth industrial revolution: epochal social changes? AI Soc. **33**(3), 379–392 (2017). https://doi.org/10.1007/s00146-017-0736-1
6. van den Ende, J., Kemp, R.: Technological transformations in history: how the computer regime grew out of existing computing regimes. Res. Policy **28** 833–851 (1999)
7. Tardieu, H., Daly, D., Esteban-Lauzán, J., Hall, J., Miller, G.: Deliberately Digi-tal. Springer, Rewriting Enterprise DNA for Enduring Success (2020). https://doi.org/10.1007/978-3-030-37955-1
8. Bharadwaj, A., El Sawy, O.A., Pavlou, P.A., Venkatraman, N.: Digital business strategy: towards a next generation of insights. MIS Quart. **37**(2), 471–482 (2013)
9. Schallmo, D.R.A., Williams, C.A.: Digital Transformation Now! Guiding the Successful Digitalization of Your Business Model. Springer Briefs in Business (2018). https://doi.org/10.1007/978-3-319-72844-5
10. Gong, C., Ribiere, V.: Developing a unified definition of digital transformation. Technovation **102**(2021), 102217 (2020). https://doi.org/10.1016/j.technovation.2020.102217
11. United Nations Industrial Development Organisation (UNIDO): A revolution in the making? Challenges and opportunities of digital production technologies for devel-oping countries (2019)

12. Liao, Y., Loures, E.R., Deschamps, F., Brezinski, G., Venancio, A.: The impact of the fourth industrial revolution: a cross-country/region comparison. Pontificia Universidafe Catolica do Parsna (2018)
13. Marwala, T.: Closing the Gap. The Fourth Industrial Revolution in Africa (2020)
14. Atkeson, A., Kehoe, P.J.: Modeling the transition to a new economy: lessons from two technological revolutions. Am. Econom. Rev. **97**(1), 64–88 (2001)
15. Kayembe, C., Nel, D.: Challenges and opportunities for education in the fourth industrial revolution. Afr. J. Public Aff. **11**(3) 79–94 (2019)
16. Schifter, C.C.: Infusing Technology into the Classroom: Continuous Practice Improvement. Temple University, USA (2008)
17. Katz, R.L.: Social and Economic Impact of Digital Transformation on the Economy. ITU GSR-17 Discussion paper (2017)
18. Moll, I.: The myth of the Fourth Industrial Revolution: Implications for teacher education'. In: F. Maringe (ed.), Higher Education in the melting pot: Emerging dis-courses of the Fourth Industrial Revolution and decolonisation (Disruptions in higher education: Impact and implication, Vol. 1, pp. 91–110, AOSIS, Cape Town (2021). https://doi.org/10.4102/aosis.2021.BK305.06
19. Pereira, A.C., Romero, F.: A review of the meanings and the implications of the industry 4.0 concept. Procedia Manufacturing **13** 1206–1214 (2017). https://doi.org/10.1016/j.promfg.2017.09.032
20. Romig, A.D., et al.: An introduction to nanotechnology policy: opportunities and constraints for emerging and established economies. Technol. Forecast. Soc. Chang. **74**(2007), 1634–1642 (2007). https://doi.org/10.1016/j.techfore.2007.04.003
21. Henriette, E., Feki, M., Boughzala, I.: The Shape of Digital Transformation: a Systematic Literature Review (2015)
22. Ensign, P.C.: Value Chain Analysis and Competitive Advantage. J. Gen. Manage. **27**(1), 18–42 (2001)
23. Recklies, D.: The Value Chain (2001)
24. Tay, S.I., Lee, T.C., Hamid, N.A.A., Ahmad, A.N.A.: An overview of industry 4.0: definition, components, and government initiatives. J. Adv. Res. Dyn. Contr. Syst. **10**(14), 1379–1387 (2018)

User-Centred Design of Machine Learning Based Internet of Medical Things (IoMT) Adaptive User Authentication Using Wearables and Smartphones

Prudence M. Mavhemwa[1(✉)], Marco Zennaro[2], Philibert Nsengiyumva[3], and Frederic Nzanywayingoma[4]

[1] African Centre of Excellence in Internet of Things, University of Rwanda, Kigali, Rwanda
pmavhemwa@gmail.com
[2] Science, Technology, and Innovation Unit, ICTP, Trieste, Italy
mzennaro@ictp.it
[3] Department of Electrical and Electronic Engineering, University of Rwanda, Kigali, Rwanda
[4] Department of Information Systems, University of Rwanda, Kigali, Rwanda
f.nzanywayingoma@ur.ac.rw

Abstract. As the world grapples with an increase in diseases including COVID-19, the Internet of Medical Things (IoMT) emerges as a complementary technology to the healthcare staff, which is constantly overburdened. Untrained users' increased online presence exposes them to cyberattack threats. Authentication is the first line of defense for protecting medical data, but existing solutions do not consider the user's context and capabilities, making them unusable for some groups of users who eventually shun them. This paper proposes a Machine Learning based adaptive user authentication framework that adapts to user profiles and context during login to determine the likelihood of the attempt being illegitimate before assigning appropriate authentication mechanisms. The proposed edge-centric framework fuses the Naive Bayes classifier and CoFRA model to determine the risk associated with a login attempt based on biometric wearable sensor data, non-biometric smartphone sensor data, and some predefined data. User backgrounds and preferences were solicited, and results showed that users despite their ICTSkills, ages, jobs, and years of experience prefer to use simple physiological biometrics for authentication. An Android App was then developed using the User-Centred design and installed on a smartphone which communicated with a PineTime smartwatch. Sensor data was used as input in calculating the risk associated with an access request to decide whether to authenticate, step up authentication, or block a request using rule and role-based access control techniques while also non-intrusively monitoring health. Once implemented, the framework is expected to improve user experience in authentication promoting the use of IoT in healthcare.

Keywords: IoMT · Healthcare · Usable-Security · Authentication · Machine Learning · Edge Computing · Smartphone

R. Silhavy and P. Silhavy (Eds.): CSOC 2023, LNNS 724, pp. 783–799, 2023.
https://doi.org/10.1007/978-3-031-35314-7_65

1 Introduction

Constant reliance on the caregiver can become frustrating at some point during a person's illness, but terminal illnesses necessitate constant medical care and monitoring. At the same time, caregivers cannot be present all of the time to monitor patients, imposing the need for self-help [1]. Internet of Things (IoT) is being used in pervasive healthcare to complement healthcare workers and caregivers world over, with International Data Corporation (IDC) estimating that more than 70% of healthcare providers in the United States are already using it [2]. IoT plays a significant role in combating infectious diseases and transforming the entire healthcare sector [3], with benefits and applications detailed in [4]. The aged population, prevalence of chronic diseases, shortage of healthcare specialists, rising medical care costs, the COVID-19 pandemic, among other factors, have contributed to the expansion of Internet of Medical Things (IoMT) as a pervasive healthcare enabler.

However, its introduction comes with several challenges compounded by the fact that the same security used to protect general IoT devices cannot secure IoMT environments because the threat landscape and malicious motives in them differ greatly [5]. Various security mechanisms have been proposed to prevent attacks on IoTs, with authentication and authorisation being the primary security concerns [6]. However, the commonly used authentication mechanisms are frequently rigid and do not consider user profiles, making them difficult to use. It is therefore critical to make it difficult for a suspicious user while making it simple for a trusted user. Usability is frequently treated as an afterthought, resulting in security solutions that are not adopted by end users, hence the need for a proper study into the balance of usability versus security on user authentication [7]. Also, a user is the weakest link who may fail to follow the rigor of security mechanisms, and as observed by Steger [8], internal misconduct accounts for a greater proportion of incidents in healthcare, making it the only industry where insiders inflict more cyber-harm.

By proposing a smartphone-based authentication solution that adapts to user profiles and context when determining which authentication mechanism to assign to them in an IoT environment, this research seeks to contribute to usable security and complement previous proposals. A plethora of research on IoT user authentication has been conducted in various parts of the world, but little has been done in relation to what may be appropriate for a specific group of users who are generally treated the same when they are different [9]. This paper will also describe a case study that will help to shape the design of an adaptive authentication prototype.

1.1 Related Work

As mentioned earlier, research is ongoing, but when it comes to adaptive authentication, the research output is not being practical enough to benefit end users. We will now look at adaptive authentication work that has been carried out by other researchers and our analysis is not limited to the medical environment. Forget et al. [10] proposed a choose your own authentication architecture (CYOA) in which users select a scheme from available alternatives based on their preferences, abilities, and usage context. The approach, however, is not dynamic thereby introducing some delay. In [11] a scheme for adjusting a smartphone's lock between voice recognition, face scan, and fingerprint based on which will be currently usable was proposed. The work focused solely on usability, leaving out security and context. Hintze et al. [12] used location-based risk assessment in conjunction with multi-modal biometrics to adjust the level of authentication required to the situational risk of unauthorized access. However, relying solely on GSM cell IDs and Wi-Fi access point MAC addresses may be difficult if third-party information is withheld. The researchers in [13] assumed that the best authentication method for wearables and nearables is the owner's biometric information. However, not all biometrics may be applicable to all users as user context affects usability, with for example, a user with body tremors may find it difficult to use fingerprint, the same with a person in a noisy environment who may be unable to use voice. In a resource-constrained environment, [14] proposed a method for continuously monitoring and analyzing user and device activities and selecting authentication or reauthentication based on the risk involved using the Naive Bayes Machine Learning algorithm. This kind of authentication minimizes attacks that occur after login but may not work well for inactive users. In [15] researchers created a risk engine that examines a user's previous login records and generates a pattern using Machine Learning to calculate the user's risk level. This approach relied heavily on previous logins and only used recall mechanisms, which hampered usability for some elderly users and dementia patients. Vhaduri and Poellabauer [16] described an implicit wearable device user authentication mechanism that used a combination of three types of biometrics: behavioral (step counts), physiological (heart rate), and hybrid (calorie burn and metabolic equivalent of task). The work was not adaptive, and the experiment only used expensive iPhone and Fitbit. He et al. [17] conducted a home IoT authentication usability study and discovered that smartphones are the most used devices to access IoT devices in the home and can closely meet user expectations. The findings concur with those of Forget et al. [10] and Batool et al. [18] who evaluated smartphone sensors and concluded that sensors embedded in smartphones and wearables enable the collection of a user's specific dataset at very low financial and computational cost. It is, however, worth noting that most previous studies were carried out in a controlled laboratory setting. They have not examined:

- investigating suspicious activity in the user authentication profile.
- feature selection that allows for precise and effective modelling of user behavior.
- the ability of authentication systems to adapt to changes in user behavior.

1.2 Contribution

In this paper, we attempt to address the above shortcomings through presenting an adaptive IoMT user authentication framework for determining authentication based on context, user profiles and available authenticators among other factors described in Sect. 2. This framework combines edge computing, machine learning, role-based access control, and rule-based access control guided by the CoFRA [19] modelling framework. We intend to develop, as part of the proposed framework, i) a contextual model that identifies a user using smartphone and wearable sensors and their pre-defined characteristics, ii) a Naive Bayes classifier that uses the contextual model and fuzzy logic to calculate the risk associated with a user's login attempt based on his/her role, and iii) rule-based access control to assign appropriate authenticators to a user based on their context risk score, usability, and availability of authenticators. We intend to use a smartphone that most users already own, as well as a low-cost PineTime smartwatch.

The paper is structured as follows: The requirements for adaptive authentication are highlighted in Sect. 2. Section 3 introduces the research method and proposed architecture, while Sect. 4 presents a case study. Section 5 discusses research findings. Section 6 concludes and gives future work.

2 Requirements for Adaptive Authentication

Most existing user authentication mechanisms are rigid and do not consider the user's capabilities, highlighting the need to classify users and tailor authentication to their needs, particularly in healthcare. Multi-factor authentication (MFA) which was once considered entirely secure in organizations with a relatively small number of users, became ineffective in instances where the risk is relatively high [20]. This necessitated the use of adaptive authentication, the requirements for which are outlined below.

Usability. This is defined in [21], and if a user correctly employs a technology, he or she will not pose a security risk. User interfaces must be usable and correspond to the user's mental model of security mechanisms and even if a system is technically secure, it is the end user who operates it and determines its usability [22].

Security. Security is also defined in [21], and because the same security controls used to protect general IoT devices cannot secure IoMT environments, a lightweight security solution must be implemented. With an unprecedented increase in online activities, there is a significant increase in security breaches, within healthcare [23].

Cost. One of the barriers to full adoption of IoMT in Sub-Saharan Africa (SSA) is cost [24]. To ensure a cost-effective solution, we intend to use the sensors of an Android smartphone and a smartwatch for context establishment and authentication.

Portability. A PineTime smartwatch will be used to supplement smartphone sensors, which will use a Low energy Bluetooth module for offloading to the smartphone. It is small enough and compatible with most android smartphones.

Low Energy Consumption. The proposed scheme's devices will run on rechargeable batteries. The smartphone will function in its normal way, and the smartwatch will be charged via USB. To save energy, wearable sensors should operate in duty cycle mode.

Low Latency. The edge-centric architecture is an appealing alternative to other models such as the cloud-centric because the collected data is first preprocessed and analyzed on the edge, allowing for time-sensitive authentication decisions to be made within the local network, reducing time consumed and congestion between the gateway and the cloud [25].

Deployability. Deployability plays a significant role in strengthening the security of IoMT. Some authentication mechanisms may offer significant protection but may not be deployable in some situations. For instance, compatibility of a mechanism with a device, in our case, on smartphones electrocardiograms may offer unique features, but cannot be deployed on smartphones that already have some built-in sensors.

3 Research Methodology

The User Centered Design (UCD) [9] methodology was used in this study which is an iterative process with four stages described below.

3.1 Specifying Usage Context

Adaptive authentication is required for all age groups, including the elderly, who may be digitally illiterate [26] and may have underlying medical conditions that will likely influence the choice of authenticators to be used.

3.2 Specify Requirements

Questionnaires were distributed to patients and medical staff to gather information about their backgrounds, skills, and needs for inclusion in the prototype design.

3.3 Producing Design Solutions

The proposed adaptive authentication scheme is classified as one of the four IoT layers shown at the bottom of Fig. 1 and described below.

Perception Layer. This layer includes biometric and non-biometric sensors that are located at the owner's wearable and smartphone.

Network Layer. This layer is made up of the edge and fog layers. The edge layer is made up of a network of sensors found in wearables and smartphones. In this case, the smartphone, which serves both senses and processes, serves as a gateway to process data before sending it to the cloud. Low energy Bluetooth will ensure communication between the smartwatch and the smartphone.

Cloud Layer. This layer performs analytics and additional processing to assist in decision-making when necessary. Google cloud services are used to store and extract geolocations and maps.

Application Layer. This layer is critical for users who require a pleasant user experience. Users will be authenticated using an app installed on their smartphone. Android Studio and Java will be used to create the app with a XAMP 3.2.4 the backend.

The proposed adaptive authentication model is based on a modified version of the Monitor Adapt Plan Execute - Knowledge with Human Machine Teaming (MAPE-K HMT) framework modified from [27]. Figure 1 depicts the detailed framework and its relationship to the IoMT architecture.

The framework was broken down to gain a better understanding of the interacting entities from the beginning to the end and Fig. 2 shows the simplified graphical context model of the framework. It was also during this phase that several risk calculating frameworks were evaluated before settling for fusing the Naïve Bayes and the CoFRA model.

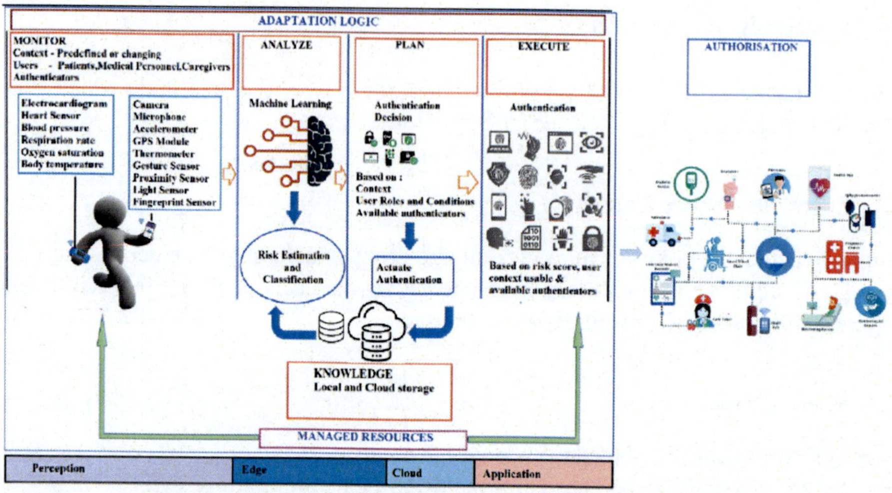

Fig. 1. Proposed framework using the MAPE-KHMT model

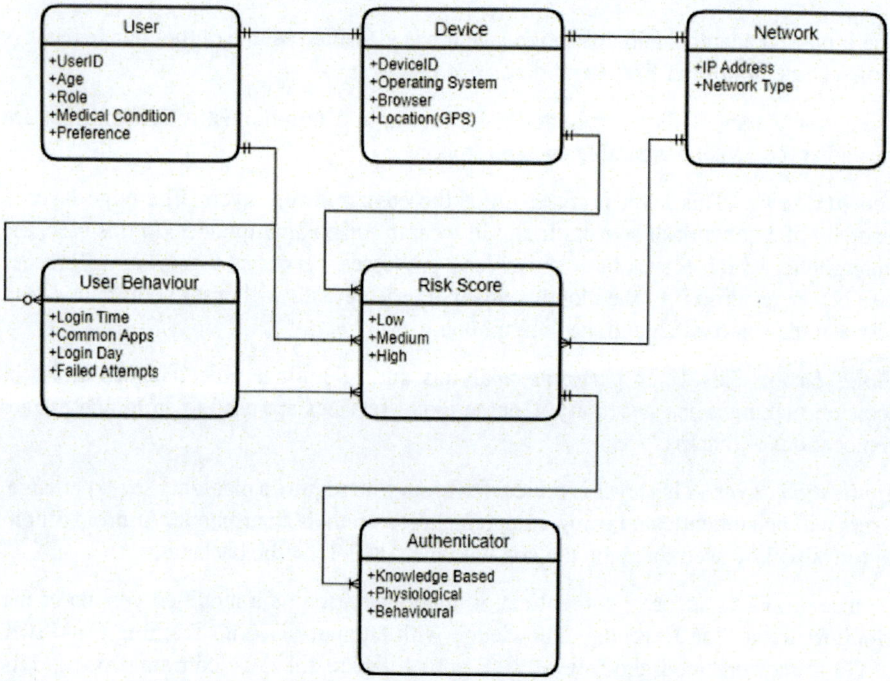

Fig. 2. Context Model showing relationships between entities

Figure 3 depicts the detailed implementation schema in which the smartwatch will communicate with the smartphone and medical devices.

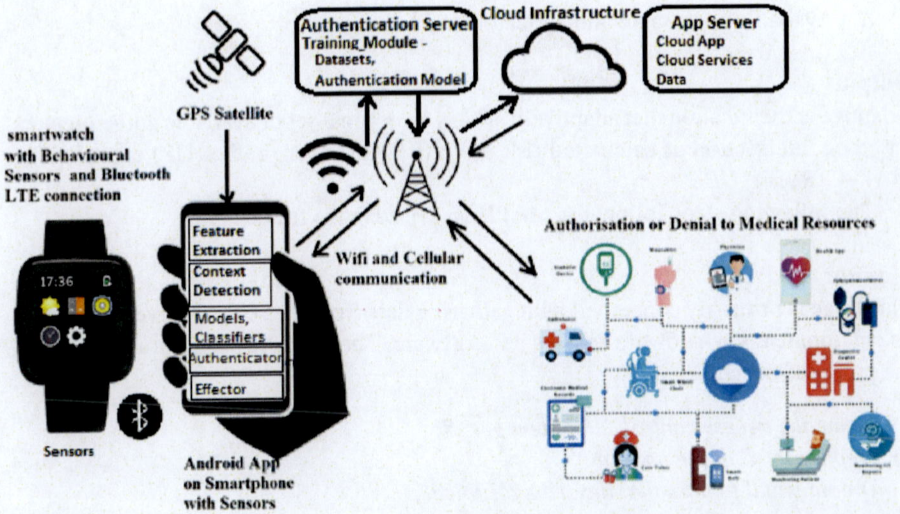

Fig. 3. Detailed Implementation Architecture

Because of the sensors shown in Fig. 4, the smartphone and smartwatch were chosen for user authentication.

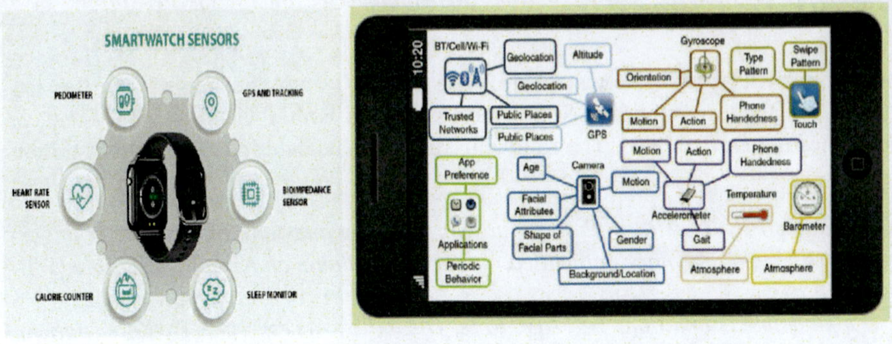

Fig. 4. Candidate sensors for authentication (Source [28, 29])

The smartphone facilitates communication between IoMT device ensuring protocol compatibility and interoperability.

Working Mechanism of Adaptive Authentication. During user authentication, an assessment of context (both fixed and variable factors) is performed to estimate the risk of the request, resulting in risk calculation for user classification. The proposed algorithm consists of the following:

Input

C – Contextual factors $\{c_1, c_2, c_3 \ldots c_n\}$ describing a user.

A – Available and usable authenticators for a user profile.

Output
Adaptive authentication that adaptively selects an optimal set of available authenticators based on intersection of calculated risk score (RS) and user profile (UP) expressed as $(AA) \rightarrow (RS) \cap (UP)$.

The following steps follow the MAPE-K_{HMT} adaptive framework.

Monitor
This stage is primarily concerned with gathering data from the self-adaptive system and its environment via hardware sensors and software. The following steps are followed in the proposed scheme.

a) *Define the access control subsystem* $s_i \in S$;
b) *Prepare set C of the possible*
   i) *contextual factors that describe a subject*
      $\forall s_i \in S: C = \{c_1, c_2, \ldots c_n\}$,
   ii. *usable authenticators available*
      $A \in A'$ *(all authenticators).*
c) *Train a mathematical model for decision-making, factoring in context, user, available authenticators, and risk score.*
   P→<C, U, A, RS> where P is user profile.
d) *Generate an adaptive authentication algorithm based on the created profile of a subject.*

Initial User and Device Registration. The user *(U)* must first register on the app installed on their smartphone (SP), which serves as both a Gateway *(GW)* and an Authenticating Device *(AD)*. The following steps are taken during the user registration phase:

1. *U chooses PIN, Password, Pattern, answer to secret question, Inputs his/her fingerprint (FP) and any other authentication factor available on AD and Wearable (WB) for storage.*
2. *U's predefined conditions i.e., Age, Role, Disability, Access-Time(T), Role (patient, nurse, doctor) are defined and captured.*
3. *U's known location(L) from Google Maps, the device fingerprint (DFP) is captured, and mapping occurs* $U \rightarrow SP$, $U \rightarrow L$, $U \rightarrow T$.
4. *Mappings are saved in the local server which is the SP memory.*

If a wearable (WB) is used, the user is mapped to it $U \rightarrow WB$.

5. Wearable (WB) is mapped to the smartphone (SP) WB → SP.

SP on receiving the registration request from WB, does the following:
Checks $U \rightarrow WB$, *if they are not related, the registration is terminated,*

otherwise. The next stage is followed.

6. Relationships are saved in the SP memory.

Adapt
Based on the monitored and expected state, environment, and any other related constraints, this phase determines whether adaptation is required.

Login Stage. Following successful registration, an authorized user *(U)* can gain access to the desired medical IoT via the adaptive authentication phase. To start the authentication process, *(U)* must open the app on their smartphone *(SP)*. The opening will trigger communication between SP, WB, and other related services. The following steps are followed during the user authentication process.

1. *First, User U uses the smartphone to open the application.*
2. *The application checks the pre-conditions associated with the user which are:*

   $(U \rightarrow SP)$ *AND* $(U \rightarrow WB)$ *AND* $(U \rightarrow L)$ *AND* $(U \rightarrow T)$.
   Risk Calculation

3. *Naïve Bayes classifier is then invoked to calculate the risk and classify user based on the degree of match of predefined conditions and contextual factors. The fuzzy score is between 0 and 1.*

   The calculation is as shown in equation (1) below.

$$P(y|x_1, \ldots\ldots\ldots x_n) = \frac{P(x_1|y)P(x_2|y)\ldots P(x_n|y)P(Y)}{P(x_1)P(x_2)\ldots\ldots.P(x_n)} \quad (1)$$

4. *The Fuzzy Score is mapped to predefined classes. Scores are shown in Table 1.*

Table 1. Risk Score classes.

Normal	Low	Medium	High
0 – 0.2	0.2 - 0.4	0.4 – 0.8	0.8 – 0.9

Plan
Rule-based authentication is planned, considering the user context, capabilities stored in the database, and available authenticators. The planned self-adaptation includes choosing

appropriate authenticators or even denying access guided by the CoFRA [19] framework shown in Fig. 2.

Execute
This is the stage at which the previously generated plan is carried out, in this case authentication using the appropriate, usable, or available authenticator and denial if attempt is deemed fraudulent as defined in [19].

5. Rule Based Authentication

While Count<=3
If U class is Normal, use normal single authenticator provided the user can use it.
Else (Low)
Use Multifactor authentication provided the user can use it.
Else (Medium)
Use Multifactor authentication AND Multifactor authentication provided the user can use it.
Else (High)
Use Multifactor authentication AND Multifactor Authentication AND Multifactor Authentication
Else
Re-authenticate.
Count ++
Do Terminate Authentication, Block User, Report

Authorisation
If the user meets all the security requirements, authorization occurs and access to a variety of medical resources required based on his or her role is granted.

Knowledge Base
The Knowledge base stores user and runtime models, as well as contextual models.

Human Machine Teaming
Human Machine Teaming refers to the interaction of the devices with the user.

4 Case Study: Adaptive User Authentication

Smartphones, smartwatches, and fitness apps are major data collectors. Figure 5 to 7 depicts IoT adaptive authentication user characteristics for our case study. As shown in Fig. 3, an Android mobile app was created to enable communication between a user's smartphone and a PineTime smartwatch. The user can be a patient, or a member of the medical staff and the patient may be in a hospital or at home. We expect our case study to help us refine our adaptive authentication framework for improved usability and security as adaptive security. A total of eighty-five (85) patients and twenty-four (24) health workers completed the pre-deployment questionnaires whose results are presented below.

5 Results and Discussion

Questionnaires gathered data that informed the design of the prototype. Furthermore, the findings were compared to previous findings and were incorporated into the Android app that was created. When compared to patients, there was an equal distribution of males and females among medical staff. Women are 26% less likely than men to use mobile internet services in low and middle-income countries [30], indicating that the gender divide has an impact on IoT adoption. This may be true for patients, but not for medical staff, who may be forced to use technology at work. Figure 5 shows analysis by age groups on both patients and medical staff.

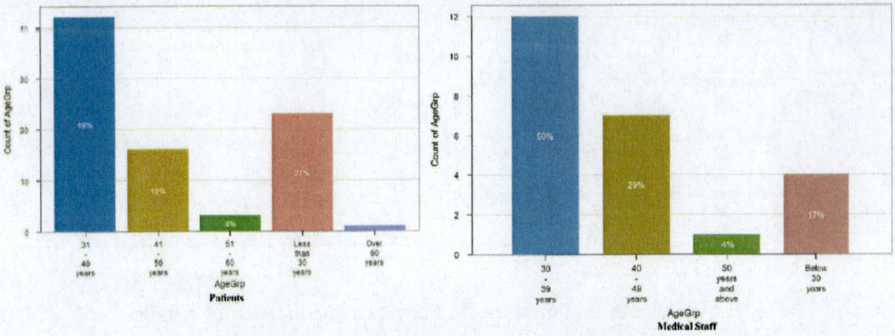

Fig. 5. Analysis of responses by age group

The 30–39 age group dominated the sample, with a smaller proportion representing the older population. This may influence the study's outcome.

Medical Conditions Medical conditions can affect an authentication candidate's cognitive, physical, and visual abilities. Only three of the elderly patients admitted to having some medical conditions in the survey. The low response rate is due in part to the small number of elderly patients among the respondents.

A question was asked to see if respondents understood the significance of protecting their medical data. Figure 6 depicts the responses of both patients and medical personnel. In terms of patient respondents, 34% of the female population and 14% of the male population were unaware of the importance of medical data security. 66% of the female population and 86% of the male population were aware of the importance of medical data security. Although there are differences in the reasons for protecting medical data, all members of the Medical Staff agreed on the importance of securing medical data.

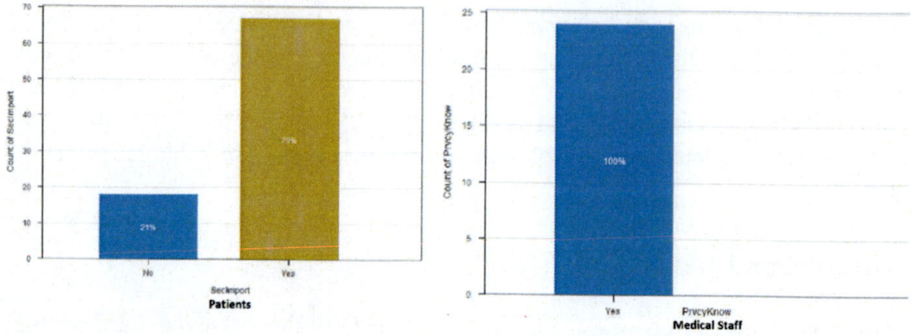

Fig. 6. Analysis of knowledge of the importance of security

An analysis of preferred smartphone authentication by job title and gender was performed. As previously stated, gender influences the ease of use of technology.

Figure 7 shows an analysis of preferred authentication on medical devices by gender on both patients and medical personnel.

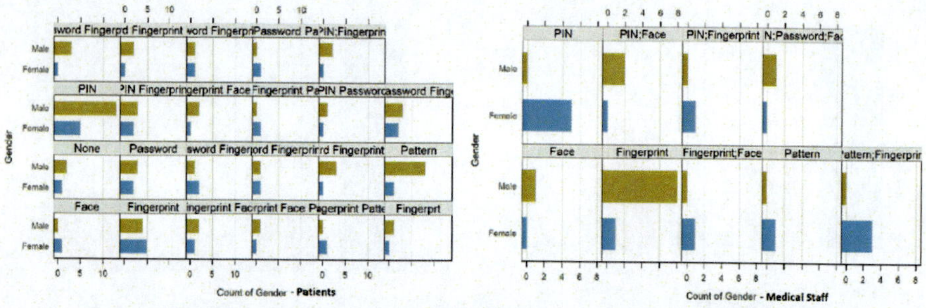

Fig. 7. Analysis of preferred medical device authentication by gender

There was no statistically significant difference between smartphone and medical device authentication among respondents.

An analysis of preferred authenticators by age group was made. PIN and fingerprint authentication were the most popular single factor authentication methods among both patients and medical staff, followed by password and none. Even when medical staff were analyzed by job, PIN and Fingerprint dominated, followed by face and multi factor combinations. Although it has technical disadvantages, PIN is preferred because it provides quick authentication [31].

According to Grindrod et al. [32], fingerprint recognition, despite being popular and widely used, has a low success rate. Although the elderly population was small, the results confirmed [31, 33] who identified reasons for the password's unpopularity among elderly users. The combination of PIN and fingerprint also dominated two-factor authentication. On multi-factor authentication, the combination of PIN, Fingerprint, and Password was dominant. This shows that despite the maturity of authentication, users still prefer the basic authentications.

All the collected data was analyzed and used to create an Android prototype, which is depicted in Fig. 8 below. Figure 9 depicts the prototype interface together with the smartwatch that was used.

Fig. 8. Code snippet of proposed prototype

Sample data collected in the laboratory environment when the prototype was tested and produced promising results is as shown in Fig. 10 below.

Verification time was done by comparing with the formula in Helkana and Snekkenes [34] and results were in reasonable range. Upon implementation we intend to do extensive evaluation of the prototype and use the usable security metrics to test its conformity to the acceptable units defined in [35]. Machine Learning Metrics will also be used for evaluation which include False Positive, False Negative, True Positive and True Negative. Energy consumption, time and space complexity will also be compared with other existing solutions.

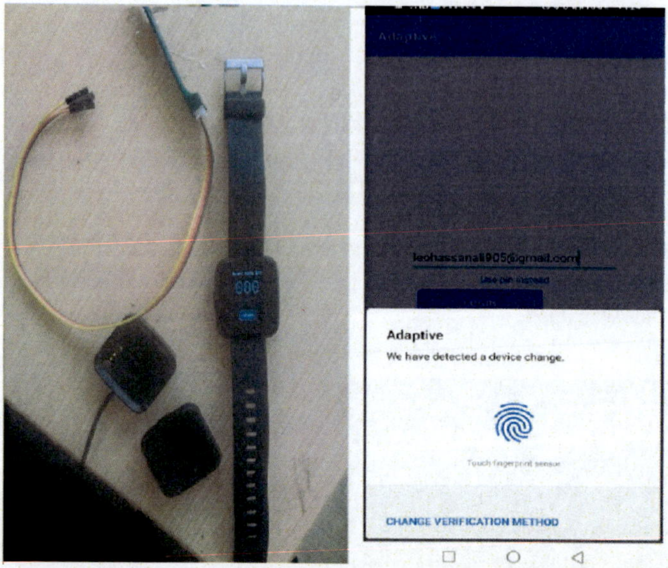

Fig. 9. Screenshots of smartwatch and mobile app

ID	DevicePIN	Latitude	Longitude	Network Type	NetworkID	Age	HasMental	HasVisual	HasPhysical	TimeTaken
1	$2y$10$h73YUkIlhFogem	-17.2898	31.3474	wifi	"Software Hub"	26	1	0	1	24.7
2	$2y$10$PRIIGmI8goABWI	-17.2898	31.3474	wifi	"Software Hub"	42	0	1	0	35.8
3	$2y$10$ZXFPynpwzvnBXS	-17.2898	31.3474	wifi	"Software Hub"	22	0	0	0	25.7
4	$2y$10$tO53xeRnYo/P3S	-17.2898	31.3476	wifi	"Software Hub"	28	0	1	0	35.8
5	$2y$10$P6kqPnsq6MHm(	-17.2901	31.3479	wifi	"Software Hub"	22	0	1	0	22.5
6	$2y$10$2myqMJx4ypdvF!	-17.2898	31.3476	wifi	"Software Hub"	26	0	1	0	24.8
7	$2y$10$zXy6hf.FcjjR323Z/	-17.2898	31.3475	wifi	"Software Hub"	26	0	0	0	17.2
8	$2y$10$pVDDOIrBfWfjhY	-17.2898	31.3477	wifi	"Software Hub"	27	0	1	0	23

Fig. 10. Sample data collected from a laboratory test

6 Conclusion and Future Work

In this paper we presented an adaptive risk based IoMT user authentication framework that uses edge computing to collect user context information and biomedical data from a wearable and a smartphone to determine the risk associated with a user's login attempt and accord the appropriate authenticators based on the user's role, services required and context. The combined use of edge computing, machine learning, role and rule-based access control is expected to create a robust adaptive user authentication system for IoMT at low cost. The authentication scheme offers several advantages, including improved security, usability, leveraging biomedical signals from wearables for health monitoring and security, high accuracy, the ability to integrate with other smart applications, and lower costs. The aim of this work is to contribute to improving security to patients and medical staff when they access various medical resources using bio data which will also monitor their health. It is expected that the ease of use will contribute to increased adherence thereby contributing to the health and well-being as enshrined in SDG number three.

Our future work will involve further development and testing the prototype of the model on young and old users in the healthcare environment and analysing it using different machine learning algorithms to see the one that offers best performance on resources.

Acknowledgements. This work was jointly supported by the African Center of Excellence in Internet of Things (ACEIoT) from College of Science and Technology, University of Rwanda, and The Regional Innovation Scholarship Fund (RSIF).

References

1. Shaikh, A.A., Gupta, N.S., Din, A., Khan, M., Artist, H.T.: Android and Internet of Things (IOT) based alzheimer care/rehabilitation system to moniter. Int. J. Innov. Res. Comput. Commun. Eng. **5**(3), 5531–5539 (2017). https://doi.org/10.15680/IJIRCCE.2017
2. Lindgren, T.: Using IoT to Fight Covid-19. *Unissu* (2020). https://www.eetasia.com/using-iot-to-fight-covid-19-pandemic/. Accessed 31 Mar 2021
3. Rahman, M.S., Peeri, N.C., Shrestha, N., Zaki, R., Haque, U., Hamid, S.H.A.: Defending against the Novel Coronavirus (COVID-19) outbreak: how can the Internet of Things (IoT) help to save the world? Heal. Policy Technol. **9**(2), 136–138 (2020). https://doi.org/10.1016/j.hlpt.2020.04.005
4. Zakaria, H., Azaliah, N., Bakar, A., Hassan, N.H., Yaacob, S.: ScienceDirect IoT security risk management model for secured practice in IoT security risk management model for secured practice in healthcare environment healthcare environment. Procedia Comput. Sci. **161**, 1241–1248 (2019). https://doi.org/10.1016/j.procs.2019.11.238
5. Hazratifard, M., Gebali, F., Mamun, M.: Using machine learning for dynamic authentication in telehealth: a tutorial. Sensors **22**(7655), 1–20 (2022)
6. Fang, H., Qi, A., Wang, X.: Fast authentication and progressive authorization in large-scale IoT: how to leverage AI for security enhancement. IEEE Netw. **34**(3), 24–29 (2020). https://doi.org/10.1109/MNET.011.1900276
7. Hayashi, E., Hong, J., Das, S., Amini, S., Oakley, I.: CASA : context - aware scalable authentication. In: Symposium on Usable Privacy and Security (SOUPS) 2013, 24–26 July 2013, Newcastle, UK, pp. 1–10 (2013)
8. Steger, A.: What Makes IoMT Devices So Difficult to Secure Against Cyberthreats. Health Magazine (2020). https://healthtechmagazine.net/article/2020/02/what-makes-iomt-devices-so-difficult-secure-perfcon. Accessed 25 Mar 2021
9. Santana-Mancilla, P.C., Anido-Rifón, L.E., Contreras-Castillo, J., Buenrostro-Mariscal, R.: Heuristic evaluation of an IoMT system for remote health monitoring in senior care. Int. J. Environ. Res. Public Health **17**(5), 1586 (2020). https://doi.org/10.3390/ijerph17051586
10. Forget, A., Chiasson, S., Biddle, R.: Choose Your Own Authentication (2015)
11. Wójtowicz, A., Joachimiak, K.: Model for adaptable context-based biometric authentication for mobile devices. Pers. Ubiquit. Comput. **20**(2), 195–207 (2016). https://doi.org/10.1007/s00779-016-0905-0
12. Hintze, D., Scholz, S., Koch, E., Mayrhofer, R.: Location-based risk assessment for mobile authentication. In: UbiComp 2016 Adjunct – Proceedings of 2016 ACM International Joint Conference on Pervasive Ubiquitous Computing. no. September 2016, pp. 85–88 (2016). https://doi.org/10.1145/2968219.2971448
13. Kumar, M.Y.T., Braeken, A., Liyanage, M.: Identity privacy preserving biometric based authentication scheme for Naked healthcare environment (2017)

14. Gebrie, M.T., Abie, H.: Risk-Based Adaptive Authentication for Internet of Things in Smart Home eHealth (2017). https://doi.org/10.1145/3129790.3129801
15. Mohammed Misbahuddin, B.D., Bhindumadhava, B.S.: Design of a Risk Based Authentication System using Machine Learning Techniques. IEEE (2017)
16. Vhaduri, S., Poellabauer, C.: Biometric-Based Wearable User Authentication During Sedentary and Non-sedentary Periods, pp. 1–4 (2018). http://arxiv.org/abs/1811.07060
17. He, W., et al.: Rethinking access control and authentication for the Home Internet of Things (IoT). In: Proceedings of the 27th USENIX Security Symposium, pp. 255–272 (2018)
18. Batool, S., Saqib, N.A., Khattack, M.K., Hassan, A.: Identification of remote IoT users using sensor data analytics. In: Arai, K., Bhatia, R. (eds.) FICC 2019. LNNS, vol. 69, pp. 328–337. Springer, Cham (2020). https://doi.org/10.1007/978-3-030-12388-8_24
19. Bumiller, A., Barais, O., Challita, S., Combemale, B., Aillery, N., Le Lan, G.: A context-driven modelling framework for dynamic authentication decisions. In: 48th Euromicro Conference on Software Engineering and Advanced Applications (SEAA), pp. 398–405 (2022). https://doi.org/10.1109/seaa56994.2022.00069
20. Jeyanthi, N., Thandeeswaran, R.: IGI Global, Security Breaches and Threat Prevention in the Internet of Things. vol. i, no. February (2017)
21. Gordieiev, O., Kharchenko, V., Vereshchak, K.: Usable Security Versus Secure Usability: an Assessment of Attributes Interaction (2017)
22. Perera, C., Ranjan, R., Wang, L., Khan, S.U., Zomaya, A.Y.: Big data privacy in the Internet of Things era. IT Prof. **17**(3), 32–39 (2015). https://doi.org/10.1109/MITP.2015.34
23. Itai Greenberg. Fifth-generation cyberattacks are here. How can the IT industry adapt? World Economic Forum (2021). https://www.weforum.org/agenda/2021/02/fifth-generation-cyberattacks/. Accessed 30 May 2022
24. Chavula, J., Phokeer, A., Feamster, N.: Insight Into Africa‘s Country-level Latencies. IEEE Africon, pp. 938–944 (2017)
25. Nizeyimana, E.: Design of a decentralized and predictive real- time framework for air pollution spikes monitoring. In: IEEE6 th International Conference on Cloud Computing and big Data Analytics, pp. 8–11 (2021)
26. Macrotrends. Sub-Saharan Africa Literacy Rate 1985–2021|MacroTrends (2021). https://www.macrotrends.net/countries/SSF/sub-saharan-africa-/literacy-rate. Accessed 12 Apr 2021
27. Cleland-huang, J., Vierhauser, M., Murphy, M.: Extending MAPE-K to support human-machine teaming. In: 17th International Symposium on Software Engineering for Adaptive and Self-Managing Systems (SEAMS '22), May 18â•fi23, 2022, PITTSBURGH, PA, USA. vol. 1, no. 1 (2022). https://doi.org/10.1145/3524844.3528054
28. Digital, T.: What are Smartwatch Sensors and How do they function? (2021). https://www.taggdigital.com/blog/what-are-smartwatch-sensors-and-how-do-they-function. Accessed 12 Mar 2023
29. Mahbub, U., Sarkar, S., Patel, V.M., Chellappa, R.: Active user authentication for smartphones: a challenge data set and benchmark results. In: 2016 IEEE 8th International Conference on Biometrics Theory, Applications and Systems (BTAS), pp. 1–8 (2016). https://doi.org/10.1109/BTAS.2016.7791155
30. GSMA Association. State of Mobile Internet Connectivity 2018. GSM Assoc., pp. 1–64 (2018)
31. Ehatisham-ul-Haq, M., et al.: Authentication of smartphone users based on activity recognition and mobile sensing. Sensors (Switzerland). **17**(9), 2043 (2017). https://doi.org/10.3390/s17092043
32. Grindrod, K., et al.: Evaluating authentication options for mobile health applications in younger and older adults. Appl. younger older adults. PLoS ONE **13**(1), e0189048 (2018). https://doi.org/10.1371/journal.pone.0189048

33. Amroun, H., Ammi, M.: Who used my smart object ? A flexible approach for the recognition of users. IEEE **3536**, 1–12. (2017). https://doi.org/10.1109/ACCESS.2017.2776098
34. Helkala, K., Snekkenes, E.: A method for ranking authentication products. In: Proceedings of the Second International Symposium on Human Aspects of Information Security & Assurance (HAISA 2008), pp 81–93 (2008)
35. Hausawi, Y.M., Allen, W.H.: Usable-security evaluation. In: Tryfonas, T., Askoxylakis, I. (eds.) HAS 2015. LNCS, vol. 9190, pp. 335–346. Springer, Cham (2015). https://doi.org/10.1007/978-3-319-20376-8_30

Automatic Searching the Neural Network Models for Time Series Classification of Small Spacecraft's Telemetry Data with Genetic Algorithms

Vadim Yu. Skobtsov[1] and Aliaksandr Stasiuk[2](✉)

[1] United Institute of Informatics Problems of the National Academy of Sciences of Belarus, Minsk 220012, Belarus
[2] EXER LABS, Denver, CO 80204, USA
alexander.stasiuk.ht@gmail.com

Abstract. The paper presents research results of using genetic algorithms (GA) for solutions to the actual problem of intelligent analysis of telemetry data from small spacecrafts in order to detect its technical states. Hybrid neural network models based on modern deep learning architectures have been researched and automatically generated by developed GA-based search method to solve the problem of multiclass classification of telemetry data, which make it possible to determine the normal and abnormal states of the spacecraft's operation. For computer analysis, we used telemetry data from onboard systems of small spacecraft (OS SS) of the AIST group of the Samara National Research University named after Academician S. P. Korolev. Neural network models were automatically built to solve the telemetry data classification problem, which slightly exceeded the classification accuracy of previously obtained hybrid neural network models based on a combination of three types of layers: 1-dimensional convolutional, recurrent and fully connected (98.8%). Also, simpler models without recurrent layers were obtained with an accuracy of 98.5%, which greatly simplifies the NN architecture and the possibilities of its use. Thus, obtained results can be considered as a confirmation of the found close to optimal solution for the certain experiment parameters.

Keywords: Genetic Algorithm · Deep Learning · Neural network architecture search. · Spacecraft's telemetry data

1 Introduction

Space telemetry is a set of technologies that allows remote collection of data about the state of onboard systems (OS) of small spacecraft (SS). Currently, with rare exceptions, telemetry is completely transferred from OS SS to the ground control complex, where telemetry data is monitored and analyzed. Timely detection and prediction of abnormal functioning of spacecraft systems allows

R. Silhavy and P. Silhavy (Eds.): CSOC 2023, LNNS 724, pp. 800–811, 2023.
https://doi.org/10.1007/978-3-031-35314-7_66

in some cases to correct their state and thereby prevent the development of an emergency situation on board, which increases the reliability and operating time of the SS. Therefore, one of the most important tasks at all stages of the life cycle of SS is the analysis of their telemetry data (TD) in terms of determining their technical state to ensure safe operation and correct control.

A large amount of data, arriving from SS and accumulating in specialized databanks, can be effectively used to improve the process of determining the technical state of the small spacecraft and its systems. The functioning data of the small spacecraft, including telemetric data, are heterogeneous irregular multidimensional data.

The volume of telemetry coming from spacecraft, due to their continuous complication, is continuously growing. The contradiction between the growing volume of telemetry data and the limited bandwidth of its transmission can be resolved new OS SS platforms if most of the telemetry analysis is performed on the spacecraft, sending only that part of it to ground control complex that the spacecraft is not able to process. In other words, instead of sending detailed and complete telemetry to Earth, it is enough to periodically send a short digest of the spacecraft technical state, and only in case of detecting an emergency situation - detailed information sufficient to analyze and predict an emergency situation. On the other hands, for existing SS platforms is important to develop effective telemetry data analysis models and tools on the ground control complex side.

Methods of machine learning, artificial intelligence and bioinspired models are currently one of the most promising and widely used approaches in data analysis of high-tech complex systems. Therefor the development of intelligent data analysis and machine learning methods and models for OS SS telemetry data is very actual problem for both cases.

The SS telemetry data are set of univariate time series that can be presented as a multivariate time series after appropriate preprocessing. In our paper, we consider the task of determining the technical state of the OS SS as a problem of multivariate time series classification. In the past two decades, time series data analysis has been considered as one of the most difficult problems in the area of intelligent data analysis and machine learning [1,2]. With the increasing availability of temporal data [3], hundreds of algorithms have been proposed since 2015 [3]. In fact, any classification problem using data that is recorded taking into account some notion of ordering can be viewed as a time series classification problem [5]. Time series are found in many real applications: data processing of electronic medical records, recognition of human activity, classification of acoustic scenes, cybersecurity, SS functioning states analysis according to TD [6–11].

Recent publications have been focused on the development of ensemble methods [12–15]. This approach stimulated the development of an ensemble of 35 COTE classifiers (Collective Of Transformation-based Ensembles) [12] and improving of COTE model, the HIVE-COTE method [13,14] - a hierarchical voting system. HIVE-COTE is currently considered as the leading time series classification algorithm among classical machine learning models when evaluating 85 datasets from the UCR/UEA archive [4]. To achieve high accuracy,

HIVE-COTE becomes extremely computationally demanding and impractical for solving real problems of intelligent analysis of big data [4]. This approach requires the training of thirty seven classifiers, as well as cross validation of each hyperparameter of these algorithms, which makes it impossible to train this approach in some situations. To emphasize this impossibility, note that one of these thirty-seven classifiers is the Shapelet transformation [16], the time complexity of which is $O(n^2 * l^4)$, where n is the number of one-dimensional time series in the dataset, and l - the length of the time series.

Having analyzed the current state-of-the-art of classical non deep classifier models, we have established the impracticality of advanced approaches in a number of cases of solving the real big data analysis problems for classifying the time series. Therefore, let us focus further on models of neural network models, which have been widely used in recent years to solve various problems of big data mining [18]. This motivated their use for solving the problems of time series analysis [5,8,19]. Earlier, we developed and researched methods for determining the state of the OS SS functioning based on unsupervised learning cluster analysis approaches. Methods of neural network clustering based on the construction of the Kohonen map and image processing methods, and streaming dynamic clustering were proposed [10,11,20].

The continuation of these results was the development and research of hybrid deep neural network models for solving the problem of binary classification of telemetry data from the Belarusian spacecraft subsystem, published in [21]. As a elaboration of the results obtained for the previous case, there was developed and researched the hybrid deep neural network model for solving the problem of multiclass classification of telemetry data from OE SS of the AIST group of the Samara National Research University named after Academician S.P. Korolev [22]. Obtained solutions showed a fairly high test classification accuracy at the level of 96–98%. The resulting neural network models were superior to the well-known deep neural networks involved in the comparative analysis, both in terms of the accuracy of the results and in terms of training and validation time.

However, it is well known that the process of developing, training, testing and comparative analysis of neural network models is very expensive both in terms of time and computing resources [5–8,18,19,21,22]. The development and research cycle can take months and involve hundreds of computational experiments. The resulting neural network solutions are often based on heuristic approaches, numerous trials and errors. Therefore, in this article, the authors present the results of solving the problem of automatic search of neural network models for the classification of time series of OS SS telemetry data using such a widespread bioinspired optimization and search method as genetic algorithm [23–25].

2 Statement of the Problem

We consider input telemetry data as time series, which can be represented as a matrix $X = (x_{ij})$ where $i - th$ row X_i is the analyzed vector of telemetry indicators at the $i-th$ moment of time, the j index corresponds to the $j-th$ indicator of the telemetry at the $i-th$ vector X_i.

Obviously that considering TD time series is M-dimensional time series $X = (X_1, X_2, ..., X_M)$, each element of which X_j is a column of the TD matrix X and at the same time a univariate time series describing the behavior of the $j-th$ telemetry indicator on an interval of discrete times $[1, T]$.

For each vector of telemetry indicators at the $i-th$ moment of time X, a label of the class $y_i \in Y$ is assigned, which characterises the SS functioning state analysed basing on the SS telemetry data or its subsystems.

We consider the case of K-class classification, since the final goal is to determine whether the analysed vector X_i of the M-dimensional time series X belongs to the failure-free, failure or fault state. In this case, the number of classes $K = 3$ and, therefore, $Y \in \{0, 1, 2\}$, where 0 denotes a failure-free state and 1, 2 - a failure states accordingly of the OS SS under analysis.

Based on the problem of the K-class classification of the M-dimensional time series as the base one, the paper investigates the problem of automatic search for neural network models as a classification problem solver. The purpose of automatic search for classification neural network models is to generate models of reasonably high classification accuracy, provided that their architecture is simplified. The condition of model simplification is important both in context of possible neural network classifier implementation on the side of OS SS and side of ground control complex. In the first case it is important from the point of view of hardware restrictions. In the second case it is important for the training time reducing of classification model.

It's well known that for GA development we should define fitness function, chromosome encoding and genetic operators for GA based search implementation [23,24]. In the next section all these basic components of our solution are represented.

3 Genetic Algorithm

3.1 Fitness Function

A fitness function in genetic algorithms defines the quality of solutions generated by the algorithm, and it is used to determine which individuals are the fittest to survive and reproduce. The fitness function assigns a fitness score to each individual based on how well they perform in solving the problem at hand. The fitness score is then used to guide the selection of individuals for reproduction.

In the context of neural networks, the accuracy of the network in predicting the output given the input can be used as the fitness function. The goal in this case is to maximize the accuracy of the network on a validation set or a test set. Thus, the fitness function would measure how well the network performs on the data set and assign higher fitness scores to individuals that produce better results. This can be used to optimize various aspects of the neural network, such as the architecture, the hyperparameters or the weights of the network.

On the other hand, performance refers to the speed at which the model can process data and generate predictions. But the peculiarity of the data makes it possible not to pay much attention to this aspect. Telemetry data arrives from spacecrafts much less frequently than the performance of the largest and slowest neural network architectures we have tried till that moment.

Considering all mentioned above fitness function for genetic algorithm in this research is the accuracy of neural network in classification task.

3.2 Selection

Selection determines which individuals will be chosen for reproduction and which ones will be eliminated from the population. The process of selection mimics the natural process of survival of the fittest, where individuals with better fitness values have a higher chance of being selected for reproduction.

The hall of fame technique was chosen as the selection mechanism to solve the problem described in this paper.

Hall of fame (HOF) is a technique used in genetic algorithms to improve the selection process. In a typical genetic algorithm, selection is performed based on fitness values, where individuals with higher fitness values are more likely to be selected for reproduction.

However, the HOF technique keeps track of the best individuals encountered during the evolution process and saves them in a separate list. These individuals are not subject to selection pressure and are allowed to remain in the population even if their fitness is lower than that of other individuals.

3.3 Chromosome Structure

Individuals in genetic algorithms can be represented in various ways, depending on the nature of the problem being solved. In genetic algorithms domain individuals often call chromosomes.

The chromosome represents the encoded solution to a particular problem. The genes in a chromosome can be of different types. In this case the chromosome is the set of genes, where gene can be a value among a variety of types or a structure. During the optimization process, genetic algorithms manipulate the genes within the chromosomes. The type of genes in a chromosome affects the way in which crossover and mutation operators are applied to generate new solutions.

Chromosome in this experiment contains 4 main genes:

- learning_rate - hyper-parameter that controls the weights of neural network with respect to the loss gradient. Possible values for this gene was restricted in reasonable limits $learning\_rate \in \{0.1^k | k \in \{1, 2, ..., 7\}\}$.
- loss_function - function that computes the distance between the space of expected output and the space of the neural network output. This gene value is a loss-function from the list developed for multiclass classification task: *poisson*, *categorical_ crossentropy*, *kl_ divergence*.

- optimizer - algorithm used to tune neural network parameters. Optimizer gene contains the predefined classical optimizers represented in the deep learning libraries - $optimizer \in \{Adam, SGD, Adagrad\}$.
- nn_architecture - architecture of the neural network. Consist of 4 types of layers: *Conv*, *GRU*, *Dense*, *Dropout*.

Also in chromosome there are some functional genes: fitness, num_params. Table 1 gives a summary of all genes in chromosome except architecture.

Table 1. Structure of Chromosome

Main genes in chromosome	
Gene	Possible values and restrictions
learning_rate	1e−1, 1e−2, 1e−3, 1e−4, 1e−5, 1e−6, 1e−7
optimizer	Adam, SGD, Adagrad
loss_function	categorical_crossentropy, poisson, kl_divergence
architecture	-
Functional genes in chromosome	
Gene	Comment
fitness	Collect fitness function value
num_params	Collect number of parameteres of neural network

Gene responsible for the architecture of neural network represented in special format is a structure which contain types of layers of neural network, order of this layers and their required parameters. This structure was also designed for convenient crossover which will be described in the next section. As a result this structure was designed as a special dataframe: rows are responsible for layers, columns are a the characteristics of this layers (see Fig. 1).

Layer	on	Conv kernel size	Conv filters	GRU units	Dense units
Conv1D	1	2	16	10	584
Conv1D	1	4	32	22	236
Conv1D	1	2	112	20	364
GRU	1	10	8	24	526
Flatten	1	10	14	16	658
Dense	1	12	126	10	56
Dense	1	6	46	22	148

Fig. 1. Example of the architecture structure

This structure has not determined size, to ensure the variety of architectures of the neural network. Therefore, the number of layers was limited as well as the number of layers of a certain type among this limit. The parameters of layers were also restricted.

3.4 Crossover

Crossover is a genetic operator used to combine the information of two parents to generate a new offspring and introduce new genetic material into the population.

Crossover for Gens Except Architecture. As was mentioned before, all genes in chromosome except architecture are simple non-structure fields with a finite number of options. For these genes, we can use the crossover approach with equiprobability selection from the values of the parent genes.(see Fig. 2).

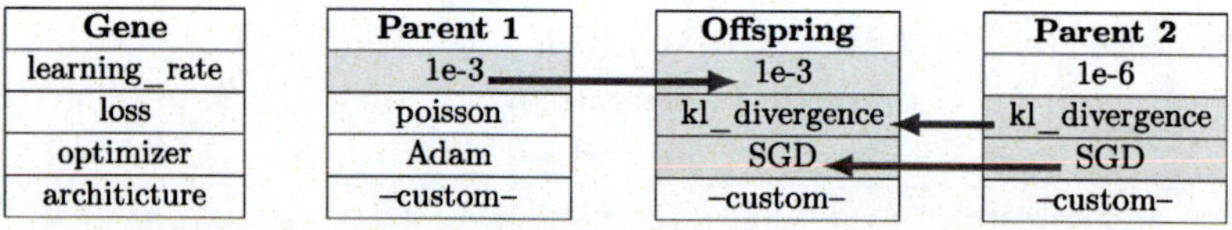

Fig. 2. Crossover process on non-architecture genes

Crossover for Architecture. For crossover between parents' architecture genes a mechanism mixing neural network layers from both parents was developed. On the first stage layers from the parents' architectures were combined into one architecture, then the layers for the offspring architecture were chosen by randomly generated binary mask (see Fig. 3).

	Layer	on	CS[1]	CF[2]	GU[3]	DU[4]
Parent 1	Conv1D	1	2	16	10	584
	GRU	1	10	8	24	526
	Flatten	1	10	14	16	658
	Dense	1	12	126	10	56
Parent 2	Conv1D	1	4	8	18	268
	Flatten	1	6	32	4	102
	Dense	1	8	26	10	344
	Dense	1	2	6	2	38
	Dense	1	2	12	20	186

*

	on
Random binary mask	1
	1
	0
	1
	0
	1
	0
	0
	1

=

	Layer	on	CS[1]	CF[2]	GU[3]	DU[4]
Offspring	Conv1D	1	2	16	10	584
	GRU	1	10	8	24	526
	Dense	1	12	126	10	56
	Flatten	1	6	32	4	102
	Dense	1	2	12	20	186

CS[1] - Conv kernel size; CF[2] - Conv filters; GU[3] - GRU units; DU[4] - Dense units;

Fig. 3. Example of the architecture crossover between two chromosomes

3.5 Mutation

Mutations are the key mechanism by which genetic algorithms introduce diversity into the population of possible solutions. Simply put, mutations are random changes made to the genetic makeup of individuals in the population, and they help to explore new regions of the search space that may contain even better solutions.

Mutations can take many different forms, depending on the specific problem being addressed and the encoding scheme used to represent individuals.

The rate at which mutations occur is an important parameter that can greatly affect the performance of a genetic algorithm. If the mutation rate is too low, the algorithm may converge prematurely to a suboptimal solution, while if the mutation rate is too high, the algorithm may lose promising solutions too quickly and fail to converge at all. As such, finding an appropriate balance between exploration and exploitation is key to effective mutation techniques in genetic algorithms.

In the experiments described in this paper the **Random Resetting** - mutation was choosen. A mutation of one of the genes occurred with a probability of **15%**. The values of *learning_rate*, *loss_function* and *optimizer* genes were assigned from the set of the possible values for this gene. The mutation of the *nn_architecture* gene was a completely new architecture generated randomly.

4 Experiment Results

Genetic algorithm was runned on Linux machine with 30 virtual cores CPU. During the entire time the algorithm was running, the CPU was loaded by 27%. Experiment takes 2778 s. Running on Apple MacBook M1 Max algorithm utilised approximately 95% of CPU and took 3586 s.

Experiment had next parameters: population size - 25, hall of fame size - 15, number of generations - 8;

Table 2 represents restriction for neural network architectures of individuals.

Table 2. Neural network architecture restrictions

Parameter	Min	Max
Number Conv layers	2	8
Number Conv filters	4	128
Conv kernel size	2	16
Number GRU layers	0	3
Number GRU units	8	32
Number Dense layers	2	10
Number Dense units	16	1024
Probability occurrence Dropout (0.2)	0.25	0.25

The best value of fitness achieved during experiment is **98.8%** (see Fig. 4a). In the end of the experiment it was noticed that the representatives with the different chromosome structures showed the results equal to maximum fitness value or the value close to maximum. These results can be considered as a confirmation of the found close to optimal solution for the certain experiment parameters.

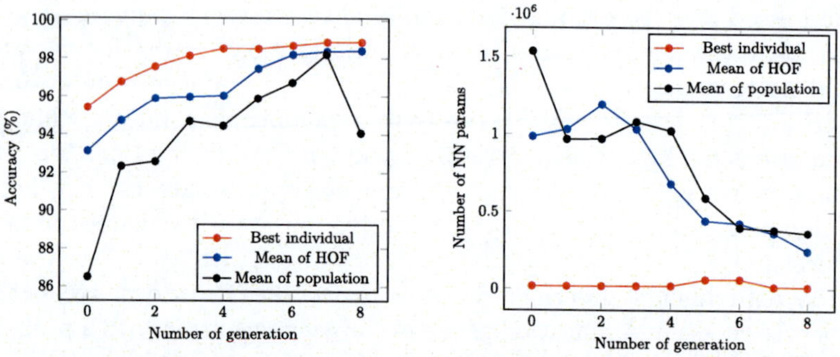

(a) Fitness of individuals

(b) Number of neural network params

Fig. 4. Genetic algorithm work process artifacts

As it can be seen on the plot with the number of neural network parameters (see Fig. 4b), during the evaluation process the value of this index reduces with each step. This confirms the idea that simpler model with fewer parameters has less capacity to memorize the training data and is forced to learn more abstract and general features that can be applied to unseen data. In contrast, a complex model with too many parameters may overfit the training data, meaning it memorizes the noise and details of the training examples instead of learning the underlying patterns.

Therefore, finding the appropriate balance between model complexity and generalization is crucial for building effective neural networks. And experiments described in this paper shows that genetic algorithms help solve this problem.

5 Conclusion

The paper presents research GA-based automatic search method of hybrid neural networks models for solutions to the actual problem of intelligent analysis of telemetry data from small spacecrafts in order to detect its technical states. Neural network models were automatically built to solve the telemetry data classification problem for real telemetry data, which slightly exceeded the classification accuracy of previously obtained hybrid neural network models [21, 22] based on a combination of three types of layers: 1-dimensional convolutional, recurrent and fully connected (98.8%). Also, simpler models without recurrent layers were obtained with an accuracy of 98.5%, which greatly simplifies the NN architecture and the possibilities of its use. The obtained results allow:

1. significantly reduce the search time, systematize and automate the construction process compared to the manual development of neural network models for the task of classifying the telemetry data;
2. the results for the previously built hybrid neural models were improved, on the other hand, simpler models were built with almost the high classification accuracy;
3. stably replicate sufficiently high classification accuracy for different models.

Thus, obtained results can be considered as a confirmation of the found close to optimal solution for the certain experiment parameters. The methods for automating the search for optimal neural networks models is promising. We are going to continue research in this direction for another bioinspired heuristic based search approaches.

Acknowledgements. The research described in this paper is partially supported by the project 9SGZ.1-222 "Develop a neural network software package for monitoring and predicting the state of the subsystems of the constellation spacecraft for the command and measurement complex" (Program of Union State "Complex-SG"), project "Behavioral and intelligent methods for detecting intrusions in telecommunication networks" (SRP "Informatics, space and security"), within the framework of research cooperation with Shaoxing Research Institute of Zhejiang University of Technology.

References

1. Yang, Q., Wu, X.: Ten challenging problems in data mining research. Int. J. Inf. Technol. Decis. Making **5**(04), 597–604 (2006). https://doi.org/10.1142/S0219622006002258
2. Ismail Fawaz, H., Forestier, G., Weber, J., Idoumghar, L., Muller, P.-A.: Deep learning for time series classification: a review. Data Min. Knowl. Disc. **33**(4), 917–963 (2019). https://doi.org/10.1007/s10618-019-00619-1
3. Silva, D.F., Giusti, R., Keogh, E., Batista, G.: Speeding up similarity search under dynamic time warping by pruning unpromising alignments. Data Min. Knowl. Disc. **32**(4), 988–1016 (2018). https://doi.org/10.1007/s10618-018-0557-y
4. Bagnall, A., Lines, J., Bostrom, A., Large, J., Keogh, E.: The great time series classification bake off: a review and experimental evaluation of recent algorithmic advances. Data Min. Knowl. Discov. **31**(3), 606–660 (2017). https://doi.org/10.1007/s10618-016-0483-9
5. Gamboa, B., Cristian, J.: Deep learning for time-series analysis. https://arxiv.org/pdf/1701.01887.pdf. Accessed 15 Nov 2020
6. Rajkomar, A., et al.: Scalable and accurate deep learning with electronic health records. npj Digit Med. **1**, 18 (2018). https://doi.org/10.1038/s41746-018-0029-1
7. Nweke, H.F., The, Y.W., Al-garadi, M.A., Alo, U.R.: Deep learning algorithms for human activity recognition using mobile and wearable sensor networks: state of the art and research challenges. Exp. Syst. Appl. **105**, 233–261 (2018). https://doi.org/10.1016/j.eswa.2018.03.056

8. Nwe, T.L, Dat, T.H., Ma, B.: Convolutional neural network with multi-task learning scheme for acoustic scene classification. In: Asia-Pacific Signal and Information Processing Association Annual Summit and Conference, pp. 1347–1350 (2017). https://doi.org/10.1109/APSIPA.2017.8282241
9. Susto, G.A., Cenedese, A., Terzi, M.: Time-series classification methods: review and applications to power systems data. In: Big Data Application in Power Systems, pp. 179–220 (2018). https://doi.org/10.1016/B978-0-12-811968-6.00009-7
10. Skobtsov, V., Novoselova, N., Arhipov, V., Potryasaev, S.: Intelligent telemetry data analysis of small satellites. In: Silhavy, R., Senkerik, R., Kominkova Oplatkova, Z., Prokopova, Z., Silhavy, P. (eds.) CSOC 2017. AISC, vol. 574, pp. 351–361. Springer, Cham (2017). https://doi.org/10.1007/978-3-319-57264-2_36
11. Skobtsov, V.Yu., Arhipau, V.I.: Neural network analysis of telemetry data of onboard equipment of spacecraft. Space Eng. Technol. **3**(34), 111–124 (2021). https://doi.org/10.33950/spacetech-2308-7625-2021-3-111-124
12. Bagnall, A., Lines, J., Hills, J., Bostrom, A.: Time-series classification with COTE: the collective of transformation-based ensembles. In: International Conference on Data Engineering, pp. 1548–1549 (2016). https://doi.org/10.1109/ICDE.2016.7498418
13. Hills, J., Lines, J., Baranauskas, E., Mapp, J., Bagnall, A.: Classification of time series by shapelet transformation. Data Min. Knowl. Disc. **28**(4), 851–881 (2013). https://doi.org/10.1007/s10618-013-0322-1
14. Bostrom, A., Bagnall, A.: Binary shapely transform for multiclass time series classification. In: Big Data Analytics and Knowledge Discovery, pp. 257–269 (2015). https://doi.org/10.1007/978-3-662-55608-5_2
15. Baydogan, M.G.: Multivariate time series classification datasets (2015). http://www.mustafabaydogan.com. Accessed 15 Nov 2020
16. Lines, J., Taylor, S., Bagnall, A.: HIVE-COTE: the hierarchical vote collective of transformation-based ensembles for time series classification. In: IEEE International Conference on Data Mining, pp. 1041–1046 (2016). https://doi.org/10.1109/ICDM.2016.0133
17. Lines, J., Taylor, S., Bagnall, A.: Time series classification with HIVE-COTE: the hierarchical vote collective of transformation-based ensembles. ACM Trans. Knowl. Discov. Data **12**(5), 52:1–52:35 (2018). https://doi.org/10.1145/3182382
18. LeCun, Y., Bengio, Y., Hinton, G.: Deep learning. Nature **521**, 436–444 (2015). https://doi.org/10.1038/nature14539
19. Wang, Z., Yan, W., Oates, T.: Time series classification from scratch with deep neural networks: a strong baseline. In: International Joint Conference on Neural Networks, pp. 1578–1585 (2017). https://doi.org/10.1109/IJCNN.2017.7966039
20. Skobtsov, V.Yu., Novoselova, N.A.: Investigation of stream clustering algorithms in solving the problem of analyzing telemetry data of small space vehicles. J. Instr. Eng. **63**(11), 1003–1011 (2020). https://doi.org/10.17586/0021-3454-2020-63-11-1003-1011
21. Skobtsov, V. Yu.: Binary classification of small satellites telemetry data based on deep learning approach. Appl. Aspects Inf. Technol. **4**(4), 299–310 (2021). https://doi.org/10.15276/aait.04.2021.1
22. Skobtsov, V.Yu., Sokolov, B.V.: Hybrid neural network models in the problem of telemetry data multiclass classification of small spacecrafts. In: Proceedings of VSU. Systems Analysis and Information Technologies, vol. 3, pp. 99–114 (2022). https://doi.org/10.17308/sait/1995-5499/2022/3/99-114
23. Holland, J.H.: Adaptation in Natural and Artificial Systems. University of Michigan Press (1975)

24. Goldberg, D.E.: Genetic Algorithms in Search, Optimization and Machine Learning, 13th edn., 432 p. Addison-Wesley Professional, 11 January 1989
25. Skobtsov, Y.A., Skobtsov, V.Y.: 13 evolutionary test generation methods for digital devices. In: Adamski, M., Barkalov, A., Węgrzyn, M. (eds.) Design of Digital Systems and Devices. LNEE, vol. 79. Springer, Heidelberg (2011). https://doi.org/10.1007/978-3-642-17545-9_13

Optimizing Data Analysis Tasks Scheduling Based on Resource Utilization Prediction

Dan Ma, Yujie Li, Huarong Xu, Mei Chen, Qingqing Liang, and Hui Li(✉)

Guizhou Engineering Laboratory for Advanced Computing and Medical Information Services, College of Computer Science and Technology, Guizhou University, Guiyang, China
cse.HuiLi@gzu.edu.cn

Abstract. In order to reduce the average response time of data analysis tasks in a decision support system and maximize the utilization of cluster resources, we proposed the Upsa algorithm for task scheduling, which is based on the prediction of resource utilization. It aims to balance the workload of analytical data processing tasks in the cluster and increase task execution efficiency. In Upsa, the random forest (RF) model is employed to accurately determine the resource utilization rate for the subsequent period to provide a rational basis for task allocation. The Upsa method can ignore a server that is about to become overloaded and assign new duties to the most suitable server. As soon as the server's resource utilization rate returns to normal levels, it will resume receiving tasks. We evaluate the effectiveness of Upsa using the TPC-H benchmark. Compared to the well-known scheduling solution Round Robin, our experimental results show that the Upsa method substantially reduces the average response time.

Keywords: Analytical Data Processing · Task Scheduling · Resource Utilization Prediction

1 Introduction

With the rapid development of informatization, a large amount of business data is generated by enterprise information systems. How to effectively obtain useful information from these massive business data for enterprise decision analysis has become a challenge for data analysts [1]. Scale out is a commonly employed technique for ensuring that the system's QoS (Quality of Services) stays stable when multiple tasks are executed concurrently. However, as the number of users and workloads grows quickly, simply increasing the number of servers will incur substantial costs [8]. Due to the immense differences in the complexity of data processing tasks, however, if the task scheduling strategy is not optimized, computing resources cannot be utilized to their utmost potential [6, 7, 11–14]. In order to effectively execute and achieve a more stable quality of service, there is an urgent need for an appropriate strategy to schedule data processing duties. The dynamic nature of the cluster's available resources and workload should be considered when scheduling tasks, as resource utilization and workload have a significant impact on the efficacy of task execution. If we are able to find a suitable method to predict the resource utilization of the server in the following period and accurately estimate the task cost, it will greatly facilitate the design of efficient task scheduling algorithms [15, 16].

R. Silhavy and P. Silhavy (Eds.): CSOC 2023, LNNS 724, pp. 812–823, 2023.
https://doi.org/10.1007/978-3-031-35314-7_67

In this study, we propose a data processing task scheduling algorithm, Upsa, based on a random forest model and task cost estimation technology, to distribute the burden equitably and reduce task response time. This method accurately predicts resource utilization by integrating the RF model and task cost estimation technology.

The structure of this paper is as follows: In Sect. 2, we discuss related work. The Sect. 3 provides an overview of our work. In Section 4, we elaborate on the proposed approach. Section 4 describes the experimental design and evaluation. The Sect. 5 concludes the paper.

2 Related Work

With the advent of the era of big data, scheduling optimization for data processing has garnered significant attention. The FCFS scheduling strategy is proposed based on the arrival order of assignments. Priority is given to the first task submitted until the task queue is vacant. FCFS is the default strategy for most simple task schedulers, and it prioritizes task equality. Sadly, since the complexity of each task and the quantity of data processed vary, so does the consumption of computing resources, resulting in an unbalanced cluster load when using the FCFS strategy.

The algorithm round-robin (RR) is utilized by process and network schedulers in computing. As the term is commonly understood, time slices (also known as time quanta) [3] are allotted to processes in equal portions and in a circular fashion, managing all processes without priority. Round-robin scheduling is straightforward, simple to implement, and devoid of deprivation. Round-robin scheduling is applicable to other scheduling issues, such as the scheduling of data packets in computer networks.

Several other researchers have developed numerous scheduling algorithms for data processing tasks. Concurrent ETL task scheduling is based on an altruistic strategy [5, 10], which chooses the optimal task execution order based on the current load of the host and the task consumption. This algorithm has more benefits than FCFS, but it can only be implemented on a single server. Han proposed a prediction-based scheduling system [4]. This strategy forecasts the execution time of tasks and assigns those with longer execution times to servers with shorter execution times in the task queue. Nodes can only execute a single task with this algorithm. Moreover, none of these algorithms are capable of scheduling real-time tasks. In general, achieving reasonable scheduling of real-time data processing tasks in a cluster environment is challenging.

3 Methodology

3.1 Overview

In this paper, we optimize the scheduling of data processing tasks using offline training (Fig. 1) and real-time task scheduling (Fig. 2). In the offline training phase, we will describe how to build a random forest prediction model to predict resource utilization for the subsequent period. During the phase of real-time scheduling, we will explain the details of Upsa.

The primary objective of this study is to accurately forecast the server's resource utilization. Resource utilization is not only related to the amount of tasks, but also to the pending tasks. Therefore, we extract features from the server's historical load information and task execution history. CPU utilization, memory consumption, the number of tasks being executed, and the cost of each task were extracted as features. This data is periodically (e.g., the default setting is 500 ms) transmitted to the scheduler via the monitoring program installed on each node. Each record is titled "load snapshot" and represents the server's state at a particular time. Once sufficient "load snapshots" are collected, we can train an effective model. The procedure is depicted in Fig. 1.

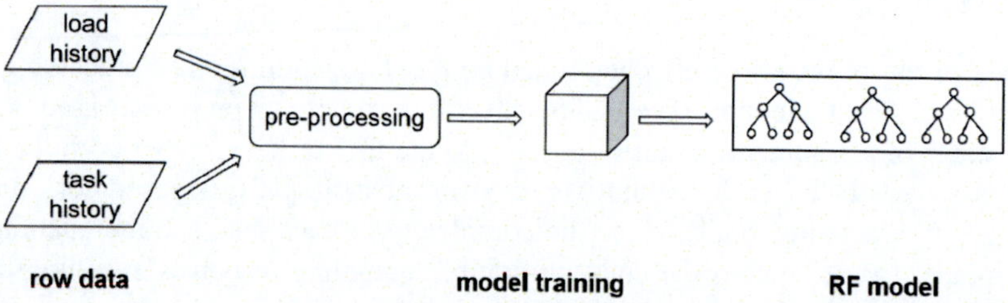

Fig. 1. Offline scheduling

Figure 2 provides an overview of Upsa's real-time scheduling. We implemented Cost calculator and Predictor in the scheduler. The cost calculator calculates the expense of each task using the Postgresql database's API. Predictor uses a random forest model [3] to forecast the future load of each cluster node, avoids overloaded servers, and returns a reasonable scheduling strategy.

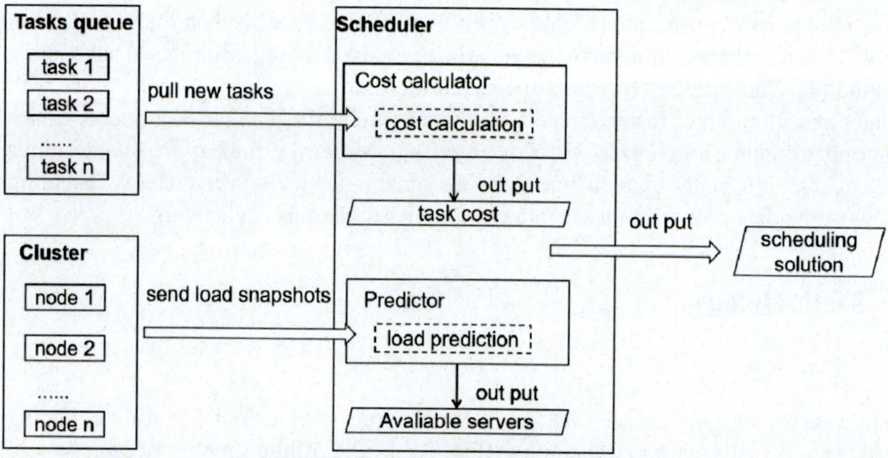

Fig. 2. Real-time task scheduling

3.2 Prediction Model

Before training the random forest, a load generator was implemented to simulate the load in the actual situation, while load samples were collected for each node. Table 1 lists the load snapshot's attributes, the attributes of I/O load is similar.

Table 1. Attributes of the CPU load snapshot.

Attributes	Description
cpu_util_T	CPU utilization at time T
cpu_util_T-t	CPU utilization at time T - t
...	
cpu_util_T-5t	CPU utilization at time T - 5t
running_task_num_T	count of running tasks at time T
running_task_num_T-t	count of running tasks at time T - t
...	
running_task_num_T-5t	count of running tasks at time T - 5t
total_task_cost_T	total cost of running task at time T
total_task_cost_T-t	total cost of running task at time T - t
...	
total_task_cost_T-5t	total cost of running task at time T - 5t
mem_usage	the usage of memory

The load generator will continuously send groups of tasks to the server at random intervals, with each group's tasks starting concurrently. Through the server-deployed monitoring system, a snapshot of the current workload is captured and preserved in a database at each moment. The procedure is depicted in Fig. 3. Note: LS means load snapshot, T represents time, and t is time increment.

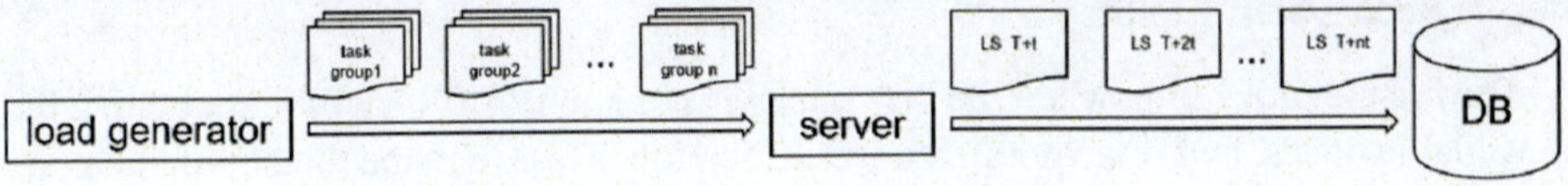

Fig. 3. Generation of training data

It is challenging to express the complex relationship between the extracted features using a generalized linear model because their relationship is complex and nonlinear. The random forest algorithm is able to handle both linear and multidimensional relationships. Therefore, we use the random forest algorithm to build the model.

The predictor will continuously receive workload samples from each node and update the node status in real-time once the model has been established. Suppose the predictor determines that the resource utilization of a node in the following period will exceed a predetermined threshold. In that case, the node's status will be unavailable until the predicted resource utilization falls below the threshold.

3.3 Task Scheduling Algorithm

After implementing the predictor and cost estimator, the scheduler is able to apply the Upsa algorithm. The fundamental concept underlying the Upsa algorithm is to assign a large Cost task to a node with a low load and a small Cost task to a node with a heavy load. And establish a limit to prevent node congestion. Figure 4 illustrates an example.

Algorithm 1: The Upsa SCHEDULING STRATEGE

```
Algorithm UPSA ()
while true do
  if taskQueue is null
    then continue
  availableServers ← getAvaliableServers()
  if availableServers.size > 0
    then
      SortByIO_PredictedValueASC(availableServers)
      excute_server ← avaliableServers[0]
      task ← taskQueue.poll()
      send(task,excute_server)
      updateServerLoad(excute_server)
end while
end Algorithm
```

The Upsa algorithm is initiated at the same time as the scheduler. The algorithm undertakes the following steps:

Step 1: The available nodes are filtered according to the procedure described in Sect. 3.2.
Step 2: Sort nodes in descending order based on their predicted resource usage.
Step 3: Obtain new tasks from the task queue.
Step 4: Sort tasks by cost in ascending order.
Sort assignments by cost in ascending order.
Step 5: Sending tasks to available servers.
Step 6: Update Sever load according to the execution of tasks.
Step 7: Repeat Step 1 to Step 6.

During the procedure, when the task queue is not empty UPSA will first obtain a list of available nodes, and if the predicted I/O utilization is less than the threshold, the node will be regarded as available. When the predicted value of I/O utilization is higher than the threshold, then continuing to assign tasks may cause resource competition and reduce execution efficiency, consequently, it will be temporarily regarded as an unavailable node until the predicted value of I/O utilization goes back to the normal level. The default threshold value is 100, implying that the node will be marked as unavailable when the predicted I/O reaches saturation.

4 Experimental Evaluation

4.1 Experiment Environment

Environment. All of our evaluations were conducted in a master-slave replication mode PostgreSQL (version 11) cluster. The cluster consists of a master node and four slave nodes, with the master node used for task scheduling and the slave nodes for task execution. The configuration is detailed in Table 2.

Table 2. Configuration of cluster.

Node Type	CPU	RAM	OS	Number
master	4 cores Intel 2.6 GHZ	8 GB	CentOS 7	1
slave	4 cores Intel 2.6 GHZ	4 GB	CentOS 7	4

Workloads. In our evaluation, we employ the TPC-H benchmark [2], which consists of a collection of business-oriented ad-hoc queries and concurrent data modifications. The database's queries and data have been selected for their industry-wide applicability. This benchmark exemplifies decision support systems that analyze large volumes of data, implement complex queries, and provide answers to crucial business questions. The TPC-H dataset generation script is executed during the evaluation to create datasets with a range of data sizes, including 256 MB, 512 MB, 1 GB, and 2 GB, which will be processed by TPC-H queries. All tasks used for experiments are randomly chosen from 76 tasks (Task1, Task2, ..., Task76), and the tasks in the simulated load will use the 19 query statements supplied by TPC-H (Q1, Q2, ..., Q19) on the aforementioned four data sets. The task uses more I/O resources at a higher rate the bigger the dataset it processes. The randomly chosen tasks represent 10% of the total tasks when processing 2 GB datasets, 30% of the tasks when processing 1 GB datasets, and 60% of the tasks when processing datasets below 1 GB to simulate the actual situation where more complex tasks are in the minority.

4.2 Evaluation

4.2.1 Predication of I/O Utilization

In this section, a total of 8826 samples were collected from TPC-H benchmark, of which 66% were used as the training set and the remainder as the test group. Three models are created for the evaluation to forecast various time periods, including 1 s, 3 s, and 5 s, respectively. I/O utilization in the upcoming seconds, in the upcoming three seconds, and in the upcoming five seconds are the three anticipated metrics. The three methods are Classification and Regression Tree (CRT), Multiple Linear Regression (MLR), and Random Forest (RF) (CART). The I/O usage ranges from 0 to 100, and Fig. 4 shows the average error of the three models for each time interval.

Figure 4 illustrates that the error increases with the length of the forecast period. When making predictions with various time horizons in preparation, the random forest model has the smallest average absolute error.

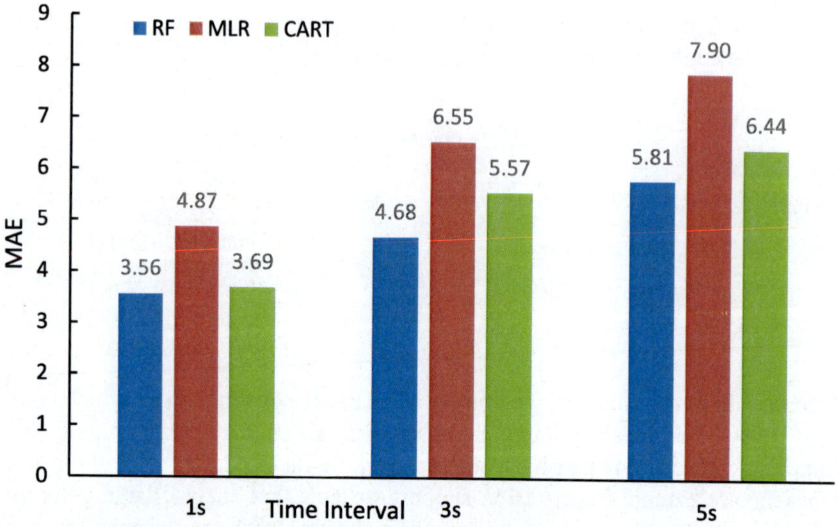

Fig. 4. MAE of RF, MLR and CART

4.2.2 Performance of Scheduling

In this study, the Round Robin (RR) policy and Upsa scheduling methods are contrasted.

Upsa: The proposed resource utilization prediction-based scheduling strategy for data processing tasks.

Round Robin: Sends the jobs, one at a time, to each execution node for processing.

The Round Robin algorithm was chosen for comparison because it is the most used task scheduling strategy, is the default policy offered by many schedulers, and performs well when used to address the issue of overloaded servers.

The experiments evaluate the effectiveness of two task scheduling strategies, UPSA and Round Robin, by sending the same set of tasks in the same amount of time and then comparing the average response time. The experiments compare UPSA and Round Robin under six distinct load conditions, each with a different degree of task tightness, where the tasks are sent within 200 s and the number of executed tasks is 100, 120, 140, 160, 180, and 200, respectively. When the time interval between two tasks is zero, the tasks are sent concurrently. During the experiment, the reaction time for each task is recorded.

In order to eliminate random errors, each load condition was executed five times and the average response time for each assignment was calculated. The experimental results obtained are depicted in Figs. 5, 6, 7 and 8, which depict the distribution of response times for the tasks when the number of tasks is 100, 120, 140, 160, 180, and 200, respectively.

From Fig. 5, 6, 7, 8, 9 and 10, it is evident that the majority of duties are completed within 5 s under both scheduling strategies. When utilizing the Upsa strategy, the number of tasks with response times greater than 5 s is significantly reduced compared to Round Robin. This can be interpreted as UPSA sharing the I/O load to a greater extent, thereby decreasing the time for tasks to queue up for I/O resources and optimizing response

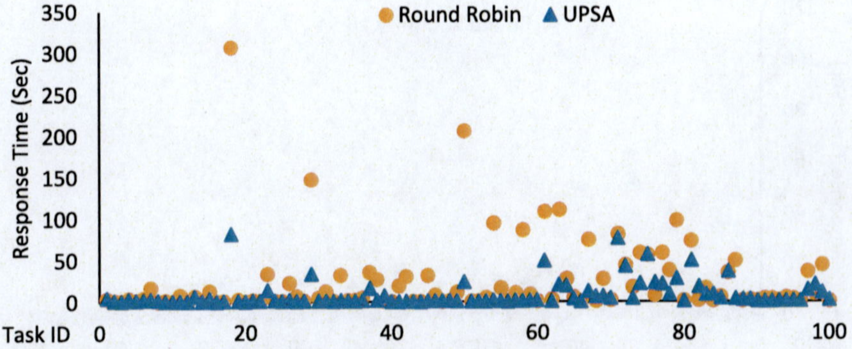

Fig. 5. Response time for 100 tasks

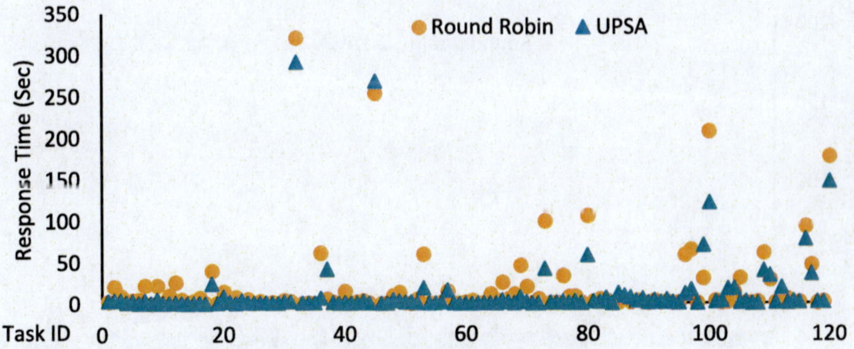

Fig. 6. Response time for 120 tasks

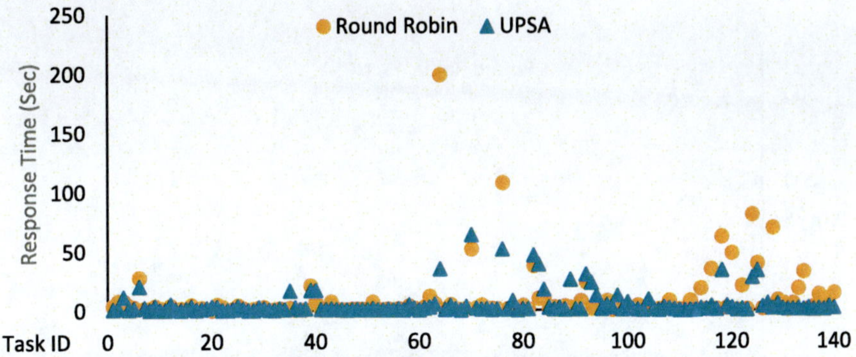

Fig. 7. Response time for 140 tasks

time. The table below compares the average response time and degree of load skewing for each load condition.

Figure 11 illustrates a comparison of the average response times of the tasks under each workload condition. Figure 12 compares the degree of burden skewing produced

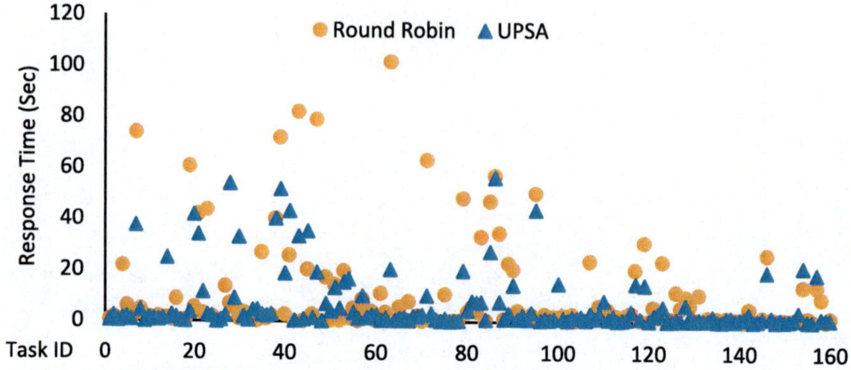

Fig. 8. Response time for 160 tasks

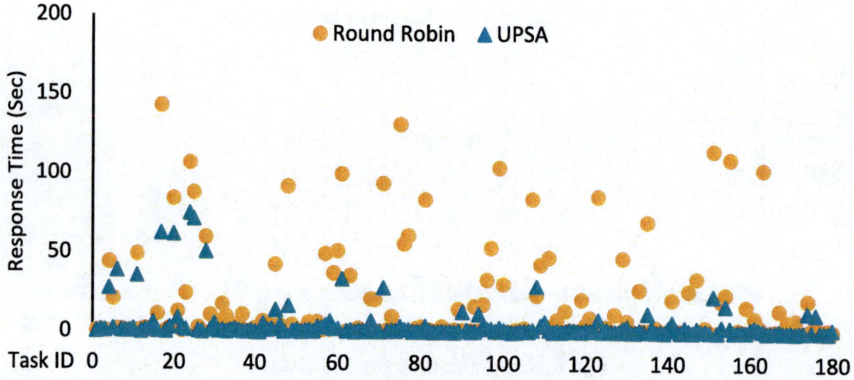

Fig. 9. Response time for 180 tasks

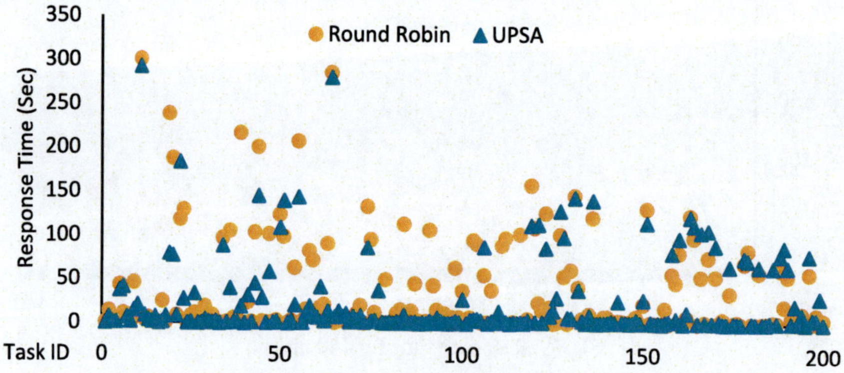

Fig. 10. Response time for 200 tasks

by the two scheduling strategies. From Figs. 11 and 12, it can be seen that the Upsa scheduling policy examined in this paper has a lower average response time and a lower load imbalance than the Round Robin policy. In the experiment, Upsa reduces the average

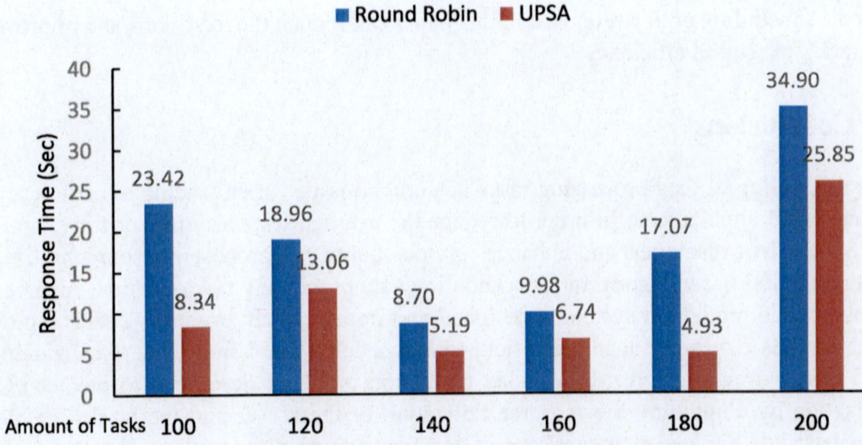

Fig. 11. Response time for 200 tasks

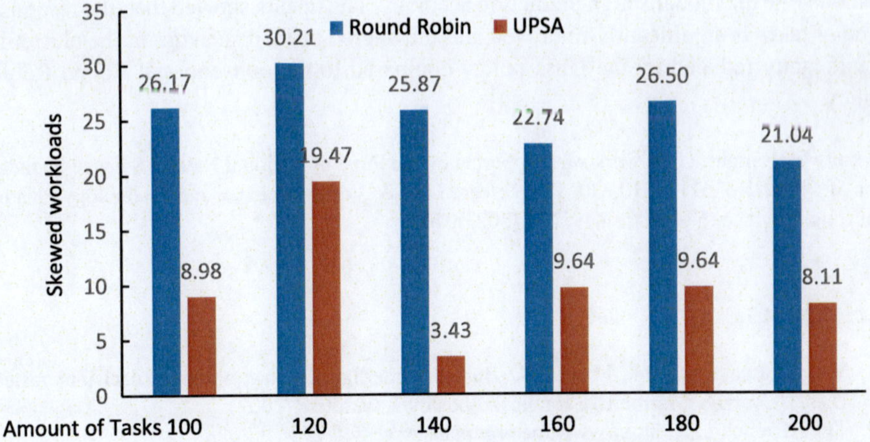

Fig. 12. Skewed workloads

response time of the task by 25% to 71% compared to Round Robin, and the optimization of the unbalanced index is in the range of 35% to 86%, demonstrating that the scheduling policy proposed in this paper can reduce the load skew and improve the task's execution efficiency.

Figure 11 illustrates a comparison of the average response times of the tasks under each workload condition. Figure 12 compares the degree of burden skewing produced by the two scheduling strategies. From Fig. 11 and Fig. 12, it can be seen that the Upsa scheduling policy examined in this paper has a lower average response time and a lower load imbalance than the Round Robin policy. In the experiment, Upsa reduces the average response time of the task by 25% to 71% compared to Round Robin, and the optimization of the unbalanced index is in the range of 35% to 86%, demonstrating that the scheduling policy proposed in this paper can reduce the load skew and improve the task's execution efficiency.

5 Conclusion

The scheduling of data processing tasks is a crucial issue for enhancing the efficiency of analytical applications. In order to reduce the average response time of tasks, optimize the use of resources, and enhance service quality, we propose a resource utilization prediction-based algorithm for scheduling data processing tasks. At first, A server resource load prediction system is designed and implemented, primarily consisting of three modules: resource load collection, task scheduling, and front-end visualization. And then, A random forest-based load prediction model is developed to predict I/O utilization by combining the resource utilization of the server and the burden of the tasks based on the load characteristics of decision support class applications. Proposed is the task scheduling policy UPSA based on capacity prediction. We conducted a series of comprehensive experiments to verify the accuracy of the load prediction model and demonstrate the model's high prediction accuracy. The results showed that, the response time of tasks is significantly improved and the degree of load skewing in the cluster is significantly reduced by the Upsa policy compared to the conventional Round Robin policy.

Acknowledgement. This work was supported by the Fund of National Natural Science Foundation of China (No. 62162010, G0112), Research Projects of the Science and Technology Plan of Guizhou Province (No. [2021]449, No. [2023]010).

References

1. Vitt, E., Luckevich, M., Misner, S.: Business intelligence: making better decisions faster. Diss. Univerza v Mariboru, Ekonomsko-poslovna fakulteta (2002)
2. TPC-H benchmark. http://www.tpc.org/tpch/. Accessed 15 Mar 2023
3. Biau, G., Scornet, E.: A random forest guided tour. TEST **25**(2), 197–227 (2016). https://doi.org/10.1007/s11749-016-0481-7
4. Han, W.B., et al.: PBS: an ETL scheduling algorithm for cluster environment. Comput. Digit. Eng. **45**(5), 793–796 (2017)
5. Tang, Li., Li, H., Chen, M., Dai, Z., Zhu, M.: Enhancing concurrent ETL task schedule with altruistic strategy. In: Silhavy, R., Silhavy, P., Prokopova, Z. (eds.) CoMeSySo 2018. AISC, vol. 859, pp. 201–212. Springer, Cham (2019). https://doi.org/10.1007/978-3-030-00211-4_19
6. Song, B., Yu, Y., Zhou, Y., Wang, Z., Du, S.: Host load prediction with long short-term memory in cloud computing. J. Supercomput. **74**(12), 6554–6568 (2017). https://doi.org/10.1007/s11227-017-2044-4
7. Chowdhury, M., Stoica, I.: Efficient coflow scheduling without prior knowledge. In: Proceedings of the 2015 ACM Conference on Special Interest Group on Data Communication, pp. 393–406 (2015)
8. Seyedhamid, M., Michael, S., Cameron, W., Sareh, P., Charles, U.: Embedding individualized machine learning prediction models for energy efficient VM consolidation within Cloud data centers. Fut. Gener. Comput. Syst. **106**, 221–233 (2020)

9. Computing Based on NSGA2. Beijing University of Posts and Telecommunications, Thesis (2018)
10. Yang, Q., Zhou, Y., Yu, Y., Yuan, J., Xing, X., Du, S.: Multi-step-ahead host load prediction using autoencoder and echo state networks in cloud computing. J. Supercomput. **71**(8), 3037–3053 (2015). https://doi.org/10.1007/s11227-015-1426-8
11. Grandl, R., Chowdhury, M., Akella, A., Ananthanarayanan, G.: Altruistic scheduling in multi-resource clusters. In: Proceedings of the 12th USENIX Symposium on Operating Systems Design and Implementation, OSDI 2016, pp. 65–80 (2016)
12. Lian, J.D., et al.: Hierarchical load balancing algorithm based on prediction mechanism. Comput. Eng. Appl. **11**, 67–71 (2015)
13. Duggan, M., Mason, K., Duggan, J., et al.: Predicting host CPU utilization in cloud computing using recurrent neural networks. In: The 12th International Conference for Internet Technology and Secured Transactions (ICITST), vol. 67–72 (2017)
14. Guerin, X.R.: Load balancing of distributed services: U.S. Patent 10,044,797[P], 7 August 2018
15. Janardhanan, D., Barrett, E.: CPU workload forecasting of machines in data centers using LSTM recurrent neural networks and ARIMA models. In: The 12th International Conference for Internet Technology and Secured Transactions (ICITST), pp. 55–60 (2017)
16. Razaque, A., Vennapusa, N.R., Soni, N., et al.: Task scheduling in cloud computing. In: IEEE Long Island Systems, Applications and Technology Conference (LISAT), pp. 1–5 (2016)

Forecasting the Sunspots Number Function in the Cycle of Solar Activity Based on the Method of Applying Artificial Neural Network

I. Krasheninnikov(✉) and S. Chumakov

Pushkov Institute of Terrestrial Magnetism, Ionosphere and Radio Wave Propagation, Russian Academy of Sciences, Moscow, Troitsk, Russia

krash@izmiran.ru

Abstract. The possibility of forecasting the sunspot number (SSN) dependency in the solar activity cycle is analyzed based on the application of the Elman artificial neural network platform to the historical data series of observatory data. A method for normalizing the initial data is proposed, i.e., constructing virtual idealized cycles using scaled coefficients in time and values of maxima in solar activity cycles. Correctness of the method is considered in the numerical simulation for the SSN time series. The intervals of changing the adaptable parameters of the neural network realization are estimated and a mathematical criterion for choosing a solution is proposed. Characteristic property of the constructed function is a significant asymmetry of the ascending and descending branches within the cycle. A forecast of the sunspots number dependence for the current 25th cycle of solar activity is presented and its general correctness is discussed in comparison with the existing results of solar activity predicting. #CSOC1120.

Keywords: Number of Sunspots · SSN · Long-Term Prediction · Solar Cycle · Artificial Neural Network

1 Introduction

Long-term forecasingt for solar activity is one of the most important factors in space weather. As one of important control indexes of geliogeophysical conditions in the Solar-Earth interaction, solar activity (SA) data in the form of sunspot number (SSN) and solar radio flux (solar radioflux - F10.7) are included in predictive ionospheric models, on the basis of which, in particular, long-term planning of the operation of ionospheric radio communication systems is carried out. Traditionally the required long-term forecasting the solar activity for the radio waves propagation purposes covers the time interval from one month to 2–3 years ahead and its most famous implementation is the generalized Waldmeier method (Combined method) [1], which is used as an official one in SIDC (Solar Influences Data Analysis Center, Brussels, EU). Mathematically, it is an extrapolation of the current data to the average forecasting scale and in complex situations, for example, as was observed at the transition from the 24th to the 25th cycle, it can lead to a significant discrepancy with the experimental data.

R. Silhavy and P. Silhavy (Eds.): CSOC 2023, LNNS 724, pp. 824–835, 2023.
https://doi.org/10.1007/978-3-031-35314-7_68

In this work, another approach is being considered, based on representing the solar cycle as a whole object for analysis and, accordingly, forecasting on the scale of the entire cycle. The quasi-periodic nature of the time dependence of the sunspots number makes it possible to apply various mathematical methods for extracting the main regularity as the basis for forecasting in observational data. In particular, this is the method of artificial neural networks (ANN) in the analysis of time series of data. The first attempts to apply ANN to long-term forecasting of solar activity were considered in [2]. Later this approach was developed in a number of works [3, 4], and it is being actively developed at the present time, for example, [5–7], as application of elements of artificial intelligence in the search for a hidden pattern in the time series of experimental data. Another direction of solving this problem is nonlinear dynamic analysis in chaos, for example, works [8, 9]. Also it should be noted studies on the possibility of predicting the SSN maximum in a cycle based on precursors in the previous cycle (cycles), in particular, based on the intensity of geomagnetic disturbances (daily Ap - index) [10, 11].

Since 2021, SWPC (Space Weather Prediction Center, Boulder, USA) is moving to a different basis in long-term forecasting of solar activity (Fig. 1), apparently based on the results of applying the neural network approach to the entire SA cycle, as the basic object in time periodic sequence of observational data. However, a number of points can be distinguished that are poorly consistent with the general properties of the solar cycle: a) a high degree of symmetry concerning the ascending and descending branches [12] and b) the incorrectness seemed on the final stage of the cycle - the descending branch does not enter in a new cycle after 2031.

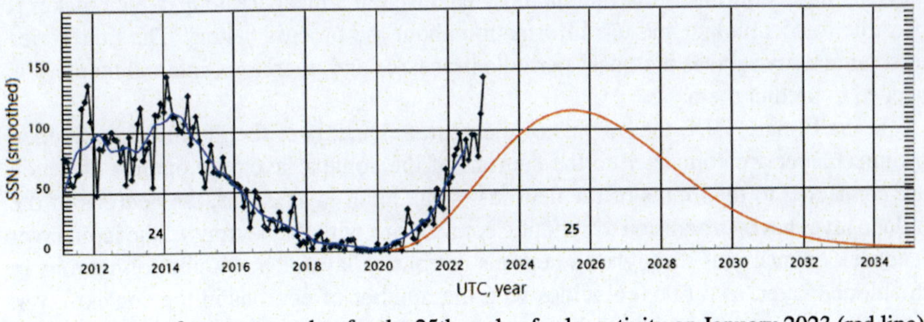

Fig. 1. Forecast of sunspot number for the 25th cycle of solar activity on January 2023 (red line) made by SWPC. The dots mark the monthly averaged SSN data, and the line marks the smoothed ones.

In this paper, based on the recurrent ANN platform we consider the possibility of long-term forecasting on the scale of the cycle of the solar activity parameter - the time function of SSN (sun spot number), using the technology of normalizing the original time series of data for analyzing and training the ANN. We evaluated the efficiency of the proposed forecasting method in numerical simulation and on the example of the previous four cycles of solar activity, from the 21st to the 24th, and a forecast is made for the current 25th cycle.

2 Elman Artificial Neural Network

Currently, there are more than 20 implementations of artificial neural networks platforms for various applications. When applied to time series, recurrent neural networks (recurrent networks) are often used. Recurrent neural networks characterized by both forward (feed forward) and reverse (feed back) propagation of information and the presence of feedback, through which the results of data processing at the previous stage are transmitted to intermediate (hidden or inner) layers of neurons. As a result, the input of the recurrent neural network at each fixed point in time is the vector of input data and the results of information processing by the network at the previous stage. Training process of such networks is based on the error backpropagation algorithm [13–15]. Also, for the problems of prediction (extrapolation) of time series, deep learning algorithms are used for deep multilayer perceptrons of direct propagation, for example, [3, 6]. In this work, as in [5], Elman's recurrent neural network [13] was used, which was developed to identify internal patterns (structures) in time series of data.

A quasi-periodic solar data structure is a suitable candidate for such a platform (Elman's recurrent neural network). The general scheme of such ANN is shown in Fig. 2. Recurrent neural networks are configurations in which the outputs of neurons of subsequent layers have synaptic connections with neurons of previous layers. This leads to the possibility of creating ANN models with memory, i.e. this models of ANN have ability to remember the process. Thus, an ANN is built, the response of which depends not only on the input signal that is currently supplied to the ANN input, but also on those signals that were processed by the neural network at previous points in time. Such an ANN has a non-linear internal memory enclosed in a feedback loop, which makes it possible to accumulate and use information about the process history. The time series of solar activity indices are quasi-periodic processes and, therefore, internal memory is needed to predict them.

In the Elman ANN, the outputs of the neural elements of the intermediate (hidden or inner) layer are connected to the neurons of the context layer, the outputs of which are connected to the inputs of the neurons of the inner layer. Thus, the neurons of the hidden layer have, in addition to synoptic connections with the neurons of the input layer feedback connections through the neurons of context layer. The number of neurons in the hidden layer, as a rule, coincides with the number of neurons in the context layer. Then the weighted sum of the i-th neural element of the intermediate layer [13, 14]:

$$S_i(t) = \sum_{j=1}^{n} w_{ji} x_j(t) + \sum_{k=1}^{m} w_{ki} c_k(t-1) - T_i, \tag{1}$$

where n is the number of neurons in the input layer or the dimension of the input data vector; $x_j(t)$, w_{ji} is the weight coefficient between the j-th neuron of the input and the i-th neuron of intermediate layer (its value determines the strength of the synoptic connection between the corresponding neurons);

m is the number of neurons of the intermediate layer;

w_{ki} is the weight coefficient between the k-th context neuron and the i-th neuron of the intermediate layer;

$c_k(t\text{-}1)$ is the output value of the k-th neuron of the inner (intermediate or hidden) layer at the previous step of calculating. His value is stored in the k-th neuron of context layer; T_i is the threshold value of the i-th neural element of the intermediate layer.

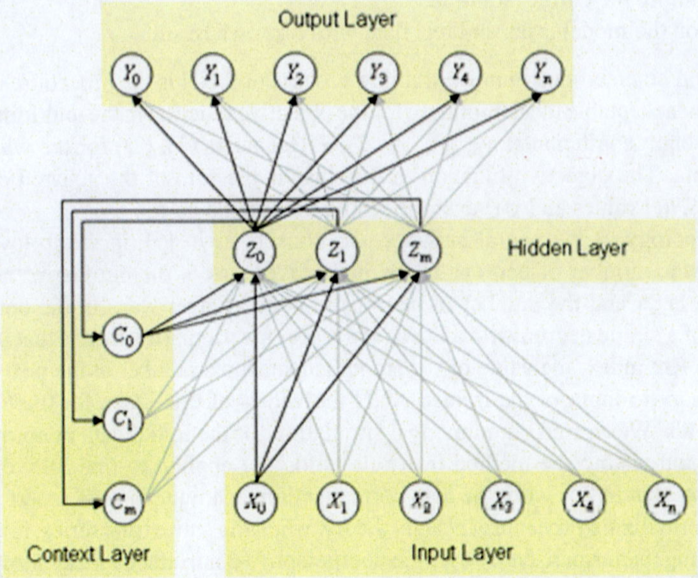

Fig. 2. Elman's ANN is a recurrent network with feedback from the hidden layer neurons.

The output value of the i-th neuron of the intermediate layer is determined as follows:

$$c_i(t) = F(S_i(t)),$$

where the function of the nonlinear transformation – F (or the activation function of neurons in the hidden layer) is usually the hyperbolic tangent or sigmoid function. The application of both functions was analyzed in the process of numerical experiments and, ultimately, the hyperbolic tangent function was chosen. In the output layer, the neural network has one neuron with a linear activation function, i.e. the value of the output neuron is a linear combination of the values of the hidden layer neurons:

$$y(t) = \sum_{k=1}^{m} v_k c_k(t) - T \tag{2}$$

where v_k is the weight coefficient between the k-th neuron of the hidden layer and the output neuron, $c_k(t)$ is the output value of the k-th neuron of the hidden layer, T is the threshold value of the output layer neuron. To train a recurrent neural network, the error backpropagation algorithm is used [13].

The general aspects of the ANN application, in addition to the choice of a working platform, include the following necessary steps:

- selection of the neural network topology (finding the optimal number of neurons in each layer);

- normalization of the initial data for the selected neural network;
- ANN training using the error backpropagation algorithm;
- experimental selection of the control parameters in the ANN training process (intervals of variation for the basic ANN parameters: training rate and training inertia) and then building the correct solution;
- testing on the model tasks and real data with a known result.

It should be noted that iterative training process of ANN is the procedure of finding the global or acceptable local minimum for the objective function in the multidimensional space of weight coefficients: w_{ji}, w_{ki}, v_k, T_i, T (formulas (1), (2)) for the whole set of training data. The objective function in this case is the sum of the squared deviations between *SSN(t)* values and *y(t)* over the whole training data set.

The topology of the neural network was chosen, as in [5], in the following configuration: the number of neurons in the input layer was 6, the number of neurons in the hidden layer and the number of neurons in context layer was 10, the output layer consisted of a single neuron with a linear activation function. This means that in order to predict the Rm index, the value of the smoothed sunspots number at the next moment, it is necessary to input of the trained ANN the values of this index for the 6 previous time intervals. We carried out numerical simulation for the initial data in normalization technique and testing the method in whole before be applied to real SSN data. Here we only note that in this work the data normalization technique is understood not as the standard normalizing procedure in ANN theory, when the entire time series, intended for ANN training, is normalized by a single coefficient. We introduced local normalization in which the sunspot number data in each individual solar activities (SA) cycle were normalized to the corresponding maximum of this solar cycle, and as a result, in each individual solar cycle, the maximum of normalized values was equal to unity.

3 Numerical Simulation

A feature of the neural network approach to time series forecasting is the problem of providing stability in constructing the correct solution - finding a global or acceptable minimum of the objective function in the iterative training process of the ANN. As a rule, it is not possible to choose control parameters of ANN training process that are universal for different cycles in an iterative training process under real data conditions or to construct an adaptive procedure for their selection. For the Elman neural network, such parameters are the speed and inertia (moment) of training in the iterative process of finding a solution [13], which can be estimated in the process of numerical experiments. For numerical experiments with our ANN model, the following representation for the monthly smoothed sunspot number (SSN) index was used to synthesize the basic function in the solar activity cycle:

$$R_m(t) = \left\{ \begin{array}{ll} A_0/2\left(1 + \sin\left(\dfrac{\pi t}{T_1} - \dfrac{\pi}{2}\right)\right), & T_1 \geq t \geq 0 \\ A_0/2\left(1 + \sin\left(-\dfrac{\pi(t - T_1)}{T_2} + \dfrac{\pi}{2}\right)\right), & T_2 \geq t \geq T_1 \end{array} \right\}, \tag{3}$$

where T_1 and T_2 are the durations of the ascending and falling branches in the cycle ($T_1 \leq T_2$), $T_0 = T_1 + T_2$ is the duration of the entire cycle, A_0 is the cycle amplitude. Parameters A_0 and T_0 were modulated by a pseudo-random sequence with $\delta A_0 / A_0 \sim 0.5$, $\delta T / T_0 \sim 0.1$ and saving the ratio $T_1/T_2 = 0.5$ for each cycle [12].

An example of cyclic time series simulating the historical series of SA data is shown in Fig. 3 (upper part of the upper panel), which shows variations in both the amplitude and duration of cycles and clearly expressed asymmetry in the temporal functional dependence of the index inside the cycle. Despite the rather high degree of idealization of the SSN cyclic sequence the direct application of the ANN technique to predict a specific cycle on the initial synthesized data has a rather low stability of the result, which is often simply incorrect from the point of view the process has to be cyclic with a minimum nearest to zero. At the stage of initial data normalization for the applying the ANN procedure, a method was developed for introducing an abstract idealized representation of a cycle that has a unit amplitude and a standard duration of 132 conditional months. That is, for each cycle, the transformation coefficients were introduced: $1 / A_0^i$ and T_N / T_0^i, where i is the cycle index, T_N- normalized cycle duration, i.e. $T_N = 132$ The lower part of the top panel in Fig. 3 displays such a normalized time dependence of the synthesized data. In addition, as in [5], thinned data with a factor of 5 was applied (in [5] it was equal to 6) and, thus, normalized data were generated for the training process of the ANN (lower part of the upper panel, Fig. 3). At the bottom part Fig. 3 shows the results of forecasting by the ANN method for the 23rd synthesized SA cycle. The base ANN forecast in the form of an idealized (abstract) cycle with thinned data is marked with empty circles against the background of the original curve (1) (lower left panel). In real time scale, from the ANN forecast with known transformation ratios, we obtain a forecast with a non-normalized (real) value of Rm - the lower right panel of Fig. 3.

It should be noted a significant increase in the stability of the result of estimating the functional dependence of SSN within the cycle with a decrease in the degrees of freedom of the problem - dividing the general process of predicting the solar cycle into partial parts: amplitude, duration and form of functional dependence from time simulated index within the cycle. The possibility of working with data in a dynamic mode has been researched, i.e. with a phase shift of the cyclic prediction interval T_0 and the prediction error in the normalized space remained at the level of the static mode. It was shown, that a fundamental feature of the Elman ANN - an existence of most appropriate values for the basic parameters of the iterative training process for the ANN is true: the optimal learning rate and the learning inertia where estimated as -0.005 and 0.5. These corresponds to the estimated intervals of changes in these parameters - [0.001 – 0.01] for the learning rate and [0.1 – 1.0] for inertia (moment) of learning given in [13].

4 Applying to Actual Data

Real SSN data have complicated structure, both in periodicity and in variations of maxima in periods; in particular, there is a weakly pronounced transition of cycles with significant variations in the minimum of solar activity. The correct operation of the ANN technique requires a strict mathematical separation of cycles - the definition of the beginning and the end of any cycle and a more clearly expressed transition process

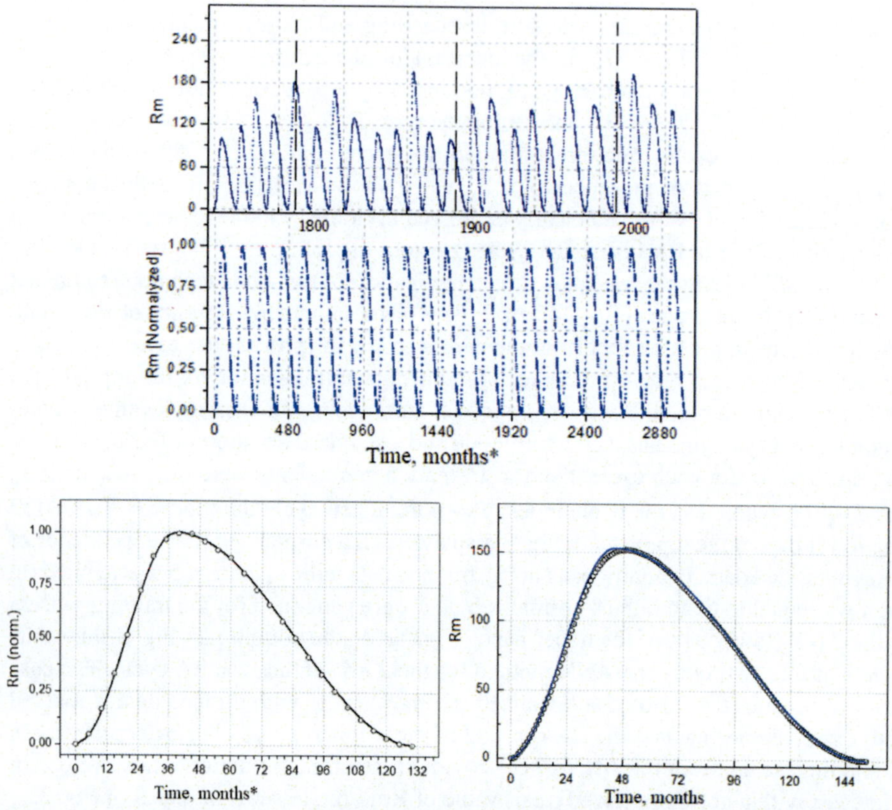

Fig. 3. Model representation of the time dependency of monthly smoothed sunspot numbers (upper panel) and prediction of the 23rd SA cycle: normalized data (lower left panel) and original simulated data (lower right panel). The solid lines represent the initial simulated curves, and the hollow circles indicate the result of the ANN prediction.

between cycles. For this purpose, we used the procedure of data approximation in the vicinity of the SA minimum at an interval of ±12 months from the current minimum SSN value (near the astronomical transition) and determined the quadratic (parabolic) dependence $R_P(t)$ and then the mathematical position of the minimum. This minimum of quadratic (parabolic) dependence was accepted for the end of the SA cycle. The point following the minimum was considered the beginning of a new SA cycle and was characterized by a positive time derivative. Further, the regularization of the time dependence $R_m(t)$ was performed near the minimum

$$R_m(t) = wR_P(t) + (1 - w)R_m(t), \;\; w(t) = \exp\left\{(t - t_{\min})^2 \Big/ \Delta t^2\right\}.$$

where $\Delta t = 6$ months, which made it possible to allocate the transition of cycles and to avoid jumps $R_m(t)$ in the docking region of the neighboring cycles (Fig. 4, top panel). At the next stage of data normalization, a transformation into a sequence of abstracted cycles with transformation coefficients was performed: $1/A_0^i$ и T_N/T_0^i, where i is the cycle index. This sequence of abstracted cycles is the training data for the ANN training. The introduction of idealized cycles with common boundary conditions - a positive time derivative at the beginning of the cycle and zero time derivative at the end of the cycle allows you to choose a solution in the variational process of finding the global minimum of the ANN objective function. A correct solution will be considered if it satisfies the general properties of normalized data at the ends of the predictive cycle (Fig. 4, lower left panel): a positive time derivative at the beginning, close to zero at the end of the cycle, and a maximum value of predictive cycle close to unity.

As an example of implicating the considered method we'll consider, the result of the prediction of the 21st SA cycle is shown in Fig. 4, which also shows the SSN dependence averaged over previous cycles (curves 1). Averaging over cycles is correct, since all normalized cycles have the same coordinate dimension. A fairly large spread of functions values in the vicinity of the SA maximum, on normalized data in cycles, is seen. What manifests itself in the values of confidence intervals (Fig. 4, lower right panel) given for several points on the averaged graph SSN. Comparison of the ANN-derived and averaged dependences for SSN shows a general characteristic asymmetry of the ascending and descending branches of the cycle and the synchronism of the curves is high enough, indicating a correct finding of the main part in the periodic process. At the same time, the ANN forecast differs from the average curve, what reflects the individual properties of a particular SA cycle.

Table 1 illustrates the final results of the analysis of solar data in the last four cycles: from the 21st to the 24th. The large variability of the parameters for the training process ANN (speed and inertia of ANN training) from cycle to cycle in the obtained solutions is clearly visible, which indicates a significant individuality of the SSN intracyclic dynamics. The numerical characteristic of the variability is given as an indicator of the efficiency of prediction (*PE*) in the cycle:

$$PE = 1 - \sum_{i=1}^{N} \left| R_p^i - R_m^i \right| \Big/ \sum_{i=1}^{N} R_m^i,$$

where R_p^i is the predicted, R_m^i is the experimental value of SSN. The lower values of the prediction efficiency for cycles 23rd and 24th are apparently associated with the "double-humped" nature of extremums R_m^i within these cycles, which manifests itself in the forecast for cycle 25th (Fig. 5). The predictive ANN curve in the normalized representation (Fig. 5, left panel) has a pronounced inflection on the descending branch, i.e. the possibility of forming an undeveloped "two-humped" peaks in the cycle is predicted. Also, like for the 21st cycle, the closeness of the ANN curve for 25th cycle is visible to the average dependence for the previous periods of the data time series. It should be noted a significant difference in the predictive curves: SWPC (marked in red in the right panel, Fig. 5) and curve predicted by ANN (highlighted in blue) in real time (Fig. 5, right panel). The maximum value for the ANN forecast is taken as 120 units and a period is taken as 132 months. (official NOAA forecast from 10 Jan 2022, https://spaceweatherarchive.com/2022/01/09/solar-cycle-25-update/). First of all, this concerns the position of the maximum value of the SSN of the SA cycle – for the ANN forecast curve, this is the end of 2023, while for the basic CWPC forecast, it is the middle of 2025. The experimental data (monthly average) are marked with hollow triangles, and for July 2022

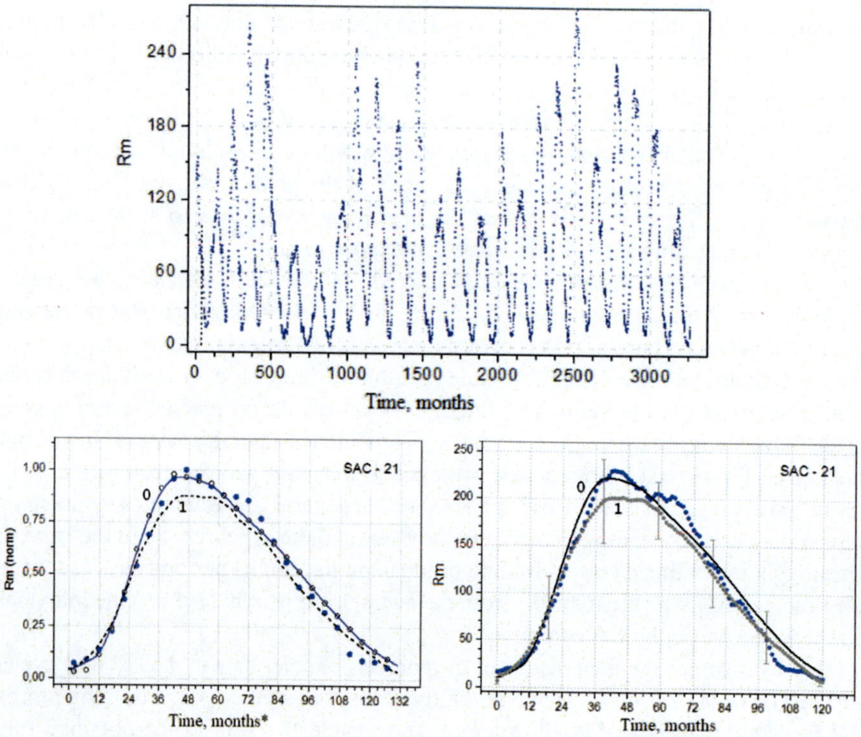

Fig. 4. Time dependency of monthly smoothed sunspot numbers (top panel) and cycle 21 prediction: normalized data (hollow circles, bottom left panel) and real data (bottom right panel). The filled circles (blue) represent observational data, the dotted line is the averaged results for previous cycles, and the solid lines are the forecast using the ANN method.

there is a significant preference for the forecast constructed by the considered method with data normalization. A large deviation from the official initial forecast is also noted in the NOAA bulletins.

Table 1. Typical parameters for applying the ANN method to SSN data.

Cycle number	Training Rate	Inertia of training	RMS (norm.)	RMS (exp.)	Prediction efficiency
21	0.005	0.75	0.06	14.97	0.91
22	0.005	0.5	0.05	12.65	0.9
23	0.0095	0.2	0.05	21.37	0.79
24	0.001	0.1	0.07	12.85	0.79
Average			0.06	15.46	0.85

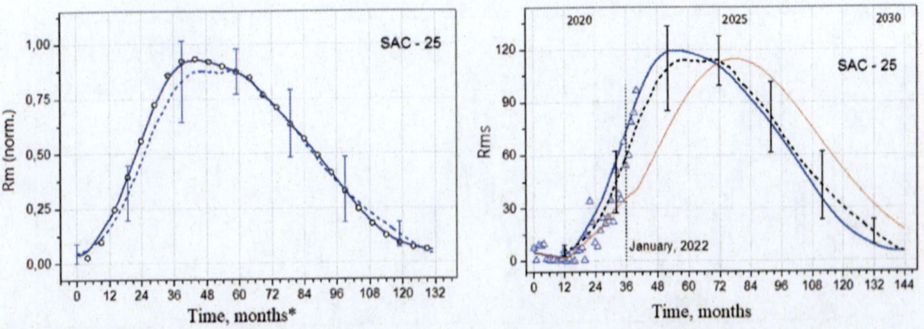

Fig. 5. Forecast of the dynamics of the sunspots number for the 25th cycle of solar activity. Normalized data (left panel): hollow circles - ANN result, dotted line with confidence intervals - averaged results for previous cycles. Real-time data (right panel): hollow triangles – current observational data (monthly average), solid line (blue) – ANN result, solid line (red) – SWPC long-term forecast (Boulder), dotted line – result of averaging over previous cycles with confidence intervals.

5 Results and Discussion

The method considered in this paper aims to predict the dynamics of SSN within the cycle, based on a decrease in the degrees of freedom of prediction as a whole, which increased the stability and, it seems, the correctness of finding a solution based on the usage of a classical recurrent ANN, in our case, Elman's ANN. In particular, the introduction of a normalized data representation space, in contrast to other approaches in the application of ANN to the long-term prediction of SA, made it possible to construct a function that has a obvious asymmetry of the ascending and descending branches within the cycle and, to some extent, reflects the features of previous SA cycles. Comparison of the predictive SWPC curves (Fig. 5) and the neural network obtained in [7] with two hidden layers of neurons, but without feedback, Fig. 6 shows great similarity of different cycles and very weak asymmetry branches within the SA cycle. It is practically absent, which is in poor agreement with the general ideas about the periodic nature of solar activity [12]. As for the maximum of the 25th cycle, the results of predictions differ greatly, from very small ~80 (applying the technique [10]) to large values of ~160, which reflects the complexity of the relationship between intrasolar processes and SSN [15] and the ambiguity of the prediction results for the same data. Thus, the deep learning method of a multilayer feed-forward neural network [7] gives an estimate of the maximum SA represented by the number of SSN of 106 units and the localization of this SA maximum in time for 2025. In work [9], based on a nonlinear dynamic data analysis, the maximum is estimated at 154 and it is expected at the beginning of 2023 (2023.02), which, to some extent, corresponds to the result of our analysis (2023.10) [16] and Fig. 5.

The proposed method for analyzing the SSN time series based on ANN has a high degree of mathematical formalization and can be the basis for both long-term forecasting on a cycle scale and for an interval of 2–3 years, using the ratio of general and partial. It seems that in both cases there is the possibility of adjusting (adaptation) to the monthly incoming SA registration data, varying the values of the predicted maximum A_0 and duration T_0 of the predicted cycle in small intervals of changes. The criterion is the

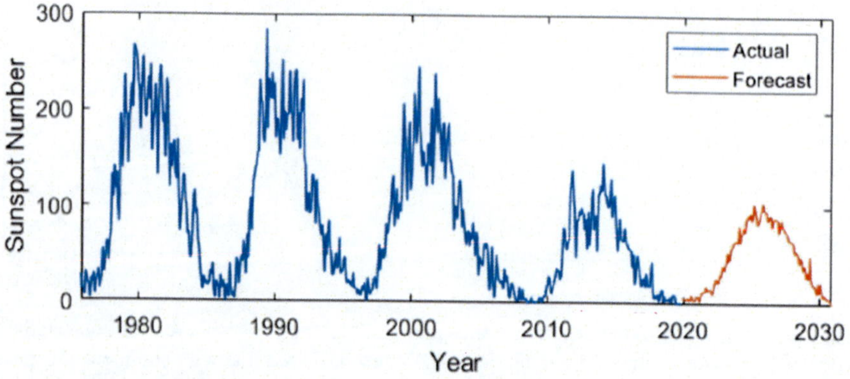

Fig. 6. Forecast for the of average monthly sunspot number for the 25th cycle of solar activity (deep neural network method)

achievement of minimum deviations of the calculated and experimental data in static and dynamic modes of operation. The software implementation of the method is based on a basis independent of commercial mathematical packages and can be improved as it is applied in practice.

6 Conclusion

Thus, based on Elman's recurrent ANN, it is shown the possibility of long-term forecasting on the scale of the whole solar activity cycle of the time function of the sunspots number in the cycle, using the above method of normalizing the initial time series of data. The proposed approach, based on decrease in the degrees of freedom of the problem, has a sufficiently high stability and forecasting efficiency, shown on the example of the previous four cycles, from the 21st to the 24th. The constructed forecast for the 25th cycle of solar activity has a higher degree of conformity with the current solar data available for the second half of 2022 than the forecast of the SWPC service. In general, the method can be considered as the basis for long-term forecasting of the sunspots number in the interval of 2–4 years and on the scale of whole cycle of solar activity as space weather indices.

References

1. Podladchikova, T., Van der Linden, R.: A Kalman filter technique for improving medium-term predictions of the sunspot number. Sol. Phys. **277**, 397–416 (2012)
2. Macpherson, K.: Neural network computation techniques applied to solar activity prediction. Adv. Space Res. **13**(9), 375–450 (1993)
3. Fessant, F., Bengio, S., Collobert, D.: On the prediction of solar activity using different neural network models. Ann. Geophys. **14**(1), 20–26 (1996)
4. Pesnell, W.D.: Solar cycle predictions (invited review). Solar Phys. **281**(1), 507 (2012). https://doi.org/10.1007/s11207-012-9997-5

5. Barkhatov, N.A., Korolev, A.V., Ponomarev, S.M., Sakharov, S.Y.: Long-term forecasting of solar activity indices using neural networks. Radiophys. Quantum Electron. **44**(9), 742–749 (2001). https://doi.org/10.1023/A:1013019328034
6. Pala, Z., Atici, R.: Forecasting sunspot time series using deep learning methods. Sol. Phys. **294**, 50 (2019). https://doi.org/10.1007/s11207-019-1434-62019
7. Benson, B., Pan, W.D., Prasad, A., Gary, G.A., Hu, Q.: Forecasting solar cycle 25 using deep neural networks. Sol. Phys. **295**(5), 1–15 (2020). https://doi.org/10.1007/s11207-020-01634-y
8. Sello, S.: Solar cycle forecasting: a nonlinear dynamics approach. Astron. Astrophys. **377**, 312–320 (2001)
9. Sarp, V., Kilcik, A., Yurchyshyn, V., Rozelot, J.P., Ozguc, A.: Prediction of solar cycle 25: A non-linear approach. MNRAS **481**, 2981–2985 (2018)
10. Thompson, R.J.: A technique for predicting the amplitude of the solar cycle. Sol. Phys. **148**(2), 383–388 (1993)
11. Hathaway, D.H., Wilson, R.M.: Geomagnetic activity indicates large amplitude for sunspot cycle 24. Geophys. Res. Lett. **33**, L18101 (2006)
12. Hathaway, D.H.: The solar cycle. Living Rev. Sol. Phys. **12**, 4 (2015). https://doi.org/10.1007/lrsp-2015-4
13. Elman, J.L.: Finding structure in time. Cogn. Sci. **14**, 179–211 (1990). https://doi.org/10.1207/s15516709cog1402_1
14. Golvko, V.A.: Neural Networks: Training, Organization and Application. BSU, Minsk, Belarus (2001)
15. Golvko, V.A., Krasnjproshin, V.V.: Neural Network Technologies for Data Processing. Classic University Edn. BSU, Minsk, Belarus (2017)
16. Nandy, D., Martens, P.C.H., Obridko, V., Dash, S., Georgieva, K.: Solar evolution and extrema: current state of understanding of long-term solar variability and its planetary impacts. Space Sci. Rev. **8**(1), 39 (2021). https://doi.org/10.1007/s11214-021-00799-7

Author Index

R. Silhavy and P. Silhavy (Eds.): CSOC 2023, LNNS 724, pp. 837–839, 2023.
https://doi.org/10.1007/978-3-031-35314-7